Catalog of Basic Functions

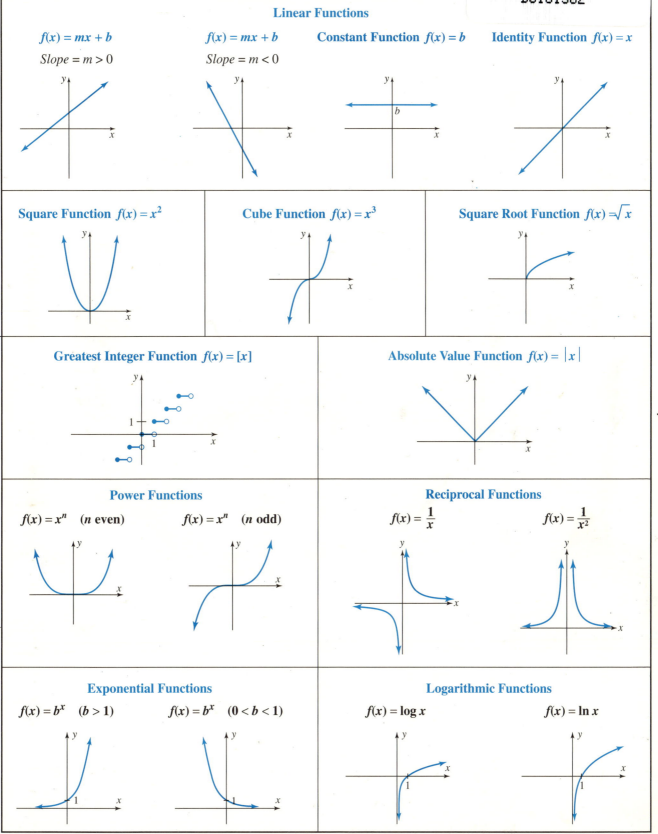

Linear Functions

$f(x) = mx + b$
$Slope = m > 0$

$f(x) = mx + b$
$Slope = m < 0$

Constant Function $f(x) = b$

Identity Function $f(x) = x$

Square Function $f(x) = x^2$

Cube Function $f(x) = x^3$

Square Root Function $f(x) = \sqrt{x}$

Greatest Integer Function $f(x) = [x]$

Absolute Value Function $f(x) = |x|$

Power Functions

$f(x) = x^n$ (*n* even)

$f(x) = x^n$ (*n* odd)

Reciprocal Functions

$f(x) = \dfrac{1}{x}$

$f(x) = \dfrac{1}{x^2}$

Exponential Functions

$f(x) = b^x$ $(b > 1)$

$f(x) = b^x$ $(0 < b < 1)$

Logarithmic Functions

$f(x) = \log x$

$f(x) = \ln x$

Contemporary College Algebra

A Graphing Approach

Second Edition

THOMAS W. HUNGERFORD
Cleveland State University
Saint Louis University

THOMSON

BROOKS/COLE

Australia • Canada • Mexico • Singapore • Spain
United Kingdom • United States

THOMSON
™
BROOKS/COLE

Sponsoring Editor: *John-Paul Ramin*
Development Editor: *Leslie Lahr*
Assistant Editor: *Katherine Brayton*
Editorial Assistant: *Darlene Amidon-Brent*
Project Manager, Editorial Production: *Janet Hill*
Technology Project Manager: *Earl Perry*
Marketing Manager: *Karin Sandberg*
Marketing Assistant: *Erin Mitchell*
Advertising Project Manager: *Bryan Vann*
Print/Media Buyer: *Barbara Britton*
Permissions Editor: *Stephanie Lee*

Production Service: *Hearthside Publishing Services/Laura Horowitz*
Text Designer: *Terri Wright*
Art Editor: *Hearthside Publishing Services*
Photo Researcher: *Terri Wright*
Copy Editor: *Barbara Willette*
Illustrator: *Hearthside Publishing Services*
Cover Designer: *Terri Wright*
Cover Printer: *Coral Graphic Services*
Compositor: *Better Graphics, Inc.*
Printer: *R. R. Donnelley/Willard*

For more information about our products, contact us at:
Thomson Learning Academic Resource Center
1-800-423-0563
For permission to use material from this text or product, submit a request online at *http://www.thomsonrights.com.*
Any additional questions about permissions can be submitted by email to thomsonrights@thomson.com.

Library of Congress Control Number: 2004110055

Student Edition: ISBN 0-534-46656-7

Instructor's Edition: ISBN 0-534-46657-5

Thomson Brooks/Cole
10 Davis Drive
Belmont, CA 94002
USA

Asia
Thomson Learning
5 Shenton Way #01-01
UIC Building
Singapore 068808

Australia/New Zealand
Thomson Learning
102 Dodds Street
Southbank, Victoria 3006
Australia

Canada
Nelson
1120 Birchmount Road
Toronto, Ontario M1K 5G4
Canada

Europe/Middle East/Africa
Thomson Learning
High Holborn House
50/51 Bedford Row
London WC1R 4LR
United Kingdom

Latin America
Thomson Learning
Seneca, 53
Colonia Polanco
11560 Mexico D.F.
Mexico

Spain/Portugal
Paraninfo
Calle/Magallanes, 25
28015 Madrid, Spain

To my wife
Mary Alice Ryan Hungerford
and my children
Anne Elizabeth Hungerford
Thomas Joseph Hungerford
who make it all worthwhile

CONTENTS

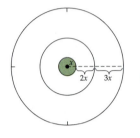
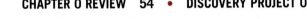
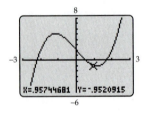
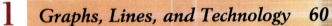
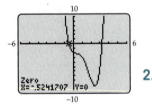

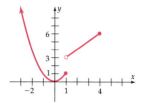

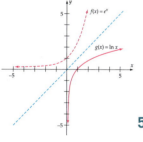

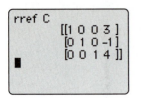

6 *Systems of Equations* 458

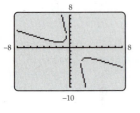

7 *Discrete Algebra* 552

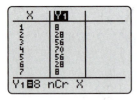

8 *Analytic Geometry* 633

PREFACE

This book is intended to provide a flexible approach to the college algebra curriculum that emphasizes real-world applications and is suitable for a variety of audiences. Mathematical concepts are presented in an informal manner that stresses meaningful motivation, careful explanations, and numerous examples, with an ongoing focus on real-world problem solving. Technology is integrated into the text and students are expected to use it to participate actively in exploring various topics from algebraic, graphical, and numerical perspectives.

Contemporary College Algebra has relatively modest mathematical prerequisites and includes a full review of basic algebra in Chapter 0. It can be used for a variety of terminal courses for students in the liberal arts, social sciences, and business, and includes a number of interesting applications such as probability and linear programming. Furthermore, all of the (non-trigonometry) topics needed in calculus are covered here in sufficient detail to prepare a student for business/social science calculus courses.*

Changes in the Second Edition

The most noticeable differences from the previous edition are the following:

Variation   An optional Section on variation has been added to Chapter 2 for those instructors who want to cover this topic.

Catalog of Functions   In response to user requests, the graphs of basic functions are now summarized in a catalog that appears in the front end-papers and at appropriate points in Chapters 3–5.

Geometry of Triangles   The Geometry Appendix now contains a full summary of the basic facts about congruent and similar triangles, including exercises.

Calculus Icons   Material that lays the groundwork for topics that are likely to appear in a calculus class is marked by **C** in the text and by ◯ in the exercises.

A variety of other changes have been made, with many examples and exercises rewritten or replaced. A few sections have been extensively rewritten to improve coverage and clarity, including graphing with technology (Section 1.2) and inverse functions (Section 5.7). Other improvements include the following:

Real-Data Applications   Significantly more exercises and examples are based on real data. All real-data examples and exercises from the previous edition have either been updated or replaced with more relevant ones.

Discovery Projects   Many of these have been replaced with more interesting topics. These investigative problems are suitable for small-group work.

Technology Tips   These have been updated, as necessary, with new tips added to clarify points that cause difficulties for students.

Artwork   In addition to an attractive multi-color format, the book now includes a number of photographs in the chapter openers and discovery projects.

---

*The companion volume *Contemporary College Algebra and Trigonometry*, which includes everything in this book and full coverage of trigonometry (beginning with triangle trigonometry) is suitable for preparing students for the standard science/engineering calculus sequence.

Mathematical and Pedagogical Features

The mathematical approach remains the same. Concepts are presented in an informal manner with careful explanations that favor student understanding over formal proof (although proofs are included when appropriate). All the helpful pedagogical features of the first edition are retained here, such as:

Cautions that alert students to common misconceptions and mistakes;

Exercises that proceed from routine drill to those requiring some thought, including graph interpretation and applied problems, as well as *Thinkers* that challenge students to "think outside the box" (most of these are not difficult—just different);

Chapter Reviews that include a list of important concepts (referenced by section and page number), a summary of important facts and formulas, and a set of review questions;

Geometry Review Appendix that summarizes frequently used facts from plane geometry, with examples and exercises.

Technology (particularly the graphing calculator) is used intelligently as an important tool for effectively developing a fuller understanding of the underlying mathematics. It is integrated into the presentation, not just an optional add-on. Several features are designed to assist students to get the most out of the technological tools available to them, including

Graphing Explorations, in which students discover and develop useful mathematics on their own;

Technology Tips that provide assistance in carrying out various procedures on specific calculators;

Calculator Investigations that encourage students to become familiar with the capabilities and limitations of graphing calculators;

Program Appendix that provides a small number of programs that are useful for updating older calculator models or for more easily carrying out procedures discussed in the text.

For the Instructor

INSTRUCTOR'S EDITION

This special version of the complete student text contains a Resource Integration Guide and a complete set of answers printed in the back of the text.
ISBN: 0-534-46657-5

TEST BANK

The Test Bank includes six tests per chapter as well as three final exams. The tests are made up of a combination of multiple-choice, free-response, true/false, and fill-in-the-blank questions.   **ISBN: 0-534-46659-1**

INSTRUCTOR'S SOLUTIONS MANUAL

The *Instructor's Solutions Manual* provides worked-out solutions to all of the problems in the text.   **ISBN: 0-534-46661-3**

TEXT-SPECIFIC VIDEOTAPES

These text-specific videotape sets, available at no charge to qualified adopters of the text, feature 10- to 20-minute problem-solving lessons that cover each section of every chapter.   **ISBN: 0-534-46794-6**

iLrn INSTRUCTOR'S VERSION

With a balance of efficiency and high-performance, simplicity and versatility, iLrn gives you the power to transform the teaching and learning experience. iLrn Instructor Version is made up of two components, iLrn Testing and iLrn Tutorial. iLrn Testing is a revolutionary, internet-ready, text-specific testing suite that allows instructors to customize exams and track student progress in an accessible, browser-based format. iLrn offers full algorithmic generation of problems and free-response mathematics. iLrn Tutorial is a text-specific, interactive tutorial software program that is delivered via the Web (at http://ilrn.com) and is offered in both student and instructor versions. Like iLrn Testing, it is browser-based, making it an intuitive mathematical guide even for students with little technological proficiency. So sophisticated, it's simple, iLrn Tutorial allows students to work with real math notation in real time, providing instant analysis and feedback. The tracking program built into the instructor version of the software enables instructors to carefully monitor student progress. The complete integration of the testing, tutorial, and course management components simplifies your routine tasks. Results flow automatically to your gradebook and you can easily communicate with individuals, sections, or entire courses.

WEBTUTOR TOOLBOX ON WEBCT AND BLACKBOARD

Preloaded with content and available free via PIN code when packaged with this text, WebTutor ToolBox for WebCT pairs all the content of this text's rich Book Companion Website with all the sophisticated course management functionality of a WebCT product. You can assign materials (including online quizzes) and have the results flow AUTOMATICALLY to your gradebook. ToolBox is ready to use as soon as you log on—or, you can customize its preloaded content by uploading images and other resources, adding Web links, or creating your own practice materials. Students only have access to student resources on the Website. Instructors can enter a PIN code for access to password-protected Instructor Resources.   **ISBN: 0-534-27488-9 WebCT, ISBN: 0-534-27489-7 Blackboard**

For the Student

STUDENT SOLUTIONS MANUAL

The *Student Solutions* manual provides worked out solutions to the odd-numbered problems in the text.   **ISBN: 0-534-46660-5**

BROOKS/COLE MATHEMATICS WEBSITE
http://mathematics.brookscole.com

The Book Companion Website offers a range of learning resources that can be used to help your students get the most from their beginning algebra class. Chapter-specific quizzes and tests allow students to assess their understanding—and receive immediate feedback. Students also have access to *Web Explorations* and an *Online Graphing Calculator Manual* that feature additional explorations of concepts.

iLRN TUTORIAL STUDENT VERSION

A complete content mastery resource offered online, iLrn Tutorial allows students to work with real math notation in real time, with unlimited practice problems, instant analysis and feedback, and streaming video to illustrate key concepts. And live, online, text-specific tutorial help is just a click away with vMentor, accessed seamlessly through iLrn Tutorial. The iLrn Tutorial access card located inside the cover of the text provides a PIN code for students to utilize when logging into iLrn.

INTERACTIVE VIDEO SKILLBUILDER CD-ROM

Think of it as portable office hours! The Interactive Video SkillBuilder CD-ROM contains more than eight hours of video instruction. The problems worked during each video lesson are shown next to the viewing screen so that students can try working them before watching the solution. To help students evaluate their progress, each section contains a 10-question Web quiz (the results can be e-mailed to the instructor) and each chapter contains a chapter test, with the answer to each problem on each test. This CD-ROM also includes MathCue tutorial and quizzing software, featuring a "SkillBuilder" that presents problems to solve and evaluates answers with step-by-step explanations; a Quiz function that enables students to generate quiz problems keyed to problem types from each section of the book; a Chapter Test that provides many problems keyed to problem types from each chapter; and a Solution Finder that allows students to enter their own basic problems and receive step-by-step help as if they were working with a tutor.   **(PACKAGED FREE WITH THIS TEXT)**

Acknowledgments

This book would not have been possible, without the diligent work of

Phil Embree, William Woods University

who checked examples and exercises for accuracy, wrote a large part of the answer section and several Discovery Projects, and provided helpful suggestions on other aspects of the text. I am also grateful to

Margaret Donlan, University of Delaware and

Sudhir Goel, Valdosta State University

who reviewed the text for accuracy and saved me from a number of embarrassments. Similar fine work was done by

Fred Safier, City College of San Francisco

who prepared the various solution manuals for the text and suggested a number of improvements.

I also want to thank my son, Thomas J. Hungerford, and Lee Windsperger, a student at St. Louis University, who assisted me with manuscript preparation.

I am grateful to the following reviewers, who provided many helpful suggestions that have improved this edition:

Kambiz Askarpour, Community College of Baltimore County at Essex; Steven B. Damelin, Georgia Southern University; Peter U. Georgakis, Santa Barbara City College; David M. George, The Ohio State University; Gregory D. Goeckel, Presbyterian College, South Carolina; Charles G. Laws, Cleveland State Community College; Mehran Mahdavi, Bowie State Univeristy; Nancy Matthews, University of Oklahoma; Annette Noble, University of Maryland Eastern Shore; Seshu Rao, Washington State University; Gary Wardall, University of Wisconsin-Green Bay; and Paul E. Yansak, Grand Valley State University

Thanks also go to reviewers from previous editions:

Deborah Adams, Jacksonville University; Kelly Bach, University of Kansas; David Blankenbaker, University of New Mexico; Bettyann Daley, University of Delaware; Margaret Donlan, University of Delaware; Patrick Dueck, Arizona State University; Betsy Farber, Bucks County Community College; Alex Feldman, Boise State University; Betty Givan, Eastern Kentucky University; William Grimes, Central Missouri State University; Frances Gulick, University of Maryland; John Hamm, University of New Mexico; Ann Lawrance, Wake Technical Community College; Charles Laws, Cleveland State Community College; Martha Lisle, Prince George's Community College; Lonnie Hass, North Dakota State University; Mathew Liu, University of Wisconsin—Stevens Point; Sergey Lvin, University of Maine; George Matthews, Onondaga Community College; Nancy Matthews, University of Oklahoma; Ruth Meyering, Grand Valley State University; William Miller, Central Michigan University; Philip Montgomery, University of Kansas; Katherine Muhs, St. Norbert College; Roger Nelson, Lewis and Clark College; Jack Porter, University of Kansas; Carolyn Robertson, University of Mississippi; Joe Rody, Arizona State University; Robert Rogers, University of New Mexico; Barbara Sausen, Fresno City College; Marvin Stick, University of Massachusetts-Lowell; Hugo Sun, California State University at Fresno; Stuart Thomas, University of Oregon; Bettie Truitt, Black Hawk College; Jan Vandever, South Dakota University; Judith Wolbert, Kettering University

I also want to thank those who have prepared various supplements for the text.

Nancy Matthews, University of Oklahoma [Test Bank]

Eric Howe, [Resource Integration Guide]

It is a pleasure to acknowledge the Brooks/Cole staff members who have been instrumental in producing this book. Thanks to

John-Paul Ramin, Sponsoring Editor

and very special thanks to

Leslie Lahr, Developmental Editor,

whose enthusiasm, energy, and good common sense have made an otherwise difficult job much easier than I had any right to expect.

I also thank

Lisa Chow, Assistant Editor

Darlene Amidon-Brent, Editorial Assistant

Karin Sandberg, Senior Marketing Manager

Earl Perry, Technology Project Manager

Janet Hill, Senior Project Manager, Editorial Production

Vernon Boes, Senior Art Director

Bryan Vann, Advertising Project Manager

I also want to thank the outside production staff.

Terri Wright of Terri Wright Design

Barbara Willette, Copyeditor

Special accolades are due to

Laura Horowitz of Hearthside Publishing Services

who coordinated the production of the book with tact, kindness, and finesse.

I especially want to thank my former colleagues at Cleveland State University, where this book was begun, for their friendship and support over the years. Thanks also go to Mike May, S. J. and my other new colleagues at Saint Louis University, where the book was finished, for the warm welcome they have given me.

Finally, I thank my wife Mary Alice, once again, for her support, understanding, and love, without which I would not long survive.

Thomas W. Hungerford

TO THE INSTRUCTOR

This book contains more than enough material for one semester. By using the chart on the facing page (and the similar ones at the beginning of each chapter that show the interdependence of sections within the chapter) you can easily design a course to fit the needs of your students and the constraints of time. When planning your syllabus, two items are worth noting:

Prerequisites Chapter 0 (Algebra Review) is a prerequisite for the entire book; it may be omitted by well-prepared classes.

Special Topics Sections with this label are usually related to the immediately preceding section and are not prerequisites for other sections of the text. I am inclined to omit most of them (particularly in a short course), but I know that a number of people consider some of them to be essential material. So feel free to include as many or as few as you wish.

This text assumes the use of technology. You and your students will need either a graphing calculator or a computer with appropriate software. All discussions of calculators in the text apply (with obvious modifications) to computer software. You should be aware of the following facts.

Minimal Requirements Current calculator models that meet the minimal requirements include TI-82 and above, HP-38 and above, and Casio 9850 and above (including FX 2.0). A few of these may also require some programs from the Program Appendix to handle certain tasks.

Technology Tips The Tips in the margin provide general information and advice, as well as listing the proper menus or keys needed to carry out procedures on specific calculators. Unless noted otherwise,

Tips for	also apply to
TI-83 Plus	TI-83, TI-82
TI-86	TI-85
TI-89	TI-92
Casio 9850	Casio 9970
HP-39+	HP-39, HP-38

You may want to explain to your students that the icon ▭ in the text indicates examples and exercises that are discussed in the Interactive Video Skillbuilder CD-ROM that accompanies the text. Finally, the symbols **C** and ⬤ indicate examples, exercises, and sections that are relevant to calculus.

INTERDEPENDENCE OF CHAPTERS

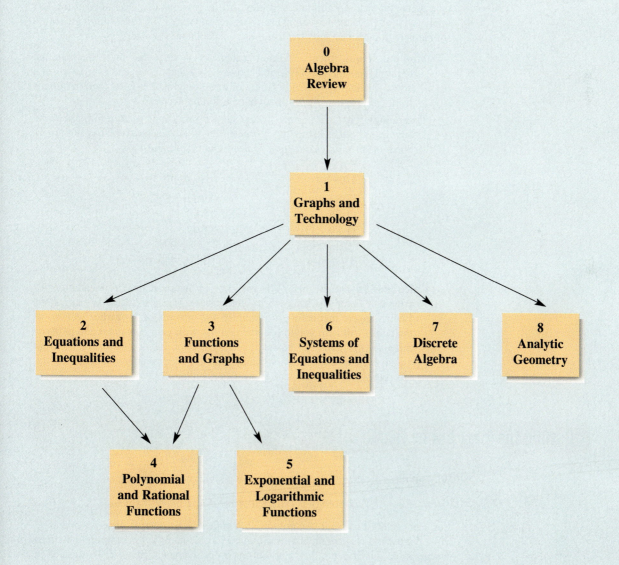

This text assumes the use of technology. You will need either a graphing calculator or a computer with appropriate software. All discussions of calculators in the text apply (with obvious modifications) to computer software. The following information should help you choose and effectively use a calculator.

Minimal Requirements Current calculator models that meet the minimal requirements include TI-82 and above, HP-38 and above, and Casio 9850 and above (including FX 2.0). A few of these may also require some programs from the Program Appendix to handle certain tasks.

Technology Tips The Tips in the margin provide general information and advice, as well as listing the proper menus or keys needed to carry out procedures on specific calculators. Unless noted otherwise,

Tips for	also apply to
TI-83 Plus	TI-83, TI-82
TI-86	TI-85
TI-89	TI-92
Casio 9850	Casio 9970
HP-39+	HP-39, HP-38

Calculator Investigations You may not be aware of the full capabilities of your calculator (or some of its limitations). The Calculator Investigations (which appear just before the exercise sets in some of the earlier sections of the book) will help you to become familiar with your calculator and to maximize the mathematical power it provides. Even if your instructor does not assign these investigations, you may want to look through them to be sure you are getting the most you can from your calculator.

Getting the Most Out of This Course

With all this talk about calculators, don't lose sight of this crucial fact:

Technology is only a *tool* for doing mathematics.

You can't build a house if you only use a hammer. A hammer is great for pounding nails, but useless for sawing boards. Similarly, a calculator is great for computations and graphing, but it is not the right tool for every mathematical task. To succeed in this course, you must develop and use your algebraic and geometric skills, your reasoning power and common sense, and you must be willing to work.

The key to success is to use all of the resources at your disposal: your instructor, your fellow students, your calculator (and its instruction manual), and this book and the Interactive Video Skillbuilder CD-ROM that accompanies it. Here are some tips for making the most of these resources.

Ask Questions   Remember the words of Hillel:

The bashful do not learn.

There is no such thing as a "dumb question" (assuming, of course, that you have attended class and read the text). Your instructor will welcome questions that arise from a serious effort on your part.

Read the Book   Not just the homework exercises, but the rest of the text as well. There is no way your instructor can possibly cover the essential topics, clarify ambiguities, explain the fine points, and answer all your questions during class time. You simply will not develop the level of understanding you need to succeed in this course and in calculus unless you read the text fully and carefully.

Be an Interactive Reader   You can't read a math book the way you read a novel or history book. You need pencil, paper, and your calculator at hand to work out the statements you don't understand and to make notes of things to ask your fellow students and/or your instructor.

Do the Graphing Explorations   When you come to a box labeled "Graphing Exploration," use your calculator as directed to complete the discussion. Typically, this will involve graphing one or more equations and answering some questions about the graphs. Doing these explorations as they arise will improve your understanding and clarify issues that might otherwise cause difficulties.

Do Your Homework   Remember that

Mathematics is not a spectator sport.

You can't expect to learn mathematics without doing mathematics, any more than you could learn to swim without getting wet. Like swimming or dancing or reading or any other skill, mathematics takes practice. Homework assignments are where you get the practice that is essential for passing this course and succeeding in subsequent mathematics courses.

CHAPTER 0

Algebra Review

Stockphoto.com/Michael Howell

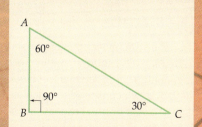

On a clear day, can you see forever?

If you are at the top of the Sears Tower in Chicago, how far can you see? In earlier centuries, the lookout on a sailing ship was posted atop the highest mast because he could see farther from there than from the deck. How much farther? These questions, and similar ones, can be answered (at least approximately) by using basic algebra and geometry. See Example 10 on page 21 and Exercise 74 on page 23.

Chapter Outline

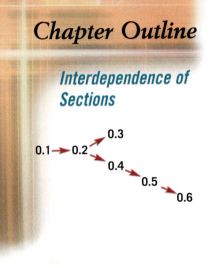
This chapter reviews the essential facts about real numbers, exponents, and the basic rules of algebra that are needed in this course and later ones. Well-prepared students may be able to skim over much of this material, but if you haven't used your algebraic skills for a while, you should review this chapter thoroughly. Your success in the rest of the course depends on your ability to use the fundamental algebraic tools presented here.

0.1 The Real Number System

You have been using **real numbers** most of your life. They include the **natural numbers** (or **positive integers**): 1, 2, 3, 4, . . . and the **integers:**

$$\ldots, -5, -4, -3, -2, -1, 0, 1, 2, 3, 4, 5, \ldots$$

A real number is said to be a **rational number** if it can be expressed as a fraction $\dfrac{r}{s}$, with r and s integers and $s \neq 0$; for instance,

$$\frac{1}{2}, \qquad -.983 = -\frac{983}{1000}, \qquad 47 = \frac{47}{1}, \qquad 8\tfrac{3}{5} = \frac{43}{5}.$$

Alternatively, rational numbers may be described as numbers that can be expressed as terminating decimals, such as $.25 = \dfrac{1}{4}$, or as nonterminating repeating decimals, in which a single digit or a block of digits repeats forever, such as

$$\frac{5}{3} = 1.66666 \cdots \qquad \text{or} \qquad \frac{58}{333} = .174174174 \cdots.$$

A real number that cannot be expressed as a fraction with integer numerator and denominator is called an **irrational number.** Alternatively, an irrational number is one that can be expressed as a nonterminating, nonrepeating decimal (no block of digits repeats forever). For example, the number π, which is used to calculate the area of a circle, is irrational.[*] More information about decimal expansions of real numbers is given in Special Topics 0.1.A.

---

[*]This fact is difficult to prove. In the past, you might have used 22/7 as π; and a calculator might display π as 3.141592654. However, these numbers are just *approximations* of π (close but not quite *equal* to π).

The real numbers are often represented geometrically as points on a **number line,** as in Figure 0–1. We shall assume that there is exactly one point on the line for every real number (and vice versa) and use phrases such as "the point 3.6" or "a number on the line." This mental identification of real numbers and points on the line is often extremely helpful.

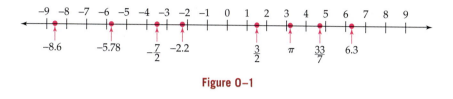

Figure 0–1

▟ ORDER

The statement $c < d$, which is read **"c is less than d,"** and the statement $d > c$ (read **"d is greater than c"**) mean exactly the same thing:

c lies to the *left* of d on the number line.

For example, Figure 0–1 shows that $-5.78 < -2.2$ and $4 > \pi$.

The statement $c \le d$, which is read **"c is less than or equal to d,"** means

Either c is less than d or c is equal to d.

Only one part of an "either . . . or" statement needs to be true for the entire statement to be true. So the statement $5 \le 10$ is true because $5 < 10$, and the statement $5 \le 5$ is true because $5 = 5$. The statement $d \ge c$ (read **"d is greater than or equal to c"**) means exactly the same thing as $c \le d$.

The statement $b < c < d$ means

$b < c$ and simultaneously $c < d$.

For example, $3 < x < 7$ means that x is a number that is strictly between 3 and 7 on the number line (greater than 3 and less than 7). Similarly, $b \le c < d$ means

$b \le c$ and simultaneously $c < d$,

and so on.

▟ ARITHMETIC

To avoid ambiguity when dealing with expressions such as $6 + 3 \times 5$, mathematicians have made the following agreement, which is also followed by your calculator.

Order of Operations

> In an expression without parentheses, multiplication and division are performed first (from left to right). Addition and subtraction are performed last (from left to right).

In light of this convention, there is only one correct way to interpret $6 + 3 \times 5$:

$6 + 3 \times 5 = 6 + 15 = 21$.   [Multiplication first, addition last]

On the other hand, if you want to "add $6 + 3$ and then multiply by 5," you must use parentheses:

$$(6 + 3) \cdot 5 = 9 \cdot 5 = 45.$$

This is an illustration of the first of two basic rules for dealing with parentheses.

Rules for Parentheses

> 1. Do all computations inside the parentheses before doing any computations outside the parentheses.
>
> 2. When dealing with parentheses within parentheses, begin with the innermost pair and work outward.

For example,

$$8 + [11 - (6 \times 3)] = 8 + (11 - 18) = 8 + (-7) = 1.$$

Inside parentheses first

We assume that you are familiar with the basic properties of real number arithmetic, particularly the following fact.

Distributive Law

> For all real numbers a, b, c,
>
> $$a(b + c) = ab + ac \qquad \text{and} \qquad (b + c)a = ba + ca.$$

The distributive law doesn't usually play a direct role in easy computations, such as $4(3 + 5)$. Most people don't say $4 \cdot 3 + 4 \cdot 5 = 12 + 20 = 32$. Instead, they mentally add the numbers in parentheses and say 4 times 8 is 32. But when symbols are involved, you can't do that, and the distributive law is essential. For example,

$$4(3 + x) = 4 \cdot 3 + 4x = 12 + 4x.$$

TECHNOLOGY TIP

To enter a negative number, such as -5, on most calculators, you must use the negation key: $(-)\,5$.

If you use the subtraction key on such calculators and enter $-\,5$, the display will read

$$\text{ANS} - 5$$

which tells the calculator to subtract 5 from the previous answer.

▉ NEGATIVE NUMBERS AND NEGATIVES OF NUMBERS

The **positive numbers** are those to the right of 0 on the number line, that is,

All numbers c with $c > 0$.

The **negative numbers** are those to the left of 0, that is,

All numbers c with $c < 0$.

The **nonnegative numbers** are the numbers c with $c \geq 0$.

The word "negative" has a second meaning in mathematics. The **negative** of a **number** c is the number $-c$. For example, the negative of 5 is -5, and the negative of -3 is $-(-3) = 3$. Thus the negative of a negative number is a positive number. Zero is its own negative, since $-0 = 0$. In summary,

Negatives

> The negative of the number c is $-c$.
>
> If c is a positive number, then $-c$ is a negative number.
>
> If c is a negative number, then $-c$ is a positive number.

TIP

To compute an expression such as $\sqrt{7^2 + 51}$ on a calculator, you must use parentheses:

$$\sqrt{\ }(7^2 + 51).$$

Without the parentheses, the calculator will compute

$$\sqrt{7^2} + 51 = 7 + 51 = 58$$

instead of the correct answer:

$$\sqrt{7^2 + 51} = \sqrt{49 + 51}$$

$$= \sqrt{100}$$

$$= 10.$$

▨ SQUARE ROOTS

The **square root** of a nonnegative real number d is defined as the nonnegative number whose square is d and is denoted $\sqrt{d}$. For instance,

$$\sqrt{25} = 5 \qquad \text{because} \qquad 5^2 = 25.$$

In the past you may have said that $\sqrt{25} = \pm 5$, since $(-5)^2$ is also 25. It is preferable, however, to have a single unambiguous meaning for the symbol $\sqrt{25}$. So in the real number system, the term "square root" and the radical symbol $\sqrt{\ }$ always denote a *nonnegative* number. To express -5 in terms of radicals, we write $-5 = -\sqrt{25}$.

Although $-\sqrt{25}$ is a real number, the expression $\sqrt{-25}$ is *not defined* in the real numbers because there is no real number whose square is -25. In fact, since the square of every real number is nonnegative,

No negative number has a square root in the real numbers.

Some square roots can be found (or verified) by hand, such as

$$\sqrt{225} = 15 \qquad \text{and} \qquad \sqrt{1.21} = 1.1.$$

Usually, however, a calculator is needed to obtain rational *approximations* of roots. For instance, we know that $\sqrt{87}$ is between 9 and 10 because $9^2 = 81$ and $10^2 = 100$. A calculator shows that $\sqrt{87} \approx 9.327379.$*

▨ ABSOLUTE VALUE

On an informal level, most students think of absolute value like this:

The absolute value of a nonnegative number is the number itself.

The absolute value of a negative number is found by "erasing the minus sign."

If $|c|$ denotes the absolute value of c, then, for example, $|5| = 5$ and $|-4| = 4$.

This informal approach is inadequate, however, for finding the absolute value of a number such as $\pi - 6$. It doesn't make sense to "erase the minus sign" here. So we must develop a more precise definition. The statement $|5| = 5$ suggests that the absolute value of a positive number ought to be the number itself. For negative numbers, such as -4, note that $|-4| = 4 = -(-4)$, that is, the absolute value of the negative number -4 is the *negative* of -4. These facts are the basis of the formal definition.

Absolute Value

The **absolute value** of a real number c is denoted $|c|$ and is defined as follows.

If $c \geq 0$, then $|c| = c$.

If $c < 0$, then $|c| = -c$.

---

*$\approx$ means "approximately equal."

EXAMPLE 1

(a) $|3.5| = 3.5$ and $|-7/2| = -(-7/2) = 7/2$.

(b) To find $|\pi - 6|$, note that $\pi \approx 3.14$, so $\pi - 6 < 0$. Hence, $|\pi - 6|$ is defined to be the *negative* of $\pi - 6$, that is,

$$|\pi - 6| = -(\pi - 6) = -\pi + 6.$$

(c) $|5 - \sqrt{2}| = 5 - \sqrt{2}$ because $5 - \sqrt{2} \geq 0$. ■

Here are the important facts about absolute value:

Properties of Absolute Value

1. $|c| \geq 0$ and $|c| > 0$ when $c \neq 0$.

2. $|c| = |-c|$

3. $|cd| = |c| \cdot |d|$

4. $\left|\dfrac{c}{d}\right| = \dfrac{|c|}{|d|}$ $(d \neq 0)$

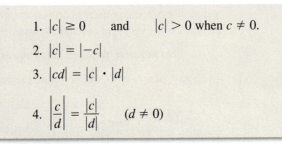

EXAMPLE 2

Here are examples of some of the properties listed in the box.

2. $|3| = 3$ and $|-3| = 3$, so $|3| = |-3|$.

3. If $c = 6$ and $d = -2$, then

$$|cd| = |6(-2)| = |-12| = 12,$$

and

$$|c| \cdot |d| = |6| \cdot |-2| = 6 \cdot 2 = 12,$$

so $|cd| = |c| \cdot |d|$.

4. If $c = -5$ and $d = 4$, then

$$\left|\frac{c}{d}\right| = \left|\frac{-5}{4}\right| = \left|-\frac{5}{4}\right| = \frac{5}{4} \quad \text{and} \quad \frac{|c|}{|d|} = \frac{|-5|}{|4|} = \frac{5}{4},$$

so $\left|\dfrac{c}{d}\right| = \dfrac{|c|}{|d|}$. ■

When c is a positive number, then $\sqrt{c^2} = c$, but when c is negative, this is *false*. For example, if $c = -3$, then

$$\sqrt{c^2} = \sqrt{(-3)^2} = \sqrt{9} = 3 \qquad (not\ -3),$$

so $\sqrt{c^2} \neq c$. In this case, however, $|c| = |-3| = 3$, so $\sqrt{c^2} = |c|$. The same thing is true for any negative number c. It is also true for positive numbers (since $|c| = c$ when c is positive). In other words,

Square Roots of Squares

For every real number c,

$$\sqrt{c^2} = |c|.$$

TECHNOLOGY

TIP

To find $|9 - 3\pi|$ on a calculator, key in

$$\text{ABS}(9 - 3\pi).$$

The ABS key is located in this menu/submenu:

TI: MATH/NUM

Casio: OPTN/NUM

HP-39+: Keyboard

When dealing with long expressions inside absolute value bars, do the computations inside first and then take the absolute value.

EXAMPLE 3

(a) $|5(2 - 4) + 7| = |5(-2) + 7| = |-10 + 7| = |-3| = 3.$
(b) $4 - |3 - 9| = 4 - |-6| = 4 - 6 = -2.$ ■

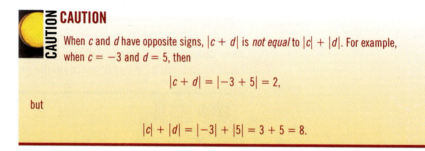

CAUTION

When c and d have opposite signs, $|c + d|$ is *not equal* to $|c| + |d|$. For example, when $c = -3$ and $d = 5$, then

$$|c + d| = |-3 + 5| = 2,$$

but

$$|c| + |d| = |-3| + |5| = 3 + 5 = 8.$$

The caution shows that $|c + d| < |c| + |d|$ when $c = -3$ and $d = 5$. In the general case, we have the following fact.

The Triangle Inequality

For any real numbers c and d,

$$|c + d| \leq |c| + |d|.$$

■■ DISTANCE ON THE NUMBER LINE

Observe that the distance from -5 to 3 on the number line is 8 units.

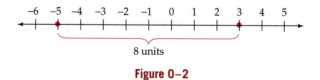

Figure 0–2

This distance can be expressed in terms of absolute value by noting that

$$|(-5) - 3| = 8.$$

That is, the distance is the *absolute value of the difference* of the two numbers. Furthermore, the order in which you take the difference doesn't matter; $|3 - (-5)|$ is also 8. This reflects the geometric fact that the distance from -5 to 3 is the same as the distance from 3 to -5. The same thing is true in the general case.

Distance on the Number Line

The distance between c and d on the number line is the number

$$|c - d| = |d - c|.$$

EXAMPLE 4

The distance from 4.2 to 9 is $|4.2 - 9| = |-4.8| = 4.8$, and the distance from 6 to $\sqrt{2}$ is $|6 - \sqrt{2}|$. ■

When $d = 0$, the distance formula shows that $|c - 0| = |c|$. Hence,

Distance to Zero

$|c|$ is the distance between c and 0 on the number line.

Algebraic problems can sometimes be solved by translating them into equivalent geometric problems. The key is to interpret statements involving absolute value as statements about distance on the number line.

EXAMPLE 5

Solve the equation $|x - 3| = 4$ geometrically.

SOLUTION   Translate the given equation into an equivalent geometric statement.

$$|x - 3| = 4$$

means

The distance between x and 3 on the number line is 4.

Figure 0–3 shows that -1 and 7 are the only numbers whose distance to 3 is 4 units:

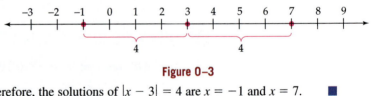

Figure 0–3

Therefore, the solutions of $|x - 3| = 4$ are $x = -1$ and $x = 7$. ■

EXAMPLE 6

Solve the equation $|x + 5| = 3$ geometrically.

SOLUTION   We rewrite it as $|x - (-5)| = 3$. In this form, it states that

The distance between x and -5 is 3 units.[*]

Figure 0–4 shows that -8 and -2 are the only two numbers whose distance to -5 is 3 units:

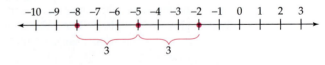

Figure 0–4

Thus, $x = -8$ and $x = -2$ are the solutions of $|x + 5| = 3$. ■

*It's necessary to rewrite the equation first because the distance formula involves the *difference* of two numbers, not their sum.

CALCULATOR

INVESTIGATIONS 0.1

1. EDIT AND REPLAY

Consider the equation $y = x^3 + 6x^2 - 5$.

(a) Find the value of y when $x = -7$ by keying in

$$(*) \qquad (-7)^3 + 6(-7)^2 - 5$$

and pressing ENTER.[†]

(b) To find the value of y when $x = 9$, without retyping the entire calculation, use the **replay** feature as follows:

Calculator	Procedure
TI	Press SECOND ENTER.
Casio	Press left or right arrow key *or* press AC and the up arrow key.
HP-39	Use up arrow key to shade the previous computation; then press COPY.

Your screen now returns to the previous calculation $(*)$. Use the left/right arrow keys and the DEL(ete) key to move through the equation and replace -7 by 9. Press ENTER, and the result of the new computation is displayed.

(c) Use the replay feature again to find the value of y when $x = 108$. On some calculators, you might need to use the INS(ert) key when replacing 9 by 108 to avoid typing over some of the computation.

(d) Press the replay keys repeatedly [on Casio, press AC and then press the up arrow repeatedly]. The calculator will display all the preceding computations, in reverse order. Go back to the first one $(*)$ and compute it again.

2. MATHEMATICAL OPERATIONS

(a) Key in each of the following, and explain why your answers are different. [See the Technology Tip on page 6.]

$$\text{ABS}(-)\,9 + 2\ \text{ENTER} \qquad \text{and} \qquad \text{ABS}\,(\,(-)\,9 + 2\,)\ \text{ENTER}$$

(b) Find INT or FLOOR in the NUM or REAL submenu of the MATH or OPTN menu. Find out what this command does when you follow it by a number and ENTER. Try all kinds of numbers (positive, negative, decimals, fractions). Answers on Casio will sometimes differ from those on other calculators.

3. SYMBOLIC CALCULATIONS

(a) To store the number 2 in memory A of a calculator, type

$$2\ \ \text{STO}\,\blacktriangleright\ \ \text{A}\ \ \ \ \text{ENTER}\ \ \ \ \ \text{[TI-85/86] or}$$

$$2\ \ \text{STO}\,\blacktriangleright\ \ \text{ALPHA}\ \ \text{A}\ \ \text{ENTER}\ \ \ \ \ \text{[TI-83+/89, HP-39+] or}$$

$$2 \rightarrow \text{ALPHA}\ \ \text{A}\ \ \text{EXE}\ \ \ \ \ \text{[Casio].}$$

If you now key in ALPHA   A   ENTER, what does the calculator display?

[†]Here and throughout the book, Casio users should read "EXE" in place of "ENTER."

(b) In a similar fashion, store the number 5 in memory B and -10 in memory C. Then, using the ALPHA keys, display this expression on the screen: B + C/A. If you press ENTER, what happens? Explain what the calculator is doing.

(c) Experiment with other expressions, such as $B^2 - 4AC$.

4. INEQUALITIES

Find out what happens when you key in each of these statements and press ENTER:

$$8 < 9, \quad 8 < 5, \quad 9 > 2, \quad 9 > 10.$$

[On TI and HP, the inequality symbols are located in the TEST(S) or INEQ (sub)menu; look on the keyboard or in the MATH menu. On Casio they are in the REL or LOGIC submenu of the PRGM menu, which is accessed from the home screen by pressing SHIFT VARS.]

✓ EXERCISES 0.1

1. Draw a number line and mark the location of each of these numbers: 0, -7, 8/3, 10, -1, -4.75, 1/2, -5, and 2.25.

2. (a) Use the π key on your calculator to find a decimal approximation of π.
   (b) Use your calculator to determine which of the following rational numbers is the best approximation of the irrational number π.

   $$\frac{22}{7}, \quad \frac{355}{113}, \quad \frac{103,993}{33,102}, \quad \frac{2,508,429,787}{798,458,000}.$$

   If your calculator says that one of these numbers equals π, it's lying. All you can conclude is that the number agrees with π for as many decimal places as your calculator can handle (usually 12–14).

In Exercises 3–14, express the given statement in symbols.

3. 12 is greater than 9.

4. 5 is less than 7.

5. -4 is greater than -8.

6. -17 is less than 14.

7. π is less than 100.

8. x is nonnegative.

9. y is less than or equal to 7.5.

10. z is greater than or equal to -4.

11. t is positive.

12. d is not greater than 2.

13. c is at most 3.

14. z is at least -17.

In Exercises 15–20, fill the blank with $<$, $=$, or $>$ so that the resulting statement is true.

15. -6 _____ -2

16. 5 _____ -3

17. 3/4 _____ .75

18. 3.1 _____ π

19. 1/3 _____ .33.

20. 2 _____ $\sqrt{2}$

The consumer price index for urban consumers (CPI-U) measures the cost of consumer goods and services such as food, housing, transportation, medical costs, etc. The table shows the yearly percentage increase in the CPI-U over a decade.[*]

Year	Percentage change
1993	3.0
1994	2.6
1995	2.8
1996	3.0
1997	2.3
1998	1.6
1999	2.2
2000	3.4
2001	2.8
2002	1.6

[*]U.S. Bureau of Labor Statistics.

In Exercises 21–25, let p denote the yearly percentage increase in the CPI-U. Find the number of years in this period which satisfied the given inequality.

21. $p \geq 2.8$ **22.** $p < 2.6$ **23.** $p > 2.3$

24. $p \leq 3.0$ **25.** $p > 3.4$

26. Galileo discovered that the period of a pendulum depends only on the length of the pendulum and the acceleration of gravity. The period T of a pendulum (in seconds) is

$$T = 2\pi \sqrt{\frac{l}{g}},$$

where l is the length of the pendulum in feet and $g \approx 32.2$ ft/sec^2 is the acceleration due to gravity. Find the period of a pendulum whose length is 4 ft.

In Exercises 27–28, use a calculator and list the given numbers in order from smallest to largest.

27. $\dfrac{189}{37}$, $\dfrac{4587}{691}$, $\sqrt{47}$, 6.735, $\sqrt{27}$, $\dfrac{2040}{523}$

28. $\dfrac{385}{177}$, $\sqrt{10}$, $\dfrac{187}{63}$, π, $\sqrt{\sqrt{85}}$, 2.9884

In Exercises 29–34, fill the blank so as to produce two equivalent statements. For example, the arithmetic statement "a is negative" is equivalent to the geometric statement "the point a lies to the left of the point 0."

Arithmetic Statement	**Geometric Statement**
29. _____	a lies c units to the right of b.
30. _____	a lies between b and c.
31. $a - b > 0$	_____
32. a is positive.	_____
33. _____	a lies to the left of b.
34. $a \geq b$	_____

In Exercises 35–42, simplify and write the given number without using absolute values.

35. $3 - |2 - 5|$ **36.** $-2 - |-2|$

37. $|6 - 4| + |-3 - 5|$ **38.** $|-6| - |6|$

39. $|(-13^2)|$ **40.** $-|-5|^2$

41. $|\pi - \sqrt{2}|$ **42.** $|\sqrt{2} - 2|$

In Exercises 43–48, fill the blank with $<$, $=$, or $>$ so that the resulting statement is true.

43. $|-2|$ ____ $|-5|$ **44.** 5 ____ $|-2|$

45. $|3|$ ____ $-|4|$ **46.** $|-3|$ ____ 0

47. -7 ____ $|-1|$ **48.** $-|-4|$ ____ 0

In Exercises 49–54, find the distance between the given numbers.

49. -3 and 4 **50.** 7 and 107

51. -7 and $15/2$ **52.** $-3/4$ and -10

53. π and 3 **54.** π and -3

55. Ann rode her bike down the Blue Ridge Parkway, beginning at Linville Falls (mile post 316.3) and ending at Mount Pisgah (mile post 408.6). Use absolute value notation to describe how far Ann rode.

56. A broker predicts that over the next six months, the price p of a particular stock will not vary from its current price of $25.75 by more than $4. Express this prediction as an inequality.

57. According to data from the Center for Science in the Public Interest, the healthy weight range for a person depends on the person's height. For example,

Height	Healthy Weight Range (lb)
5 ft 8 in.	143 ± 21
6 ft 0 in.	163 ± 26

Express each of these ranges as an absolute value inequality in which x is the weight of the person.

The wind-chill factor, shown in the table, calculates how a given temperature feels to a person's skin when the wind is taken into account. For example, the table shows that a temperature of 20° in a 40 mph wind feels like $-1°$.[*]

				Wind (mph)					
Calm	**5**	**10**	**15**	**20**	**25**	**30**	**35**	**40**
40	36	34	32	30	29	28	28	27
30	25	21	19	17	16	15	14	13
20	13	9	6	4	3	1	0	−1
10	1	−4	−7	−9	−11	−12	−14	−15
0	−11	−16	−19	−22	−24	−26	−27	−29
−10	−22	−28	−32	−35	−37	−39	−41	−43
−20	−34	−41	−45	−48	−51	−53	−55	−57
−30	−46	−53	−58	−61	−64	−67	−69	−71
−40	−57	−66	−71	−74	−78	−80	−82	−84

Temperature (°F)

In Exercises 58–60, find the absolute value of the difference of the two given wind-chill factors. For example, the difference between the wind-chill at 30° with a 15 mph wind and one at $-10°$ with a 10 mph wind is $|19 - (-28)| = 47°$ or, equivalently, $|-28 - 19| = 47°$.

58. 10° with a 25 mph wind and 20° with a 20 mph wind

_____

[*]Table from the Joint Action Group for Temperature Indices, 2001.

59. 30° with a 10 mph wind and 10° with a 30 mph wind

60. −30° with a 5 mph wind and 0° with a 10 mph wind

61. The use of digital devices (cell phones, DVD players, PCs, etc.) continues to grow. The graph shows the approximate number of digital devices (in billions) in use worldwide from 2000 to 2004.[*]

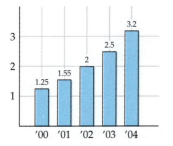

In what years was the following statement true:

$$|x - 2{,}000{,}000{,}000| \geq 500{,}000{,}000,$$

where x is the number of digital devices in use in that year?

62. At Statewide Insurance, each department's expenses are reviewed monthly. A department can fail to pass the budget variance test in a category if either (i) the absolute value of the difference between actual expenses and the budget is more than $500 or (ii) the absolute value of the difference between the actual expenses and the budget is more than 5% of the budgeted amount. Which of the following items fail the budget variance test? Explain your answers.

Item	Budgeted Expense ($)	Actual Expense ($)
Wages	220,750	221,239
Overtime	10,500	11,018
Shipping and Postage	530	589

In Exercises 63–68, write the given expression without using absolute values.

63. $|t^2|$

64. $|u^2 + 2|$

---

[*]Data, estimates, and projections from IDC.

65. $|(-3 - y)^2|$

66. $|-2 - y^2|$

67. $|b - 3|$ if $b \geq 3$

68. $|a - 5|$ if $a < 5$

In Exercises 69–74, express the given geometric statement about numbers on the number line algebraically, using absolute values.

69. The distance from x to 5 is less than 4.

70. x is more than 6 units from c.

71. x is at most 17 units from −4.

72. x is within 3 units of 7.

73. c is closer to 0 than b is.

74. x is closer to 1 than to 4.

In Exercises 75–78, translate the given algebraic statement into a geometric statement about numbers on the number line.

75. $|x - 3| < 2$

76. $|x - c| > 6$

77. $|x + 7| \leq 3$

78. $|u + v| \geq 2$

In Exercises 79–84, use the geometric approach explained in the text to solve the given equation.

79. $|x| = 1$

80. $|x| = \dfrac{3}{2}$

81. $|x - 2| = 1$

82. $|x + 3| = 2$

83. $|x + \pi| = 4$

84. $\left|x - \dfrac{3}{2}\right| = 5$

Thinkers

In Exercises 85 and 86, explain why the given statement is true for any numbers c and d. [Hint: Look at the boxes on pages 5 and 6.]

85. $|(c - d)^2| = c^2 - 2cd + d^2$

86. $\sqrt{9c^2 - 18cd + 9d^2} = 3|c - d|$

87. Explain why the statement $|a| + |b| + |c| > 0$ is algebraic shorthand for "at least one of the numbers a, b, c, is different from zero."

88. Find an algebraic shorthand version of the statement "the numbers a, b, c, are all different from zero."

O.1.A   *SPECIAL TOPICS* Decimal Representation of Real Numbers

Every rational number can be expressed as a terminating or repeating decimal. For instance, $3/4 = 0.75$. To express $15/11$ as a decimal, divide the numerator by the denominator:

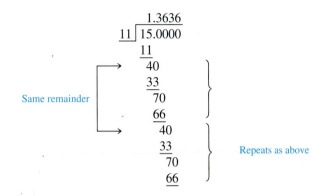

Since the remainder at the first step (namely, 4) occurs again at the third step, it is clear that the division process goes on forever with the two-digit block "36" repeating over and over in the quotient $15/11 = 1.3636363636 \cdots$.

The method used in the preceding example can be used to express any rational number as a decimal. During the division process, some remainder *necessarily repeats*. If the remainder at which this repetition starts is 0, the result is a repeating decimal ending in zeros—that is, a terminating decimal (for instance, $.75000 \cdots = .75$). If the remainder at which the repetition starts is nonzero, then the result is a nonterminating repeating decimal, as in the example above.

Conversely, there is a simple method for converting any repeating decimal into a rational number.

TECHNOLOGY TIP

To convert repeating decimals to fractions automatically on TI, use FRAC in this menu/submenu:

TI-83+: MATH

TI-86: MATH/MISC

On HP-39+, select FRACTION number format in the MODE menu; then enter the decimal.

On Casio FX 2.0, press the *a b/c* key after a calculation has been run to change the answer to a mixed number.*

On other Casio models, use the FRAC program in the Program Appendix.

EXAMPLE 1

Write $d = .272727 \cdots$ as a rational number.

SOLUTION   Assuming that the usual rules of arithmetic hold, we see that

$$100d = 27.272727 \cdots \qquad \text{and} \qquad d = .272727 \cdots.$$

Now subtract d from $100d$:

$$
\begin{array}{r}
100d = 27.272727 \cdots \\
-d = -.272727 \cdots \\
\hline
99d = 27
\end{array}
$$

Dividing both sides of this last equation by 99 shows that $d = 27/99 = 3/11$. ∎

---

*The calculator denotes $12\frac{3}{4}$, for example, by 12⌐3⌐4 and denotes 15/29 by 15⌐29.

▚ IRRATIONAL NUMBERS

Many nonterminating decimals are *nonrepeating* (that is, no block of digits repeats forever), such as .202002000200002 · · · (where after each 2 there is one more zero than before). Although the proof is too long to give here, it is in fact true that every nonterminating and nonrepeating decimal represents an *irrational* real number. Conversely every irrational number can be expressed as a nonterminating and nonrepeating decimal (no proof to be given here).

A typical calculator can hold only the first 10–14 digits of a number in decimal form. Consequently, a calculator can contain the *exact* value only of those rational numbers whose decimal expansion terminates after 10–14 places. It must *approximate* all other real numbers.

Since every real number is either a rational number or an irrational one, the preceding discussion can be summarized as follows.

Decimal Representation

1. Every real number can be expressed as a decimal.

2. Every decimal represents a real number.

3. The terminating decimals and the nonterminating repeating decimals are the rational numbers.

4. The nonterminating, nonrepeating decimals are the irrational numbers.

CALCULATOR INVESTIGATIONS 0.1.A

1. FRAC KEY

If your calculator has a FRAC key or fraction conversion program (see the Program Appendix), test its limitations by entering each of the following numbers and then pressing the FRAC key.

(a) .058823529411   (b) .0588235294117

(c) .058823529411724   (d) .0588235294117985

Which of your answers are correct? [*Hint:* Use Exercise 25 below to find the decimal expansion of 1/17.]

✓ EXERCISES 0.1.A 📼

In Exercises 1–6, express the given rational number as a repeating decimal.

1. 7/9   **2.** 2/13   **3.** 23/14

4. 19/88   **5.** 1/19 (long)   **6.** 9/11

In Exercises 7–13, express the given repeating decimal as a fraction.

7. .373737 · · ·   **8.** .929292 · · ·

9. 76.63424242 · · · [*Hint:* Consider 10,000d − 100d, where d = 76.63424242 · · · .]

10. 13.513513 · · · [*Hint:* Consider 1000d − d, where d = 13.513513 · · · .]

11. .135135135 · · · [*Hint:* See Exercise 10.]

12. .33030303 · · ·   **13.** 52.31272727 · · ·

14. If two real numbers have the same decimal expansion through three decimal places, how far apart can they be on the number line?

In Exercises 15–22, state whether a calculator can express the given number exactly.

15. 2/3   **16.** 7/16   **17.** 1/64   **18.** 1/22

19. 3π/2   **20.** π − 3   **21.** 1/.625   **22.** 1/.16

Thinkers

23. Use the methods in Exercises 7–13 to show that both .74999 ··· and .75000 ··· are decimal expansions of 3/4. [Every terminating decimal can also be expressed as a decimal ending in repeated 9's. It can be proved that these are the only real numbers with more than one decimal expansion.]

Finding remainders with a calculator.

24. If you use long division to divide 369 by 7, you obtain

$$
\begin{array}{r}
52 \quad \leftarrow \text{Quotient} \\
\text{Divisor} \rightarrow \quad 7\,\overline{)369} \quad \leftarrow \text{Dividend} \\
\underline{35} \\
19 \\
\underline{14} \\
5 \quad \leftarrow \text{Remainder}
\end{array}
$$

If you use a calculator to find $369 \div 7$, the answer is displayed as 52.71428571. Observe that the integer part of this calculator answer, 52, is the quotient when you do the problem by long division. The usual "checking procedure" for long division shows that

$$7 \cdot 52 + 5 = 369 \quad \text{or, equivalently} \quad 369 - 7 \cdot 52 = 5.$$

Thus, the remainder is

$$\text{Dividend} - (\text{divisor})\left(\begin{array}{c}\text{integer part of} \\ \text{calculator answer}\end{array}\right).$$

Use this method to find the quotient and remainder in these problems:
(a) $5683 \div 9$ (b) $1,000,000 \div 19$
(c) $53,000,000 \div 37$

In Exercises 25–30, find the decimal expansion of the given rational number. All these expansions are too long to fit in a calculator but can be readily found by using the hint in Exercise 25.

25. 1/17 [*Hint:* The first part of dividing 1 by 17 involves working this division problem: $1,000,000 \div 17$. The method of Exercise 24 shows that the quotient is 58,823 and the remainder is 9. Thus the decimal expansion of 1/17 begins .058823, and the next block of digits in the expansion will be the quotient in the problem $9,000,000 \div 17$. The remainder when 9,000,000 is divided by 17 is 13, so the next block of digits in the expansion of 1/17 is the quotient in the problem $13,000,000 \div 17$. Continue in this way until the decimal expansion repeats.]

26. 3/19   **27.** 1/29   **28.** 3/43   **29.** 283/47

30. 768/59

31. (a) Show that there are at least as many irrational numbers (nonrepeating decimals) as there are terminating decimals. [*Hint:* With each terminating decimal associate a nonrepeating decimal.]
(b) Show that there are at least as many irrational numbers as there are repeating decimals. [*Hint:* With each repeating decimal, associate a nonrepeating decimal by inserting longer and longer strings of zeros: for instance, with .11111111 ··· associate the number .101001000100001 ····.]

0.2   Integral Exponents

Exponents provide a convenient shorthand for certain products. If c is a real number, then c^2 denotes cc and c^3 denotes ccc. More generally, for any positive integer n,

$$c^n \text{ denotes the product } ccc \cdot \cdot \cdot c \qquad (n \text{ factors}).$$

In this notation, c^1 is just c, so we usually omit the exponent 1.

EXAMPLE 1

$3^4 = 3 \cdot 3 \cdot 3 \cdot 3 = 81$ and

$$(-2)^5 = (-2)(-2)(-2)(-2)(-2) = -32.$$

For every positive integer n, $0^n = 0 \cdot \cdot \cdot 0 = 0$. ∎

EXAMPLE 2

To find $(2.4)^9$, use the $\wedge$ (or a^b or x^y) key on your calculator:

$$2.4 \wedge 9 \text{ ENTER}$$

which produces the (approximate) answer 2641.80754.   ■

CAUTION

Be careful with negative bases. For instance, if you want to compute $(-12)^4$, which is a positive number, but you key in $(-)$ 12 $\wedge$ 4 ENTER, the calculator will interpret this as $-(12^4)$ and produce a negative answer. To get the correct answer, you must key in the parentheses:

$$((-)12) \wedge 4 \text{ ENTER}.$$

Because exponents are just shorthand for multiplication, it is easy to determine the rules they obey. For instance,

$$c^3 c^5 = (ccc)(ccccc) = c^8, \qquad \text{that is,} \qquad c^3 c^5 = c^{3+5}.$$

The same thing works in general.

Multiplication with Exponents

To multiply c^m by c^n, add the exponents:

$$c^m \cdot c^n = c^{m+n}.$$

For division, we have, for example,

$$\frac{c^7}{c^4} = \frac{ccccccc}{cccc} = ccc = c^3, \qquad \text{that is,} \qquad \frac{c^7}{c^4} = c^{7-4}.$$

In general, we have the following.

Division with Exponents

To divide c^m by c^n, subtract the exponents:

$$\frac{c^m}{c^n} = c^{m-n}.$$

Finally, note that

$$(c^2)^3 = (cc)^3 = (cc)(cc)(cc) = c^6, \qquad \text{that is,} \qquad (c^2)^3 = c^{2 \cdot 3},$$

which suggests the following rule.

Power of a Power

To find a power of a power $(c^m)^n$, multiply the exponents.

$$(c^m)^n = c^{mn}.$$

EXAMPLE 3

(a) To find $4^2 \cdot 4^7$, we note that this is multiplication and *add* the exponents:
$4^2 \cdot 4^7 = 4^{2+7} = 4^9$.

(b) To divide $2^8/2^3$, we *subtract* the exponents: $2^8/2^3 = 2^{8-3} = 2^5$.

(c) To compute $(3^5)^4$, which is a power of a power, we *multiply* the exponents:
$(3^5)^4 = 3^{5 \cdot 4} = 3^{20}$.   ■

When an exponent applies to the product or quotient of two numbers, use the definitions carefully. For instance,

$$(7 \cdot 19)^3 = (7 \cdot 19)(7 \cdot 19)(7 \cdot 19) = 7 \cdot 7 \cdot 7 \cdot 19 \cdot 19 \cdot 19 = 7^3 \cdot 19^3$$

In the general case, we have the following.

Product to a Power

To find $(cd)^n$, apply the exponent to each term in parentheses.

$$(cd)^n = c^n d^n.$$

CAUTION

A common mistake is to write an expression such as $(2x)^5$ incorrectly as $2x^5$, that is, to apply the exponent only to the second term in parentheses. Using the property in the box, however, we see that

$$(2x)^5 = 2^5 x^5 = 32x^5.$$

A similar conclusion holds for division.

Quotient to a Power

To find $(c/d)^n$, apply the exponent to both numerator and denominator:

$$\left(\frac{c}{d}\right)^n = \frac{c^n}{d^n}.$$

For example,

$$\left(\frac{x}{7}\right)^2 = \frac{x}{7} \cdot \frac{x}{7} = \frac{x^2}{7^2} = \frac{x^2}{49}.$$

CAUTION

The preceding rules do *not* apply to situations such as $2^4 \cdot 3^5$ or $4^7/5^2$, where powers of two different numbers are being multiplied or divided.

Next we consider exponents other than positive integers. The symbol c^0 has no obvious meaning (what would it mean to multiply c by itself 0 times?). On the other hand, consider the rule for multiplying with exponents. For this rule to hold with a zero exponent, we would have to have

$$c^5 c^0 = c^{5+0} = c^5.$$

On the other hand, we know that

$$c^5 \cdot 1 = c^5,$$

so it is reasonable to *define* c^0 to be 1.

Zero Exponent

If $c \neq 0$, then c^0 is defined to be the number 1.

Similarly, for the multiplication rule to hold for negative exponents we would have to have, for example,

$$c^5 c^{-5} = c^{5-5} = c^0 = 1.$$

But we know that

$$c^5 \cdot \frac{1}{c^5} = 1,$$

so it is also reasonable to define c^{-5} to be $1/c^5$. More generally, if $c \neq 0$ and n is a positive integer, we have the following.

Negative Exponents

c^{-n} is defined to be the number $\dfrac{1}{c^n}$.

Note that 0^0 and negative powers of 0 are not defined (negative powers of 0 would involve division by 0).

EXAMPLE 4

$6^{-3} = 1/6^3 = 1/216$ and $(-2)^{-5} = 1/(-2)^5 = -1/32$.
A calculator shows that $(.287)^{-12} \approx 3{,}201{,}969.857$.   ∎

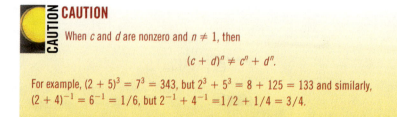

CAUTION

When c and d are nonzero and $n \neq 1$, then

$$(c + d)^n \neq c^n + d^n.$$

For example, $(2 + 5)^3 = 7^3 = 343$, but $2^3 + 5^3 = 8 + 125 = 133$ and similarly, $(2 + 4)^{-1} = 6^{-1} = 1/6$, but $2^{-1} + 4^{-1} = 1/2 + 1/4 = 3/4$.

Because of the way zero and negative exponents are defined,

The multiplication, division, and power rules for exponents in the boxes above hold for all integer exponents (positive, negative, or zero),

as illustrated in the following examples.

EXAMPLE 5

(a) By the multiplication and division rules,

$$\pi^{-5}\pi^2 = \pi^{-5+2} = \pi^{-3} = 1/\pi^3 \quad \text{and} \quad x^9/x^4 = x^{9-4} = x^5.$$

(b) The power of a power rule shows that $(5^{-3})^2 = 5^{(-3)\cdot 2} = 5^{-6}$.

(c) The power rule for quotients and the power of a power rule show that

$$\left(\frac{x^2}{3}\right)^4 = \frac{(x^2)^4}{3^4} = \frac{x^8}{81}.$$

(d) By the negative exponent rule,

$$\frac{1}{x^{-5}} = \frac{1}{1/x^5} = x^5. \quad \blacksquare$$

The exponent rules can often be used to simplify complicated expressions.

EXAMPLE 6

Simplify:

(a) $(2x^2y^3z)^4$ (b) $(r^{-3}s^2)^{-2}$ (c) $\dfrac{x^5(y^2)^3}{(x^2y)^2}$

SOLUTION

(a)   Product to a power: $(2x^2y^3z)^4 = 2^4(x^2)^4(y^3)^4z^4$

     Power of a power: $= 16x^8y^{12}z^4$

(b)   Product to a power: $(r^{-3}s^2)^{-2} = (r^{-3})^{-2}(s^2)^{-2}$

     Power of a power: $= r^6s^{-4} = \dfrac{r^6}{s^4}$

(c)   Power of a power: $\dfrac{x^5(y^2)^3}{(x^2y)^2} = \dfrac{x^5y^6}{(x^2y)^2}$

     Product to a power: $= \dfrac{x^5y^6}{(x^2)^2y^2}$

     Power of a power: $= \dfrac{x^5y^6}{x^4y^2}$

     Division with exponents: $= x^{5-4}y^{6-2} = xy^4 \quad \blacksquare$

Using the exponent laws with negative exponents is usually more efficient than first converting to positive exponents. That conversion, if necessary, can always be done at the end.

EXAMPLE 7

Simplify and express without negative exponents:

$$\frac{a^{-2}(b^2c^3)^{-2}}{(a^{-3}b^{-5})^2c}.$$

SOLUTION   Product to a power:

$$\frac{a^{-2}(b^2c^3)^{-2}}{(a^{-3}b^{-5})^2c} = \frac{a^{-2}(b^2)^{-2}(c^3)^{-2}}{(a^{-3})^2(b^{-5})^2c}$$

Power of a power:

$$= \frac{a^{-2}b^{-4}c^{-6}}{a^{-6}b^{-10}c}$$

Division with exponents:

$$= a^{-2-(-6)}b^{-4-(-10)}c^{-6-1}$$

Negative exponent:

$$= a^4b^6c^{-7} = \frac{a^4b^6}{c^7}$$ ∎

▪▪ SCIENTIFIC NOTATION

A positive real number is said to be in **scientific notation** when it is written in the form

$$a \times 10^n \qquad \text{where } 1 \le a < 10 \text{ and } n \text{ is an integer.}$$

EXAMPLE 8

Write each of these numbers in scientific notation:

(a) 356   (b) 1,564,000   (c) .072   (d) .00000087

SOLUTION   In each case, move the decimal point to the left or right to obtain a number between 1 and 10; then write the original number in scientific notation as follows.

(a) $356 = 3.56 \times 100 = 3.56 \times 10^2$ [Decimal point is moved 2 places to the *left*, and 10 is raised to the power of 2.]

(b) $1,564,000 = 1.564 \times 1,000,000 = 1.564 \times 10^6$ [Decimal point is moved 6 places to the *left*, and 10 is raised to the 6th power.]

(c) $.072 = 7.2 \times \dfrac{1}{100} = 7.2 \times 10^{-2}$ [Decimal point is moved 2 places to the *right*, and 10 is raised to −2.]

(d) $.00000087 = 8.7 \times \dfrac{1}{10,000,000} = 8.7 \times 10^{-7}$ [Decimal point is moved 7 places to the *right*, and 10 is raised to −7.]   ∎

Scientific notation is useful for computations with very large or very small numbers.

EXAMPLE 9

$$(.00000002)(4,300,000,000) = (2 \times 10^{-8})(4.3 \times 10^9)$$
$$= 2(4.3)10^{-8+9} = (8.6)10^1 = 86.$$ ∎

When calculators display numbers in scientific notation, they omit the 10 and use an E to indicate the exponent. For instance,

$$7.235 \times 10^{-12} \qquad \text{is displayed as} \qquad 7.235 \text{ E} -12.$$

To enter a number in scientific notation, for example, 5.6×10^{73}, type in 5.6, then press the EE key (labeled EEX or EXP or 10^x on some calculators) and type in

73. Calculators automatically switch to scientific notation whenever a number is too large or too small to be displayed in the standard way. If you try to enter a number with more digits than the calculator can handle, such as 45,000,000,333,222,111, a typical calculator will approximate it using scientific notation as 4.500000033 E 16, that is, as 45,000,000,330,000,000.

EXAMPLE 10

Suppose you are located h feet above the ground. Because of the curvature of the earth, the maximum distance you can see is approximately d miles, where

$$d = \sqrt{1.5h + (3.587 \times 10^{-8})h^2}.$$

How far can you see from the 500-ft-high Smith Tower in Seattle and from the 1454-ft-high Sears Tower in Chicago?

SOLUTION　For the Smith Tower, substitute 500 for h in the formula, and use your calculator:

$$d = \sqrt{1.5(500) + (3.587 \times 10^{-8})500^2} \approx 27.4 \text{ miles.}$$

For the Sears Tower, you can see almost 47 miles because

$$d = \sqrt{1.5(1454) + (3.587 \times 10^{-8})1454^2} \approx 46.7 \text{ miles.} \quad \blacksquare$$

EXERCISES 0.2

In Exercises 1–12, evaluate the expression.

1. $(-6)^2$

2. -6^2

3. $5 + 4(3^2 + 2^3)$

4. $(-3)2^2 + 4^2 - 1$

5. $\dfrac{(-3)^2 + (-2)^4}{-2^2 - 1}$

6. $\dfrac{(-4)^2 + 2}{(-4)^2 - 7} + 1$

7. $2^4 - 2^7$

8. $3^3 - 3^{-7}$

9. $(2^{-2} + 2)^2$

10. $(3^{-1} - 3^3)^2$

11. $\dfrac{1}{2^3} + \dfrac{1}{2^{-4}}$

12. $3^2\left(\dfrac{1}{3} + \dfrac{1}{3^{-2}}\right)$

In Exercises 13–28, simplify the expression. Each letter represents a nonzero real number and should appear at most once in your answer.

13. $x^2 \cdot x^3 \cdot x^5$

14. $y \cdot y^4 \cdot y^6$

15. $(.03)y^2 \cdot y^7$

16. $(1.3)z^3 \cdot z^5$

17. $(2x^2)^3 3x$

18. $(3y^3)^4 5y^2$

19. $(3x^2y)^2$

20. $(2xy^3)^3$

21. $(2w)^3(3w)(4w)^2$

22. $(3d)^4(2d)^2(5d)$

23. $a^{-2}b^3a^3$

24. $c^4d^5c^{-3}$

25. $(2x)^{-2}(2y)^3(4x)$

26. $(3x)^{-3}(2y)^{-2}(2x)$

27. $(2x^2y)^0(3xy)$

28. $(3x^2y^4)^0$

In Exercises 29–32, express the given number as a power of 2.

29. $(64)^2$

30. $\left(\dfrac{1}{8}\right)^3$

31. $(2^4 \cdot 16^{-2})^3$

32. $\left(\dfrac{1}{2}\right)^{-8}\left(\dfrac{1}{4}\right)^4\left(\dfrac{1}{16}\right)^{-3}$

In Exercises 33–44, simplify and write the given expression without negative exponents. All letters represent nonzero real numbers.

33. $\dfrac{x^4(x^2)^3}{x^3}$

34. $\left(\dfrac{z^2}{t^3}\right)^4 \cdot \left(\dfrac{z^3}{t}\right)^5$

35. $\left(\dfrac{e^6}{c^4}\right)^2 \cdot \left(\dfrac{c^3}{e}\right)^3$

36. $\left(\dfrac{x^7}{y^6}\right)^2 \cdot \left(\dfrac{y^2}{x}\right)^4$

37. $\left(\dfrac{a^6}{b^{-4}}\right)^2$

38. $\left(\dfrac{x^{-2}}{y^{-2}}\right)^2$

39. $\left(\dfrac{c^5}{d^{-3}}\right)^{-2}$

40. $\left(\dfrac{x^{-1}}{2y^{-1}}\right)\left(\dfrac{2y}{x}\right)^{-2}$

41. $\dfrac{(a^{-3}b^2c)^{-2}}{(ab^{-2}c^3)^{-1}}$

42. $\dfrac{(-2cd^2e^{-1})^3}{(5c^{-3}de)^{-2}}$

43. $(c^{-1}d^{-2})^{-3}$

44. $[(x^2y^{-1})^2]^{-3}$

In Exercises 45–50, determine the sign of the given number without calculating the product.

45. $(-2.6)^3(-4.3)^{-2}$

46. $(4.1)^{-2}(2.5)^{-3}$

47. $(-1)^9(6.7)^5$

48. $(-4)^{12}6^9$

49. $(-3.1)^{-3}(4.6)^{-6}(7.2)^7$

50. $(45.8)^{-7}(-7.9)^{-9}(-8.5)^{-4}$

In Exercises 51–54, r, s, and t are positive integers and a, b, and c are nonzero real numbers. Simplify and write the given expression without negative exponents.

51. $\dfrac{3^{-r}}{3^{-s-r}}$

52. $\dfrac{4^{-(t+1)}}{4^{2-t}}$

53. $\left(\dfrac{a^6}{b^{-4}}\right)^t$

54. $\dfrac{c^{-t}}{(6b)^{-s}}$

In Exercises 55–60, express the given numbers (based on 2003 estimates) in scientific notation.

55. Population of the world: 6,275,000,000

56. Population of the United States: 290,300,000

57. Mean distance from the earth to Pluto: 5,910,000,000,000 meters

58. Radius of a hydrogen atom: .00000000001 meter

59. Width of a DNA double helix: .000000002 meter

60. Diameter of an electron: .0000000000004 centimeter

In Exercises 61–64, express the given number in normal decimal notation.

61. Speed of light in a vacuum: 2.9979×10^8 miles per second

62. Average distance from the earth to the sun: 1.50×10^{11} meters

63. Electron charge: 1.602×10^{-27} coulomb

64. Proton mass: 1.6726×10^{-19} kilogram

In Exercises 65–72, express your answer in scientific notation.

64. One light-year is the distance light travels in a 365-day year. The speed of light is about 186,282.4 miles per second.

(a) How long is 1 light-year (in miles)?

(b) Light from the North Star takes 680 years to reach the earth. How many miles is the North Star from the earth?

66. The gross federal debt was 6412 billion dollars in February 2003, when the U.S. population was approximately 289.2 million people.

(a) Express the debt and population in scientific notation.

(b) At that time, what was each person's share of the debt?

67. If a single gene has an average weight of 10^{-17} gram, how many genes would you expect in an organism having cells with a DNA weight of 10^{-12} gram?

68. Johannes Kepler discovered that the ratio R^3/T^2 is the same for the moon and other objects that orbit the earth, where R is the object's average distance from the earth and T is the object's orbital period. The moon is 240,000 miles from the earth and makes one complete orbit every 27 days. Use Kepler's discovery to find the orbital period of a satellite orbiting the earth at a distance of 900,000 miles.

69. Assume that the earth is a sphere with a radius of 6400 kilometers. What is the surface area of the earth in square kilometers? [*Hint:* The surface area S of a sphere of radius r is given by $S = 4\pi r^2$.]

70. If the circumference of the earth at the equator is 25,000 miles, how many inches is it around the equator?

71. Assume that light travels at 1.86×10^5 miles per sec. Sound travels at 1.1×10^3 feet per sec. Elizabeth is at a rock concert sitting 100 ft from the band. Her friend back home—1000 miles away—is listening to the concert on the radio (radio waves travel at the speed of light). Who hears the music first?

72. Australia has 7,617,930 square kilometers of land area, and in 2005, it is estimated to have 19,547,000 people, Singapore has 622.6 square kilometers of land area, and in 2005, it is estimated to have 4,553,000 people.

(a) Express Australia's and Singapore's populations and land areas in scientific notation.

(b) What is the population density (people per square kilometer) for Australia? For Singapore?

73. After five years an investment of $1000 at interest rate r, compounded annually, will be worth $1000(1 + r)^5$, where r is written as a decimal. Use a calculator to fill in this table.

r	$1000(1 + r)^5$
$3\frac{1}{2}\%$	
5%	
$7\frac{1}{2}\%$	
10%	

74. Suppose you are k miles (not feet) above the ground. The radius of the earth is approximately 3960 miles, and the point where your line of sight meets the earth is perpendicular to the radius of the earth at that point, as shown in the figure.

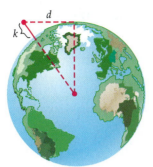

(a) Use the Pythagorean Theorem (see the Geometry Review Appendix) to show that

$$d = \sqrt{(3960 + k)^2 - 3960^2}.$$

(b) Show that the equation in part (a) simplifies to $d = \sqrt{7920k + k^2}$.

(c) If you are h feet above the ground, then you are $h/5280$ miles high. Why? Use this fact and the equation in part (b) to obtain the formula used in Example 10.

Errors to Avoid

In Exercises 75–78, give a numerical example to show that the statement is false for some numbers.

75. $a^r a^s = a^{rs}$

76. $a^r b^s = (ab)^{r+s}$

77. $c^{-r} = -c^r$

78. $\dfrac{c^r}{c^s} = c^{r/s}$

0.3   Roots, Radicals, and Rational Exponents

The square root of 5 is the positive number $\sqrt{5}$ whose square is 5. Similarly, the fourth root of 5, denoted $\sqrt[4]{5}$, is the positive number whose fourth power is 5, and in the general case:

Even Roots

> If n is an even positive integer and c is any nonnegative real number, then
>
> $\sqrt[n]{c}$ denotes the nonnegative number whose nth power is c.*
>
> $\sqrt[n]{c}$ is called the **nth root** of c.

For example, $\sqrt[4]{81} = 3$ because $3^4 = 81$ and $\sqrt[6]{64} = 2$ because $2^6 = 64$. When $n = 2$, we shall continue to use the ordinary radical $\sqrt{}$, rather than $\sqrt[2]{}$. Since even powers of nonzero numbers are positive, no negative number can have a fourth root, sixth root, etc.

The odd roots (third, fifth, etc.) of a number c are defined similarly, except that c need not be nonnegative.

Odd Roots

> If n is an odd positive integer and c is any real number, then
>
> $\sqrt[n]{c}$ denotes the number whose nth power is c.
>
> $\sqrt[n]{c}$ is called the **nth root** of c.

---

*We are assuming for now that there is exactly one such number, a fact that will be verified when we study equations and polynomial graphs later in the book.

EXAMPLE 1

Find $\sqrt[3]{-27}$, $\sqrt[3]{64}$, and $\sqrt[5]{32}$.

SOLUTION $\sqrt[3]{-27} = -3$ because $(-3)^3 = -27$ and $\sqrt[3]{64} = 4$ because $4^3 = 64$, as you can easily verify. Similarly, $\sqrt[5]{32} = 2$ because $2^5 = 32$. ■

EXAMPLE 2

Show that $\sqrt[3]{64} = \sqrt[3]{16} \cdot \sqrt[3]{4}$ without using a calculator.

SOLUTION   By the exponent rules and the definition of cube root,

$$(\sqrt[3]{16} \cdot \sqrt[3]{4})^3 = (\sqrt[3]{16})^3(\sqrt[3]{4})^3 = 16 \cdot 4 = 64.$$

Thus, $\sqrt[3]{16} \cdot \sqrt[3]{4}$ is the number whose cube is 64. But $\sqrt[3]{64}$ is the number whose cube is 64. Therefore, $\sqrt[3]{64} = \sqrt[3]{16} \cdot \sqrt[3]{4}$. ■

Since $64 = 16 \cdot 4$, Example 2 illustrates the first of the following facts, which apply to all real numbers c and d that have nth roots.

Products and Quotients of Roots

$$\sqrt[n]{cd} = \sqrt[n]{c} \cdot \sqrt[n]{d} \qquad \text{and} \qquad \sqrt[n]{\frac{c}{d}} = \frac{\sqrt[n]{c}}{\sqrt[n]{d}}$$

EXAMPLE 3

Simplify:

(a) $\sqrt{63}$ (b) $\sqrt{12} - \sqrt{75}$ (c) $\dfrac{2 + 3\sqrt{20}}{2}$

SOLUTION

(a) Factor 63 as the product of 9 (a perfect square) and 7 and then use the product of roots property.

$$\sqrt{63} = \sqrt{9 \cdot 7} = \sqrt{9}\sqrt{7} = 3\sqrt{7}$$

(b) Look for perfect squares.

Factor perfect squares out of 12 and 75: $\sqrt{12} - \sqrt{75} = \sqrt{4 \cdot 3} - \sqrt{25 \cdot 3}$

Product of roots: $= \sqrt{4}\sqrt{3} - \sqrt{25}\sqrt{3}$

$$= 2\sqrt{3} - 5\sqrt{3} = -3\sqrt{3}$$

(c)

Factor perfect square out of 20:
$$\frac{2 + 3\sqrt{20}}{2} = \frac{2 + 3\sqrt{4 \cdot 5}}{2}$$

Product of roots:
$$= \frac{2 + 3\sqrt{4}\sqrt{5}}{2}$$

$$= \frac{2 + 3 \cdot 2\sqrt{5}}{2} = \frac{2 + 6\sqrt{5}}{2}$$

Factor out 2 and simplify:
$$= \frac{2(1 + 3\sqrt{5})}{2} = 1 + 3\sqrt{5} \quad \blacksquare$$

EXAMPLE 4

Simplify:

(a) $\sqrt[3]{40k^4}$ (b) $\sqrt[3]{\dfrac{y^4}{8x^6}}$

SOLUTION

(a) Note that $8k^3$ is a factor of $40k^4$ and that $8k^3 = (2k)^3$ is a perfect cube.

Factor out $8k^3$:
$$\sqrt[3]{40k^4} = \sqrt[3]{8k^3 \cdot 5k}$$

Product of roots:
$$= \sqrt[3]{8k^3}\sqrt[3]{5k}$$

$$= \sqrt[3]{(2k)^3}\sqrt[3]{5k} = 2k\sqrt[3]{5k}$$

(b)

Quotient of roots:
$$\sqrt[3]{\frac{y^4}{8x^6}} = \frac{\sqrt[3]{y^4}}{\sqrt[3]{8x^6}}$$

Factor out perfect cubes:
$$= \frac{\sqrt[3]{y^3 y}}{\sqrt[3]{2^3 (x^2)^3}}$$

Product of roots:
$$= \frac{\sqrt[3]{y^3}\sqrt[3]{y}}{\sqrt[3]{2^3}\sqrt[3]{(x^2)^3}} = \frac{y\sqrt[3]{y}}{2x^2} \quad \blacksquare$$

CAUTION

When c and d are nonzero, then

$$\sqrt[n]{c + d} \neq \sqrt[n]{c} + \sqrt[n]{d}.$$

For example,

$$\sqrt[3]{8 + 27} = \sqrt[3]{35} \approx 3.271, \quad \text{but} \quad \sqrt[3]{8} + \sqrt[3]{27} = 2 + 3 = 5.$$

Similarly, the expression $\sqrt{4 + x^2}$ is *NOT* equal to $2 + x$.

Finally, we note that nth roots of nth powers behave in the same way as square roots when n is even, but not when n is odd.

nth Roots of ***nth Powers***	$$\sqrt[n]{c^n} = |c| \qquad \text{for } n \text{ even};$$ $$\sqrt[n]{c^n} = c \qquad \text{for } n \text{ odd}.$$

For example, $\sqrt[4]{(-2)^4} = \sqrt[4]{16} = 2 = |-2|$ and $\sqrt[5]{(-2)^5} = \sqrt[5]{-32} = -2$.

▪▪ RATIONAL EXPONENTS

nth roots provide a natural way to define fractional exponents. Suppose, for example, that we want to define $5^{1/2}$ in such a way that the exponent rules continue to hold. Then we would certainly want

$$(5^{1/2})^2 = 5^{\frac{1}{2}\cdot 2} = 5^1 = 5.$$

But we know that

$$\left(\sqrt{5}\right)^2 = 5.$$

So it is reasonable to define $5^{1/2}$ to be the number $\sqrt{5}$. Similarly, for any real number c and positive integer n, we have the following.

Reciprocal Exponents	$c^{1/n}$ is defined to be the number $\sqrt[n]{c}$ (provided that it exists).

EXAMPLE 5

Use a calculator to approximate

(a) $\sqrt[5]{40}$ (b) $\sqrt[11]{225}$.

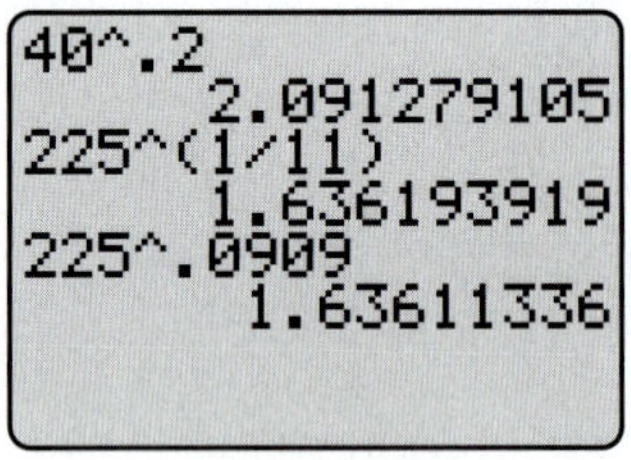

```
40^.2
         2.091279105
225^(1/11)
         1.636193919
225^.0909
         1.63611336
```

Figure 0–5

SOLUTION   Although most calculators have an nth root function (usually on a submenu of the MATH menu), it's easier to use fractional exponents.

(a) Since $\sqrt[5]{40} = 40^{1/5}$ and $1/5 = .2$, we compute $40^{.2}$, as in Figure 0–5.

(b) To approximate $\sqrt[11]{225} = 225^{1/11}$, note that $1/11$ has an infinite decimal expansion, $1/11 = .09090909\cdots$. To ensure the highest degree of accuracy, key in $225 \wedge (1 \div 11)$. If you round off the decimal expansion of $1/11$, say to $.0909$, you won't get the same answer, as Figure 0–5 shows. ■

The next step is to give a meaning to fractional exponents for any fraction, not just those of the form $1/n$. If possible, they should be defined in such a way that the various exponent rules continue to hold. Consider, for example, how $4^{3/2}$ might be defined. The exponent $3/2$ can be written as either

$$3 \cdot \left(\frac{1}{2}\right) \qquad \text{or} \qquad \left(\frac{1}{2}\right) \cdot 3$$

If the power of a power property $(c^m)^n = c^{mn}$ is to hold, we might define $4^{3/2}$ as either $(4^3)^{1/2}$ or $(4^{1/2})^3$. The result is the same in both cases:

$$(4^3)^{1/2} = 64^{1/2} = \sqrt{64} = 8 \qquad \text{and} \qquad (4^{1/2})^3 = (\sqrt{4})^3 = 2^3 = 8.$$

It can be proved that the same thing is true in the general case, which leads to this definition.

Rational Exponents

> If c is a real number and t/k is a rational number in lowest terms with positive denominator, then
>
> $$c^{t/k} \text{ is defined to be the number } \sqrt[k]{c^t} \qquad \text{(provided that it exists).}$$

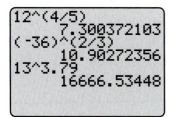

Figure 0–6

TECHNOLOGY TIP

TI-89 and Casio FX 2.0 will produce real number answers to computations like $(-8)^{2/3}$ if you select "Real" in the appropriate menu/submenu:

TI-89: MODE/COMPLEX FORMAT

Casio FX 2.0: SET-UP/COMPLEX MODE

Fractional powers can sometimes be computed by hand, for example,

$$8^{2/3} = \sqrt[3]{8^2} = \sqrt[3]{64} = 4.$$

Most of the time, however, we approximate them with a calculator, as shown in Figure 0–6. Note that decimal exponents now have a meaning, since they are simply a special case of fractional exponents; for instance, $13^{3.79} = 13^{379/100}$.

CAUTION

We know that

$$(-8)^{2/3} = \sqrt[3]{(-8)^2} = \sqrt[3]{64} = 4$$

but entering $(-8)^{2/3}$ on some calculators produces either an error message or a complex number (which the calculator indicates by an ordered pair or an expression involving i). If this happens on your calculator, you can get the correct answer by keying in one of

$$[(-8)^2]^{1/3} \qquad \text{or} \qquad [(-8)^{1/3}]^2$$

each of which is equal to $(-8)^{2/3}$.

Rational exponents were defined in a way that guaranteed that one of the familiar properties of exponents remains valid. In fact, all of the exponent properties developed for integer exponents are valid for rational exponents, as summarized here and illustrated in Examples 6–8.

Exponent Laws

> Let c and d be any real numbers and let r and s be any rational numbers in lowest terms for which each of the following exists. Then
>
> 1. $c^r c^s = c^{r+s}$
>
> 2. $\dfrac{c^r}{c^s} = c^{r-s} \quad (c \neq 0)$
>
> 3. $(c^r)^s = c^{rs}$
>
> 4. $(cd)^r = c^r d^r$
>
> 5. $\left(\dfrac{c}{d}\right)^r = \dfrac{c^r}{d^r} \quad (d \neq 0)$
>
> 6. $c^{-r} = \dfrac{1}{c^r}$

EXAMPLE 6

Simplify: $(x^{5/2}y^4)(xy^{7/4})$.

SOLUTION

Group like terms together: $(x^{5/2}y^4)(xy^{7/4}) = (x^{5/2}x)(y^4y^{7/4})$

Multiplication with exponents (law 1): $= x^{5/2+1}y^{4+7/4}$

$\dfrac{5}{2} + 1 = \dfrac{7}{2}$ and $4 + \dfrac{7}{4} = \dfrac{23}{4}$: $= x^{7/2}y^{23/4}$ ■

CAUTION

The exponent laws deal only with products and quotients. There are no analogous properties for sums. In particular, if both c and d are nonzero, then

$(c + d)^r$ is **not** equal to $c^r + d^r$.

EXAMPLE 7

Compute the product $x^{1/2}(x^{3/4} - x^{3/2})$.

SOLUTION

Distributive law: $x^{1/2}(x^{3/4} - x^{3/2}) = x^{1/2}x^{3/4} - x^{1/2}x^{3/2}$

Multiplication with exponents (law 1): $= x^{1/2+3/4} - x^{1/2+3/2}$

$\dfrac{1}{2} + \dfrac{3}{4} = \dfrac{5}{4}$ and $\dfrac{1}{2} + \dfrac{3}{2} = 2$: $= x^{5/4} - x^2$ ■

EXAMPLE 8

Simplify $(8r^{3/4}s^{-3})^{2/3}$, and express it without negative exponents.

SOLUTION

Product to a power (law 4): $(8r^{3/4}s^{-3})^{2/3} = 8^{2/3}(r^{3/4})^{2/3}(s^{-3})^{2/3}$

Power of a power (law 3): $= 8^{2/3}r^{(3/4)(2/3)}s^{(-3)(2/3)}$

$\dfrac{3}{4} \cdot \dfrac{2}{3} = \dfrac{1}{2}$ and $(-3)\dfrac{2}{3} = -2$: $= 8^{2/3}r^{1/2}s^{-2}$

Negative exponents (law 6): $= \dfrac{8^{2/3}r^{1/2}}{s^2}$

$8^{2/3} = 4$ (as shown on page 27): $= \dfrac{4r^{1/2}}{s^2}$ or $\dfrac{4\sqrt{r}}{s^2}$ ■

✓ EXERCISES 0.3

In Exercises 1–18, simplify the expression without using a calculator.

1. $\sqrt{80}$ **2.** $\sqrt{96}$

3. $\sqrt{147}$ **4.** $\sqrt{405}$

5. $\sqrt{6}\sqrt{12}$ **6.** $\sqrt{8}\sqrt{96}$

7. $\dfrac{-6 + \sqrt{99}}{15}$ **8.** $\dfrac{5 - \sqrt{175}}{10}$

9. $\sqrt{50} - \sqrt{72}$ **10.** $\sqrt{75} + \sqrt{192}$

11. $5\sqrt{20} - \sqrt{45} + 2\sqrt{80}$

12. $\sqrt[3]{40} + 2 \cdot \sqrt[3]{135} - 5 \cdot \sqrt[3]{320}$

13. $\sqrt{16a^8b^{-2}}$ **14.** $\sqrt{24x^6y^{-4}}$

15. $\dfrac{\sqrt{c^2d^6}}{\sqrt{4c^3d^{-4}}}$ **16.** $\dfrac{\sqrt{a^{-10}b^{-12}}}{\sqrt{a^{14}d^{-4}}}$

17. $\sqrt[3]{27a^9b^{-3}}$ **18.** $\sqrt[3]{40x^6y^{-6}}$

In Exercises 19–22, fill the blank with one of $<, =, >$ so that the resulting statement is true.

19. $\sqrt[3]{4} + \sqrt[3]{2}$ ___ $\sqrt[3]{6}$ **20.** $\sqrt{11} - \sqrt{2}$ ___ 3

21. $\sqrt[5]{7}\sqrt[5]{11}$ ___ $\sqrt[5]{77}$ **22.** $\sqrt[3]{35}\sqrt[4]{35}$ ___ $\sqrt[7]{35}$

In Exercises 23–28, write the given expression without using radicals.

23. $\sqrt[3]{a^2 + b^2}$ **24.** $\sqrt[4]{a^{-3} - b^3}$ **25.** $\sqrt[4]{\sqrt[4]{a^3}}$

26. $\sqrt{\sqrt[3]{a^3b^4}}$ **27.** $\sqrt[5]{t}\sqrt{16t^5}$ **28.** $\sqrt{x}\left(\sqrt[3]{x^2}\right)\left(\sqrt[4]{x^3}\right)$

In Exercises 29–36, simplify the expression.

29. $(25k^2)^{3/2}(16k^{1/3})^{3/4}$

30. $(3x^{2/3})(4y^{3/4})(2x^{-1/3})(3y^{13/4})$

31. $\sqrt{x^7} \cdot x^{5/2} \cdot x^{-3/2}$ **32.** $(x^{1/2}y^3)(x^0y^7)^{-2}$

33. $\dfrac{(x^2)^{1/3}(y^2)^{2/3}}{3x^{2/3}y^2}$ **34.** $\dfrac{(c^{1/2})^3(d^3)^{1/2}}{(c^3)^{1/4}(d^{1/4})^3}$

35. $\dfrac{(7a)^2(5b)^{3/2}}{(5a)^{3/2}(7b)^4}$ **36.** $\dfrac{(6a)^{1/2}\sqrt{ab}}{a^2b^{3/2}}$

In Exercises 37–42, compute and simplify.

37. $x^{1/2}(x^{2/3} - x^{4/3})$ **38.** $x^{1/2}(3x^{3/2} + 2x^{-1/2})$

39. $(x^{1/2} + y^{1/2})(x^{1/2} - y^{1/2})$

40. $(x^{1/3} + y^{1/2})(2x^{1/3} - y^{3/2})$

41. $(x + y)^{1/2}[(x + y)^{1/2} - (x + y)]$

42. $(x^{1/3} + y^{1/3})(x^{2/3} - x^{1/3}y^{1/3} + y^{2/3})$

In Exercises 43–46, assume that you have invested $2000 at 5.5% interest compounded annually. The value of your investment after x years is $2000(1.055^x)$. Find the value of your investment (to the nearest penny) at the end of the given period.

43. Five years and six months

44. 21 months

45. Eight years and seven months

46. 12 years and one month

In Exercises 47–50, use the equation $y = 92.8935 \cdot x^{.6669}$, which gives the approximate distance y (in millions of miles) from the sun to a planet that takes x earth years to complete one orbit of the sun. Find the distance from the sun to the planet whose orbit time is given.

47. Mercury (.24 years)   **48.** Mars (1.88 years)

49. Saturn (29.46 years)   **50.** Pluto (247.69 years)

Between 1790 and 1860 the population y of the United States (in millions) in year x was given by $y = 3.9572(1.0299^x)$, where $x = 0$ corresponds to 1790. In Exercises 51–54, find the U.S. population in the given year.

51. 1800   **52.** 1817   **53.** 1845   **54.** 1859

The consumer price index for U.S. cities (CPI-U) stood at 100 in 1982–1984. In recent years, it can be approximated by

$$102.68x^{.2249} \qquad (x \geq 7)$$

where $x = 7$ corresponds to 1997.[*] In Exercises 55–58, use this fact to approximate the CPI-U in the given year.

55. First half of 2000 ($x = 10$)

56. Last half of 2001 ($x = 11.5$)

57. Last half of 2004

58. First half of 2006

59. The output Q of an industry depends on labor L and capital C according to the equation

$$Q = L^{1/4}C^{3/4}.$$

(a) Use a calculator to determine the output for the following resource combinations.

L	C	$Q = L^{1/4}C^{3/4}$
10	7	
20	14	
30	21	
40	28	
60	42	

(b) When you double both labor and capital what happens to the output? When you triple both labor and capital what happens to the output?

60. Do Exercise 59 when the equation relating output to resources is $Q = L^{1/4}C^{1/2}$.

61. Do Exercise 59 when the equation relating output to resources is $Q = L^{1/2}C^{3/4}$.

62. In Exercises 59–61, how does the sum of the exponents on L and C affect the increase in output?

63. Here are some of the reasons why restrictions are necessary when defining fractional powers of a negative number.
(a) Explain why the equations $x^2 = -4$, $x^4 = -4$, $x^6 = -4$, etc., have no real solutions. Hence we cannot define $c^{1/2}$, $c^{1/4}$, $c^{1/6}$ when $c = -4$.
(b) Since $1/3$ is the same as $2/6$, it should be true that $c^{1/3} = c^{2/6}$, that is, that $\sqrt[3]{c} = \sqrt[6]{c^2}$. Show that this is false when $c = -8$.

64. Use a calculator to find a *six*-place decimal approximation of $(311)^{-4.2}$. Explain why your answer cannot possibly be the number $(311)^{-4.2}$.

---

[*]Based on data from the U.S. Bureau of Labor Statistics.

65. Each of the following formulas can be used to approximate square roots:

(i) $\sqrt{a^2 + b} \approx a + \dfrac{b}{2a + 1}$

(ii) $\sqrt{a^2 + b} \approx a + \dfrac{b}{2a}$

(iii) $\sqrt{a^2 + b} \approx a + \dfrac{b}{2a} - \dfrac{b^2}{8a^3}$

(a) Use each of the formulas to find an approximation of $\sqrt{10}$. [*Hint:* $10 = 9 + 1$.] Use a calculator to see which approximation is best.

(b) Do the same for $\sqrt{19}$.

Errors to Avoid

In Exercises 66–69, give a numerical example to show that the given statement is false *for some numbers.*

66. $\sqrt[3]{a + b} = \sqrt[3]{a} + \sqrt[3]{b}$

67. $\sqrt{c^2 + d^2} = c + d$

68. $\sqrt{8a} = 4\sqrt{a}$

69. $\sqrt{a}(\sqrt[3]{a}) = \sqrt[4]{a}$

0.4   Polynomials

Expressions such as

$$b + 3c^2, \quad 3x^2 - 5x + 4, \quad \sqrt{x^3 + z}, \quad \text{and} \quad \dfrac{x^3 + 4xy - \pi}{x^2 + xy}$$

are called **algebraic expressions.** The letters in algebraic expressions represent numbers. A letter that represents a number whose value remains unchanged throughout the discussion is called a **constant.** Constants are usually denoted by letters near the beginning of the alphabet. A letter that may represent *any* number is called a **variable.** Letters near the end of the alphabet usually denote variables.

EXAMPLE 1

In the expression $3x + 5 + c$, it is understood that x is a variable and c is a constant. Evaluate this expression when $x = 4$ and when $x = 1/3$.

SOLUTION   When $x = 4$, then

$$3x + 5 + c = 3(4) + 5 + c = 12 + 5 + c = 17 + c.$$

When $x = 1/3$, then

$$3x + 5 + c = 3\left(\dfrac{1}{3}\right) + 5 + c = 1 + 5 + c = 6 + c. \quad \blacksquare$$

The most common algebraic expressions in this book are **polynomials,** such as

$$4x^3 - 6x^2 + \dfrac{1}{2}x \quad \text{or} \quad y^{15} + y^{10} + 7 \quad \text{or} \quad 3z - \pi \quad \text{or} \quad 12.$$

in which the exponents on the variable are required to be nonnegative integers. Thus, $x^{1/2} + 5$ and $x^2 + x^{-1} + 2$ are *not* polynomials, but $\frac{1}{2}z - 6$ is a polynomial.

As the preceding examples illustrate, a polynomial is a sum of **terms** of the form

$$(\text{constant}) \times (\text{nonnegative integer power of the variable}).$$

We assume that $x^0 = 1$, $y^0 = 1$, etc., so terms such as 7 and 12 may be considered to be $7y^0$ and $12x^0$.

Any letter may be used for the variable in a polynomial, but we shall usually use x. The constants that appear in each term of a polynomial are called the **coefficients** of the polynomial. The coefficient of x^0 is called the **constant term** of the polynomial.

EXAMPLE 2

Identify the coefficients and the constant term of the polynomial:

(a) $3x^2 - x + 4$ (b) $5x^3 + x - 2$.

SOLUTION

(a) The coefficients are 3, -1, and 4, and the constant term is 4.

(b) The coefficients are 5, 0, 1 and -2 because the polynomial can be written $5x^3 + 0x^2 + x - 2$. The constant term is -2. ∎

A polynomial that consists of a constant term, such as 12, is called a **constant polynomial.** The **zero polynomial** is the constant polynomial 0. The **degree** of a polynomial is the *exponent* of the highest power of x that appears with *nonzero* coefficient and the nonzero coefficient of this highest power of x is the **leading coefficient** of the polynomial. For example,

Polynomial	Degree	Leading Coefficient	Constant Term
$6x^7 + 4x^3 + 5x^2 - 7x + 10$	7	6	10
$-x^4 + 2x^3 + \frac{1}{2}$	4	-1	$\frac{1}{2}$
x^3	3	1	0
12	0	12	12

The degree of the zero polynomial is *not defined* since no exponent of x occurs with nonzero coefficient. First-degree polynomials are often called **linear polynomials.** Second- and third-degree polynomials are called **quadratics** and **cubics,** respectively.

■■ POLYNOMIAL ARITHMETIC

The usual rules of arithmetic are valid for polynomials and other algebraic expressions.

EXAMPLE 3

Use the distributive law to *combine like terms;* for instance,

$$3x + 5x - 2x = (3 + 5 - 2)x = 6x.$$

In practice, you do the middle part in your head and simply write $3x + 5x - 2x = 6x$. ∎

EXAMPLE 4

Compute: $(2x^3 - 6x^2 + x) + (4x^5 + 5x^3 - 3x - 7)$.

SOLUTION

Eliminate parentheses:	$= 2x^3 - 6x^2 + x + 4x^5 + 5x^3 - 3x - 7$
Group like terms:	$= 4x^5 + (2x^3 + 5x^3) - 6x^2 + (x - 3x) - 7$
Combine like terms:	$= 4x^5 + 7x^3 - 6x^2 - 2x - 7$ ∎

Be careful when parentheses are preceded by a minus sign. For instance, the expression $-(b - 3)$ means $(-1)(b - 3)$, and hence, by the distributive law,

$$-(b - 3) = (-1)(b - 3) = (-1)b + (-1)(-3) = -b + 3.$$

A common mistake is to write $-(b - 3) = -b - 3$, rather than the correct answer $-b + 3$. Consequently, we have the following rules.

Rules for Eliminating Parentheses

Parentheses preceded by a plus sign (or no sign) may be deleted.

Parentheses preceded by a minus sign may be deleted *if* the sign of every term within the parentheses is changed.

EXAMPLE 5

Compute: $(4x^3 - 5x^2 + 7x - 2) - (2x^3 + x^2 - 6x - 8)$.

SOLUTION

Eliminate parentheses:	$= 4x^3 - 5x^2 + 7x - 2 - 2x^3 - x^2 + 6x + 8$
Group like terms:	$= (4x^3 - 2x^3) + (-5x^2 - x^2) + (7x + 6x) + (-2 + 8)$
Combine like terms:	$= 2x^3 - 6x^2 + 13x + 6$ ∎

To multiply polynomials, use the distributive law repeatedly, as shown in the following examples. The net result is to *multiply every term in the first polynomial by every term in the second.*

EXAMPLE 6

Compute: $(y - 2)(3y^2 - 7y + 4)$.

SOLUTION   We first apply the distributive law, treating $(3y^2 - 7y + 4)$ as a single number:

$$(y - 2)(3y^2 - 7y + 4) = y(3y^2 - 7y + 4) - 2(3y^2 - 7y + 4)$$

Distributive law: $= 3y^3 - 7y^2 + 4y - 6y^2 + 14y - 8$

Regroup: $= 3y^3 - \underbrace{7y^2 - 6y^2} + \underbrace{4y + 14y} - 8$

Combine like terms: $= 3y^3 - \quad 13y^2 \quad + \quad 18y \quad - 8.$ ■

EXAMPLE 7

Compute: $(2x - 5y)(3x + 4y)$.

SOLUTION   Treat $3x + 4y$ as a single number and use the distributive law.

$$(2x - 5y)(3x + 4y) = 2x(3x + 4y) - 5y(3x + 4y)$$

Distributive law: $= 2x \cdot 3x + 2x \cdot 4y + (-5y) \cdot 3x + (-5y) \cdot 4y$

Multiply: $= 6x^2 + \underbrace{8xy - 15xy} - 20y^2$

Combine like terms: $= 6x^2 \quad - 7xy \quad - 20y^2.$ ■

Observe the pattern in the second line of Example 7 and its relationship to the terms being multiplied:

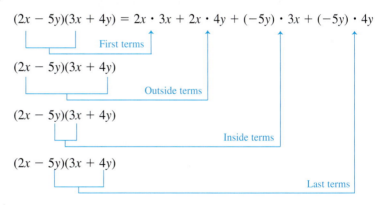

CAUTION

The FOIL method can be used only when multiplying two expressions that each have two terms.

This pattern is easy to remember by using the acronym FOIL (**F**irst, **O**utside, **I**nside, **L**ast). The FOIL Method makes it easy to find products such as this one mentally, without the necessity of writing out the intermediate steps.

EXAMPLE 8

$$(3x + 2)(x + 5) = 3x^2 + 15x + 2x + 10 = 3x^2 + 17x + 10.$$ ■

First   Outside   Inside   Last

Polynomials arise in a variety of situations, as illustrated in the following example.

EXAMPLE 9

Four squares, each measuring x by x inches, are cut from the corners of a rectangular sheet of cardboard that measures 30 by 22 inches, and the flaps are folded up to form an open-top box, as shown in Figure 0–7.

(a) Express the length and width of the box as polynomials in x.

(b) Express the volume of the box as a polynomial.

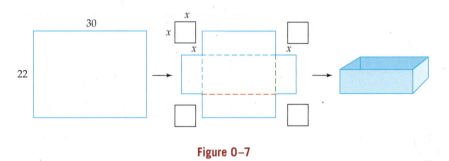

Figure 0–7

SOLUTION

(a) The red dashed line in Figure 0–7 is the length of the resulting box. The cardboard is 30 inches long. Cutting a piece x inches long from each end means that the length of the red dashed line is

$$30 - x - x = 30 - 2x.$$

Similarly, the green dashed line in Figure 0–7 is the width of the resulting box. The cardboard is 22 inches wide, so the green dashed line is $22 - 2x$ inches long. Hence, the box has length $30 - 2x$ and width $22 - 2x$.

(b) When the flaps are folded up to form the box, the height of the box is x inches. Therefore,

$$\text{Volume} = \text{length} \times \text{width} \times \text{height}$$
$$= (30 - 2x)(22 - 2x)x$$
$$= (660 - 104x + 4x^2)x$$
$$= 4x^3 - 104x^2 + 660x. \quad \blacksquare$$

✓ EXERCISES 0.4

In Exercises 1–50, perform the indicated operations and simplify your answer.

1. $x + 7x$

2. $5w + 7w - 3w$

3. $6a^2b + (-8b)a^2$

4. $-6x^3\sqrt{t} + 7x^3\sqrt{t} - 15x^3\sqrt{t}$

5. $(x^2 + 2x + 1) - (x^3 - 3x^2 + 4)$

6. $\left(u^4 - (-3)u^3 + \dfrac{u}{2} + 1\right) + \left(u^4 - 2u^3 + 5 - \dfrac{u}{2}\right)$

7. $\left(u^4 - (-3)u^3 + \dfrac{u}{2} + 1\right) - \left(u^4 - 2u^3 + 5 - \dfrac{u}{2}\right)$

8. $(6a^2b + 3a\sqrt{c} - 5ab\sqrt{c}) + (-6ab^2 - 3ab + 6ab\sqrt{c})$

9. $(4z - 6z^2w - (-2)z^3w^2) + (8 - 6z^2w - zw^3 + 4z^3w^2)$

10. $(x^5y - 2x + 3xy^3) - (-2x - x^5y + 2xy^3)$

11. $(9x - x^3 + 1) - [2x^3 + (-6)x + (-7)]$

12. $(x - \sqrt{y} - z) - (x + \sqrt{y} - z) - (\sqrt{y} + z - x)$

13. $(x^2 - 3xy) - (x + xy) - (x^2 + xy)$

14. $2x(x^2 + 2)$ **15.** $(-5y)(-3y^2 + 1)$

16. $x^2y(xy - 6xy^2)$

17. $3ax(4ax - 2a^2y + 2ay)$ **18.** $2x(x^2 - 3xy + 2y^2)$

19. $6z^3(2z + 5)$ **20.** $-3x^2(12x^6 - 7x^5)$

21. $3ab(4a - 6b + 2a^2b)$ **22.** $(-3ay)(4ay - 5y)$

23. $(x + 1)(x - 2)$ **24.** $(x + 2)(2x - 5)$

25. $(-2x + 4)(-x - 3)$ **26.** $(y - 6)(2y + 2)$

27. $(y + 3)(y + 4)$ **28.** $(w - 2)(3w + 1)$

29. $(3x + 7)(-2x + 5)$ **30.** $(ab + 1)(a - 2)$

31. $(y - 3)(3y^2 + 4)$ **32.** $(y + 8)(y - 8)$

33. $(x + 4)(x - 4)$ **34.** $(3x - y)(3x + y)$

35. $(4a + 5b)(4a - 5b)$ **36.** $(x + 6)^2$

37. $(y - 11)^2$ **38.** $(2x + 3y)^2$

39. $(5x - b)^2$

40. $(2s^2 - 9y)(2s^2 + 9y)$ **41.** $(4x^3 - y^4)^2$

42. $(4x^3 - 5y^2)(4x^3 + 5y^2)$ **43.** $(-3x^2 + 2y^4)^2$

44. $(c - 2)(2c^2 - 3c + 1)$

45. $(2y + 3)(y^2 + 3y - 1)$

46. $(x + 2y)(2x^2 - xy + y^2)$

47. $(5w + 6)(-3w^2 + 4w - 3)$

48. $(5x - 2y)(x^2 - 2xy + 3y^2)$

49. $2x(3x + 1)(4x - 2)$ **50.** $3y(-y + 2)(3y + 1)$

In Exercises 51–56, perform the indicated multiplication and simplify your answer if possible.

51. $\left(\sqrt{x} + 5\right)\left(\sqrt{x} - 5\right)$

52. $\left(2\sqrt{x} + \sqrt{2y}\right)\left(2\sqrt{x} - \sqrt{2y}\right)$

53. $\left(3 + \sqrt{y}\right)^2$ **54.** $\left(7w - \sqrt{2x}\right)^2$

55. $\left(1 + \sqrt{3x}\right)\left(x + \sqrt{3}\right)$

56. $\left(2y + \sqrt{3}\right)\left(\sqrt{5y} - 1\right)$

In Exercises 57–62, compute the product and arrange the terms of your answer according to decreasing powers of x, with each power of x appearing at most once.
Example: $(ax + b)(4x - c) = 4ax^2 + (4b - ac)x - bc$.

57. $(ax + b)(3x + 2)$ **58.** $(4x - c)(dx + c)$

59. $(ax + b)(bx + a)$ **60.** $rx(3rx + 1)(4x - r)$

61. $(x - a)(x - b)(x - c)$ **62.** $(2dx - c)(3cx + d)$

In Exercises 63–66, fill the blanks, using the Box Method of multiplying polynomials, which is illustrated here.

    To multiply $(2x + 3)(5x - 7)$, form a 2-by-2 box, with row and column labels as shown. Then enter the products of each row label by each column label; for instance, the entry 15x in the lower left corner is the product of 5x and 3.

	$5x$	-7
$2x$		
3		

	$5x$	-7
$2x$	$10x^2$	$-14x$
3	$15x$	-21

The sum of the entries in the box is the product:

$$10x^2 + 15x - 14x - 21 = 10x^2 + x - 21.$$

63. $(3x + 2)(4x - 5) = $ _____

	$3x$	____
____	$12x^2$	
-5		-10

64. $(4x - 7)($_____$) = 12x^2 - x - 35$

	$4x$	-7
____	$12x^2$	
____		-35

65. $(2x + 5)(4x^2 - 10x + 25) = $ _____

	$4x^2$	$-10x$	25
$2x$			
5			

66. $(4x - 3)($_____$) = 64x^3 - 27$

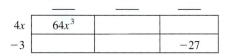

	____	____	____
$4x$	$64x^3$		
-3			-27

67. Four squares, each measuring x by x inches, are cut from the corners of a rectangular sheet of cardboard that measures 20 by 15 inches and the flaps are folded up to form an open-top box, as in Example 9.
  (a) Express the length and width of the resulting box as polynomials in x.
  (b) Express the volume of the resulting box as a polynomial.
  (c) Find the volume of the box when $x = 1$, when $x = 2$, and when $x = 4$.

68. Do Exercise 67 when the sheet of cardboard measures 40 by 40 inches.

69. A gutter is to be constructed by folding up the edge of a 16-inch-wide piece of metal, as shown in the figure.

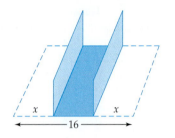

(a) Express the area of the cross section of the gutter as a polynomial in x.

(b) Fill in the table.

x	1	2	3	4	5	6	7	8
Area of cross section								

(c) Which value of x appears to produce the largest cross-sectional area (and hence the largest volume of water in the gutter)?

In Exercises 70–72, express the area of the shaded region as a polynomial in x.

70.

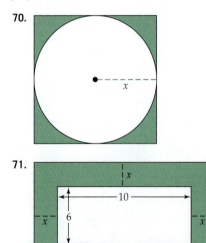

71.

72.

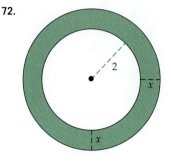

Thinkers

In Exercises 73 and 74, explain algebraically why each of these parlor tricks always works.

73. Write down a nonzero number. Add 1 to it and square the result. Subtract 1 from the original number and square the result. Subtract this second square from the first one. Divide by the number with which you started. The answer is 4.

74. Write down a positive number. Add 4 to it. Multiply the result by the original number. Add 4 to this result and then take the square root. Subtract the number with which you started. The answer is 2.

75. Invent a similar parlor trick in which the answer is always the number with which you started.

76. Show that there do *not* exist real numbers c and d such that

$$x^2 + 1 = (x + c)(x + d).$$

0.5   Factoring

Factoring is the reverse of multiplication. We begin with a product and find the factors that multiply together to produce this product. Factoring skills are necessary to simplify expressions, to do arithmetic with fractional expressions, and to solve equations and inequalities.

The first general rule for factoring is as follows.

Common Factors

If there is a common factor in every term of the expression, factor out the common factor of highest degree.

EXAMPLE 1

In $4x^6 - 8x$, for example, each term contains a factor of $4x$, so that

$$4x^6 - 8x = 4x(x^5 - 2).$$

Similarly, the common factor of highest degree in $x^3y^2 + 2xy^3 - 3x^2y^4$ is xy^2, and

$$x^3y^2 + 2xy^3 - 3x^2y^4 = xy^2(x^2 + 2y - 3xy^2). \quad \blacksquare$$

You can greatly increase your factoring proficiency by learning to recognize multiplication patterns that appear frequently. Here are the most common ones.

Quadratic Factoring Patterns

Difference of Squares $\qquad u^2 - v^2 = (u + v)(u - v)$

Perfect Squares $\qquad\qquad u^2 + 2uv + v^2 = (u + v)^2$

$$u^2 - 2uv + v^2 = (u - v)^2$$

EXAMPLE 2

(a) $x^2 - 9y^2$ can be written $x^2 - (3y)^2$, a difference of squares. Therefore, $x^2 - 9y^2 = (x + 3y)(x - 3y)$.

(b) $y^2 - 7 = y^2 - \left(\sqrt{7}\right)^2 = \left(y + \sqrt{7}\right)\left(y - \sqrt{7}\right)$.*

(c) $36r^2 - 64s^2 = (6r)^2 - (8s)^2 = (6r + 8s)(6r - 8s)$

$$= 2(3r + 4s)2(3r - 4s) = 4(3r + 4s)(3r - 4s). \quad \blacksquare$$

EXAMPLE 3

Factor: $4x^2 - 36x + 81$.

SOLUTION   Since the first and last terms of $4x^2 - 36x + 81$ are perfect squares, we try to use the perfect square pattern with $u = 2x$ and $v = 9$:

$$4x^2 - 36x + 81 = (2x)^2 - 36x + 9^2$$

$$= (2x)^2 - 2 \cdot 2x \cdot 9 + 9^2 = (2x - 9)^2. \quad \blacksquare$$

---

*When a polynomial has integer coefficients, we normally look only for factors with integer coefficients. But when it is easy to find other factors, as here, we shall do so.

Cubic Factoring
Patterns

Difference of Cubes	$u^3 - v^3 = (u - v)(u^2 + uv + v^2)$
Sum of Cubes	$u^3 + v^3 = (u + v)(u^2 - uv + v^2)$
Perfect Cubes	$u^3 + 3u^2v + 3uv^2 + v^3 = (u + v)^3$
	$u^3 - 3u^2v + 3uv^2 - v^3 = (u - v)^3$

EXAMPLE 4

Factor:

(a) $x^3 - 125$ (b) $x^3 + 8y^3$ (c) $x^3 - 12x^2 + 48x - 64$

SOLUTION

(a) $x^3 - 125 = x^3 - 5^3 = (x - 5)(x^2 + 5x + 5^2)$
$$= (x - 5)(x^2 + 5x + 25).$$

(b) $x^3 + 8y^3 = x^3 + (2y)^3 = (x + 2y)[x^2 - x \cdot 2y + (2y)^2]$
$$= (x + 2y)(x^2 - 2xy + 4y^2).$$

(c) $x^3 - 12x^2 + 48x - 64 = x^3 - 12x^2 + 48x - 4^3$
$$= x^3 - 3x^2 \cdot 4 + 3x \cdot 4^2 - 4^3$$
$$= (x - 4)^3. \quad \blacksquare$$

When none of the multiplication patterns applies, other techniques must be used. The next example illustrates a method that works for quadratic polynomials with leading coefficient 1.

EXAMPLE 5

Factor: $x^2 + 9x + 18$.

SOLUTION   We must find integers b and d such that

$$x^2 + 9x + 18 = (x + b)(x + d)$$
$$= x^2 + dx + bx + bd$$
$$= x^2 + (b + d)x + bd.$$

If the polynomials are equal, then their coefficients are equal term by term, so

$$9 = b + d \quad \text{and} \quad 18 = bd.$$

In particular, b and d are integer factors of 18. The possibilities are summarized in this table.

Factors b, d of 18	Sum $b + d$
$1 \cdot 18$	$1 + 18 = 19$
$2 \cdot 9$	$2 + 9 = 11$
$3 \cdot 6$	$3 + 6 = 9$
$(-1)(-18)$	$-1 + (-18) = -19$
$(-2)(-9)$	$-2 + (-9(= -11$
$(-3)(-6)$	$-3 + (-6) = -9$

Note that we really didn't have to list the negative factors, since their sums are also negative and we must have $b + d = 9$. There is just one choice with sum 9, namely, 3 and 6. So we have this factorization (which you should verify by multiplying out the right-hand side):

$$x^2 + 9x + 18 = (x + 3)(x + 6). \qquad \blacksquare$$

EXAMPLE 6

Factor: $x^2 + 3x - 10$.

SOLUTION   We must find factors $x + b$ and $x + d$ such that $bd = -10$ and $b + d = 3$. The table shows the possibilities.

Factors b, d of -10	Sum $b + d$
$1(-10)$	$1 + (-10) = -9$
$(-1)(10)$	$-1 + 10 = 9$
$2(-5)$	$2 + (-5) = -3$
$(-2)(5)$	$-2 + 5 = 3$

The only factors with product -10 and sum 3 are -2 and 5. So the correct factorization is

$$x^2 + 3x - 10 = (x - 2)(x + 5). \qquad \blacksquare$$

Although we used tables in Examples 5 and 6 to summarize all the possibilities, you can often do this without writing out a table by mentally checking the various possibilities. The same idea works for factoring quadratic polynomials in which the leading coefficient is not 1, as shown in the next example.

EXAMPLE 7

Factor: $6x^2 + 11x + 4$.

SOLUTION   We must find factors $ax + b$ and $cx + d$ such that

$$6x^2 + 11x + 4 = (ax + b)(cx + d)$$
$$= acx^2 + adx + bcx + bd$$
$$= acx^2 + (ad + bc)x + bd.$$

Once again, the coefficients must be the same, term by term, on both sides. In particular, we must have

$$6 = ac \qquad \text{and} \qquad 4 = bd.$$

There are quite a few possibilities, as shown here.

$ac = 6$	a	± 1	± 2	± 3	± 6
	c	± 6	± 3	± 2	± 1

$bd = 4$	b	± 1	± 2	± 4
	d	± 4	± 2	± 1

Guided by the possibilities listed in the tables, we use trial and error to find the right factors. Typically, it might go like this.

$a = 1, c = 6$ and $b = 1, d = 4$:

$(ax + b)(cx + d) = (x + 1)(6x + 4) = 6x^2 + 10x + 4$ Incorrect

$a = 2, c = 3$ and $b = 4, d = 1$:

$(ax + b)(cx + d) = (2x + 4)(3x + 1) = 6x^2 + 14x + 4$ Incorrect

$a = 2, c = 3$ and $b = 1, d = 4$:

$(ax + b)(cx + d) = (2x + 1)(3x + 4) = 6x^2 + 11x + 4$ Correct   ■

Substitution and appropriate factoring patterns can sometimes be used to factor expressions involving exponents larger than 2.

EXAMPLE 8

Factor: $x^4 - 2x^2 - 3$.

SOLUTION   The idea is to put the polynomial in quadratic form. Note that $x^4 = (x^2)^2$ and let $u = x^2$. Then

$$x^4 - 2x^2 - 3 = (x^2)^2 - 2x^2 - 3$$
$$= u^2 - 2u - 3 = (u + 1)(u - 3)$$
$$= (x^2 + 1)(x^2 - 3)$$
$$= (x^2 + 1)(x + \sqrt{3})(x - \sqrt{3}). \quad ■$$

EXAMPLE 9

Factor:
(a) $x^6 - y^6$ (b) $x^8 - 1$.

SOLUTION   Once again, we want to put each polynomial in quadratic form.

(a) Note that $x^6 = (x^3)^2$ and $y^6 = (y^3)^2$. Let $u = x^3$ and $v = y^3$. Then

$$x^6 - y^6 = (x^3)^2 - (y^3)^2$$
$$= u^2 - v^2 = (u + v)(u - v) \qquad \text{[Difference of squares]}$$
$$= (x^3 + y^3)(x^3 - y^3)$$
$$= (x + y)(x^2 - xy + y^2)(x^3 - y^3) \qquad \text{[Sum of cubes]}$$
$$= (x + y)(x^2 - xy + y^2)(x - y)(x^2 + xy + y^2). \quad \text{[Difference of cubes]}$$

(b) Note that $x^8 = (x^4)^2$ and let $u = x^4$. Then

$$x^8 - 1 = (x^4)^2 - 1 = u^2 - 1 = (u + 1)(u - 1) \qquad \text{[Difference of squares]}$$
$$= (x^4 + 1)(x^4 - 1).$$

Observe that $x^4 = (x^2)^2$ and let $v = x^2$. Then

$$x^8 - 1 = (x^4 + 1)(x^4 - 1)$$
$$= [(x^2)^2 + 1][(x^2)^2 - 1]$$
$$= (v^2 + 1)(v^2 - 1)$$
$$= (v^2 + 1)(v + 1)(v - 1) \qquad \text{[Difference of squares]}$$
$$= (x^4 + 1)(x^2 + 1)(x^2 - 1)$$
$$= (x^4 + 1)(x^2 + 1)(x + 1)(x - 1). \qquad \text{[Difference of squares]} \qquad ■$$

Occasionally, a polynomial can be factored by regrouping.

EXAMPLE 10

Factor: $3x^3 + 3x^2 + 2x + 2$.

SOLUTION   Write the polynomial as

$$(3x^3 + 3x^2) + (2x + 2)$$

and factor a common factor out of each expression in parentheses:

$$3x^2(x + 1) + 2(x + 1).$$

Now use the distributive law (considering $x + 1$ as a single quantity).

$$3x^2(x + 1) + 2(x + 1) = (3x^2 + 2)(x + 1) \qquad ■$$

✓ EXERCISES 0.5

In Exercises 1–50, factor the expression.

1. $x^2 - 4$
2. $x^2 + 6x + 9$
3. $9y^2 - 25$
4. $y^2 - 4y + 4$
5. $81x^2 + 36x + 4$
6. $4x^2 - 12x + 9$
7. $5 - x^2$
8. $1 - 36u^2$
9. $49 + 28z + 4z^2$
10. $25u^2 - 20uv + 4v^2$
11. $x^4 - y^4$
12. $x^2 - 1/9$
13. $x^2 + x - 6$
14. $y^2 + 11y + 30$
15. $z^2 + 4z + 3$
16. $x^2 - 8x + 15$
17. $y^2 + 5y - 36$
18. $z^2 - 9z + 14$
19. $x^2 - 6x + 9$
20. $4y^2 - 81$
21. $x^2 + 7x + 10$
22. $w^2 - 6w - 16$
23. $x^2 + 11x + 18$
24. $x^2 + 3xy - 28y^2$
25. $3x^2 + 4x + 1$
26. $4y^2 + 4y + 1$
27. $2z^2 + 11z + 12$
28. $10x^2 - 17x + 3$

29. $9x^2 - 72x$
30. $4x^2 - 4x - 3$
31. $10x^2 - 8x - 2$
32. $7z^2 + 23z + 6$
33. $8u^2 + 6u - 9$
34. $2y^2 - 4y + 2$
35. $4x^2 + 20xy + 25y^2$
36. $63u^2 - 46uv + 8v^2$
37. $x^3 - 125$
38. $y^3 + 64$
39. $x^3 + 6x^2 + 12x + 8$
40. $y^3 - 3y^2 + 3y - 1$
41. $8 + x^3$
42. $27 - t^3$
43. $x^3 + 1$
44. $x^3 - 1$
45. $8x^3 - y^3$
46. $(x - 1)^3 + 1$
47. $x^6 - 64$
48. $x^5 - 8x^2$
49. $y^4 + 7y^2 + 10$
50. $81 - y^4$

In Exercises 51–56, factor by regrouping and using the distributive law (as in Example 10).

51. $x^2 - yz + xz - xy$
52. $x^6 - 2x^4 - 8x^2 + 16$

53. $a^3 - 2b^2 + 2a^2b - ab$ **54.** $u^2v - 2w^2 - 2uvw + uw$

55. $x^3 + 4x^2 - 8x - 32$ **56.** $z^8 - 5z^7 + 2z - 10$

In Exercises 57–60, use the hint in Exercise 57 to factor the fourth-degree polynomial.

57. $x^4 + x^2 + 1$ [*Hint:* Add and subtract x^2, regroup, and factor to obtain

$$x^4 + x^2 + 1 + x^2 - x^2 = (x^4 + x^2 + 1 + x^2) - x^2$$

$$= (x^4 + 2x^2 + 1) - x^2 = (x^2 + 1)^2 - x^2.$$

Note that the last polynomial is a difference of squares and hence, can be factored.]

58. $x^4 + 3x^2 + 4$ **59.** $x^4 + 5x^2 + 9$

60. $x^4 + 7x^2 + 16$

61. (a) On the basis of the factoring pattern in Exercises 57–60, guess how each of these polynomials factors.

$$x^4 + 9x^2 + 25 \quad \text{and} \quad x^4 + 11x^2 + 36$$

   (b) Verify your guesses in part (a) by multiplying out the factors.

62. The difference of squares and difference of cubes patterns show that

$$x^2 - 1 = (x - 1)(x + 1) \quad \text{and}$$

$$x^3 - 1 = (x - 1)(x^2 + x + 1).$$

(a) Verify that $x^4 - 1 = (x - 1)(x^3 + x^2 + x + 1)$.

(b) On the basis of the preceding factoring patterns, write each of $x^5 - 1$ and $x^6 - 1$ as a product of $x - 1$ and another factor. Verify your answer by multiplying out the two factors.

(c) Based on these results, how do you think that $x^n - 1$ (with n a positive integer) can be written as a product of $x - 1$ and another factor?

Errors to Avoid

In Exercises 63–70, find a numerical example to show that the given statement is false. Then find the mistake in the statement and correct it.

Example: The statement $-(b + 2) = -b + 2$ is false when $b = 5$, since $-(5 + 2) = -7$ but $-5 + 2 = -3$. The mistake is the sign on the 2. The correct statement is $-(b + 2) = -b - 2$.

63. $3(y + 2) = 3y + 2$

64. $x - (3y + 4) = x - 3y + 4$

65. $(x + y)^2 = x + y^2$

66. $(2x)^3 = 2x^3$

67. $y + y + y = y^3$

68. $(a - b)^2 = a^2 - b^2$

69. $(x - 3)(x - 2) = x^2 - 5x - 6$

70. $(a + b)(a^2 + b^2) = a^3 + b^3$

0.6   Fractional Expressions

Quotients of algebraic expressions are called **fractional expressions.** A quotient of two polynomials, such as

$$\frac{x^2 - 2}{x - 1}, \quad \frac{1}{x^3 + 4x^2 - x}, \quad \text{or} \quad \frac{3x - 7}{x^3(x + 1)}$$

is called a **rational expression.** Throughout this section, we assume that all denominators are nonzero.

The basic rules for dealing with fractional expressions are essentially the same as those for ordinary numerical fractions. In particular, the usual cancellation property holds.

Cancellation

If $k \neq 0$, then $\dfrac{ka}{kb} = \dfrac{a}{b}$.

EXAMPLE 1

Simplify:

(a) $\dfrac{36}{56}$ (b) $\dfrac{x^4 - 1}{x^2 + 1}$.

SOLUTION

(a) Factor numerator and denominator: $\dfrac{36}{56} = \dfrac{9 \cdot 4}{7 \cdot 8} = \dfrac{9 \cdot 4}{7 \cdot 2 \cdot 4}$

Cancel the common factor 4: $= \dfrac{9}{7 \cdot 2} = \dfrac{9}{14}$.

(b) The same procedure works for the rational expression.

Factor the numerator: $\dfrac{x^4 - 1}{x^2 + 1} = \dfrac{(x^2 + 1)(x^2 - 1)}{x^2 + 1}$

Cancel the common factor $x^2 + 1$: $= \dfrac{x^2 - 1}{1} = x^2 - 1$ ∎

From arithmetic, we know that

$$\dfrac{2}{4} = \dfrac{3}{6}.$$

Note that the cross-products are equal: $2 \cdot 6 = 4 \cdot 3$. The same thing is true in the general case.

Equality Rule

$$\dfrac{a}{b} = \dfrac{c}{d} \qquad \text{exactly when} \qquad ad = bc.$$

Proof   Suppose

$$\dfrac{a}{b} = \dfrac{c}{d}.$$

Multiply both sides by bd: $\dfrac{a}{b} \cdot bd = bd \cdot \dfrac{c}{d}$

Use cancellation: $ad = bc$

Conversely, if $ad = bc$, then dividing both sides by bd and cancelling shows that $\dfrac{a}{b} = \dfrac{c}{d}$.   ∎

EXAMPLE 2

We know that

$$\dfrac{x^2 + 2x}{x^2 + x - 2} = \dfrac{x}{x - 1}$$

because the cross-products are equal:

$$(x^2 + 2x)(x - 1) = x^3 + x^2 - 2x \qquad \text{and} \qquad (x^2 + x - 2)x = x^3 + x^2 - 2x.$$

∎

A fraction is in **lowest terms** if its numerator and denominator have no common factors except ± 1.

EXAMPLE 3

Express $\dfrac{x^2 + x - 6}{x^2 - 3x + 2}$ in lowest terms.

SOLUTION   Proceed as we did in Example 1: Factor numerator and denominator, and cancel common factors.

$$\frac{x^2 + x - 6}{x^2 - 3x + 2} = \frac{(x - 2)(x + 3)}{(x - 2)(x - 1)} = \frac{x + 3}{x - 1} \quad \blacksquare$$

To add or subtract two fractions with the same denominator, simply add or subtract the numerators.

$$\frac{a}{b} + \frac{c}{b} = \frac{a + c}{b} \qquad \text{and} \qquad \frac{a}{b} - \frac{c}{b} = \frac{a - c}{b}$$

EXAMPLE 4

$$\frac{7x^2 + 2}{x^2 + 3} - \frac{4x^2 + 2x - 5}{x^2 + 3} = \frac{(7x^2 + 2) - (4x^2 + 2x - 5)}{x^2 + 3}$$

$$= \frac{7x^2 + 2 - 4x^2 - 2x + 5}{x^2 + 3}$$

$$= \frac{3x^2 - 2x + 7}{x^2 + 3}. \quad \blacksquare$$

To add or subtract fractions with different denominators, you must use a **common denominator,** that is, replace the given fractions by equivalent ones that have the same denominator. For instance, to add $1/3$ and $1/4$, we use the common denominator 12 (the product of 3 and 4):

$$\frac{1}{3} = \frac{4}{12} \qquad \text{and} \qquad \frac{1}{4} = \frac{3}{12},$$

so that

$$\frac{1}{3} + \frac{1}{4} = \frac{4}{12} + \frac{3}{12} = \frac{4 + 3}{12} = \frac{7}{12}.$$

EXAMPLE 5

Compute: $\dfrac{2x + 1}{3x} - \dfrac{x^2 - 2}{x - 1}$.

SOLUTION   We use the product of the denominators $3x(x - 1)$ as the common denominator. To rewrite the first fraction in terms of this denominator, we must find the right-side numerator here.

$$\frac{2x + 1}{3x} = \frac{?}{3x(x - 1)}$$

Note that to get from the left-side denominator to the right-side one, you multiply by $x - 1$. To keep the fractions equal, you must multiply the left-hand numerator by $x - 1$ also:

$$\frac{2x + 1}{3x} = \frac{(2x + 1)(x - 1)}{3x(x - 1)}$$

A similar procedure works with the other fraction.

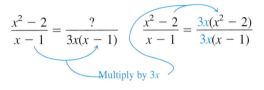

Now that the fractions have the same denominator, it's easy to subtract them.

$$\frac{2x + 1}{3x} - \frac{x^2 - 2}{x - 1} = \frac{(2x + 1)(x - 1)}{3x(x - 1)} - \frac{3x(x^2 - 2)}{3x(x - 1)}$$

$$= \frac{(2x + 1)(x - 1) - 3x(x^2 - 2)}{3x(x - 1)}$$

$$= \frac{2x^2 - x - 1 - 3x^3 + 6x}{3x^2 - 3x}$$

$$= \frac{-3x^3 + 2x^2 + 5x - 1}{3x^2 - 3x} \qquad \blacksquare$$

Although the product of the denominators can always be used as a common denominator, it's usually more efficient to use the **least common denominator,** that is, the smallest quantity that can be the denominator of both fractions. The next example illustrates an easy way to find the least common denominator when dealing with numbers.

EXAMPLE 6

Find the least common denominator for $\dfrac{1}{100}$ and $\dfrac{1}{120}$.

SOLUTION   The two denominators can be factored as

$$100 = 2^2 \cdot 5^2 \qquad \text{and} \qquad 120 = 2^3 \cdot 3 \cdot 5.$$

The distinct factors that appear in both numbers are 2, 3, and 5. For each factor, take the highest power to which it appears in either number (namely, 2^3, 3^1, and 5^2). The product of these highest powers is the least common denominator.

$$2^3 \cdot 3 \cdot 5^2 = 600$$

Note that $600 = 100 \cdot 6$ and $600 = 120 \cdot 5$, so

$$\frac{1}{100} = \frac{1 \cdot 6}{100 \cdot 6} = \frac{6}{600} \qquad \text{and} \qquad \frac{1}{120} = \frac{1 \cdot 5}{120 \cdot 5} = \frac{5}{600}. \qquad \blacksquare$$

The technique of Example 6 can also be used to find the least common denominator of rational expressions.

EXAMPLE 7

Find the least common denominator for

$$\frac{1}{x^2 + 2x + 1}, \quad \frac{5x}{x^2 - x}, \quad \text{and} \quad \frac{3x - 7}{x^4 + x^3}$$

and express each fraction in terms of this common denominator.

SOLUTION   Begin by factoring each denominator completely.

$$x^2 + 2x + 1 = (x + 1)^2 \qquad x^2 - x = x(x - 1) \qquad x^4 + x^3 = x^3(x + 1)$$

The distinct factors are x, $x + 1$, and $x - 1$. The least common denominator is

$$x^3(x + 1)^2(x - 1).$$

Therefore,

$$\frac{1}{(x + 1)^2} = \frac{1}{(x + 1)^2} \cdot \frac{x^3(x - 1)}{x^3(x - 1)} = \frac{x^3(x - 1)}{x^3(x + 1)^2(x - 1)}$$

$$\frac{5x}{x(x - 1)} = \frac{5x}{x(x - 1)} \cdot \frac{x^2(x + 1)^2}{x^2(x + 1)^2} = \frac{5x^3(x + 1)^2}{x^3(x + 1)^2(x - 1)}$$

$$\frac{3x - 7}{x^3(x + 1)} = \frac{3x - 7}{x^3(x + 1)} \cdot \frac{(x + 1)(x - 1)}{(x + 1)(x - 1)} = \frac{(3x - 7)(x + 1)(x - 1)}{x^3(x + 1)^2(x - 1)}. \quad \blacksquare$$

EXAMPLE 8

Compute $\dfrac{1}{z} + \dfrac{3z}{z + 1} - \dfrac{z^2}{(z + 1)^2}$.

SOLUTION   We use the least common denominator $z(z + 1)^2$ and express each fraction in terms of this denominator.

$$\frac{1}{z} = \frac{1}{z} \cdot \frac{(z + 1)^2}{(z + 1)^2} = \frac{(z + 1)^2}{z(z + 1)^2}$$

$$\frac{3z}{z + 1} = \frac{3z}{z + 1} \cdot \frac{z(z + 1)}{z(z + 1)} = \frac{3z^2(z + 1)}{z(z + 1)^2}$$

$$\frac{z^2}{(z + 1)^2} = \frac{z^2}{(z + 1)^2} \cdot \frac{z}{z} = \frac{z^3}{z(z + 1)^2}$$

Then

$$\frac{1}{z} + \frac{3z}{z + 1} - \frac{z^2}{(z + 1)^2} = \frac{(z + 1)^2}{z(z + 1)^2} + \frac{3z^2(z + 1)}{z(z + 1)^2} - \frac{z^3}{z(z + 1)^2}$$

$$= \frac{(z + 1)^2 + 3z^2(z + 1) - z^3}{z(z + 1)^2}$$

$$= \frac{z^2 + 2z + 1 + 3z^3 + 3z^2 - z^3}{z(z + 1)^2}$$

$$= \frac{2z^3 + 4z^2 + 2z + 1}{z(z + 1)^2}. \quad \blacksquare$$

Multiplication of fractional expressions is done in the same way as multiplication of numerical fractions.

Multiplication of Fractional Expressions

If a, b, c, and d are algebraic expressions, then

$$\frac{a}{b} \cdot \frac{c}{d} = \frac{ac}{bd}.$$

In other words, to multiply fractional expressions, multiply the corresponding numerators and denominators.

EXAMPLE 9

Multiply $\dfrac{x^2 - 1}{x^2 + 2} \cdot \dfrac{3x - 4}{x + 1}$ and simplify your answer.

SOLUTION

$$\frac{x^2 - 1}{x^2 + 2} \cdot \frac{3x - 4}{x + 1} = \frac{(x^2 - 1)(3x - 4)}{(x^2 + 2)(x + 1)}$$

$$= \frac{(x - 1)(x + 1)(3x - 4)}{(x^2 + 2)(x + 1)} = \frac{(x - 1)(3x - 4)}{x^2 + 2}. \quad \blacksquare$$

In Example 9, we multiplied and then factored and canceled. You can also factor and cancel before multiplying, as illustrated in the next example.

EXAMPLE 10

$$\frac{x^2 + 6x + 8}{x + 3} \cdot \frac{x^2 + x - 6}{x^2 + 3x + 2} = \frac{(x + 2)(x + 4)}{x + 3} \cdot \frac{(x + 3)(x - 2)}{(x + 2)(x + 1)}$$

$$= \frac{(x + 4)(x - 2)}{x + 1}$$

$$= \frac{x^2 + 2x - 8}{x + 1} \quad \blacksquare$$

Division of fractional expressions is also done in the same manner as division of numerical fractions.

Division of Fractional Expressions

If a, b, c, and d are algebraic expressions, then

$$\frac{a}{b} \div \frac{c}{d} = \frac{a}{b} \cdot \frac{d}{c} = \frac{ad}{bc}.$$

That is, to divide fractional expressions, invert the divisor and multiply.

48   CHAPTER 0   **Algebra Review**

EXAMPLE 11

$$\frac{x^2 + x - 2}{x^2 - 6x + 9} \div \frac{x^2 - 1}{x - 3} = \frac{x^2 + x - 2}{x^2 - 6x + 9} \cdot \frac{x - 3}{x^2 - 1}$$

$$= \frac{(x + 2)(x - 1)}{(x - 3)(x - 3)} \cdot \frac{x - 3}{(x - 1)(x + 1)}$$

$$= \frac{x + 2}{(x - 3)(x + 1)}. \quad \blacksquare$$

Division problems can also be written in fractional form. For instance, 8/2 means 8 ÷ 2. Similarly, a compound fraction such as

$$\frac{\dfrac{x + 1}{x^2 - 4}}{\dfrac{3x + 2}{x + 2}}$$

means

$$\frac{x + 1}{x^2 - 4} \div \frac{3x + 2}{x + 2}, \quad \text{which is equivalent to} \quad \frac{x + 1}{x^2 - 4} \cdot \frac{x + 2}{3x + 2}.$$

Consequently, the basic rule for simplifying a compound fraction is:

Invert the denominator and multiply it by the numerator.

EXAMPLE 12

Compute:

(a) $\dfrac{\dfrac{16y^2z}{8yz^2}}{\dfrac{yz}{6y^3z^3}}$ (b) $\dfrac{\dfrac{y^2}{y + 2}}{y^3 + y}$

SOLUTION

(a) $\dfrac{16y^2z/8yz^2}{yz/6y^3z^3} = \dfrac{16y^2z}{8yz^2} \cdot \dfrac{6y^3z^3}{yz} = \dfrac{\overset{2}{16} \cdot 6 \cdot y^5z^4}{8y^2z^3}$

$$= 2 \cdot 6 \cdot y^{5-2}z^{4-3} = 12y^3z$$

(b) $\dfrac{\dfrac{y^2}{y + 2}}{y^3 + y} = \dfrac{y^2}{y + 2} \cdot \dfrac{1}{y^3 + y} = \dfrac{y^2}{(y + 2)(y^3 + y)}$

$$= \dfrac{y^2}{(y + 2)y(y^2 + 1)}$$

$$= \dfrac{y}{(y + 2)(y^2 + 1)} \quad \blacksquare$$

C

EXAMPLE 13

Simplify the compound fraction

$$\frac{\dfrac{1}{x+h} - \dfrac{1}{x}}{h}.$$

SOLUTION   First write the numerator as a single fraction, then invert and multiply.

$$\frac{\dfrac{1}{x+h} - \dfrac{1}{x}}{h} = \frac{\dfrac{x}{x(x+h)} - \dfrac{x+h}{x(x+h)}}{h}$$

$$= \frac{\dfrac{x-(x+h)}{x(x+h)}}{h} = \frac{\dfrac{-h}{x(x+h)}}{h}$$

$$= \frac{-h}{x(x+h)} \cdot \frac{1}{h} = \frac{-1}{x(x+h)} \qquad \blacksquare$$

▪▪ RATIONALIZING NUMERATORS AND DENOMINATORS

When dealing with fractions in the days before calculators, it was customary to *rationalize the denominators,* that is, write equivalent fractions with no radicals in the denominator, because this made many computations easier. With calculators, of course, there is no computational advantage to rationalizing denominators. Nevertheless, rationalizing denominators or numerators is sometimes needed to simplify expressions and to derive useful formulas.

EXAMPLE 14

Rationalize the denominators of

(a) $\dfrac{7}{\sqrt{5}}$ (b) $\dfrac{2}{3+\sqrt{6}}$

SOLUTION

(a) The key is to multiply the fraction by 1, with 1 written as a radical fraction.

$$\frac{7}{\sqrt{5}} = \frac{7}{\sqrt{5}} \cdot 1 = \frac{7}{\sqrt{5}} \cdot \frac{\sqrt{5}}{\sqrt{5}} = \frac{7\sqrt{5}}{5}$$

(b) The same technique works here.

Multiply by 1:

$$\frac{2}{3 + \sqrt{6}} = \frac{2}{3 + \sqrt{6}} \cdot 1$$

Rewrite 1 as a radical fraction:

$$= \frac{2}{3 + \sqrt{6}} \cdot \frac{3 - \sqrt{6}}{3 - \sqrt{6}}$$

Multiply:

$$= \frac{2(3 - \sqrt{6})}{(3 + \sqrt{6})(3 - \sqrt{6})}$$

Use the multiplication pattern
$(a + b)(a - b) = a^2 - b^2$ in the denominator:

$$= \frac{6 - 2\sqrt{6}}{3^2 - (\sqrt{6})^2}$$

Simplify:

$$= \frac{6 - 2\sqrt{6}}{9 - 6} = \frac{6 - 2\sqrt{6}}{3} \quad \blacksquare$$

*C

EXAMPLE 15

Assume that $h \neq 0$ and rationalize the numerator of

$$\frac{\sqrt{x + h} - \sqrt{x}}{h};$$

that is, write an equivalent fraction with no radicals in the numerator.

SOLUTION   Begin by multiplying the fraction by 1, with 1 written as a suitable radical fraction.

$$\frac{\sqrt{x + h} - \sqrt{x}}{h} = \frac{\sqrt{x + h} - \sqrt{x}}{h} \cdot 1$$

$$= \frac{\sqrt{x + h} - \sqrt{x}}{h} \cdot \frac{\sqrt{x + h} + \sqrt{x}}{\sqrt{x + h} + \sqrt{x}}$$

Use the multiplication pattern
$(a + b)(a - b) = a^2 - b^2$:

$$= \frac{(\sqrt{x + h})^2 - (\sqrt{x})^2}{h(\sqrt{x + h} + \sqrt{x})} = \frac{x + h - x}{h(\sqrt{x + h} + \sqrt{x})}$$

$$= \frac{h}{h(\sqrt{x + h} + \sqrt{x})}$$

Cancel h:

$$= \frac{1}{\sqrt{x + h} + \sqrt{x}} \quad \blacksquare$$

---

* **C** and ⬤ indicate examples, exercises, and sections that are relevant to calculus.

✓ EXERCISES 0.6

In Exercises 1–10, express the fraction in lowest terms.

1. $\dfrac{63}{49}$ **2.** $\dfrac{121}{33}$

3. $\dfrac{13 \cdot 27 \cdot 22 \cdot 10}{6 \cdot 4 \cdot 11 \cdot 12}$ **4.** $\dfrac{x^2 - 4}{x + 2}$

5. $\dfrac{x^2 - x - 2}{x^2 + 2x + 1}$ **6.** $\dfrac{z + 1}{z^3 + 1}$

7. $\dfrac{a^2 - b^2}{a^3 - b^3}$ **8.** $\dfrac{x^4 - 3x^2}{x^3}$

9. $\dfrac{(x + c)(x^2 - cx + c^2)}{x^4 + c^3 x}$ **10.** $\dfrac{x^4 - y^4}{(x^2 + y^2)(x^2 - xy)}$

In Exercises 11–28, perform the indicated operations.

11. $\dfrac{3}{7} + \dfrac{2}{5}$ **12.** $\dfrac{7}{8} - \dfrac{5}{6}$ **13.** $\left(\dfrac{19}{7} + \dfrac{1}{2}\right) - \dfrac{1}{3}$

14. $\dfrac{1}{a} - \dfrac{2a}{b}$ **15.** $\dfrac{c}{d} + \dfrac{3c}{e}$ **16.** $\dfrac{r}{s} + \dfrac{s}{t} + \dfrac{t}{r}$

17. $\dfrac{b}{c} - \dfrac{c}{b}$ **18.** $\dfrac{a}{b} + \dfrac{2a}{b^2} + \dfrac{3a}{b^3}$ **19.** $\dfrac{1}{x + 1} - \dfrac{1}{x}$

20. $\dfrac{1}{2x + 1} + \dfrac{1}{2x - 1}$

21. $\dfrac{1}{x + 4} + \dfrac{2}{(x + 4)^2} - \dfrac{3}{x^2 + 8x + 16}$

22. $\dfrac{1}{x} + \dfrac{1}{xy} + \dfrac{1}{xy^2}$ **23.** $\dfrac{1}{x} - \dfrac{1}{3x - 4}$

24. $\dfrac{3}{x - 1} + \dfrac{4}{x + 1}$ **25.** $\dfrac{1}{x + y} + \dfrac{x + y}{x^3 + y^3}$

26. $\dfrac{6}{5(x - 1)(x - 2)^2} + \dfrac{x}{3(x - 1)^2(x - 2)}$

27. $\dfrac{1}{4x(x + 1)(x + 2)^3} - \dfrac{6x + 2}{4(x + 1)^3}$

28. $\dfrac{x + y}{(x^2 - xy)(x - y)^2} - \dfrac{2}{(x^2 - y^2)^2}$

In Exercises 29–40, multiply and express in lowest terms.

29. $\dfrac{3}{4} \cdot \dfrac{12}{5} \cdot \dfrac{10}{9}$ **30.** $\dfrac{10}{45} \cdot \dfrac{6}{14} \cdot \dfrac{1}{2}$ **31.** $\dfrac{3a^2 c}{4ac} \cdot \dfrac{8ac^3}{9a^2 c^4}$

32. $\dfrac{6x^2 y}{2x} \cdot \dfrac{y}{21xy}$ **33.** $\dfrac{7x}{11y} \cdot \dfrac{66y^2}{14x^3}$ **34.** $\dfrac{ab}{c^2} \cdot \dfrac{cd}{a^2 b} \cdot \dfrac{ad}{bc^2}$

35. $\dfrac{3x + 9}{2x} \cdot \dfrac{8x^2}{x^2 - 9}$ **36.** $\dfrac{4x + 16}{3x + 15} \cdot \dfrac{2x + 10}{x + 4}$

37. $\dfrac{5y - 25}{3} \cdot \dfrac{y^2}{y^2 - 25}$ **38.** $\dfrac{6x - 12}{6x} \cdot \dfrac{8x^2}{x - 2}$

39. $\dfrac{u}{u - 1} \cdot \dfrac{u^2 - 1}{u^2}$ **40.** $\dfrac{t^2 - t - 6}{t^2 - 6t + 9} \cdot \dfrac{t^2 + 4t - 5}{t^2 - 25}$

In Exercises 41–52, compute the quotient and express in lowest terms.

41. $\dfrac{5}{12} \div \dfrac{4}{14}$ **42.** $\dfrac{\dfrac{100}{52}}{\dfrac{27}{26}}$ **43.** $\dfrac{uv}{v^2 w} \div \dfrac{uv}{u^2 v}$

44. $\dfrac{3x^2 y}{(xy)^2} \div \dfrac{3xyz}{x^2 y}$ **45.** $\dfrac{\dfrac{x + 3}{x + 4}}{\dfrac{2x}{x + 4}}$ **46.** $\dfrac{\dfrac{(x + 2)^2}{(x - 2)^2}}{\dfrac{x^2 + 2x}{x^2 - 4}}$

47. $\dfrac{\dfrac{(c + d)^2}{c^2 - d^2}}{cd}$ **48.** $\dfrac{\dfrac{1}{x} - \dfrac{3}{2}}{\dfrac{2}{x - 2} + \dfrac{5}{x}}$

49. $\dfrac{\dfrac{6}{y} - 3}{1 - \dfrac{1}{y - 1}}$ **50.** $\dfrac{\dfrac{1}{3x} - \dfrac{1}{4y}}{\dfrac{5}{6x^2} + \dfrac{1}{y}}$

51. $\dfrac{\dfrac{1}{(x + h)^2} - \dfrac{1}{x^2}}{h}$ **52.** $(x^{-1} + y^{-1})^{-1}$

In Exercises 53–58, rationalize the denominator and simplify your answer.

53. $\dfrac{3}{\sqrt{8}}$ **54.** $\dfrac{2}{\sqrt{6}}$

55. $\dfrac{3}{2 + \sqrt{12}}$ **56.** $\dfrac{1 + \sqrt{3}}{5 + \sqrt{10}}$

57. $\dfrac{2}{\sqrt{x} + 2}$ **58.** $\dfrac{\sqrt{x}}{\sqrt{x} - \sqrt{c}}$

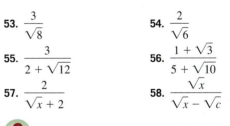

In Exercises 59–62, rationalize the numerator and simplify your answer.

59. $\dfrac{\sqrt{x + h + 1} - \sqrt{x + 1}}{h}$

60. $\dfrac{2\sqrt{x + h + 3} - 2\sqrt{x + 3}}{h}$

61. $\dfrac{\sqrt{(x + h)^2 + 1} - \sqrt{x^2 + 1}}{h}$

62. $\dfrac{\sqrt{(x+h)^2 - (x+h)} - \sqrt{x^2 - x}}{h}$

In Exercises 63–66, use the following formula, which gives the monthly payment required to pay off in n months a loan of P dollars at an annual interest rate i compounded monthly. The interest rate i is expressed as a decimal (for instance, 9% = .09).

$$R = \dfrac{Pi}{12 - 12\left[1 + \dfrac{i}{12}\right]^{-n}}$$

63. (a) If you get a car loan of $12,000 for four years at 9%, what is your monthly payment? [*Hint:* 4 years = 48 months.]

   (b) How much interest will you pay on this loan? [*Hint:* The total you pay is 48 × (monthly payment); the difference between this and $12,000 is the interest.]

64. What would your monthly payment be in Exercise 63 if you wanted to pay off the loan in three years?

65. You buy a house for $80,000. Your parents give you $8000 for a down payment and you get a 30-year mortgage for $72,000 at an interest rate of 7.5%.

   (a) Find your monthly payment.

   (b) Find the total amount of interest you will pay over 30 years.

66. Do Exercise 65 for a 15-year mortgage (same amount and interest rate).

In Exercises 67–70, a dartboard is shown. Assuming that every dart actually hits the board, find the probability that the dart will land in the shaded area. This probability is the quotient

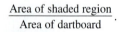

$$\dfrac{\text{Area of shaded region}}{\text{Area of dartboard}}.$$

67.

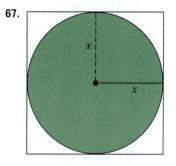

68.

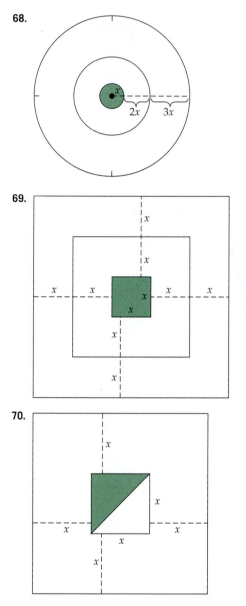

69.

70.

71. The concentration of a drug in a person's bloodstream *t* hours after its injection is modeled by the rational expression.

$$\dfrac{3t}{4t^2 + 5}$$

   (a) What is the concentration after $2\frac{1}{2}$ hours?

   (b) Complete the following table:

t (hours)	$\frac{1}{2}$	1	$1\frac{1}{2}$	2	$2\frac{1}{2}$	3
concentration						

   (c) About what time is the concentration the highest?

(d) What happens to the concentration of the drug as time goes on?

72. The rational expression

$$\frac{.62x^3 - 5.95x^2 + 91.7x + 124}{x}$$

gives the average cost per case of producing x cases of pencils.

(a) What is the average cost of producing 3 cases?

(b) Complete the following table:

x	1	2	3	4	5	6	7	8	9	10
Average cost										

(c) How many cases should be produced to minimize average cost?

Errors to Avoid

In Exercises 73–79, find a numerical example to show that the given statement is false. Then find the mistake in the statement and correct it.

73. $\dfrac{1}{a} + \dfrac{1}{b} = \dfrac{1}{a + b}$

74. $\dfrac{x^2}{x^2 + x^6} = 1 + x^3$

75. $\left(\dfrac{1}{\sqrt{a} + \sqrt{b}}\right)^2 = \dfrac{1}{a + b}$

76. $\dfrac{r + s}{r + t} = 1 + \dfrac{s}{t}$

77. $\dfrac{x}{2 + x} = \dfrac{1}{2}$

78. $\dfrac{x}{\dfrac{1}{y}} = \dfrac{1}{xy}$

79. $(\sqrt{x} + \sqrt{y})\dfrac{1}{\sqrt{x} + \sqrt{y}} = x + y$

Chapter 0 Review

IMPORTANT FACTS & FORMULAS

- $|c - d|$ = distance from c to d on the number line.
- Laws of Exponents:

$$c^r c^s = c^{r+s} \qquad (cd)^r = c^r d^r$$

$$\frac{c^r}{c^s} = c^{r-s} \qquad \left(\frac{c}{d}\right)^r = \frac{c^r}{d^r}$$

$$(c^r)^s = c^{rs} \qquad c^{-r} = \frac{1}{c^r}$$

REVIEW QUESTIONS

1. Fill the blanks with one of the symbols $<$, $=$, or $>$ so that the resulting statement is true.
   (a) 142 ____ $|-51|$
   (b) $\sqrt{2}$ ____ $|-2|$
   (c) -1000 ____ $\dfrac{1}{10}$
   (d) $|-2|$ ____ $-|6|$
   (e) $|u - v|$ ____ $|v - u|$ where u and v are fixed real numbers.

2. List two real numbers that are *not* rational numbers.

3. Express in symbols:
   (a) y is negative, but greater than -10;
   (b) x is nonnegative and not greater than 10.

4. Express in symbols:
   (a) $c - 7$ is nonnegative;
   (b) .6 is greater than $|5x - 2|$.

5. Express in symbols:
   (a) x is less than 3 units from -7 on the number line;

(b) y is farther from 0 than x is from 3 on the number line.

6. Simplify: $|b^2 - 2b + 1|$

7. Solve: $|x - 5| = 3$

8. Solve: $|x + 2| = 4$

9. Solve: $|x + 3| = \dfrac{5}{2}$

10. Solve: $|x - 5| = 2$

11. (a) $|\pi - 7| =$ _____
    (b) $|\sqrt{23} - \sqrt{3}| =$ _____

12. If c and d are real numbers with $c \neq d$ what are the possible values of $\dfrac{c - d}{|c - d|}$?

13. A number x is one of the five numbers A, B, C, D, E on the number line shown in the figure.

If x satisfies *both* $A \le x < D$ and $B < x \le E$, then
(a) $x = A$ (b) $x = B$
(c) $x = C$ (d) $x = D$
(e) $x = E$

14. Which one of these statements is *always* true for any real number x, y?
    (a) $|2x| = 2x$ (b) $\sqrt{x^2} = x$
    (c) $|x - y| = x - y$ (d) $|x - y| = |y - x|$
    (e) $|x - y| = |x| + |y|$

15. Express $0.282828 \cdots$ as a fraction.

16. Express $.362362362 \cdots$ as a fraction.

In Questions 17–20, simplify the expression and use only positive exponents in your answer.

17. $(c^3 d^{-3} e^{-1})^5$

18. $(c^3 d^2)^3 (c^{-1} d)^{-2}$

19. $\dfrac{(8u^5)^2 2^{-4} u^{-3}}{2u^8}$

20. $\dfrac{(2x^2)^{-1} yz^3}{x^2 (yz)^3}$

21. Express in scientific notation:
    (a) 12,320,000,000,000,000   (b) .0000000000789

22. Express in decimal notation:
    (a) 4.78×10^8 (b) 6.53×10^{-9}

In Questions 23–28, simplify the expression.

23. $\sqrt{\sqrt[3]{c^{12}}}$

24. $(\sqrt[3]{4c^3 d^2})^3 (c\sqrt{d})^2$

25. $(a^{-2/3} b^{2/5})(a^3 b^6)^{4/3}$

26. $\dfrac{(3c)^{3/5} (2d)^{-2} (4c)^{1/2}}{(4c)^{1/5} (2d)^4 (2c)^{-3/2}}$

27. $(u^{1/4} - v^{1/4})(u^{1/4} + v^{1/4})$

28. $c^{3/2}(2c^{1/2} + 3c^{-3/2})$

In Questions 29 and 30, simplify and write the expression without radicals or negative exponents:

29. $\dfrac{\sqrt[3]{6c^4 d^{14}}}{\sqrt[3]{48c^{-2} d^2}}$

30. $\dfrac{(8u^5)^{1/4} 2^{-1} u^{-3}}{2u^8}$

31. Let a and b be positive real numbers. Which of these statements is *false*?
    (a) $a^r a^s = a^{r+s}$ (b) $\sqrt{a}\sqrt{b} = \sqrt{ab}$
    (c) $\sqrt[3]{a^3} = a$ (d) $\sqrt{a + b} = \sqrt{a} + \sqrt{b}$
    (e) $\sqrt{\sqrt{a}} = \sqrt[4]{a}$

32. The length L of an animal is related to its surface area S by the equation $L = \left(\dfrac{S}{a}\right)^{1/2}$, where a is a constant that depends on the type of animal. Suppose one animal has twice the surface area of another animal of the same type. How much longer is the first animal than the second?

In Questions 33–40, perform the indicated operations and simplify your answer.

33. $(-2p^3 - 5p + 7) + (-4p^2 + 8p + 2)$

34. $(-4y^2 - 3y + 8) - (2y^2 - 6y - 2)$

35. $(3z + 5)(4z^2 - 2z + 1)$

36. $(2k + 3)(4k^3 - 3k^2 + k)$

37. $(6k - 1)(2k - 3)$

38. $(8r + 3)(r - 1)$

39. $3x(2x - 2)^2$

40. $(\sqrt{x} + 2y)(\sqrt{x} - y)$

41. Suppose the cost of making x cartons of videotapes is
$$.02x^3 + x^2 - 2x + 520 \qquad (0 \le x \le 65),$$
and the revenue from selling them is
$$-.0027x^4 + .27x^3 - 5.2x^2 + 68x.$$
Find a polynomial that expresses the profit on the sale of x cartons.

(Exercises continue on next page.)

42. The volume V of the frustum of a square pyramid, as shown in the figure, is given by

$$V = \frac{h(a^2 + ab + b^2)}{3}.$$

Suppose the side of the base b is twice as long as the side of the top a.
(a) Find a formula for the volume in terms of a and h.
(b) Find a formula for the volume in terms of b and h.

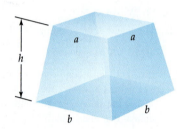

In Questions 43–54, factor the given expression.

43. $x^2 + 5x + 4$ **44.** $x^2 - 4x + 3$

45. $x^2 - 2x - 8$ **46.** $x^2 - 10x + 25$

47. $4x^2 - 9$ **48.** $49y^2 - \frac{1}{4}$

49. $4x^2 - 4x - 15$ **50.** $6x^2 - x - 2$

51. $3x^3 + 5x^2 - 2x$ **52.** $2x^3 + 4x^2 - 6x$

53. $x^4 - 1$ **54.** $x^3 - 8$

55. Which of these statements is *true*?
(a) $(4x)^3 = 4x^3$
(b) $(x + y)^2 = x^2 + y^2$
(c) $(a + b)(a^2 + b^2) = a^3 + b^3$
(d) $x - (2y + 6) = x - 2y + 6$
(e) None of the above

56. Which of these statements is *false*?
(a) $3x - (2x - 7) = x + 7$
(b) $x^3 - 27 = (x - 3)(x^2 + 3x + 9)$
(c) $\left(2\sqrt{u}\right)^4 = 16u^2$
(d) $\left(3\sqrt[4]{v}\right)^2 = 9v^2$
(e) None of the above

In Questions 57–70, compute and simplify.

57. $\dfrac{2a + 4a^2}{2a}$

58. $\dfrac{b^2 - b - 6}{b + 2}$

59. $\dfrac{4x - 7}{8x} - \dfrac{3x - 5}{12x}$

60. $\dfrac{3y + 6}{y} + \dfrac{y - 2}{3}$

61. $1 - \dfrac{x}{x - y}$

62. $\dfrac{2x - 3}{x - 2} + \dfrac{x}{x + 2}$

63. $\dfrac{x^2 - 3x + 2}{x^2 - 4} \cdot \dfrac{x + 2}{x - 3}$

64. $\dfrac{9x - 18}{6x + 12} \cdot \dfrac{3x + 6}{15x - 30}$

65. $\dfrac{12r + 24}{36r - 36} \div \dfrac{6r + 12}{8r - 8}$

66. $\dfrac{4a + 12}{2a - 10} \div \dfrac{a^2 - 9}{a^2 - a - 20}$

67. $\dfrac{6r - 18}{9r^2 + 6r - 24} \cdot \dfrac{12r - 16}{4r - 12}$

68. $\dfrac{k^2 - k - 6}{k^2 + k - 12} \cdot \dfrac{k^2 + 3k - 4}{k^2 + 2k - 3}$

69. $\dfrac{1 + \dfrac{1}{x}}{1 - \dfrac{1}{x}}$

70. $\dfrac{2 - \dfrac{2}{y}}{2 + \dfrac{2}{y}}$

71. Rationalize the numerator and simplify:

$$\frac{\sqrt{2x + 2h + 1} - \sqrt{2x + 1}}{h}.$$

72. Rationalize the denominator:

$$\frac{5}{\sqrt{x} - 3}.$$

DISCOVERY PROJECT O   Real Numbers in the Real World

Mathematical calculations arise throughout our technological society. They may range from a simple calculation of the sales tax to the extremely complex logic and calculation used in modern aircraft navigation systems. We mostly ignore the presence of these calculations and simply do what computer software tells us to do. When the software is written, however, decisions must be made to account for the fact that measurement and commerce do not use nonrepeating nonterminating decimals; in other words, people use only rational numbers, not the whole family of real numbers. In fact, in most cases we are restricted to a very small selection of noninteger rational numbers.

1. Suppose you are constructing a triangular display case with a right angle and that you want the legs of the triangle to be 35 cm each. How long should the long side be? Answer this question using the Pythagorean Theorem and assume that your measuring device is accurate to the nearest millimeter.

2. Supose you are constructing a triangular display case with a right angle and that you want the legs of the triangle to be 14 inches each. If your measuring device is accurate to the nearest sixteenth of an inch, how long should the third side be?

   In your answers to Questions 1 and 2, you had to approximate an irrational number, such as $\sqrt{70}$, and round your answer to meet the required tolerance (nearest millimeter or nearest sixteenth of an inch). In this case, rounding does not cause any difficulty, but when a rounded value is used in subsequent calculations, problems can arise. One situation in which such difficulties occur is in compound interest calculations, when amounts are rounded to the nearest penny. For example, suppose you borrow $50 at an interest rate of 1.5%, compounded monthly. At the end of the first month, the amount of interest is

$$.015 \times 50 = .75$$

so that you now owe $50.75. At the end of the second month, you owe interest on this entire amount, namely,

$$.015 \times 50.75 = .76125.$$

This can be rounded down to .76 or up to .77. Depending on which choice is made, you owe

$$\$50.75 + .76 = \$51.51 \qquad \text{or} \qquad \$50.75 + .77 = \$51.52.$$

Either way, there is a problem: The balance due is either less than or greater than the amount you actually owe. A difference of a penny might not matter now, but after a year or so, the difference may be several cents. Even that won't seriously affect you, but for a bank that has thousands of loans, the difference can be a substantial amount of money.

3. Complete the following three tables, which track the outstanding balance on a $50 loan (with deferred payment) at an interest rate of 1.5%, compounded monthly (a common rate for cash advances on a credit card).

(a) Round interest amounts down to the next penny.

Month	Old Balance	Interest	New Balance
1	$50.00	$0.75	$50.75
2	$50.75	$0.76	$51.51
3	$51.51		
4			
5			
6			
7			
8			
9			
10			
11			
12			

(b) Round interest amounts up to the next penny.

Month	Old Balance	Interest	New Balance
1	$50.00	$0.75	$50.75
2	$50.75	$0.77	$51.52
3	$51.52		
4			
5			
6			
7			
8			
9			
10			
11			
12			

(c) Round to the nearest penny [.014 rounds to .01; .015 rounds to .02].

Month	Old Balance	Interest	New Balance
1	$50.00	$0.75	$50.75
2	$50.75	$0.76	$51.51
3	$51.51	$0.77	
4			
5			
6			
7			
8			
9			
10			
11			
12			

(d) How do the results in parts (a)–(c) compare? Which method is best for you, as a consumer? Which one is best for the bank, and how much do they stand to make if they have 250,000 such accounts?

4. As we shall see in Section 5.2, the formula for determining the amount owed at interest rate r per period, after n periods, is $A = P(1 + r)^n$. In our case, $r = .015$ and $n = 12$. How does the answer from the formula compare with your results in Question 3?

Graphs, Lines, and Technology

AP/Wide World Photos

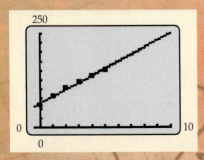

Are you on the Web?

The number of Internet users in the United States increased significantly from 1998 to 2003. Suppose this trend continues and you have data for the years 1998–2003. Then you can use technology to construct a linear model to predict the number of Internet users in future years. See Exercise 21 on page 109.

Chapter Outline

This chapter begins with the essential facts about the coordinate plane and graphs that will be used throughout the text. The last half of the chapter reviews the basic properties of straight lines and shows how technology can be used to construct linear models of real-life situations.

1.1   The Coordinate Plane

Just as real numbers are identified with points on the number line, ordered *pairs* of real numbers can be identified with points in the plane. To do this, draw two number lines in the plane, one vertical and one horizontal, as in Figure 1–1. The horizontal line is usually called the **x-axis,** and the vertical line the **y-axis,** but other letters may be used if desired. The point where the axes intersect is the **origin.** The axes divide the plane into four regions, called **quadrants,** that are numbered as in Figure 1–1.

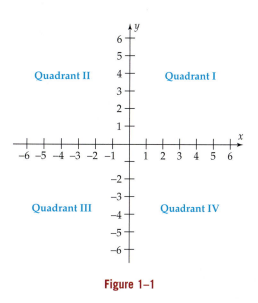

Figure 1–1

*Section 1.3 may be covered immediately after Section 1.1 by omitting the Graphing Exploration in the text and those exercises that require graphing technology.

If P is a point in the plane, draw vertical and horizontal lines through P to the coordinate axes, as shown in Figure 1–2. These lines intersect the x-axis at some number c and the y-axis at d. We say that P has **coordinates** (c, d). The number c is the **x-coordinate** of P, and d is the **y-coordinate** of P. The plane is said to have a **rectangular (or Cartesian) coordinate system.**

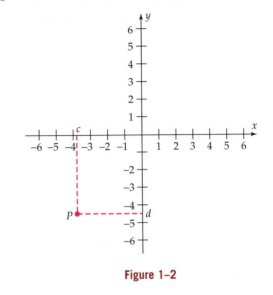

Figure 1–2

EXAMPLE 1

Plot the points $(4, -3)$, $(-5, 2)$, $(2, -5)$, $(0, 13/3)$, $(2, 2)$, and $(-4, -1.2)$.

SOLUTION   You can think of the coordinates of a point as directions for locating it. To find $(4, -3)$, for instance, start at the origin and move 4 units to the right along the x-axis (positive direction), then move 3 units downward (negative direction), as shown in Figure 1–3. To find $(-5, 2)$, start at the origin and move 5 units to the left along the x-axis (negative direction), then move 2 units upward (positive direction). The other points are plotted similarly, as shown in Figure 1–3. ■

CAUTION

The coordinates of a point are an *ordered* pair. Figure 1–3 shows that the point P with coordinates $(-5, 2)$ is quite different from the point Q with coordinates $(2, -5)$. The same numbers (2 and -5) occur in both cases, but in *different order*.

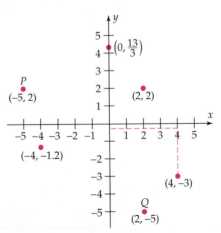

Figure 1–3

EXAMPLE 2

The following table, from the U.S. Department of Education shows the maximum Pell Grant for college students in selected years.

Year	1990	1992	1994	1996	1998	2000	2002	2004
Amount	2300	2400	2300	2470	3000	3300	4000	4050

One way to represent this data graphically is to represent each year's maximum by a point; for instance (1990, 2300) and (2002, 4000). Alternatively, to avoid using large numbers, we can let x be the number of years since 1990, so that $x = 0$ is 1990 and $x = 12$ is 2002. We can also list the dollar amounts in *hundreds,* so the points for 1990 and 2002 are (0, 23) and (12, 40). Plotting all the data in this way leads to the **scatter plot** in Figure 1–4. Connecting these data points with line segments produces the **line graph** in Figure 1–5. ■

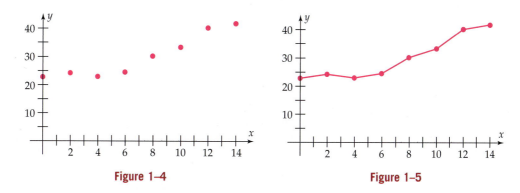

| Figure 1–4 | Figure 1–5 |

▪▪ THE DISTANCE FORMULA

We shall often identify a point with its coordinates and refer, for example, to the point (2, 3). When dealing with several points simultaneously, it is customary to label the coordinates of the first point (x_1, y_1), the second point (x_2, y_2), the third point (x_3, y_3), and so on.[*] Once the plane is coordinatized, it is easy to compute the distance between any two points.

The Distance Formula

The distance between points (x_1, y_1) and (x_2, y_2) is
$$\sqrt{(x_1 - x_2)^2 + (y_1 - y_2)^2}.$$

Before proving the distance formula, we shall see how it is used.

---

[*]"x_1" is read "x-one" or "x-sub-one"; it is a *single symbol* denoting the first coordinate of the first point, just as c denotes the first coordinate of (c, d). Analogous remarks apply to y_1, x_2, and so on.

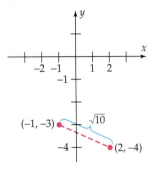

Figure 1–6

EXAMPLE 3

To find the distance between the points $(-1, -3)$ and $(2, -4)$ in Figure 1–6, substitute $(-1, -3)$ for (x_1, y_1) and $(2, -4)$ for (x_2, y_2) in the distance formula:

$$\text{Distance formula:} \quad \text{Distance} = \sqrt{(x_1 - x_2)^2 + (y_1 - y_2)^2}$$

$$\text{Substitute:} \quad = \sqrt{(-1 - 2)^2 + (-3 - (-4))^2}$$

$$\text{Simplify:} \quad = \sqrt{(-3)^2 + (-3 + 4)^2}$$

$$= \sqrt{9 + 1} = \sqrt{10}.$$

The order in which the points are used in the distance formula doesn't make a difference. If we substitute $(2, -4)$ for (x_1, y_1) and $(-1, -3)$ for (x_2, y_2), we get the same answer:

$$\sqrt{[2 - (-1)]^2 + [-4 - (-3)]^2} = \sqrt{3^2 + (-1)^2} = \sqrt{10}. \quad \blacksquare$$

EXAMPLE 4

In a Cubs game at Wrigley Field, a fielder catches the ball near the right-field corner and throws it to second base. The right-field corner is 353 feet from home plate along the right-field foul line. If the fielder is 5 feet from the outfield wall and 5 feet from the foul line, how far must he throw the ball?

SOLUTION   Imagine that the playing field is placed on the coordinate plane, with home plate at the origin and the right-field foul line along the positive x-axis, as shown in Figure 1–7 (not to scale).

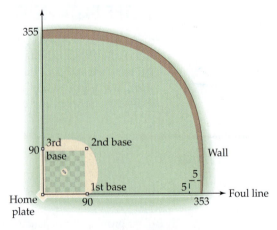

Figure 1–7

Since the four bases form a square whose sides measure 90 ft each, second base has coordinates $(90, 90)$. The fielder is located 5 feet from the wall, so his x-coordinate is $353 - 5 = 348$. His y-coordinate is 5, since he is 5 feet from the foul line. Therefore the distance he throws is the distance from (348, 5) to (90, 90), which can be found as follows.

Distance formula: $\quad\text{Distance} = \sqrt{(x_1 - x_2)^2 + (y_1 - y_2)^2}$

Substitute: $\qquad\qquad = \sqrt{(348 - 90)^2 + (5 - 90)^2}$

Simplify: $\qquad\qquad = \sqrt{258^2 + (-85)^2}$

$\qquad\qquad\qquad = \sqrt{73{,}789} \approx 271.6 \text{ feet.}$

Therefore, he must throw about 272 feet.   ■

Proof of the Distance Formula   Figure 1–8 shows typical points P and Q in the plane. We must find length d of line segment PQ.

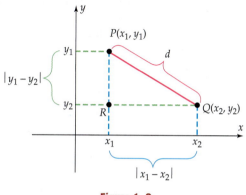

Figure 1–8

As shown in Figure 1–8, the length of RQ is the same as the distance from x_1 to x_2 on the x-axis (number line), namely, $|x_1 - x_2|$. Similarly, the length of PR is the same as the distance from y_1 to y_2 on the y-axis, namely, $|y_1 - y_2|$. According to the Pythagorean Theorem[*] the length d of PQ is given by

$$(\text{Length } PQ)^2 = (\text{Length } RQ)^2 + (\text{Length } PR)^2$$

$$d^2 = |x_1 - x_2|^2 + |y_1 - y_2|^2.$$

Since $|c|^2 = |c| \cdot |c| = |c^2| = c^2$ (because $c^2 \geq 0$), this equation becomes

$$d^2 = (x_1 - x_2)^2 + (y_1 - y_2)^2.$$

Since the length d is nonnegative, we must have

$$d = \sqrt{(x_1 - x_2)^2 + (y_1 - y_2)^2}.\qquad ■$$

The distance formula can be used to prove the following useful fact (see Exercise 68).

The Midpoint Formula

The midpoint of the line segment from (x_1, y_1) to (x_2, y_2) is

$$\left(\frac{x_1 + x_2}{2}, \frac{y_1 + y_2}{2} \right).$$

*See the Geometry Review Appendix.

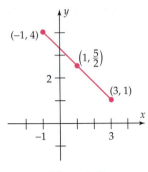

Figure 1–9

EXAMPLE 5

To find the midpoint of the segment joining $(-1, 4)$ and $(3, 1)$, use the formula in the box with $x_1 = -1$, $y_1 = 4$, $x_2 = 3$, and $y_2 = 1$. The midpoint is

$$\left(\frac{x_1 + x_2}{2}, \frac{y_1 + y_2}{2}\right) = \left(\frac{-1 + 3}{2}, \frac{4 + 1}{2}\right) = \left(1, \frac{5}{2}\right),$$

as shown in Figure 1–9. ■

▉▉ GRAPHS

A **graph** is a set of points in the plane. Some graphs are based on data points, such as Figures 1–4 and 1–5. Other graphs arise from equations, as follows. A **solution** of an equation in variables x and y is a pair of numbers such that the substitution of the first number for x and the second for y produces a true statement. For instance, $(3, -2)$ is a solution of $5x + 7y = 1$ because

$$5 \cdot 3 + 7(-2) = 1,$$

and $(-2, 3)$ is *not* a solution because $5(-2) + 7 \cdot 3 \neq 1$. The **graph of an equation** in two variables is the set of points in the plane whose coordinates are solutions of the equation. Thus the graph is a *geometric picture of the solutions.*

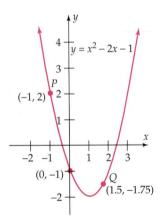

Figure 1–10

EXAMPLE 6

The graph of $y = x^2 - 2x - 1$ is shown in Figure 1–10. Verify that the coordinates of P and Q actually are solutions of the equation.

SOLUTION   For the point $(-1, 2)$, substitute -1 for x in the equation:

$$y = x^2 - 2x - 1$$

$$y = (-1)^2 - 2(-1) - 1 = 1 + 2 - 1 = 2.$$

Therefore, $(-1, 2)$ is a solution. Similarly, substituting $x = 1.5$ shows that

$$y = x^2 - 2x - 1 = 1.5^2 - 2(1.5) - 1 = 2.25 - 3 - 1 = -1.75.$$

Hence, $(1.5, -1.75)$ is also a solution.   ■

▉▉ CIRCLES

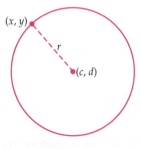

Figure 1–11

If (c, d) is a point in the plane and r a positive number, then the **circle with center (c, d) and radius r** consists of all points (x, y) that lie r units from (c, d), as shown in Figure 1–11. According to the distance formula, the statement that "the distance from (x, y) to (c, d) is r units" is equivalent to

$$\sqrt{(x - c)^2 + (y - d)^2} = r.$$

Squaring both sides shows that (x, y) satisfies this equation:

$$(x - c)^2 + (y - d)^2 = r^2.$$

Reversing the procedure shows that any solution (x, y) of this equation is a point on the circle. Therefore, we have the following.

Circle Equation

The circle with center (c, d) and radius r is the graph of

$$(x - c)^2 + (y - d)^2 = r^2.$$

We say that $(x - c)^2 + (y - d)^2 = r^2$ is the **equation of the circle** with center (c, d) and radius r.

EXAMPLE 7

Identify the graph of the equation $(x - 4)^2 + (y - 2)^2 = 9$.

SOLUTION   Since $9 = 3^2$, we can write the equation as

$$(x - 4)^2 + (y - 2)^2 = 3^2.$$

Now the equation is of the form shown in the box above, with $c = 4$, $d = 2$ and $r = 3$. So the graph is a circle with center $(4, 2)$ and radius 3, as shown in Figure 1–12. ■

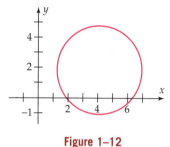

Figure 1–12

EXAMPLE 8

Find the equation of the circle with center $(-3, 2)$ and radius 2 and sketch its graph.

SOLUTION   Here the center is $(c, d) = (-3, 2)$ and the radius is $r = 2$, so the equation of the circle is

$$(x - c)^2 + (y - d)^2 = r^2$$
$$[x - (-3)]^2 + (y - 2)^2 = 2^2$$
$$(x + 3)^2 + (y - 2)^2 = 4.$$

Its graph is shown in Figure 1–13. ■

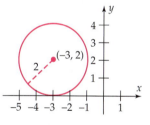

Figure 1–13

EXAMPLE 9

Find the equation of the circle with center $(3, -1)$ that passes through $(2, 4)$ and sketch its graph.

SOLUTION   We must first find the radius. Since $(2, 4)$ is on the circle, the radius is the distance from $(2, 4)$ to $(3, -1)$ as shown in Figure 1–14, namely,

$$\sqrt{(2 - 3)^2 + (4 - (-1))^2} = \sqrt{1 + 25} = \sqrt{26}.$$

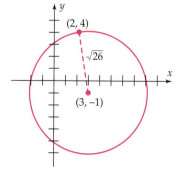

Figure 1–14

The equation of the circle with center at $(3, -1)$ and radius $\sqrt{26}$ is

$$(x - 3)^2 + (y - (-1))^2 = (\sqrt{26})^2$$
$$(x - 3)^2 + (y + 1)^2 = 26$$
$$x^2 - 6x + 9 + y^2 + 2y + 1 = 26$$
$$x^2 + y^2 - 6x + 2y - 16 = 0. \qquad ■$$

When the center of a circle of radius r is at the origin $(0, 0)$, its equation takes a simpler form.

Circle at the Origin

The circle with center $(0, 0)$ and radius r is the graph of

$$x^2 + y^2 = r^2.$$

Proof   Substitute $c = 0$ and $d = 0$ in the equation for the circle with center (c, d) and radius r.

$$(x - c)^2 + (y - d)^2 = r^2$$
$$(x - 0)^2 + (y - 0)^2 = r^2$$
$$x^2 + y^2 = r^2. \qquad ■$$

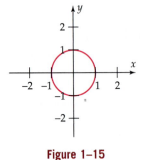

Figure 1–15

EXAMPLE 10

Letting $r = 1$ shows that the graph of $x^2 + y^2 = 1$ is the circle of radius 1 centered at the origin, as shown in Figure 1–15. This circle is called the **unit circle.** ■

✓ EXERCISES 1.1

1. Find the coordinates of points $A–I$.

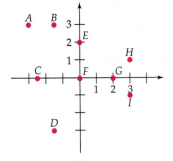

In Exercises 2–5, find the coordinates of the point P.

2. P lies 4 units to the left of the y-axis and 5 units below the x-axis.

3. P lies 3 units above the x-axis and on the same vertical line as $(-6, 7)$.

4. P lies 2 units below the x-axis, and its x-coordinate is three times its y-coordinate.

5. P lies 4 units to the right of the y-axis, and its y-coordinate is half its x-coordinate.

In Exercises 6–8, sketch a scatter plot and a line graph of the given data.

6. Tuition and fees at four-year public colleges in the fall of each year are shown in the table[*]. Let $x = 0$ correspond to 1995.

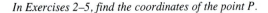

*The College Board.

Year	Tuition and Fees
1995	2811
1996	2975
1997	3111
1998	3247
1999	3362
2000	3487
2001	3754

7. The table shows projected sales of personal digital video recorders.[*] Let $x = 0$ correspond to 2000, and measure y in thousands.

Year	Number Sold
2000	257,000
2001	129,000
2002	143,000
2003	214,000
2004	315,000
2005	485,000

8. The maximum yearly contribution to an individual retirement account (IRA) is now $3000. It will change to $4000 in 2005 and change to $5000 in 2008. Assuming 3% inflation, however, the picture is somewhat different. The table shows the maximum IRA contribution in fixed 2003 dollars. Let $x = 0$ correspond to 2000.

Year	Maximum Contribution
2003	3000
2004	2910
2005	3764
2006	3651
2007	3541
2008	4294

9. (a) If the first coordinate of a point is greater than 3 and its second coordinate is negative, in what quadrant does it lie?
   (b) What is the answer in part (a) if the first coordinate is less than 3?

10. In what quadrant(s) does a point lie if the product of its coordinates is
   (a) positive?   (b) negative?

11. (a) Plot the points $(3, 2)$, $(4, -1)$, $(-2, 3)$, and $(-5, -4)$.
   (b) Change the sign of the y-coordinate in each of the points in part (a), and plot these new points.
   (c) Explain how the points (a, b) and $(a, -b)$ are related graphically. [*Hint:* What are their relative positions with respect to the x-axis?]

12. (a) Plot the points $(5, 3)$, $(4, -2)$, $(-1, 4)$, and $(-3, -5)$.
   (b) Change the sign of the x-coordinate in each of the points in part (a), and plot these new points.
   (c) Explain how the points (a, b) and $(-a, b)$ are related graphically. [*Hint:* What are their relative positions with respect to the y-axis?]

In Exercises 13–20, find the distance between the two points and the midpoint of the segment joining them.

13. $(-3, 5)$, $(2, -7)$ **14.** $(2, 4)$, $(1, 5)$

15. $(1, -5)$, $(2, -1)$ **16.** $(-2, 3)$, $(-3, 2)$

17. $(\sqrt{2}, 1)$, $(\sqrt{3}, 2)$ **18.** $(-1, \sqrt{5})$, $(\sqrt{2}, -\sqrt{3})$

19. (a, b), (b, a) **20.** (s, t), $(0, 0)$

21. Find the perimeter of the shaded area in the figure.

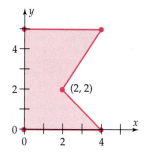

22. What is the perimeter of the triangle with vertices $(1, 1)$, $(5, 4)$, and $(-2, 5)$?

In Exercises 23–24, show that the three points are the vertices of a right triangle, and state the length of the hypotenuse. [You may assume that a triangle with sides of lengths a, b, c is a right triangle with hypotenuse c provided that $a^2 + b^2 = c^2$.]

23. $(0, 0)$, $(1, 1)$, $(2, -2)$

24. $(3, -2)$, $(0, 4)$, $(-2, 3)$

25. Sales of organically grown foods in the United States were $7.8 billion in 2000 and are projected to be $16 billion in 2004.[†]
   (a) Represent this data graphically by two points.
   (b) Find the midpoint of the line segment joining these two points.
   (c) How might this midpoint be interpreted? What assumptions, if any, are needed to make this interpretation?

---

26. Suppose a baseball playing field is placed on the coordinate plane, as in Example 4.
(a) Find the coordinates of first and third base.
(b) If the left fielder is at the point (50, 325), how far is he from first base?
(c) How far is the left fielder in part (b) from the right fielder, who is at the point (280, 20)?

27. A standard football field is 100 yards long and $53\frac{1}{3}$ yards wide. The quarterback, who is standing on the 10-yard line, 20 yards from the left sideline, throws the ball to a receiver who is on the 45-yard line, 5 yards from the right sideline, as shown in the figure.
(a) How long was the pass? [*Hint:* Place the field in the first quadrant of the coordinate plane, with the left sideline on the y-axis and the goal line on the x-axis. What are the coordinates of the quarterback and the receiver?]
(b) A player is standing halfway between the quarterback and the receiver. What are his coordinates?

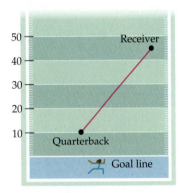

28. How far is the quarterback in Exercise 27 from a player who is on the 50-yard line, halfway between the sidelines?

In Exercises 29–34, determine whether the point is on the graph of the given equation.

29. $(1, -2)$; $3x - y - 5 = 0$

30. $(2, -1)$; $x^2 + y^2 - 6x + 8y = -15$

31. $(6, 2)$; $3y + x = 12$

32. $(1, -2)$; $3x + y = 12$

33. $(3, 4)$; $(x - 2)^2 + (y + 5)^2 = 4$

34. $(1, -1)$; $\dfrac{x^2}{2} + \dfrac{y^2}{3} = 1$

In Exercises 35–38, find the equation of the circle with given center and radius r.

35. $(-3, 4)$; $r = 2$

36. $(-2, -1)$; $r = 3$

37. $(0, 0)$; $r = \sqrt{2}$

38. $(5, -2)$; $r = 1$

In Exercises 39–42, sketch the graph of the equation. Label the x- and y-intercepts.

39. $(x - 5)^2 + (y + 2)^2 = 5$

40. $(x + 6)^2 + y^2 = 4$

41. $(x + 1)^2 + (y - 3)^2 = 9$

42. $(x - 2)^2 + (y - 4)^2 = 1$

In Exercises 43–50, find the equation of the circle.

43. Center (2, 2); passes through the origin.

44. Center $(-1, -3)$; passes through $(-4, -2)$.

45. Center (1, 2); intersects x-axis at -1 and 3.

46. Center (3, 1); diameter 2.

47. Center $(-5, 4)$; tangent (touching at one point) to the x-axis.

48. Center $(2, -6)$; tangent to the y-axis.

49. Endpoints of diameter are (3, 3) and $(1, -1)$.

50. Endpoints of diameter are $(-3, 5)$ and $(7, -5)$.

51. One diagonal of a square has endpoints $(-3, 1)$ and $(2, -4)$. Find the endpoints of the other diagonal.

52. Find the vertices of all possible squares with this property: Two of the vertices are (2, 1) and (2, 5). [*Hint:* There are three such squares.]

53. Find a number x such that (0, 0), (3, 2), and $(x, 0)$ are the vertices of an isosceles triangle, neither of whose two equal sides lie on the x-axis.

54. Do Exercise 53 if one of the two equal sides lies on the positive x-axis.

55. The graph, which is based on data from the Actuarial Society of South Africa and assumes no changes in current behavior, shows the projected new cases of AIDS in South Africa (in millions) in coming years ($x = 0$ corresponds to 2000).
(a) Estimate the number of new cases in 2010.
(b) Estimate the year in which the largest number of new cases will occur. About how many new cases will there be in that year?
(c) In what years will the number of new cases be below 7,000,000?

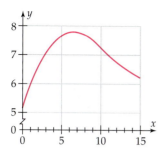

56. The graph shows the total number of alcohol-related car crashes in Ohio at a particular time of the day for the years 1991–2000.[*] Time is measured in hours after midnight. During what periods is the number of crashes
(a) below 5000?
(b) above 15,000?

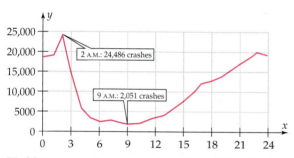

57. Many companies are changing their traditional employee pension plans to so-called cash balance plans. The graph shows pension accrual by age for two hypothetical plans.[†]

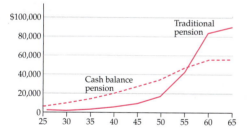

(a) Assuming that you can take your accrued pension benefits in cash when you leave the company before retirement, for what age group is the cash balance plan better?
(b) At what age is the accrued amount the same for either type of pension plan?
(c) If you remain with the company until retirement, how much better off are you with a traditional instead of cash balance plan?

*The Cleveland *Plain Dealer*
†Data from Steve J. Kopp and Lawrence W. Sher, *The Pension Forum,* Vol 11, No. 1. Graph from "What if a Pension Shift Hit Lawmakers Too?" by M. W. Walsh, *New York Times*, March 9, 2003. Copyright © 2003 The New York Times Co. Reprinted with permission.

58. In an ongoing consumer confidence survey, respondents are asked two questions: Are jobs plentiful? Are jobs hard to get? The graph shows the percentage of people answering "yes" to each question over the years.[‡]

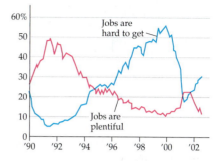

(a) In what year did the most people feel that jobs were plentiful? In that year, approximately what percentage of people felt that jobs were hard to get?
(b) In what year did the most people feel that jobs were hard to get? In that year approximately what percentage of people felt that jobs were plentiful?
(c) In what years was the percentage of those who thought jobs were plentiful the same as the percentage of those that thought jobs were hard to get?

In Exercises 59–62, determine which of graphs A, B, C best describes the given situation.

59. You have a job that pays a fixed salary for the week. The graph shows your salary.

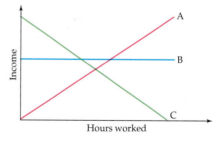

60. You have a job that pays an hourly wage. The graph shows your salary.

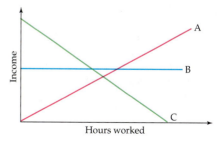

‡Data for The Conference Board. Graph from "Tight U.S. Job Market Adds to Jitters Among Consumers." by A. Berenson, *The New York Times,* March 1, 2003. Copyright © 2003 The New York Times Co. Reprinted with permission.

61. You take a ride on a Ferris wheel. The graph shows your distance from the ground.*

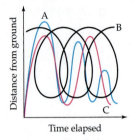

62. Alison's wading pool is filled with a hose by her big sister Emily, and Alison plays in the pool. When they are finished, Emily empties the pool. The graph shows the water level of the pool.

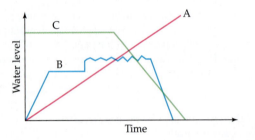

63. Show that the midpoint M of the hypotenuse of a right triangle is equidistant from the vertices of the triangle. [*Hint:* Place the triangle in the first quadrant of the plane, with right angle at the origin so that the situation looks like the figure.]

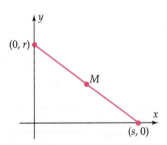

64. Show that the diagonals of a parallelogram bisect each other. [*Hint:* Place the parallelogram in the first quadrant with a vertex at the origin and one side along the x-axis so that the situation looks like the figure.]

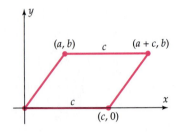

Thinkers

65. For each nonzero real number k, the graph of $(x - k)^2 + y^2 = k^2$ is a circle. Describe all possible such circles.

66. Suppose every point in the coordinate plane is moved 5 units straight up.
   (a) To what point does each of these points go: $(0, -5), (2, 2), (5, 0), (5, 5), (4, 1)$?
   (b) Which points go to each of the points in part (a)?
   (c) To what point does (a, b) go?
   (d) To what point does $(a, b - 5)$ go?
   (e) What point goes to $(-4a, b)$?
   (f) What points go to themselves?

67. Let (c, d) be any point in the plane with $c \neq 0$. Prove that (c, d) and $(-c, -d)$ lie on the same straight line through the origin, on opposite sides of the origin, the same distance from the origin. [*Hint:* Find the midpoint of the line segment joining (c, d) and $(-c, -d)$.]

68. *Proof of the Midpoint Formula* Let P and Q be the points (x_1, y_1) and (x_2, y_2), respectively, and let M be the point with coordinates

$$\left(\frac{x_1 + x_2}{2}, \frac{y_1 + y_2}{2} \right).$$

Use the distance formula to compute the following:
   (a) The distance d from P to Q;
   (b) The distance d_1 from M to P;
   (c) The distance d_2 from M to Q.
   (d) Verify that $d_1 = d_2$.
   (e) Show that $d_1 + d_2 = d$. [*Hint:* Verify that $d_1 = \frac{1}{2}d$ and $d_2 = \frac{1}{2}d$.]
   (f) Explain why parts (d) and (e) show that M is the midpoint of PQ.

1.2 Graphs and Graphing Technology

The traditional method of graphing an equation "by hand" is as follows: Construct a table of values with a reasonable number of entries, plot the corresponding points, and use whatever algebraic or other information is available to make an educated guess about the rest.

EXAMPLE 1

The graph of $y = x^2$ consists of all points (x, x^2), where x is a real number. You can easily construct a table of values and plot the corresponding points, as in Figure 1–16. These points suggest that the graph looks like the one in Figure 1–17, which is obtained by connecting the plotted points and extending the graph upward. ∎

x	$y = x^2$
-2.5	6.25
-2	4
-1.5	2.25
-1	1
$-.5$	.25
0	0
$.5$	.25
1	1
1.5	2.25
2	4
2.5	6.25

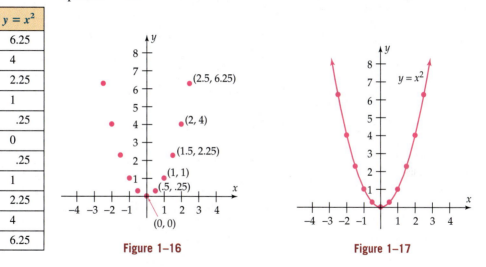

Figure 1–16　　　　　**Figure 1–17**

▨ GRAPHING WITH TECHNOLOGY

A graphing calculator or computer graphing program graphs in the same way you would graph by hand, but with much greater speed: It plots 95 or more points and simultaneously connects them with line segments. These graphs are generally quite accurate, but no technology is perfect. Algebra and geometry may be needed to interpret misleading (or incorrect) screen images. The next example illustrates the basics of graphing with technology.

EXAMPLE 2

Graph the equation　$2x^3 - 8x - 2y + 4 = 0$

(a) using a calculator;

(b) using a computer graphing program.

SOLUTION   (a) For calculator graphing, proceed as follows.

Step 1   *Set the **viewing window**—the portion of the coordinate plane that will appear on the screen.* Since we don't know yet where the graph lies, we'll try the window with $-10 \leq x \leq 10$ and $-10 \leq y \leq 10$. We can change it later if necessary. Press the WINDOW (or RANGE or V-WINDOW or PLOT SET-UP) key,* and enter the appropriate numbers, as in Figure 1–18 (which shows a TI-83+; other calculators are similar). Then the calculator will display the portion of the plane shown in Figure 1–19.

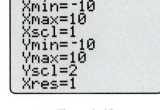

Figure 1–18

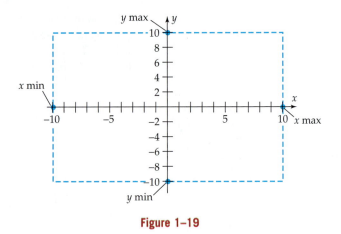

Figure 1–19

Setting $X\text{scl} = 1$ in Figure 1–18 puts the tick marks 1 unit apart on the x-axis; similarly, $Y\text{scl} = 2$ puts the tick marks 2 units apart on the y-axis.[†]

Step 2   *Enter the equation in the calculator.* To do this, you must first solve the equation for y:

$$2x^3 - 8x - 2y + 4 = 0$$

Rearrange terms: $$-2y = -2x^3 + 8x - 4$$

Divide by -2: $$y = x^3 - 4x + 2.$$

Now call up the **equation memory** by pressing Y= on TI (or SYMB on HP-39+ or GRAPH (main menu) on Casio). Use the

---

*On TI-86, press GRAPH first, then WINDOW will appear as a menu choice.
[†]$X\text{scl}$ is labeled "X scale" on Casio and "$X\text{tick}$" on HP-39. Some calculators do not have an $X\text{res}$ setting. On those that do, it should normally be set at 1 (or at "detail" on HP-39).

> **TECHNOLOGY TIP**
> When entering an equation for graphing, use the "variable" key rather than the ALPHA *X* key. It has a label such as *X*, T, *θ* or *X*, T, *θ*, *n* or *X*, *θ*, T or x-VAR. On TI-89, however, use the *x* key on the keyboard.

Technology Tip in the margin to enter the equation, as shown in Figure 1–20.

Step 3 *Graph the equation.* Press GRAPH (or PLOT or DRAW) to obtain Figure 1–21.

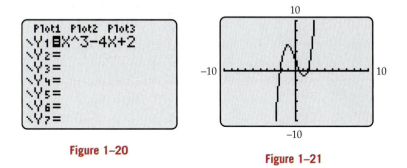

Figure 1–20

Figure 1–21

Because of the limited resolution of a calculator screen, the graph appears to consist of short adjacent line segments rather than a smooth unbroken curve.

Step 4 *If necessary, adjust the viewing window for a better view.* The graph in Figure 1–21 is scrunched up in the middle of the screen. So we change the viewing window (Figure 1–22) and press GRAPH again to obtain Figure 1–23.

> **TECHNOLOGY TIP**
> On TI-83+/86 you may get an error message when you press GRAPH if one of Plot 1, Plot 2, or Plot 3, at the top of the Y= menu (see Figure 1–20) is shaded. In this case, move the cursor over the shading and press ENTER to remove it.

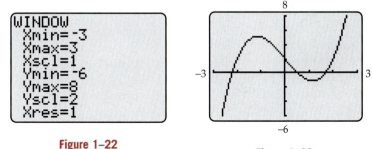

Figure 1–22

Figure 1–23

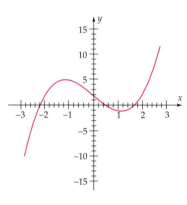

(b) A typical computer graphing program requires the same information as does a calculator, namely, the equation to be graphed and the viewing window in which to graph it. The procedures for entering this information vary widely, so check your instruction manual or help key index. For example, the following command in Maple produces Figure 1–24, in which the range of *y*-values was chosen automatically by Maple.

plot(*x*^3 −4*x + 2, x = −3..3);

To duplicate Figure 1–23, we specify the range of *y*-values and the number of tick marks on each axis:

plot(x^3 −4*x + 2, x = −3..3, y = −6..8,
xtickmarks = 6, ytickmarks = 6);

Figure 1–24

The result is Figure 1–25.

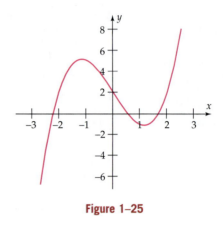

Figure 1–25

As is typical of computer-generated graphs, this one appears smooth and connected, as it should. ∎

> **NOTE**   An equation stays in a calculator's equation memory until you delete it. When several equations are in the memory, you must turn "on" those you want graphed and turn "off" those you don't want graphed, as explained in the Technology Tip in the margin.

TECHNOLOGY TIP

An equation is "on" if its equal sign is shaded or if there is a check mark next to it. Only equations that are "on" are graphed when you press GRAPH.

To turn an equation "on" or "off" on TI-83+, move the cursor over the equal sign and press ENTER. On other calculators, move the cursor to the equation and press SELECT or CHECK.

▟ GRAPHING TOOLS: THE TRACE FEATURE

A calculator obtains a graph by plotting points and simultaneously connecting them. To see which points were actually plotted, press TRACE, and a flashing cursor appears on the graph. Use the left and right arrow keys to move the cursor along the graph. The coordinates of the point the cursor is on appear at the bottom of the screen. Figure 1–26 illustrates this for the graph of $y = x^3 - 4x + 2$.

TECHNOLOGY TIP

If TRACE is not on the keyboard, it will appear on screen after you press GRAPH (or PLOT or DRAW).

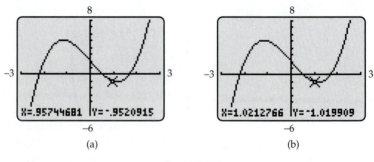

(a)   (b)

Figure 1–26

The trace cursor displays only the points that the calculator actually plotted. For instance, $(1, -1)$ is on the graph of $y = x^3 - 4x + 2$, as you can easily verify, but was not one of the points the calculator plotted to produce Figure 1–26. So the trace lands on two nearby points that *were* plotted but skips $(1, -1)$.

TECHNOLOGY

TIP

If ZOOM is not on the keyboard, it will appear on screen after you press GRAPH (or PLOT or DRAW).

To set the zoom factors, look for FACT, ZFACT, or (SET) FACTORS in the ZOOM menu (or its MEMORY submenu).

▪▪ GRAPHING TOOLS: ZOOM IN/OUT

The ZOOM-OUT and ZOOM-IN keys on the ZOOM menu make it easy to increase or decrease the size of the viewing window with a single keystroke. Figure 1–27 shows the graph of $y = x^3 - 2x^2 + 1$ in two windows. Window (b) was obtained from window (a) by zooming in at the origin by a factor of 4.

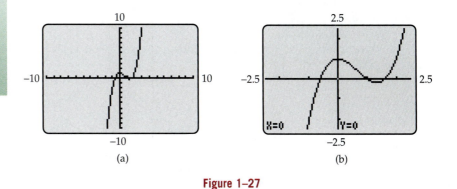

Figure 1–27

The calculator automatically changes the range of x- and y-values when zooming, but does not change the Xscl or Yscl settings. This may cause occasional viewing problems.

NOTE

The heading GRAPHING EXPLORATION indicates that you are to use your calculator or computer as directed to complete the discussion.

GRAPHING EXPLORATION

Graph $y = x^3 - 2x^2 + 1$ in the same window as Figure 1–27(a). Then zoom *out* from the origin by a factor of 4. Can you read the tick marks on the axes? Change the settings to Xscl $= 5$ and Yscl $= 5$, and regraph. Can you read them now?

▪▪ GRAPHING TOOLS: SPECIAL VIEWING WINDOWS

The ZOOM menu also has keys that enable you to obtain frequently used viewing windows in a single keystroke, including most or all of the following.

1. The **standard viewing window** (also labeled **Zstandard** or **Zstd** or **Default**) has $-10 \le x \le 10$ and $-10 \le y \le 10$.

2. A **decimal window** (also labeled **Zdecimal** or **Zdecm** or **Init**) is one in which the horizontal distance between two adjacent pixels on the screen is .1. The size of the decimal window depends on the width of your calculator screen.

GRAPHING EXPLORATION

Graph $y = x^4 - 2x^2 - 1$ in the standard window. Use the TRACE key, and watch the values of the x-coordinates. Now regraph by using DECIMAL in the ZOOM menu. Then use the TRACE key again. How do the x-coordinates change at each step? Finally, look in WINDOW to find the dimensions of your decimal window.

3. In a **square window,** a one-unit segment on the *x*-axis has the same length on the screen as a one-unit segment on the *y*-axis. Because calculator screens are wider than they are high, the *y*-axis in a square window must be shorter than the *x*-axis (see the Technology Tip in the margin). The next example illustrates the usefulness of square windows.

EXAMPLE 3

Graph the circle $x^2 + y^2 = 9$ on a calculator.

SOLUTION   First, we solve the equation for *y:*

$$y^2 = 9 - x^2$$

$$y = \sqrt{9 - x^2} \quad \text{or} \quad y = -\sqrt{9 - x^2}.$$

Graphing both of these equations on the same screen will produce the graph of the circle. In the standard viewing window (Figure 1–28), however, the graph doesn't look like a circle.

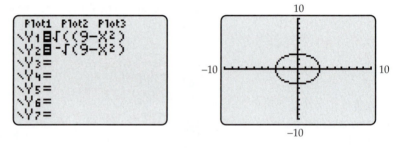

Figure 1–28

The reason is that this is *not* a square window (the 20 one-unit segments on the *y*-axis occupy less space than the 20 one-unit segments on the *x*-axis, which means that the ones on the *x*-axis are a tad longer than the ones on the *y*-axis). So we press SQUARE (or ZSQUARE or ZOOMSQR) on the ZOOM menu and obtain Figure 1–29. The size of the *x*-axis has been changed to produce a square window in which the circle looks round, as it should.* ∎

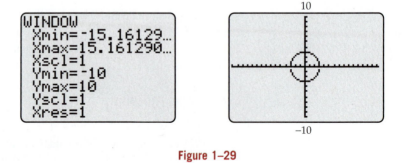

Figure 1–29

*On some calculators, the SQUARE key changes the *y*-axis length instead of the *x*-axis. The result in either case is a square window.

Although any convenient viewing window is usually OK, you should use a square window when you want circles to look round and perpendicular lines to look perpendicular.

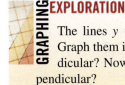 **EXPLORATION**

The lines $y = .5x$ and $y = -2x + 2$ are perpendicular (why?). Graph them in the standard viewing window. Do they look perpendicular? Now graph them in a square window. Do they look perpendicular?

The method used to graph the circle $x^2 + y^2 = 4$ in Example 3 can be used to graph any equation that can be solved for y.

EXAMPLE 4

To graph $12x^2 - 4y^2 + 16x + 12 = 0$, solve the equation for y:

$$4y^2 = 12x^2 + 16x + 12$$
$$y^2 = 3x^2 + 4x + 3$$
$$y = \pm\sqrt{3x^2 + 4x + 3}.$$

Every point on the graph of the equation is on the graph of either

$$y = \sqrt{3x^2 + 4x + 3} \qquad \text{or} \qquad y = -\sqrt{3x^2 + 4x + 3}.$$

 EXPLORATION

Graph the previous two equations on the same screen. The result will be the graph of the original equation. ■

▪▪ GRAPHING TOOLS: THE MAXIMUM/MINIMUM FINDER

Many graphs have peaks and valleys (for instance, see Figure 1–27 on page 77 or Figure 1–30). A calculator's maximum finder or minimum finder can locate these points with a high degree of accuracy, as illustrated in the next example.

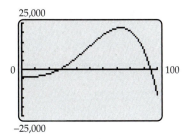

Figure 1–30

EXAMPLE 5

The Cortopassi Computer Company can produce a maximum of 100,000 computers a year. Their annual profit is given by

$$y = -.003x^4 + .3x^3 - x^2 + 5x - 4000,$$

where y is the profit (in thousands of dollars) from selling x thousand computers. Use graphical methods to estimate how many computers should be sold to make the largest possible profit.

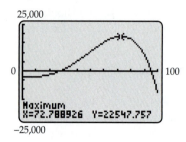

Figure 1–31

SOLUTION We first choose a viewing window. The number x of computers is nonnegative, and no more than 100,000 can be produced, so that $0 \le x \le 100$ (because x is measured in thousands). The profit y may be positive or negative (the company could lose money). So we try a window with $-25,000 \le y \le 25,000$ and obtain Figure 1–30. For each point on the graph,

The x-coordinate is the number of thousands of computers produced;

The y-coordinate is the profit (in thousands) on that number of computers.

The largest possible profit occurs at the point with the largest y-coordinate, that is, the highest point in the window. The maximum finder on a TI-83+ (see the Technology Tip in the margin) produced Figure 1–31. Since x and y are measured in thousands, we see that making about 72,789 computers results in a maximum profit of about $22,547,757. ■

TECHNOLOGY TIP

The graphical maximum finder is in the following menu/submenu:

TI-83+: CALC

TI-86/89: GRAPH/MATH

Casio: GRAPH/G-SOLVE

HP-39+: PLOT/FNC

It is labeled MAXIMUM, MAX, FMAX or EXTREMUM.

On some TI calculators, you must first select a left (or lower) bound, meaning an x-value to the left of the highest point, and a right (or upper) bound, meaning an x-value to its right, and make an initial guess. On other calculators, you may have to move the cursor near the point you are seeking.

GRAPHING EXPLORATION

Graph $y = .3x^3 + .8x^2 - 2x - 1$ in the window with $-5 \le x \le 5$ and $-5 \le y \le 5$. Use your maximum finder to approximate the coordinates of the highest point to the left of the y-axis. Then use your minimum finder (in the same menu) to approximate the coordinates of the *lowest* point to the right of the y-axis. How do these answers compare with the ones you get by using the trace feature?

COMPLETE GRAPHS

A viewing window is said to display a **complete graph** if it shows all the important features of the graph (peaks, valleys, intercepts, etc.) and suggests the general shape of the portions of the graph that aren't in the window. Many different windows may show a complete graph. It's usually best to use a window that is small enough to show as much detail as possible.

In later chapters, we shall develop algebraic facts that will enable us to know when certain graphs are complete. For the present, however, the best you may be able to do is try several different windows to see which, if any, appear to display a complete graph.

EXAMPLE 6

Sketch a complete graph of

$$y = .007x^5 - .2x^4 + 1.332x^3 - .004x^2 + 10.$$

SOLUTION Four different viewing windows for this graph are shown in Figure 1–32. Graph (a) (the standard window) is certainly not complete, since it shows no points to the right of the y-axis. Graph (b) is not complete, since it indicates that parts of the graph lie outside the window.

TIP

Most calculators have an "auto scaling" feature. Once the range of *x*-values has been set, the calculator selects a viewing window that includes all the points on the graph whose *x*-coordinates are in the chosen range.

It is in the ZOOM menu (VIEWS menu on HP-39+) and is called ZOOMFIT, ZFIT, or AUTO(-SCALE).

This feature can eliminate some guesswork but may produce a window so large that it hides some of the features of the graph.

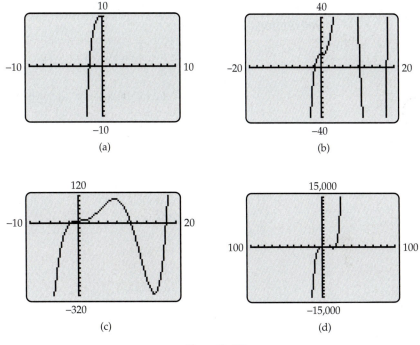

(a) (b) (c) (d)

Figure 1–32

Graph (d) tends to confirm what graph (c) suggests: that the graph keeps climbing sharply forever as you move to the right and that it keeps falling sharply forever as you move to the left. Because of its large scale, however, graph (d) doesn't show the features of the graph near the origin. So we conclude that graph (c) is probably a complete graph, since it shows important features (twists and turns) near the origin, as well as suggesting the shape of the graph farther out. ■

TIP

When you get a blank screen, press TRACE and use the left/right arrow keys. The coordinates of points on the graph will be displayed at the bottom of the screen, even though they aren't in the current viewing window. Use these coordinates as a guide for selecting a viewing window in which the graph does appear.

EXAMPLE 7

If you graph $y = -2x^3 + 26x^2 + 18x + 50$ in the standard viewing window, you get a blank screen (try it!). In such cases, you can usually find at least one point on the graph by setting $x = 0$ and determining the corresponding value of y (the y-intercept of the graph). If $x = 0$ here, then $y = 50$, so the point $(0, 50)$ is on the graph. Consequently, the y-axis of our viewing window should extend well beyond 50. An alternative method of finding some points on the graph is in the Technology Tip in the margin.

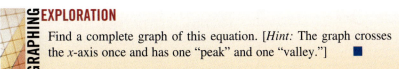

GRAPHING EXPLORATION

Find a complete graph of this equation. [*Hint:* The graph crosses the x-axis once and has one "peak" and one "valley."] ■

As a general rule, you should follow the directions in the following box when graphing equations.

Graphing Conventions

1. Unless directed otherwise, use technology for graphing.

2. Complete graphs are required unless a viewing window is specified or the context of a problem indicates that a partial graph is acceptable.

3. If the directions say "obtain the graph," "find the graph," or "graph the equation," you need not actually draw the graph on paper. For review purposes, however, it may be helpful to record the viewing window used.

4. The directions "sketch the graph" mean "draw the graph on paper, indicating the scale on each axis." This may involve simply copying the display on the screen, or it may require work if the display is misleading.

CALCULATOR INVESTIGATIONS 1.2

1. TICK MARKS

(a) Set Xscl = 1 so that adjacent tick marks on the x-axis are one unit apart. Find the largest range of x values such that the tick marks on the x-axis are clearly distinguishable and appear to be equally spaced.

(b) Do part (a) with y in place of x.

2. VIEWING WINDOWS

Look in the ZOOM menu to find out how many built-in viewing windows your calculator has. Take a look at each one.

3. MAXIMUM/MINIMUM FINDERS

Use your minimum finder to approximate the x-coordinates of the lowest point on the graph of $y = x^3 - 2x + 5$ in the window with $0 \leq x \leq 5$ and $-3 \leq y \leq 8$. The correct answer is

$$x = \sqrt{\frac{2}{3}} \approx .816496580928.$$

How good is your approximation?

4. SQUARE WINDOWS

Find a square viewing window on your calculator that has $-10 \leq x \leq 10$.

5. DOT GRAPHING MODE

To see which points your calculator actually graphs (without any connecting line segments), change the graphing mode. Select DOT (or DRAW DOT or PLOT) in the appropriate menu/submenu:

TI-83+: MODE   TI-89: Y= / STYLE

TI-86: GRAPH/FORMAT   Casio: SETUP/DRAWTYPE

On HP-39+, uncheck "connect" on the second page of the PLOT SETUP menu.

Now graph $y = .5x^3 - 2x^2 + 1$ in the standard window. Try some other equations as well.

✓ EXERCISES 1.2

Exercises 1–4 are representations of calculator screens. State the approximate coordinates of the points P and Q.

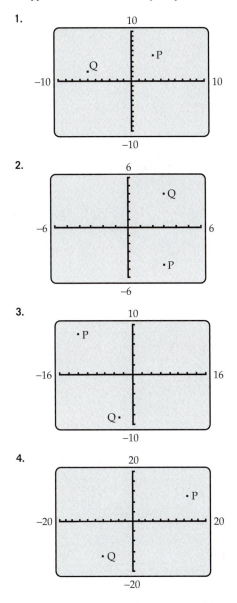

1.

2.

3.

4.

In Exercises 5–10, graph the equation by hand by plotting no more than six points and filling in the rest of the graph as best you can. Then use the calculator to graph the equation and compare the results.

5. $y = |x - 2|$

6. $y = \sqrt{x + 5}$

7. $y = x^2 - x$

8. $y = x^2 + x + 1$

9. $y = x^3 + 1$

10. $y = \dfrac{1}{x}$

In Exercises 11–16, find the graph of the equation in the standard window.

11. $3 + y = .5x$

12. $y - 2x = 4$

13. $y = x^2 - 5x + 2$

14. $y = .3x^2 + x - 4$

15. $y = .2x^3 + .1x^2 - 4x + 1$

16. $y = .2x^4 - .2x^3 - 2x^2 - 2x + 5$

In Exercises 17–22, determine which of the following viewing windows gives the best view of the graph of the given equation.

(a) $-10 \le x \le 10$; $-10 \le y \le 10$

(b) $-5 \le x \le 25$; $0 \le y \le 20$

(c) $-10 \le x \le 10$; $-100 \le y \le 100$

(d) $-20 \le x \le 15$; $-60 \le y \le 250$

(e) None of a, b, c, d gives a complete graph.

17. $y = 18x - 3x^2$

18. $y = 4x^2 + 80x + 350$

19. $y = \dfrac{1}{3}x^3 - 25x + 100$

20. $y = x^4 + x - 5$

21. $y = x^2 + 50x + 625$

22. $y = .01(x - 15)^4$

23. A toy rocket is shot straight up from ground level and then falls back to earth; wind resistance is negligible. Use your calculator to determine which of the following equations has a graph whose portion above the *x*-axis provides the most plausible model of the path of the rocket.

(a) $y = .1(x - 3)^3 - .1x^2 + 5$

(b) $y = -x^4 + 16x^3 - 88x^2 + 192x$

(c) $y = -16x^2 + 117x$

(d) $y = .16x^2 - 3.2x + 16$

(e) $y = -(.1x - 3)^6 + 600$

24. Monthly profits at DayGlo Tee Shirt Company appear to be given by the equation

$$y = -.00027(x - 15{,}000)^2 + 60{,}000,$$

where *x* is the number of shirts sold that month and *y* is the profit. DayGlo's maximum production capacity is 15,000 shirts per month.

(a) If you plan to graph the profit equation, what range of *x* values should you use? [*Hint:* You can't make a negative number of shirts.]

(b) The president of DayGlo wants to motivate the sales force (who are all in the profit-sharing plan) so he asks you to prepare a graph that shows DayGlo's profits increasing *dramatically* as sales increase. Using the profit equation and the *x* range from part (a), what viewing window is suitable?

(c) The City Council is talking about imposing more taxes. The president asks you to prepare a graph showing that DayGlo's profits are essentially flat. Using the profit equation and the *x* range from part (a), what viewing window is suitable?

In each of the applied situations in Exercises 25–28, find an appropriate viewing window for the equation (that is, a window that includes all the points relevant to the problem, but does not include large regions that are not relevant to the problem, and has easily readable tick marks on the axes). Explain why you chose this window. See the Hint in Exercise 24(a).

25. Beginning in 1905 the deer population in a region of Arizona rapidly increased because of a lack of natural predators. Eventually food resources were depleted to such a degree that the deer population completely died out. In the equation $y = -.125x^5 + 3.125x^4 + 4000$, y is the number of deer in year x, where $x = 0$ corresponds to 1905.

26. A cardiac test measures the concentration y of a dye x seconds after a known amount is injected into a vein near the heart. In a normal heart,

$$y = -.006x^4 + .14x^3 - .053x^2 + 179x.$$

27. The concentration of a certain medication in the bloodstream at time x hours is approximated by the equation

$$y = \frac{375x}{.1x^3 + 50},$$

where y is measured in milligrams per liter. After two days, the medication has no effect.

28. A winery can produce x barrels of red wine and y barrels of white wine, where

$$y = \frac{200,000 - 50x}{2000 + x}.$$

In Exercises 29–31, use the viewing windows found in Exercises 25–27.

29. (a) Use the trace feature to estimate the year when the deer population in Exercise 25 first reached 40,000. In what later year was it also 40,000?
    (b) When was the deer population at its maximum? Approximately how many deer were there at that time?

30. (a) In the cardiac test of Exercise 26, use the trace feature to estimate the time at which all dye is gone from the body.
    (b) At what time during the test was the concentration of the dye the greatest?

31. (a) When was the concentration of the medication in Exercise 27 at its peak?
    (b) Use the trace feature to determine approximately when the concentration dropped below 10 milligrams per liter and stayed below that level.

In Exercises 32–36, use your algebraic knowledge to state whether or not the two equations have the same graph. Confirm your answer by graphing the equations in the standard window.

32. $y = |x| - 4$ and $y = |x - 4|$

33. $y = |x + 3|$ and $y = |x| + 3$

34. $y = \sqrt{x^2 + 6x + 9}$ and $y = |x + 3|$

35. $y = \sqrt{x^2}$ and $y = |x|$

36. $y = \sqrt{x^2 + 9}$ and $y = x + 3$

37. (a) Confirm the accuracy of the factorization

$$x^2 - 5x + 6 = (x - 2)(x - 3)$$

graphically. [*Hint:* Graph $y = x^2 - 5x + 6$ and $y = (x - 2)(x - 3)$ on the same screen. If the factorization is correct the graphs will be identical, which means that you will see only a single graph on the screen.]
    (b) Show graphically that $(x + 5)^2 \neq x^2 + 5^2$. [*Hint:* Graph $y = (x + 5)^2$ and $y = x^2 + 5^2$ on the same screen. If the graphs are different, then the two expressions cannot be equal.]

True or False   *In Exercises 38–40, use the technique of Exercise 37 to determine graphically whether the given statement is possibly true or definitely false. [We say "possibly true" because two graphs that appear identical on a calculator screen may actually differ by small amounts or at places not shown in the window.]*

38. $x^3 - 7x - 6 = (x + 1)(x + 2)(x - 3)$

39. $(1 - x)^6 = 1 - 6x + 15x^2 - 20x^3 + 15x^4 - 6x^5 + x^6$

40. $x^5 - 8x^4 + 16x^3 - 5x^2 + 4x - 20 =$
$$(x - 2)^2(x - 5)(x^2 + x + 1)$$

41. (a) Graph $y = x^3 - 2x^2 + x - 2$ in the standard window.
    (b) Use the trace feature to show that the portion of the graph with $0 \leq x \leq 1.5$ is not actually horizontal. [*Hint:* All the points on a horizontal segment must have the same y-coordinate. (Why?)]
    (c) Find a viewing window that clearly shows that the graph is not horizontal when $0 \leq x \leq 1.5$.

42. (a) Graph $y = \dfrac{1}{x^2 + 1}$ in the standard window.
    (b) Does the graph appear to stop abruptly part-way along the x-axis? Use the trace feature to explain why this happens. [*Hint:* In this viewing window, each pixel represents a rectangle that is approximately .32 units high.]
    (c) Find a viewing window with $-10 \leq x \leq 10$, which shows a complete graph that does not fade into the x-axis.

In Exercises 43–52, use the techniques of Examples 3 and 4 to graph the equation in a suitable square viewing window.

43. $x^2 + y^2 = 16$ **44.** $y^2 = x + 2$

45. $3x^2 + 2y^2 = 48$

46. $25(x - 5)^2 + 36(y + 4)^2 = 900$

47. $(x - 4)^2 + (y + 2)^2 = 25$

48. $9x^2 + 4y^2 = 36$ **49.** $4x^2 - 9y^2 = 36$

50. $9y^2 - x^2 = 9$ **51.** $9x^2 + 5y^2 = 45$

52. $x = y^2 - 2$

In Exercises 53–54, use zoom-in or a maximum/minimum finder to determine the highest and lowest point on the graph in the given window.

53. $y = .4x^3 - 3x^2 + 4x + 3$
($0 \le x \le 5$ and $-5 \le y \le 5$)

54. $y = .07x^5 - .3x^3 + 1.5x^2 - 2$
($-3 \le x \le 2$ and $-6 \le y \le 6$)

In Exercises 55–58, find an appropriate viewing window for the graph of the equation (which may take some experimentation) and use a maximum/minimum finder to answer the question.

55. The population y of New Orleans (in thousands) in year x of the 20th century is approximated by

$$y = .000046685x^4 - .0108x^3 + .7194x^2 - 9.2426x + 305$$
$$(0 \le x \le 100),$$

where $x = 0$ corresponds to 1900. According to this model, in what year was the population largest?

56. The NASDAQ stock index between March 2000 and March 2002 is roughly approximated by

$$y = 11.19x^2 - 470x + 6358 \qquad (3 \le x \le 27),$$

where y is the index in month x, with $x = 3$ corresponding to March 2000. Determine the months during this period when the NASDAQ reached its high and its low. What were the approximate values of the NASDAQ at these times?

57. The number y of subscribers to direct broadcast satellite TV (in millions) in year x can be approximated by the equation

$$y = -.0255x^3 + .57x^2 - .65x - 10.24 \qquad (8 \le x \le 15),$$

where $x = 8$ corresponds to 1998.[*] In what year was the number of subscribers the largest?

58. The following equation gives the approximate number of thefts at Cleveland businesses each hour of the day:

$$y = .00357x^4 - .3135x^3 + 6.87x^2 - 38.3x + 118.4$$
$$(0 \le x \le 23),$$

where x is measured in hours after midnight.[†]
   (a) About how many thefts occurred around 5 A.M.?
   (b) When did the largest number of thefts occur?

59. Four squares, each measuring x by x inches, are cut from the corners of a rectangular sheet of cardboard that measures 30 by 22 inches and the flaps are folded up to form an open-top box, as shown in Example 9 of Section 0.4. In that exercise you found a polynomial that expresses the volume of the resulting box.
   (a) Graph $y =$ volume polynomial in the window with $0 \le x \le 11$ and $0 \le y \le 1300$.
   (b) Explain what the coordinates of each point on the graph represent.
   (c) Find the value of x for which the volume of the box is as large as possible.

60. Do Exercise 59 when the sheet of cardboard measures 40 by 40 inches. [*Hint:* You will need a different viewing window.]

In Exercises 61–66, obtain a complete graph of the equation by trying various viewing windows. List a viewing window that produces this complete graph. (Many correct answers are possible; consider your answer to be correct if your window shows all the features in the window given in the answer section.)

61. $y = 7x^3 + 35x + 10$

62. $y = x^3 - 5x^2 + 5x - 6$

63. $y = \sqrt{x^2 - x}$

64. $y = 1/x^2$

65. $y = -.1x^4 + x^3 + x^2 + x + 50$

66. $y = .002x^5 + .06x^4 - .001x^3 + .04x^2 - .2x + 15$

---

*Based on data and projections in *Newsweek*, December 23, 2002.

[†]Cleveland Police Department; based on data for 2002.

1.3 Lines

When you move from a point P to a point Q on a line,[*] two numbers are involved, as illustrated in Figure 1–33:

(i) The vertical distance you move (the **change in y**);

(ii) The horizontal distance you move (the **change in x**).

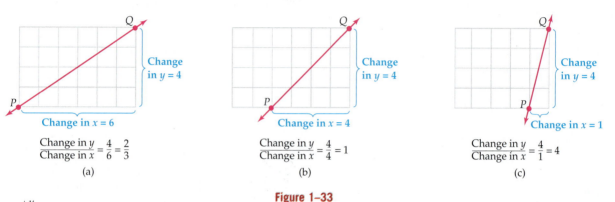

(a)

$$\frac{\text{Change in } y}{\text{Change in } x} = \frac{4}{6} = \frac{2}{3}$$

(b)

$$\frac{\text{Change in } y}{\text{Change in } x} = \frac{4}{4} = 1$$

(c)

$$\frac{\text{Change in } y}{\text{Change in } x} = \frac{4}{1} = 4$$

Figure 1–33

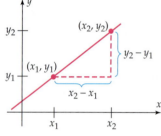

Figure 1–34

The number $\dfrac{\text{change in } y}{\text{change in } x}$ measures the steepness of the line: the steeper the line, the larger the number. In Figure 1–33, the grid allowed us to measure the change in y and the change in x. When the coordinates of P and Q are given, then

The change in y is the difference of the y-coordinates of P and Q;

The change in x is the difference of the x-coordinates of P and Q;

as shown in Figure 1–34. Consequently, we have the following definition.

Slope of a Line

> If (x_1, y_1) and (x_2, y_2) are points with $x_1 \neq x_2$, then the **slope** of the line through these points is the number
>
> $$\frac{\text{change in } y}{\text{change in } x} = \frac{y_2 - y_1}{x_2 - x_1}.$$

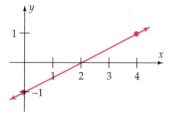

Figure 1–35

EXAMPLE 1

To find the slope of the line through $(0, -1)$ and $(4, 1)$ (see Figure 1–35), we apply the formula in the previous box with $x_1 = 0$, $y_1 = -1$ and $x_2 = 4$, $y_2 = 1$.

$$\text{Slope} = \frac{y_2 - y_1}{x_2 - x_1} = \frac{1 - (-1)}{4 - 0} = \frac{2}{4} = \frac{1}{2}.$$

---

[*]In this section, "line" means "straight line," and movement is from left to right.

The order of the points makes no difference; if you use (4, 1) for (x_1, y_1) and (0, −1) for (x_2, y_2), you obtain the same number.

$$\text{Slope} = \frac{y_2 - y_1}{x_2 - x_1} = \frac{-1 - 1}{0 - 4} = \frac{-2}{-4} = \frac{1}{2}. \quad \blacksquare$$

CAUTION

When finding slopes, you must subtract the *y*-coordinates and *x*-coordinates in the same order. With the points (3, 4) and (1, 8), for instance, if you use 8 − 4 in the numerator, you must use 1 − 3 in the denominator (*not* 3 − 1).

■ SLOPE-INTERCEPT FORM

A nonvertical line intersects the *y*-axis at a point with coordinates (0, *b*) for some number *b* (because every point on the *y*-axis has first coordinate 0). The number *b* is called the **y-intercept** of the line. For example, the line in Figure 1–35 has *y*-intercept −1 because it crosses the *y*-axis at (0, −1).

Let *L* be a nonvertical line with slope *m* and *y*-intercept *b*. Then (0, *b*) is a point on *L*. Let (*x*, *y*) be any other point on *L*. Using points (0, *b*) and (*x*, *y*) to compute the slope of *L* (see Figure 1–36), we have

$$\text{Slope of } L = \frac{y - b}{x - 0}.$$

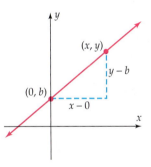

Figure 1–36

Since the slope of *L* is *m*, this equation becomes

$$m = \frac{y - b}{x}$$

Multiply both sides by *x*: $mx = y - b$

Rearrange terms: $y = mx + b$

Thus, the coordinates of any point on *L* satisfy the equation $y = mx + b$. So we have the following fact.

Slope-Intercept Form

The line with slope *m* and *y*-intercept *b* is the graph of the equation

$$y = mx + b.$$

EXAMPLE 2

Show that the graph of $2y - 5x = 2$ is a straight line. Find its slope, and graph the line.

SOLUTION   We begin by solving the equation for *y*.

Add 5*x* to both sides: $2y = 5x + 2$

Divide both sides by 2: $y = 2.5x + 1.$

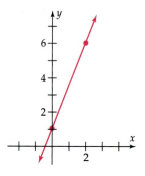

Figure 1–37

The equation now has the form in the preceding box, with $m = 2.5$ and $b = 1$. Therefore, its graph is the line with slope 2.5 and y-intercept 1.

Since the y-intercept is 1, the point $(0, 1)$ is on the graph. To find another point on the line, choose a value for x, say, $x = 2$, and compute the corresponding value of y.

$$y = 2.5x + 1 = 2.5(2) + 1 = 6.$$

Hence, $(2, 6)$ is on the line. Plotting the line through $(0, 1)$ and $(2, 6)$ produces Figure 1–37. ■

Although we don't need a calculator to graph lines, the calcuator makes it easy to see some essential properties of "slope."

GRAPHING EXPLORATION

Using the standard viewing window, graph the following equations on the same screen and answer the questions below.

$$y_1 = .5x, \qquad y_2 = x, \qquad y_3 = 3x, \qquad y_4 = 7x.$$

What are the slopes of these lines?

Which line rises least steeply (from left to right)? Which one rises most steeply?

Which line has the smallest slope? Which has the largest slope?

How is the slope of the line related to how steeply it rises?

Now graph the following equations on the same screen:

$$y_5 = 2, \qquad y_6 = -x + 2, \qquad y_7 = -2.5x + 2, \qquad y_8 = -5x + 2.$$

What are the slopes of these lines?

Which line falls least steeply (from left to right)? Which one falls most steeply?

How is the slope of the line related to how steeply it falls?

Your answers to the questions in the preceding Graphing Exploration should indicate that slope measures the steepness of the line, as summarized here:

Properties
of Slope

The slope of a nonvertical line is a number m that measures how steeply the line rises or falls.

If $m > 0$, the line rises from left to right; the larger m is, the more steeply the line rises.

If $m = 0$, the line is horizontal.

If $m < 0$, the line falls from left to right; the larger $|m|$ is, the more steeply the line falls.

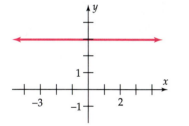

Figure 1–38

EXAMPLE 3

Describe and sketch the graph of the equation $y = 3$.

SOLUTION   We can write $y = 3$ as $y = 0x + 3$. So its graph is a line with slope 0, which means that the line is horizontal, and has y-intercept 3, which means that the line crosses the y-axis at 3. This is sufficient information to obtain the graph in Figure 1–38.   ■

Example 3 is an illustration of this fact.

Horizontal Lines

> The horizontal line with y-intercept b is the graph of the equation
> $$y = b.$$

We have seen that the geometric interpretation of slope is that it measures the *steepness* or direction of the line. The next two examples show that slope can also be interpreted as a *rate of change*.

* Ⓒ

EXAMPLE 4

According to the Kelley Blue Book, a Chevrolet Prizm that is now worth \$13,500 will be worth \$8500 in four years, provided that it is in good condition with average mileage. Assuming linear depreciation, what will the car be worth in two years? At what rate is the car depreciating?

SOLUTION   Linear depreciation means that the equation that gives the value y of the car in year x is linear. So the equation is of the form $y = mx + b$ for some constants m and b. Since the car is worth \$13,500 now ($x = 0$), we have

$$y = mx + b$$
$$13,500 = m \cdot 0 + b$$
$$b = 13,500,$$

so the equation is $y = mx + 13,500$. Since the car is worth \$8,500 after four years (that is, $y = 8500$ when $x = 4$), we have

$$y = mx + 13,500$$
$$8500 = m \cdot 4 + 13,500$$

Subtract 13,500 from both sides: $-5000 = 4m$

Divide both sides by 4: $m = -\dfrac{5000}{4} = -1250.$

Consequently, the depreciation equation is $y = -1250x + 13,500$. The value of the car after two years ($x = 2$) is

$$y = -1250(2) + 13,500 = \$11,000.$$

---

* Ⓒ and ⚪ indicate examples, exercises, and sections that are relevant to calculus.

The equation $y = -1250x + 13,500$ shows that the car depreciates $1250 each year (that is, each time x changes by 1). In other words, its value changes at the *rate* of -1250 per year. This rate is the *slope* of the graph of the depreciation equation. ■

EXAMPLE 5

A factory that makes can openers has fixed costs (for building, fixtures, machinery, etc.) of $26,000. The variable cost (materials and labor) for making one can opener is $2.75.

(a) What is the total cost of making 1000 can openers? 20,000? 40,000?

(b) What is the average cost per can opener in each case?

SOLUTION

(a) Since each can opener costs $2.75, the variable costs for making x can openers is $2.75x$. The total cost y of making x can openers is

$$y = \text{variable costs} + \text{fixed costs} = 2.75x + 26,000.$$

The cost of making 1000 can openers is

$$y = 2.75x + 26,000 = 2.75(1000) + 26,000 = \$28,750.$$

Similarly, the cost of making 20,000 can openers is

$$y = 2.75(20,000) + 26,000 = \$81,000,$$

and the cost of 40,000 is

$$y = 2.75(40,000) + 26,000 = \$136,000.$$

The cost equation $y = 2.75x + 26,000$ shows that the cost of making x can openers increases at the rate of $2.75 per can opener. This rate of change is the slope of the graph of the cost equation.

(b) The average cost per can opener in each case is the total cost divided by the number of can openers. So the average cost per can opener is as follows.

For 1000: $28,750/1000 = $28.75 per can opener;

For 20,000: $81,000/20,000 = $4.05 per can opener;

For 40,000: $136,000/40,000 = $3.40 per can opener.   ■

Examples 4 and 5 illustrate the following fact.

Linear Rate of Change

The slope m of the line with equation

$$y = mx + b$$

is the rate of change of y with respect to x.

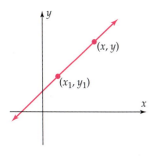

Figure 1–39

▪▪ POINT-SLOPE FORM

Suppose the line L passes through the point (x_1, y_1) and has slope m. Let (x, y) be any other point on L. Using the points (x_1, y_1) and (x, y) to compute the slope m of L (see Figure 1–39), we have

$$\frac{y - y_1}{x - x_1} = \text{slope of } L$$

$$\frac{y - y_1}{x - x_1} = m$$

Multiply both sides by $x - x_1$: $y - y_1 = m(x - x_1)$.

Thus, the coordinates of every point on L satisfy the equation

$$y - y_1 = m(x - x_1),$$

and we have this fact:

Point-Slope Form

> The line with slope m through the point (x_1, y_1) is the graph of the equation
> $$y - y_1 = m(x - x_1).$$

C

EXAMPLE 6

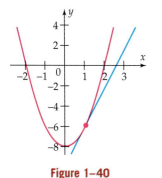

Figure 1–40

The tangent line to the graph of $y = 2x^2 - 8$ at the point $(1, -6)$ is shown in Figure 1–40. In calculus, it is shown that this line passes through $(1, -6)$ and has slope 4. Find the equation of the tangent line.

SOLUTION   Substitute 4 for m and $(1, -6)$ for (x_1, y_1) in the point-slope equation.

$$y - y_1 = m(x - x_1)$$

$$y - (-6) = 4(x - 1) \qquad \text{[point-slope form]}$$

$$y + 6 = 4x - 4$$

$$y = 4x - 10 \qquad \text{[slope-intercept form]} \quad ▪$$

EXAMPLE 7

The annual out-of-pocket spending (per person) on doctors and clinical services was approximately \$105 in 1997. According to projections from the Health Care

Financing Committee, this cost is expected to rise linearly to $196 in 2010, as indicated in Figure 1–41.

(a) Find an equation that gives the out-of-pocket cost y in year x.

(b) Use this equation to estimate the out-of-pocket costs in 2004 and 2006.

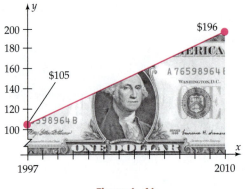

Figure 1–41

SOLUTION

(a) Let $x = 0$ correspond to 1997, so that $x = 13$ corresponds to 2010. Then the given information can be represented by the points $(0, 105)$ and $(13, 196)$. We must find the equation of the line through these points. Its slope is

$$\frac{196 - 105}{13 - 0} = \frac{91}{13} = 7.$$

Now we use the slope 7 and one of the points $(0, 105)$ or $(13, 196)$ to find the equation of the line. It doesn't matter which point, since both lead to the same equation.

$$y - y_1 = m(x - x_1) \qquad\qquad y - y_1 = m(x - x_1)$$
$$y - 105 = 7(x - 0) \qquad\qquad y - 196 = 7(x - 13)$$
$$y = 7x + 105 \qquad\qquad y - 196 = 7x - 91$$
$$\qquad\qquad\qquad\qquad\qquad\qquad y = 7x + 105.$$

(b) Since 2004 corresponds to $x = 7$, the projected out-of-pocket costs in 2004 are

$$y = 7x + 105 = 7 \cdot 7 + 105 = \$154.$$

The costs in 2006 ($x = 9$) are

$$y = 7x + 105 = 7 \cdot 9 + 105 = \$168. \qquad\blacksquare$$

▪▪ VERTICAL LINES AND GENERAL FORM

The preceding discussion does not apply to vertical lines, whose equations have a different form from those examined earlier.

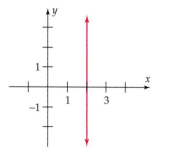

Figure 1–42

EXAMPLE 8

Every point on the vertical line in Figure 1–42 has first coordinate 2. Thus, every point on the line satisfies $x + 0y = 2$. Thus, the line is the graph of the equation $x = 2$. If you try to compute the slope of this line, say, using $(2, 1)$ and $(2, 4)$, you obtain $\dfrac{4 - 1}{2 - 2} = \dfrac{3}{0}$, which is not defined. ■

Example 8 illustrates these facts.

Vertical Lines

The vertical line with x-intercept c is the graph of the equation

$$x = c.$$

The slope of this line is undefined.

By rearranging terms, if necessary, the equation of any line can be written in the **general form** $Ax + By = C$ for some constants A, B, and C. For instance,

This equation	can be written as
$y = 4x - 10$	$4x - 1y = 10$
$y - 196 = 7(x - 13)$	$-7x + 1y = 105$
$y = 3$	$0x + 1y = 3$
$x = 2$	$1x + 0y = 2$

In summary, we have the following.

General Form

Every line is the graph of an equation of the form

$$Ax + By = C,$$

where A and B are not both zero.

■ PARALLEL AND PERPENDICULAR LINES

The slope of a line measures how steeply it rises or falls. Since parallel lines rise or fall equally steeply, the following fact should be plausible (see Exercises 64 and 65 for a proof).

Parallel Lines

Two nonvertical lines are parallel exactly when they have the same slope.

EXAMPLE 9

Find the equation of the line L through $(2, -1)$ that is parallel to the line M whose equation is $3x - 2y + 6 = 0$.

SOLUTION   First find the slope of M by rewriting its equation in slope-intercept form.

$$3x - 2y + 6 = 0$$
$$-2y = -3x - 6$$
$$y = \frac{3}{2}x + 3$$

Therefore M has slope $3/2$. The parallel line L must have the same slope, $3/2$. Since $(2, -1)$ is on L, we can use the point-slope form to find its equation.

$$y - y_1 = m(x - x_1)$$
$$y - (-1) = \frac{3}{2}(x - 2) \qquad \text{[point-slope form]}$$
$$y + 1 = \frac{3}{2}x - 3$$
$$y = \frac{3}{2}x - 4 \qquad \text{[slope-intercept form]} \quad \blacksquare$$

Two lines that meet in a right angle (90° angle) are said to be **perpendicular.** As you might suspect, there is a close relationship between the slopes of two perpendicular lines.

Perpendicular
Lines

Two nonvertical lines are perpendicular exactly when the product of their slopes is -1.

A proof of this fact is outlined in Exercise 66.

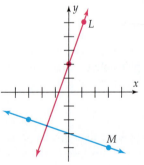

Figure 1–43

EXAMPLE 10

In Figure 1–43, the line L through $(0, 2)$ and $(1, 5)$ appears to be perpendicular to the line M through $(-3, -2)$ and $(3, -4)$. We can confirm this fact by computing the slopes of these lines.

$$\text{Slope } L = \frac{5 - 2}{1 - 0} = 3 \quad \text{and} \quad \text{Slope } M = \frac{-4 - (-2)}{3 - (-3)} = \frac{-2}{6} = -\frac{1}{3}.$$

Since $3(-1/3) = -1$, the lines L and M are perpendicular. $\quad \blacksquare$

> ⚠ **CAUTION**
> Perpendicular lines may not appear to be perpendicular on a calculator screen unless you use a square window (see Exercise 30).

✓ EXERCISES 1.3 📼💿

1. For which of the line segments in the figure is the slope
(a) largest?   (b) smallest?
(c) largest in absolute value?   (d) closest to zero?

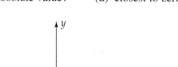

2. The doorsill of a campus building is 5 ft above ground level. To allow wheelchair access, the steps in front of the door are to be replaced by a straight ramp with constant slope 1/12, as shown in the figure. How long must the ramp be? [The answer is *not* 60 ft.]

In Exercises 3–6, find the slope and y-intercept of the line whose equation is given.

3. $2x - y + 5 = 0$ **4.** $3x + 4y = 7$

5. $3(x - 2) + y = 7 - 6(y + 4)$

6. $2(y - 3) + (x - 6) = 4(x + 1) - 2$

In Exercises 7–10, find the slope of the line through the given points.

7. $(1, 2); (3, 7)$ **8.** $(-1, -2); (2, -1)$

9. $(1/4, 0); (3/4, 2)$ **10.** $(\sqrt{2}, -1); (2, -9)$

11. Let L be a nonvertical straight line through the origin. L intersects the vertical line through $(1, 0)$ at a point P. Show that the second coordinate of P is the slope of L.

12. On one graph, sketch five line segments, not all meeting at a single point, whose slopes are five different positive numbers. Do this in such a way that the left-hand line segment has the largest slope, the second line segment from the left has the next largest slope, and so on.

In Exercises 13–16, find the equation of the line with slope m that passes through the given point.

13. $m = 1;$ $(3, 5)$ **14.** $m = 2;$ $(-2, 1)$

15. $m = -1;$ $(6, 2)$ **16.** $m = 0;$ $(-4, -5)$

In Exercises 17–20, find the equation of the line through the given points.

17. $(0, -5)$ and $(-3, -2)$ **18.** $(4, 3)$ and $(2, -1)$

19. $(4/3, 2/3)$ and $(1/3, 3)$ **20.** $(6, 7)$ and $(6, 15)$

In Exercises 21–24, determine whether the line through P and Q is parallel or perpendicular to the line through R and S or neither.

21. $P = (2, 5), Q = (-1, -1)$ and $R = (4, 2), S = (6, 1)$

22. $P = (0, 3/2), Q = (1, 1)$ and $R = (2, 7), S = (3, 9)$

23. $P = (-3, 1/3), Q = (1, -1)$ and $R = (2, 0),$
   $S = (4, -2/3)$

24. $P = (3, 3), Q = (-3, -1)$ and $R = (2, -2), S = (4, -5)$

In Exercises 25–26, determine whether the lines whose equations are given are parallel, perpendicular, or neither.

25. $2x + y - 2 = 0$ and $4x + 2y + 18 = 0$

26. $3x + y - 3 = 0$ and $6x + 2y + 17 = 0$

27. Use slopes to show that the points $(-5, -2)$, $(-3, 1)$, $(3, 0)$, and $(5, 3)$ are the vertices of a parallelogram.

28. Use slopes to show that the points $(-4, 6)$, $(-1, 12)$, and $(-7, 0)$ all lie on the same straight line.

29. Use slopes to determine if $(9, 6)$, $(-1, 2)$, and $(1, -3)$ are the vertices of a right triangle.

30. (a) Show that the lines $y = 2x + 4$ and $y = -.5x - 3$ are perpendicular.
(b) Graph the lines in part (a), using the standard viewing window. Do the lines look perpendicular?
(c) Find a viewing window in which the lines in part (a) appear to be perpendicular.

In Exercises 31–38, find an equation for the line satisfying the given conditions.

31. Through $(-2, 1)$ with slope 3.

32. y-intercept -7 and slope 1.

33. Through $(2, 3)$ and parallel to $3x - 2y = 5$.

34. Through $(1, -2)$ and perpendicular to $y = 2x - 3$.

35. x-intercept 5 and y-intercept -5.

36. Through $(-5, 2)$ and parallel to the line through $(1, 2)$ and $(4, 3)$.

37. Through $(-1, 3)$ and perpendicular to the line through $(0, 1)$ and $(2, 3)$.

38. y-intercept 3 and perpendicular to $2x - y + 6 = 0$.

If P is a point on a circle with center C, then the tangent line to the circle at P is the straight line through P that is perpendicular to the radius CP. In Exercises 39–42, find the equation of the tangent line to the circle at the given point.

39. $x^2 + y^2 = 25$ at $(3, 4)$ [*Hint:* Here C is $(0, 0)$ and P is $(3, 4)$; what is the slope of radius CP?]

40. $x^2 + y^2 = 169$ at $(-5, 12)$

41. $(x - 1)^2 + (y - 3)^2 = 5$ at $(2, 5)$

42. $(x + 3)^2 + (y - 4)^2 = 10$ at $(-2, 1)$

43. Let A, B, C, D be nonzero real numbers. Show that the lines $Ax + By + C = 0$ and $Ax + By + D = 0$ are parallel.

44. Let L be a line that is neither vertical nor horizontal and which does not pass through the origin. Show that L is the graph of $\dfrac{x}{a} + \dfrac{y}{b} = 1$, where a is the x-intercept and b is the y-intercept of L.

45. According to the March 31, 2001, issue of *The Economist,* worldwide automobile production was about 54 million vehicles in 2000 and is expected to be 63 million vehicles

in 2008. This production growth is approximately linear.
(a) Let $x = 0$ correspond to 2000. Find a linear equation that gives the number y of vehicles (in millions) produced in year x.
(b) Estimate the number of vehicles produced in 2003.

46. Carbon dioxide (CO_2) concentration is measured regularly at the Mauna Loa observatory in Hawaii. The mean annual concentration in parts per million in various years is given in the table.[*]

Year	Concentration (ppm)
1981	339.9
1986	347.2
1996	362.6
2001	370.9

(a) Do the data points lie on a single line? How do you know?
(b) Write a linear equation that uses the data from 1981 and 2001 to model CO_2 concentration over time. What does this model say the concentration should have been in 1986? Does the model overestimate or underestimate the concentration?
(c) Write a linear equation that uses the data from 1996 and 2001 to model CO_2 concentration over time. What does this model say the concentration should have been in 1981? Does the model overestimate or underestimate the concentration?
(d) What do the two models say about the concentration in the year 2007? Which model do you think is the more accurate? Why?

47. The consumer price index (CPI), which measures the cost of a typical package of consumer goods, stood at 144.5 in 1993 and 184.2 in 2003.
(a) Find a linear equation that approximates the CPI y in year x, with $x = 0$ corresponding to 1993.
(b) Estimate the CPI in 2007.
(c) At what yearly rate is the CPI increasing?

48. At sea level, water boils at 212°F. At a height of 1100 feet, water boils at 210°F. The relationship between boiling point and height is linear.
(a) Find an equation that gives the boiling point y of water at a height of x feet.
Find the boiling point of water in each of the following cities (whose altitudes are given).
(b) Cincinnati, OH (550 feet)
(c) Springfield, MO (1300 feet)
(d) Billings, MT (3120 feet)
(e) Flagstaff, AZ (6900 feet)

---

[*]C. D. Keeling and T. P. Whorf, Scripps Institution of Oceanography.

49. A Chrysler Concorde LXi that was worth $24,000 in 2002 will be worth $13,000 in 2006. Assuming linear depreciation, what is the rate of depreciation? What will the car be worth in 2009?

50. The age-adjusted death rate from heart disease was 586.8 per 100,000 people in 1950 and 257.9 per 100,000 in 2000.[*]
(a) Assuming that the rate decreased linearly, find an equation that gives the number y of deaths per 100,000 in year x, with $x = 0$ corresponding to 1950.
(b) Use this equation to estimate the death rate in 1995 and 2007. [For comparison purposes, the actual rate was 296.3 in 1995.]
(c) Is this equation likely to remain accurate long into the future? [*Hint:* What does it give for a death rate in 2040?]

51. Total health care expenditures were approximately $696 billion in 1990 and $1300 billion in 2000.[*] The growth in health care costs was approximately linear.
(a) Find an equation that gives the approximate health care costs y in year x (with $x = 0$ corresponding to 1990).
(b) Use the equation in part (a) to estimate the health care costs in 1996 and 2006.

52. The profit p (in thousands of dollars) on x thousand units of a specialty item is $p = .6x - 14.5$. The cost c of manufacturing x items is given by $c = .8x + 14.5$.
(a) Find an equation that gives the revenue r from selling x items.
(b) How many items must be sold for the company to break even (that is, for revenue to equal cost)?

53. A publisher has fixed costs of $110,000 for a mathematics text. The variable costs are $50 per book. The book sells for $72. Find equations that give
(a) The cost c of making x books
(b) The revenue r from selling x books
(c) The profit p from selling x books
(d) What is the publisher's break-even point (see Exercise 52(b))?

54. At the factory in Example 5, the cost of producing x can openers was given by $y = 2.75x + 26,000$.
(a) Write an equation that gives the average cost per can opener when x can openers are produced.
(b) How many can openers should be made to have an average cost of $3 per can opener?

55. The Whismo Hat Company has fixed costs of $50,000 and variable costs of $8.50 per hat.
(a) Find an equation that gives the total cost y of producing x hats.
(b) What is the average cost per hat when 20,000 are made? 50,000? 100,000?

56. Suppose the cost of making x TV sets is given by $y = 145x + 120,000$.
(a) Write an equation that gives the average cost per set when x sets are made.
(b) How many sets should be made in order to have an average cost per set of $175?

57. There were about 1 million subscribers to digital cable TV in 1998. This number is expected to increase linearly to 31 million in 2005.[†]
(a) Give an equation that gives the approximate number of subscribers y (in millions) in year x, where $x = 0$ corresponds to 1998.
(b) How many subscribers were there in 2002? If this trend continues, how many will there be in 2008?

58. The poverty level income for a family of four was $9287 in 1981. Because of inflation and other factors, the poverty level rose approximately linearly to $18,556 in 2002.[‡]
(a) At what rate is the poverty level increasing?
(b) Estimate the poverty level in 1990 and 2007.

59. A 75-gallon water tank is being emptied. The graph shows the amount of water in the tank after x minutes.
(a) At what rate is the tank emptying during the first 2 minutes? During the next 3 minutes? During the last minute?

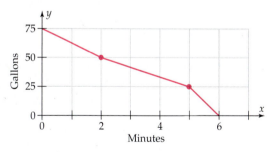

(b) Suppose the tank is emptied at a constant rate of 10 gallons per minute. Draw the graph that shows the amount of water after x minutes. What is the equation of the graph?

---

[*]Center for Health Statistics, Department of Health and Human Services.

[†]*Newsweek,* December 23, 2002.
[‡]U.S. Census Bureau.

Use the graph and the following information for Exercises 60–62. Rocky is an "independent" ticket dealer who markets choice tickets for Los Angeles Lakers' home games. (California currently has no laws against scalping.) Each graph shows how many tickets will be demanded by buyers at a particular price. For instance, when the Lakers play the Chicago Bulls, the graph shows that at a price of $160, no tickets are demanded. As the price (y-cooordinate) gets lower, the number of tickets demanded (x-coordinate) increases.

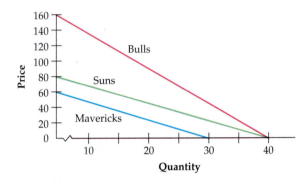

60. Write a linear equation that relates the quantity x of tickets demanded at price y when the Lakers play the
(a) Dallas Mavericks.
(b) Phoenix Suns.
(c) Chicago Bulls.
[*Hint:* In each case, use the points where the graph crosses the two axes to determine its slope.]

61. Use the equations from Exercise 60 to find the number of tickets Rocky would sell at a price of $40 for a game against the
(a) Mavericks.   (b) Bulls.

62. Suppose Rocky has 20 tickets to sell. At what price could he sell them all when the Lakers play the
(a) Mavericks?   (b) Suns?

63. According to the Center of Science in the Public Interest, the maximum healthy weight for a person who is 5 feet 5 inches tall is 150 pounds, and the maximum healthy weight for someone 6 feet 3 inches tall is 200 pounds. The relationship between weight and height here is linear.
(a) Find a linear equation that gives the maximum healthy weight y for a person whose height is x inches over 4 feet 10 inches. (Thus, $x = 0$ corresponds to 4 feet 10 inches, $x = 2$ to 5 feet, etc.)
(b) What is the maximum healthy weight for a person whose height is 5 feet? 6 feet?
(c) How tall is a person who is at a maximum healthy weight of 220 pounds?

Thinkers

64. Prove that nonvertical parallel lines L and M have the same slope, as follows. Suppose M lies above L and choose two points (x_1, y_1) and (x_2, y_2) on L.
(a) Let P be the point on M with first coordinate x_1. Let b denote the vertical distance from P to (x_1, y_1). Show that the second coordinate of P is $y_1 + b$.
(b) Let Q be the point on M with first coordinate x_2. Use the fact that L and M are parallel to show that the second coordinate of Q is $y_2 + b$.
(c) Compute the slope of L using (x_1, y_1) and (x_2, y_2). Compute the slope of M using the points P and Q. Verify that the two slopes are the same.

65. Show that two nonvertical lines with the same slope are parallel. [*Hint:* The equations of distinct lines with the same slope must be of the form $y = mx + b$ and $y = mx + c$ with $b \neq c$ (why?). If (x_1, y_1) were a point on both lines, its coordinates would satisfy both equations. Show that this leads to a contradiction and conclude that the lines have no point in common.]

66. This exercise provides a proof of the statement about slopes of perpendicular lines in the box on page 94. First, assume that L and M are nonvertical perpendicular lines that both pass through the origin. L and M intersect the vertical line $x = 1$ at the points $(1, k)$ and $(1, m)$ respectively, as shown in the figure.

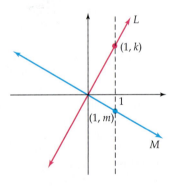

(a) Use $(0, 0)$ and $(1, k)$ to show that L has slope k. Use $(0, 0)$ and $(1, m)$ to show that M has slope m.
(b) Use the distance formula to compute the length of each side of the right triangle with vertices $(0, 0)$, $(1, k)$, and $(1, m)$.
(c) Use part (b) and the Pythagorean Theorem to find an equation involving k, m, and various constants. Show that this equation simplifies to $km = -1$. This proves one half of the statement.

(d) To prove the other half, assume that $km = -1$ and show that L and M are perpendicular as follows. You may assume that a triangle whose sides a, b, c satisfy $a^2 + b^2 = c^2$ is a right triangle with hypotenuse c. Use this fact and do the computation in part (b) in reverse (starting with $km = -1$) to show that the triangle with vertices $(0, 0)$, $(1, k)$ and $(1, m)$ is a right triangle, so L and M are perpendicular.

(e) Finally, to prove the general case when L and M do not intersect at the origin, let L_1 be a line through the origin that is parallel to L and M_1 a line through the origin that is parallel to M. Then L and L_1 have the same slope and M and M_1 have the same slope (why?). Use this fact and parts (a)–(d) to prove that L is perpendicular to M exactly when $km = -1$.

67. Show that the diagonals of a square are perpendicular. [*Hint:* Place the square in the first quadant of the plane, with one vertex at the origin and sides on the positive axes. Label the coordinates of the vertices appropriately.]

1.4   Linear Models*

People working in business, medicine, agriculture, and other fields often need to make judgments based on past data. For instance, a stock analyst might use the past profits of a company to estimate next year's profits, or a doctor might use data on previous patients to determine the ideal dosage of a drug for a new patient. In such situations, the available data can sometimes be used to construct a **mathematical model,** such as an equation or graph, that approximates the likely outcome in cases that are not included in the data. In this section, we consider applications in which the data can be modeled by a linear equation.

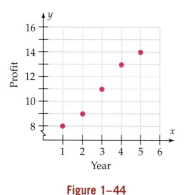

Figure 1–44

EXAMPLE 1

The profits of the General Electric Company (in billions of dollars) over a five-year period are shown in the table.[†]

Year	1997	1998	1999	2000	2001
Profit	8	9	11	13	14

Let $x = 1$ correspond to 1997, $x = 2$ to 1998, and so on. Then the data points given by the table are $(1, 8)$, $(2, 9)$, $(3, 11)$, $(4, 13)$, and $(5, 14)$. They are plotted in Figure 1–44, which shows that they lie roughly in a straight line. One way to

---

*This section is optional. It will be used only in clearly identifiable exercises or (sub)sections of the text that can be omitted by those not interested.
[†]GE Annual Report 2001, March 2002. The profits (net earnings) are rounded to the nearest billion.

construct a linear model is to use the line determined by two of the data points. Here are two such models:

Points: (1, 8) and (5, 14)

Slope of line: $\dfrac{14 - 8}{5 - 1} = \dfrac{6}{4} = 1.5$

Equation: $y - 8 = 1.5(x - 1)$

$$y = 1.5x - 1.5 + 8$$

$$y = 1.5x + 6.5.$$

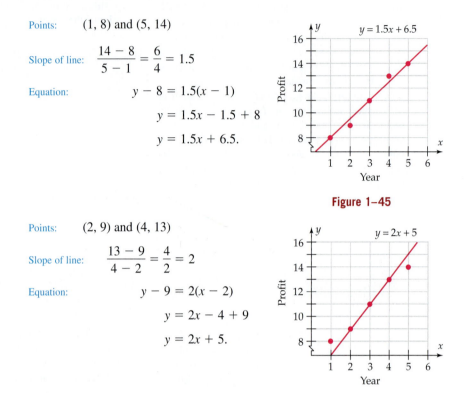

Figure 1–45

Points: (2, 9) and (4, 13)

Slope of line: $\dfrac{13 - 9}{4 - 2} = \dfrac{4}{2} = 2$

Equation: $y - 9 = 2(x - 2)$

$$y = 2x - 4 + 9$$

$$y = 2x + 5.$$

Figure 1–46

Choosing two data points is not the only way to construct a linear model. Figure 1–47 shows another model for the data, $y = 1.6x + 6.2$, that was obtained by a process that will be described later. It is difficult to determine visually which of these models is best. The first two models (Figures 1–45 and 1–46) each go through three of the data points but miss two data points by a noticeable amount. The third model (Figure 1–47) is very close to all the data points, although it passes through only one of them.

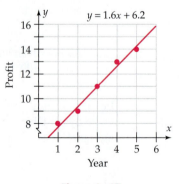

Figure 1–47

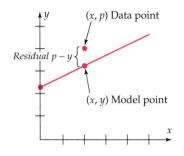

Figure 1–48

To determine the best model, we compute for each one the difference between the actual profit p and the amount y given by the model. If the data point is (x, p) and the corresponding point on the line is (x, y), then the difference $p - y$ measures the error in the model for that particular value of x. The number $p - y$ is called a **residual.** As shown in Figure 1–48, the residual $p - y$ is the vertical distance from the data point to the line (positive when the data point is above the line, negative when it is below the line, and 0 when it is on the line). The following tables compute the residuals for the three models, as well as some additional information that is discussed below.

$y = 1.5x + 6.5$			
Data Point (x, p)	Model Point (x, y)	Residual $p - y$	Squared Residual $(p - y)^2$
(1, 8)	(1, 8)	0	0
(2, 9)	(2, 9.5)	−.5	.25
(3, 11)	(3, 11)	0	0
(4, 13)	(4, 12.5)	.5	.25
(5, 14)	(5, 14)	0	0
		Sum: 0	*Sum:* .5

$y = 2x + 5$			
Data Point (x, p)	Model Point (x, y)	Residual $p - y$	Squared Residual $(p - y)^2$
(1, 8)	(1, 7)	1	1
(2, 9)	(2, 9)	0	0
(3, 11)	(3, 11)	0	0
(4, 13)	(4, 13)	0	0
(5, 14)	(5, 15)	−1	1
		Sum: 0	*Sum:* 2

$y = 1.6x + 6.2$			
Data Point (x, p)	Model Point (x, y)	Residual $p - y$	Squared Residual $(p - y)^2$
(1, 8)	(1, 7.8)	.2	.04
(2, 9)	(2, 9.4)	−.4	.16
(3, 11)	(3, 11)	0	0
(4, 13)	(4, 12.6)	.4	.16
(5, 14)	(5, 14.2)	−.2	.04
		Sum: 0	*Sum:* .4

The tables show that the sum of the residuals in each case is 0. This indicates that positive and negative errors cancel each other out (suggesting that the models are reasonable), but it doesn't help us to decide whether one model is better than another.

So we try another accuracy measure: Compute the sum of the *squares* of the residuals, as shown in the last columns of the preceding tables. Now there is no canceling, and each model has a different sum. Using this number as a measure of accuracy has the effect of emphasizing large errors (those with absolute value greater than 1) because the square is greater than the residual and minimizing small errors (those with absolute value less than 1) because the square is less than the residual. So a smaller sum means that the line has less overall error and fits the data better. By this measure, the best of the three models is the third one, $y = 1.6x + 6.2$, because the sum of the squares of its residuals is the smallest. ■

The following fact, which requires multivariate calculus for its proof, shows that there is always a best possible model for linear data.

Linear Regression Theorem

For any set of data points, there is one and only one line for which the sum of the squares of the residuals is as small as possible. This line is called the **least squares regression line.**

The computational process for finding the least squares regression line is called **linear regression.** The linear regression formulas are quite complicated and can be tiresome to use with a large data set. Fortunately, however, linear regression routines are built into most calculators and are also available in spreadsheet and other computer programs. A calculator's linear regression feature was used to obtain the third model in Example 1. According to the Linear Regression Theorem, it is the best possible linear model for that data.

EXAMPLE 2

The following table from the U.S. Census Bureau shows the poverty level for a family of four in selected years (families whose income is below this level are considered to be in poverty).

Year	1995	1997	1998	1999	2000	2001	2002
Income	$15,569	16,400	16,660	17,029	17,603	18,267	18,556

(a) Use linear regression to find an equation that models this data.

(b) Use the equation to estimate the poverty level in 1996 and 2006.

SOLUTION

(a) Let $x = 0$ correspond to 1990, and write the income in thousands (for instance, 15.569 in place of 15,569). Then the data points are (5, 15.569), (7, 16.400), . . . , (12, 18.556). We display the calculator's statistics editor and enter the data points: x-coordinates in the first list and y-coordinates (in the same order) in the second list, as in Figure 1–49.* Using the statistical plotting feature to plot the data points, we obtain Figure 1–50, which shows that the data is approximately linear.

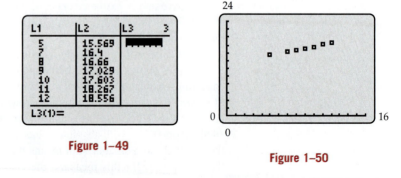

Figure 1–49

Figure 1–50

*Consult the Technology Tip at the end of this section for details on how to carry out this and subsequent steps.

Now we use the linear regression feature to obtain Figure 1–51, which shows that the equation of the least squares regression line is

$$y = .43675x + 13.2865.$$

When you do this on most calculators, you can simultaneously store the regression equation in the equation memory so that it can be graphed along with the data points, as in Figure 1–52.

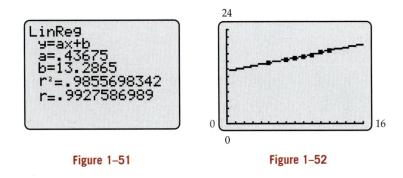

Figure 1–51   Figure 1–52

Figure 1–52 suggests that the regression line provides a good model of the data. This is also indicated in Figure 1–51 by the number r, as is explained after the example.

(b) The year 1996 corresponds to $x = 6$. If $x = 6$, then

$$y = .43675(6) + 13.2865 \approx 15.907.$$

So the regression model estimates the poverty level at \$15,907 in 1996. The actual level in that year was \$16,036, a difference of only \$129. Similarly, the model predicts the poverty level in 2006 ($x = 16$) to be about \$20,275 because

$$y = .43675(16) + 13.2865 \approx 20.275.$$

Because 2006 lies outside of the range of the data points, this figure might not prove to be as accurate as the estimate for 1996, which lies within the data range. ■

Figure 1–51 contains a number r (and its square) in addition to the coefficients of the regression line. The number r, which is called the **correlation coefficient,** is a statistical measure of how well the regression line fits the data. It is always between -1 and 1. The closer the absolute value of r is to 1, the better the fit. For instance, the regression line in Example 2 is an excellent fit because $r \approx .99$, as shown in Figure 1–51. When $|r| = 1$, the fit is perfect: All the data points are on the regression line. Conversely, a regression coefficient near 0 indicates a poor fit.

EXAMPLE 3

The table shows the number of accidental deaths per 100,000 population in the United States in selected years.[*]

Year	Death Rate	Year	Death Rate
1910	84.4	1970	56.2
1920	71.2	1980	46.5
1930	80.5	1990	36.9
1940	73.4	1995	35.5
1950	60.3	2000	35.6
1960	52.1	2001	34.3

Find a linear model for this data, and graph it. If the model fits the data well, use it to estimate the accidental death rate in 2006.

SOLUTION   We let $x = 0$ correspond to 1900 and enter the data points in the statistical editor of a calculator. Then we proceed as above to find the least squares regression line (Figure 1–53), which is graphed in Figure 1–54.

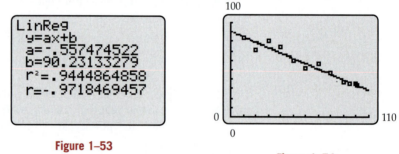

Figure 1–53

Figure 1–54

The correlation coefficient $r \approx -.97$ indicates that the model fits the data well, and the graph confirms this. To estimate the rate in 2006, substitute $x = 106$ in the regression equation:

$$y = -.557474522(106) + 90.23133279 \approx 31.1.$$

In 2006, the accidental death rate should be about 31.1 per 100,000 population. ∎

The correlation coefficient always has the same sign as the slope of the least squares regression line. So when r is negative, as in Example 3, the regression line moves downward from left to right (Figure 1–54). In other words, as x increases, y decreases. In such cases, we say that the data has a **negative correlation.** When r is positive, as in Example 2, the regression line slopes upward from left to right (Figure 1–52), and we say that the data has a **positive correlation:** As x increases, y also increases. When r is close to 0 (regardless of sign), we say that there is **no correlation.**

*National Center for Health Statistics, U.S. Department of Health and Human Services.

EXAMPLE 4

The number of unemployed people in the labor force (in millions) for 1988–2003 is as follows.[*]

Year	Unemployed	Year	Unemployed	Year	Unemployed
1988	6.701	1993	8.940	1998	6.210
1989	6.528	1994	7.996	1999	5.880
1990	7.047	1995	7.404	2000	5.692
1991	8.628	1996	7.236	2001	6.801
1992	9.613	1997	6.739	2002	8.378
				2003	9.018

Is a linear equation a good model for this data?

SOLUTION After entering the data as two lists in the statistics editor (with $x = 0$ corresponding to 1980), you can test it graphically or analytically.

Graphical: Plotting the data points (Figure 1–55) shows that they do not form a linear pattern (unemployment tends to rise and fall).

Analytical: Linear regression (Figure 1–56) produces an equation whose correlation coefficient is $r \approx -.08$, a number close to 0. This indicates that there is no correlation, that is, the regression line is a very poor fit for the data.

Therefore, a linear equation is not a good model for this data. ∎

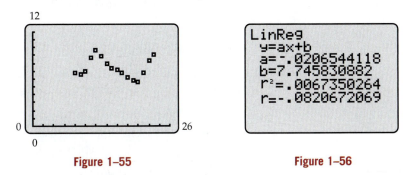

Figure 1–55   Figure 1–56

GRAPHING EXPLORATION

Enter the data from Example 4 in the statistics editor of your calculator. Graph the data points. Graph the least squares regression line on the same screen to see how poorly it fits the data.

The following Technology Tip may be helpful for learning how to use the regression feature of your calculator.

[*]U.S. Bureau of Labor Statistics; 2003 data is for the first quarter.

TECHNOLOGY TIP

Most calculators allow you to store three or more statistics graphs (identified by number); the following directions assume that the first one is used.

To call up the statistics editor, use these commands:

TI-83+: STAT   EDIT   (lists are L_1, L_2, . . .);

TI-85/86: STAT   EDIT   (built-in lists are x-stat, y-stat; it is usually better to create your own lists; we use L1 and L2 as list names here);[*]

TI-89: APPS   DATA/MATRIX EDITOR   NEW; then choose DATA as the TYPE, enter a VARIABLE name (we use L here), and key in ENTER (lists are C_1, C_2, . . .);

Casio 9850: STAT   (lists are List 1, List 2, . . .);

HP-39+: APLET   STATISTICS   (lists are C_1, C_2, . . .).

To graph the data points that have been entered as lists (x-coordinates in the first list, corresponding y-coordinates in the second), choose an appropriate viewing window on TI calculators. On TI-85, use STAT   DRAW   SCAT   to plot the points determined by the lists currently chosen in the STAT   EDIT screen. On other calculators, use these commands to enter the setup screen:

TI-83+: STAT-PLOT (on keyboard); choose PLOT 1;

TI-86: STAT; choose PLOT and PLOT 1;

TI-89: from the Data Editor, choose PLOT-SETUP (F_2) and DEFINE (F_1);

Casio: STAT   GRPH   SET;

HP-39+: APLET   STATISTICS   PLOT-SETUP   and   SYMB.

Key in the appropriate information [on/off (TI-83+/86), graph number (Casio and HP), lists to be used, graph type (scatter plot), the mark to be used for data points, and, on HP-39+, the viewing window]. Then press GRAPH (on TI) or GPH 1 (on Casio) or PLOT (on HP-39+).

To produce the least squares regression line, store its equation as y_1 in the equation memory, and graph it, use these commands:

TI-83+: STAT   CALC   LinReg ($ax + b$) L_1, L_2, Y_1;

TI-85: STAT   CALC   L1, L2   LINR; then press STREG and enter y_1 as the name;

TI-86: STAT   CALC   Lin R   L1, L2, y_1;

TI-89: From the Data Editor, choose CALC (F_5); enter LIN REG as CALCULATION TYPE, C_1 as x, C_2 as y; and choose $y_1(x)$ in STORE REGEQ;

Then press GRAPH or DRAW to obtain the graph of the equation and the data points (assuming that PLOT 1 has not been turned off). On HP-39+, after plotting the data points, use MENU FIT to graph the regression line and SYMB to see its equation. On Casio 9850, after plotting the data points, press X for the equation of the regression line and DRAW for its graph. On Casio FX 2.0, press F4 (CALC), then F2 (LINEAR).

---

[*]When statistical computations are run on TI-85/86, the lists used are automatically copied into the x-stat and y-stat lists, replacing whatever was there before. So use the x-stat and y-stat lists only if you don't want to save them. See your instruction manual to find out how to create new lists in the statistics editor.

✓ EXERCISES 1.4

In Exercises 1–4, two linear models are given for the data. For each model,
(a) Find the residuals and their sum;
(b) Find the sum of the squares of the residuals;
(c) Determine which model is the better fit.

1. The weekly amount spent on advertising and the weekly sales revenue of a small store over a five-week period are shown in the table. Two models are $y = x$ and $y = .5x + 1.5$.

Advertising Expenditures x (in hundreds of dollars)	1	2	3	4	5
Sales Revenue y (in thousands of dollars)	2	2	3	3	5

2. The table gives the consumer price index (CPI) in March of selected years.[*] Two models are $y = 4.1x + 171.4$ and $y = 4.3x + 171.2$, where $x = 0$ corresponds to 2000.

Year	2000	2001	2002	2003
CPI	171.2	176.2	178.8	184.2

3. The projected numbers of amended federal income tax returns (in millions) in selected years are shown in the table.[†] The two models are

$$y = .17x + 1.4 \qquad \text{and} \qquad y = .18x + 138,$$

where $x = 0$ corresponds to 1990.

Year	1990	1995	2000	2005
Amended Returns	1.4	2.2	3.2	4

4. Advertising expenditures in the United States (in billions of dollars) in selected years are shown in the table.[‡] Two models are $y = 10.6x + 122.6$ and $y = 9.3x + 129$, where $x = 0$ corresponds to 1990.

Year	1990	1995	2000	2001
Amount	129	161	244	231

In Exercises 5–8, determine whether the given scatter plot of the data indicates that there is a positive correlation, a negative correlation, or very little correlation.

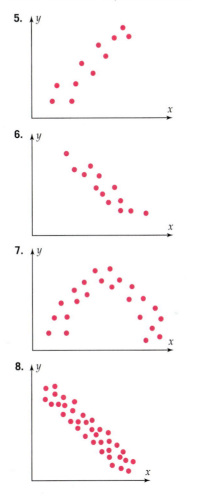

5.

6.

7.

8.

In Exercises 9–14, construct a scatter diagram for the data and answer these questions: (a) Does the data appear to be linear? (b) If so, is there a positive or negative correlation?

9. U.S. consumers owe more than ever. The table shows total outstanding credit card debt (in billions of dollars) in recent years.[§] Let $x = 0$ correspond to 1990.

Year	1997	1999	2000	2001	2002
Debt	531	598	667	700	729

---

[*]Bureau of Labor Statistics, U.S. Department of Labor.
[†]Internal Revenue Service.
[‡]*Advertising Age.*

[§]Federal Reserve.

10. The U.S. gross domestic product (GDP) is the total value of all goods and services produced in the United States. The table shows the GDP in billions of 1996 dollars.[*] Let $x = 0$ correspond to 1990.

Year	1992	1994	1996	1998	2000	2002
GDP	6880.0	7347.7	7813.2	8508.9	9191.4	9439.9

11. In mid-2002, Grange Life Insurance advertised the following monthly premium rates for a $50,000 term policy for a female nonsmoker. Let x represent age and y the monthly premium.

Age	30	35	40	45	50
Premium	6.48	6.52	6.78	7.88	9.93

Age	55	60	65	70
Premium	12.69	16.71	23.67	36.79

12. The vapor pressure y of water depends on the temperature x, as given in the table.

Temperature (°C)	Pressure (mm Hg)
0	4.6
10	9.2
20	17.5
30	31.8
40	55.3
50	92.5
60	149.4
70	233.7
80	355.1
90	525.8
100	760

13. The table shows the U.S. Bureau of Census population data for St. Louis, Missouri. Let $x = 0$ correspond to 1950.

Year	Population
1950	856,796
1970	622,236
1980	452,801
1990	396,685
2000	348,189

14. The table shows the U.S. disposable income (personal income less personal taxes) in billions of dollars.[†] Let $x = 0$ correspond to 1990.

Year	Disposal Personal Income
1995	5422.6
1997	5968.2
1999	6627.4
2000	7120.2
2001	7393.2
2002	7815.5
2003	8019.6

15. The table[‡] gives the annual U.S. consumption (in millions of pounds) of beef and poultry.

(a) Make scatter plots for both beef and poultry consumption, using the actual years (1996, 1997, etc.) as x in each case.

(b) Without graphing, use your knowledge of slopes to determine which of the following equations models beef consumption and which one models poultry consumption. Confirm your answer by graphing.

$$y_1 = 763.1x - 1,496,436 \qquad y_2 = 335.3x - 643,550$$

Year	Beef	Poultry
1996	25,861	26,760
1997	25,611	27,261
1998	26,305	27,821
1999	26,937	29,584
2000	27,377	30,031
2001	27,022	30,087

16. The table gives the median weekly earnings of full-time workers 25 years and older by the amount of education.[§]

(a) Make four scatter plots, one for each educational group, using $x = 0$ to correspond to 1997.

(b) Four linear models are given here. Match each model with the appropriate data set.

$$y_1 = 19.6x + 538 \qquad y_2 = 13.5x + 321$$

$$y_3 = 33.3x + 787 \qquad y_4 = 14.97x + 462$$

---

[*]Bureau of Economic Analysis, U.S. Department of Commerce.

[†]Bureau of Economic Analysis; data for 2003 is for the first quarter only.
[‡]U.S. Department of Agriculture.
[§]Bureau of Labor Statistics, U.S. Department of Labor.

Year	No High School Diploma	High School Graduate	Some College	College Graduate
1997	$321	$461	$535	$779
1998	$337	$479	$558	$821
1999	$346	$490	$580	$860
2000	$360	$506	$598	$896
2001	$378	$520	$621	$924
2002	$388	$538	$631	$943

In Exercises 17–24, use linear regression to find the requested linear model.

17. Expenditures for research and development at Intel Corporation (in billions of dollars) are shown in the table.*

Year	1994	1996	1998	2000	2001	2002
Expenditures	1.1	1.8	2.5	3.9	3.8	4.0

(a) Make a scatter plot of the data, with $x = 0$ corresponding to 1990.
(b) Find a linear model for the data.
(c) Use the model to estimate the expenditures in 1997 and 1999. [For comparison, the actual expenditures were $2.3 billion and $3.1 billion, respectively.]
(d) Assuming that the model remains accurate after 2002, estimate the expenditures in 2005.

18. The table shows the median time (in months after the application has been made) for the U.S. Food and Drug Administration to approve a new drug.†

Year	Median Time to Approval
1993	26.9
1994	22.1
1995	18.7
1996	17.8
1997	15.0
1998	12.0
1999	13.8
2000	12.0
2001	14.0
2002	15.3

(a) Make a scatter plot of the data, with $x = 0$ corresponding to 1990.
(b) Find a linear model for the data.
(c) What are the limitations of this model? [*Hint:* What does it say about approval time in 2011?]

19. National expenditures on health care (in billions of dollars) in various years are shown in the table.‡

Year	1990	1995	1997	1999
Amount	699.4	993.7	1092.4	1228.5

Year	2000	2001	2002	2003
Amount	1316.2	1403.6	1495.5	1590.4

(a) Find a linear model for this data, with $x = 0$ corresponding to 1990.
(b) Interpret the meaning of the slope and the y-intercept.
(c) If the model remains accurate in the future, what will health care expenditures be in 2007?

20. The table shows the estimated number (in millions) of people ages 15–24 living with HIV/AIDS.§

Year	2001	2002	2003	2004	2005
People with HIV/AIDS	12	13.5	14.5	15.05	17

(a) Find a linear model for this data (with $x = 0$ corresponding to 2000).
(b) Assuming the model remains accurate, predict the number of people in this age group who will be living with HIV/AIDS in 2008 and 2010.

21. The table shows the number of Internet users in the United States (in millions) in recent years.¶

Year	1998	1999	2000	2001	2002	2003
Users	85	102	117	143	158	162

(a) Find a linear model for this data (with $x = 0$ corresponding to 1998), and use it to predict the number of Internet users in 2009.
(b) For how long do you think this model might remain accurate? [*Note:* The U.S. population was approximately 294 million in 2003 and is increasing by about 3 million per year.

---

‡Data and projections from the U.S. Health Care Financing Administration.
§Kaiser Family Foundation, UNICEF, Census Bureau.
¶U.S. Department of Commerce; Jupiter Media Metrix.

---

*Intel Corporation, 2002 Annual Report.
†U.S. Food and Drug Administration; priority approvals not included.

22. The projected number of new cases of Alzheimer's disease (in thousands) in the United States in selected years is shown in the table.[*]

Year	2000	2010	2020	2030	2040	2050
New Cases	400	467	489	600	800	956

(a) Find a linear model for this data, with $x = 0$ corresponding to 2000, and use it to estimate the number of new cases in the years 2005, 2015, and 2025.
(b) Use the model to predict the years in which the number of new cases will be 750,000 and 1,000,000.

23. The table shows the maximum Pell Grant for college students in recent years.[†]

Year	1995	1996	1997	1998	1999
Maximum Grant	$2340	$2470	$2700	$3000	$3125

Year	2000	2001	2002	2003
Maximum Grant	$3300	$3750	$4000	$4050

(a) Find a linear model for this data, with $x = 0$ corresponding to 1990.
(b) According to this model, when will the maximum Pell Grant reach $5000?

24. During the current decade, the amount of money each American spends annually on prescription drugs (in addition to the amounts paid by insurance) is projected to increase, as shown in the table.[‡]

Year	2000	2002	2004	2006	2008	2010
Amount	$143	$175	$207	$231	$287	$351

(a) Find a linear model for this data, with $x = 0$ corresponding to 2000.
(b) Use the model to estimate the amount each person will spend on prescription drugs in 2005, 2007, and 2011.
(c) According to this model, in what year will each American spend $400 on prescription drugs?

---

[*]L. Hebert, L. Beckett, P. Scherr, D. Evans, "Annual Incidence of Alzheimer Disease in the United States Projected to the Years 2000 Through 2050," in *Alzheimer Disease and Associated Disorders,* Vol. 15, No. 4, 2001.
[†]U.S. Department of Education.

[‡]Centers for Medicare and Medicaid Services.

Chapter 1 Review

IMPORTANT

CONCEPTS

FACTS & FORMULAS

- *Distance Formula:* The distance from (x_1, y_1) to (x_2, y_2) is

$$\sqrt{(x_1 - x_2)^2 + (y_1 - y_2)^2}.$$

- *Midpoint Formula:* The midpoint of the line segment from (x_1, y_1) to (x_2, y_2) is

$$\left(\frac{x_1 + x_2}{2}, \frac{y_1 + y_2}{2}\right).$$

- Equation of the circle with center (c, d) and radius r.

$$(x - c)^2 + (y - d)^2 = r^2.$$

- The slope of the line through (x_1, y_1) and (x_2, y_2) (where $x_1 \neq x_2$) is

$$\frac{y_2 - y_1}{x_2 - x_1}.$$

- The equation of the line with slope m and y-intercept b is

$$y = mx + b.$$

- The equation of the line through (x_1, y_1) with slope m is

$$y - y_1 = m(x - x_1).$$

- Two nonvertical lines are parallel exactly when they have the same slope.
- Two nonvertical lines are perpendicular exactly when the product of their slopes is -1.

1. Find the distance from $(1, -2)$ to $(4, 5)$.

2. Find the distance from $(3/2, 4)$ to $(3, 5/2)$.

3. Find the distance from (c, d) to $(c - d, c + d)$.

4. Find the midpoint of the line segment from $(-4, 7)$ to $(9, 5)$.

5. Find the midpoint of the line segment from (c, d) to $(2d - c, c + d)$.

6. Find the equation of the circle with center $(-3, 4)$ that passes through the origin.

7. (a) If $(1, 1)$ is on a circle with center $(2, -3)$, what is the radius of the circle?
   (b) Find the equation of the circle in part (a).

8. Sketch the graph of $3x^2 + 3y^2 = 12$.

9. Sketch the graph of $(x - 5)^2 + y^2 - 9 = 0$.

10. Find the equation of the circle of radius 4 whose center is the midpoint of the line segment joining $(3, 5)$ and $(-5, -1)$.

11. Which of statements (a)–(d) are descriptions of the circle with center $(0, -2)$ and radius 5?
    (a) The set of points (x, y) that satisfy $|x| + |y + 2| = 5$.
    (b) The set of all points whose distance from $(0, -2)$ is 5.
    (c) The set of all points (x, y) such that $x^2 + (y + 2)^2 = 5$.
    (d) The set of all points (x, y) such that $\sqrt{x^2 + (y + 2)^2} = 5$.

12. If the equation of a circle is $3x^2 + 3(y - 2)^2 = 12$, which of the following statements is true?
    (a) The circle has diameter 3.
    (b) The center of the circle is $(2, 0)$.
    (c) The point $(0, 0)$ is on the circle.
    (d) The circle has radius $\sqrt{12}$.
    (e) The point $(1, 1)$ is on the circle.

13. The table shows the number of Americans (in millions) who snowboarded each year from 1994 to 2000.[*]

Year	1994	1995	1996	1997	1998	1999	2000
Number	2	2.7	3.7	2.5	3.5	3.2	4.2

Sketch a scatter plot and a line graph for this data, letting $x = 0$ correspond to 1990.

14. The table shows the average speed (mph) of the winning car in the Indianapolis 500 race in selected years.[†]

Year	1990	1992	1994	1996	1998	2000	2002
Speed	186	134	161	148	145	168	166

Sketch a scatter plot and a line graph for this data, letting $x = 0$ correspond to 1990.

In Questions 15–20,
(a) Determine which of the viewing windows a–e shows a complete graph of the equation.
(b) For each viewing window that does not show a complete graph, explain why.
(c) Find a viewing window that gives a "better" complete graph than windows a–e (meaning that the window is small enough to show as much detail as possible, yet large enough to show a complete graph).

(a) Standard viewing window
(b) $-10 \le x \le 10$, $-200 \le y \le 200$
(c) $-20 \le x \le 20$, $-500 \le y \le 500$
(d) $-50 \le x \le 50$, $-50 \le y \le 50$
(e) $-1000 \le x \le 1000$, $-1000 \le y \le 1000$

15. $y = .2x^3 - .8x^2 - 2.2x + 6$

16. $y = x^3 - 11x^2 - 25x + 275$

17. $y = x^4 - 7x^3 - 48x^2 + 180x + 200$

18. $y = x^3 - 6x^2 - 4x + 24$

19. $y = .03x^5 - 3x^3 + 69.12x$

20. $y = .00000002x^6 - .0000014x^5 - .00017x^4 + .0107x^3 + .2568x^2 - 12.096x$

21. According to data from the U.S. Department of Labor, the number y of unemployed people in the labor force (in millions) from 1988 to 2003 can be approximated by the equation

$$y = (5.55 \times 10^{-4})x^4 - .0208x^3 + .1393x^2 + 1.8466x - 9.23,$$

where $x = 8$ corresponds to 1988.
(a) In what year between 1988 and 2003 was unemployment the lowest?
(b) In what year between 1988 and 2000 was unemployment the highest?
(c) In what year between 2000 and 2003 was unemployment the highest?

*National Sporting Goods Association, TransWorld Snowboarding Business.
†Indianapolis Motor Speedway Hall of Fame and Museum.

22. Data and projections from the U.S. National Center for Educational Statistics show that the number y of children enrolled in public elementary schools (in millions) between 1965 and 2005 can be approximated by the equation

$$y = (-4.28 \times 10^{-5})x^4 + .00356x^3 - .0819x^2 + .4135x + 31.08,$$

where $x = 0$ corresponds to 1965. According to this model, when was enrollment the lowest? The highest?

In Questions 23 and 24, determine graphically whether the statement could possibly be true or is definitely false.

23. $6x^4 - 7x^3 + 8x^2 - 7x + 2 =$
$$(2x - 1)(3x + 2)(x^2 - 1)$$

24. $x^7 - 2x^6 - 6x^5 + 8x^4 + 17x^3 - 6x^2 - 20x - 8 =$
$$(x - 2)^3(x + 1)^4$$

In Questions 25–28, sketch a complete graph of the equation.

25. $y = |x - 10| - 3$

26. $y = -|x + 8| + 12$

27. $y = x^2 + 10$

28. $y = x^3 - 7x + 24$

29. (a) What is the y-intercept of the graph of
$$y = x - \frac{x - 2}{5} + \frac{3}{5}?$$
(b) What is the slope of this line?

30. Find the equation of the line passing through $(1, 3)$ and $(2, 5)$.

31. Find the equation of the line passing through $(2, -1)$ with slope 3.

32. (a) Find the y-intercept of the line $2x + 3y - 4 = 0$.
(b) Find the equation of the line through $(1, 3)$ that has the same y-intercept as the line in part (a).

33. Find the equation of the line through $(-4, 5)$ that is parallel to the line through $(1, 3)$ and $(-4, 2)$.

34. Sketch the graph of the line $3x + y - 1 = 0$.

35. As a balloon is launched from the ground, the wind is blowing it due east. The conditions are such that the balloon is ascending along a straight line with slope $1/5$. After 1 hour the balloon is 5000 ft vertically above the ground. How far east has the balloon blown?

36. The point (u, v) lies on the line $y = 5x - 10$. What is the slope of the line passing through (u, v) and the point $(0, -10)$?

In Questions 37–41, determine whether the statement is true or false.

37. The graph of $x = 5y + 6$ has y-intercept 6.

38. The graph of $2y - 8 = 3x$ has y-intercept 4.

39. The lines $3x + 4y = 12$ and $4x + 3y = 12$ are perpendicular.

40. Slope is not defined for horizontal lines.

41. The line in the figure has positive slope.

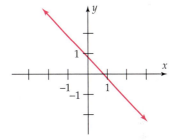

42. The line in the figure above does not pass through the third quadrant.

43. The y-intercept of the line in the figure above is negative.

44. Consider the *slopes* of the lines shown in the figure below. Which line has the slope with the largest *absolute value*?

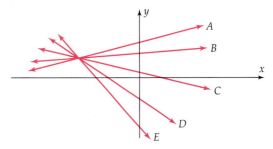

45. Which of the following lines rises most steeply from left to right?
(a) $y = -4x - 10$
(b) $y = 3x + 4$
(c) $20x + 2y - 20 = 0$
(d) $4x = y - 1$
(e) $4x = 1 - y$

46. Which of the following lines is *not* perpendicular to the line $y = x + 5$?
(a) $y = 4 - x$
(b) $y + x = -5$
(c) $4 - 2x - 2y = 0$
(d) $x = 1 - y$
(e) $y - x = \dfrac{1}{5}$

47. Which of the following lines does *not* pass through the third quadrant?
(a) $y = x$
(b) $y = 4x - 7$
(c) $y = -2x - 5$
(d) $y = 4x + 7$
(e) $y = -2x + 5$

48. What is the y-intercept of the line $2x - 3y + 5 = 0$?

49. Average life expectancy increased linearly from 70.8 years for a person born in 1970 to 76.8 years for a person born in 2000.
  (a) Find a linear equation that gives the average life expectancy y of a person born in year x, with $x = 0$ corresponding to 1970.
  (b) Use the equation in part (a) to estimate the average life expectancy of a person born in 1984.
  (c) Assuming the equation remains valid, in what year will the average life expectancy be 80 years for people born in that year?

50. The population of San Diego, California, grew in an approximately linear fashion from 334,400 in 1950 to 1,223,400 in 2000.
  (a) Find a linear equation that gives the population y of San Diego (in thousands) in year x, with $x = 0$ corresponding to 1950.
  (b) Estimate the population of San Diego in 1995.
  (c) Assuming the equation remains accurate, when will San Diego's population reach 1.5 million people?

In Exercises 51–54, match the given information with the graph, and determine the slope of the graph.

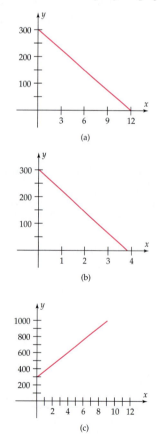

(a)

(b)

(c)

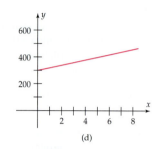

(d)

51. A salesman is paid $300 per week plus $75 for each unit sold.

52. A person is paying $25 per week to repay a $300 loan.

53. A gold coin that was purchased for $300 appreciates $20 per year.

54. A CD player that was purchased for $300 depreciates $80 per year.

55. The table shows monthly premiums (as of 2002) for a $100,000 term life insurance policy from Grange Life Insurance for a male smoker.

Age	Premium
25	$20.30
30	$20.39
35	$20.39
40	$22.05
45	$28.61
50	$41.65
55	$60.90
60	$85.58
65	$132.91

  (a) Make a scatter plot of the data, using x for age and y for premiums.
  (b) Does the data appear to be linear?

56. For which of the following scatter plots would a linear model be reasonable? Which sets of data show positive correlation, and which show negative correlation?

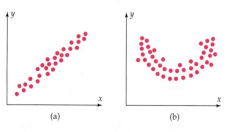

(a)   (b)

REVIEW QUESTIONS

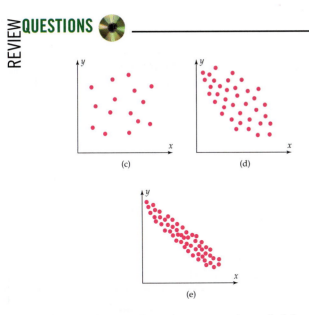

(c)   (d)

(e)

In Questions 57–62, use linear regression to find the required linear models.

57. The estimated revenue generated by prepaid telephone calling cards (in billions of dollars) is shown in the table.[*]

Year	1998	1999	2000	2001	2002	2003	2004
Revenue	2.6	3.0	3.1	3.4	3.6	4.1	4.6

(a) Make a scatter plot of the data, with $x = 8$ corresponding to 1998.
(b) Use linear regression to find a model for the data.
(c) If this model remains accurate, what will be the revenue generated in 2006 ($x = 16$)?

58. The amount spent on marketing prescription drugs (in billions of dollars) in recent years is shown in the table.[†]

Year	1996	1997	1998	1999	2000	2001
Amount	9.3	11	12.5	13.8	15.8	19.0

(a) Find a linear model for the data, with $x = 0$ corresponding to 1996.
(b) Estimate the amount spent on marketing in 2002.

59. The table shows the average hourly earnings of production workers in manufacturing.[‡]

Year	1993	1995	1997	1999	2001	2003
Hourly Earnings	$11.74	$12.37	$13.17	$13.90	$14.84	$15.58

(a) Find a linear model for the data, with $x = 0$ corresponding to 1990.
(b) Use the model to estimate the average hourly earnings in 1995 and 2001. How do the estimates of the model compare with the actual figures?
(c) Estimate the average hourly earnings in 2004.

60. The table shows the winning times (in minutes) for men's 1500-meter freestyle swimming at the Olympics in selected years.

Year	Time
1912	22.00
1924	20.11
1936	19.23
1948	19.31
1960	17.33
1972	15.88
1984	15.09
1996	14.94
2000	14.81

(a) Find a linear model for this data, with $x = 0$ corresponding to 1900.
(b) The Olympic record of 14.72 minutes was set in 1992 by Kieren Perkins of Australia. How accurately did your model estimate his time?
(c) How long is this model likely to remain accurate? Why?

[*]AMC Atlantic.
[†]IMS Health.

[‡]Bureau of Labor Statistics, U.S. Department of Labor; 2003 data is for the first third of the year.

61. The table shows the total amount of charitable giving (in billions of dollars) in the United States in recent years.[*]

Year	Charitable Giving
1993	116.5
1994	119.2
1995	124.0
1996	138.6
1997	157.1
1998	174.8
1999	199.0
2000	210.9
2001	212.0

(a) Find a linear model for the data, with $x = 0$ corresponding to 1990.
(b) Estimate the amount of charitable giving in 2004 and 2007.
(c) According to the model, in what year will charitable giving reach $330 billion?

62. The table shows, for selected states, the percent of high school students in the class of 2002 who took the SAT and the average SAT math score.[†]

State	Students Who Took SAT (%)	Average Math Score
Connecticut	83	509
Delaware	69	500
Georgia	65	491
Idaho	18	541
Indiana	62	503
Iowa	5	602
Montana	23	547
Nevada	34	518
New Jersey	82	513
New Mexico	14	543
North Dakota	4	610
Ohio	27	540
Pennsylvania	72	500
South Carolina	59	493
Washington	54	529

(a) Make a scatter plot of average SAT math score y and percent x of students who took the SAT.
(b) Find a linear model for the data.
(c) What is the slope of your linear model? What does this mean in the context of the problem?
(d) Here are the data on four additional states. How well does the model match the actual figures for these states?

State	Students Who Took SAT (%)	Average Math Score
Alaska	52	519
Arizona	36	523
Hawaii	53	520
Oklahoma	8	562

---

[*]*Statistical Abstract of the United States, 2002.*
[†]The College Board.

DISCOVERY PROJECT 1   Supply and Demand

Claus Meyer/Black Star Publishing/PictureQuest

Economists study the forces at play as buyers and sellers interact in what we call the **market** for a product. Because their money is limited, consumers tend to buy less of a particular product if the price is higher and more if the price is lower. This is the essence of the **demand curve,** which is just a depiction of the relationship between the price of an item and the maximum quantity that people will buy at that price. The demand curve represents the aggregate demand of all consumers in a market.

Manufacturers, however, have other considerations. Production costs and limited resources (materials, labor, technology, and capital, for instance) factor into the decision to produce at different levels. Ultimately, as the ones supplying (selling) the product, manufacturers want to sell more of them if the price is high and less of them if the price is low. This is the essence of the **supply curve,** a depiction of the relationship between the price of an item and the number of those items the manufacturers are willing to sell at that price. Like the demand curve, the supply curve represents the aggregate of all suppliers of a given product.

Here is a simple example of these ideas. The table below shows the quantity demanded for a hypothetical product at several different prices. You can easily verify that the equation $p = 5.85 - .005q$ describes this relationship.

Price p (in dollars)	Quantity q (in thousands)
3.25	520
3.00	570
2.75	620
2.50	670
2.25	720

Suppose the supply curve for this product is given by the equation $p = .01q - 3.75$ and that the producer would like to set the price of the product at \$2.80. To find the supply and demand at this price we must solve the appropriate equation for q when $p = 2.80$.[*]

Supply

$$.01q - 3.75 = p$$

$$.01q - 3.75 = 2.80$$

$$.01q = 6.55$$

$$q = \frac{6.55}{.01} = 655$$

Demand

$$5.85 - .005q = p$$

$$5.85 - .005q = 2.80$$

$$-.005q = -3.05$$

$$q = \frac{-3.05}{-.005} = 610.$$

---

[*]We use algebra to solve the equation, but you could also use graphical means or an equation solver. The technological methods are often the best choice for nonlinear supply and demand equations.

These results pose a problem: At a price of $2.80, the producer would be willing to sell 655,000 items, but the demand curve says that only 610,000 items would be demanded by consumers at this price. In technical terms, there is a **surplus** because supply exceeds demand.

You can readily verify that the situation is reversed when the price is $2.40. Solving the supply and demand equations when $p = 2.40$ shows that consumer demand is 690,000 items, but producers are willing to sell only 615,000. In this case, there is a **shortage** because demand exceeds supply.

When there is a surplus, sellers are inclined to lower the price because storing unsold items (or having to discard perishable ones) is expensive. When there is a shortage, consumers are willing to pay more to get a product they want. These tendencies push the market toward the **equilibrium point,** at which buyers and sellers agree on both price and quantity. This occurs when

Quantity demanded = Quantity supplied.

The price p at this point is called the **equilibrium price,** the price at which supply equals demand.

The equilibrium point can be found graphically by finding the intersection point of the supply curve and the demand curve. Because economists normally put price (p) on the vertical axis and quantity (q) on the horizontal axis, we follow the same convention here (using X in place of q and Y in place of p when graphing on a calculator). Figure 1 shows that the equilibrium point for our example is (640, 2.65), which means that the equilibrium price is $2.65 and that 640,000 items will be demanded and supplied at that price.

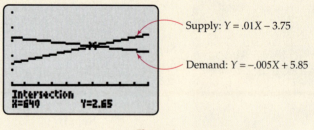

Supply: $Y = .01X - 3.75$

Demand: $Y = -.005X + 5.85$

Intersection
X=640   Y=2.65

Figure 1

Supply and demand curves provide quantities that correspond to a limited collection of feasible prices at a given point in time. So the corresponding equations are valid only in this range and might not be valid outside of it. In our example, the equations are valid when $520 \leq q \leq 720$, which determines the viewing window in Figure 1.

When these ideas are used to make business decisions, the first step is to determine the supply and demand equations. This is often done by using statistical data from a trial market, as in the following exercises.

1. An economist obtained the following data about the demand for oranges.

Price per pound (dollars)	0.93	0.90	0.84	0.70	0.65	0.63	0.60
Quantity demanded (in 100,000 tons)	10	10.6	11.5	13.8	14.7	15.1	15.5

Find a linear demand equation of the form $p = mq + b$ either by using linear regression *or* by using the first and last data points [(10, .93) and (15.5, .60)]. If you use regression, round the coefficients to four decimal places.

2. Use the model you found in Exercise 1 to find the quantity demanded at a price of 75¢ per pound.

3. The economist in Exercise 1 also obtained this data about the supply of oranges.

Price per pound (dollars)	0.20	0.36	0.45	0.55	0.65	0.75	1.00
Quantity supplied (in 100,000 tons)	10	11.4	12.4	13.5	14.6	15.5	18

Find a linear supply equation of the form $p = mq + b$ either by using linear regression *or* by using the first and last data points, as in Exercise 1.

4. Use the model you found in Exercise 3 to find the price at which producers are willing to supply 1,300,000 tons of oranges.

5. Find the equilibrium point for the situation in Exercises 1 and 3. What is the equilibrium price? How many oranges will be supplied/demanded at this price?

6. Here are the equations of the curves for the sale of apples:

$$p = .06q - .2 \quad \text{and} \quad p = -.04q + .7,$$

where p is in dollars and q is in 100,000 tons.

(a) Which curve is supply and which one is demand? How can you tell?

(b) Find the equilibrium point. What is the equilibrium price? How many apples will be supplied/demanded at this price?

7. Graph $p = .28$ together with the demand and supply curves of Exercise 6. Is this price above or below the equilibrium price? Would this price lead to a surplus, a shortage, or neither? Explain.

Equations and Inequalities

Peter Menzel/Stock, Boston Inc./PictureQuest

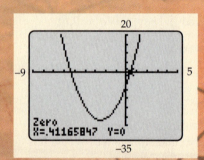

How deep is that well?

Measuring the height of a cliff or the depth of a well directly can be very difficult. It can sometimes be done indirectly by using sound waves and solving an appropriate equation. The solution of many other practical problems in construction, quality control, economics, and a variety of other fields also depends on solving equations and inequalities. See Exercise 83 on page 151, Exercise 42 on page 168, and Exercise 51 on page 169.

Chapter Outline

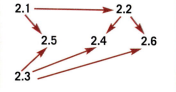
Dealing effectively with real-world problems often involves solving equations and inequalities. The emphasis in the first part of this chapter is on algebraic methods of solving equations. Then we examine graphical and numerical solution methods that are necessary when algebraic methods are inadequate. Finally, we consider both algebraic and graphical methods of solving inequalities.

2.1 First-Degree Equations and Applications

Two equations are said to be **equivalent** if they have the same solutions. For example, $3x + 2 = 17$ and $x - 2 = 3$ are equivalent because 5 is the only solution of each one. The usual strategy for solving equations algebraically is to use the following principles to transform a given equation into an equivalent one whose solutions are known.

Basic Principles for Solving Equations

> Performing any of the following operations on an equation produces an equivalent equation:
>
> 1. Add or subtract the same quantity on both sides of the equation.
>
> 2. Multiply or divide both sides of the equation by the same *nonzero* quantity.

The Basic Principles apply to all equations. In this section, however, we shall deal only with **first-degree equations,** which are those that involve only constants and the first power of the variable. Solving most first-degree equations is quite straightforward.

EXAMPLE 1

Solve: $3x - 6 = 7x + 4$.

SOLUTION   We use the Basic Principles to transform this equation into an equivalent one whose solution is obvious.

$$3x - 6 = 7x + 4$$

Add 6 to both sides: $3x = 7x + 10$

Subtract $7x$ from both sides: $-4x = 10$

Divide both sides by -4: $x = \dfrac{10}{-4} = -\dfrac{5}{2}$

Since $-5/2$ is the only solution of this last equation, $-5/2$ is the only solution of the original equation, $3x - 6 = 7x + 4$. ■

EXAMPLE 2

Solve: $\dfrac{13t - 7}{2} = \dfrac{2t - 4}{3}$.

SOLUTION   The first step is to eliminate the fractions by multiplying both sides by their common denominator 6.

$$6\left(\frac{13t - 7}{2}\right) = 6\left(\frac{2t - 4}{3}\right)$$

Simplify: $3(13t - 7) = 2(2t - 4)$

Multiply out both sides: $39t - 21 = 4t - 8$

Subtract $4t$ from both sides: $35t - 21 = -8$

Add 21 to both sides: $35t = 13$

Divide both sides by 35: $t = \dfrac{13}{35}$ ■

EXAMPLE 3

The surface area S of the rectangular box in Figure 2–1 is given by

$$2lh + 2lw + 2wh = S.$$

Solve this equation for h.

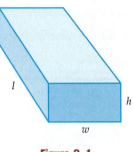

Figure 2–1

SOLUTION   Treat h as the variable and all the other letters as constants. First, get all the terms involving the variable h on one side of the equation and everything else on the other side.

$$2lh + 2lw + 2wh = S$$

Sutract $2lw$ from both sides: $2lh + 2wh = S - 2lw$

Factor out h on the left side: $(2l + 2w)h = S - 2lw$

Divide both sides by $(2l + 2w)$: $h = \dfrac{S - 2lw}{2l + 2w}$ ■

The second Basic Principle applies only to *nonzero* quantities. Multiplying both sides of an equation by a quantity involving the variable (which might be zero for some values) may lead to an **extraneous solution,** a number that does not satisfy the original equation. To avoid errors in such situations, always *check your solutions in the original equation.*

EXAMPLE 4

Solve: $\dfrac{x-3}{x-7} = \dfrac{3x-17}{x-7}$.

SOLUTION   Begin by multiplying both sides by $x-7$ to eliminate the fractions.

$$\frac{x-3}{x-7}(x-7) = \frac{3x-17}{x-7}(x-7)$$

Simplify both sides:	$x - 3 = 3x - 17$
Add 3 to both sides:	$x = 3x - 14$
Subtract $3x$ from both sides:	$-2x = -14$
Divide both sides by -2:	$x = 7$

Substituting 7 for x in the original equation yields

$$\frac{7-3}{7-7} = \frac{3 \cdot 7 - 17}{7-7} \qquad \text{or, equivalently,} \qquad \frac{4}{0} = \frac{4}{0}.$$

Since division by 0 is not defined, 7 is *not* a solution of the original equation. So this equation has no solutions.   ■

EXAMPLE 5

Solve: $\dfrac{5}{x+6} = \dfrac{3}{x-3} + \dfrac{1}{2x-6}$.

SOLUTION   We begin by multiplying both sides by a common denominator to eliminate the fractions. Note that the third denominator here is $2x - 6 = 2(x - 3)$. Hence, $2(x - 3)(x + 6)$ is a common denominator for all three fractions. Multiplying both sides by this common denominator, we have

$$\frac{5}{x+6} \cdot 2(x-3)(x+6) = \frac{3}{x-3} \cdot 2(x-3)(x+6) + \frac{1}{2x-6} \cdot 2(x-3)(x+6)$$

Cancel like factors:	$5 \cdot 2(x-3) = 3 \cdot 2(x+6) + (x+6)$
Multiply out both sides:	$10(x-3) = 6(x+6) + x + 6$
	$10x - 30 = 6x + 36 + x + 6$
Combine like terms:	$10x - 30 = 7x + 42$
Subtract $7x$ from both sides:	$3x - 30 = 42$
Add 30 to both sides:	$3x = 72$
Divide both sides by 3:	$x = 24.$

By substituting 24 for x in the original equation, it can be seen that 24 *is* a solution.

$$\frac{5}{24+6} = \frac{5}{30} = \frac{1}{6} \quad \text{and} \quad \frac{3}{24-3} + \frac{1}{2 \cdot 24 - 6} = \frac{3}{21} + \frac{1}{42} = \frac{1}{6} \quad \blacksquare$$

Algebraic methods are normally used with first-degree equations, but there are cases when the use of a calculator is advisable.

EXAMPLE 6

Solve: $3.69x + \dfrac{21}{14.77} = 4.53$.

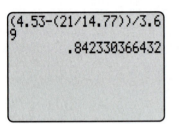

Figure 2–2

SOLUTION   We proceed as before, but to avoid round-off errors in the intermediate steps, we do all the algebra first.

Subtract $21/14.77$ from both sides: $3.69x = 4.53 - \dfrac{21}{14.77}$

Divide both sides by 3.69: $x = \dfrac{4.53 - \dfrac{21}{14.77}}{3.69}$

Now use the calculator, as in Figure 2–2, to find the solution $x \approx .8423$. Alternatively, this equation could be solved as in the Graphing Exploration below. $\blacksquare$

GRAPHING EXPLORATION

Solve the equation in Example 6 by finding the point where the graphs of $y = 3.69x + 21/14.77$ and $y = 4.53$ intersect. How does your answer compare with the one in Figure 2–2?

▉ APPLICATIONS

Real-life problem situations are usually described verbally. To solve such problems, you must interpret this verbal information and express it in mathematical language (typically as an equation or inequality). The following guidelines may be helpful.

Setting Up
Applied Problems

1. *Read* the problem carefully and determine what is asked for.

2. *Label* the unknown quantities by letters (variables) and, if appropriate, draw a picture of the situation.

3. *Translate* the verbal statements in the problem and the relationships between the known and unknown quantities into mathematical language.

4. *Consolidate* the mathematical information into an equation in one variable that can be solved or an equation in two variables that can be graphed in order to determine at least one of the unknown quantities.

EXAMPLE 7   Measurement

What is the height of a rectangular box with a square base measuring 8 inches on a side and a volume of 928 cubic inches.

SOLUTION   *Read:* We know the volume and the dimensions of the base and must find the height.
Label: Draw a diagram with h denoting the height (Figure 2–3).
Translate:

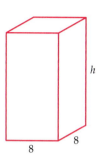

Figure 2–3

English Language	Mathematical Language
Height	h
Volume (Length $\times$ Width $\times$ Height)	$8 \times 8 \times h$
Volume is 928 cu in.	$8 \times 8 \times h = 928$
	$64h = 928$

Consolidate: We need only solve the equation $64h = 928$.

Divide both sides by 64: $h = \dfrac{928}{64} = 14.5$ in.

Therefore, the box measures 8 by 8 by 14.5 inches high.   ■

EXAMPLE 8   Counting

A box of nickels and quarters contains 52 coins that are worth a total of $10. How many nickels and how many quarters are in the box?

SOLUTION   *Read:* We are asked to find the number of nickels and the number of quarters.
Label: Let x be the number of nickels and y the number of quarters.
Translate:

English Lanuage	Mathematical Language
Number of nickels	x
Number of quarters	y
Total number of coins is 52	$x + y = 52$
Value of the nickels	$.05x$
Value of the quarters	$.25y$
Value of collection is $10	$.05x + .25y = 10$

Consolidate: When you have two variables one technique is to express one variable in terms of the other and use this to obtain an equation in a single variable. In this case, we can solve $x + y = 52$ for y.

(∗)
$$y = 52 - x$$

and substitute the result in the other equation.

$$.05x + .25y = 10$$

$$.05x + .25(52 - x) = 10$$

Now solve this last equation.

Multiply out left side: $$.05x + 13 - .25x = 10$$

Combine like terms: $$-.20x + 13 = 10$$

Subtract 13 from both sides: $$-.20x = -3$$

Divide both sides by $-.20$: $$x = \frac{-3}{-.20} = 15$$

Therefore, there are 15 nickels. From equation (∗), the number of quarters is

$$y = 52 - x$$

$$y = 52 - 15 = 37. \quad ■$$

Recall that 4% is written as the decimal .04 and that "4% of 227" means ".04 *times* 227." For example, if $227 is deposited in a savings account that pays 4% annual interest, then the interest after one year is 4% of $227, that is, .04(227) = $9.08. This is an example of the basic rule of annual simple interest:

Interest = Rate × Amount.

EXAMPLE 9   Interest

How much money must be invested at 9% to produce $128.25 interest annually?

SOLUTION   *Read and Label:* We are asked to find amount x that should be invested.
Translate and Consolidate: The basic rule of simple interest tells us that

$$\text{Rate} \times \text{Amount} = \text{Interest}$$

$$9\% \text{ of } x = 128.25$$

$$.09x = 128.25$$

Divide both sides by .09: $$x = \frac{128.25}{.09} = \$1425. \quad ■$$

EXAMPLE 10   Interest

A high-risk stock pays dividends at a rate of 12% per year, and a savings account pays 6% interest per year. How much of a $9000 investment should be put in the stock and how much in the savings account to obtain a return of 8% per year on the total investment?

SOLUTION   *Label:* Let x be the amount invested in stock. Then the rest of the $9000, namely, $(9000 - x)$ dollars, goes in the savings account.
Translate and Consolidate: We want the total return on $9000 to be 8%, so we have

$$\begin{pmatrix} \text{Return on } x \text{ dollars} \\ \text{of stock at 12\%} \end{pmatrix} + \begin{pmatrix} \text{Return on } (9000 - x) \\ \text{dollars of savings at 6\%} \end{pmatrix} = 8\% \text{ of } \$9000$$

$$(12\% \text{ of } x \text{ dollars}) + (6\% \text{ of } (9000 - x) \text{ dollars}) = 8\% \text{ of } \$9000$$

$$.12x + .06(9000 - x) = .08(9000)$$
$$.12x + .06(9000) - .06x = .08(9000)$$
$$.12x + 540 - .06x = 720$$
$$.12x - .06x = 720 - 540$$
$$.06x = 180$$
$$x = \frac{180}{.06} = 3000.$$

Therefore, $3000 should be invested in stock and $(9000 - 3000) = \$6000$ in the savings account. If this is done, the total return will be 12% of $3000 ($360) plus 6% of $6000 ($360), a total of $720, which is precisely 8% of $9000. ∎

EXAMPLE 11   **Mixture**

A car radiator contains 12 quarts of fluid, 20% of which is antifreeze. How much fluid should be drained and replaced with pure antifreeze so that the resulting mixture is 50% antifreeze?

SOLUTION   Let x be the number of quarts of fluid to be replaced by pure antifreeze.[*] When x quarts are drained, there are $12 - x$ quarts of fluid left in the radiator, 20% of which is antifreeze. So we have

$$\begin{pmatrix} \text{Amount of antifreeze} \\ \text{in radiator after} \\ \text{draining } x \text{ quarts} \\ \text{of fluid} \end{pmatrix} + \begin{pmatrix} x \text{ quarts of} \\ \text{antifreeze} \end{pmatrix} = \begin{pmatrix} \text{Amount of} \\ \text{antifreeze in} \\ \text{final mixture} \end{pmatrix}$$

$$20\% \text{ of } (12 - x) \quad + \quad x \quad = \quad 50\% \text{ of } 12$$

$$.20(12 - x) + x = .50(12)$$
$$2.4 - .2x + x = 6$$
$$-.2x + x = 6 - 2.4$$
$$.8x = 3.6$$
$$x = \frac{3.6}{.8} = 4.5.$$

---

[*]Hereafter, we omit the headings Label, Translate, etc.

Therefore, 4.5 quarts should be drained and replaced with pure antifreeze. ■

The basic formula for problems involving distance and a uniform rate of speed is

$$\textbf{Distance} = \textbf{Rate} \times \textbf{Time.}$$

For instance, if you drive at a rate of 55 mph for 2 hours, you travel a distance of $55 \cdot 2 = 110$ miles.

EXAMPLE 12   **Distance/Rate**

A car leaves Chicago traveling at an average speed of 64 kilometers per hour toward Decatur, which is 239 kilometers away. An hour later, a second car leaves Decatur for Chicago on the same road, traveling at an average speed of 86 km/h. When will the cars meet?

SOLUTION   Let t be the number of hours the first car has traveled. Since the second car left an hour later, it will have traveled $t - 1$ hours. When the cars meet, the distances they have traveled will total 239 km, as shown in Figure 2–4.

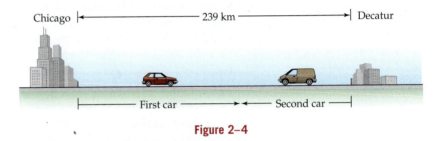

Figure 2–4

We now translate this information into an equation.

$$\begin{pmatrix}\text{Distance traveled}\\\text{by first car}\end{pmatrix} + \begin{pmatrix}\text{Distance traveled}\\\text{by second car}\end{pmatrix} = 239$$

$$\text{Rate} \cdot \text{Time} \quad + \quad \text{Rate} \cdot \text{Time} \quad = 239$$

$$64t \quad + \quad 86(t - 1) \quad = 239$$

$$64t + 86t - 86 = 239$$

$$64t + 86t = 239 + 86$$

$$150t = 325$$

$$t = \frac{325}{150} = 2\frac{1}{6} \text{ hours} \quad ■$$

> **EXAMPLE 13**   **Work**

Tom can paint the fence in 6 hours, and Huck can paint it in 4 hours. If they work together, how long will it take to paint the fence?

SOLUTION   Since Tom can paint the entire fence in 6 hours, he can paint $\frac{1}{6}$ of it in 1 hour. Similarly, Huck can paint $\frac{1}{4}$ of the fence in 1 hour. If t is the number of hours it takes to paint the fence when they work together, then together they can paint $1/t$ of the fence in 1 hour. Therefore,

$$\begin{pmatrix} \text{Part of fence} \\ \text{painted by Tom} \\ \text{in 1 hour} \end{pmatrix} + \begin{pmatrix} \text{Part of fence} \\ \text{painted by Huck} \\ \text{in 1 hour} \end{pmatrix} = \begin{pmatrix} \text{Part of fence} \\ \text{painted by both} \\ \text{in 1 hour} \end{pmatrix}$$

$$\frac{1}{6} + \frac{1}{4} = \frac{1}{t}$$

$$12t\left(\frac{1}{6} + \frac{1}{4}\right) = 12t\left(\frac{1}{t}\right)$$

$$2t + 3t = 12$$

$$5t = 12$$

$$t = \frac{12}{5} = 2.4 \text{ hours.} \quad \blacksquare$$

✓ EXERCISES 2.1

In Exercises 1–30, solve the equation.

1. $2x = 3$

2. $4x = 0$

3. $-x = 5$

4. $2 - x = 7$

5. $3x + 2 = 26$

6. $\dfrac{y}{5} - 3 = 14$

7. $3x + 2 = 9x + 7$

8. $-7(t + 2) = 3(4t + 1)$

9. $\dfrac{3y}{4} - 6 = y + 2$

10. $2(1 + x) = 3x + 5$

11. $\dfrac{1}{2}(2 - z) = \dfrac{3}{2} + 6z$

12. $\dfrac{4y}{5} + \dfrac{5y}{4} = 1$

13. $2x - \dfrac{x - 5}{4} = \dfrac{3x - 1}{2} + 1$

14. $2\left(\dfrac{x}{3} + 1\right) + 5\left(\dfrac{4x}{3} - 2\right) = 2$

15. $\dfrac{2x - 1}{2x + 1} = \dfrac{1}{4}$

16. $\dfrac{2x}{x - 3} = 1 + \dfrac{6}{x - 3}$

17. $\dfrac{1}{2t} - \dfrac{2}{5t} = \dfrac{1}{10t} - 1$

18. $\dfrac{1}{2} + \dfrac{2}{y} = \dfrac{1}{3} + \dfrac{3}{y}$

19. $\dfrac{2x - 7}{x + 4} = \dfrac{5}{x + 4} - 2$

20. $\dfrac{z + 4}{z + 5} = \dfrac{-1}{z + 5}$

21. $\dfrac{6x}{3x - 1} - 5 = \dfrac{2}{3x - 1}$

22. $1 + \dfrac{1}{x} = \dfrac{2 + x}{x} + 1$

23. $(x + 2)^2 = (x + 3)^2$

24. $x^2 + 3x - 7 = (x + 4)(x - 4)$

25. $21.31 + 41.29x = 17.51x - 8.17$

26. $2.37x + 3.1288 = 6.93x - 2.48$

27. $18.923y - 15.4228 = 10.003y + 18.161$

28. $6.31(x - 3.53) = 5.42(x + 1.07) - 21.1584$

29. $18.34x - \dfrac{14.21}{3} = \dfrac{33.41}{3} - 46.82x$

30. $\dfrac{423.1}{51.71} + \dfrac{53.19x}{21.72} = x$

In Exercises 31–38, solve the given equation for the indicated variable.

31. $x = 3y - 5$ for y **32.** $5x - 2y = 1$ for x

33. $\frac{1}{4}x - \frac{1}{3}y - 6 = 0$ for y

34. $3x + \frac{1}{2}y - z = 4$ for z

35. $A = \frac{h}{2}(b + c)$ for b

36. $V = \pi b^2 c$ for c

37. $V = \frac{\pi d^2 h}{4}$ for h **38.** $\frac{1}{r} = \frac{1}{s} + \frac{1}{t}$ for r

39. Find the number c such that $3x - 6 = c$ and $2 - x = 1$ have the same solutions.

40. If $1/3$ is a solution of $cx + d = 0$, then how are c and d related?

In a simple model of the economy (by John Maynard Keynes), equilibrium between national output and national expenditures is given by the equilibrium equation

$$Y = C + I + G + (X - M),$$

where Y is the national income, C is consumption (which depends on national income), I is the amount of investment, G is government spending, X is exports, and M is imports. In Exercises 41 and 42, solve the equilibrium equation for Y under the given conditions.

41. $C = 120 + .9Y$, $M = 20 + .2Y$, $I = 140$, $G = 150$, and $X = 60$

42. $C = 60 + .85Y$, $M = 35 + .2Y$, $I = 95$, $G = 145$, and $X = 50$

43. According to data from the U.S. Department of Defense and the Congressional Research Service, the estimated annual cost y (in thousands of dollars) per active duty service member in the U.S. armed forces in year x is given by the equation

$$6000x - 2000y + 136,000 = 0 \quad (4 \le x \le 18),$$

where $x = 4$ corresponds to 1994. Find the year in which the estimated annual cost per service member is
(a) $92,000   (b) $110,000.

44. The number of pounds of tomatoes y consumed per person in the United States in year x is given by

$$5x - 22y + 286 = 0,$$

where $x = 0$ corresponds to 1980.[*] Find the year in which tomato consumption per person is
(a) 15.5 lb   (b) 18 lb

45. Data from the Centers for Medicare and Medical Services suggests that the estimated annual cost per person y for prescription drugs in the United States in year x is given by

$$104x - 5y = -715,$$

where $x = 0$ corresponds to 2000. Find the year in which the estimated annual cost is
(a) $247   (b) $351.

46. DVD sales in the United States were $1 billion in 1999 and were expected to be $9 billion in 2004.[†]
(a) Find a linear equation that models this data, with $x = 9$ corresponding to 1999. [*Hint:* See Section 1.3.]
(b) In what year were DVD sales $6 billion?

47. Online retail sales of home products in the United States were $7.5 billion in January 2002 and are projected to be $45 billion in January 2008.[‡]
(a) Find a linear equation that models this data, with $x = 2$ corresponding to 2002.
(b) In what year are sales expected to be $30 billion?

48. According to data from Merrill Lynch, AOL had 25 million subscribers in 2001 and is expected to have 30 million in 2005.
(a) Find a linear equation that models this data, with $x = 1$ corresponding to 2001.
(b) In what year will AOL have 28 million subscribers?

In Exercises 49–52, a problem situation is given.
(a) Decide what is being asked for and label the unknown quantities.
(b) Translate the verbal statements in the problem and the relationships between the known and unknown quantities into mathematical language, using a table as in Examples 7 and 8. You need not find an equation to be solved.

49. A student has exam scores of 88, 62, and 79. What score does he need on the fourth exam to have an average of 80?

English Language	Mathematical Language
Score on fourth exam	
Sum of scores on four exams	
Average of scores on four exams	

---

*U.S. Department of Agriculture.
[†]Alexander and Associates.
[‡]Forester Research.

50. What number when added to 30 is 11 times the number?

English Language	Mathematical Language
The number	
The number added to 30	
11 times the number	

51. A rectangular garden has a perimeter of 270 ft and its length is twice its width. What are the dimensions of the garden?

English Language	Mathematical Language
Length of garden	
Width of garden	
Perimeter of garden	
The length is twice the width	

52. How many gallons of a 12% salt solution should be combined with 10 gallons of an 18% salt solution to obtain a 16% solution?

English Language	Mathematical Language
Gallons of 12% solution	
Total gallons of mixture	
Amount of salt in 10 gallons of the 18% solution	
Amount of salt in the 12% solution	
Amount of salt in the mixture	

In Exercises 53–56, set up the problem by labeling the unknowns, translating the given information into mathematical language, and finding an equation that will lead to a solution of the problem. You need not solve this equation.

53. A worker gets an 8% pay raise and now makes $1593 per month. What was the worker's salary before the raise?

54. A student has exam scores of 64, 82, and 91. What is her average? What score must she get on the fourth exam to raise her average by 3 points?

55. If 9 is added to a certain number, the result is 1 less than 3 times the number. What is the number?

56. A merchant has 5 pounds of mixed nuts that cost $30. She wants to add peanuts that cost $1.50 per pound and cashews that cost $4.50 per pound to obtain 50 pounds of a mixture that costs $2.90 per pound. How many pounds of peanuts are needed?

In the remaining exercises, solve the applied problem.

57. An item on sale costs $13.41. All merchandise was marked down 25% for the sale. What was the original price of the item?

58. A discount store sells a jacket for $28, which is 20% below the list price. What is the list price?

59. Inflation caused the price of a workbook to increase by 5%. The new price is $15.12. What was the old price?

60. Lawnmowers are priced at $300 during April. In May the price is rasied by 20%. In September, the price is reduced by 20%. What is the final price? Explain why it is *not* $300.

61. You have already invested $550 in a stock with an annual return of 11%. How much of an additional $1100 should be invested at 12% and how much at 6% so that the total annual return on the entire $1650 is 9%?

62. $25,000 is placed in the bank, part in an account earning 10% interest and part in an account earning 15% interest. The annual interest paid on the $25,000 is $3350. How much money is in the 10% account?

63. If you borrow $500 from a credit union at 12% annual interest and $250 from a bank at 18% annual interest, what is the *effective annual interest rate* (that is, what single rate of interest on $750 results in the same total amount of interest)?

64. Matt's credit card bill this month is $2534. He has $4000 in his savings account which earns 0.5% interest monthy. The credit card company has a service charge of 1.75% monthly on the unpaid balance.
  (a) Assuming that there is no other activity on either account for the next 30-day period, how much of the credit card bill should Matt pay off with his savings so that the interest he earns on the balance in his savings will equal the amount of the service charge on the remainder of the credit card bill?
  (b) What will be the new balance on the credit card bill?
  (c) What will be the service charge on this bill?
  (d) What will be the new balance in Matt's savings account?
  (e) What will be the interest accrued on the new savings balance?

65. A student has an average of 72 on four exams. If the final exam is to count double, what score does she need to have an average of 80?

66. A student scored 63 and 81 on two exams, each of which accounts for 30% of the course grade. The final exam accounts for 40% of the course grade. What score must she get on the final exam to have an average grade of 76?

67. One alloy is one part copper to three parts tin. A second alloy is one part copper to four parts tin. How much of the second alloy should be combined with 24 pounds of the first alloy to obtain a new alloy that is two parts copper to seven parts tin?

68. A chemist has 10 mL of a 30% acid solution. How many milliliters of pure acid must be added in order to obtain a 50% acid solution?

69. A radiator contains 8 quarts of fluid, 40% of which is antifreeze. How much fluid should be drained and replaced with pure antifreeze so that the new mixture is 60% antifreeze?

70. Emily has 10 pounds of premium coffee that cost $7.50 per pound. She wants to mix this with ordinary coffee that costs $4 per pound to make a mixture that costs $5.50 per pound. How much ordinary coffee must be used? What will the weight of the final mixture be?

71. Two cars leave a gas station at the same time, one traveling north and the other south. The northbound car travels at 50 mph. After 3 hours the cars are 345 miles apart. How fast is the southbound car traveling?

72. A train leaves New York for Boston, 200 miles away, at 3 P.M. and averages 75 mph. Another train leaves Boston for New York on an adjacent set of tracks at 5 P.M. and averages 45 mph. At what time will the trains meet?

73. A car passes a certain point at 1 P.M. and continues along at a constant speed of 64 kilometers per hour. A second car passes the point at 2 P.M. and goes on at a constant speed of 88 kilometers per hour, following the same route as the first car. At what time will the second car catch up with the first one?

74. In a motorcycle race the winner averaged 100 mph and finished 15 minutes ahead of the loser, who averaged 95 mph. How long was the race (in miles) and what was the winner's time?

75. An airplane flew with the wind for 2.5 hours and returned the same distance against the wind in 3.5 hours. If the cruising speed of the plane was a constant 360 mph, how fast was the wind blowing? [*Hint:* If the wind speed is r miles per hour, then the plane travels at $(360 + r)$ mph with the wind and at $(360 - r)$ mph against the wind.]

76. A motorboat goes 15 miles downstream at its top speed and then turns around and returns 15 miles upstream at top speed. The trip upstream took twice as long as the trip downstream. If the boat's top speed is 10 mph in still water, how fast is the current? [See Exercise 75.]

77. If Charlie and Nick work together, they can paint the house in 20 hours. Charlie can do the job alone in 36 hours. How long would it take Nick to do the job alone?

78. Using two lawnmowers and working together, Tom and Anne can mow the lawn in 36 minutes. It takes Anne 90 minutes if she mows the entire lawn herself. How long would it take Tom to do the job if he worked alone?

79. One pipe can fill a pool in 50 minutes. Another pipe takes 75 minutes. How long does it take for both pipes together to fill the pool? [*Hint:* This is a "work problem" in disguise: How much of the pool does the first pipe fill in 1 minute?]

80. One pipe can fill a tank in 4 hours, a second pipe can fill it in 10 hours, and a third pipe can fill it in 12 hours. The pipes are connected to the same tank. How long does it take to fill the tank if all three pipes are used together? [See Exercise 79.]

81. (With apologies to Mrs. Morgan of *How Green Was My Valley*) The water faucet can fill a tub in 15 minutes. But there is a hole in the tub through which a full tub of water can drain out in 18 minutes. If the tub is empty and the faucet is turned on, how long will it take to fill the tub?

82. The length of a rectangle is 15 inches more than the width. If the perimeter is 398 inches, what are the dimensions of the rectangle?

83. The perimeter of a rectangle is 160 meters. If a new rectangle is formed by halving the length of each of one pair of opposite sides and increasing the remaining pair of sides by 30 meters each, then the new rectangle also has a perimeter of 160 meters. What were the dimensions of the original rectangle?

84. The perimeter of a rectangular garden is 72 feet. If two such gardens were placed side by side they would form a square. What are the dimensions of the garden?

Thinkers

85. Insurance on overseas shipment of goods is charged at the rate of 4% of the total value of the shipment (which is the sum of the value of the goods, the freight charges, *and* the cost of the insurance). The freight charges for $30,000 worth of goods are $4500. How much does the insurance cost?

86. A squirrel and a half eat a nut and a half in a day and a half. How many nuts do six squirrels eat in six days?

87. Diophantus of Alexandria, a third-century mathematician, lived one-sixth of his life in childhood, one-twelfth in his youth, and one-seventh as a bachelor. Then he married and five years later had a son. The son died four years before Diophantus at half the age Diophantus was when himself died. How long did Diophantus live?

OMIT

2.1.A   *SPECIAL TOPICS* Variation

If a plane flies 400 mph for t hours, then it travels a distance of $400t$ miles. So the distance d is related to the time t by the equation $d = 400t$. We say that d *varies directly as t* and that the *constant of variation* is 400. Direct variation also occurs in other settings and is formally defined as follows.

Direct Variation

Suppose the quantities u and v are related by the equation

$$v = ku,$$

where k is a nonzero constant. We say that v **varies directly as u** or that v is **directly proportional to u.** In this case, k is called the **constant of variation** or the **constant of proportionality.**

EXAMPLE 1

When you swim underwater, the pressure p varies directly with the depth d at which you swim. At a depth of 20 feet, the pressure is 8.6 pounds per square inch. What is the pressure at 65 feet deep?

SOLUTION   The variation equation is $p = kd$ for some constant k. We first use the given information to find k. Since $p = 8.6$ when $d = 20$, we have

$$p = kd$$

$$8.6 = k(20)$$

$$k = \frac{8.6}{20} = .43.$$

Therefore, the variation equation is $p = .43d$. To find the pressure at 65 feet, substitute 65 for d in the equation:

$$p = .43d = .43(65) = 27.95 \text{ pounds per square inch.} \quad \blacksquare$$

The basic idea in direct variation is that the two quantities grow or shrink together. For instance, when $v = 3u$, then as u gets large, so does v. But two quantities can also be related in such a way that one grows as the other shrinks or vice versa. For example, if $v = 5/u$, then when u is large (say, $u = 500$), v is small: $v = 5/500 = .01$. This is an example of *inverse variation.*

Inverse Variation

Suppose the quantities u and v are related by the equation

$$v = \frac{k}{u},$$

where k is a nonzero constant. We say that v **varies inversely as u** or that v **is inversely proportional to u.** Once again, k is called the **constant of variation** or the **constant of proportionality.**

EXAMPLE 2

According to one of Parkinson's laws, the amount of time the Math Department spends discussing an item in its budget is inversely proportional to the cost of the item. If the department spent 40 minutes discussing the purchase of a copying machine for $450, how much time will it spend discussing a $100 appropriation for the annual picnic?

SOLUTION   Let t denote time (in minutes) and c the cost of an item. Then the variation equation is

$$t = \frac{k}{c}$$

for some constant k. We know that $t = 40$ when $c = 450$. Substituting these numbers in the equation, we see that

$$40 = \frac{k}{450}$$

$$k = 40 \cdot 450 = 18{,}000.$$

So the variation equation is

$$t = \frac{18{,}000}{c}.$$

To find the time spent discussing the picnic, let $c = 100$. Hence,

$$t = \frac{18{,}000}{100} = 180 \text{ minutes } (= 3 \text{ hours!}). \quad \blacksquare$$

The terminology of variation also applies to situations involving powers of variables, as illustrated in the table below.

Example	Terminology	General Case*
$y = 7x^4$	y varies directly as the fourth power of x. y is directly proportional to the fourth power of x. (Constant of variation is 7.)	$v = ku^n$
$A = \pi r^2$	A varies directly as the square of r. A is directly proportional to the square for r. (Constant of variation is π.)	
$y = \dfrac{10}{x^3}$	y varies inversely as the cube of x. y is inversely proportional to the cube of x. (Constant of variation is 10.)	$v = \dfrac{k}{u^n}$
$W = \dfrac{3.48 \times 10^9}{d^2}$	W varies inversely as the square of d. W is inversely proportional to the square of d. (Constant of variation is 3.48×10^9.)	

*k is a nonzero constant; u and v are variables.

EXAMPLE 3

The density of light I falling on an object (measured in lumens per square foot) is inversely proportional to the square of the distance d from the light source. If an object 2 feet from a standard 100-watt light bulb receives 34.82 lumens per square foot, how much light falls on an object 10 feet from this bulb?

SOLUTION   The variation equation is

$$I = \frac{k}{d^2}$$

for some constant k. Since $I = 34.82$, when $d = 2$, we have

$$34.82 = \frac{k}{2^2}$$

$$k = 34.82(4) = 139.28.$$

Therefore, the variation equation is

$$I = \frac{139.28}{d^2}.$$

So the light density at 10 feet is

$$I = \frac{139.28}{10^2} = 1.3928 \text{ lumens per square foot.}\quad\blacksquare$$

In many situations, variation involves more than two variables. Typical terminology in such cases is illustrated in the following table.

Example	Terminology
$I = 100rt$	I **varies jointly** as r and t. I is **jointly proportional** to r and t.
$H = .06sd^3$	H varies jointly as s and the cube of d. H is jointly proportional to s and the cube of d.
$P = \dfrac{3T}{V}$	P varies directly as T and inversely as V. P is directly proportional to T and inversely proportional to V.
$S = \dfrac{2.5wt^2}{\ell}$	S varies directly as w and the square of t and inversely as ℓ. S is directly proportional to w and the square of t and inversely proportional to ℓ.

EXAMPLE 4

The electrical resistance R of wire (of uniform material) varies directly as the length L and inversely as the square of the diameter d. A 2-meter-long piece of wire with a diameter of 4.4 millimeters has a resistance of 500 ohms. What diameter wire should be used if a 10-meter piece is to have a resistance of 1300 ohms?

SOLUTION   The variation equation is

$$R = \frac{kL}{d^2}$$

for some constant k. So we substitute the known information ($R = 500$ when $L = 2$ and $d = 4.4$) and solve for k.

$$R = \frac{kL}{d^2}$$

$$500 = \frac{k \cdot 2}{(4.4)^2}$$

$$k = \frac{500(4.4)^2}{2} = 4840$$

Hence, the equation is

$$R = \frac{4840L}{d^2}.$$

We must find d when $L = 10$ and $R = 1300$.

$$1300 = \frac{4840 \cdot 10}{d^2}$$

$$1300d^2 = 48400$$

$$d^2 = \frac{48400}{1300} \approx 37.2308$$

Since the diameter is positive, we must have $d \approx \sqrt{37.2308} \approx 6.1$ mm.   ■

✓ EXERCISES 2.1.A

In Exercises 1–6, express each geometric formula as a statement about variation by filling in the blanks in this sentence:

_____ varies _____ as _____;
the constant of variation is _____.

1. Area of a circle of radius r: $A = \pi r^2$.

2. Volume of a sphere of radius r: $V = \frac{4}{3}\pi r^3$.

3. Area of a rectangle of length l and width w: $A = lw$.

4. Area of a triangle of base b and height h: $A = \frac{bh}{2}$.

5. Volume of a right circular cone of base radius r and height
h: $V = \frac{\pi r^2 h}{3}$.

6. Volume of a triangular cylinder of base b, height h, and
length l: $V = \frac{1}{2}bhl$.

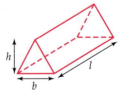

In Exercises 7–12, find the variation equation. Use k for the constant of variation.

7. a varies inversely as b.

8. r is proportional to t.

9. z varies jointly as x, y, and w.

10. The weight w of an object varies inversely as the square of the distance d from the object to the center of the earth.

11. The distance d one can see to the horizon varies directly as the square root of the height h above sea level.

12. The pressure p exerted on the floor by a person's shoe heel is directly proportional to the weight w of the person and inversely proportional to the square of the width r of the heel.

In Exercises 13–22, express the given statement as an equation and find the constant of variation.

13. v varies directly as u; $v = 8$ when $u = 2$.

14. v is directly proportional to u; $v = .4$ when $u = .8$.

15. v varies inversely as u; $v = 8$ when $u = 2$.

16. v is inversely proportional to u; $v = .12$ when $u = .1$.

17. t varies jointly as r and s; $t = 24$ when $r = 2$ and $s = 3$.

18. B varies inversely as u and v; $B = 4$ when $u = 1$ and $v = 3$.

19. w varies jointly as x and y^2; $w = 96$ when $x = 3$ and $y = 4$.

20. p varies directly as the square of z and inversely as r; $p = 32/5$ when $z = 4$ and $r = 10$.

21. T varies jointly as p and the cube of v and inversely as the square of u; $T = 24$ when $p = 3$, $v = 2$, and $u = 4$.

22. D varies jointly as the square of r and the square of s and inversely as the cube of t; $D = 18$ when $r = 4$, $s = 3$, and $t = 2$.

In Exercises 23–30, use an appropriate variation equation to find the required quantity.

23. If r varies directly as t and $r = 6$ when $t = 3$, find r when $t = 2$.

24. If r is directly proportional to t and $r = 4$ when $t = 2$, find t when $r = 2$.

25. If b varies inversely as x and $b = 9$ when $x = 3$, find b when $x = 12$.

26. If b is inversely proportional to x and $b = 10$ when $x = 4$, find x when $b = 12$.

27. Suppose w is directly proportional to the sum of u and the square of v. If $w = 200$ when $u = 1$ and $v = 7$, then find u when $w = 300$ and $v = 5$.

28. Suppose z varies jointly as x and y. If $z = 30$ when $x = 5$ and $y = 2$, then find x when $z = 45$ and $y = 3$.

29. Suppose r varies inversely as s and t. If $r = 12$ when $s = 3$ and $t = 1$, then find r when $s = 6$ and $t = 2$.

30. Suppose u varies jointly as r and s and inversely as t. If $u = 1.5$ when $r = 2$, $s = 3$, and $t = 4$, then find r when $u = 27$, $s = 9$, and $t = 5$.

31. A resident of Michigan whose federal adjusted gross income (AGI) was $24,000 dollars paid state income tax of $1008 in 2001. Use the fact that state income tax varies directly with the AGI to determine the state tax paid by someone with an income of $39,000.

32. By experiment, you discover that the amount of water that comes from your garden hose varies directly with the water pressure. A pressure of 10 pounds per square inch is needed to produce a flow of 3 gallons per minute.
   (a) What pressure is needed to produce a flow of 4.2 gallons per minute?
   (b) If the pressure is 5 pounds per square inch, what is the flow rate?

33. According to Hooke's law, the distance d that a spring stretches when an object is attached to it is directly proportional to the weight of the object. Suppose that the string stretches 15.75 inches when a 7-pound weight is attached. What weight is required to stretch the spring 27 inches?

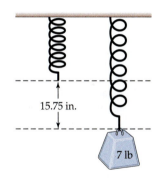

15.75 in.

7 lb

34. In a thunderstorm, there is a gap between the time you see the lightning and the time you hear the thunder. Your distance from the storm is directly proportional to the time interval between the lightning and the thunder. If you hear thunder from a storm that is 1.36 miles away 6.6 seconds after you see lightning, how far away is a storm for which the time gap between lightning and thunder is 12 seconds?

35. At a fixed temperature, the pressure of an enclosed gas is inversely proportional to its volume. The pressure is 50 kilograms per square centimeter when the volume is 200 cubic centimeters. If the gas is compressed to 125 cubic centimeters, what is the pressure?

36. The electrical resistance in a piece of wire of a given length and material varies inversely as the square of the diameter of the wire. If a wire of diameter .01 cm has a resistance of .4 ohm, what is the resistance of a wire of the same length and material but with diameter .025 cm?

37. The distance traveled by a falling object (subject only to gravity, with wind resistance ignored) is directly proportional to the square of the time it takes to fall that far. If an object falls 100 feet in 2.5 seconds, how far does it fall in 5 seconds?

38. The weight of an object varies inversely with the square of its distance from the center of the earth. Assume that the radius of the earth is 3960 miles. If an astronaut weighs 180 pounds on the surface of the earth, what does he weigh when traveling 300 miles above the surface of the earth?

39. The weight of a cylindrical can of glop varies jointly as the height and the square of the base radius. The weight is 250 ounces when the height is 20 inches and the base radius is 5 inches. What is the height when the weight is 960 ounces and the base radius is 8 inches?

40. The force of the wind blowing directly on a flat surface varies directly with the area of the surface and the square of the velocity of the wind. A 10 mile an hour (mph) wind blowing on a wooden gate that measures 3 by 6 feet exerts a force of 15 pounds. What is the force on a 1.5 by 2 foot traffic sign from a 50 mph wind?

41. The force needed to keep a car from skidding on a circular curve varies inversely as the radius of the curve and jointly as the weight of the car and the square of the speed. It takes 1500 kilograms of force to keep a 1000-kilogram car from skidding on a curve of radius 200 meters at a speed of 50 kilometers per hour. What force is needed to keep the same car from skidding on a curve of radius 320 meters at 100 kilometers per hour?

42. Boyle's law states that the volume of a given mass of gas varies directly as the temperature and inversely as the pressure. If a gas has a volume of 14 cubic inches when the temperature is $40°$ and the pressure is 280 pounds per square inch, what is the volume at a temperature of $50°$ and pressure of 175 pounds per square inch?

43. The maximum safe load that a rectangular beam can support varies directly with the width w and the square of the height h, and inversely with the length l of the beam. A 6-foot-long beam that is 2 inches high and 4 inches wide has a maximum safe load of 1000 pounds.
    (a) What is the maximum safe load for a 10-foot-long beam that is 4 inches high and 4 inches wide?
    (b) How long should a 4 inch by 4 inch beam be to safely support a 6000-pound load?

44. The period of a pendulum is the time it takes for the pendulum to make one complete swing (going both left and right) and return to its starting point. The period P varies directly with the *square root* of the length l of the pendulum.
    (a) Write the variation equation that describes this situation, using k for the constant of variation.
    (b) What happens to the period if the length of the pendulum is quadrupled?

2.2 Quadratic Equations and Applications

A **second-degree,** or **quadratic, equation** is an equation that can be written in the form

$$ax^2 + bx + c = 0$$

for some constants a, b, c with $a \neq 0$. There are several algebraic techniques for solving such equations. We begin with the **factoring method,** which makes use of this property of real numbers:

Zero Products

If a product of real numbers is zero, then at least one of the factors is zero; in other words

If $cd = 0$, then $c = 0$ or $d = 0$.

EXAMPLE 1

Solve: $7x^2 - 5x - 2 = 0$.

SOLUTION   We first factor the left-hand side.

$$7x^2 - 5x - 2 = 0$$

$$(7x + 2)(x - 1) = 0$$

When a product of real numbers is 0, than at least one of the factors must be 0. So this equation is equivalent to

$$7x + 2 = 0 \quad \text{or} \quad x - 1 = 0$$

$$7x = -2 \qquad\qquad x = 1.$$

$$x = -2/7$$

Therefore, the exact solutions are $-2/7$ and 1. ■

EXAMPLE 2

Solve: $3x^2 + x = 10$.

SOLUTION   We first rearrange the terms to make one side 0 and then factor.

$$3x^2 + x = 10$$

Subtract 10 from each side: $3x^2 + x - 10 = 0$

Factor left-hand side: $(3x - 5)(x + 2) = 0$

Since the product on the left-hand side is 0, one of its factors must be 0. Hence,

$$3x - 5 = 0 \quad \text{or} \quad x + 2 = 0$$

$$3x = 5 \qquad\qquad x = -2.$$

$$x = 5/3$$

The solutions are -2 and $5/3$. ■

CAUTION
To guard against mistakes check your solutions by substituting each one in the *original* equation to make sure it really *is* a solution.

▪▪ THE QUADRATIC FORMULA

The solutions of $x^2 = 7$ are the numbers whose square is 7. Although 7 has just one square root, there are *two* numbers whose square is 7, namely, $\sqrt{7}$ and $-\sqrt{7}$. So the solutions of $x^2 = 7$ are $\sqrt{7}$ and $-\sqrt{7}$, or in abbreviated form $\pm\sqrt{7}$. A similar argument works with any positive number d in place of 7:

The solutions of $x^2 = d$ are $\sqrt{d}$ and $-\sqrt{d}$.

The same reasoning enables us to solve other equations.

EXAMPLE 3

Solve: $(z - 2)^2 = 5$.

SOLUTION　The equation says that $z - 2$ is a number whose square is 5. Since there are only two numbers whose square is 5, namely, $\sqrt{5}$ and $-\sqrt{5}$, we must have

$$z - 2 = \sqrt{5} \qquad \text{or} \qquad z - 2 = -\sqrt{5}$$
$$z = \sqrt{5} + 2 \qquad\qquad\qquad z = -\sqrt{5} + 2.$$

In compact notation, the solutions of the equation are $\pm\sqrt{5} + 2$.　■

Let a, b, c be any real numbers with $a \neq 0$. We shall solve the equation

$$ax^2 + bx + c = 0.$$

We begin by noting a fact that will be needed in the solution process. By multiplying out the right-hand side, you should verify that

$(*)$
$$x^2 + \frac{b}{a}x + \left(\frac{b}{2a}\right)^2 = \left(x + \frac{b}{2a}\right)^2.$$

Now solve the equation as follows.[*]

$$ax^2 + bx + c = 0$$

Divide both sides by a:
$$x^2 + \frac{b}{a}x + \frac{c}{a} = 0$$

Subtract $\frac{c}{a}$ from both sides:
$$x^2 + \frac{b}{a}x = -\frac{c}{a}$$

Add $\left(\frac{b}{2a}\right)^2$ to both sides:
$$x^2 + \frac{b}{a}x + \left(\frac{b}{2a}\right)^2 = \left(\frac{b}{2a}\right)^2 - \frac{c}{a}$$

Factor the left-hand side of this last equation by using equation $(*)$ above:

$$\left(x + \frac{b}{2a}\right)^2 = \left(\frac{b}{2a}\right)^2 - \frac{c}{a}$$

Find common denominator for right-hand side:
$$\left(x + \frac{b}{2a}\right)^2 = \frac{b^2}{4a^2} - \frac{c}{a} = \frac{b^2 - 4ac}{4a^2}$$

This last equation says that $x + \frac{b}{2a}$ is a number whose square is $\frac{b^2 - 4ac}{4a^2}$.

So we must have

$$x + \frac{b}{2a} = \sqrt{\frac{b^2 - 4ac}{4a^2}} \qquad \text{or} \qquad x + \frac{b}{2a} = -\sqrt{\frac{b^2 - 4ac}{4a^2}}$$

or, in more compact notation,

$$x + \frac{b}{2a} = \pm\sqrt{\frac{b^2 - 4ac}{4a^2}} = \pm\frac{\sqrt{b^2 - 4ac}}{2a}.$$

---

[*]Don't be put off by all the letters here—the algebra/arithmetic is all easy. Also, don't worry about how anyone thought to do these steps; just verify that each one follows from the preceding one.

Adding $\dfrac{-b}{2a}$ to both ends shows that

$$x = \frac{-b}{2a} \pm \frac{\sqrt{b^2 - 4ac}}{2a} = \frac{-b \pm \sqrt{b^2 - 4ac}}{2a}.$$

We have proved a major result.

The Quadratic Formula

> The solutions of the quadratic equation $ax^2 + bx + c = 0$ are
> $$x = \frac{-b \pm \sqrt{b^2 - 4ac}}{2a}.$$

You should memorize the quadratic formula.

EXAMPLE 4

Solve $x^2 + 3 = -8x$.

SOLUTION   Rewrite the equation as $x^2 + 8x + 3 = 0$, and apply the quadratic formula with $a = 1$, $b = 8$, and $c = 3$.

$$x = \frac{-b \pm \sqrt{b^2 - 4ac}}{2a} = \frac{-8 \pm \sqrt{8^2 - 4 \cdot 1 \cdot 3}}{2 \cdot 1}$$

$$= \frac{-8 \pm \sqrt{52}}{2} = \frac{-8 \pm \sqrt{4 \cdot 13}}{2}$$

$$= \frac{-8 \pm 2\sqrt{13}}{2} = -4 \pm \sqrt{13}$$

The equation has two distinct real solutions: $-4 + \sqrt{13}$ and $-4 - \sqrt{13}$.   ■

EXAMPLE 5

Solve $x^2 - 194x + 9409 = 0$.

SOLUTION   Use a calculator and the quadratic formula with $a = 1$, $b = -194$, and $c = 9409$.

$$x = \frac{-b \pm \sqrt{b^2 - 4ac}}{2a} = \frac{-(-194) \pm \sqrt{(-194)^2 - 4 \cdot 1 \cdot 9409}}{2 \cdot 1}$$

$$= \frac{194 \pm \sqrt{37636 - 37636}}{2} = \frac{194 \pm 0}{2} = 97$$

Thus, 97 is the only solution of the equation.   ■

EXAMPLE 6

Solve $2x^2 + x + 3 = 0$.

SOLUTION   Using the quadratic formula with $a = 2$, $b = 1$, and $c = 3$, we have

$$x = \frac{-b \pm \sqrt{b^2 - 4ac}}{2a} = \frac{-1 \pm \sqrt{1^2 - 4 \cdot 2 \cdot 3}}{2 \cdot 2} = \frac{-1 \pm \sqrt{1 - 24}}{4}$$

$$= \frac{-1 \pm \sqrt{-23}}{4}.$$

Since $\sqrt{-23}$ is not a real number, this equation has *no real solutions* (that is, no solutions in the real number system).   ■

EXAMPLE 7

If an object is thrown upward, dropped, or thrown downward and travels in a straight line subject only to gravity (with wind resistance ignored), the height h of the object above the ground (in feet) after t seconds is given by

$$h = -16t^2 + v_0 t + h_0,$$

where h_0 is the height of the object when $t = 0$ and v_0 is the initial velocity at time $t = 0$. The value of v_0 is taken as positive if the object moves upward and negative if it moves downward. If a baseball is thrown down from the top of a 640-foot-high building with an initial velocity of 52 feet per second, how long does it take to reach the ground?

SOLUTION   In this case, v_0 is -52 and h_0 is 640, so that the height equation is

$$h = -16t^2 - 52t + 640.$$

The object is on the ground when $h = 0$, so we must solve the equation

$$0 = -16t^2 - 52t + 640.$$

Using the quadratic formula and a calculator, we see that

$$t = \frac{-(-52) \pm \sqrt{(-52)^2 - 4(-16)(640)}}{2(-16)} = \frac{52 \pm \sqrt{43,664}}{-32} \approx \begin{cases} -8.15 \\ \text{or} \\ 4.90. \end{cases}$$

Only the positive answer makes sense in this case. So it takes about 4.9 seconds for the baseball to reach the ground.   ■

■■ THE DISCRIMINANT

The expression $b^2 - 4ac$ in the quadratic formula is called the **discriminant.** As the last four examples illustrate, the discriminant determines the *number* of real solutions of the equation $ax^2 + bx + c = 0$.

Real Solutions of a
Quadratic Equation

Discriminant $b^2 - 4ac$	Number of Real Solutions of $ax^2 + bx + c = 0$	Example
> 0	Two distinct real solutions	$x^2 + 8x + 3 = 0$
$= 0$	One real solution	$x^2 - 194x + 9409 = 0$
< 0	No real solutions	$2x^2 + x + 3 = 0$

The quadratic formula and a calculator can be used to solve any quadratic equation with nonnegative discriminant. Experiment with your calculator to find the most efficient sequence of keystrokes for doing this.

EXAMPLE 8

TECHNOLOGY TIP

Most calculators have built-in polynomial equation solvers that will solve quadratic and other polynomial equations. See Calculator Investigation 1 at the end of this section. A quadratic formula program for other calculators is in the Program Appendix.

Use a calculator to solve $3.2x^2 + 15.93x - 7.1 = 0$.

First compute $\sqrt{b^2 - 4ac} = \sqrt{15.93^2 - 4(3.2)(-7.1)}$ and store the result in memory D. Then the solutions of the equation are given by

$$x = \frac{-b \pm \sqrt{b^2 - 4ac}}{2a} = \frac{-15.93 \pm D}{2(3.2)}.$$

They are easily approximated on a calculator.

$$\frac{-15.93 + D}{2(3.2)} \approx .411658467 \quad \text{and} \quad \frac{-15.93 - D}{2(3.2)} \approx -5.389783467.$$

Remember that these answers are *approximations,* so they might not check exactly when substituted in the original equation. ■

▓▓ APPLICATIONS

The guidelines for setting up applied problems in Section 2.1 are equally useful in situations involving quadratic equations. However, setting up a problem is only half the job. You must then solve the equation you have obtained, getting exact answers whenever possible. Finally, you must

> **Interpret your answers in terms of the original problem. Do they make sense? Do they satisfy the required conditions?**

In particular, an equation may have several solutions, some of which may not make sense in the context of the problem. For instance, distance can't be negative, the number of people in a room cannot be a proper fraction, etc.

EXAMPLE 9

A rectangle of area 24.5 square inches is twice as long as it is wide. What are its dimensions?

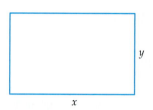

Figure 2–5

SOLUTION   Figure 2–5 shows the situation. The given information can be translated into mathematical terms as follows:

English Language	Mathematical Language
Length and width	x and y
Length is twice the width	$x = 2y$
Area (length × width)	xy
Area is 24.5 sq. in.	$xy = 24.5$

Now substitute $x = 2y$ into the area equation and solve it.

$$xy = 24.5$$
$$(2y)y = 24.5$$
$$2y^2 = 24.5$$
$$y^2 = 12.25$$
$$y = \pm\sqrt{12.25} = \pm 3.5$$

Since the width is a positive number, $y = 3.5$ is the only relevant solution. Therefore, the rectangle has width 3.5 and length $x = 2y = 2(3.5) = 7$.   ■

EXAMPLE 10

John and Kathy want to put a cement walk of uniform width around a rectangular garden that measures 24 by 40 feet. They have enough cement to cover 660 square feet. How wide should the walk be to use up all the cement?

SOLUTION   Let x denote the width of the walk (in feet) and draw a picture of the situation (Figure 2–6).

The length of the outer rectangle is $40 + 2x$ (the garden length plus walks on each end) and its width is $24 + 2x$.

$$\left(\begin{array}{c}\text{Area of outer}\\ \text{rectangle}\end{array}\right) - \left(\begin{array}{c}\text{Area of}\\ \text{garden}\end{array}\right) = \text{Area of walk}$$

$$\text{Length} \cdot \text{Width} - \text{Length} \cdot \text{Width} = 660$$

$$(40 + 2x)(24 + 2x) - 40 \cdot 24 = 660$$
$$960 + 128x + 4x^2 - 960 = 660$$
$$4x^2 + 128x - 660 = 0$$

Figure 2–6

Dividing both sides by 4 and applying the quadratic formula yields

$$x^2 + 32x - 165 = 0$$

$$x = \frac{-32 \pm \sqrt{(32)^2 - 4 \cdot 1 \cdot (-165)}}{2 \cdot 1}$$

$$x = \frac{-32 \pm \sqrt{1684}}{2} \approx \begin{cases} 4.5183 \\ \text{or} \\ -36.5183. \end{cases}$$

Only the positive solution makes sense in the context of this problem. The walk should be approximately 4.5 feet wide. ■

EXAMPLE 11

A stone is dropped from the top of a cliff. Five seconds later, the sound of the stone hitting the ground is heard. Use the fact that the speed of sound is 1100 feet per second and that the distance traveled by a falling object in t seconds is $16t^2$ feet to determine the height of the cliff.

SOLUTION   If t is the time it takes the rock to fall to the ground, then the height of the cliff is $16t^2$ feet. The total time for the rock to fall and the sound of its hitting ground to return to the top of the cliff is 5 seconds. So the sound must take $5 - t$ seconds to return to the top, as indicated in Figure 2–7. Therefore,

$$\begin{pmatrix} \text{Distance rock falls} \\ \text{in } t \text{ seconds} \end{pmatrix} = \text{Height of cliff} = \begin{pmatrix} \text{Distance sound} \\ \text{travels upward} \\ \text{in } 5 - t \text{ seconds} \end{pmatrix}$$

$$16t^2 = \text{Rate} \cdot \text{Time}$$

$$16t^2 = 1100(5 - t)$$

$$16t^2 = 5500 - 1100t$$

$$16t^2 + 1100t - 5500 = 0.$$

The quadratic formula and a calculator show that

$$t = \frac{-1100 \pm \sqrt{(1100)^2 - 4 \cdot 16 \cdot (-5500)}}{2 \cdot 16}$$

$$= \frac{-1100 \pm \sqrt{1,562,000}}{32} \approx \begin{cases} 4.6812 \\ \text{or} \\ -73.4312. \end{cases}$$

Since the time must be a number between 0 and 5, we see that $t \approx 4.6812$ seconds. Therefore the height of the cliff is

$$16(4.6812)^2 \approx 350.6 \text{ feet.} \quad ■$$

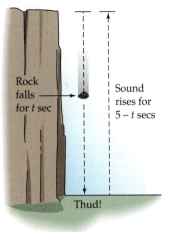

Rock falls for t sec

Sound rises for $5 - t$ secs

Thud!

Figure 2–7

EXAMPLE 12

A pilot wants to make the 840-mile round trip from Cleveland to Peoria and back in 5 hours flying time. Going to Peoria, there will be a headwind of 30 mph, that is, a wind opposite to the direction the plane is flying. It is estimated that on the return trip to Cleveland, there will be a 40-mph tailwind (in the direction the plane is flying). At what constant speed should the plane be flown?

SOLUTION   Let x be the engine speed of the plane. On the trip to Peoria, the actual speed will be $x - 30$ mph because of the headwind.

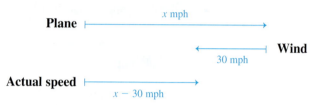

Plane — x mph

Wind — 30 mph

Actual speed — $x - 30$ mph

The distance from Cleveland to Peoria is 420 miles (half the round trip). So we have

$$\text{Rate} \cdot \text{Time} = \text{Distance}$$

$$\text{Time} = \frac{\text{Distance}}{\text{Rate}} = \frac{420}{x - 30}.$$

On the return trip, the actual speed will be $x + 40$ because of the tailwind.

Plane — x mph   **Wind** — 40 mph

Actual speed — $x + 40$ mph

So we have

$$\text{Rate} \cdot \text{Time} = \text{Distance}$$

$$\text{Time} = \frac{\text{Distance}}{\text{Rate}} = \frac{420}{x + 40}.$$

Therefore,

$$5 = \left(\begin{array}{l}\text{Time from Cleveland} \\ \text{to Peoria}\end{array}\right) + \left(\begin{array}{l}\text{Time from Peoria} \\ \text{to Cleveland}\end{array}\right)$$

$$5 = \frac{420}{x - 30} + \frac{420}{x + 40}.$$

Multiplying both sides by the common denominator $(x - 30)(x + 40)$ and simplifying, we have

$$5(x - 30)(x + 40) = \frac{420}{x - 30} \cdot (x - 30)(x + 40) + \frac{420}{x + 40} \cdot (x - 30)(x + 40)$$

$$5(x - 30)(x + 40) = 420(x + 40) + 420(x - 30)$$

$$(x - 30)(x + 40) = 84(x + 40) + 84(x - 30)$$

$$x^2 + 10x - 1200 = 84x + 3360 + 84x - 2520$$

$$x^2 - 158x - 2040 = 0$$

$$(x - 170)(x + 12) = 0$$

$$x - 170 = 0 \quad \text{or} \quad x + 12 = 0$$

$$x = 170 \qquad\qquad x = -12.$$

Obviously, the negative solution doesn't apply. Since we multiplied both sides by a quantity involving the variable, we must check that 170 actually is a solution of the original equation. It is; so the plane should be flown at a speed of 170 mph. ■

INVESTIGATIONS 2.2

1. POLYNOMIAL EQUATION SOLVERS

If you have one of the calculators mentioned below, use its polynomial equation solver to solve the equation $3.2x^2 + 15.93x - 7.1 = 0$. How do the answers you obtain compare with those in Example 8?

	Solver Name and Location	*Special Syntax (if needed)*
TI-86	POLY on Keyboard	For "order," enter the degree of the polynomial.
TI-89	SOLVE in ALGEBRA menu	SOLVE $(3.2x^2 + 15.93x - 7.1 = 0, x)$
Casio	EQUATION in main menu	
HP-39+	POLYROOT in POLYNOM submenu of MATH menu	POLYROOT([3.2, 15.93, −7.1])

Note: TI-83+ and TI-89 users can download a free polynomial solver similar to the one on TI-86 [http://education.ti.com/apps].

Use the same technique to solve the following equations.

(a) $x^2 + 2x - 4 = 0$

(b) $x^2 + 5x + 2 = 0$

✓ EXERCISES 2.2

In Exercises 1–12, solve the equation by factoring.

1. $x^2 - 8x + 15 = 0$

2. $x^2 + 5x + 6 = 0$

3. $x^2 - 5x = 14$

4. $x^2 + x = 20$

5. $2y^2 + 5y - 3 = 0$

6. $3t^2 - t - 2 = 0$

7. $4t^2 + 9t + 2 = 0$

8. $9t^2 + 2 = 11t$

9. $3u^2 + u = 4$

10. $5x^2 + 26x = -5$

11. $12x^2 + 13x = 4$

12. $18x^2 = 23x + 6$

In Exercises 13–22, use the quadratic formula to solve the equation.

13. $x^2 - 4x + 1 = 0$

14. $x^2 - 2x - 1 = 0$

15. $x^2 + 6x + 7 = 0$

16. $x^2 + 4x - 3 = 0$

17. $x^2 + 6 = 2x$

18. $x^2 + 11 = 6x$

19. $4x^2 - 4x = 7$

20. $4x^2 - 4x = 11$

21. $4x^2 - 8x + 1 = 0$

22. $2t^2 + 4t + 1 = 0$

In Exercises 23–28, find the number of real solutions of the equation by computing the discriminant.

23. $x^2 + 4x + 1 = 0$

24. $4x^2 - 4x - 3 = 0$

25. $9x^2 = 12x + 1$

26. $9t^2 + 15 = 30t$

27. $25t^2 + 49 = 70t$

28. $49t^2 + 5 = 42t$

In Exercises 29–32, solve the equation and check your answers. [Hint: First, eliminate fractions by multiplying both sides by a common denominator.]

29. $1 - \dfrac{3}{x} = \dfrac{40}{x^2}$

30. $\dfrac{4x^2 + 5}{3x^2 + 5x - 2} = \dfrac{4}{3x - 1} - \dfrac{3}{x + 2}$

31. $\dfrac{2}{x^2} - \dfrac{5}{x} = 4$

32. $\dfrac{x}{x - 1} + \dfrac{x + 2}{x} = 3$

In Exercises 33–42, solve the equation by any method.

33. $x^2 + 9x + 18 = 0$

34. $3t^2 - 11t - 20 = 0$

35. $4x(x + 1) = 1$

36. $25y^2 = 20y + 1$

37. $2x^2 = 7x + 15$

38. $2x^2 = 6x + 3$

39. $t^2 + 4t + 13 = 0$

40. $5x^2 + 2x = -2$

41. $\dfrac{7x^2}{3} = \dfrac{2x}{3} - 1$

42. $25x + \dfrac{4}{x} = 20$

In Exercises 43 and 44, use a calculator and the quadratic formula to find approximate solutions of the equation.

43. $4.42x^2 - 10.14x + 3.79 = 0$

44. $8.06x^2 + 25.8726x - 25.047256 = 0$

45. The atmospheric pressure a (in pounds per square foot) at height h thousand feet above sea level is approximately

$$a = .8315h^2 - 73.93h + 2116.1.$$

  (a) Find the atmospheric pressure at sea level and at the top of Mount Everest, the tallest mountain in the world (29,035 feet*). [Remember that h is measured in thousands.]

  (b) The atmospheric pressure at the top of Mount Rainier is 1223.43 pounds per square foot. How high is Mount Rainier?

46. Data from the U.S. Department of Health and Human Services indicates that the cumulative number N of reported cases of AIDS in the United States in year x can be approximated by the equation

$$N = 3362.1x^2 - 17{,}270.3x + 24{,}043,$$

where $x = 0$ corresponds to 1980. In what year did the total reach 550,000?

47. According to data from the U.S. Census Bureau, the population P of Cleveland, Ohio (in thousands) in year x can be approximated by $P = .08x^2 - 13.08x + 927$, where $x = 0$ corresponds to 1950. In what year in the past was the population about 804,200?

48. According to data from the National Highway Traffic Safety Administration, the driver fatality rate D in automobile accidents per 1000 licensed drivers every 100 million miles can be approximated by the equation

$$D = .0031x^2 - .291x + 7.1,$$

where x is the age of the driver.

  (a) For what ages is the driver fatality rate approximately 1 death per 1000 drivers every 100 million miles?

  (b) For what ages is the rate three times greater than in part (a).

In Exercises 49–51, find a number k such that the given equation has exactly one real solution.

49. $x^2 + kx + 25 = 0$

50. $x^2 - kx + 49 = 0$

51. $kx^2 + 8x + 1 = 0$

---

*Based on measurements in 1999 by climbers sponsored by the Boston Museum of Science and the National Geographic Society, using satellite-based technology.

In Exercises 52–54, the discriminant of the equation $ax^2 + bx + c = 0$ (with a, b, c integers) is given. Use it to determine whether or not the solutions of the equation are rational numbers.

52. $b^2 - 4ac = 25$

53. $b^2 - 4ac = 0$

54. $b^2 - 4ac = 72$

In Exercises 55–58, solve the equation for the indicated letter. State any conditions that are necessary for your answer to be valid.

55. $E = mc^2$ for c **56.** $V = 4\pi r^2$ for r

57. $A = \pi rh + \pi r^2$ for r **58.** $d = -16t^2 + vt$ for t

59. Find a number k such that 4 and 1 are the solutions of $x^2 - 5x + k = 0$.

60. Find the error in the following "proof" that $6 = 3$.

$$x = 3$$

Multiply both sides by x:	$x^2 = 3x$
Subtract 9 from both sides:	$x^2 - 9 = 3x - 9$
Factor each side:	$(x - 3)(x + 3) = 3(x - 3)$
Divide both sides by $x - 3$:	$x + 3 = 3$
Because $x = 3$:	$3 + 3 = 3$
Oops!	$6 = 3.$

In Exercises 61–64, label the unknown quantities and fill in the rest of the right-hand column of the table. Then use the information in the table to obtain an equation in one variable that will provide the solution to the problem. Solve the equation and interpret the answers.

61. A rectangle has perimeter 45 cm and an area of 112.5 sq cm. What are the dimensions?

English Language	Mathematical Language
Length of rectangle	
Width of rectangle	
Perimeter of rectangle	
Perimeter is 45	
Area of rectangle	
Area is 112.5.	

62. A triangle has an area of 96 sq in. and its height is two-thirds of its base. What are the base and height of the triangle?

English Language	Mathematical Language
Base of triangle	
Height of triangle	
Height is two-thirds of base.	
Area of triangle	
Area is 96.	

63. A rectangular window has 5.7 square feet of glass. Its length is 2.3 ft longer than its width. What are the dimensions of the window?

English Language	Mathematical Language
Length of window	
Width of window	
Area of window	
Area of window is 5.7.	
Length is 2.3 ft longer than width.	

64. The area of a circular window (in square feet) is two and half times the length of the diameter. What is the radius of the window?

English Language	Mathematical Language
Radius of circle	
Area of circle	
Diameter of circle	
Two and a half times the diameter	
Relationship of diameter and radius	

65. The two legs of a right triangle differ in length by 1 cm and the hypotenuse is 1 cm longer than the longer leg. How long is each side?

66. The area of a triangle is 20 square feet. If the base is 2 feet longer than the height, find the height.

67. The radius of a circle is 8 cm. By what amount must the radius be decreased in order to decrease the area of the circle by 48π square centimeters?

68. A right triangle has a hypotenuse of length 12.054 meters. The sum of the lengths of the other two sides is 16.96 meters. How long is each side?

69. A 13-foot-long ladder leans on a wall, as shown in the figure. The bottom of the ladder is 5 feet from the wall. If the bottom is pulled out 3 feet farther from the wall, how far does the top of the ladder move down the wall? [*Hint:* Draw pictures of the right triangle formed by the ladder, the ground, and the wall before and after the ladder is moved. In each case, use the Pythagorean Theorem to find the distance from the top of the ladder to the ground.]

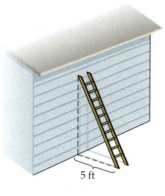

5 ft

70. A 15-foot-long pole leans against a wall. The bottom is 9 feet from the wall. How much farther should the bottom be pulled away from the wall so that the top moves the same amount down the wall?

71. A concrete walk of uniform width is to be built around a circular pool, as shown in the figure. The radius of the pool is 12 meters and enough concrete is available to cover 52π square meters. If all the concrete is to be used, how wide should the walk be?

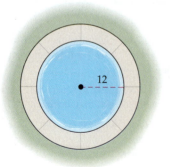

12

72. Mary Alice has 92 one-foot-square tiles that will be laid around the edges of a 12-by-15-foot room. A rectangular rug that is 3 feet longer than it is wide is to be placed over the center area where there are no tiles. To the nearest quarter foot, find the dimensions of the smallest rug that will cover the untiled part of the floor. [Assume that all the tiles are used and that none of them are split.]

73. A corner lot has dimensions of 25 by 40 yards. The city plans to take a strip of uniform width along the two sides bordering the streets in order to widen these roads. To the nearest tenth of a yard, how wide should the strip be if the remainder of the lot is to have an area of 844 square yards?

74. A box with a square base and no top is to be made from a square sheet of aluminum by cutting out 3-inch squares from each corner and folding up the sides as shown in the figure at the bottom of the page. If the box is to have a volume of 48 cubic inches, what size should the piece of aluminum be?

75. A box with no top is to be made by taking a rectangular aluminum sheet 8 by 10 inches and cutting a square of the same size out of each corner and folding up the sides. If the area of the base is to be 24 square inches, what size squares should be cut out?

76. A box with no top is to be made by taking a rectangular piece of aluminum and cutting a square of the same size out of each corner and folding up the sides. The box is to be 2 inches deep. The length of its base is to be 5 inches more than the width. If the volume is to be 352 cubic inches, what size should the piece of aluminum be?

77. It requires 53.4 square inches of metal to make a cylindrical can whose height is 4.25 inches. To the nearest hundredth, what is the radius of the can? [*Hint:* The formula for the surface area S of a right circular cylinder of radius r and height h is $S = 2\pi r^2 + 2\pi rh$.]

78. Two trains leave the same city at the same time, one going north and the other east. The northbound train travels 20 mph faster than the eastbound one. If they are 300 miles apart after 5 hours, what is the speed of each train?

79. A student leaves the university at noon, bicycling south at a constant rate. At 12:30 a second student leaves the same point and heads west, bicycling 7 mph faster than the first student. At 2 P.M. they are 30 miles apart. How fast is each one going?

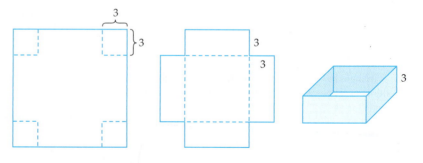

80. To get to work Sam jogs 3 kilometers to the train, then rides the remaining 5 kilometers. If the train goes 40 kilometers per hour faster than Sam's constant rate of jogging and the entire trip takes 30 minutes, how fast does Sam jog?

81. In 1991, Rick Mears won the Indianapolis 500 (mile) race. His speed (rounded to the nearest whole number) was 100 mph faster than the speed of Ray Harroun, who won the race in 1911. Mears took 3.74 hours less than Harroun. What was Mears's speed (to the nearest whole number)?

82. A canoeist paddles at a constant rate and takes 2 hours longer to go 12 kilometers upstream than to go the same distance downstream. If the current runs at 3 kilometers per hour, how long would a 12-kilometer trip take in still water?

83. Tom drops a rock into a well and 3 seconds later hears the sound of its splash. How deep is the well? [Make the same assumptions about falling rocks and sound as in Example 11.]

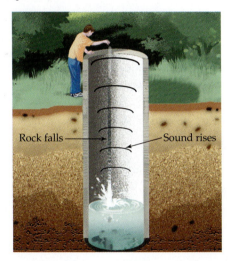

Rock falls ——→   ←—— Sound rises

84. A group of homeowners are to share equally in the $210 cost of repairing a bus-stop shelter near their homes. At the last moment two members of the group decide not to participate and this raises the share of each remaining person by $28. How many people were in the group at the beginning?

85. A woman and her son working together can paint a garage in 4 hours and 48 minutes. The woman working alone can paint it in 4 hours less than the son would take to do it alone. How long would it take the son to paint the garage alone?

86. When each works alone Barbara can do a job in 3 hours less time than John. When they work together it takes them 2 hours to complete the job. How long does it take each one working alone to do the job?

In Exercises 87–90, use the height equation in Example 7. Note that an object that is dropped (rather than thrown downward) has initial velocity $v_0 = 0$.

87. How long does it take a baseball to reach the ground if it is dropped from the top of a 640-foot-high building? Compare with Example 7.

88. You are standing on a cliff that is 200 feet high. How long will it take a rock to reach the ground if
 (a) you drop it?
 (b) you throw it downward at an initial velocity of 40 feet per second?
 (c) How far does the rock fall in 2 seconds if you throw it downward with an initial velocity of 40 feet per second?

89. A rocket is fired straight up from ground level with an initial velocity of 800 feet per second.
 (a) How long does it take the rocket to rise 3200 feet?
 (b) When will the rocket hit the ground?

90. A rocket loaded with fireworks is to be shot vertically upward from ground level with an initial velocity of 200 feet per second. When the rocket reaches a height of 400 feet on its upward trip the fireworks will be detonated. How many seconds after liftoff will this take place?

Thinkers

91. Two ferryboats leave opposite sides of a lake at the same time. They pass each other when they are 800 meters from the nearest shore. When it reaches the opposite side, each boat spends 30 minutes at the dock and then starts back. This time the boats pass each other when they are 400 meters from the nearest shore. Assuming that each boat travels at the same speed in both directions, how wide is the lake between the two ferry docks?

92. Charlie was crossing a narrow bridge. When he was halfway across, he saw a truck on the opposite side of the bridge, 200 yards away and heading toward him. He turned and ran back. The truck continued at the same speed and missed him by a hair. If Charlie had tried to cross the bridge, the truck would have hit him 3 yards before he reached the other end. How long is the bridge?

2.3   Solving Equations Graphically and Numerically

Algebraic techniques are effective for solving linear and quadratic equations, as we saw in the preceding sections. For many other equations, however, graphical and numerical approximation methods are the only practical alternative. These methods are based on a connection between graphs and equations that we now examine.

In the coordinate plane, all points on the x-axis have second coordinate 0. Consequently, when a graph intersects the x-axis, the intersection point has coordinates of the form $(a, 0)$ for some real number a. The number a is called an **x-intercept** of the graph. For example, the graph of $y = x^2 + x - 2$ in Figure 2–8 has x-intercepts at -2 and 1 because it intersects the x-axis at $(-2, 0)$ and $(1, 0)$. To say that $(1, 0)$ is on the graph of $y = x^2 + x - 2$ means that $x = 1$ and $y = 0$ satisfy the equation:

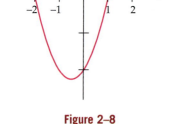

$$y = x^2 + x - 2$$
$$0 = 1^2 + 1 - 2.$$

In other words, the x-intercept 1 is a *solution* of the equation

$$x^2 + x - 2 = 0.$$

Figure 2–8

Similarly, the other x-intercept, -2, is also a solution of $x^2 + x - 2 = 0$ because $(-2)^2 + (-2) - 2 = 0$. The same argument works in the general case.

Solutions and Intercepts

> The real solutions of a one-variable equation of the form
>
> $$\text{Expression in } x = 0$$
>
> are the x-intercepts of the graph of the two-variable equation
>
> $$y = \text{Expression in } x.$$

Therefore, to solve an equation, you need only find the x-intercepts of its graph, as illustrated in the next example.

EXAMPLE 1

Solve the equation $x^4 - 4x^3 + 3x^2 - x = 2$ graphically.

SOLUTION   We first rewrite the equation so that one side is 0.

$$x^4 - 4x^3 + 3x^2 - x - 2 = 0.$$

As we have just seen, the solutions of this equation are the x-intercepts of the graph of

$$y = x^4 - 4x^3 + 3x^2 - x - 2.$$

TECHNOLOGY TIP

The graphical root finder is labeled ROOT or ZERO or X-INCPT and is in this menu/submenu:

TI-83+: CALC

TI-86/89: GRAPH/MATH

Casio: DRAW/G-SOLV

HP-39+: PLOT/FCN

On most TI calculators, you must select a left and a right bound (*x*-values to the left and right of the intercept) and make an initial guess.

These *x*-intercepts may be found by graphing the equation and using the graphical root finder, as in Figure 2–9.* (See the Technology Tip in the margin.)

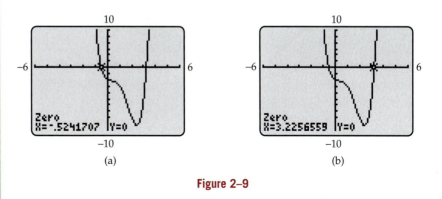

(a)   (b)

Figure 2–9

Therefore, the approximate solutions of the original equation are $x \approx -.5242$ and $x \approx 3.2257$. ▪

EXAMPLE 2

Solve: $x^3 - 4x^2 - 2x + 5 = 0$.

SOLUTION   We graph $y = x^3 - 4x^2 - 2x + 5$ (Figure 2–10) and see that it has three *x*-intercepts. So the equation has three real solutions. A root finder shows that one solutions is $x \approx -1.1926$ (Figure 2–11). ▪

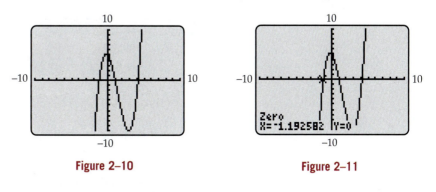

Figure 2–10   **Figure 2–11**

GRAPHING EXPLORATION

Use your graphical root finder to find the other two solutions (*x*-intercepts) of the equation in Example 2.

When an equation can be solved algebraically, this may be preferable to using graphical methods.

---

*Unless stated otherwise in the examples and exercises of this section, you may assume that the given viewing window includes all *x*-intercepts of the graph.

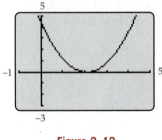

Figure 2–12

EXAMPLE 3

How many solutions does this equation have?

$$x^2 - 4.2x + 4.4 = 0.$$

SOLUTION   The graph of $y = x^2 - 4.2x + 4.4$ in Figure 2–12 suggests that the equation has just one solution (x-intercept). However, the discriminant of the equation is

$$b^2 - 4ac = (-4.2)^2 - 4(1)(4.4) = .04.$$

Since the discriminant is positive, the equation has *two* solutions. The moral of this story is: Be careful when using a graph to determine the number of solutions of an equation.   ■

GRAPHING EXPLORATION

Find a viewing window that clearly shows that the graph of $y = x^2 - 4.2x + 4.4$ has two x-intercepts, that is, that the equation $x^2 - 4.2x + 4.4 = 0$ has two real solutions. Then find the solutions.

▪▪ THE INTERSECTION METHOD

The next examples illustrate an alternative graphical method, which is sometimes more convenient for solving equations.

EXAMPLE 4

Solve: $|x^2 - 4x - 3| = x^3 + x - 6$.

SOLUTION   Let $y_1 = |x^2 - 4x - 3|$ and $y_2 = x^3 + x - 6$, and graph both equations on the same screen (Figure 2–13). Consider the point where the two graphs intersect. Since it is on the graph of y_1, its second coordinate is $|x^2 - 4x - 3|$, and since it is also on the graph of y_2, its second coordinate is $x^3 + x - 6$. So for this number x, we must have $|x^2 - 4x - 3| = x^3 + x - 6$. In other words, *the x-coordinate of the intersection point is the solution of the equation.*

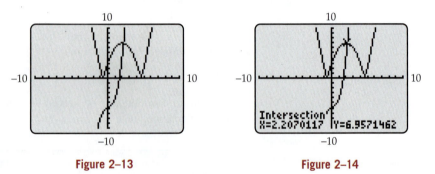

Figure 2–13   **Figure 2–14**

This coordinate can be approximated by using a graphical intersection finder (see the Technology Tip in the margin), as shown in Figure 2–14. Therefore, the solution of the original equation is $x \approx 2.207$. ■

EXAMPLE 5

Solve: $x^2 - 2x - 6 = \sqrt{2x + 7}$.

SOLUTION   We graph $y_1 = x^2 - 2x - 6$ and $y_2 = \sqrt{2x + 7}$ on the same screen. Figure 2–15 shows that there are two solutions (intersection points). According to an intersection finder (Figure 2–16), the positive solution is $x \approx 4.3094$. ■

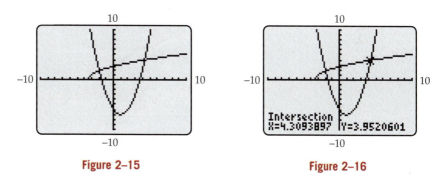

Figure 2–15   **Figure 2–16**

GRAPHING EXPLORATION

Use a graphical intersection finder to approximate the negative solution of the equation in Example 5.

EXAMPLE 6

According to data from the U.S. Census Bureau, the approximate population y (in millions) of Detroit is given by

$$y = (-6.985 \times 10^{-7})x^4 + (9.169 \times 10^{-5})x^3 - .00373x^2 + .0268x + 1.85,$$

where $x = 0$ corresponds to 1950. The approximate population of San Diego during the same period is given by

$$y = (-6.216 \times 10^{-6})x^3 + (4.569 \times 10^{-4})x^2 + .01035x + .346.$$

(a) Approximately when did the two cities have the same population?

(b) In what year was the population of Detroit about 1,500,000?

SOLUTION

(a) The populations of the two cities are equal when

$$(-6.985 \times 10^{-7})x^4 + (9.169 \times 10^{-5})x^3 - .00373x^2 + .0268x + 1.85$$
$$= (-6.216 \times 10^{-6})x^3 + (4.569 \times 10^{-4})x^2 + .01035x + .346.$$

We solve this equation by graphing the left side as y_1 and the right side as y_2 and finding the intersection point, as in Figure 2–17. Rounding the x-coordinate to the nearest year ($x \approx 38$), we see that the populations were equal around 1988.

(b) To find when the population of Detroit was 1,500,000 (that is, 1.5 million), we must solve the equation

$$(-6.985 \times 10^{-7})x^4 + (9.169 \times 10^{-5})x^3 - .00373x^2 + .0268x + 1.85 = 1.5.$$

Since the left side of this equation has already been graphed as y_1, we need only graph $y_3 = 1.5$ on the same screen and find the intersection point of y_1 and y_3, as in Figure 2–18. The solution $x \approx 20.45$ shows that Detroit had a population of 1.5 million in 1970.   ∎

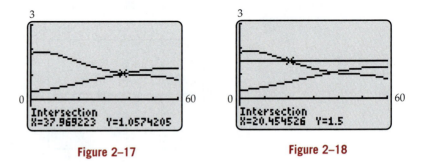

Figure 2–17   **Figure 2–18**

▪▪ NUMERICAL METHODS

In addition to graphical tools for solving equations, calculators and computer algebra systems have **equation solvers** that can find or approximate the solutions of most equations. On most calculators, the solver finds one solution at a time. You must enter the equation and an initial guess and possibly an interval in which to search. A few calculators (such as TI-89) and most computer systems have **one-step equation solvers** that will find or approximate all the solutions in a single step. Check you calculator instruction manual or your computer's help menu for directions on using these solvers.

Several calculators and many computer algebra systems also have **polynomial solvers**, one-step solvers designed specifically for polynomial equations; you need only enter the degree and coefficients of the polynomial. A few of these (such as Casio 9850) are limited to equations of degree 2 and 3. Directions for using polynomial solvers on calculators are in Calculator Investigation 1 at the end of Section 2.2. Computer users should check the help menu for the proper syntax.

EXAMPLE 7

When asked to search the interval $-10 \le x \le 10$ with an initial guess of 3, the equation solver on a TI-83+ found one solution of

$$4x^5 - 12x^3 + 8x - 1 = 0,$$

TIP

The equation solver is labeled SOLVE or SOLVER. It is on the keyboard or in this menu:

TI-83+: MATH

TI-89: ALGEBRA

HP-39+: APLET

Casio: EQUA (Main Menu)

The syntax varies, so check your instruction manual.

namely, $x \approx .128138691376$ (Figure 2–19). The POLY solver on a TI-86 produced that solution and four others (Figure 2–20) as did the downloadable POLY solver for TI-83+.

Figure 2–19   **Figure 2–20**

On Maple, the command

$$\text{fsolve}(4*x\char`\^5 - 12*x\char`\^3 + 8*x - 1 = 0, x);$$

also produced all five solutions of the equation. ■

■■ STRATEGIES FOR SPECIAL CASES

Because of various technological shortcomings, some equations are easier to solve if an indirect approach is used.

EXAMPLE 8

Solve $\sqrt{x^4 + x^2 - 2x - 1} = 0$.

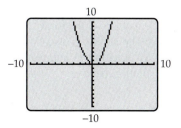

Figure 2–21

SOLUTION   The graph of $y = \sqrt{x^4 + x^2 - 2x - 1}$ in Figure 2–21 does not even appear to touch the x-axis (although it actually does touch at two places). Consequently, some solvers and graphical root finders will return an error message (check yours). Some computer solvers (such as "fsolve" on Maple) produce only one solution, unless told to search a specific interval. Even if your root finder or solver can handle this equation, it may fail in other similar situations. So the best approach is to use the following fact:

The only number whose square root is 0 is 0 itself.

Thus, $\sqrt{x^4 + x^2 - 2x - 1} = 0$ exactly when $x^4 + x^2 - 2x - 1 = 0$. So you need only solve the polynomial equation $x^4 + x^2 - 2x - 1 = 0$, which is easily done on any calculator or computer. In fact, if you have a polynomial or one-step solver, you can find all the solutions at once, which is faster than solving the original equation with a root finder or other solvers.

GRAPHING EXPLORATION

If you have a polynomial or one-step solver, use it to find all the real solutions of $x^4 + x^2 - 2x - 1 = 0$ at once. Otherwise, graph $y = x^4 + x^2 - 2x - 1$ and use the graphical root finder to obtain the solutions one at a time. Verify that the real solutions of this equation, and hence, of the original one, are $x \approx -.40469$ and $x \approx 1.18413$. ■

The technique used in Example 8 is recommended for all equations of the form

$$\sqrt{\text{Expression in } x} = 0:$$

Set the expression under the radical equal to 0 and solve.

EXAMPLE 9

Solve $\dfrac{7x - 4}{9x^2 - 9x + 2} = 0.$

SOLUTION   If you graph

$$y = \frac{7x - 4}{9x^2 - 9x + 2},$$

you may get "garbage," as in Figure 2–22. You could experiment with other viewing windows, but it's easier to use this fact:

**A fraction is 0 exactly when its numerator is 0 and
its denominator is nonzero.**

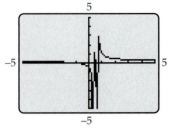

Figure 2–22

To find where the numerator is 0, we need only solve $7x - 4 = 0$, which is easily done.

Add 4 to both sides.	$7x = 4$
Divide both sides by 7:	$x = 4/7$

You can readily verify that $x = 4/7$ does not make the denominator 0. Hence the solution of the original equation is $4/7$. ■

The technique of Example 9 is recommended for equations of the form "fraction = 0": Set the numerator equal to 0, and solve. The solutions that do not make the denominator 0 are the solutions of the original equation. A number that makes *both* numerator and denominator 0 is *not* a solution because $0/0$ is not defined.

▪▪ CHOOSING A SOLUTION METHOD

We have seen that equations can be solved by algebraic, graphical, and numerical methods. Each method has both advantages and disadvantages, as summarized in the table on the next page.

The choice of solution method is up to you. In the rest of this book, we normally use algebraic means for solving linear and quadratic equations because this is often the fastest and most accurate method. Naturally, any such equation can also be solved graphically or numerically (and you might want to do that as a check against errors). Except in special cases, some of which are considered in Section 2.4, graphical and numerical methods will normally be used for more complicated equations.

Solution Method	Advantages	Possible Disadvantages
Algebraic	Produces exact solutions. Easiest method for most linear and quadratic equations.	May be difficult or impossible to use with complicated equations.
Graphical **Root Finder or** **Intersection Finder**	Works well for a large variety of equations. Gives visual picture of the location of the solutions.	Solutions may be approximations. Finding a useable viewing window may take a lot of time.
Numerical **Equation Solvers**		Solutions may be approximations.
Polynomial solver	Fast and easy.	Works only for polynomial equations.
One-step solver	Fast and easy.	May miss some solutions or be unable to solve certain equations.
Other solvers		May require a considerable amount of work to find the particular solution you want.

✓ **EXERCISES 2.3**

In Exercises 1–6, determine graphically the number of solutions of the equation, but don't solve the equation. You may need a viewing window other than the standard one to find all the x-intercepts.

1. $x^5 + 5 = 3x^4 + x$

2. $x^3 + 5 = 3x^2 + 24x$

3. $x^7 - 10x^5 + 15x + 10 = 0$

4. $x^5 + 36x + 25 = 13x^3$

5. $x^4 + 500x^2 - 8000x = 16x^3 - 32{,}000$

6. $6x^5 + 80x^3 + 45x^2 + 30 = 45x^4 + 86x$

In Exercises 7–20, use graphical approximation (a root finder or an intersection finder) to find a solution of the equation in the given interval.

7. $x^3 + 4x^2 + 10x + 15 = 0; \quad -3 < x < -2$

8. $x^3 + 9 = 3x^2 + 6x; \quad 1 < x < 2$

9. $x^4 + x - 3 = 0; \quad x > 0$

10. $x^5 + 5 = 3x^4 + x; \quad x < 1$

11. $\sqrt{x^4 + x^3 - x - 3} = 0; \quad x > 0$

12. $\sqrt{8x^4 - 14x^3 - 9x^2 + 11x - 1} = 0; \quad x < 0$

13. $\sqrt{\frac{2}{5}x^5 + x^2 - 2x} = 0; \quad x > 0$

14. $\sqrt{x^4 + x^2 - 3x + 1} = 0; \quad 0 < x < 1$

15. $x^2 = \sqrt{x + 5}; \quad -2 < x < -1$

16. $\sqrt{x^2 - 1} - \sqrt{x + 9} = 0; \quad 3 < x < 4$

17. $\dfrac{2x^5 - 10x + 5}{x^3 + x^2 - 12x} = 0; \quad x < 0$

18. $\dfrac{3x^5 - 15x + 5}{x^7 - 8x^5 + 2x^2 - 5} = 0; \quad x > 1$

19. $\dfrac{x^3 - 4x + 1}{x^2 + x - 6} = 0; \quad x > 1$

20. $\dfrac{4}{x + 2} - \dfrac{3}{x + 1} = 0; \quad x > 0$ [*Hint:* Write the left side as a single fraction.]

In Exercises 21–34, use algebraic, graphical, or numerical methods to find all real solutions of the equation, approximating when necessary.

21. $2x^3 - 4x^2 + x - 3 = 0$

22. $6x^3 - 5x^2 + 3x - 2 = 0$

23. $x^5 - 6x + 6 = 0$

24. $x^3 - 3x^2 + x - 1 = 0$

25. $10x^5 - 3x^2 + x - 6 = 0$ **26.** $\frac{1}{4}x^4 - x - 4 = 0$

27. $2x - \frac{1}{2}x^2 - \frac{1}{12}x^4 = 0$

28. $\frac{1}{4}x^4 + \frac{1}{3}x^2 + 3x - 1 = 0$

29. $\frac{5x}{x^2 + 1} - 2x + 3 = 0$ **30.** $\frac{2x}{x + 5} = 1$

31. $|x^2 - 4| = 3x^2 - 2x + 1$ **32.** $|x^3 + 2| = 5 + x - x^2$

33. $\sqrt{x^2 + 3} = \sqrt{x - 2} + 5$

34. $\sqrt{x^3 + 2} = \sqrt{x + 5} + 4$

In Exercises 35–40, find an exact solution of the equation in the given interval. (For example, if the graphical approximation of a solution begins .3333, check to see whether 1/3 is the exact solution. Similarly, $\sqrt{2} \approx 1.414$; so if your approximation begins 1.414, check to see whether $\sqrt{2}$ is a solution.)

35. $3x^3 - 2x^2 + 3x - 2 = 0;\quad 0 < x < 1$

36. $4x^3 - 3x^2 - 3x - 7 = 0;\quad 1 < x < 2$

37. $12x^4 - x^3 - 12x^2 + 25x - 2 = 0;\quad 0 < x < 1$

38. $8x^5 + 7x^4 - x^3 + 16x - 2 = 0;\quad 0 < x < 1$

39. $4x^4 - 13x^2 + 3 = 0;\quad 1 < x < 2$

40. $x^3 + x^2 - 2x - 2 = 0;\quad 1 < x < 2$

Exercises 41–44 deal with exponential, logarithmic, and trigonometric equations, which will be dealt with in later chapters. If you are familiar with these concepts, solve each equation graphically or numerically.

41. $10^x - \frac{1}{4}x = 28$ **42.** $x + \sin\left(\frac{1}{2}x\right) = 4$

43. $\ln x - x^2 + 3 = 0$ **44.** $e^x - 6x = 5$

45. According to data from the U.S. Department of Education, the average cost y of tuition and fees at four-year public colleges and universities in year x is approximated by

$$y = .325x^3 - 7.624x^2 + 220.276x + 2034,$$

where $x = 0$ corresponds to 1990. If this model continues to be accurate, in what year will tuition and fees reach $5000?

46. Use the information in Example 6 to determine the year in which the population of San Diego reached 1.1 million people.

47. According to data from the U.S. Department of Health and Human Services, the cumulative number y of AIDS cases (in thousands) as of year x is approximated by

$$y = .1006x^4 - 2.581x^3 + 19.282x^2 + 23.731x + 188$$
$$(0 \le x \le 11),$$

where $x = 0$ corresponds to 1990.
(a) When did the cumulative number of cases reach 750,000?
(b) If this model remains accurate after 2001, in what year does the cumulative number of cases reach one million?

48. (a) How many real solutions does the equation

$$.2x^5 - 2x^3 + 1.8x + k = 0$$

have when $k = 0$?
(b) How many real solutions does it have when $k = 1$?
(c) Is there a value of k for which the equation has just one real solution?
(d) Is there a value of k for which the equation has no real solutions?

2.4   Polynomial, Radical, and Absolute Value Equations

In this section, we explore both algebraic and graphical techniques for solving various kinds of equations, beginning with polynomial equations of degree greater than 2. Although graphical and numerical approximation methods usually work best for such equations, some of them can be solved algebraically.

EXAMPLE 1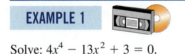

Solve: $4x^4 - 13x^2 + 3 = 0$.

SOLUTION   Substitute u for x^2 and solve the resulting quadratic equation.

$$4x^4 - 13x^2 + 3 = 0$$
$$4(x^2)^2 - 13x^2 + 3 = 0$$
$$4u^2 - 13u + 3 = 0$$
$$(u - 3)(4u - 1) = 0$$

$$u - 3 = 0 \quad \text{or} \quad 4u - 1 = 0$$
$$u = 3 \qquad\qquad 4u = 1$$
$$u = \frac{1}{4}$$

Since $u = x^2$, we see that

$$x^2 = 3 \qquad \text{or} \qquad x^2 = \frac{1}{4}$$

$$x = \pm\sqrt{3} \qquad\qquad x = \pm\frac{1}{2}.$$

Hence, the original equation has four solutions: $-\sqrt{3}, \sqrt{3}, -1/2, 1/2$.   ■

EXAMPLE 2

Solve: $x^4 - 4x^2 + 1 = 0$.

SOLUTION   Let $u = x^2$; then

$$x^4 - 4x^2 + 1 = 0$$
$$u^2 - 4u + 1 = 0.$$

The quadratic formula shows that

$$u = \frac{-(-4) \pm \sqrt{(-4)^2 - 4 \cdot 1 \cdot 1}}{2 \cdot 1} = \frac{4 \pm \sqrt{12}}{2}$$

$$= \frac{4 \pm \sqrt{4 \cdot 3}}{2} = \frac{4 \pm 2\sqrt{3}}{2} = 2 \pm \sqrt{3}.$$

Since $u = x^2$, we have the equivalent statements

$$x^2 = 2 + \sqrt{3} \qquad \text{or} \qquad x^2 = 2 - \sqrt{3}$$
$$x = \pm\sqrt{2 + \sqrt{3}} \qquad\qquad x = \pm\sqrt{2 - \sqrt{3}}.$$

Therefore the original equation has four solutions.   ■

EXAMPLE 3

A rectangular box with a square base and no top is to have a volume of 20,000 cubic cm. If each side of the base can be no longer than 40 cm and the surface area of the box is 4000 square cm, what are its dimensions?

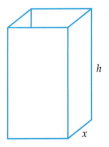

Figure 2–23

SOLUTION   Let x denote the length of each side of the base and let h denote the height, as in Figure 2–23. We have these translations.

English Language	Mathematical Language
Length, width, and height	x, x, and h
Volume (length $\times$ width $\times$ height)	x^2h
Volume is 20,000 cu cm	$x^2h = 20{,}000$
Area of base (length $\times$ width)	x^2
Area of each side (length $\times$ height)	xh
Surface area of box (base and 4 sides)	$x^2 + 4xh$
Surface area is 4000 sq cm	$x^2 + 4xh = 4000$

Since there are two variables, we solve the volume equation for h,

$$x^2h = 20{,}000$$

$$h = \frac{20{,}000}{x^2},$$

and substitute this result into the surface area equation.

$$x^2 + 4xh = 4000$$

$$x^2 + 4x\left(\frac{20{,}000}{x^2}\right) = 4000$$

$$x^2 + \frac{80{,}000}{x} = 4000$$

$$x^3 + 80{,}000 = 4000x$$

$$x^3 - 4000x + 80{,}000 = 0.$$

Substitution and factoring seem unlikely to work here, but we have two other ways to solve the equation.

Graphical: The graph of $y = x^3 - 4000x + 80{,}000$ in Figure 2–24 shows that the equation has three solutions. The negative one and the one near 50 do not apply here because the side of the base must be positive and less than 40 cm long. The third solution, $x \approx 23.069$, is easily found (Figure 2–25).

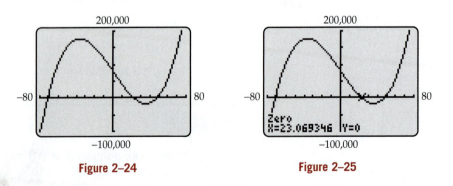

Figure 2–24   **Figure 2–25**

Numerical: The polynomial solver on a TI-86 or the "fsolve" command on Maple produces the three solutions (Figure 2–26). As above, only one of them is applicable here, namely, $x \approx 23.069$.

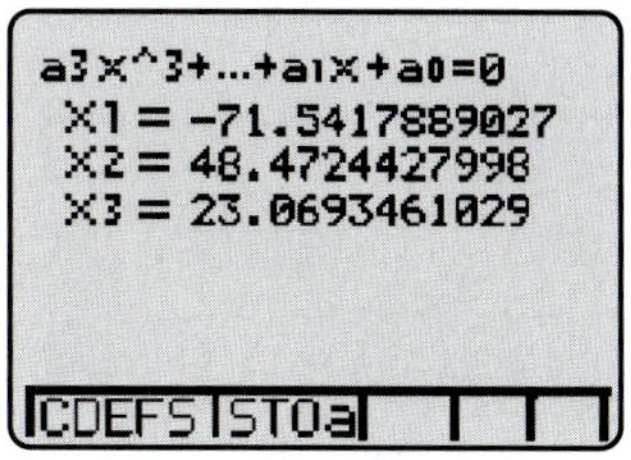

```
a3x^3+...+a1x+a0=0
X1= -71.5417889027
X2= 48.4724427998
X3= 23.0693461029


CDEFS STO3
```

```
MAPLE

> fsolve(x^3 – 4000*x + 80000 = 0, x);
                    –71.5417889, 23.0693461, 48.4724428
```

Figure 2–26

Therefore, the height is

$$h = \frac{20,000}{x^2} \approx \frac{20,000}{23.069^2} \approx 37.581.$$

So the approximate dimensions of the box are 23.069 by 23.069 by 37.581 cm.

■■ RADICAL EQUATIONS

The algebraic solution of equations involving radicals depends on this fact: If two quantities are equal, say,

$$x - 2 = 3,$$

then their squares are also equal.

$$(x - 2)^2 = 9.$$

Thus,

Every solution of $x - 2 = 3$ is also a solution of $(x - 2)^2 = 9$.

But *be careful:* This works only in one direction. For instance, -1 is a solution of $(x - 2)^2 = 9$ but not of $x - 2 = 3$. This is an example of the Power Principle.

Power Principle

If both sides of an equation are raised to the same positive integer power, then every solution of the original equation is also a solution of the new equation. But the new equation may have solutions that are *not* solutions of the original one.

Consequently, if you raise both sides of an equation to a power, you must *check your solutions* in the *original* equation. Graphing provides a quick way to eliminate most extraneous solutions. But only an algebraic computation can confirm an exact solution.

EXAMPLE 4

Solve: $5 + \sqrt{3x - 11} = x$.

SOLUTION We first rearrange terms to get the radical expression alone on one side.

$$\sqrt{3x - 11} = x - 5.$$

Then we square both sides and solve the resulting equation.

$$(\sqrt{3x - 11})^2 = (x - 5)^2$$
$$3x - 11 = x^2 - 10x + 25$$
$$0 = x^2 - 13x + 36$$
$$0 = (x - 4)(x - 9)$$
$$x - 4 = 0 \quad \text{or} \quad x - 9 = 0$$
$$x = 4 \qquad\qquad x = 9.$$

If these numbers are solutions of the original equation, they should be x-intercepts of the graph of $y = \sqrt{3x - 11} - x + 5$ (why?). But the graph of f in Figure 2–27 doesn't have an x-intercept at $x = 4$, so 4 is not a solution of the original equation. The graph appears to show that $x = 9$ is a solution of the original equation, and direct calculation confirms this.

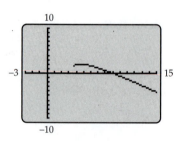

Figure 2–27

Left side:	Right side:
$5 + \sqrt{3x - 11}$	x
$5 + \sqrt{3 \cdot 9 - 11}$	9
$5 + \sqrt{16}$	
9	

Hence, 9 is the only solution of the original equation. ∎

EXAMPLE 5

Solve: $\sqrt{2x - 3} - \sqrt{x + 7} = 2$.

SOLUTION We first rearrange terms so that one side contains only a single radical term.

$$\sqrt{2x - 3} = \sqrt{x + 7} + 2$$

Then we square both sides and simplify.

$$(\sqrt{2x - 3})^2 = (\sqrt{x + 7} + 2)^2$$
$$2x - 3 = (\sqrt{x + 7})^2 + 2 \cdot 2 \cdot \sqrt{x + 7} + 2^2$$
$$2x - 3 = x + 7 + 4\sqrt{x + 7} + 4$$
$$x - 14 = 4\sqrt{x + 7}$$

Now we square both sides and solve the resulting equation.

$$(x - 14)^2 = (4\sqrt{x + 7})^2$$
$$x^2 - 28x + 196 = 4^2 \cdot (\sqrt{x + 7})^2$$
$$x^2 - 28x + 196 = 16(x + 7)$$
$$x^2 - 28x + 196 = 16x + 112$$
$$x^2 - 44x + 84 = 0$$
$$(x - 2)(x - 42) = 0$$
$$x - 2 = 0 \quad \text{or} \quad x - 42 = 0$$
$$x = 2 \qquad\qquad x = 42$$

Substituting 2 and 42 in the left side of the original equation shows that

$$\sqrt{2 \cdot 2 - 3} - \sqrt{2 + 7} = \sqrt{1} - \sqrt{9} = 1 - 3 = -2$$
$$\sqrt{2 \cdot 42 - 3} - \sqrt{42 + 7} = \sqrt{81} - \sqrt{49} = 9 - 7 = 2.$$

Therefore, 42 is the only solution of the equation. ■

EXAMPLE 6

Nancy Hoffman, who is standing at point A on the bank of a 2.5-kilometer-wide river, wants to reach point B, 15 kilometers downstream on the opposite bank. She plans to row to a point C on the opposite shore and then run to B, as shown in Figure 2–28. She can row at a rate of 4 kilometers per hour and can run at 8 kilometers per hour. If her trip is to take 3 hours, how far from B should she land?

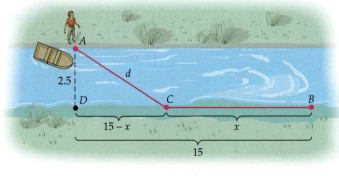

Figure 2–28

SOLUTION   Let x be the distance that Nancy must run from C to B. Using the basic formula for distance, we have

$$\text{Rate} \times \text{Time} = \text{Distance}$$

$$\text{Time} = \frac{\text{Distance}}{\text{Rate}} = \frac{x}{8}.$$

Similarly, the time required to row distance d is

$$\text{Time} = \frac{\text{Distance}}{\text{Rate}} = \frac{d}{4}.$$

Since $15 - x$ is the distance from D to C, the Pythagorean Theorem applied to right triangle ADC shows that

$$d^2 = (15 - x)^2 + 2.5^2 \quad \text{or, equivalently,} \quad d = \sqrt{(15 - x)^2 + 6.25}.$$

Therefore, the total time for the trip is given by

$$T(x) = \text{Rowing time} + \text{Running time} = \frac{d}{4} + \frac{x}{8} = \frac{\sqrt{(x - 15)^2 + 6.25}}{4} + \frac{x}{8}.$$

If the trip is to take 3 hours, then $T(x) = 3$, and we must solve the equation

$$\frac{\sqrt{(x - 15)^2 + 6.25}}{4} + \frac{x}{8} = 3.$$

GRAPHING EXPLORATION

Using the viewing window with $0 \le x \le 15$ and $-2 \le y \le 2$, graph the function

$$f(x) = \frac{\sqrt{(x - 15)^2 + 6.25}}{4} + \frac{x}{8} - 3$$

and use a root finder to find its x-intercept (the solution of the equation).

This Graphing Exploration shows that Nancy should land approximately 6.74 kilometers from B to make the trip in 3 hours. ■

■■ ABSOLUTE VALUE EQUATIONS

If c is a real number, then by the definition of absolute value, $|c|$ is either c or $-c$ (whichever one is positive). This fact can be used to solve absolute value equations algebraically.

EXAMPLE 7

To solve $|3x - 4| = 8$, apply the fact stated above with $c = 3x - 4$. Then $|3x - 4|$ is either $3x - 4$ or $-(3x - 4)$, so

$$
\begin{array}{lll}
3x - 4 = 8 & \quad\text{or}\quad & -(3x - 4) = 8 \\
3x = 12 & & -3x + 4 = 8 \\
x = 4 & & -3x = 4 \\
& & x = -4/3.
\end{array}
$$

So there are two possible solutions of the original equation $|3x - 4| = 8$. You can readily verify that both 4 and $-4/3$ actually are solutions. ■

EXAMPLE 8

Solve: $|x + 4| = 5x - 2$.

SOLUTION   The left side of the equation is either $x + 4$ or $-(x + 4)$ (why?). Hence,

$$x + 4 = 5x - 2 \qquad \text{or} \qquad -(x + 4) = 5x - 2$$

$$-4x + 4 = -2 \qquad\qquad\qquad -x - 4 = 5x - 2$$

$$-4x = -6 \qquad\qquad\qquad\qquad -6x = 2$$

$$x = \frac{-6}{-4} = \frac{3}{2} \qquad\qquad x = \frac{2}{-6} = -\frac{1}{3}.$$

We must check each of these possible solutions in the original equation,

$$|x + 4| = 5x - 2.$$

We see that $x = 3/2$ is a solution, since

$$\left|\frac{3}{2} + 4\right| = \frac{11}{2} \qquad \text{and} \qquad 5\left(\frac{3}{2}\right) - 2 = \frac{11}{2}.$$

However, $x = -1/3$ is not a solution, since

$$\left|-\frac{1}{3} + 4\right| = \frac{11}{3} \qquad \text{but} \qquad 5\left(-\frac{1}{3}\right) - 2 = -\frac{11}{3}.$$

You can graphically confirm the fact that $x = -1/3$ is not a solution by graphing $y = |x + 4|$ and $y = 5x - 2$ on the same screen (Figure 2–29). Since there is no intersection point with a negative x-coordinate, $x = -1/3$ cannot be a solution.   ■

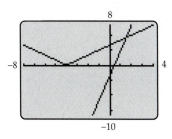

Figure 2–29

EXERCISES 2.4

In Exercises 1–8, find all real solutions of each equation exactly.

1. $y^4 - 7y^2 + 6 = 0$ **2.** $x^4 - 2x^2 + 1 = 0$

3. $x^4 - 2x^2 - 35 = 0$ **4.** $x^4 - 2x^2 - 24 = 0$

5. $2y^4 - 9y^2 + 4 = 0$ **6.** $6z^4 - 7z^2 + 2 = 0$

7. $10x^4 + 3x^2 = 1$ **8.** $6x^4 - 7x^2 = 3$

In Exercises 9–28, find all real solutions of each equation. Find exact solutions when possible and approximate ones otherwise.

9. $\sqrt{x + 2} = 3$ **10.** $\sqrt{x - 7} = 4$

11. $\sqrt{4x + 9} = 5$ **12.** $\sqrt{3x - 2} = 7$

13. $\sqrt[3]{5 - 11x} = 3$ [*Hint:* Cube both sides.]

14. $\sqrt[3]{6x - 10} = 2$ **15.** $\sqrt[3]{x^2 - 1} = 2$

16. $(x + 1)^{2/3} = 4$ **17.** $\sqrt{x^2 - x - 1} = 1$

18. $\sqrt{x^2 - 5x + 4} = 2$ **19.** $\sqrt{x + 7} = x - 5$

20. $\sqrt{x + 5} = x - 1$ **21.** $\sqrt{3x^2 + 7x - 2} = x + 1$

22. $\sqrt{4x^2 - 10x + 5} = x - 3$

23. $\sqrt[3]{x^3 + x^2 - 4x + 5} = x + 1$

24. $\sqrt[3]{x^3 - 6x^2 + 2x + 3} = x - 1$

25. $\sqrt{5x + 6} = 3 + \sqrt{x + 3}$

26. $\sqrt{3y + 1} - 1 = \sqrt{y + 4}$

27. $\sqrt{2x - 5} = 1 + \sqrt{x - 3}$

28. $\sqrt{x - 3} + \sqrt{x + 5} = 4$

In Exercises 29–32, assume that all letters represent positive numbers and solve each equation for the required letter.

29. $A = \sqrt{1 + \dfrac{a^2}{b^2}}$ for b **30.** $T = 2\pi\sqrt{\dfrac{m}{g}}$ for g

31. $K = \sqrt{1 - \dfrac{x^2}{u^2}}$ for u **32.** $R = \sqrt{d^2 + k^2}$ for d

In Exercises 33–41, find all real solutions of each equation.

33. $|2x + 3| = 9$ **34.** $|3x - 5| = 7$

35. $|6x - 9| = 0$ **36.** $|4x - 5| = 9$

37. $|2x + 3| = 4x - 1$ **38.** $|3x - 2| = 5x + 4$

39. $|x - 3| = x$ **40.** $|2x - 1| = 2x + 1$

41. $|x^2 + 4x - 1| = 4$

42. In statistical quality control, one needs to find the proportion of the product that is not acceptable. The upper and lower control limits are found by solving the following equation (in which $\bar{p}$ is the mean percent defective and n is the sample size) for CL.

$$|CL - \bar{p}| = 3\sqrt{\frac{\bar{p}(1 - \bar{p})}{n}}.$$

Find the control limits when $\bar{p} = .02$ and $n = 200$.

43. What are the dimensions of a rectangle whose diagonal is 130 cm long and whose area is 6000 square centimeters?

44. The dimensions of a rectangular box are consecutive integers. If the box has volume 13,800 cubic centimeters, what are its dimensions?

45. The surface area S of the right circular cone in the figure is given by $S = \pi r\sqrt{r^2 + h^2}$. What radius should be used to produce a cone of height 5 inches and surface area 100 square inches?

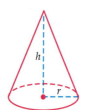

46. What is the radius of the base of a cone whose surface area is 18π square centimeters and whose height is 4 cm?

47. Find the radius of the base of a conical container whose height is 1/3 of the radius and whose volume is 180 cubic inches. [*Note:* The volume of a cone of radius r and height h is $\pi r^2 h/3$.]

48. The surface area of the right square pyramid in the figure at the top of the next column is given by $S = b\sqrt{b^2 + 4h^2}$. If the pyramid has height 10 feet and surface area 100 square feet, what is the length of a side b of its base?

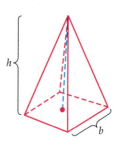

49. A rope is to be stretched at uniform height from a tree to a 35-foot-long fence, which is 20 feet from the tree, and then to the side of a building at a point 30 feet from the fence, as shown in the figure. If 63 feet of rope is to be used, how far from the building wall should the rope meet the fence?

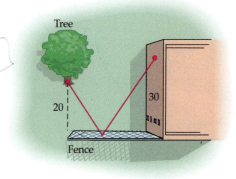

50. Anne is standing on a straight road and wants to reach her helicopter, which is located 2 miles down the road from her, a mile from the road in a field (see figure). She can run 5 miles per hour on the road and 3 miles per hour in the field. She plans to run down the road, then cut diagonally across the field to reach the helicopter. Where should she leave the road to reach the helicopter in exactly 42 minutes (.7 hour)?

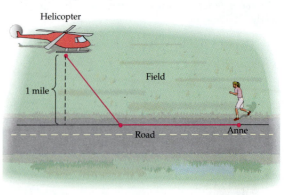

51. A power plant is located on the bank of a river that is $\frac{1}{2}$ mile wide. Wiring is to be laid across the river and then along the shore to a substation 8 miles downstream, as shown in the figure. It costs $12,000 per mile for underwater wiring and $8000 per mile for wiring on land. If $72,000 is to be spent on the project, how far from the substation should the wiring come to shore?

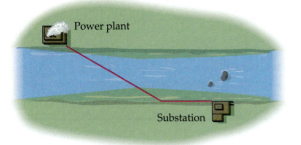

52. A homemade loaf of bread turns out to be a perfect cube. Five slices of bread, each .6 in. thick, are cut from one end of the loaf. The remainder of the loaf now has a volume of 235 cubic inches. What were the dimensions of the original loaf?

53. A rectangular bin with an open top and volume of 38.72 cubic feet is to be built. The length of its base must be twice the width, and the bin must be at least 3 feet high.

Material for the base of the bin costs $12 per square foot, and material for the sides costs $8 per square foot. If it costs $538.56 to build the bin, what are its dimensions?

Thinker

54. One corner of an 8.5-by-11-inch piece of paper is folded over to the opposite side, as shown in the figure. The area of the darkly shaded triangle at the lower left is 6 square inches and we want to find the length x.

(a) Take a piece of paper this size and experiment. Approximately, what is the largest value x could have (and still have the paper look like the figure)? With this value of x, what is the approximate area of the triangle? Try some other possibilities.

(b) Now find an exact answer by constructing and solving a suitable equation. Explain why one of the solutions to the equation is not an answer to this problem.

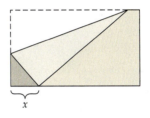

2.5   Linear Inequalities

Inequalities such as

$$5x + 3 \le 7x + 6 \quad \text{or} \quad 4 < 3 - 5x < 18$$

are called **linear inequalities.** They may be solved by algebraic or graphical means, both of which are discussed here.

We say that two inequalities are **equivalent** if they have the same solutions. The usual strategy for solving inequalities algebraically is to use the following Basic Principles to transform a given inequality into an equivalent one whose solutions are known.

Basic Principles for Solving Inequalities

Performing any of the following operations on an inequality produces an equivalent inequality:

1. Add or subtract the same quantity on both sides of the inequality.
2. Multiply or divide both sides of the inequality by the same *positive* quantity.
3. Multiply or divide both sides of the inequality by the same *negative* quantity and *reverse the direction of the inequality.*

Note Principle 3 carefully. It says, for example, that when you multiply both sides of $-3 < 5$ by -2, you must reverse the direction of the inequality to have a true statement:

$$-3 < 5$$

Reverse direction: $(-2)(-3) > (-2)(5)$

$$6 > -10$$

Failure to do this is probably the most common algebraic error when solving inequalities.

The best way to solve most linear inequalities is to use algebra. For those interested, however, graphical solution methods are also given in many of the following examples. They can be omitted without any loss of continuity.

EXAMPLE 1

Solve: $3x + 1 < 5$.

SOLUTION   *Algebraic Method:*   We use the Basic Principles to transform this inequality into an equivalent one whose solutions are obvious.

$$3x + 1 < 5$$

Subtract 1 from both sides: $3x < 4$

Divide both sides by 3: $x < 4/3$

Therefore, the solutions are all real numbers that are less than $4/3$, as shown on the number line in Figure 2–30.

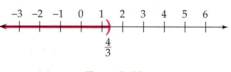

Figure 2–30

Graphical Method: One approach is to change the inequality to an equivalent one with 0 on one side.

$$3x + 1 < 5$$

Subtract 5 from both sides: $3x - 4 < 0.$

Now graph $y_1 = 3x - 4$ (Figure 2–31) and note how the graph is related to this last inequality:

When a point (x, y_1) lies below the x-axis, then its y-coordinate is negative, which means that

$$y_1 < 0, \quad \text{or, equivalently,} \quad 3x - 4 < 0.$$

Thus, the solutions of the inequality are the x-coordinates of all points on the graph that lie below the x-axis. Figure 2–31 shows that these are the points to the

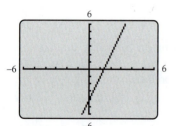

Figure 2–31

left of the *x*-intercept of the graph. The calculator's root finder shows that the *x*-intercept is approximately 1.333 (see Figure 2–32). So the (approximate) solutions of this inequality are all *x* such that $x < 1.333$. Since the decimal expansion of 4/3 begins 1.333, this answer is consistent with the one obtained algebraically. ■

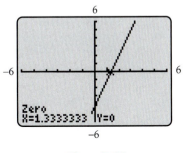

Figure 2–32

EXAMPLE 2

Solve: $x + 2 \leq 8x + 4$.

SOLUTION   *Algebraic Method:*

$$x + 2 \leq 8x + 4$$

Subtract 2 from both sides: $\qquad\qquad x \leq 8x + 2$

Subtract 8*x* from both sides: $\qquad\qquad -7x \leq 2$

Divide both sides by −7 and reverse
the direction of the inequality: $\qquad\qquad x \geq -2/7.$

Therefore, the solutions are all real numbers greater than or equal to $-2/7$. They are shown on the number line in Figure 2–33.

Figure 2–33

Graphical Method: The inequality can be solved by using the method of Example 1. A second method is to graph $y_1 = x + 2$ and $y_2 = 8x + 4$ on the same screen and note how points on the two graphs are related:

When the point (x, y_1) lies below the point (x, y_2), then the *y*-coordinate of (x, y_1) is smaller than the *y*-coordinate of (x, y_2), which means that

$$y_1 \leq y_2, \qquad \text{or, equivalently,} \qquad x + 2 \leq 8x + 4.$$

Thus, the solutions of the inequality are the *x*-coordinates of all points for which the graph of y_1 lies below or on the graph of y_2. Figure 2–34 shows that the graph

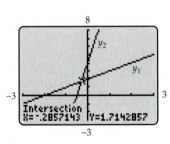

Figure 2–34

of y_1 lies below the graph of y_2 to the right of the point where they intersect. The calculator's intersection finder shows that this point is approximately $(-.285714, 1.771429)$. Points to the right of the intersection point have larger x-coordinates. So the approximate solutions of the inequality are all x such that $x \geq -.285714$. Verify that this answer is consistent with the one obtained algebraically by finding the decimal expansion of $-2/7$.   ∎

EXAMPLE 3

Solve: $2 \leq 3x + 5 < 2x + 11$.

SOLUTION   By definition, the solutions of $2 \leq 3x + 5 < 2x + 11$ are the numbers that are solutions of *both* of these inequalities:

$$2 \leq 3x + 5 \qquad \text{and} \qquad 3x + 5 < 2x + 11.$$

Each of these inequalities can be solved by the methods used above. For the first one we have:

$$2 \leq 3x + 5$$

Subtract 5 from both sides: $-3 \leq 3x$

Divide both sides by 3: $-1 \leq x.$

The second inequality is solved similarly:

$$3x + 5 < 2x + 11$$

Subtract 5 from both sides: $3x < 2x + 6$

Subtract $2x$ from both sides: $x < 6.$

The solutions of the original inequality are the numbers that satisfy *both* $-1 \leq x$ *and* $x < 6$, that is, all x with $-1 \leq x < 6$, as shown in Figure 2–35.   ∎

> **CAUTION**
>
> All inequality signs in an inequality should point in the same direction. *Don't* write things like $4 < x > 2$ or $-3 \geq x < 5$.

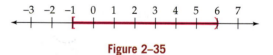

Figure 2–35

> **GRAPHING EXPLORATION**
>
> Graphically confirm the results in Example 3 as follows. Rewrite the first inequality $2 \leq 3x + 5$ as $0 \leq 3x + 3$ and graph $y_1 = 3x + 3$. The solutions are the x-coordinates of points that lie on or *above* the x-axis. (Why?) Solve the second inequality $3x + 5 < 2x + 11$ by finding the x-coordinates of the points where the graph of $y_2 = 3x + 5$ lies below the graph of $y_3 = 2x + 11$.

EXAMPLE 4

Solve: $-1 < 4 - 3x < 7$.

SOLUTION   When solving an inequality like this, in which the variable appears only in the middle part, you can proceed as follows:

$$-1 < 4 - 3x < 7$$

Subtract 4 from each part: $$-5 < -3x < 3$$

Divide each part by -3 and reverse the direction of the inequalities: $$\frac{5}{3} > x > -1.$$

Reading this last inequality from right to left, we see that the solutions are all real numbers x such that $-1 < x < 5/3$.

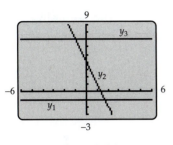

Figure 2–36

GRAPHING EXPLORATION

Confirm that these are the solutions by graphing $y_1 = -1$, $y_2 = 4 - 3x$, and $y_3 = 7$ on the same screen, as in Figure 2–36, and using an intersection finder to determine the x-coordinates of points on the graph of y_2 that lie above the graph of y_1 and below the graph of y_3. ■

EXAMPLE 5

You can rent a car for $40 per day plus 5¢ per mile, or you can pay $10 per day plus 30¢ per mile. How many miles must you drive in one day for the first plan to be cheaper?

SOLUTION   Let x be the number of miles driven in a day. We must solve the following inequality.

Cost of driving x miles with plan 1 $<$ Cost of driving x miles with plan 2

$$\$40 + 5¢ \text{ per mile} < \$10 + 30¢ \text{ per mile}$$

$$40 + .05x < 10 + .30x$$

We will solve it algebraically. (It can also be solved graphically.)

$$40 + .05x < 10 + .30x$$

Subtract 40 from both sides: $$.05x < -30 + .30x$$

Subtract $.30x$ from both sides: $$-.25x < -30$$

Divide both sides by $-.25$ and *reverse the direction* of the inequality: $$x > \frac{-30}{-.25}$$

Simplify right-hand side: $$x > 120$$

Therefore plan 1 is cheaper when you drive more than 120 miles. ■

▉▉ ABSOLUTE VALUE INEQUALITIES

Recall that if r is a real number, then $|r|$ is the distance from r to 0 on the number line. Thus, the inequality $|r| \leq 5$ states that the distance from r to 0 is 5 units or less. A glance at the number line in Figure 2–37 shows that these are the numbers r such that $-5 \le r \le 5$.

Figure 2–37

Similarly, the numbers r such that $|r| \ge 5$ are those whose distance to 0 is 5 or more units, that is, the numbers r with $r \le -5$ or $r \ge 5$. This argument works with any positive number k in place of 5 and proves the following facts (which are also true with $<$ and $>$ in place of $\le$ and $\ge$).

Absolute Value Inequalities

Let k be a positive number and r any real number.

$$|r| \le k \quad \text{is equivalent to} \quad -k \le r \le k.$$

$$|r| \ge k \quad \text{is equivalent to} \quad r \le -k \quad \text{or} \quad r \ge k.$$

EXAMPLE 6

Solve: $|3x - 5| \le 4$.

SOLUTION   *Algebraic Method:*   Apply the first fact in the preceding box, with $3x - 5$ in place of r and 4 in place of k, to obtain this equivalent inequality:

$$-4 \le 3x - 5 \le 4$$

Add 5 to each part: $1 \le 3x \le 9$

Divide each part by 3: $\dfrac{1}{3} \le x \le 3.$

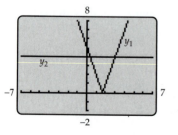

Figure 2–38

Therefore, the solutions of the original inequality are all real numbers x such that $1/3 \le x \le 3$.

Graphical Method: One way to solve the inequality is to graph $y_1 = |3x - 5|$ and $y_2 = 4$ on the same screen (Figure 2–38) and determine the x-coordinates of the points where the graph of y_1 lies on or below the graph of y_2. Alternatively, you can rewrite the original inequality as

$$|3x - 5| - 4 \le 0$$

and graph $y_3 = |3x - 5| - 4$ (Figure 2–39). The solutions of the inequality are the x-coordinates of points on the graph of y_3 that lie on or below the x-axis.

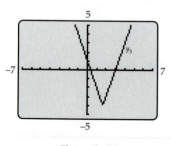

GRAPHING EXPLORATION

Use one of the graphical methods just described to solve the inequality. ■

Figure 2–39

EXAMPLE 7

Solve: $|5x + 2| > 3$.

SOLUTION   Apply the second fact in the box before Example 6, with $5x + 2$ in place of r, 3 in place of k, and $>$ in place of $\geq$. This produces the equivalent statement

$$5x + 2 < -3 \quad \text{or} \quad 5x + 2 > 3$$
$$5x < -5 \qquad\qquad 5x > 1$$
$$x < -1 \quad \text{or} \quad x > \frac{1}{5}.$$

Therefore, the solution of the original inequality consists of the numbers that are either less than -1 or greater than $1/5$.

GRAPHING EXPLORATION

Graphicaly confirm that this solution is correct. Use either one of the methods described in Example 6. ■

EXERCISES 2.5

In Exercises 1–12, solve the inequality algebraically. Confirm your answer graphically.

1. $2x + 4 \leq 7$
2. $3x - 5 > -6$
3. $3 - 5x < 13$
4. $2 - 3x < 11$
5. $6x + 3 \leq x - 5$
6. $5x + 3 \leq 2x + 7$
7. $5 - 7x < 2x - 4$
8. $5 - 3x > 7x - 3$
9. $2 < 3x - 4 < 8$
10. $1 < 5x + 6 < 9$
11. $0 < 5 - 2x \leq 11$
12. $-4 \leq 7 - 3x < 0$

In Exercises 13–18, solve the inequality exactly (no decimal approximations) by any means.

13. $2x + 7(3x - 2) < 2(x - 1)$
14. $x + 3(x - 5) \geq 3x + 2(x + 1)$
15. $\dfrac{x + 1}{2} - 3x \leq \dfrac{x + 5}{3}$
16. $\dfrac{x - 1}{4} + 2x \geq \dfrac{2x - 1}{3} + 2$
17. $2x + 3 \leq 5x + 6 < -3x + 7$
18. $4x - 2 < x + 8 < 9x + 1$

19. Women's weekly earnings as a percentage of men's earnings are approximated by the equation

$$y = .6522x + 62.5 \qquad (0 \leq x \leq 23),$$

where $x = 0$ corresponds to 1979.[*] In what years were women's earnings less than 75% of men's earnings?

20. The number of stockbrokers in the United States (in thousands) is approximated by

$$y = 70x/3 + 394 \qquad (3 \leq x \leq 13),$$

where $x = 3$ corresponds to 1993.[†] In what years were there more than 600,000 stockbrokers?

In Exercises 21–24, a, b, c, and d are positive constants. Solve the inequality for x.

21. $ax - b < c$
22. $d - cx > a$
23. $0 < x - c < a$
24. $-d < x - c < d$

---

[*]Bureau of Labor Statistics.
[†]National Association of Securities Dealers.

25. The average temperature on the planet Mercury is 333°F. The temperature t on Mercury can be as much as 507°F above or 633°F below the average. Write the range of temperatures on Mercury as an inequality.

26. If you were single in 2003 and had a taxable income I between \$28,400 and \$68,800, then your federal income tax T was

$$27\% \text{ of (income} - 28,400) + 3960.$$

(a) Write this income bracket as an inequality.

(b) Write the tax range for this income bracket as an inequality.

27. A company has fixed overhead costs of \$350 per day and variable costs of \$7.50 per unit for every unit produced. During December, the total daily cost T varied from \$3150 to \$4575 per day.

(a) Write an equation that expresses T in terms of the number x of units produced.

(b) Write an inequality that expresses the variability of the total daily cost in December.

(c) Write an inequality that gives the lowest and highest number of units produced daily during December.

28. Using statistical quality control methods, the management of a large hotel chain determines that the percentage of rooms that are not ready when the guest checks in ranges from 0.16% to 0.44%. When the room is not ready at check-in, a guest receives the first night free. The chain has 20,000 rooms, each of which costs \$85 per night. Let x be the number of rooms that are not ready at check-in and C the cost to the hotel chain for rooms not ready at check-in.

(a) Write an equation that relates C and x.

(b) Write an inequality that gives the range of values for x.

(c) Write an inequality that gives the range of values of C.

In Exercises 29 and 30, read the solution of the inequality from the given graph.

29. $3 - 2x < .8x + 7$

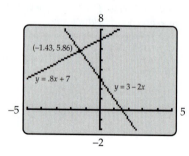

30. $8 - |7 - 5x| > 3$

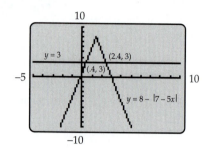

In Exercises 31–34, match the equation or inequality with the graph of its solution on the number line.

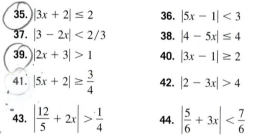

31. $|x - 17| = 7$

32. $|x - 17| \le 7$

33. $|x - 17| \ge -7$

34. $|x - 17| > 7$

In Exercises 35–44, solve the inequality exactly (no decimal approximations).

35. $|3x + 2| \le 2$

36. $|5x - 1| < 3$

37. $|3 - 2x| < 2/3$

38. $|4 - 5x| \le 4$

39. $|2x + 3| > 1$

40. $|3x - 1| \ge 2$

41. $|5x + 2| \ge \dfrac{3}{4}$

42. $|2 - 3x| > 4$

43. $\left|\dfrac{12}{5} + 2x\right| > \dfrac{1}{4}$

44. $\left|\dfrac{5}{6} + 3x\right| < \dfrac{7}{6}$

45. A broker predicts that over the next six months, the price p of Gigantica stock will not vary from its current price of \$25$\frac{3}{4}$ by more than \$4. Use absolute value to write this prediction as an inequality.

46. About two-thirds of people taking a standard IQ test will score within 12 points of 100. Write an absolute value inequality that describes this situation. Be sure to state what the variable represents.

47. In 2003, the cost of a taxi in New York city was \$2 plus 30 cents for each 1/5 of a mile.[*] How far could you travel for at least \$6 but no more than \$12?

---

[*]Provided that the taxi's speed is at least 8 mph during the trip, which we shall assume is the case.

48. The graphs of the revenue and cost functions for a manufacturing firm are shown in the figure.
  (a) What is the break-even point?
  (b) Shade in the region representing profit.
  (c) What does the *y*-intercept of the cost graph represent? Why is the *y*-intercept of the revenue graph 0?

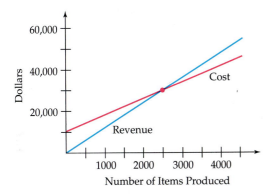

49. One freezer costs $623.95 and uses 90 kilowatt-hours (kwh) of electricity each month. A second freezer costs $500 and uses 100 kwh of electricity each month. The expected life of each freezer is 12 years. What is the minimum electric rate (in *cents* per kwh) for which the 12-year total cost (purchase price + electricity costs) will be less for the first freezer?

50. A business executive leases a car for $300 per month. She decides to lease another brand for $250 per month but has to pay a penalty of $1000 for breaking the first lease. How long must she keep the second car to come out ahead?

51. One salesperson is paid a salary of $1000 per month plus a commission of 2% of her total sales. A second salesperson receives no salary but is paid a commission of 10% of her total sales. What dollar amount of sales must the second salesperson have in order to earn more per month than the first?

52. A Gas Guzzler SUV has a 26-gallon gas tank and gets 12 miles per gallon. If it travels more than 210 miles and runs out of gas, what possible amounts of gas were in the tank at the beginning of the trip?

53. If $5000 is invested at 8%, how much more should be invested at 10% to guarantee a total annual interest income between $800 and $940?

54. How many gallons of a 12% salt solution should be added to 10 gallons of an 18% salt solution to produce a solution whose salt content is between 14 and 16%?

2.6   Polynomial and Rational Inequalities

Polynomial and rational inequalities, such as

$$2x^3 - 15x < x^2, \quad \text{or} \quad 2x^2 + 3x - 4 \le 0, \quad \text{or} \quad \frac{x}{x-1} > -6,$$

may be solved by algebraic or graphical methods, or by a combination of the two. The Basic Principles for solving inequalities (page 169) are valid for all inequalities and will be used frequently. Whenever possible, we use algebra to obtain exact solutions. When algebraic methods are too difficult or unavailable, approximate graphical solutions will be found.

To illustrate the recommended graphical solution method, we consider these inequalities:

$$2x^3 - x^2 - 15x < 0 \quad \text{and} \quad 2x^3 - x^2 - 15x > 0.$$

To solve either one, we graph $y_1 = 2x^3 - x^2 - 15x$, as in Figure 2–40. A point (x, y_1) on the graph lies *below* the *x*-axis when its *y*-coordinate is negative, that is, when

$$y_1 < 0 \quad \text{or, equivalently,} \quad 2x^3 - x^2 - 15x < 0.$$

Thus, the solutions of $2x^3 - x^2 - 15x < 0$ are the *x*-coordinates of all points on

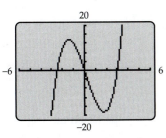

Figure 2–40

the graph that lie *below* the x-axis. Similarly, the point (x, y_1) lies above the x-axis when its y-coordinate is positive, that is, when

$$y_1 > 0 \qquad \text{or, equivalently,} \qquad 2x^3 - x^2 - 15x > 0.$$

So the solutions of $2x^3 - x^2 - 15x > 0$ are the x-coordinates of all points on the graph that lie *above* the x-axis.

To determine precisely which points are above and which are below the x-axis, we must find the x-intercepts of the graph (the points where the graph intersects the x-axis). This should be done algebraically if possible and by graphical approximation otherwise.

EXAMPLE 1

Solve these inequalities:

(a) $2x^3 - x^2 - 15x < 0$ (b) $2x^3 - x^2 - 15x > 0$.

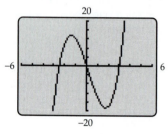

Figure 2–40

SOLUTION   Figure 2–40 shows the graph of $y_1 = 2x^3 - x^2 - 15x$. To find its x-intercepts, we note that these are the points of the graph that are *on* the x-axis, that is, the points (x, y_1) with $y_1 = 0$. So we solve the equation $y_1 = 0$.

$$2x^3 - x^2 - 15x = 0$$

Factor out x:
$$x(2x^2 - x - 15) = 0$$

Factor other term:
$$x(2x + 5)(x - 3) = 0$$

$$x = 0 \quad \text{or} \quad 2x + 5 = 0 \quad \text{or} \quad x - 3 = 0$$

$$x = -5/2 \qquad\qquad x = 3.$$

Therefore, the x-intercepts are $-5/2$, 0, and 3. Now we can read the solutions of the two inequalities from the graph.

(a) Figure 2–40 shows that the graph lies below the x-axis when

$$x < -5/2 \qquad \text{or} \qquad 0 < x < 3.$$

Therefore, the numbers satisfying either of these conditions are the solutions of $2x^3 - x^2 - 15x < 0$.

(b) Figure 2–40 also shows that the graph lies above the x-axis when

$$-5/2 < x < 0 \qquad \text{or} \qquad x > 3.$$

The solutions of $2x^3 - x^2 - 15x > 0$ are all numbers x satisfying either of these conditions.   ■

EXAMPLE 2

The solutions of $2x^2 + 3x - 4 \le 0$ are the numbers x for which the graph of $y = 2x^2 + 3x - 4$ lies on or below the x-axis (Figure 2–41). The x-intercepts

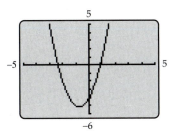

Figure 2–41

of the graph can be found by using the quadratic formula to solve the equation $y = 0$.

$$2x^2 + 3x - 4 = 0$$

$$x = \frac{-3 \pm \sqrt{3^2 - 4 \cdot 2(-4)}}{2 \cdot 2} = \frac{-3 \pm \sqrt{41}}{4}.$$

Figure 2–41 shows that the graph lies below the x-axis between the two x-intercepts. Therefore, the solutions to the inequality are all numbers x such that

$$\frac{-3 - \sqrt{41}}{4} \le x \le \frac{-3 + \sqrt{41}}{4}. \qquad \blacksquare$$

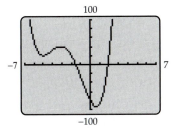

Figure 2–42

EXAMPLE 3

Solve: $x^4 + 10x^3 + 21x^2 + 8 > 40x + 88$.

SOLUTION   We begin by rewriting the inequality in an equivalent form:

$$x^4 + 10x^3 + 21x^2 - 40x - 80 > 0.$$

The graph of $y = x^4 + 10x^3 + 21x^2 - 40x - 80$ in Figure 2–42 has two x-intercepts, one between -2 and -1, and the other near 2. Finding the x-intercepts algebraically would involve solving a fourth-degree equation. So we approximate them graphically.

GRAPHING EXPLORATION

Use a root finder to show that the approximate x-intercepts are -1.53 and 1.89.

Therefore, the approximate solutions of the inequality (the numbers x for which the graph is above the x-axis) are all numbers x such that $x < -1.53$ or $x > 1.89$. $\blacksquare$

CAUTION

Do not attempt to write the solutions in Example 3, namely, $x < -1.53$ or $x > 1.89$, as a single inequality. If you do, the result will be a *nonsense* statement, such as $-1.53 > x > 1.89$ (which says, among other things, that $-1.53 > 1.89$).

■■ FACTORABLE INEQUALITIES

The preceding examples show that solving a polynomial inequality depends only on knowing the x-intercepts of a graph and the places where it is above or below the x-axis. When the polynomial can be completely factored, technology isn't necessary to determine this information.

EXAMPLE 4

Solve: $(x + 15)(x - 2)^6(x - 10) \leq 0$.

SOLUTION The x-intercepts of the graph of $y = (x + 15)(x - 2)^6(x - 10)$ are the solutions of $y = 0$. They are easily read from the factored form.

$$y = 0$$
$$(x + 15)(x - 2)^6(x - 10) = 0$$

$$x + 15 = 0 \quad \text{or} \quad x - 2 = 0 \quad \text{or} \quad x - 10 = 0$$
$$x = -15 \qquad\qquad x = 2 \qquad\qquad x = 10.$$

We need only determine the places where the graph is below the x-axis. To do this without technology, note that the three x-intercepts divide the x-axis into four intervals.

$$x < -15, \qquad -15 < x < 2, \qquad 2 < x < 10, \qquad x > 10.$$

Consider, for example, the interval with $2 < x < 10$. As in previous examples, the graph is an unbroken curve that crosses the x-axis only at the x-intercepts. Since there are no x-intercepts between 2 and 10, the graph must lie *entirely above or entirely below* the x-axis on that interval. To determine which one, choose any number between 2 and 10, say $x = 4$. Then

$$y = (x + 15)(x - 2)^6(x - 10) = (4 + 15)(4 - 2)^6(4 - 10) = 19(2^6)(-6).$$

We don't need to finish the computation to see that y is negative when $x = 4$. Therefore, the point with x-coordinate 4 lies *below* the x-axis. Since one point between 2 and 10 lies below the x-axis, the entire graph must be below the x-axis between 2 and 10.

The location of the graph on the other intervals can also be determined algebraically by choosing a "test number" in each interval, as summarized in the following chart.

Interval	$x < -15$	$-15 < x < 2$	$2 < x < 10$	$x > 10$
Test number in this interval	-20	0	4	11
Value of y at test number	$(-5)(-22)^6(-30)$	$15(-2)^6(-10)$	$19(2^6)(-6)$	$26(9^6)(1)$
Sign of y at test number	$+$	$-$	$-$	$+$
Graph	Above x-axis	Below x-axis	Below x-axis	Above x-axis

The last line of the chart shows that the graph is below the x-axis when $-15 < x < 2$ and when $2 < x < 10$. Since the graph touches the x-axis at -15, 2, and 10, the solutions of the inequality (numbers for which the graph is *on* or below the x-axis) are all numbers x such that $-15 \leq x \leq 10$. ■

The procedures used in Examples 1–4 may be summarized as follows.

Solving Polynomial Inequalities

1. To solve an inequality in one of these forms

$$y > 0, \qquad y \geq 0, \qquad y < 0, \qquad y \leq 0,$$

where y is a polynomial in x, consider the graph of y.

2. Determine the x-intercepts of the graph by solving the equation $y = 0$. Find exact solutions if possible (as in Examples 1, 2, and 4). Otherwise, approximate the solutions (as in Example 3).

3. Use a calculator graph (as in Examples 1–3) or a sign chart (as in Example 4) to determine the intervals where the graph is above or below the x-axis.

4. Use the preceding information to read off the solutions of the inequality.

▪▪ RATIONAL INEQUALITIES

Inequalities involving rational expressions may be solved in much the same way as polynomial inequalities, with one difference. Unlike a polynomial graph that can cross the x-axis only at an x-intercept, the graph of a rational expression may change from one side of the x-axis to the other at its x-intercepts *or* at undefined points (where its denominator is 0).

EXAMPLE 5

Solve: $\dfrac{x}{x-1} > -6$.

SOLUTION   There are three ways to solve this inequality.

Graphical Method: The fastest way to get approximate solutions is to replace the given inequality by an equivalent one,

$$\frac{x}{x-1} + 6 > 0,$$

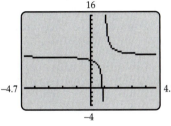

Figure 2–43

and graph the equation $y = \dfrac{x}{x-1} + 6$, as in Figure 2–43.[*] The equation is undefined when $x = 1$ (why?), so there is no point on the graph there. The graph lies above the x-axis everywhere except between the x-intercept and the undefined point at $x = 1$. Using a root finder, we determine that the x-intercept is approximately .857. So the approximate solutions of the inequality are all numbers x such that

$$x < .857 \qquad \text{or} \qquad x > 1.$$

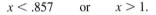

[*]If you don't use a decimal window, as we did, your graph may show an erroneous vertical line at $x = 1$.

Algebraic/Graphical Method: Proceed as before, but write the left-hand side as a single fraction before graphing.

$$\frac{x}{x-1} + 6 > 0$$

$$\frac{x}{x-1} + \frac{6(x-1)}{x-1} > 0$$

$$\frac{x}{x-1} + \frac{6x-6}{x-1} > 0$$

$$\frac{7x-6}{x-1} > 0$$

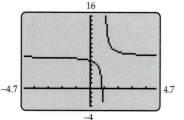

Figure 2–43

When the equation $y = \dfrac{x}{x-1} + 6$ is rewritten as $y = \dfrac{7x-6}{x-1}$, we obtain the same graph (Figure 2–43), but now its x-intercept can be determined exactly. The x-intercept occurs when $y = 0$, which happens when the numerator is 0:

$$7x - 6 = 0$$

$$x = 6/7.$$

So the x-intercept is $6/7$ (whose decimal expansion begins .857), and the exact solutions of the original inequality are those numbers x such that

$$x < 6/7 \qquad \text{or} \qquad x > 1.$$

Algebraic Method: Begin as above by rewriting the given inequality as

$$\frac{7x-6}{x-1} > 0.$$

The numerator of the fraction is 0 when $x = 6/7$ and the denominator is 0 when $x = 1$. The numbers $6/7$ and 1 divide the x-axis into three intervals. Use test numbers and a sign chart, instead of graphing, to determine the location of the graph of $y = \dfrac{7x-6}{x-1}$ on each interval.[*]

Interval	$x < 6/7$	$6/7 < x < 1$	$x > 1$
Test number in this interval	0	.9	2
Value of y at test number	$\dfrac{7 \cdot 0 - 6}{0 - 1}$	$\dfrac{7(.9) - 6}{.9 - 1}$	$\dfrac{7 \cdot 2 - 6}{2 - 1}$
Sign of y at test number	+	−	+
Graph	Above x-axis	Below x-axis	Above x-axis

The last line of the chart shows that the solutions of the original inequality (the numbers x for which the graph is above the x-axis) are all such that $x < 6/7$ or $x > 1$. ■

---

[*]The justification for this approach is the same as in Example 4. The graph can change from one side of the x-axis to the other only at an x-intercept (where the numerator is 0) or at an undefined point (where the denominator is 0).

CAUTION

Don't treat rational inequalities as if they are equations, as in this *incorrect* "solution" of the preceding example:

$$\frac{x}{x-1} > -6$$

$$x > -6(x-1) \qquad \text{[Both \textit{sides} multiplied by } x-1]$$

$$x > -6x + 6$$

$$7x > 6$$

$$x > \frac{6}{7}.$$

According to this, the inequality has no negative solutions and $x = 1$ is a solution, but as we saw in Example 5, *every* negative number is a solution and $x = 1$ is not.[*]

The algebraic technique of writing the left side of an inequality as a single fraction is useful when the resulting numerator and denominator have low degree, so that setting them equal to 0 produces exact solutions (x-intercepts and undefined points). It can be omitted, however, when these solutions must be approximated.

■■ APPLICATIONS

EXAMPLE 6

A computer store has determined that the cost C of ordering and storing x laser printers is given by

$$C = 2x + \frac{300,000}{x}.$$

If the delivery truck can bring at most 450 printers per order, how many printers should be ordered at a time to keep the cost below $1600?

SOLUTION   To find the values of x that make C less than 1600, we must solve the inequality

$$2x + \frac{300,000}{x} < 1600 \qquad \text{or equivalently,} \qquad 2x + \frac{300,000}{x} - 1600 < 0.$$

We shall solve this inequality graphically, although it can also be solved algebraically. In this context, the only solutions that make sense are those between

---

*The source of the error is multiplying by $x - 1$, which is negative for some values of x and positive for others. So you must consider two cases and reverse the direction of the inequality when $x - 1$ is negative.

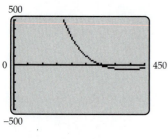

Figure 2–44

0 and 450. So we choose the viewing window in Figure 2–44 and graph

$$y = 2x + \frac{300{,}000}{x} - 1600.$$

Figure 2–44 shows that the desired solutions (numbers for which the graph is below the x-axis) are all numbers between the x-intercept and 450. A root finder shows that the x-intercept is $x \approx 300$. In fact, this is the exact x-intercept because $y = 0$ when $x = 300$, as you can easily verify. Therefore, to keep costs under \$1600, x printers should be ordered each time, with $300 < x \le 450$. ∎

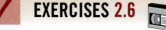

EXERCISES 2.6

In Exercises 1–20, solve the inequality exactly.

1. $x^2 - 4x + 3 \le 0$
2. $x^2 - 7x + 10 \le 0$

3. $x^2 + 9x + 14 \ge 0$
4. $x^2 + 8x + 15 \ge 0$

5. $x^2 \ge 9$
6. $x^2 \ge 16$

7. $6 + x - x^2 \le 0$
8. $4 - 3x - x^2 \ge 0$

9. $6x^2 - x \ge 2$
10. $4x^2 + 11x + 6 \ge 0$

11. $x^2 + 11x + 6 > 0$
12. $x^2 - 9x + 6 > 0$

13. $x^2 - 5x < 5$
14. $x^2 + 6x < 4$

15. $x^2 + 4x + 5 > 0$
16. $x^2 + 3x + 4 < 0$

17. $x^3 - x \ge 0$
18. $x^3 + 2x^2 + x > 0$

19. $x^3 - 2x^2 - 3x < 0$
20. $x^4 - 14x^3 + 48x^2 \ge 0$

In Exercises 21–46, solve the inequality. Find exact solutions when possible and approximate ones otherwise.

21. $(x + 7)(x + 3)(x - 5)^2 \ge 0$

22. $(x - 5)(x + 4)(x - 3)(x + 2) \ge 0$

23. $(x + 4)^2(x + 1)(x - 3)^2 \le 0$

24. $(x - 6)^3(x + 2)^3(2x - 1) \le 0$

25. $x^3 + 6x^2 + 6x > 0$

26. $x^3 + 2x^2 < 4x$

27. $x^4 - 3x^3 \le 5x^2$

28. $x^4 + 3x^2 \ge 7x^3$

29. $x^3 - 2x^2 - 5x + 7 \ge 2x + 1$

30. $x^4 - 6x^3 + 2x^2 < 5x - 2$

31. $2x^4 + 3x^3 < 2x^2 + 4x - 2$

32. $x^5 + 5x^4 > 4x^3 - 3x^2 + 2$

33. $x^4 - 5x^3 \le 35x^2 + 4x - 9$

34. $x^4 + 5x^2 + 27x \ge 8x^3 + 6$

35. $4x^5 + 17x^4 - 12x^3 < 85x^2 + 60x - 4$

36. $x^5 + x^4 + 206x + 12 > 34x^3 + 69x^2$

37. $\dfrac{3x + 1}{2x - 4} > 0$
38. $\dfrac{2x - 1}{5x + 3} \ge 0$

39. $\dfrac{x^2 + x - 2}{x^2 - 2x - 3} < 0$
40. $\dfrac{2x^2 + x - 1}{x^2 - 4x + 4} \ge 0$

41. $\dfrac{x - 2}{x - 1} < 1$
42. $\dfrac{-x + 5}{2x + 3} \ge 2$

43. $\dfrac{x - 3}{x + 3} \le 5$
44. $\dfrac{2x + 1}{x - 4} > 3$

45. $\dfrac{x^3 - 3x^2 + 5x - 29}{x^2 - 7} > 3$

46. $\dfrac{x^4 - 3x^3 + 2x^2 + 2}{x - 2} > 15$

47. The Euro became the common currency of most of Europe in 1999. The value of one Euro in U.S. dollars is approximated by the equation

$$y = .08x^2 - 1.765x + 10.575 \qquad (9 \le x \le 13),$$

where $x = 9$ corresponds to 1999. When was one Euro worth less than one dollar?

48. Actual and projected Medicare costs (in trillions of dollars) are approximated by

$$y = .0011x^2 + .00138x + .05,$$

where $x = 0$ corresponds to 1980.[*] For what period are Medicare costs expected to be below one trillion dollars?

---

[*]Congressional Budget Office.

Use the following information for Exercises 49 and 50. If appropriate steps are taken to reduce emissions of carbon dioxide and other greenhouse gases (to prevent the west Antarctic ice sheet from disintegrating), then the amount of such gases (in billions of tons per year) during the next three centuries will be approximated by

$$y = (-1.6443 \times 10^{-8})x^4 + (1.073 \times 10^{-5})x^3 - .00235x^2$$
$$+ .132x + 8.396,$$

where $x = 0$ corresponds to 2000.[†]

49. During what years will emission be less than 10 billion tons per year?

50. During what years will emissions be below five billion tons per year?

51. Find all pairs of numbers that satisfy these two conditions: Their sum is 20 and the sum of their squares is less than 362.

52. The length of a rectangle is 6 inches longer than its width. What are the possible widths if the area of the rectangle is at least 667 square inches?

53. It costs a craftsman $5 in materials to make a medallion. He has found that if he sells the medallions for $50 - x$ dollars each, where x is the number of medallions produced each week, then he can sell all that he makes. His fixed costs are $350 per week. If he wants to sell all he makes and show a profit each week, what are the possible numbers of medallions he should make?

54. A retailer sells file cabinets for $80 - x$ dollars each, where x is the number of cabinets she receives from the supplier each week. She pays $10 for each file cabinet and has fixed costs of $600 per week. How many file cabinets should she order from the supplier each week to guarantee that she makes a profit?

In Exercises 55–58, you will need the formula for the height h of an object above the ground at time t seconds:
$h = -16t^2 + v_0t + h_0$; *this formula was explained on page 142.*

55. A toy rocket is fired straight up from ground level with an initial velocity of 80 feet per second. During what time interval will it be at least 64 feet above the ground?

56. A projectile is fired straight up from ground level with an initial velocity of 72 feet per second. During what time interval is it at least 37 feet above the ground?

57. A ball is dropped from the roof of a 120-foot-high building. During what time period will it be strictly between 56 and 39 feet above the ground?

58. A ball is thrown straight up from a 40-foot-high tower with an initial velocity of 56 feet per second.
(a) During what time interval is the ball at least 8 feet above the ground?
(b) During what time interval is the ball between 53 feet and 80 feet above the ground?

Thinkers

In Exercises 59–62, solve the inequality.

59. $x^4 - 50x^3 + 125x^2 + 250x - 8 > 0$

60. $.1x^4 + 6x^3 + 22x^2 - 153x - 16 < 0$

61. $\dfrac{2x^2 + 6x - 8}{2x^2 + 5x - 3} < 1$

62. $\dfrac{2x^2}{x^2 + x - 2} > 1.8$

[†]Based on data from *Nature*.

Chapter 2 Review

IMPORTANT FACTS & FORMULAS

- *Quadratic Formula:* If $a \neq 0$, then the solutions of $ax^2 + bx + c = 0$ are

$$x = \frac{-b \pm \sqrt{b^2 - 4ac}}{2a}.$$

- If $a \neq 0$, then the number of real solutions of $ax^2 + bx + c = 0$ is 0, 1, or 2, depending on whether the discriminant $b^2 - 4ac$ is negative, zero, or positive.
- The solutions of $|x| \leq k$ are all numbers x such that $-k \leq x \leq k$.
- The solutions of $|x| \geq k$ are all numbers x such that $x \leq -k$ or $x \geq k$.

REVIEW QUESTIONS

1. Solve for x: $2\left(\dfrac{x}{5} + 7\right) - 3x = \dfrac{x + 2}{5} - 4$

2. Solve for t: $\dfrac{t + 1}{t - 1} = \dfrac{t + 2}{t - 3}$

3. Solve for x: $\dfrac{3}{x} - \dfrac{2}{x - 1} = \dfrac{1}{2x}$

4. Solve for z: $2 - \dfrac{1}{z - 2} = \dfrac{z - 3}{z - 2}$

5. Solve for r: $Q = \dfrac{b - a}{2r}$

6. Solve for R: $\dfrac{v}{t} = 2\pi Rh$

7. The median hourly pay for women is approximated by

$$y = .095x + 8.97 \qquad (x \geq 3),$$

where $x = 3$ corresponds to 1973. In what year will the median pay be $13?

8. Bert weighs 10 times as much as the Thanksgiving turkey, and Sally weighs 7 times as much as the turkey. If Bert is 48 pounds heavier than Sally, how much does Bert weigh?

REVIEW QUESTIONS

9. A jeweler wants to make a 1-ounce ring consisting of gold and silver, using $200 worth of metal. If gold costs $600 per ounce and silver $50 per ounce, how much of each metal should she use?

10. A printer is on sale for 15% less than the list price. The sale price, plus a 5% shipping charge, totals $210. What is the list price?

11. Mike can do a job in 5 hours, and Larry can do the same job in 4 hours. How long will it take them to do the job together?

12. A car leaves the city traveling at 54 mph. One-half hour later, a second car leaves from the same place and travels at 63 mph along the same road. How long will it take for the second car to catch up with the first?

13. A 12-foot-long rectangular board is cut in two pieces so that one piece is 4 times as long as the other. How long is the bigger piece?

14. George owns 200 shares of stock, 40% of which are in the computer industry. How many more shares must he buy in order to have 50% of his total shares in computers?

15. If x and y are directly proportional and $x = 12$ when $y = 36$, then what is y when $x = 2$?

16. Driving time varies inversely as speed. If it takes 3 hours to drive to the beach at an average speed of 48 mph, how long will it take to drive home at an average speed of 54 mph?

17. Find the constant of variation when T varies directly as the square of R and inversely as S if $T = .6$ when $R = 3$ and $S = 15$.

18. The statement "r varies directly as s and the square root of t and inversely as the cube of x" means that for some constant k,

(a) $r = kstx^3$
(b) $rx^3 = ks\sqrt{t}$
(c) $r = \dfrac{kx^3\sqrt{t}}{s}$
(d) $k = \dfrac{rs\sqrt{t}}{x^3}$
(e) $kx^3s = k\sqrt{t}$

19. Solve for x: $3x^2 - 2x + 5 = 0$

20. Solve for x: $5x^2 + 6x = 7$

21. Solve for y: $3y^2 - 2y = 5$

22. Find the *number* of real solutions of the equation $20x^2 + 12 = 31x$.

23. Find the *number* of real solutions of the equation
$$\frac{1}{3x + 6} = \frac{2}{x + 2}.$$

24. For what value of k does the equation $kt^2 + 5t + 2 = 0$ have exactly one real solution for t?

25. The population P (in thousands) of St. Louis, Missouri, can be approximated by
$$P = .11x^2 - 15.95x + 864,$$
where $x = 0$ corresponds to 1950. When did the population first drop below 450,000?

26. The median sales price S (in thousands of dollars) for a single-family home in the United States Midwest since 1990 can be approximated by $S = .159x^2 + 3.342x + 75.185$, where $x = 0$ corresponds to 1990.* When did the median price reach $131,200?

27. Do there exist two real numbers whose sum is 2 and whose product is 2? Justify your answer.

28. Find two consecutive integers, the sum of whose squares is 481.

29. The size of a computer monitor is its diagonal measurement, as shown in the figure. Assume that the height of the screen is three-fourths of its width.
(a) Find the viewing area of a 21-inch monitor.
(b) Find the viewing area of a 14-inch monitor.
(c) How many times larger is the viewing area on a 21-inch monitor than on a 14-inch one?

30. The radius of a circle is 10 inches. By how many inches should the radius be increased so that the area increases by 5π square inches?

In Questions 31–36, find a solution of the equation that lies in the given interval.

31. $x^4 + x^3 - 10x^2 = 8x + 16;$ $(x \geq 0)$

32. $2x^4 + x^3 - 2x^2 + 6x + 2 = 0;$ $(x < -1)$

---

*National Association of Realtors

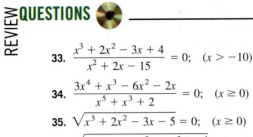

33. $\dfrac{x^3 + 2x^2 - 3x + 4}{x^2 + 2x - 15} = 0; \quad (x > -10)$

34. $\dfrac{3x^4 + x^3 - 6x^2 - 2x}{x^5 + x^3 + 2} = 0; \quad (x \geq 0)$

35. $\sqrt{x^3 + 2x^2 - 3x - 5} = 0; \quad (x \geq 0)$

36. $\sqrt{1 + 2x - 3x^2 + 4x^3 - x^4} = 0; \quad (-5 < x < 5)$

37. According to data from the U.S. Department of Justice, the total number of prisoners y in state and federal prisons (in thousands) can be approximated by the equation

$$y = .074x^4 - 1.91x^3 + 13.75x^2 + 37.94x + 738.613,$$

where $x = 0$ corresponds to 1990. In what year did the prison population first reach 1,300,000?

38. Since 1995, the amount spent on federal student assistance (such as Pell Grants and other programs) can be approximated by

$$y = .1366x^3 - 3.3107x^2 + 27.434x - 35.924,$$

where $x = 5$ corresponds to 1995 and y is in billions of dollars. In what year did student assistance reach $50 billion?

In Questions 39–42, find all real solutions of the equation. Do not approximate.

39. $x^4 - 11x^2 + 18 = 0$ **40.** $x^6 - 4x^3 + 4 = 0$

41. $|3x - 1| = 4$ **42.** $|2x - 1| = x + 4$

In Questions 43–50, find all real solutions of the equation.

43. $x^4 - 2x^2 - 15 = 0$ **44.** $x^4 - x^2 = 6$

45. $x^6 + 7x^3 - 8 = 0$ **46.** $3y^7 - 3y^5 - 15y^3 = 0$

47. $\sqrt{x - 1} = 2 - x$ **48.** $\sqrt[3]{1 - t^2} = -2$

49. $\sqrt{x + 1} + \sqrt{x - 1} = 1$

50. $\sqrt{3x - 1} + \sqrt{x} = 2$ **51.** Solve for s: $t = \sqrt{\dfrac{2s}{g}}$

In Questions 52–56, solve the inequality.

52. $-3(x - 4) \leq 5 + x$ **53.** $-4 < 2x + 5 < 9$

54. $-3 < 8x + 5 < 13$ **55.** $|3x + 2| \geq 2$

56. $\left|\dfrac{y + 2}{3}\right| \geq 5$

57. On which intervals is $\dfrac{2x - 1}{3x + 1} < 1$?

58. On which intervals is $\dfrac{2}{x + 1} < x$?

59. Let C = consumer spending and I = income. Suppose $C = 90 + \dfrac{3}{4}I$. Express the range of incomes (I) over which consumer spending exceeds income.

60. Solve for x: $x^2 + x > 12$.

61. Solve for x: $(x - 1)^2(x^2 - 1)x \leq 0$.

62. If $0 < r \leq s - t$, then which of these statements is *false*?
(a) $s \geq r + t$ (b) $t - s \leq -r$
(c) $-r \geq s - t$ (d) $\dfrac{s - t}{r} > 0$
(e) $s - r \geq t$

63. If $\dfrac{x + 3}{2x - 3} > 1$, then which of these statements is *true*?
(a) $\dfrac{x - 3}{2x + 3} < -1$ (b) $\dfrac{2x - 3}{x + 3} < -1$
(c) $\dfrac{3 - 2x}{x + 3} > 1$ (d) $2x + 3 < x - 3$
(e) None of the preceding statements.

64. Solve: $2x - 3 \leq 5x + 9 < -3x + 4$.

In Questions 65–72, solve the inequality.

65. $|2 - 5x| \geq 2$ **66.** $x^2 + x - 20 > 0$

67. $\dfrac{x - 2}{x + 4} \leq 3$

68. $(x + 1)^2(x - 3)^4(x + 2)^3(x - 7)^5 > 0$

69. $\dfrac{x^2 + x - 9}{x + 3} < 1$ **70.** $\dfrac{x^2 - x - 6}{x - 3} > 1$

71. $\dfrac{x^2 - x - 5}{x^2 + 2} > -2$ **72.** $\dfrac{x^4 - 3x^2 + 2x - 3}{x^2 - 4} < -1$

73. Social Security costs y (in trillions of dollars) can be approximated by

$$y = .00127x^2 + .0096x + .3 \qquad (0 \leq x \leq 23),$$

where $x = 0$ corresponds to 1980. During what years will costs be between one and two trillion dollars per year?

DISCOVERY PROJECT 2   Inequalities in Yes/No Decisions

David Young Wolff/PhotoEdit

In this chapter, you learned techniques for determining the values for which an inequality is true. In many practical applications, it is not necessary to know what is happening for every value of the variable. In fact, in most situations, we are only interested in a very limited range of values for the variable. Consider the construction of sports facilities, for example. An important consideration in the construction of the roofs of such facilities is whether or not the ball used is likely to hit the ceiling during play of the game.

1. A new field house is to be constructed on a college campus to support the very successful volleyball program. The roof design has the lights suspended from the ceiling at a height of 46 feet above the floor. When the ball is hit so that it rises high, its height h after t seconds is typically given by a formula such as $h = -16t^2 + 50t + 5$. Assuming this formula to be universally valid, can a shot hit the lights? [*Hint:* Are there any positive values of t for which $h > 46$?]

A similar question can be posed for more complicated paths.

2. A particuarly well-hit baseball will leave the bat at 80 miles per hour with an upward angle of 30°. Its height h above the ground s feet from the bat is given by

$$h = \frac{-9s^2}{9744} + \frac{19s}{30} + 3.$$

The height k of the new stadium roof along the right field line, s feet from home plate, is given by

$$k = \frac{2\sqrt{-4s^2 + 1500s + 250{,}000}}{5} - 80.$$

Will the ball hit the ceiling before crossing out of play 385 feet from home plate? In other words, is the height of the ball always less than the height of the ceiling?

3. A movie director wants to film a scene in which a speeding car evades a police car by crossing the state line first. The car passes a stopped police officer, who immediately begins the chase. The officer's position is given by $p = 17t^{1.5}$, where t is measured in seconds. The speeder drives at a constant 80 miles per hour, giving it a position of $s = \dfrac{352t}{3}$ after t seconds. If the speeder reaches the border in 45 seconds, will the police car catch up?

Functions and Graphs

Spencer Grant/PhotoEdit

Looking for a house?

If you buy a house, you will probably need a mortgage. If you can get a low interest rate, your monthly payments will be lower (or, alternatively, you can afford a more expensive house). The timing of your purchase can make a difference because mortgage interest rates constantly fluctuate. In mathematical terms, rates are a function of time. The graph of this function provides a picture of how interest rates change. See Exercise 47 on page 226.

191

Chapter Outline

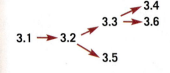
The concept of a function and functional notation are central to modern mathematics and its applications. In this chapter, you will be introduced to functions and operations on functions, learn how to use functional notation, and develop skill in constructing and interpreting graphs of functions.

3.1 Functions

To understand the origin of the concept of function, it may help to consider some real-life situations in which one numerical quantity depends on, corresponds to, or determines another.

EXAMPLE 1

The amount of state income tax Louisiana residents pay depends on their income. The way that the income determines the tax is given by the following tax law.[*]

Income		Tax
At Least	**But Less Than**	
0	$4,500	0
$4,500	$12,500	2% of amount over $4,500
$12,500	$25,000	$162.50 + 4% of amount over $12,500
$25,000		$662.50 + 6% of amount over $25,000

---

[*]2003 rates for a single person with one exemption and no deductions; actual tax amount may vary slightly from this formula when tax tables are used.

EXAMPLE 2

The graph in Figure 3–1 shows the temperatures in Cleveland, Ohio, on April 11, 2001, as recorded by the U.S. Weather Bureau at Hopkins Airport. The graph indicates the temperature that corresponds to each given time.   ■

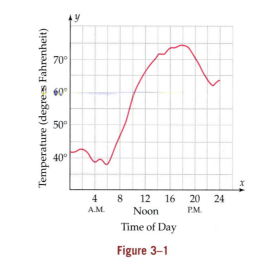

Figure 3–1

EXAMPLE 3

Suppose a rock is dropped straight down from a high place. Physics tells us that the distance traveled by the rock in t seconds is $16t^2$ feet. So the distance depends on the time.   ■

These examples share several common features. Each involves two sets of numbers, which we can think of as inputs and outputs. In each case, there is a rule by which each input determines an output, as summarized here.

	Set of Inputs	Set of Outputs	Rule
Example 1	All incomes	All tax amounts	The tax law
Example 2	Hours since midnight	Temperatures during the day	Time/temperature graph
Example 3	Seconds elapsed after dropping the rock	Distance rock travels	Distance $= 16t^2$

Each of these examples may be mentally represented by an idealized calculator that has a single operation key: A number is entered [*input*], the rule key is pushed [*rule*], and an answer is displayed [*output*]. The formal definition of function incorporates these common features (input/rule/output), with a slight change in terminology.

Functions

A **function** consists of

> A set of inputs (called the **domain**);
>
> A **rule** by which each input determines one and only one output;
>
> A set of outputs (called the **range**).

The phrase "one and only one" in the definition of the rule of a function might need some clarification. In Example 2, for each time of day (input), there is one and only one temperature (output). But it is quite possible to have the same temperature (output) at different times (inputs). In general,

> **For each input, the rule of a function determines exactly one output. But different inputs may produce the same output.**

Although real-world situations, such as Examples 1–3, are the motivation for functions, much of the emphasis in mathematics courses is on the functions themselves, independent of possible interpretations in specific situations, as illustrated in the following examples.

EXAMPLE 4

Could either of the following be the table of values of a function?

(a)

Input	−4	−2	0	2	4
Output	21	7	1	3	7

(b)

Input	3	2	1	3	5
Output	4	0	2	6	9

SOLUTION In table (a), two different inputs (−2 and 4) produce the same output, but that's OK because each input produces exactly one output. So table (a) could be the table of values of a function. In table (b), however, the input 3 produces two different outputs (4 and 6), so this table cannot possibly represent a function. ■

EXAMPLE 5

For each real number s that is not an integer, let $[s]$ denote the *integer* that is closest to s on the *left* side of s on the number line; if s is itself an integer, we define $[s] = s$. Some examples are shown in Figure 3–2.

$$[-4.7] = -5, \quad [-3] = -3, \quad [-1.5] = -2,$$

$$[0] = 0, \quad \left[\frac{5}{3}\right] = 1, \quad [\pi] = 3.$$

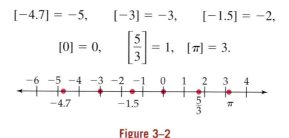

Figure 3–2

The **greatest integer function** is the function whose domain is the set of all real numbers, whose range is the set of integers, and whose rule is

For each input (real number) x, the output is the integer $[x]$. ■

TECHNOLOGY

TIP

The greatest integer function is denoted INT or FLOOR on TI and HP, and INTG on Casio. It is in this menu/submenu:

TI: MATH/NUM

HP-39+: MATH/REAL

Casio: OPTN/NUM

▪▪ FUNCTIONS DEFINED BY EQUATIONS

Equations in two variables are *not* the same things as functions. However, many equations can be used to define functions.

EXAMPLE 6

The equation $4x - 2y^3 + 5 = 0$ can be solved uniquely for y.

$$2y^3 = 4x + 5$$

$$y^3 = 2x + \frac{5}{2}$$

$$y = \sqrt[3]{2x + \frac{5}{2}}$$

If a number is substituted for x in this equation, then exactly one value of y is produced. So we can define a function whose domain is the set of all real numbers and whose rule is

The input x produces the output $\sqrt[3]{2x + 5/2}$.

In this situation, we say that the equation defines **y as a function of x.**

The original equation can also be solved for x.

$$4x = 2y^3 - 5$$

$$x = \frac{2y^3 - 5}{4}$$

Now if a number is substituted for y, exactly one value of x is produced. So we can think of y as the input and the corresponding x as the output and say that the equation defines **x as a function of y.** ■

EXAMPLE 7

If you solve the equation

$$y^2 - x + 1 = 0$$

for y, you obtain

$$y^2 = x - 1$$
$$y = \pm\sqrt{x - 1}.$$

This equation does *not* define y as a function of x because, for example, the input $x = 5$ produces two outputs: $y = \pm 2$. ■

EXAMPLE 8

A group of students drives from Cleveland to Seattle, a distance of 2350 miles, at an average speed of 52 mph.

(a) Express their distance from Cleveland as a function of time.

(b) Express their distance from Seattle as a function of time.

SOLUTION

(a) Let t denote the time traveled in hours after leaving Cleveland, and let D be the distance from Cleveland at time t. Then the equation that expresses D as a function of t is

$$D = \text{Distance traveled in } t \text{ hours at 52 mph} = 52t.$$

(b) At time t, the car has traveled $52t$ miles of the 2350-mile journey, so the distance K remaining to Seattle is given by $K = 2350 - 52t$. This equation expresses K as a function of t. ■

Graphing calculators are designed to deal with equations that define y as a function of x. Calculators can evaluate such functions (that is, produce the outputs from various inputs). One method is illustrated in Calculator Investigation 1 at the end of this section, and another in the next example.

EXAMPLE 9

The equation $y = x^3 - 2x + 3$ defines y as a function of x. Use the table feature of a calculator* to find the outputs for each of the following inputs.

(a) $-3, -2, -1, 0, 1, 2, 3, 4, 5$ (b) $-5, -11, 8, 7.2, -.44$

---

*TI-85 does not have a built-in table feature, but a program to proivde one is in the Program Appendix.

TECHNOLOGY

TIP

To find the table setup screen, look for TBLSET (or RANG or NUM SETUP) on the keyboard or in the TABLE menu.

The increment is called ΔTBL on TI (and PITCH or NUM-STEP on others).

The table type is called INDPNT on TI, and NUMTYPE on HP-39+.

SOLUTION

(a) To use the table feature, we first enter $y = x^3 - 2x + 3$ in the equation memory, say, as y_1. Then we call up the setup screen (see the Technology Tip in the margin and Figure 3–3) and enter the *starting number* (-3), the *increment* (the amount the input changes for each subsequent entry, which is 1 here), and the *table type* (AUTO, which means the calculator will compute all the outputs at once).* Then press TABLE to obtain the table in Figure 3–4. To find values that don't appear on the screen in Figure 3–4, use the up and down arrow keys to scroll through the table.

(b) With an apparently random list of inputs, as here, we change the table type to ASK (or USER or BUILD YOUR OWN).† Then key in each value of x, and hit ENTER. This produces the table one line at a time, as in Figure 3–5. ■

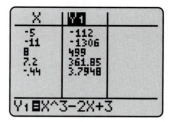

Figure 3–3   **Figure 3–4**   **Figure 3–5**

CALCULATOR EXPLORATION

Construct a table of values for the function in Example 9 that shows the outputs for these inputs: 2, 2.4, 2.8, 3.2, 3.6, and 4. What is the increment here?

EXAMPLE 10

The revenues y of MTV in year x can be approximated by the equation

$$y = .257x^3 + 11.6x^2 + 4.5x + 177.5 \qquad (0 \le x \le 14),$$

where $x = 0$ corresponds to 1987 and y is in millions of dollars.‡ So the revenue y is a function of the year x. In what year did revenues first exceed one billion dollars?

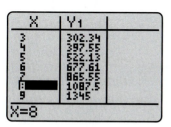

Figure 3–6

SOLUTION   Since y is in millions, one billion dollars corresponds to $y = 1000$. Make a table of values for the function (Figure 3–6). It shows that the revenue was approximately \$865,550,000 in 1994 ($x = 7$) and \$1,087,500,000 in 1995 ($x = 8$). ■

---

*There is no table type selection on Casio, but you must enter a maximum value for x.
†Casio users should follow the directions in Calculator Investigation 2 at the end of this section. This type of table is not available on TI-85, but there is an effective alternative (see Calculator Investigation 3).
‡Based on data from MTV Networks.

INVESTIGATIONS 3.1

1. FUNCTION EVALUATION

The equation $y = x^3 - 4x + 1$ defines y as a function of x. On calculators other than HP-39+, the function may be evaluated as follows.

(a) Enter this equation as y_1 in the equation memory, and return to the home screen. To evaluate y at $x = 5$, store 5 in memory X*, where X* denotes the "variable" key that is used to enter the equation y_1 in the function memory. Then choose y_1, which is found in the following menu/submenu:

TI-83+: VARS/Y-VARS

Casio: VARS/GRPH [choose Y and type in 1]

TI-86/89: Type in y_1 on the keyboard.

Now press ENTER and the calculator displays the value of y at $x = 5$.

(b) Evaluate y at the following values of x: 3.5, 2, −1, −4.7.

(c) Enter the equation $y = x^2 + x - 3$ as y_2 in the equation memory. Evaluate this function at $x = 7$ and $x = -5.5$.

2. USER-CONTROLLED TABLES ON CASIO

To construct the table in Example 9(b), proceed as follows.

(a) Enter the equation $y = x^3 - 2x + 3$ in the equation memory. This can be done by selecting either TABLE or GRAPH in the MAIN menu.

(b) Return to the MAIN menu and select LIST. Enter the numbers at which you want to evaluate this function [in this case, −5, −11, 8, 7.2, −.44] as List 1.

(c) Return to the MAIN menu and select TABLE. Then press SET-UP [that is, 2nd MENU] and select LIST as the Variable; on the LIST menu, choose List 1. Press EXIT and then press TABL to produce the table.

(d) Use the up/down arrow key to scroll through the table. If you change an entry in the X column, the corresponding y_1 value will automatically change. Find the value of y_1 at $x = 11$ and $x = -6$.

3. FUNCTION EVALUATION ON TI-85/86

The equation $y = x^2 - 3.7x + 4.5$ defines y as a function of x.

(a) Enter this equation as y_1 in the equation memory, and return to the home screen by pressing 2nd QUIT. To determine the value of y when $x = 29$, key in EVAL 29 ENTER (you will find EVAL in the MISC submenu of the MATH menu). Use this method to find the value of y when $x = .45$ and 611.

(b) Enter two more equations in the equation memory: $y_2 = x^3 - 6$ and $y_3 = 5x + 2$. Now when you key in EVAL 29 ENTER, the value of all three functions at $x = 29$ will be displayed as a list, which you can scroll through by using the arrow keys.

(c) In the equation memory, turn off y_2, and return to the home screen. If you use EVAL now, what happens?

4. CUSTOM MENU ON TI-85/86

If you press CUSTOM, the calculator displays a menu that may be blank. You may fill in any entries you want, so that frequently used operations (such as EVAL and FRAC) may be accessed quickly from the CUSTOM menu, rather than searching for them in varius menus and submenus. Check your instruction manual for details.

✓ EXERCISES 3.1

In Exercises 1–4, determine whether or not the given table could possibly be a table of values of a function. Give reasons for your answer.

1.

Input	−2	0	3	1	−5
Output	2	3	−2.5	2	14

2.

Input	−5	3	0	−3	5
Output	7	3	0	5	−3

3.

Input	−5	1	3	−5	7
Output	0	2	4	6	8

4.

Input	1	−1	2	−2	3
Output	1	−2	±5	−6	8

Exercises 5–10 deal with the greatest integer function of Example 5, which is given by the equation $y = [x]$. Compute the following values of the function:

5. $[6.75]$

6. $[.75]$

7. $[−4/3]$

8. $[5/3]$

9. $[−16.0001]$

10. Does the equation $y = [x]$ define x as a function of y? Give reasons for your answer.

In Exercises 11–18, determine whether the equation defines y as a function of x or defines x as a function of y.

11. $y = 3x^2 − 12$

12. $y = 2x^4 + 3x^2 − 2$

13. $y^2 = 4x + 1$

14. $5x − 4y^4 + 64 = 0$

15. $3x + 2y = 12$

16. $y − 4x^3 − 14 = 0$

17. $x^2 + y^2 = 9$

18. $y^2 − 3x^4 + 8 = 0$

In Exercises 19–22, each equation defines y as a function of x. Create a table that shows the values of the function for the given values of x.

19. $y = x^2 + x − 4$; $x = −2, −1.5, −1, \ldots, 3, 3.5, 4$.

20. $y = x^3 − x^2 + 4x + 1$; $x = 3, 3.1, 3.2, \ldots, 3.9, 4$

21. $y = \sqrt{4 − x^2}$; $x = −2, −1.2, −.04, .04, 1.2, 2$

22. $y = |x^2 − 5|$; $x = −8, −6, \ldots, 8, 10, 12$

Exercises 23–26 refer to the tax law in Example 1.

23. Find the output (tax amount) that is produced by each of the following inputs (incomes):

$500 $1509 $3754

$6783 $12,500 $55,342

24. Find four different numbers in the domain of this function that produce the same output (number in the range).

25. Explain why your answer in Exercise 24 does *not* contradict the definition of a function (in the box on page 194).

26. Is it possible to do Exercise 24 if all four numbers in the domain are required to be greater than 4500? Why or why not?

27. The amount of postage required to mail a first-class letter is determined by its weight. In this situation, is weight a function of postage? Or vice versa? Or both?

28. Could the following statement ever be the rule of a function?

> For input x, the output is the number whose square is x.

Why or why not? If there is a function with this rule, what is its domain and range?

29. (a) Use the following chart to make two tables of values (one for an average man and one for an average woman) in which the inputs are the number of drinks per hour and the outputs are the corresponding blood alcohol contents.[*]

Blood alcohol content

A look at the number of drinks consumed and blood alcohol content in one hour under optimum conditions:

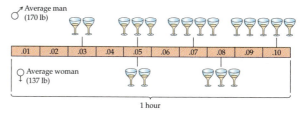

(b) Does each of these tables define a function? If so, what are the domain and range of each function? [Remember that you can have part of a drink.]

30. The table on the next page relates a large-framed woman's height to the weight at which the woman should live longest.[†] In this situation, is weight a function of height? Is height a function of weight? Justify your answer.

---

[*]National Highway Traffic Safety Administration. Art by AP/Amy Kranz.
[†]Metropolitan Life Insurance Company.

Height (in shoes)	Weight (in pounds, in indoor clothing)
4'10"	118–131
4'11"	120–134
5'0"	122–137
5'1"	125–140
5'2"	128–143
5'3"	131–147
5'4"	134–151
5'5"	137–155
5'6"	140–159
5'7"	143–163
5'8"	146–167
5'9"	149–170
5'10"	152–173
5'11"	155–176
6'0"	158–179

31. The chart shows part of a body mass index (BMI) table.[*]
Using only the data in the chart for a 5-foot-tall person,
(a) Is BMI a function of weight? Why?
(b) Is weight a function of BMI? Why?
(c) Do parts (a) and (b) for a person 5 ft 10 inches tall.

	Height	
Weight (lb)	5 ft	5ft 10 in
125	24	18
130	25	19
135	26	19
140	27	20
145	28	21
150	29	22
155	30	22
160	31	23
165	32	24
170	33	24
175	34	25
180	35	26
185	36	27
190	37	27
195	38	28
200	39	29

32. The prime rate is the rate that large banks charge their best corporate customers for loans. The graph shows how the prime rate charged by a particular bank has varied in recent years.[†] Answer the following questions by reading the graph as best you can.
(a) What was the prime rate in January 1998? In January 2002? In early 2003?
(b) In what time periods was the prime rate 6%?
(c) On the basis of the data provided by this graph, can the prime rate be considered a function of time? Can time be considered a function of the prime rate?

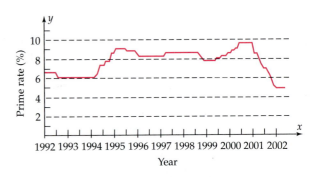

33. Find an equation that expresses the area A of a circle as a function of its
(a) radius r (b) diameter d

34 Find an equation that expresses the area of a square as a function of its
(a) side x (b) diagonal d

35. A box with a square base of side x is four times higher than it is wide. Express the volume V of the box as a function of x.

36. The surface area of a cylindrical can of radius r and height h is $2\pi r^2 + 2\pi rh$. If the can is twice as high as the diameter of its top, express its surface area S as a function of r.

37. Suppose you drop a rock from the top of a 400-foot-high building. Express the distance D from the rock to the ground as a function of time t. What is the range of this function? [*Hint:* See Example 3.]

38. A bicycle factory has weekly fixed costs of $26,000. In addition, the material and labor costs for each bicycle are $125. Express the total weekly cost C as a function of the number x of bicycles that are made.

---

*BMI is the body weight in kilograms divided by the height in meters squared. A person with a BMI of 25–29 is considered overweight, and one with a BMI of 30 or more is considered obese.

†Federal Reserve Board.

39. The table below shows the percentage of single-parent families in various years.*

Year	Percent
1960	12.8
1970	13.2
1980	17.5
1990	20.8
2000	23.2

(a) The equation

$$y = (2.0284 \times 10^{-7})x^4 - (2.0237 \times 10^{-5})x^3 \\ + (6.584 \times 10^{-4})x^2 - .00434x + .1282,$$

in which $x = 0$ corresponds to 1960, defines y as a function of x. Make a table of values that includes the x-values corresponding to the years in the Census Bureau table.

(b) How do the values of y in your table compare with the percentages in the Census Bureau table? Does this equation seem to provide a reasonable model of the Census Bureau data?

(c) Use the equation to estimate the percentage of single-parent families in 1995 and 2005.

(d) Assuming that this model remains reasonably accurate, in what year will 30% of families be single-parent families?

40. The table shows the amount spent on student scholarships (in millions of dollars) by Oberlin College in recent years. [1995 indicates the school year 1995–96, and so on.]

Year	Scholarships
1995	19.8
1996	22.0
1997	25.7
1998	27.5
1999	28.7
2000	31.1
2001	34.3

(a) Use linear regression to find an equation that expresses the amount of scholarships y as a function of the year x, with $x = 0$ corresponding to 1995.

(b) Assuming that the function in part (a) remains accurate, estimate the amount spent on scholarships in 2004.

Thinkers

41. Consider the function whose rule uses a calculator as follows: "Press COS, and then press LN; then enter a number in the domain, and press ENTER."† Experiment with this function, then answer the following questions. You may not be able to prove your answers—just make the best estimate you can based on the evidence from your experiments.

(a) What is the largest set of real numbers that could be used for the domain of this function? [If applying the rule to a number produces an error message or a complex number, that number cannot be in the domain.]

(b) Using the domain in part (a), what is the range of this function?

42. Do Exercise 41 for the function whose rule is "Press 10^x, and then press TAN; then enter a number in the domain, and press ENTER."

43. The *integer part* function has the set of all real numbers (written as decimals) as its domain. The rule is "For each input number, the output is the part of the number to the left of the decimal point." For instance, the input 37.986 produces the output 37, and the input -1.5 produces the output -1. On most calculators, the integer part function is denoted "iPart." On calculators that use "Intg" or "Floor" for the greatest integer function, the integer part function is denoted by "INT."

(a) For each nonnegative real number input, explain why both the integer part function and the greatest integer function (see Example 5) produce the same output.

(b) For which negative numbers do the two functions produce the same output?

(c) For which negative numbers do the two functions produce different outputs?

---

*U. S. Census Bureau.

†You don't need to know what these keys mean to do this exercise.

3.2   Functional Notation

Functional notation is a convenient shorthand language that facilitates the analysis of mathematical problems involving functions. It arises from real-life situations, such as the following.

EXAMPLE 1

In Section 3.1, we saw that the 2003 Louisiana state income tax rates (for a single person with one exemption and no deductions) were as follows.

Income		Tax
At Least	But Less Than	
0	$4,500	0
$4,500	$12,500	2% of amount over $4,500
$12,500	$25,000	$162.50 + 4% of amount over $12,500
$25,000		$662.50 + 6% of amount over $25,000

Let I denote income, and write $T(I)$ (read "T of I") to denote the amount of tax on income I. In this shorthand language, $T(7500)$ denotes "the tax on an income of 7500." The sentence "The tax on an income of $7500 is $60" is abbreviated as $T(7500) = 60$. Similarly, $T(35,000) = 1262.5$ says that the tax on an income of $35,000 is 1262.50. There is nothing that forces us to use the letters T and I here.

> **Any choice of letters will do, provided that we make clear what is meant by these letters.**   ■

EXAMPLE 2

Recall that a falling rock travels $16t^2$ feet after t seconds. Let $d(t)$ stand for the phrase "the distance the rock has traveled after t seconds." Then the sentence "The distance the rock has traveled after t seconds is $16t^2$ feet" can be abbreviated as $d(t) = 16t^2$. For instance,

$$d(1) = 16 \cdot 1^2 = 16$$

means "the distance the rock has traveled after 1 second is 16 feet," and

$$d(4) = 16 \cdot 4^2 = 256$$

means "the distance the rock has traveled after 4 seconds is 256 feet."   ■

Functional notation is easily adapted to mathematical settings, in which the particulars of time, distance, etc., are not mentioned. Suppose a function is given. Denote the function by f and let x denote a number in the domain. Then

> **$f(x)$ denotes the output produced by input x.**

CAUTION

The parentheses in $d(t)$ in Example 2 do *not* denote multiplication as in the algebraic equation $3(a + b) = 3a + 3b$. The entire symbol $d(t)$ is part of a *shorthand language.* In particular

$d(1 + 4)$ is *not* equal to $d(1) + d(4)$.

Example 2 shows that $d(1) = 16$ and $d(4) = 256$, so $d(1) + d(4) = 16 + 256 = 272$. But $d(1 + 4)$ is "the distance traveled after $1 + 4$ seconds," that is, the distance after 5 seconds, namely, $16 \cdot 5^2 = 400$. In general,

Functional notation is a convenient shorthand for phrases and sentences in the English language. It is *not* the same as ordinary algebraic notation.

For example, $f(6)$ is the output produced by the input 6. The sentence

"*y* is the output produced by input *x* according to the rule of the function *f*"

is abbreviated

$$y = f(x),$$

which is read "*y* equals *f* of *x*." The output $f(x)$ is sometimes called the **value** of the function *f* at *x*.

In actual practice, functions are seldom presented in the style of domain, rule, range, as they have been here. Usually, you will be given a phrase such as "the function $f(x) = \sqrt{x^2 + 1}$." This should be understood as a set of directions:

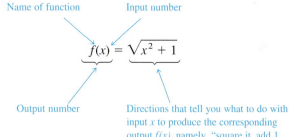

Name of function   Input number

$$f(x) = \sqrt{x^2 + 1}$$

Output number   Directions that tell you what to do with input *x* to produce the corresponding output *f(x)*, namely, "square it, add 1, and take the square root of the result."

TECHNOLOGY TIP

Functional notation can be used on calculators other than TI-85 and Casio. If the function is y_1 in the function memory, then

$$y_1(5) \quad \text{ENTER}$$

gives the value of the function at $x = 5$.

You can key in y_1 from the keyboard on TI-86/89 and HP-39+. On TI-83+, y_1 is in the FUNCTION submenu of the Y-VARS menu.

Warning: Keying in $y_1(5)$ on TI-85 or Casio will produce an answer, but it usually will be wrong.

For example, to find $f(3)$, the output of the function *f* for input 3, simply replace *x* by 3 in the formula.

$$f(x) = \sqrt{x^2 + 1}$$
$$f(3) = \sqrt{3^2 + 1} = \sqrt{10}.$$

Similarly, replacing *x* by -5 and 0 shows that

$$f(-5) = \sqrt{(-5)^2 + 1} = \sqrt{26} \qquad \text{and} \qquad f(0) = \sqrt{0^2 + 1} = 1.$$

EXAMPLE 3

The expression

$$h(x) = \frac{x^2 + 5}{x - 1}$$

defines the function *h* whose rule is

$$\text{For input } x, \text{ the output is the number } \frac{x^2 + 5}{x - 1}.$$

Find each of the following:

$$h(\sqrt{3}), \qquad h(-2), \qquad h(-a), \qquad h(r^2 + 3), \qquad h(\sqrt{c} + 2).$$

SOLUTION   To find $h(\sqrt{3})$ and $h(-2)$, replace *x* by $\sqrt{3}$ and -2, respectively, in the rule of *h*.

$$h(\sqrt{3}) = \frac{(\sqrt{3})^2 + 5}{\sqrt{3} - 1} = \frac{8}{\sqrt{3} - 1} \qquad \text{and} \qquad h(-2) = \frac{(-2)^2 + 5}{-2 - 1} = -3.$$

The value of the function h at any quantity, such as $-a$, $r^2 + 3$, etc., can be found by using the same procedure: *Replace x in the formula for $h(x)$ by that quantity.*

$$h(-a) = \frac{(-a)^2 + 5}{-a - 1} = \frac{a^2 + 5}{-a - 1}$$

$$h(r^2 + 3) = \frac{(r^2 + 3)^2 + 5}{(r^2 + 3) - 1} = \frac{r^4 + 6r^2 + 9 + 5}{r^2 + 2} = \frac{r^4 + 6r^2 + 14}{r^2 + 2}$$

$$h(\sqrt{c + 2}) = \frac{(\sqrt{c + 2})^2 + 5}{\sqrt{c + 2} - 1} = \frac{c + 2 + 5}{\sqrt{c + 2} - 1} = \frac{c + 7}{\sqrt{c + 2} - 1} \quad \blacksquare$$

When functional notation is used in expressions such as $f(-x)$ or $f(x + h)$, the same basic rule applies: Replace x in the formula by the *entire* expression in parentheses.

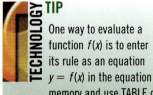

TECHNOLOGY TIP

One way to evaluate a function $f(x)$ is to enter its rule as an equation $y = f(x)$ in the equation memory and use TABLE or (on TI-85/86) EVAL; see Example 9 or Calculator Investigation 3 in Section 3.1.

EXAMPLE 4

If $f(x) = x^2 + x - 2$, then

$$f(-x) = (-x)^2 + (-x) - 2 = x^2 - x - 2.$$

Note that in this case, $f(-x)$ is *not* the same as $-f(x)$, because $-f(x)$ is the negative of the number $f(x)$, that is,

$$-f(x) = -(x^2 + x - 2) = -x^2 - x + 2. \quad \blacksquare$$

***C**

EXAMPLE 5

If $f(x) = x^2 - x + 2$ and $h \neq 0$, find

(a) $f(x + h)$ (b) $f(x + h) - f(x)$ (c) $\dfrac{f(x + h) - f(x)}{h}$.

SOLUTION

(a) Replace x by $x + h$ in the rule of the function.

$$f(x) = x^2 - x + 2$$

$$f(x + h) = (x + h)^2 - (x + h) + 2 = x^2 + 2xh + h^2 - x - h + 2$$

(b) By part (a),

$$f(x + h) - f(x) = [(x + h)^2 - (x + h) + 2] - [x^2 - x + 2]$$
$$= [x^2 + 2xh + h^2 - x - h + 2] - [x^2 - x + 2]$$
$$= x^2 + 2xh + h^2 - x - h + 2 - x^2 + x - 2$$
$$= 2xh + h^2 - h.$$

(c) By part (b), we have

$$\frac{f(x + h) - f(x)}{h} = \frac{2xh + h^2 - h}{h} = \frac{h(2x + h - 1)}{h} = 2x + h - 1. \quad \blacksquare$$

---

* **C** and ⬤ indicate examples, exercises, and sections that are relevant to calculus.

If f is a function, then the quantity $\dfrac{f(x+h)-f(x)}{h}$, as in Example 5(c), is called the **difference quotient** of f. Difference quotients, whose significance is explained in Special Topics 3.6.A, play an important role in calculus.

C

EXAMPLE 6

Compute and simplify the difference quotient for the function $f(x) = x^3 + 2x$.

SOLUTION

$$\frac{f(x+h)-f(x)}{h} = \frac{\overbrace{[(x+h)^3 + 2(x+h)]}^{f(x+h)} - \overbrace{[x^3 + 2x]}^{f(x)}}{h}$$

$$= \frac{[x^3 + 3x^2h + 3xh^2 + h^3 + 2x + 2h] - [x^3 + 2x]}{h}$$

$$= \frac{x^3 + 3x^2h + 3xh^2 + h^3 + 2x + 2h - x^3 - 2x}{h}$$

$$= \frac{3x^2h + 3xh^2 + h^3 + 2h}{h}$$

$$= \frac{h(3x^2 + 3xh + h^2 + 2)}{h} = 3x^2 + 3xh + h^2 + 2. \quad \blacksquare$$

As the preceding examples illustrate, functional notation is a specialized shorthand language. Treating it as ordinary algebraic notation can lead to mistakes.

EXAMPLE 7

CAUTION

Common Mistakes with Functional Notation

Each of the following statements may be FALSE:

1. $f(a+b) = f(a) + f(b)$
2. $f(a-b) = f(a) - f(b)$
3. $f(ab) = f(a)f(b)$
4. $f(ab) = af(b)$
5. $f(ab) = f(a)b$

Here are examples of three of the errors listed in the Caution box.

1. If $f(x) = x^2$, then

$$f(3+2) = f(5) = 5^2 = 25.$$

But

$$f(3) + f(2) = 3^2 + 2^2 = 9 + 4 = 13.$$

So $f(3+2) \neq f(3) + f(2)$.

3. If $f(x) = x + 7$, then

$$f(3 \cdot 4) = f(12) = 12 + 7 = 19.$$

But

$$f(3)f(4) = (3+7)(4+7) = 10 \cdot 11 = 110.$$

So $f(3 \cdot 4) \neq f(3)f(4)$.

5. If $f(x) = x^2 + 1$, then

$$f(2 \cdot 3) = (2 \cdot 3)^2 + 1 = 36 + 1 = 37.$$

But

$$f(2) \cdot 3 = (2^2 + 1)3 = 5 \cdot 3 = 15.$$

So $f(2 \cdot 3) \neq f(2) \cdot 3.$ ■

▦ DOMAINS

When the rule of a function is given by a formula, as in Examples 3–7, its domain (set of inputs) is determined by the following convention.

Domain Convention

Unless specific information to the contrary is given, the domain of a function f includes every real number (input) for which the rule of the function produces a real number as output.

Thus, the domain of a polynomial function such as $f(x) = x^3 - 4x + 1$ is the set of all real numbers, since $f(x)$ is defined for every value of x. In cases in which applying the rule of a function leads to division by zero or to the square root of a negative number, however, the domain may not consist of all real numbers.

EXAMPLE 8

Find the domain of the function given by

(a) $k(x) = \dfrac{x^2 - 6x}{x - 1}$

(b) $f(u) = \sqrt{u + 2}.$

SOLUTION

(a) When $x = 1$, the denominator of $\dfrac{x^2 - 6x}{x - 1}$ is 0, and the fraction is not defined. When $x \neq 1$, however, the denominator is nonzero, and the fraction *is* defined. Therefore, the domain of the function k consists of all real numbers *except* 1.

(b) Since negative numbers do not have real square roots, $\sqrt{u + 2}$ is a real number only when $u + 2 \geq 0$, that is, when $u \geq -2$. Therefore, the domain of f consists of all real numbers greater than or equal to -2.   ■

EXAMPLE 9

A **piecewise-defined** function is one whose rule includes several formulas, such as

$$f(x) = \begin{cases} 2x + 3 & \text{if } x < 4 \\ x^2 - 1 & \text{if } 4 \leq x \leq 10. \end{cases}$$

Find each of the following.

(a) $f(-5)$ (b) $f(8)$ (c) $f(k)$

(d) The domain of f.

SOLUTION

(a) Since $-5 < 4$, the first part of the rule applies.

$$f(-5) = 2(-5) + 3 = -7.$$

(b) Since 8 is between 4 and 10, the second part of the rule applies.

$$f(8) = 8^2 - 1 = 63.$$

(c) We cannot find $f(k)$ unless we know whether $k < 4$ or $4 \le k \le 10$.

(d) The rule of f gives no directions when $x > 10$, so the domain of f consists of all real numbers x with $x \le 10$. ■

EXAMPLE 10

Use Example 1 to write the rule of the piecewise-defined function T that gives the Louisiana state income tax $T(x)$ on an income of x dollars.

SOLUTION   By translating the information in the table in Example 1 into functional notation, we obtain

$$T(x) = \begin{cases} 0 & \text{if } 0 \le x < 4500 \\ .02(x - 4500) & \text{if } 4500 \le x < 12{,}500 \\ 162.50 + .04(x - 12{,}500) & \text{if } 12{,}500 \le x < 25{,}000 \\ 662.50 + .06(x - 25{,}000) & \text{if } x \ge 25{,}000. \end{cases}$$ ■

■■ APPLICATIONS

The domain convention does not always apply when dealing with applications. Consider, for example, the distance function for falling objects, $d(t) = 16t^2$ (see Example 2). Since t represents time, only nonnegative values of t make sense here, even though the rule of the function is defined for all values of t. Analogous comments apply to other applications.

A real-life situation may lead to a function whose domain does not include all the numbers for which the rule of the function is defined.

EXAMPLE 11

A glassware factory has fixed expenses (mortgage, taxes, machinery, etc.) of $12,000 per week. It costs 80 cents to make one cup (labor, materials, shipping). A cup sells for $1.95. At most 18,000 cups can be manufactured and sold each week.

(a) Express the weekly revenue as a function of the number x of cups made.

(b) Express the weekly costs as a function of x.

(c) Find the domain and the rule of the weekly profit function.

SOLUTION

(a) If $R(x)$ is the weekly revenue from selling x cups, then

$$R(x) = \text{(price per cup)} \times \text{(number sold)}$$

$$R(x) = 1.95x.$$

(b) If $C(x)$ is the weekly cost of manufacturing x cups, then

$$C(x) = \text{(cost per cup)} \times \text{(number sold)} + \text{(fixed expenses)}$$

$$C(x) = .80x + 12{,}000.$$

(c) If $P(x)$ is the weekly profit from selling x cups, then

$$P(x) = \text{Revenue} - \text{Cost}$$

$$P(x) = R(x) - C(x)$$

$$P(x) = 1.95x - (.80x + 12{,}000) = 1.95x - .80x - 12{,}000$$

$$P(x) = 1.15x - 12{,}000.$$

Although this rule is defined for all real numbers x, the domain of the function P consists of the possible number of cups that can be made each week. Since you can make only whole cups and the maximum production is 18,000, the domain of P consists of all integers from 0 to 18,000. ■

EXAMPLE 12

Let P be the profit function in Example 11.

(a) What is the profit from selling 5000 cups? From 14,000 cups?

(b) What is the break-even point?

SOLUTION

(a) We evaluate the function $P(x) = 1.15x - 12{,}000$ at the required values of x.

$$P(5000) = 1.15(5000) - 12{,}000 = -\$6250$$

$$P(14{,}000) = 1.15(14{,}000) - 12{,}000 = \$4100$$

Thus, sales of 5000 cups produce a loss of $6250, while sales of 14,000 produce a profit of $4100.

(b) The break-even point occurs when revenue equals costs (that is, when profit is 0). So we set $P(x) = 0$ and solve for x.

$$1.15x - 12{,}000 = 0$$

$$1.15x = 12{,}000$$

$$x = \frac{12{,}000}{1.15} \approx 10{,}434.78$$

Thus, the break-even point occurs between 10,434 and 10,435 cups. There is a slight loss from selling 10,434 cups and a slight profit from selling 10,435. ■

✓ EXERCISES 3.2 📼

In Exercises 1 and 2, find the indicated values of the function by hand and by using the table feature of a calculator (or the EVAL key on TI-85/86). If your answers do not agree with each other or with those at the back of the book, you are either making algebraic mistakes or incorrectly entering the function in the equation memory.

1. $f(x) = \dfrac{x-3}{x^2+4}$

(a) $f(-1)$ (b) $f(0)$ (c) $f(1)$ (d) $f(2)$ (e) $f(3)$

2. $g(x) = \sqrt{x+4} - 2$

(a) $g(-2)$ (b) $g(0)$ (c) $g(4)$ (d) $g(5)$ (e) $g(12)$

Exercises 3–24 refer to these three functions:

$$f(x) = \sqrt{x+3} - x + 1 \qquad g(t) = t^2 - 1$$

$$h(x) = x^2 + \frac{1}{x} + 2$$

In each case, find the indicated value of the function.

3. $f(0)$ | **4.** $f(1)$
5. $f(\sqrt{2})$ | **6.** $f(\sqrt{2} - 1)$
7. $f(-2)$ | **8.** $f(-3/2)$
9. $h(3)$ | **10.** $h(-4)$
11. $h(3/2)$ | **12.** $h(\pi + 1)$
13. $h(a + k)$ | **14.** $h(-x)$
15. $h(2 - x)$ | **16.** $h(x - 3)$
17. $g(3)$ | **18.** $g(-2)$
19. $g(0)$ | **20.** $g(x)$
21. $g(s + 1)$ | **22.** $g(1 - r)$
23. $g(-t)$ | **24.** $g(t + h)$

25. If $f(x) = x^3 + cx^2 + 4x - 1$ for some constant c and $f(1) = 2$, find c. [*Hint:* Use the rule of f to compute $f(1)$.]

26. If $g(x) = \sqrt{cx - 4}$ and $g(8) = 6$, find c.

27. If $f(x) = \dfrac{dx - 5}{x - 3}$ and $f(4) = 3$, find d.

28. If $g(x) = \dfrac{dx + 1}{x + 2}$ and $g(3) = .2$, find d.

In Exercises 29–32, find the values of x for which $f(x) = g(x)$.

29. $f(x) = 2x^2 + 4x - 3$; $g(x) = x^2 + 12x + 7$

30. $f(x) = 2x^2 + 12x - 14$; $g(x) = 7x - 2$

31. $f(x) = 3x^2 - x + 5$; $g(x) = x^2 - 2x + 26$

32. $f(x) = 2x^2 - x + 1$; $g(x) = x^2 - 4x + 4$

C

In Exercises 33–42, assume that $h \neq 0$. Compute and simplify the difference quotient

$$\frac{f(x+h) - f(x)}{h}.$$

33. $f(x) = x + 1$ | **34.** $f(x) = -10x$
35. $f(x) = 3x + 7$ | **36.** $f(x) = x^2$
37. $f(x) = x - x^2$ | **38.** $f(x) = x^3$
39. $f(x) = \sqrt{x}$ | **40.** $f(x) = 1/x$
41. $f(x) = x^2 + 3$ | **42.** $f(x) = x^2 + 2x + 1$

43. In each part, compute $f(a)$, $f(b)$, and $f(a + b)$ and determine whether the statement "$f(a + b) = f(a) + f(b)$" is true or false for the given function.

(a) $f(x) = x^2$ (b) $f(x) = 3x$

44. In each part, compute $g(a)$, $g(b)$, and $g(ab)$ and determine whether the statment "$g(ab) = g(a) \cdot g(b)$" is true or false for the given function.

(a) $g(x) = x^3$ (b) $g(x) = 5x$

45. If $f(x) = \begin{cases} x^2 + 2x & \text{if } x < 2 \\ 3x - 5 & \text{if } 2 \le x \le 20, \end{cases}$ find

(a) the domain of f

(b) $f(-3)$ (c) $f(-1)$ (d) $f(2)$ (e) $f(7/3)$

46. If $g(x) = \begin{cases} 2x - 3 & \text{if } x < -1 \\ |x| - 5 & \text{if } -1 \le x \le 2 \\ x^2 & \text{if } x > 2, \end{cases}$ find

(a) the domain of g

(b) $g(-2.5)$ (c) $g(-1)$ (d) $g(2)$ (e) $g(4)$

47. In a certain state, the sales tax $T(p)$ on an item of price p dollars is 5% of p. Which of the following formulas give the correct sales tax in all cases?

(i) $T(p) = p + 5$
(ii) $T(p) = 1 + 5p$
(iii) $T(p) = p/20$
(iv) $T(p) = p + (5/100)p = p + .05p$
(v) $T(p) = (5/100)p = .05p$

48. Let T be the sales tax function of Exercise 47 and find $T(3.60)$, $T(4.80)$, $T(60)$, an $T(0)$.

In Exercises 49–62, determine the domain of the function according to the usual convention.

49. $f(x) = x^2$

50. $g(x) = \dfrac{1}{x^2} + 2$

51. $h(t) = |t| - 1$

52. $k(u) = \sqrt{u}$

53. $k(x) = |x| + \sqrt{x - 1}$

54. $h(x) = \sqrt{(x + 1)^2}$

55. $g(u) = \dfrac{|u|}{u}$

56. $h(x) = \dfrac{\sqrt{x - 1}}{x^2 - 1}$

57. $g(y) = [-y]$

58. $f(t) = \sqrt{-t}$

59. $g(u) = \dfrac{u^2 + 1}{u^2 - u - 6}$

60. $f(t) = \sqrt{4 - t^2}$

61. $f(x) = -\sqrt{9 - (x - 9)^2}$

62. $f(x) = \sqrt{-x} + \dfrac{2}{x + 1}$

63. Give an example of two different functions f and g that have all of the following properties:

$$f(-1) = 1 = g(-1) \quad \text{and} \quad f(0) = 0 = g(0)$$
$$\text{and} \quad f(1) = 1 = g(1).$$

64. Give an example of a function g with the property that $g(x) = g(-x)$ for every real number x.

In Exercises 65–68, the rule of a function f is given. Write an algebraic formula for f(x).

65. Double the input, subtract 5, and take the square root of the result.

66. Square the input, multiply by 3, and subtract the result from 8.

67. Cube the input, add 6, and divide the result by 5.

68. Take the square root of the input, add 7, divide the result by 8, and add this result to the original input.

69. The table shows the 2004 federal income tax rates for a single person.

| Taxable Income | | Tax |
Over	But Not Over	
0	$7150	10% of income
$7150	$29,050	$715 + 15% of amount over $7150
$29,050	$70,350	$4,000 + 25% of amount over $29,050
$70,350	$146,750	$14,325 + 28% of amount over $70,350
$146,750	$319,100	$35,717 + 33% of amount over $146,750
$319,100		$92,592.50 + 35% of amount over $319,100

(a) Write the rule of a piecewise-defined function T such that $T(x)$ is the tax due on a taxable income of x dollars.

(b) Find $T(24{,}000)$, $T(35{,}000)$, and $T(100{,}000)$.

70. According to the National Center for Health Statistics, the number of marriages in the United States was 1,523,000 in 1960, 2,443,000 in 1990, and 2,327,000 in 2001. Here are three functions that model this data (where $x = 0$ corresponds to 1960 and $f(x)$ is in thousands):

$$f(x) = 15.9x + 1822$$
$$f(x) = -1.2x^2 + 65.8x + 1600$$
$$f(x) = .02x^3 - 2.4x^2 + 84.2x + 1526.$$

(a) Make a table or otherwise evaluate each function at the values of x corresponding to 1960, 1990, and 2001. Determine which function provides the best model for the given data.

(b) Use the function determined in part (a) to estimate the number of marriages in 2000.

71. Data and projections from the International Trade Administration indicate that the number of foreign tourists visiting the United States was 43.3 million in 1995 and 50.9 million in 2000 and will be 56.3 million in 2005. This data can be modeled by the following functions (where $x = 5$ corresponds to 1995 and $g(x)$ is in millions):

$$g(x) = 1.1x + 39$$
$$g(x) = -.005x^2 + 1.1x + 44$$
$$g(x) = .07x^3 - 2x^2 + 20x - 14.8.$$

(a) Make a table or otherwise evaluate each function at the values of x corresponding to 1995, 2000, and 2005. Determine which function provides the best model for the given data.

(b) Use the function determined in part (a) to estimate the number of foreign tourists in 2006.

72. Suppose a car travels at a constant rate of 55 mph for two hours and travels 45 mph thereafter. Find the rule of the function that expresses the distance traveled as a function of time.

73. Jack and Jill are salespeople in the suit department of a clothing store. Jack is paid $200 per week plus $5 for each suit he sells, whereas Jill is paid $10 for every suit she sells.

(a) Let $f(x)$ denote Jack's weekly income, and let $g(x)$ denote Jill's weekly income from selling x suits. Find the rules of the functions f and g.

(b) Use algebra or a table to find: $f(20)$ and $g(20)$; $f(35)$ and $g(35)$; $f(50)$ and $g(50)$.

(c) If Jack sells 50 suits a week, how many must Jill sell to have the same income as Jack?

74. A potato chip factory has a daily overhead from salaries and building costs of $1800. The cost of ingredients and packaging to produce a pound of potato chips is 50¢. A pound of potato chips sells for $1.20. Show that the factory's daily profit is a function of the number of pounds of potato chips sold and find the rule of this function. (Assume that the factory sells all the potato chips it produces each day.)

75. A rectangular region of 6000 sq ft is to be fenced in on three sides with fencing costing $3.75 per foot and on the fourth side with fencing costing $2.00 per foot. Express the cost of the fence as a function of the length x of the fourth side.

76. A box with a square base measuring $t \times t$ ft is to be made of three kinds of wood. The cost of the wood for the base is 85¢ per square foot; the wood for the sides costs 50¢ per square foot, and the wood for the top costs $1.15 per square foot. The volume of the box is to be 10 cubic feet. Express the total cost of the box as a function of the length t.

77. Suppose that the width and height of the box in the figure are equal and that the sum of the length and the girth is 108 (the maximum size allowed by the post office).

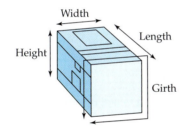

(a) Express the length y as a function of the width x. [*Hint:* Use the girth.]
(b) Express the volume V of the box as a function of the width x. [*Hint:* Find a formula for the volume and use part (a).]

78. A person who needs crutches can determine the correct length as follows: a 50-inch-tall person needs a 38-inch-long crutch. For each additional inch in the person's height, add .72 inch to the crutch length.
(a) If a person is y inches taller than 50 inches, write an expression for the proper crutch length.
(b) Write the rule of a function f such that $f(x)$ is the proper crutch length (in inches) for a person who is x inches tall. [*Hint:* Replace y in your answer to part (a) with an expression in x. How are x and y related?]

79. Average tuition and fees in private four-year colleges in recent years were as follows.[*]

Year	Tuition and Fees
1996	$15,605
1997	$16,552
1998	$17,229
1999	$18,340
2000	$19,307
2001	$20,143

(a) Use linear regression to find the rule of a function f that gives the approximate average tuition in year x, where $x = 0$ corresponds to 1990.
(b) Find $f(7), f(9)$, and $f(11)$. How do they compare with the actual figures?
(c) Use f to estimate tuition and fees in 2005.

80. The number of U.S. commercial radio stations whose primary format is top 40 hits has been increasing in recent years, as shown in the table.[†]

Year	Number of Stations
1995	318
1996	333
1997	358
1998	379
1999	401
2001	468
2002	474

(a) Use linear regression to find the rule of a function g that gives the number of top-40 stations in year x, where $x = 0$ corresponds to 1990.
(b) Find $g(7)$ and $g(11)$. How do they compare with the actual figures?
(c) Data for the year 2000 is missing. Estimate the number of stations in 2000.
(d) Assuming that this function remains accurate, estimate the number of stations in 2006.

*U.S. National Center for Education Statistics.
†*World Almanac and Book of Facts, 2003.*

3.3 Graphs of Functions

The graph of a function f is the graph of the equation $y = f(x)$. Hence,

The graph of the function f consists of the points $(x, f(x))$ for every number x in the domain of f.

When the rule of a function is given by an algebraic formula, the graph is easily obtained with technology. However, machine-generated graphs can sometimes be incomplete or misleading. So the emphasis here is on using your algebraic knowledge *before* reaching for a calculator. Doing so will often tell you that a calculator is inappropriate or help you to interpret screen images when a calculator is used.

Some functions appear so frequently that you should memorize the shapes of their graphs, which are easily obtained by hand-graphing or by using technology. Regardless of how you first find these graphs, you should be able to reproduce them without looking them up or resorting to technology. These basic graphs are summarized in the **catalog of functions** at the end of this section. A title on an example (such as "linear functions" in Example 1) indicates a function that is in the catalog.

EXAMPLE 1   **Linear Functions**

The graph of a function of the form

$$f(x) = mx + b \qquad \text{(with m and b constants)}$$

is the graph of the equation $y = mx + b$. As we saw in Section 1.3, the graph is a straight line with slope m and y-intercept b that can easily be obtained by hand. Some typical linear functions are graphed in Figure 3–7 (some of them have special names). On a calculator, these graphs will usually look a bit bumpy and jagged.   ■

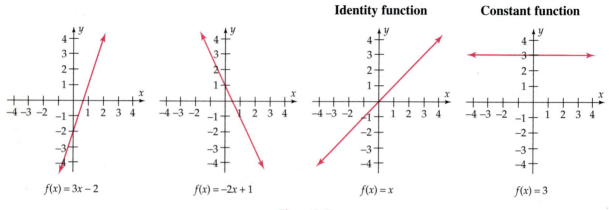

$f(x) = 3x - 2$ \qquad $f(x) = -2x + 1$ \qquad **Identity function** $f(x) = x$ \qquad **Constant function** $f(x) = 3$

Figure 3–7

EXAMPLE 2   **Square and Cube Functions**

Figure 3–8 shows the graphs of $f(x) = x^2$ and $g(x) = x^3$. They can be obtained by plotting points (as was done for $y = x^2$ in Example 1 of Section 1.2) or by using technology. ■

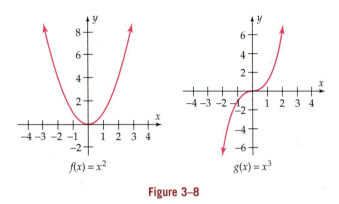

$f(x) = x^2$ $g(x) = x^3$

Figure 3–8

EXAMPLE 3   **Square Root Function**

The graph of $f(x) = \sqrt{x}$ in Figure 3–9 is easily found. Why does it lie entirely in the first quadrant? ■

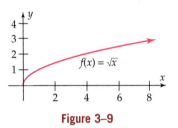

Figure 3–9

▪▪ ▪ STEP FUNCTIONS

The greatest integer function $f(x) = [x]$ was introduced in Example 5 of Section 3.1. It can easily be graphed by hand, as in the next example.

EXAMPLE 4   **Greatest Integer Function**

The function $f(x) = [x]^*$ can easily be graphed by hand by considering the values of the function between each two consecutive integers. For instance,

x	$-2 \leq x < -1$	$-1 \leq x < 0$	$0 \leq x < 1$	$1 \leq x < 2$	$2 \leq x < 3$
$[x]$	-2	-1	0	1	2

Thus, between $x = -2$ and $x = -1$, the value of $f(x) = [x]$ is always -2, so the graph there is a horizontal line segment, all of whose points have second coordinate -2. The rest of the graph is obtained similarly (Figure 3–10). An open circle in Figure 3–10 indicates that the endpoint of the segment is *not* on the graph, whereas a closed circle indicates that the endpoint is on the graph. ■

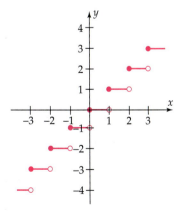

Figure 3–10

A function whose graph consists of horizontal line segments, such as Figure 3–10, is called a **step function.** Graphing step functions with reasonable

---

*See Example 5 in Section 3.1.

accuracy on a calculator requires some care. Even then, some features of the graph might not be shown.

GRAPHING EXPLORATION

Graph the greatest integer function $f(x) = [x]$ on your calculator (see the Technology Tip on page 195). Does your graph look like Figure 3–10, or does it include vertical segments? Now change the graphing mode of your calculator to "dot" rather than "connected" (see the Technology Tip in the margin), and graph again. How does this graph compare with Figure 3–10? Can you tell from the graph which endpoints are included and which are excluded?

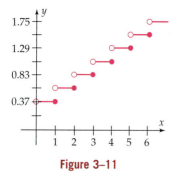

Figure 3–11

EXAMPLE 5

At this writing, first-class postage rates are 37¢ for the first ounce and 23¢ for each additional ounce or fraction thereof. Verify that the postage $P(x)$ for a letter weighing x ounces is given by $P(x) = .37 - .23[1 - x]$. For instance, the postage for a 2.5-ounce letter is

$$P(2.5) = .37 - .23[1 - 2.5] = .37 - .23(-2) = .83$$

Although the rule of P makes sense for all real numbers, the domain of the function consists of positive numbers (why?). The graph of P is in Figure 3–11. ∎

▪▪ PIECEWISE-DEFINED FUNCTIONS

Piecewise-defined functions were introduced in Example 9 of Section 3.2. Graphing them correctly requires some care.

EXAMPLE 6

The graph of the piecewise-defined function

$$f(x) = \begin{cases} x^2 & \text{if } x \le 1 \\ x + 2 & \text{if } 1 < x \le 4 \end{cases}$$

is made up of *parts* of two graphs, corresponding to the different parts of the rule of the function:

$x \le 1$ For these values of x, the graph of f coincides with the graph of $y = x^2$, which was sketched in Figure 3–8 (page 213).

$1 < x \le 4$ For these values of x, the graph of f coincides with the graph of $y = x + 2$, which is a straight line.

Therefore, we must graph

$$y = x^2 \quad \text{when } x \le 1 \qquad \text{and} \qquad y = x + 2 \quad \text{when } 1 < x \le 4.$$

Combining these partial graphs produces the graph of *f* in Figure 3–12.   ■

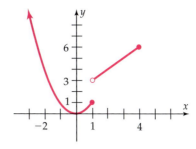

Figure 3–12

Piecewise-defined functions can be graphed on a calculator, provided that you use the correct syntax. Once again, however, the screen does not show which endpoints are included or excluded from the graph.

TECHNOLOGY TIP

Inequality symbols can be found in the following menu/submenu:

TI-83+/86: TEST

TI-89: MATH/TEST

HP-39+: MATH/TESTS

GRAPHING EXPLORATION

Use the Technology Tip in the margin and the directions here to graph the function *f* of Example 6 on a calculator. On TI-83+/86 and HP-39+, graph these two equations on the same screen:

$$y_1 = \frac{x^2}{(x \le 1)}$$

$$y_2 = \frac{(x + 2)}{(x > 1)(x \le 4)}.$$

On TI-89, graph the following equations on the same screen (the symbol | is on the keyboard; "and" is in the TESTS submenu of the MATH menu):

$$y_1 = x^2 | x \le 1$$

$$y_2 = x + 2 | x > 1 \quad \text{and} \quad x \le 4.$$

To graph *f* on Casio, with the viewing window of Figure 3–12, graph these equations on the same screen (including commas and square brackets):

$$y_1 = x^2, [-6, 1]$$

$$y_2 = x + 2, [1, 4].$$

How does your graph compare with Figure 3–12?

EXAMPLE 7   **Absolute Value Function**

The absolute value function $f(x) = |x|$ is also a piecewise-defined function, since by definition,

$$|x| = \begin{cases} x & \text{if } x \ge 0 \\ -x & \text{if } x < 0. \end{cases}$$

Its graph can be obtained by drawing the part of the line $y = x$ to the right of the origin and the part of the line $y = -x$ to the left of the origin (Figure 3–13) or by graphing $y = \text{ABS } x$ on a calculator (Figure 3–14).   ■

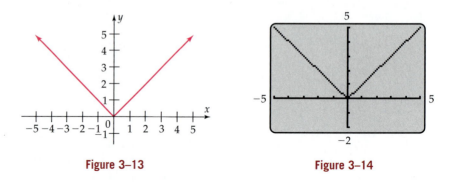

Figure 3–13   Figure 3–14

▪▪ LOCAL MAXIMA AND MINIMA

The graph of a function may include some peaks and valleys (Figure 3–15). A peak is not necessarily the highest point on the graph, but it is the highest point in its neighborhood. Similarly, a valley is the lowest point in the neighborhood but not necessarily the lowest point on the graph.

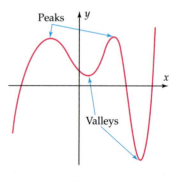

Figure 3–15

More formally, we say that a function f has a **local maximum** at $x = c$ if the graph of f has a peak at the point $(c, f(c))$. This means that all nearby points $(x, f(x))$ have smaller y-coordinates, that is,

$$f(x) \le f(c) \qquad \text{for all } x \text{ near } c.$$

Similarly, a function has a **local minimum** at $x = d$ provided that

$$f(x) \ge f(d) \qquad \text{for all } x \text{ near } d.$$

In other words, the graph of f has a valley at $(d, f(d))$, since all nearby points $(x, f(x))$ have larger y-coordinates.

Calculus is usually needed to find the exact location of local maxima and minima (the plural forms of maximum and minimum). However, they can be accurately approximated by the maximum finder or minimum finder of a calculator.

EXAMPLE 8

The graph of $f(x) = x^3 - 1.8x^2 + x + 1$ in Figure 3–16 does not appear to have any local maxima or minima. However, if you use the trace feature to move along the flat segment to the right of the y-axis, you find that the y-coordinates increase, then decrease, then increase (try it!). To see what's really going on, we change viewing windows (Figure 3–17) and see that the function actually has a local maximum and a local minimum (Figure 3–18). The calculator's minimum finder shows that the local minimum occurs when $x \approx .7633$. ∎

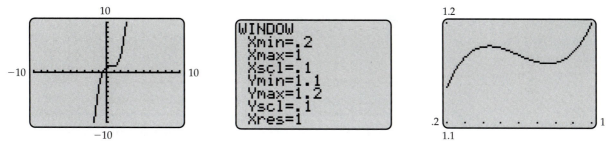

Figure 3–16   **Figure 3–17**   **Figure 3–18**

GRAPHING EXPLORATION

Graph the function in Example 8 in the viewing window of Figure 3–18. Use the maximum finder to approximate the location of the local maximum.

EXAMPLE 9

U.S. military troop levels (in millions) during 1980–2003 can be approximated by the function

$$g(x) = (2.063 \times 10^{-5})x^4 - (5.9 \times 10^{-4})x^3$$
$$- (4.88 \times 10^{-4})x^2 + .047x + 2.052,$$

where $x = 0$ corresponds to 1980. During this period, when were troop levels at their highest?

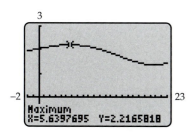

Figure 3–19

SOLUTION   We graph the function and use the maximum finder to determine that the local maximum occurs when $x \approx 5.640$ and $g(5.640) \approx 2.217$, as shown in Figure 3–19. Therefore, troop levels reached their highest level (about 2,217,000) in late 1985. ∎

▉▉ INCREASING AND DECREASING FUNCTIONS

A function is said to be **increasing on an interval** if its graph always rises as you move from left to right over the interval. It is **decreasing on an interval** if its graph always falls as you move from left to right over the interval. A function is said to be **constant on an interval** if its graph is horizontal over the interval.

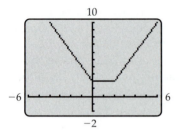

Figure 3–20

EXAMPLE 10

The graph of $f(x) = |x| + |x - 2|$ in Figure 3–20 suggests that f is decreasing when $x < 0$, increasing when $x > 2$, and constant between 0 and 2. You can confirm that the function is actually constant between 0 and 2 by using the trace feature to move along the graph there (the y-coordinates remain the same, as they should on a horizontal segment). For an algebraic proof that f is constant between 0 and 2, see Exercise 18. ■

CAUTION

A horizontal segment on a calculator graph does not always mean that the function is constant there. There may be **hidden behavior,** as was the case in Example 8. When in doubt, use the trace feature to see whether the y-coordinates remain constant as you move along the "horizontal" segment, or change the viewing window.

EXAMPLE 11

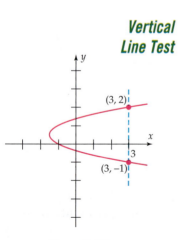

Figure 3–21

On what (approximate) intervals is the function $g(x) = .5x^3 - 3x$ increasing or decreasing?

SOLUTION   The (complete) graph of g in Figure 3–21 shows that g has a local maximum at P and a local minimum at Q. The maximum and minimum finders show that the approximate coordinates of P and Q are

$$P = (-1.4142, 2.8284) \quad \text{and} \quad Q = (1.4142, -2.8284).$$

Therefore, f is increasing when $x < -1.4142$ and when $x > 1.4142$. It is decreasing when $-1.4142 < x < 1.4142$. ■

▪▪ THE VERTICAL LINE TEST

The following fact, which distinguishes graphs of functions from other graphs, can also be used to interpret some calculator-generated graphs.

Vertical Line Test

The graph of a function $y = f(x)$ has this property:

No vertical line intersects the graph more than once.

Conversely, any graph with this property is the graph of a function.

Figure 3–22

To see why this is true, consider Figure 3–22, in which the graph intersects the vertical line at two points. If this were the graph of a function f, then we would have $f(3) = 2$ [because $(3, 2)$ is on the graph] *and* $f(3) = -1$ [because $(3, -1)$ is on the graph]. This means that the input 3 produces two different outputs, which is impossible for a function. Therefore, Figure 3–22 is not the graph of a function. A similar argument works in the general case.

Care must be used when applying the Vertical Line Test to a calculator graph, as the following example illustrates.

EXAMPLE 12

The rule $g(x) = x^{15} + 2$ does give a function (each input produces exactly one output), but its graph in Figure 3–23 appears to be a vertical line near $x = 1$. To see why this is not actually true, change the viewing window. Figure 3–24 shows that the graph is not vertical between $x = 1$ and $x = 1.2$. This detail is lost in Figure 3–23 because all the x values in Figure 3–24 occupy only a single pixel width in Figure 3–23. ■

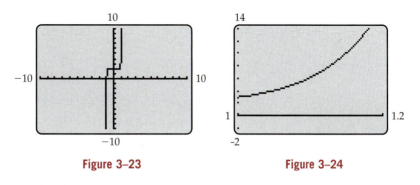

Figure 3–23   Figure 3–24

EXPLORATION

GRAPHING

Find a viewing window that shows that the graph of the function g in Example 12 is not actually vertical near $x = -1$.

■■ GRAPH READING

Until now, we have concentrated on translating statements into functional notation and functional notation into graphs. It is just as important, however, to be able to translate graphical information into equivalent statements in English or functional notation.

EXAMPLE 13

The entire graph of a function f is shown in Figure 3–25. Find the domain and range of f.

SOLUTION   The graph of f consists of all points of the form $(x, f(x))$. Thus, the first coordinates of points on the graph are the inputs (numbers in the domain of f), and the second coordinates are the outputs (the numbers in the range of f). Figure 3–25 shows that the first coordinates of points on the graph all satisfy $-1 \leq x \leq 1$, so these numbers are the domain of f. Similarly, the range of f consists of all numbers y such that $0 \leq y \leq \pi$, because these are the second coordinates of points on the graph. ■

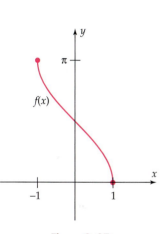

Figure 3–25

EXAMPLE 14

The consumer confidence level reflects people's feelings about their employment opportunities and income prospects. Let $C(t)$ be the consumer confidence level at time t (with $t = 0$ corresponding to 1970) and consider the graph of the function C in Figure 3–26.*

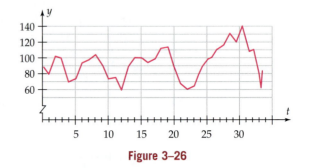

Figure 3–26

(a) How did the consumer confidence level vary in the 1980s?

(b) What was the lowest level of consumer confidence during the 1990s?

(c) During what time periods was the confidence level above 110?

SOLUTION

(a) The 1980s correspond to the interval $10 \le t \le 20$, so we consider the part of the graph that lies between the vertical lines $t = 10$ and $t = 20$. The second coordinates of these points range from approximately 60 to 114. So the consumer confidence level varied from a low of 60 to a high of 114 during the 1980s.

(b) The 1990s correspond to the interval $20 \le t \le 30$. Figure 3–26 shows that the graph has local minimums at $t = 22$ and $t = 29$. The lowest of these two points is the one at $t = 22$. Hence, the lowest level of consumer confidence in the 1990s occurred at the beginning of 1992.

(c) We must find the values of t for which the graph lies above the horizontal line through 110. Figure 3–26 shows that this occurs approximately when $17.5 \le t \le 19.5$ and when $26.5 \le t \le 31.5$. Thus, the confidence level was above 110 from the middle of 1987 to the middle of 1989 and from the middle of 1996 to the middle of 2001. ■

EXAMPLE 15

Use the graphs of the functions g and h in Figure 3–27 to find:

(a) All numbers x such that $h(x) < 0$;

---

*The consumer confidence level is scaled to be 100 in 1985.

(b) All numbers x with $-3 \le x \le 3$ such that $g(x) = 2$;

(c) The largest interval over which g is increasing and h is deceasing and $h(x) \ge g(x)$ for every x in the interval.

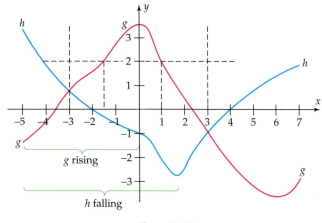

Figure 3–27

SOLUTION

(a) The graph of h consists of all points $(x, h(x))$. The numbers such that $h(x) < 0$ correspond to the points on the graph with negative y-coordinates, that is, points that lie below the x-axis. The graph of h shows that these points are the ones with $-2 < x < 4$.

(b) The graph of g consists of the points $(x, g(x))$. The points on the graph that lie *between* the vertical lines through -3 and 3 and *on* the horizontal line through 2 have x-coordinates satisfying $-3 \le x \le 3$ and y-coordinate $g(x) = 2$. Figure 3–27 shows that the only such points are $(-1.5, 2)$ and $(1, 2)$. So the answer is $x = -1.5$ and $x = 1$.

(c) Figure 3–27 shows that the only place where the graph of h is falling *and* the graph of g is rising is where $-5 \le x \le 0$. Now $h(x) \ge g(x)$ when the point $(x, h(x))$ lies above the point $(x, g(x))$. This occurs when $-5 \le x \le -3$. ■

▦ CATALOG OF BASIC FUNCTIONS

As we noted at the beginning of this section, there are a number of functions whose graphs you should know by heart. The ones in this section are listed in the table on the next page; others will be added as we go along. The entire catalog appears on the inside front cover of this book.

CATALOG OF BASIC FUNCTIONS—PART 1

Linear Functions

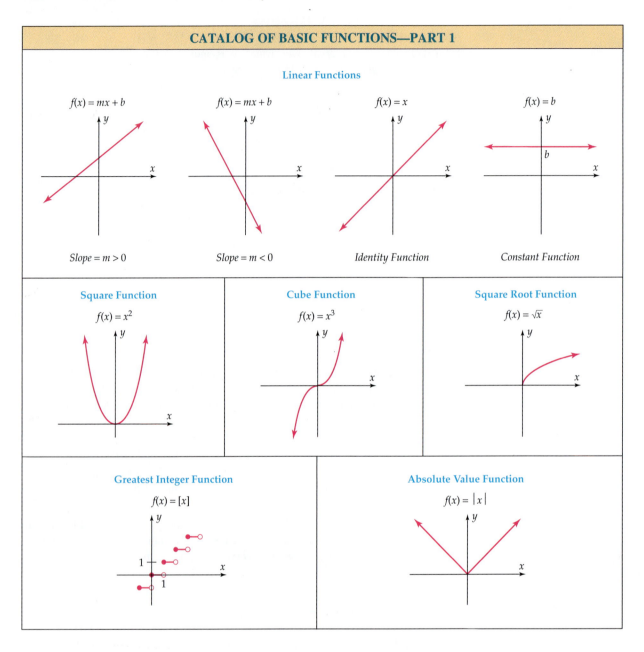

$f(x) = mx + b$

Slope = m > 0

$f(x) = mx + b$

Slope = m < 0

$f(x) = x$

Identity Function

$f(x) = b$

Constant Function

Square Function

$f(x) = x^2$

Cube Function

$f(x) = x^3$

Square Root Function

$f(x) = \sqrt{x}$

Greatest Integer Function

$f(x) = [x]$

Absolute Value Function

$f(x) = |x|$

EXERCISES 3.3

In Exercises 1–4, state whether or not the graph is the graph of a function. If it is, find $f(3)$.

1.

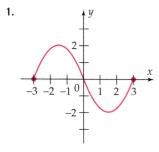

2.

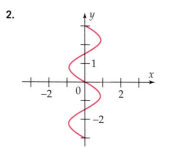

3.

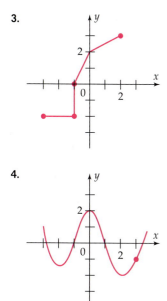

4.

In Exercises 5–12, sketch the graph of the function, being sure to indicate which endpoints are included and which ones are excluded.

5. $f(x) = 2[x]$

6. $f(x) = -[x]$

7. $g(x) = [-x]$ [This is *not* the same function as in Exercise 6.]

8. $h(x) = [x] + [-x]$

9. $f(x) = \begin{cases} x^2 & \text{if } x \geq -1 \\ 2x + 3 & \text{if } x < -1 \end{cases}$

10. $g(x) = \begin{cases} |x| & \text{if } x < 1 \\ -3x + 4 & \text{if } x \geq 1 \end{cases}$

11. $k(u) = \begin{cases} -2u - 2 & \text{if } u < -3 \\ u - [u] & \text{if } -3 \leq u \leq 1 \\ 2u^2 & \text{if } u > 1 \end{cases}$

12. (a) Sketch the graph of the following function, which gives the approximate percentage increase in health insurance premiums in year x, where $x = 0$ corresponds to 1988.[*]

$$f(x) = \begin{cases} 6x + 12 & \text{if } 0 \leq x \leq 1 \\ -2.5x + 20.5 & \text{if } 1 < x \leq 8 \\ 2x - 15.5 & \text{if } 8 < x \leq 14. \end{cases}$$

For example, $x = 5$ corresonds to 1993 and $f(5) = 8$, which means that premiums increased 8% in 1993.

(b) Does your graph show that there was a time interval during this period when premiums were decreasing? Explain your answer.

13. At this writing, first-class postage rates are 37¢ for the first ounce plus 23¢ for each additional ounce or fraction thereof. Assume that each first-class letter carries one 37¢ stamp and as many 23¢ stamps as are necessary. Then the *number* of stamps required for a first-class letter is a function of the weight of the letter in ounces. Call this function the *postage stamp function.*

(a) Describe the rule of the postage stamp function algebraically.

(b) Sketch the graph of the postage stamp function.

(c) Sketch the graph of the function whose rule is

$$f(x) = p(x) - [x],$$

where p is the postage stamp function.

14. A common mistake is to graph the function f in Example 6 by graphing both $y = x^2$ and $y = x + 2$ on the same screen (with no restrictions on x). Explain why this graph could not possibly be the graph of a function.

[*]Based on data from the Health Research and Education Trust of the Kaiser Family Foundation.

In Exercises 15–17, (a) Use the fact that the absolute value function is piecewise-defined (see Example 7) to write the rule of the given function as a piecewise-defined function whose rule does not include any absolute value bars. (b) Graph the function.

15. $f(x) = |x| + 2$

16. $g(x) = |x| - 4$

17. $g(x) = |x + 3|$

18. Show that the function $f(x) = |x| + |x - 2|$ is constant when $0 \leq x \leq 2$. [*Hint:* Use the definition of absolute value (see Example 7) to compute $f(x)$ when $0 \leq x \leq 2$.]

In Exercises 19–24, find the approximate location of all local maxima and minima of the function.

19. $f(x) = x^3 - x$

20. $g(t) = -\sqrt{16 - t^2}$

21. $h(x) = \dfrac{x}{x^2 + 1}$

22. $k(x) = x^3 - 3x + 1$

23. $f(x) = x^3 - 1.8x^2 + x + 2$

24. $g(x) = 2x^3 + x^2 + 1$

In Exercises 25–26, find the approximate intervals on which the function whose graph is shown is increasing and those on which it is decreasing.

25.

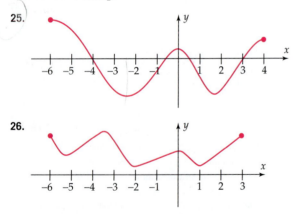

26.

In Exercises 27–32, find the approximate intervals on which the function is increasing, those on which it is decreasing, and those on which it is constant.

27. $f(x) = |x - 1| - |x + 1|$

28. $g(x) = |x - 1| + |x + 2|$

29. $f(x) = -x^3 - 8x^2 + 8x + 5$

30. $f(x) = x^4 - .7x^3 - .6x^2 + 1$

31. $g(x) = .2x^4 - x^3 + x^2 - 2$

32. $g(x) = x^4 + x^3 - 4x^2 + x - 1$

In Exercises 33–34, sketch the graph of a function f that satisfies all of the given conditions. The function whose graph you sketch need not be given by an algebraic formula.

33. (i)   The domain of f consists of all real numbers x with $-2 \leq x \leq 4$.

   (ii)   The range of f consists of all real numbers y with $-5 \leq y \leq 6$.

   (iii) $f(-1) = f(3)$

   (iv) $f\left(\frac{1}{2}\right) = 0$

34. (i) $f(-1) = 2$

   (ii) $f(x) \geq 2$ when $-1 \leq x < 1/2$

   (iii) $f(x)$ starts decreasing when $x = 1$.

   (iv) $f(0) = 3$ and $f(3) = 3$

   (v) $f(x)$ starts increasing when $x = 5$.

35. Find the dimensions of the rectangle with perimeter 100 inches and largest possible area, as follows.

(a) Use the figure to write an equation in x and z that expresses the fact that the perimeter of the rectangle is 100.

(b) The area A of the rectangle is given by $A = xz$. (Why?) Write an equation that expresses A as a function of x. [*Hint:* Solve the equation in part (a) for z, and substitute the result in the area equation.]

(c) Graph the function in part (b), and find the value of x that produces the largest possible value of A. What is z in this case?

36. Find the dimensions of the rectangle with area 240 square inches and smallest possible perimeter, as follows.

(a) Using the figure for Exercise 35, write an equation for the perimeter P of the rectangle in terms of x and z.

(b) Write an equation in x and z that expresses the fact that the area of the rectangle is 240.

(c) Write an equation that expresses P as a function of x. [*Hint:* Solve the equation in part (b) for z, and substitute the result in the equation of part (a).]

(d) Graph the function in part (c), and find the value of x that produces the smallest possible value of P. What is z in this case?

37. Find the dimensions of a box with a square base that has a volume of 867 cubic inches and the smallest possible surface area, as follows.

(a) Write an equation for the surface area S of the box in terms of x and h. [Be sure to include all four sides, the top, and the bottom of the box.]

(b) Write an equation in x and h that expresses the fact that the volume of the box is 867.

(c) Write an equation that expresses S as a function of x. [*Hint:* Solve the equation in part (b) for h, and substitute the result in the equation of part (a).]

(d) Graph the function in part (c), and find the value of x that produces the smallest possible value of S. What is h in this case?

38. Find the radius r and height h of a cylindrical can with a surface area of 60 square inches and the largest possible volume, as follows.

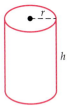

(a) Write an equation for the volume V of the can in terms of r and h.

(b) Write an equation in r and h that expresses the fact that the surface area of the can is 60. [*Hint:* Think of cutting the top and bottom off the can; then cut the side of the can lengthwise and roll it out flat; it's now a rectangle. The surface area is the area of the top and bottom plus the area of this rectangle. The length of the rectangle is the same as the circumference of the original can (why?).]

(c) Write an equation that expresses V as a function of r. [*Hint:* Solve the equation in part (b) for h, and substitute the result in the equation of part (a).]

(d) Graph the function in part (c), and find the value of r that produces the largest possible value of V. What is h in this case?

39. Match each of the following functions with the graph that best fits the situation.

(a) The phases of the moon as a function of time;

(b) The demand for a product as a function of its price;

(c) The height of a ball thrown from the top of a building as a function of time;

(d) The distance a woman runs at constant speed as a function of time;

(e) The temperature of an oven turned on and set to 350° as a function of time.

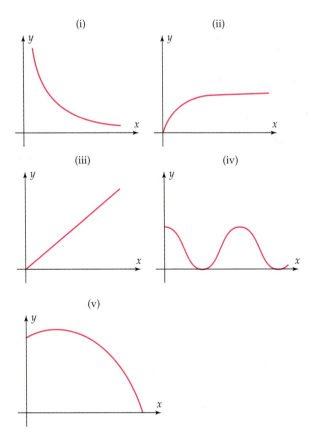

In Exercises 40–42, sketch a plausible graph of the given function. Label the axes and specify a reasonable domain and range.

40. The distance from the top of your head to the ground as you jump on a trampoline is a function of time.

41. The amount you spend on gas each week is a function of the number of gallons you put in your car.

42. The temperature of an oven that is turned on, set to 350°, and 45 minutes later turned off is a function of time.

43. A bacteria population in a laboratory culture contains about three million bacteria at 8 A.M. The culture grows very rapidly until noon, when a bactericide is introduced and the bacteria population plunges. By 4 P.M., the bacteria have adapted to the bactericide, and the culture slowly increases in population until 10 P.M., when the culture is accidentally destroyed by the cleanup crew. Let $g(t)$ denote the bacteria population at time t, and draw a plausible graph of the function g. [Many correct answers are possible.]

44. A plane flies from Austin, Texas, to Cleveland, Ohio, a distance of 1200 miles. Let f be the function whose rule is $f(t) = $ distance (in miles) from Austin at time t hours. Draw a plausible graph of f under the given circumstances. [There are many possible correct answers for each part.]
   (a) The flight is nonstop and takes less than 4 hours.
   (b) Bad weather forces the plane to land in Dallas (about 200 miles from Austin), remain overnight (for 8 hours), and continue the next day.
   (c) The flight is nonstop, but owing to heavy traffic, the plane must fly in a holding pattern over Cincinnati (about 200 miles from Cleveland) for an hour before going on to Cleveland.

In Exercises 45–46, the graph of a function f is shown. Find and label the given points on the graph.

45. (a) $(k, f(k))$
   (b) $(-k, f(-k))$
   (c) $(k, -f(k))$

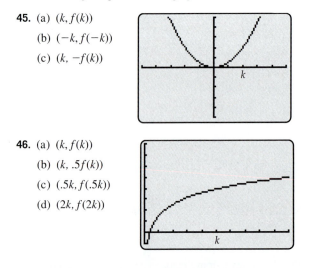

46. (a) $(k, f(k))$
   (b) $(k, .5f(k))$
   (c) $(.5k, f(.5k))$
   (d) $(2k, f(2k))$

47. The graph of the function f, whose rule is $f(x) = $ average interest rate on a 30-year fixed-rate mortgage in year x, is shown in the figure.* Use it to answer these questions (reasonable approximations are OK).
   (a) $f(1980) = ?$ (b) $f(2000) = ?$ (c) $f(2003) = ?$
   (d) In what year between 1990 and 1996 were rates the lowest? The highest?
   (e) During what three-year period were rates changing the fastest? How do you determine this from the graph?

*Federal Home Mortgage Corporation.

48. The figure shows the graph of the function g whose rule is $g(x) = $ the Pentagon's budget in year x (in billions of 2001 dollars, adjusted for inflation), with $x = 0$ corresponding to 1975.† Use it to answer the following questions.

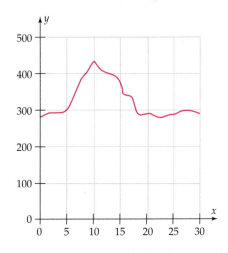

   (a) Over what intervals was this function approximately constant?
   (b) Over what intervals was this function decreasing?
   (c) When did the Pentagon budget reach a local maximum?

Exercises 49–58 deal with the function g whose entire graph is shown in the figure.

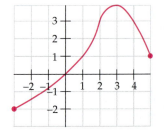

49. What is the domain of g?

50. What is the range of g?

51. If $t = 1.5$, then $g(2t) = ?$

52. If $t = 1.5$, then $2g(t) = ?$

53. If $y = 2$, then $g(y + 1.5) = ?$

54. If $y = 2$, then $g(y) + g(1.5) = ?$

55. If $y = 2$, then $g(y) + 1.5 = ?$

56. For what values of x is $g(x) < 0$?

57. For what values of z is $g(z) = 1$?

58. For what values of z is $g(z) = -1$?

†Department of Defense; figures for 2002–2005 are estimates.

Exercises 59–64 deal with the function f whose entire graph is shown in the figure.

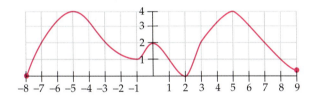

59. What is the domain of f ?

60. What is the range of f ?

61. Find all numbers x such that $f(x) = 2$.

62. Find all numbers x such that $f(x) > 2$.

63. Find all numbers x such that $f(x) = f(7)$.

64. Find two numbers x such that $f(x - 2) = 4$.

65. Let $F(x) =$ the U.S. federal debt in year x, and let $p(x) =$ the federal debt as a percent of the gross domestic product in year x. The graphs of these functions appear below.[*] Explain why the graph of F is increasing after 1996, while the graph of p is decreasing during that period.

Gross Federal Debt

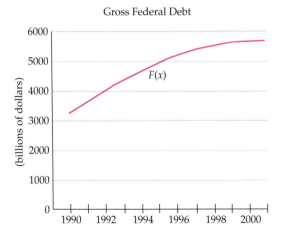

Federal Debt as a Percent of Gross Domestic Product

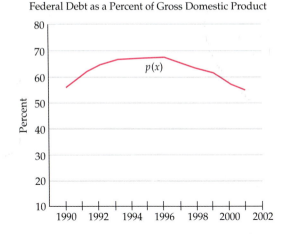

66. The annual percentage changes in various consumer price indexes (CPIs) are shown in the figure.[†] Use it to answer the following questions. In each case, explain how you got your answer from the graph.

(a) Did the CPI for medical care increase or decrease from 1990 to 1996?

(b) During what time intervals was the CPI for fuel oil increasing?

(c) If the CPI for fuel oil stood at 91 at the beginning of 1999, approximately what was it at the beginning of 2000?

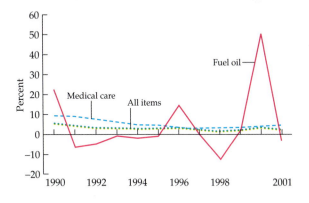

---

[*]Graphs prepared by U.S. Census Bureau, based on data from the U.S. Office of Management and Budget.

[†]Graph prepared by U.S. Census Bureau, based on data from the Bureau of Labor Statistics, U.S. Department of Labor.

For Exercises 67–70, use the Cleveland temperature graph from Example 2 of Section 3.1, which is reproduced below. Let T(x) denote the temperature at time x hours after midnight.

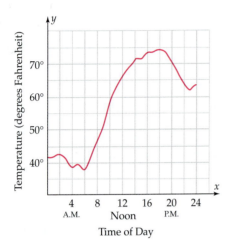

67. Find $T(4)$ and $T(7 + 11)$.

68. Is $T(8)$ larger than, equal to, or less than $T(16)$?

69. Find an eight-hour period in which $T(x) > 65°$ for all x in this period. [There are many possible correct answers.]

70. Determine whether the following statements are true or false.
(a) $T(4 \cdot 3) = T(4) \cdot T(3)$
(b) $T(4 \cdot 3) = 4 \cdot T(3)$
(c) $T(4 + 14) = T(4) + T(14)$

Exercises 71–76 deal with the two functions f and g whose entire graphs are shown in the figure.

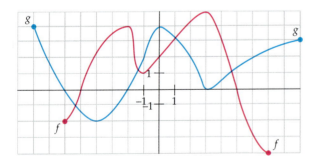

71. What is the domain of f? The domain of g?

72. What is the range of f? The range of g?

73. Find all numbers x with $-3 \le x \le 1$ such that
$$f(x) = 2.$$

74. For how many values of x is it true that $f(x) = g(x)$?

75. Find all intervals over which both functions are defined, f is decreasing, and g is increasing.

76. Find all intervals over which g is decreasing.

Exercises 77–80 deal with this situation: The owners of the Melville & Pluth Hammer Factory have determined that both their weekly manufacturing expenses and their weekly sales income are functions of the number of hammers manufactured each week. The figure shows the graphs of these two functions.

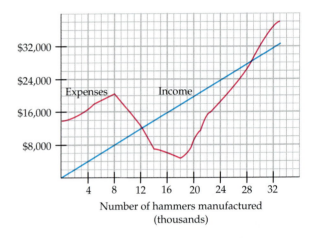

Number of hammers manufactured (thousands)

77. Use careful measurement on the graph and the fact that profit = income − expenses to determine the weekly profit if 5000 hammers are manufactured.

78. Do the same if 10,000, 14,000, 18,000, or 22,000 hammers are manufactured.

79. What is the smallest number of hammers that can be manufactured each week without losing money?

80. What is the largest number of hammers that can be manufactured without losing money?

In Exercises 81–82, express the length h as a function of x.

81.

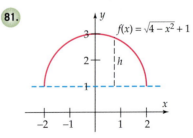

82.

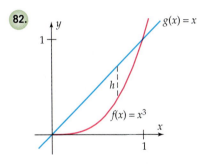

83. The table shows total spending for wireless communication services (in billions of dollars) over a five year period.*

Year	Spending
1999	45.1
2000	59.2
2001	72.9
2002	87.6
2003	103.3

(a) Make a scatter plot of the data, with $x = 0$ corresponding to 1999.
(b) Use linear regression to find a function that models this data. Assume that the model remains accurate.
(c) What is spending in 2005?
(d) When will spending reach \$200 billion?

84. The percentage of adults in the United States who smoke has been decreasing, as shown in the table.[†]

Year	Percent Who Smoke
1980	33.2
1983	32.1
1987	28.8
1991	25.7
1995	24.7
1999	23.5
2001	22.8

(a) Make a scatter plot of the data, with $x = 0$ corresponding to 1980.
(b) Use linear regression to find a function that models this data.
(c) Use the model to estimate the percentage of smokers in 1985 and 2005. [For comparison purposes, the actual figure for 1985 is 30.1%]
(d) If this model remains accurate, when will less than 15% of adults smoke?
(e) According to this model, will smoking ever disappear entirely? If so, when?

Thinkers

85. A jogger begins her daily run from her home. The graph shows her distance from home at time t minutes. The graph shows, for example, that she ran at a slow but steady pace for 10 minutes, then increased her pace for 5 minutes, all the time moving farther from home. Describe the rest of her run.

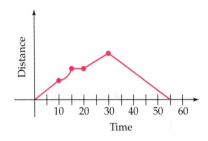

86. The graph shows the speed (in mph) at which a driver is going at time t minutes. Describe his journey.

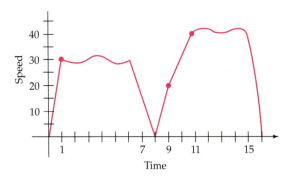

---

*New York Times 2003 Almanac.
[†]Centers for Disease Control and Prevention.

3.3.A   *SPECIAL TOPICS* Parametric Graphing

OMIT

Most of the functions we have seen up to now can be described by equations in which y is a function of x and, hence, are easily graphed on a calculator. When you must graph an equation that expresses x as a function of y, a different calculator graphing technique is needed.

EXAMPLE 1

Graph $x = y^3 - 3y^2 - 4y + 7$.

SOLUTION   Let t be any real number. If $y = t$, then

$$x = y^3 - 3y^2 - 4y + 7 = t^3 - 3t^2 - 4t + 7.$$

Thus, the graph consists of all points (x, y) such that

$$x = t^3 - 3t^2 - 4t + 7 \qquad \text{and} \qquad y = t \quad (t \text{ any real number}).$$

When written in this form, the equation can be graphed as follows.

Change the graphing mode to **parametric mode** (see the Technology Tip in the margin), and enter the equations

$$x_{1t} = t^3 - 3t^2 - 4t + 7$$

$$y_{1t} = t$$

in the equation memory.* Next set the viewing window so that

$$-6 \le t \le 6, \qquad -10 \le x \le 10, \qquad -6 \le y \le 6,$$

as partially shown in Figure 3–28 (scroll down to see the rest). We use the same range for t and y here because $y = t$. Note that we must also set "t-step" (or "t pitch"), which determines how much t changes each time a point is plotted. A t-step between .05 and .15 usually produces a relatively smooth graph in a reasonable amount of time. Finally, pressing GRAPH (or PLOT or DRAW) produces the graph in Figure 3–29. ■

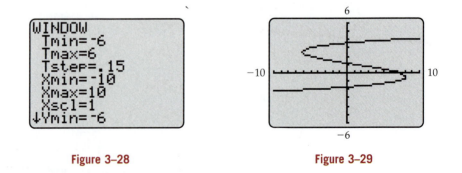

Figure 3–28   **Figure 3–29**

---

*On some calculators, x_{1t} is denoted x_{t1} or $x_1(T)$, and similarly for y_{1t}.

As is illustrated in Example 1, the underlying idea of **parametric graphing** is to express both *x* and *y* as functions of a third variable *t*. The equations that define *x* and *y* are called **parametric equations,** and the variable *t* is called the **parameter.** Example 1 illustrates just one of the many applications of parametric graphing. It can also be used to graph curves that are not graphs of a single equation in *x* and *y*.

EXAMPLE 2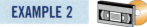

Graph the curve given by

$$x = t^2 - t - 1 \quad \text{and} \quad y = t^3 - 4t - 6 \quad (-2 \le t \le 3).$$

SOLUTION   Using the standard viewing window, we obtain the graph in Figure 3–30. Note that the graph crosses over itself at one point and that it does not extend forever to the left and right but has endpoints.

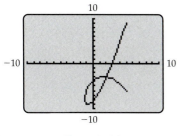

Figure 3–30

GRAPHING EXPLORATION

Graph these same parametric equations, but set the range of *t* values so that $-4 \le t \le 4$. What happens to the graph? Now change the range of *t* values so that $-10 \le t \le 10$. Find a viewing window large enough to show the entire graph, including endpoints. ∎

Any function of the form $y = f(x)$ can be expressed in terms of parametric equations and graphed that way. For instance, to graph $f(x) = x^2 + 1$, let $x = t$ and $y = f(t) = t^2 + 1$.

✓ EXERCISES 3.3.A

In Exercises 1–6, use parametric graphing. Find a viewing window that shows a complete graph of the equation.

1. $x = y^3 + 5y^2 - 4y - 5$

2. $\sqrt[3]{y^2 - y - 1} - x + 2 = 0$

3. $xy^2 + xy + x = y^3 - 2y^2 + 4$
  [*Hint:* First solve for *x*.]

4. $2y = xy^2 + 180x$ **5.** $x - \sqrt{y} + y^2 + 8 = 0$

6. $y^2 - x - \sqrt{y + 5} + 4 = 0$

In Exercises 7–12, find a viewing window that shows a complete graph of the curve determined by the parametric equations.

7. $x = 3t^2 - 5 \quad \text{and} \quad y = t^2 \quad (-4 \le t \le 4)$

8. The Zorro curve: $x = .1t^3 - .2t^2 - 2t + 4 \quad \text{and} \quad y = 1 - t \quad (-5 \le t \le 6)$

9. $x = t^2 - 3t + 2 \quad \text{and} \quad y = 8 - t^3 \quad (-4 \le t \le 4)$

10. $x = t^2 - 6t \quad \text{and} \quad y = \sqrt{t + 7} \quad (-5 \le t \le 9)$

11. $x = 1 - t^2 \quad \text{and} \quad y = t^3 - t - 1 \quad (-4 \le t \le 4)$

12. $x = t^2 - t - 1 \quad \text{and} \quad y = 1 - t - t^2$

Thinkers

13. Graph the curve given by

$$x = (t^2 - 1)(t^2 - 4)(t + 5) + t + 3$$

$$y = (t^2 - 1)(t^2 - 4)(t^3 + 4) + t - 1$$

$$(-2.5 \le t \le 2.5)$$

How many times does this curve cross itself?

14. Use parametric equations to describe a curve that crosses itself more times than the curve in Exercise 13. [Many correct answers are possible.]

3.4   Graphs and Transformations

In this section, we shall see that when the rule of a function is algebraically changed in certain ways, so as to produce a new function, then the graph of the new function can be obtained from the graph of the original function by a simple geometric transformation. The same format will be used for each topic:

First   You will be asked to assemble some evidence by doing a graphing exploration.

Next   General conclusions deduced from the evidence will be summarized in the boxes.

Last   There may be some additional examples or explorations.

■■ VERTICAL SHIFTS

GRAPHING EXPLORATION

Using the standard viewing window, graph these three functions on the same screen:

$$f(x) = x^2 \qquad g(x) = x^2 + 5 \qquad h(x) = x^2 - 7,$$

and answer these questions:

Do the graphs of *g* and *h* look very similar to the graph of *f* in *shape*?

How do their vertical positions differ?

Where would you predict that the graph of $k(x) = x^2 - 9$ is located relative to the graph $f(x) = x^2$, and what is its shape?

Confirm your prediction by graphing *k* on the same screen as *f, g,* and *h.*

The results of this Exploration should make the following statements plausible.

Vertical Shifts

Let *f* be a function and *c* a positive constant.

The graph of $g(x) = f(x) + c$ is the graph of *f* shifted *c* units upward.

The graph of $h(x) = f(x) - c$ is the graph of *f* shifted *c* units downward.

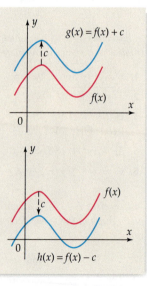

EXAMPLE 1

A calculator was used to obtain a complete graph of $f(x) = .04x^3 - x - 3$ in Figure 3–31. The graph of

$$h(x) = f(x) - 4 = (.04x^3 - x - 3) - 4 = .04x^3 - x - 7$$

is the graph of f shifted 4 units downward, as shown in Figure 3–32.

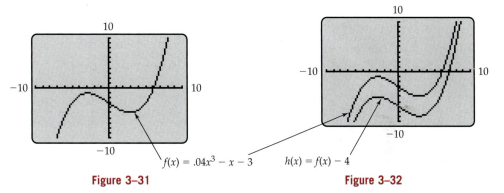

$f(x) = .04x^3 - x - 3$ $h(x) = f(x) - 4$

Figure 3–31   **Figure 3–32**

Although it may appear that the graph of h is closer to the graph of f at the outer edges of Figure 3–32 than in the center, this is an optical illusion. The *vertical* distance between the graphs is always 4 units.

EXPLORATION

Use the trace feature of your calculator as follows to confirm that the vertical distance is always 4:*

Move the cursor to any point on the graph of f, and note its coordinates.

Use the down arrow to drop the cursor to the graph of h, and note the coordinates of the cursor in its new position.

The x-coordinates will be the same in both cases, and the new y-coordinate will be 4 less than the original y-coordinate. ■

▪▪ HORIZONTAL SHIFTS

EXPLORATION

Using the standard viewing window, graph these three functions on the same screen:

$$f(x) = 2x^3 \qquad g(x) = 2(x + 6)^3 \qquad h(x) = 2(x - 8)^3,$$

and answer these questions:

Do the graphs of g and h look very similar to the graph of f in *shape*?

How do their horizontal positions differ?

Where would you predict that the graph of $k(x) = 2(x + 2)^3$ is located relative to the graph of $f(x) = 2x^3$, and what is its shape?

Confirm your prediction by graphing k on the same screen as f, g, and h.

---

*The trace cursor can be moved vertically from graph to graph by using the up and down arrows.

The results of this Exploration should make the following statements plausible.

Horizontal
Shifts

Let f be a function and c a positive constant.

The graph of $g(x) = f(x + c)$ is the graph of f shifted horizontally c units to the left.

The graph of $h(x) = f(x - c)$ is the graph of f shifted horizontally c units to the right.

To see why the first statement is true, suppose $c = 4$ and $g(x) = f(x + 4)$. Then the value of g at x is the same as the value of f at $x + 4$, which is 4 units to the right of x on the horizontal axis. So the graph of f is the graph of g shifted 4 units to the *right,* which means that the graph of g is the graph of f shifted 4 units to the *left.* An analogous argument works for the second statement in the box.

TECHNOLOGY TIP

If the function f of Example 2 is entered as $y_1 = x^2 - 7$, then the functions g and h can be entered as $y_2 = y_1(x + 5)$ and $y_3 = y_1(x - 4)$ on calculators other than TI-85 and Casio.

EXAMPLE 2

In some cases, shifting the graph of a function f horizontally may produce a graph that overlaps the graph of f. For instance, a complete graph of $f(x) = x^2 - 7$ is shown in red in Figure 3–33. The graph of

$$g(x) = f(x + 5) = (x + 5)^2 - 7 = x^2 + 10x + 25 - 7 = x^2 + 10x + 18$$

is the graph of f shifted 5 units to the left, and the graph of

$$h(x) = f(x - 4) = (x - 4)^2 - 7 = x^2 - 8x + 16 - 7 = x^2 - 8x + 9$$

is the graph of f shifted 4 units to the right, as shown in Figure 3–33. ■

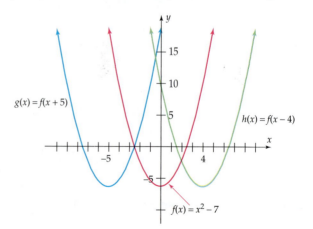

Figure 3–33

██ EXPANSIONS AND CONTRACTIONS

TECHNOLOGY

TIP

On most calculators, you can graph both functions in the Exploration at the same time by keying in

$$y = \{1, 3\}(x^2 - 4).$$

GRAPHING

EXPLORATION

In the viewing window with $-5 \le x \le 5$ and $-15 \le y \le 15$, graph the following functions on the same screen: $Y_1 = f(x)$ and $Y_2 = g(x)$.

$$f(x) = x^2 - 4 \qquad g(x) = 3f(x) = 3(x^2 - 4).$$

The table of values in Figure 3–34 shows that the y-coordinates on the graph of $Y_2 = g(x)$ are always 3 times the y-coordinates of the corresponding points on the graph of $Y_1 = f(x)$. To translate this into visual terms, imagine that the graph of f is nailed to the x-axis at its intercepts (± 2). The graph of g is then obtained by "stretching" the graph of f away from the x-axis (with the nails holding the x-intercepts in place) by a factor of 3, as shown in Figure 3–35.

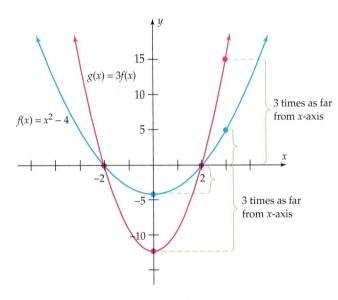

X	Y₁	Y₂
-3	5	15
-2	0	0
-1	-3	-9
0	-4	-12
1	-3	-9
2	0	0
3	5	15

X=3

Figure 3–34

Figure 3–35

GRAPHING

EXPLORATION

In the viewing window with $-5 \le x \le 5$ and $-5 \le y \le 10$, graph these functions on the same screen:

$$f(x) = x^2 - 4 \qquad h(x) = \frac{1}{4}(x^2 - 4).$$

Your screen should suggest that the graph of h is the graph of f "shrunk" vertically toward the x-axis by a factor of $1/4$.

Analogous facts are true in the general case.

Expansions and Contractions

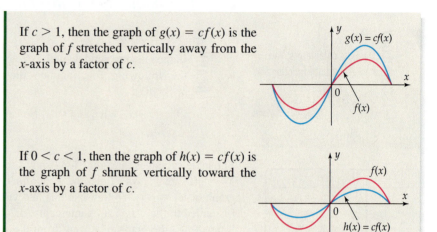

If $c > 1$, then the graph of $g(x) = cf(x)$ is the graph of f stretched vertically away from the x-axis by a factor of c.

If $0 < c < 1$, then the graph of $h(x) = cf(x)$ is the graph of f shrunk vertically toward the x-axis by a factor of c.

⠿ REFLECTIONS

GRAPHING EXPLORATION

In the standard viewing window, graph these functions on the same screen.

$$f(x) = .04x^3 - x \qquad g(x) = -f(x) = -(.04x^3 - x).$$

By moving your trace cursor from graph to graph, verify that for every point on the graph of f, there is a point on the graph of g with the same first coordinate that is on the opposite side of the x-axis, the same distance from the x-axis.

This Exploration shows that the graph of g is the mirror image (reflection) of the graph of f, with the x-axis being the (two-way) mirror. The same thing is true in the general case.

Reflections

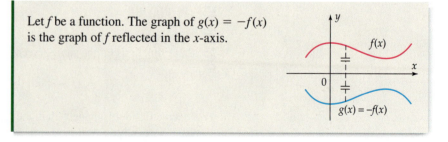

Let f be a function. The graph of $g(x) = -f(x)$ is the graph of f reflected in the x-axis.

EXAMPLE 3

If $f(x) = x^2 - 3$, then the graph of

$$g(x) = -f(x) = -(x^2 - 3)$$

is the reflection of the graph of f in the x-axis, as shown in Figure 3–36.   ■

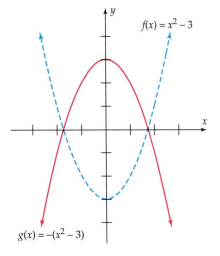

$f(x) = x^2 - 3$

$g(x) = -(x^2 - 3)$

Figure 3–36

EXPLORATION

In the standard viewing window, graph the following functions on the same screen.

$$f(x) = \sqrt{5x + 10} \qquad \text{and} \qquad h(x) = f(-x) = \sqrt{5(-x) + 10}.$$

Reflect carefully: How are the two graphs related to the y-axis? Now graph the following two functions on the same screen.

$$f(x) = x^2 + 3x - 3$$

$$h(x) = f(-x) = (-x)^2 + 3(-x) - 3 = x^2 - 3x - 3.$$

Are the graphs of f and h related in the same way as the first pair?

This Exploration shows that the graph of h in each case is the mirror image (reflection) of the graph of f, with the y-axis as the mirror. The same thing is true in the general case.

Reflections

Let f be a function. The graph of $h(x) = f(-x)$ is the graph of f reflected in the y-axis.

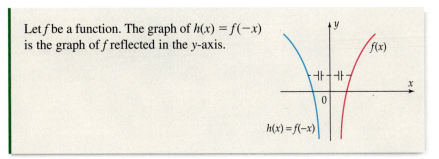

$f(x)$

$h(x) = f(-x)$

Other algebraic operations and their graphical effects are considered in Exercises 43–50.

■■ COMBINING TRANSFORMATIONS

The transformations described above may be used in sequence to analyze the graphs of functions whose rules are algebraically complicated.

EXAMPLE 4

To understand the graph of $g(x) = 2(x - 3)^2 - 1$, note that the rule of g may be obtained from the rule of $f(x) = x^2$ in three steps:

$$f(x) = x^2 \xrightarrow{\text{Step 1}} (x - 3)^2 \xrightarrow{\text{Step 2}} 2(x - 3)^2 \xrightarrow{\text{Step 3}} 2(x - 3)^2 - 1 = g(x).$$

Step 1 shifts the graph of f horizontally 3 units to the right; step 2 stretches the resulting graph away from the x-axis by a factor of 2; step 3 shifts this graph 1 unit downward, thus producing the graph of g in Figure 3–37. ■

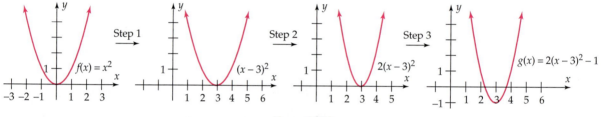

Figure 3–37

INVESTIGATIONS 3.4

Investigations 1 and 2 show that, because of its small screen size, a calculator might not always display clearly what it should.

1. Graph $f(x) = x^2 - x - 6$ in the standard viewing window. Describe verbally what the graph of $h(x) = f(x - 1000)$ should look like (see the box on page 234). Now find an appropriate viewing window, and graph h. Can you find a viewing window that clearly displays both the graph of f and the graph of h?

2. Graph $f(x) = x^2 - x - 6$ in the standard viewing window. Describe verbally what the graph of $g(x) = 1000f(x)$ should look like (see the box on page 236). Now find an appropriate viewing window, and graph g. Can you find a viewing window that clearly displays both the graph of f and the graph of g?

✓ EXERCISES 3.4

In Exercises 1–8, use the catalog of functions at the end of Section 3.3 and information from this section to match each function with its graph, which is one of those shown here.

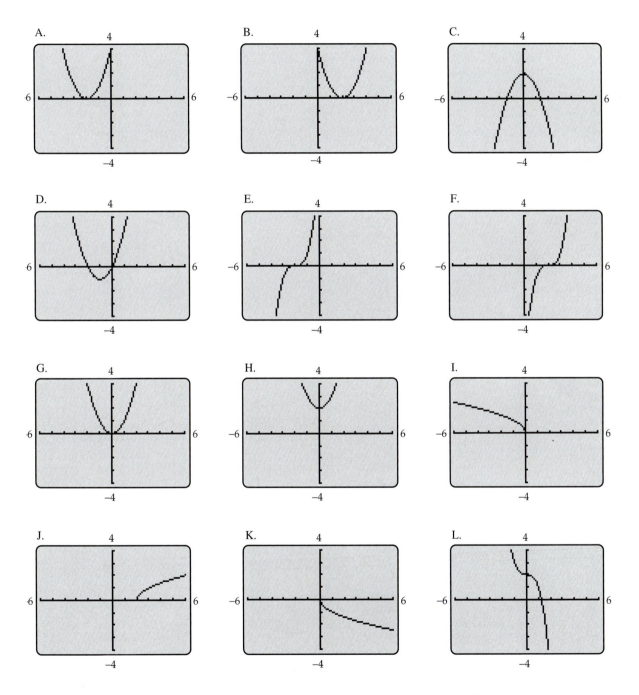

1. $f(x) = x^2 + 2$

2. $f(x) = \sqrt{x - 2}$

3. $g(x) = (x - 2)^3$

4. $g(x) = (x - 2)^2$

5. $f(x) = -\sqrt{x}$

6. $f(x) = (x + 1)^2 - 1$

7. $g(x) = -x^2 + 2$

8. $g(x) = -x^3 + 2$

In Exercises 9–12, use the catalog of functions and information from this section (but not a calculator) to sketch the graph of the function.

9. $f(x) = |x - 2|$

10. $(g)|x| = |x| - 2$

11. $g(x) = -|x|$

12. $f(x) = |x + 2| - 2$

In Exercises 13–16, find a single viewing window that shows complete graphs of the functions f, g, and h.

13. $f(x) = .25x^3 - 9x + 5;$ $g(x) = f(x) + 15;$
$h(x) = f(x) - 20$

14. $f(x) = \sqrt{x^2 - 9} - 5;$ $g(x) = 3f(x);$
$h(x) = .5f(x)$

15. $f(x) = |x^2 - 5|;$ $g(x) = f(x + 8);$
$h(x) = f(x - 6)$

16. $f(x) = .125x^3 - .25x^2 - 1.5x + 5;$
$g(x) = f(x) - 5;$ $h(x) = 5 - f(x)$

In Exercises 17 and 18, find complete graphs of the functions f and g in the same viewing window.

17. $f(x) = \dfrac{4 - 5x^2}{x^2 + 1};$ $g(x) = -f(x)$

18. $f(x) = x^4 - 4x^3 + 2x^2 + 3;$ $g(x) = f(-x)$

In Exercises 19–22, describe a sequence of transformations that will transform the graph of the function f into the graph of the function g.

19. $f(x) = x^2 + x;$ $g(x) = (x - 3)^2 + (x - 3) + 2$

20. $f(x) = x^2 + 5;$ $g(x) = (x + 2)^2 + 10$

21. $f(x) = \sqrt{x^3 + 5};$ $g(x) = -\dfrac{1}{2}\sqrt{x^3 + 5} - 6$

22. $f(x) = \sqrt{x^4 + x^2 + 1};$ $g(x) = 10 - \sqrt{4x^4 + 4x^2 + 4}$

In Exercises 23–26, write the rule of a function g whose graph can be obtained from the graph of the function f by performing the transformations in the order given.

23. $f(x) = x^2 + 2;$ shift the graph horizontally 5 units to the left and then vertically upward 4 units.

24. $f(x) = x^2 - x + 1;$ reflect the graph in the x-axis, then shift it vertically upward 3 units.

25. $f(x) = \sqrt{x};$ shift the graph horizontally 6 units to the right, stretch it away from the x-axis by a factor of 2, and shift it vertically downward 3 units.

26. $f(x) = \sqrt{-x};$ shift the graph horizontally 3 units to the left, then reflect it in the x-axis, and shrink it toward the x-axis by a factor of $1/2$.

27. Let $f(x) = x^2 + 3x$, and let $g(x) = f(x) + 2$.
(a) Write the rule of $g(x)$.
(b) Find the difference quotients of $f(x)$ and $g(x)$. How are they related?

28. Let $f(x) = x^2 + 5$, and let $g(x) = f(x - 1)$.
(a) Write the rule of $g(x)$ and simplify.
(b) Find the difference quotients of $f(x)$ and $g(x)$.
(c) Let $d(x)$ denote the difference quotient of $f(x)$. Show that the difference quotient of $g(x)$ is $d(x - 1)$.

In Exercises 29–31, use the graph of the function f in the figure to sketch the graph of the function g.

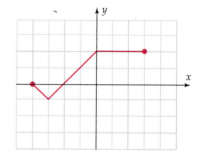

29. $g(x) = f(x) + 3$

30. $g(x) = f(x) - 1$

31. $g(x) = 3f(x)$

32. The graph shows the approximate annual net revenue of Intel (in billions of dollars) over a 10-year period, with $x = 0$ corresponding to 1993.[*] Sketch the revenue graph corresponding to each of the following conditions:

(a) Revenue was $5 billion smaller each year.
(b) Revenue was 25% larger each year.

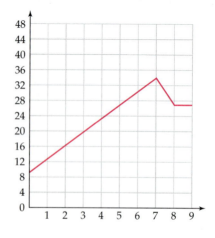

---

*Based on data in the 2002 Annual report of Intel Corporation.

In Exercises 33–36, use the graph of the function f in the figure to sketch the graph of the function h.

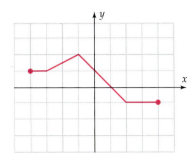

33. $h(x) = -f(x)$,

34. $h(x) = -4f(x)$

35. $h(x) = f(-x)$

36. $h(x) = f(-x) + 2$

In Exercises 37–40, use the graph of the function f in the figure to sketch the graph of the function g.

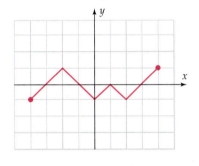

37. $g(x) = f(x + 3)$

38. $g(x) = f(x - 2)$

39. $g(x) = f(x - 2) + 3$

40. $g(x) = 2 - f(x)$

41. Graph $f(x) = -|x - 3| - |x - 17| + 20$ in the window with $0 \le x \le 20$ and $-2 \le y \le 12$. Think of the *x*-axis as a table and the graph as a side view of a fast-food carton placed upside down on the table (the flat part of the graph is the bottom of the carton). Find the rule of a function *g* whose graph (in this viewing window) looks like another fast-food carton, which has been placed right side up on top of the first one.

42. A factory has a linear cost function $c(x) = ax + b$, where *b* represents fixed costs and *a* represents the variable costs (labor and materials) of making one item, both in thousands of dollars. If property taxes (part of the fixed costs) are increased by \$28,000 per year, what effect does this have on the graph of the cost function?

In Exercises 43–45, assume $f(x) = (.2x)^6 - 4$. Use the standard viewing window to graph the functions f and g on the same screen.

43. $g(x) = f(2x)$ **44.** $g(x) = f(3x)$ **45.** $g(x) = f(4x)$

46. On the basis of the results of Exercises 43–45, describe the transformation that transforms the graph of a function $f(x)$ into the graph of the function $f(cx)$, where *c* is a constant with $c > 1$. [*Hint:* How are the two graphs related to the *y*-axis? Stretch your mind.]

In Exercises 47–49, assume $f(x) = x^2 - 3$. Use the standard viewing window to graph the functions f and g on the same screen.

47. $g(x) = f\left(\dfrac{1}{2}x\right)$ **48.** $g(x) = f\left(\dfrac{1}{3}x\right)$

49. $g(x) = f\left(\dfrac{1}{4}x\right)$

50. On the basis of the results of Exercises 47–49, describe the transformation that transforms the graph of a function $f(x)$ into the graph of the function $f(cx)$, where *c* is a constant with $0 < c < 1$. [*Hint:* How are the two graphs related to the *y*-axis?]

3.4.A *SPECIAL TOPICS* Symmetry

A graph is **symmetric with respect to the y-axis** if the part of the graph on the right side of the y-axis is the mirror image of the part on the left side of the y-axis (with the y-axis being the mirror), as shown in Figure 3–38.

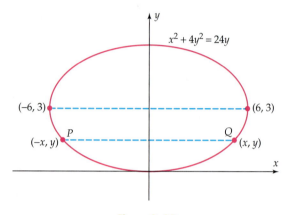

Figure 3–38

Each point P on the left side of the graph has a mirror image point Q on the right side of the graph, as indicated by the dashed lines. Note that:

> Their second coordinates are the same (P and Q are on the same side of the x-axis and the same distance from it);

> Their first coordinates are negatives of each other (P and Q lie on opposite sides of the y-axis and the same distance from it).

Thus, a graph is symmetric with respect to the y-axis provided that

> Whenever (x, y) is on the graph, then $(-x, y)$ is also on it.

In algebraic terms, this means that replacing x by $-x$ in the equation leads to the same number y. In other words, replacing x by $-x$ produces an equivalent equation.

EXAMPLE 1

Replacing x by $-x$ in the equation $y = x^4 - 5x^2 + 3$ produces

$$(-x)^4 - 5(-x)^2 + 3,$$

which is the same equation because $(-x)^2 = x^2$ and $(-x)^4 = x^4$. Therefore, the graph is symmetric with respect to the y-axis.

GRAPHING EXPLORATION

Confirm this fact by graphing the equation. ■

▟ *x*-AXIS SYMMETRY

A graph is **symmetric with respect to the *x*-axis** if the part of the graph above the *x*-axis is the mirror image of the part below the *x*-axis (the *x*-axis being the mirror), as shown in Figure 3–39.

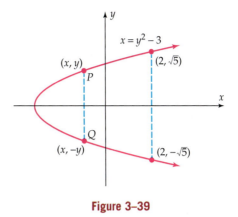

Figure 3–39

Using Figure 3–39 and argument analogous to the one preceding Example 1, we see that a graph is symmetric with respect to the *x*-axis provided that

Whenever (x, y) is on the graph, then $(x, -y)$ is also on it.

In algebraic terms, this means that replacing y by $-y$ in the equation leads to the same number x. In other words, replacing y by $-y$ produces an equivalent equation.

EXAMPLE 2

Replacing y by $-y$ in the equation $y^2 = 4x - 12$ produces $(-y)^2 = 4x - 12$, which is the same equation, so the graph is symmetric with respect to the *x*-axis.

GRAPHING EXPLORATION

Confirm this fact by graphing the equation. To do this, note that every point on the graph of $y^2 = 4x - 12$ is also on the graph of either $y = \sqrt{4x - 12}$ or $y = -\sqrt{4x - 12}$. Each of these latter equations defines a function; graph them both on the same screen. ■

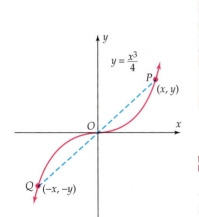

Figure 3–40

▟ ORIGIN SYMMETRY

A graph is **symmetric with respect to the origin** if a straight line through the origin and any point P on the graph also intersects the graph at a point Q such that the origin is the midpoint of segment PQ, as shown in Figure 3–40.

Using Figure 3–40, we can also describe symmetry with respect to the origin in terms of coordinates and equations (as proved in Exercise 34):

Whenever (x, y) is on the graph, then $(-x, -y)$ is also on it.

In algebraic terms, this means that replacing x by $-x$ and y by $-y$ in the equation produces an equivalent equation.

EXAMPLE 3

Replacing x by $-x$ and y by $-y$ in the equation $y = \dfrac{x^3}{10} - x$ yields

$$-y = \frac{(-x)^3}{10} - (-x), \qquad \text{that is,} \qquad -y = \frac{-x^3}{10} + x.$$

This equation is equivalent to

$$y = \frac{x^3}{10} - x,$$

since it can be obtained from it by multiplying by -1. Therefore, the graph of

$$y = \frac{x^3}{10} - x$$

is symmetric with respect to the origin.

GRAPHING EXPLORATION

Confirm this fact by graphing the equation.   ■

Here is a summary of the various tests for symmetry:

Symmetry Tests

Symmetry with Respect to	Coordinate Test for Symmetry	Algebraic Test for Symmetry
y-axis	(x, y) on graph implies $(-x, y)$ on graph.	Replacing x by $-x$ produces an equivalent equation.
x-axis	(x, y) on graph implies $(x, -y)$ on graph.	Replacing y by $-y$ produces an equivalent equation.
origin	(x, y) on graph implies $(-x, -y)$ on graph.	Replacing x by $-x$ and y by $-y$ produces an equivalent equation.

■■ EVEN AND ODD FUNCTIONS

For *functions,* the algebraic description of symmetry takes a different form. A function f whose graph is symmetric with respect to the y-axis is called an **even function.** To say that the graph of $y = f(x)$ is symmetric with respect to the *y*-axis means that replacing *x* by $-x$ produces the same *y* value. In other words, the function takes the same value at both *x* and $-x$. Therefore,

Even Functions

> A function *f* is even provided that
>
> $$f(x) = f(-x) \text{ for every number } x \text{ in the domain of } f.$$
>
> The graph of an even function is symmetric with respect to the *y*-axis.

For example, $f(x) = x^4 + x^2$ is even because

$$f(-x) = (-x)^4 + (-x)^2 = x^4 + x^2 = f(x).$$

Thus, the graph of *f* is symmetric with respect to the *y*-axis, as you can easily verify with your calculator (do it!).

Except for zero functions ($f(x) = 0$ for every *x* in the domain), *the graph of a function is never symmetric with respect to the x-axis*. The reason is the vertical line test: The graph of a function never contains two points with the same first coordinate. If both (5, 3) and (5, -3), for instance, were on the graph, this would say that $f(5) = 3$ and $f(5) = -3$, which is impossible when *f* is a function.

A function whose graph is symmetric with respect to the origin is called an **odd function.** If both (*x*, *y*) and ($-x$, $-y$) are on the graph of such a function *f*, then we must have both

$$y = f(x) \qquad \text{and} \qquad -y = f(-x),$$

so $f(-x) = -y = -f(x)$. Therefore,

Odd Functions

> A function *f* is odd provided that
>
> $$f(-x) = -f(x) \text{ for every number } x \text{ in the domain of } f.$$
>
> The graph of an odd function is symmetric with respect to the origin.

For example, $f(x) = x^3$ is an odd function because

$$f(-x) = (-x)^3 = -x^3 = -f(x).$$

Hence, the graph of *f* is symmetric with respect to the origin (verify this with your calculator).

✓ EXERCISES 3.4.A

In Exercises 1–4, find the graph of the equation. If the graph is symmetric with respect to the x-axis, the y-axis, or the origin, say so.

1. $y = x^2 + 2$

2. $x = (y - 3)^2$

3. $y = x^3 + 2$

4. $y = (x + 2)^3$

In Exercises 5–14, determine whether the given function is even, odd, or neither.

5. $f(x) = 4x$

6. $k(t) = -5t$

7. $f(x) = x^2 - |x|$

8. $h(u) = |3u|$

9. $k(t) = t^4 - 6t^2 + 5$

10. $f(x) = x(x^4 - x^2) + 4$

11. $f(t) = \sqrt{t^2 - 5}$ **12.** $h(x) = \sqrt{7 - 2x^2}$

13. $f(x) = \dfrac{x^2 + 2}{x - 7}$ **14.** $g(x) = \dfrac{x^2 + 1}{x^2 - 1}$

In Exercises 15–18, determine algebraically whether or not the graph of the given equation is symmetric with respect to the x-axis.

15. $x^2 - 6x + y^2 + 8 = 0$ **16.** $x^2 + 8x + y^2 = -15$

17. $x^2 - 2x + y^2 + 2y = 2$ **18.** $x^2 - x + y^2 - y = 0$

In Exercises 19–24, determine whether the given graph is symmetric with respect to the y-axis, the x-axis, or the origin.

19.

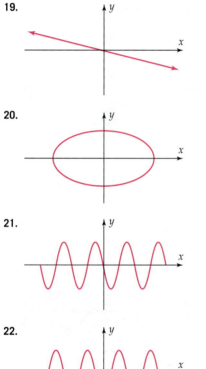

20.

21.

22.

23.

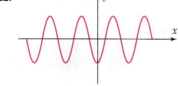

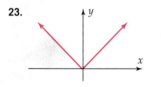

24.

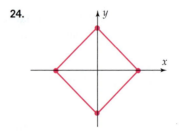

In Exercises 25–28, complete the graph of the given function, assuming that it satisfies the given symmetry condition.

25. Even

26. Even

27. Odd

28. Odd

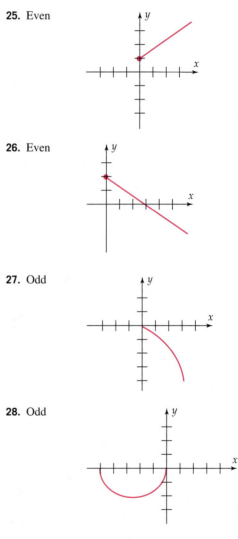

29. (a) Draw some coordinate axes, and plot the points $(0, 1)$, $(1, -3)$, $(-5, 2)$, $(-3, 5)$, $(2, 3)$, and $(4, 1)$.

(b) Suppose the points in part (a) lie on the graph of an *even* function f. Plot the points $(0, f(0))$, $(-1, f(-1))$, $(5, f(5))$, $(3, f(3))$, $(-2, f(-2))$, and $(-4, f(-4))$.

30. Draw the graph of an *even* function that includes the points $(0, -3)$, $(-3, 0)$, $(2, 0)$, $(1, -4)$, $(2.5, -1)$, $(-4, 3)$, and $(-5, 3)$.*

31. (a) Plot the points $(0, 0)$; $(2, 3)$; $(3, 4)$; $(5, 0)$; $(7, -3)$; $(-1, -1)$; $(-4, -1)$; $(-6, 1)$.

(b) Suppose the points in part (a) lie on the graph of an *odd* function f. Plot the points $(-2, f(-2))$; $(-3, f(-3))$; $(-5, f(-5))$; $(-7, f(-7))$; $(1, f(1))$; $(4, f(4))$; $(6, f(6))$.

(c) Draw the graph of an odd function f that includes all the points plotted in parts (a) and (b).*

32. Draw the graph of an odd function that includes the points $(-3, 5)$, $(-1, 1)$, $(2, -6)$, $(4, -9)$, and $(5, -5)$.*

33. Show that any graph that has two of the three types of symmetry (x-axis, y-axis, origin) necessarily has the third type also.

34. Use the midpoint formula to show that $(0, 0)$ is the midpoint of the segment joining (x, y) and $(-x, -y)$. Conclude that the coordinate test for symmetry with respect to the origin (page 244) is correct.

*There are many correct answers.

3.5   Operations on Functions

We now examine ways in which two or more given functions can be used to create new functions. If f and g are functions, then their **sum** is the function h defined by the rule

$$h(x) = f(x) + g(x).$$

For example, if $f(x) = 3x^2 + x$ and $g(x) = 4x - 2$, then

$$h(x) = f(x) + g(x) = (3x^2 + x) + (4x - 2) = 3x^2 + 5x - 2.$$

Instead of using a different letter h for the sum function, we shall usually denote it by $f + g$. Thus, the sum $f + g$ is defined by the rule

$$(f + g)(x) = f(x) + g(x).$$

This rule is *not* just a formal manipulation of symbols. If x is a number, then so are $f(x)$ and $g(x)$. The plus sign in $f(x) + g(x)$ is addition of *numbers,* and the result is a number. But the plus sign in $f + g$ is addition of *functions,* and the result is a new function.

The **difference** $f - g$ is the function defined by the rule

$$(f - g)(x) = f(x) - g(x).$$

The domain of the sum and difference functions is the set of all real numbers that are in both the domain of f and the domain of g.

EXAMPLE 1

If $f(x) = \sqrt{9 - x^2}$ and $g(x) = \sqrt{x - 2}$, find the rules of the functions $f + g$ and $f - g$ and their domains.

SOLUTION   We have

$$(f + g)(x) = f(x) + g(x) = \sqrt{9 - x^2} + \sqrt{x - 2}$$
$$(f - g)(x) = f(x) - g(x) = \sqrt{9 - x^2} - \sqrt{x - 2}.$$

The domain of f consists of all x such that $9 - x^2 \geq 0$ (so that the square root will be defined), that is, all x with $-3 \leq x \leq 3$. Similarly, the domain of g consists of all x such that $x \geq 2$. The domain of $f + g$ and $f - g$ consists of all real numbers in both the domain of f and the domain of g, namely, all x such that $2 \leq x \leq 3$. ∎

The **product** and **quotient** of functions f and g are the functions defined by the rules

$$(fg)(x) = f(x)g(x) \qquad \text{and} \qquad \left(\frac{f}{g}\right)(x) = \frac{f(x)}{g(x)}.$$

The domain of fg consists of all real numbers in both the domain of f and the domain of g. The domain of f/g consists of all real numbers x in both the domain of f and the domain of g such that $g(x) \neq 0$.

EXAMPLE 2

If $f(x) = \sqrt{3x}$ and $g(x) = x^2 - 1$, find the rules of the functions fg and f/g and their domains.

SOLUTION The rules are

$$(fg)(x) = f(x)g(x) = \sqrt{3x}\,(x^2 - 1) = \sqrt{3x}\,x^2 - \sqrt{3x}$$

$$\left(\frac{f}{g}\right)(x) = \frac{f(x)}{g(x)} = \frac{\sqrt{3x}}{x^2 - 1}.$$

The domain of fg consists of all numbers x in both the domain of f (all nonnegative real numbers) and the domain of g (all real numbers), that is, all $x \geq 0$. The domain of f/g consists of all these x for which $g(x) \neq 0$, that is, all nonnegative real numbers *except $x = 1$*. ∎

If c is a real number and f is a function, then the product of f and the constant function $g(x) = c$ is usually denoted cf. For example, if the function

$$f(x) = x^3 - x + 2,$$

and $c = 5$, then $5f$ is the function given by

$$(5f)(x) = 5 \cdot f(x) = 5(x^3 - x + 2) = 5x^3 - 5x + 10.$$

■■ COMPOSITION OF FUNCTIONS

Another way of combining functions is illustrated by the function $h(x) = \sqrt{x^3}$. To compute $h(4)$, for example, you first find $4^3 = 64$ and then take the square root $\sqrt{64} = 8$. So the rule of h may be rephrased as follows:

First apply the function $f(x) = x^3$,

Then apply the function $g(t) = \sqrt{t}$ to the result.

The same idea can be expressed in functional notation like this:

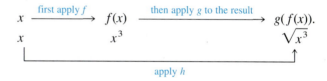

So the rule of h may be written as $h(x) = g(f(x))$, where $f(x) = x^3$ and $g(t) = \sqrt{t}$. We can think of h as being made up of two simpler functions f and g, or we can think of f and g being "composed" to create the function h. Both viewpoints are useful.

EXAMPLE 3

Suppose $f(x) = 4x^2 + 1$ and $g(t) = \dfrac{1}{t+2}$. Define a new function h whose rule is "first apply f; then apply g to the result." In functional notation,

$$x \xrightarrow{\text{first apply } f} f(x) \xrightarrow{\text{then apply } g \text{ to the result}} g(f(x)).$$

So the rule of the function h is $h(x) = g(f(x))$. Evaluating $g(f(x))$ means that whenever t appears in the formula for $g(t)$, we must replace it by $f(x) = 4x^2 + 1$.

$$h(x) = g(f(x)) = \frac{1}{f(x) + 2} = \frac{1}{(4x^2 + 1) + 2} = \frac{1}{4x^2 + 3}. \qquad \blacksquare$$

The function h in Example 3 is an illustration of the following definition.

Composite Functions

> Let f and g be functions. The **composite function** of f and g is defined as follows.
>
> For input x, the output is $g(f(x))$.
>
> This composite function is denoted $g \circ f$.

The symbol "$g \circ f$" is read "g circle f" or "f followed by g." (Note the order carefully; the functions are applied *right* to *left*.) So the rule of the composite function is

$$(g \circ f)(x) = g(f(x)).$$

EXAMPLE 4

If $f(x) = 2x + 5$ and $g(t) = 3t^2 + 2t + 4$, then find

$$(f \circ g)(2), \qquad (g \circ f)(-1), \qquad (g \circ f)(5).$$

SOLUTION

$$(f \circ g)(2) = f(g(2)) = f(3 \cdot 2^2 + 2 \cdot 2 + 4) = f(20) = 2 \cdot 20 + 5 = 45.$$

Similarly,

$$(g \circ f)(-1) = g(f(-1)) = g(2(-1) + 5) = g(3) = 3 \cdot 3^2 + 2 \cdot 3 + 4 = 37.$$

The value of a composite function can also be computed as follows.

$$(g \circ f)(5) = g(f(5)) = 3(f(5)^2) + 2(f(5)) + 4 = 3(15^2) + 2(15) + 4 = 709$$

■

The domain of $g \circ f$ is determined by this convention.

Domain of $g \circ f$

> The domain of the composite function $g \circ f$ is the set of all real numbers x such that x is in the domain of f and $f(x)$ is in the domain of g.

EXAMPLE 5

Find the rule and domain of $g \circ f$, when $f(x) = \sqrt{x}$ and $g(t) = t^2 - 5$.

SOLUTION

$$(g \circ f)(x) = g(f(x)) = (f(x))^2 - 5 = (\sqrt{x})^2 - 5 = x - 5.$$

Although $x - 5$ is defined for every real number x, the domain of $g \circ f$ is *not* the set of all real numbers. The domain of g is the set of all real numbers, but the function $f(x) = \sqrt{x}$ is defined only when $x \geq 0$. So the domain of $g \circ f$ is the set of nonnegative real numbers.   ■

TECHNOLOGY TIP

Evaluating composite functions is easy on calculators other than TI-85 and most Casio calculators. If the functions are entered in the equation memory as $y_1 = g(x)$ and $y_2 = h(x)$ (with f in place of y on HP-39+), then keying in $y_2(y_1(5))$ ENTER produces the number $h(g(5))$.

On TI-85 and most Casio calculators, this syntax does *not* produce $h(g(5))$; it produces

$$h(x) \cdot g(x) \cdot 5$$

for whatever number is stored in the x-memory.

EXAMPLE 6

Write the function $h(x) = \sqrt{3x^2 + 1}$ in two different ways as the composite of two functions.

SOLUTION   Let $f(x) = 3x^2 + 1$ and $g(x) = \sqrt{x}.^*$ Then

$$(g \circ f)(x) = g(f(x)) = g(3x^2 + 1) = \sqrt{3x^2 + 1} = h(x).$$

Similarly, h is also the composite $j \circ k$, where $j(x) = \sqrt{x + 1}$ and $k(x) = 3x^2$:

$$(j \circ k)(x) = j(k(x)) = j(3x^2) = \sqrt{3x^2 + 1} = h(x).$$ ■

EXAMPLE 7

If $k(x) = (x^2 - 2x + \sqrt{x})^3$, then k is $g \circ f$, where $f(x) = x^2 - 2x + \sqrt{x}$ and $g(x) = x^3$ because

$$(g \circ f)(x) = g(f(x)) = g(x^2 - 2x + \sqrt{x}) = (x^2 - 2x + \sqrt{x})^3 = k(x).$$ ■

---

*Now that you have the idea of composite functions, we'll use the same letter for the variable in both functions.

As you might have noticed, there are two possible ways to form a composite function from two given functions. If f and g are functions, we can consider either

$$(g \circ f)(x) = g(f(x)),$$ [the composite of f and g]

$$(f \circ g)(x) = f(g(x)).$$ [the composite of g and f]

The *order is important,* as we shall now see.

$g \circ f$ and $f \circ g$ usually are *not* the same function.

EXAMPLE 8

If $f(x) = x^2$ and $g(x) = x + 3$, then

$$(g \circ f)(x) = g(f(x)) = g(x^2) = x^2 + 3,$$

but

$$(f \circ g)(x) = f(g(x)) = f(x + 3) = (x + 3)^2 = x^2 + 6x + 9.$$

Obviously, $g \circ f \neq f \circ g$, since, for example, they have different values at $x = 0$. ∎

CAUTION

Don't confuse the product function fg with the composite function $f \circ g$ (g followed by f). For instance, if $f(x) = 2x^2$ and $g(x) = x - 3$, then the product fg is given by

$$(fg)(x) = f(x)g(x) = 2x^2(x - 3) = 2x^3 - 6x^2.$$

It is *not* the same as the composite $f \circ g$ because

$$(f \circ g)(x) = f(g(x)) = f(x - 3) = 2(x - 3)^2 = 2x^2 - 12x + 18.$$

▪▪ APPLICATIONS

Composition of functions arises in applications involving several functional relationships simultaneously. In such cases, one quantity may have to be expressed as a function of another.

EXAMPLE 9

A circular puddle of liquid is evaporating and slowly shrinking in size. After t minutes, the radius r of the puddle measures $\dfrac{18}{2t + 3}$ inches; in other words, the radius is a function of time. The area A of the puddle is given by $A = \pi r^2$, that is, area is a function of the radius r. We can express the area as a function of time by substituting $r = \dfrac{18}{2t + 3}$ in the area equation:

$$A = \pi r^2 = \pi \left(\frac{18}{2t + 3} \right)^2.$$

This amounts to forming the composite function $f \circ g$, where $f(r) = \pi r^2$ and

$$g(t) = \frac{18}{2t + 3}.$$

$$(f \circ g)(t) = f(g(t)) = f\left(\frac{18}{2t + 3}\right) = \pi\left(\frac{18}{2t + 3}\right)^2$$

When area is expressed as a function of time, it is easy to compute the area of the puddle at any time. For instance, after 12 minutes, the area of the puddle is

$$A = \pi\left(\frac{18}{2t + 3}\right)^2 = \pi\left(\frac{18}{2 \cdot 12 + 3}\right)^2 = \frac{4\pi}{9} \approx 1.396 \text{ square inches.} \quad \blacksquare$$

✓ EXERCISES 3.5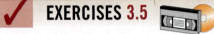

In Exercises 1–4, find $(f + g)(x)$, $(f - g)(x)$, and $(g - f)(x)$.

1. $f(x) = -3x + 2$, $g(x) = x^3$
2. $f(x) = x^2 + 2$, $g(x) = -4x + 7$
3. $f(x) = 1/x$, $g(x) = x^2 + 2x - 5$
4. $f(x) = \sqrt{x}$, $g(x) = x^2 + 1$

In Exercises 5–8, find $(fg)(x)$, $(f/g)(x)$, and $(g/f)(x)$.

5. $f(x) = -3x + 2$, $g(x) = x^3$
6. $f(x) = 4x^2 + x^4$, $g(x) = \sqrt{x^2 + 4}$
7. $f(x) = x^2 - 3$, $g(x) = \sqrt{x - 3}$
8. $f(x) = \sqrt{x^2 - 1}$, $g(x) = \sqrt{x - 1}$

In Exercises 9–12, find the domains of fg and f/g.

9. $f(x) = x^2 + 1$, $g(x) = 1/x$
10. $f(x) = x + 2$, $g(x) = \dfrac{1}{x + 2}$
11. $f(x) = \sqrt{4 - x^2}$, $g(x) = \sqrt{3x + 4}$
12. $f(x) = 3x^2 + x^4 + 2$, $g(x) = 4x - 3$

In Exercises 13–16, find the indicated values, where

$$g(t) = t^2 - t \quad \text{and} \quad f(x) = 1 + x.$$

13. $g(f(0))$
14. $(f \circ g)(3)$
15. $g(f(2) + 3)$
16. $f(2g(1))$

In Exercises 17–20, find $(g \circ f)(3)$, $(f \circ g)(1)$, and $(f \circ f)(0)$.

17. $f(x) = 3x - 2$, $g(x) = x^2$
18. $f(x) = |x + 2|$, $g(x) = -x^2$
19. $f(x) = x$, $g(x) = -3$
20. $f(x) = x^2 - 1$, $g(x) = \sqrt{x}$

In Exercises 21–24, find the rule of the function $f \circ g$, the domain of $f \circ g$, the rule of $g \circ f$, and the domain of $g \circ f$.

21. $f(x) = x^2$, $g(x) = x + 3$
22. $f(x) = -3x + 2$, $g(x) = x^3$
23. $f(x) = 1/x$, $g(x) = \sqrt{x}$
24. $f(x) = \dfrac{1}{2x + 1}$, $g(x) = x^2 - 1$

In Exercises 25–28, find the rules of the functions ff and $f \circ f$.

25. $f(x) = x^3$
26. $f(x) = (x - 1)^2$
27. $f(x) = 1/x$
28. $f(x) = \dfrac{1}{x - 1}$

In Exercises 29–32, verify that $(f \circ g)(x) = x$ and $(g \circ f)(x) = x$ for every x.

29. $f(x) = 9x + 2$, $g(x) = \dfrac{x - 2}{9}$
30. $f(x) = \sqrt[3]{x - 1}$, $g(x) = x^3 + 1$
31. $f(x) = \sqrt[3]{x} + 2$, $g(x) = (x - 2)^3$
32. $f(x) = 2x^3 - 5$, $g(x) = \sqrt[3]{\dfrac{x + 5}{2}}$

Exercises 33 and 34 refer to the function f whose graph is shown in the figure.

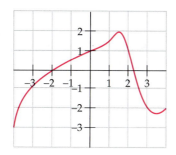

35.

x	$(g \circ f)(x)$
1	4
2	
3	5
4	
5	

36.

t	$(f \circ g)(t)$
1	
2	2
3	
4	
5	

37.

x	$(f \circ f)(x)$
1	
2	
3	3
4	
5	

38.

t	$(g \circ g)(t)$
1	
2	
3	
4	4
5	

33. Let g be the composite function $f \circ f$, that is,

$$g(x) = (f \circ f)(x) = f(f(x)).$$

Use the graph of f to fill in the following table (approximate where necessary).

x	f(x)	$g(x) = f(f(x))$
−4		
−3		
−2	0	1
−1		
0		
1		
2		
3		
4		

34. Use the information obtained in Exercise 33 to sketch the graph of the function g.

In Exercises 35–38, fill the blanks in the given table. In each case the values of the functions f and g are given by these tables:

x	f(x)
1	3
2	5
3	1
4	2
5	3

t	g(t)
1	5
2	4
3	4
4	3
5	2

C *In Exercises 39–44, write the given function as the composite of two functions, neither of which is the identity function, as in Examples 6 and 7. (There may be more than one way to do this.)*

39. $f(x) = \sqrt[3]{x^2 + 2}$

40. $g(x) = \sqrt{x + 3} - \sqrt[3]{x + 3}$

41. $h(x) = (7x^3 - 10x + 17)^7$

42. $k(x) = \sqrt[3]{(7x - 3)^2}$

43. $f(x) = \dfrac{1}{3x^2 + 5x - 7}$

44. $g(t) = \dfrac{3}{\sqrt[3]{t - 3}} + 7$

45. If f is any function and I is the identity function, what are $f \circ I$ and $I \circ f$?

46. Give an example of a function f such that

$$f\left(\frac{1}{x}\right) \neq \frac{1}{f(x)}.$$

In Exercises 47 and 48, graph both $f \circ g$ and $g \circ f$ on the same screen. Use the graphs to determine whether $f \circ g$ is the same function as $g \circ f$.

47. $f(x) = x^5 - x^3 - x$; $g(x) = x - 2$

48. $f(x) = x^3 + x$; $g(x) = \sqrt[3]{x - 1}$

C *In Exercises 49–52, find $g \circ f$, and find the difference quotient of the function $g \circ f$.*

49. $f(x) = x + 3$; $g(x) = x^2 + 1$

50. $f(x) = 2x$; $g(x) = x^2 + x$

51. $f(x) = x + 1$; $g(x) = 1/x$

52. $f(x) = x^3$; $g(x) = x + 2$

53. (a) What is the area of the puddle in Example 9 after one day? After a week? After a month?

(b) Does the puddle ever totally evaporate? Is this realistic? Under what circumstances might this area function be an accurate model of reality?

54. In a laboratory culture, the number $N(d)$ of bacteria (in thousands) at temperature d degrees Celsius is given by the function

$$N(d) = \frac{-90}{d + 1} + 20 \quad (4 \le d \le 32).$$

The temperature $D(t)$ at time t hours is given by the function $D(t) = 2t + 4 \quad (0 \le t \le 14)$.

(a) What does the composite function $N \circ D$ represent?

(b) How many bacteria are in the culture after 4 hours? After 10 hours?

55. A certain fungus grows in a circular shape. Its diameter after t weeks is $6 - \dfrac{50}{t^2 + 10}$ inches.

(a) Express the area covered by the fungus as a function of time.

(b) What is the area covered by the fungus when $t = 0$? What area does it cover at the end of 8 weeks?

(c) When is its area 25 square inches?

56. Tom left point P at 6 A.M. walking south at 4 mph. Anne left point P at 8 A.M. walking west at 3.2 mph.

(a) Express the distance between Tom and Anne as a function of the time t elapsed since 6 A.M.

(b) How far apart are Tom and Anne at noon?

(c) At what time are they 35 miles apart?

57. As a weather balloon is inflated, its radius increases at the rate of 4 centimeters per second. Express the volume of the balloon as a function of time and determine the volume of the balloon after 4 seconds. [*Hint:* The volume of a sphere of radius r is $4\pi r^3/3$.]

58. Express the surface area of the weather balloon in Exercise 57 as a function of time. [*Hint:* The surface area of a sphere of radius r is $4\pi r^2$.]

59. Phil, who is 6 feet tall, walks away from a streetlight that is 15 feet high at a rate of 5 feet per second, as shown in the figure. Express the length s of Phil's shadow as a function of time. [*Hint:* First use similar triangles to express s as a function of the distance d from the streetlight to Phil.]

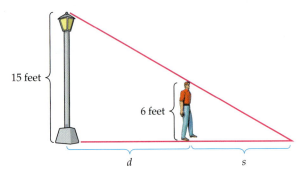

15 feet

6 feet

d s

C 3.6   Rates of Change

OMIT

Rates of change play a central role in the analysis of many real-world situations. To understand the basic ideas involved in rates of change, we take another look at the falling rock from Sections 3.1 and 3.2. We saw that when the rock is dropped from a high place, then the distance it travels (ignoring wind resistance) is given by the function

$$d(t) = 16t^2$$

with distance $d(t)$ measured in feet and time t in seconds. The following table shows the distance the rock has fallen at various times.

Time t	0	1	2	3	3.5	4	4.5	5
Distance $d(t)$	0	16	64	144	196	256	324	400

To find the distance the rock falls from time $t = 1$ to $t = 3$, we note that at the end of three seconds, the rock has fallen $d(3) = 144$ feet, whereas it had fallen only $d(1) = 16$ feet at the end of one second. So during this time interval, the rock traveled

$$d(3) - d(1) = 144 - 16 = 128 \text{ feet.}$$

The distance traveled by the rock during other time intervals can be found similarly.

Time Interval	Distance Traveled
$t = 1$ to $t = 4$	$d(4) - d(1) = 256 - 16 = 240$
$t = 2$ to $t = 3.5$	$d(3.5) - d(2) = 196 - 64 = 132$
$t = 2$ to $t = 4.5$	$d(4.5) - d(2) = 324 - 64 = 260$

The same procedure works in general.

The distance traveled from time $t = a$ to time $t = b$ is $d(b) - d(a)$ feet.

In the preceding chart, the length of each time interval can be computed by taking the difference between the two times. For example, from $t = 1$ to $t = 4$ is a time interval of length $4 - 1 = 3$ seconds. Similarly, the interval from $t = 2$ to $t = 3.5$ is of length $3.5 - 2 = 1.5$ seconds, and in general,

The time interval from $t = a$ to $t = b$ is an interval of $b - a$ seconds.

Since Distance = Average speed × Time,

$$\text{Average speed} = \frac{\text{Distance traveled}}{\text{Time interval}}.$$

Hence, the average speed over the time interval from $t = a$ to $t = b$ is

$$\text{Average speed} = \frac{\text{Distance traveled}}{\text{Time interval}} = \frac{d(b) - d(a)}{b - a}.$$

For example, to find the average speed from $t = 1$ to $t = 4$, apply the preceding formula with $a = 1$ and $b = 4$.

$$\text{Average speed} = \frac{d(4) - d(1)}{4 - 1} = \frac{256 - 16}{4 - 1} = \frac{240}{3} = 80 \text{ ft per second.}$$

Similarly, the average speed from $t = 2$ to $t = 4.5$ is

$$\frac{d(4.5) - d(2)}{4.5 - 2} = \frac{324 - 64}{4.5 - 2} = \frac{260}{2.5} = 104 \text{ ft per second.}$$

The units in which average speed is measured here (feet per second) indicate the number of units of distance traveled during each unit of time, that is, the *rate of change* of distance (feet) with respect to time (seconds). The preceding discussion can be summarized by saying that the average speed (rate of change of distance with respect to time) as time changes from $t = a$ to $t = b$ is given by

$$\text{Average speed} = \text{Average rate of change}$$

$$= \frac{\text{Change in distance}}{\text{Change in time}} = \frac{d(b) - d(a)}{b - a}.$$

Although speed is the most familiar example, rates of change play a role in many other situations as well, as illustrated in Examples 1–3 below. Consequently, we define the average rate of change of any function as follows.

Average Rate of Change

Let f be a function. The **average rate of change of $f(x)$** with respect to x as x changes from a to b is the number

$$\frac{\text{Change in } f(x)}{\text{Change in } x} = \frac{f(b) - f(a)}{b - a}.$$

EXAMPLE 1

A large heavy-duty balloon is being filled with water. Its approximate volume (in gallons) is given by

$$V(x) = \frac{x^3}{55},$$

where x is the radius of the balloon (in inches). Find the average rate of change of the volume of the balloon as the radius increases from 5 to 10 inches.

SOLUTION

$$\frac{\text{Change in volume}}{\text{Change in radius}} = \frac{V(10) - V(5)}{10 - 5} \approx \frac{18.18 - 2.27}{10 - 5} = \frac{15.91}{5}$$

$$= 3.182 \text{ gallons per inch.} \quad \blacksquare$$

EXAMPLE 2

According to U.S. Census Bureau data, the population of Cleveland, Ohio, can be approximated by

$$f(x) = (4.063 \times 10^{-5})x^4 - .00587x^3 - .0274x^2 + 21.8x + 377$$
$$(0 \leq x \leq 100),$$

where $x = 0$ corresponds to 1900 and $f(x)$ is in thousands. Find the average rate of change of the population during the following time periods.

(a) 1900–1950   (b) 1960–2000

SOLUTION

(a) Now 1900 corresponds to $x = 0$ and 1950 to $x = 50$, so

$$\text{Average rate of change} = \frac{f(50) - f(0)}{50 - 0} \approx \frac{918.69 - 377}{50} \approx 10.83.$$

Since $f(x)$ is measured in thousands, the population was increasing at an average rate of approximately 10,830 people per year during this period.

(b) During this period ($x = 60$ to $x = 100$), the average rate of change was

$$\frac{f(100) - f(60)}{100 - 60} \approx \frac{476 - 845}{40} \approx -9.23.$$

The rate is negative, which means that the population was *decreasing* at an average rate of approximately 9,230 people per year during this period. ■

EXAMPLE 3

Figure 3–41 is the graph of the temperature function f during a particular day; $f(x)$ is the temperature at x hours after midnight. What is the average rate of change of the temperature (a) from 4 A.M. to noon? (b) from 3 P.M. to 8 P.M.?

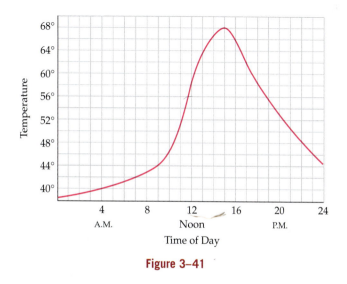

Figure 3–41

SOLUTION

(a) The graph shows that the temperature at 4 A.M. is $f(4) = 40°$ and the temperature at noon is $f(12) = 58°$. The average rate of change of temperature is

$$\frac{\text{Change in temperature}}{\text{Change in time}} = \frac{f(12) - f(4)}{12 - 4} = \frac{58 - 40}{12 - 4} = \frac{18}{8}$$

$$= 2.25° \text{ per hour.}$$

The rate of change is positive because the temperature is increasing at an average rate of 2.25° per hour.

(b) Now 3 P.M. corresponds to $x = 15$ and 8 P.M. to $x = 20$. The graph shows that $f(15) = 68°$ and $f(20) = 53°$. Hence, the average rate of change of temperature is

$$\frac{\text{Change in temperature}}{\text{Change in time}} = \frac{f(20) - f(15)}{20 - 15} = \frac{53 - 68}{20 - 15} = \frac{-15}{5}$$

$$= -3° \text{ per hour.}$$

The rate of change is negative because the temperature is decreasing at an average rate of 3° per hour. ■

▉▉ GEOMETRIC INTERPRETATION OF AVERAGE RATE OF CHANGE

If P and Q are points on the graph of a function f, then the straight line determined by P and Q is called a **secant line.** Figure 3–42 shows the secant line joining the points $(4, 40)$ and $(12, 58)$ on the graph of the temperature function f of Example 3.

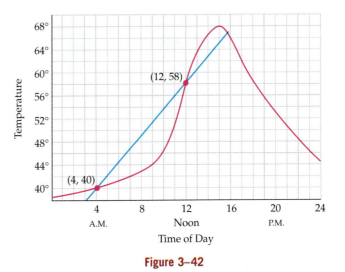

Figure 3–42

Using the points $(4, 40)$ and $(12, 58)$, we see that the slope of this secant line is

$$\frac{58 - 40}{12 - 4} = \frac{18}{8} = 2.25.$$

To say that $(4, 40)$ and $(12, 58)$ are on the graph of f means that $f(4) = 40$ and $f(12) = 58$. Thus,

$$\text{Slope of secant line} = 2.25 = \frac{58 - 40}{12 - 4} = \frac{f(12) - f(4)}{12 - 4}$$

$$= \text{Average rate of change as } x \text{ goes from 4 to 12.}$$

The same thing happens in the general case.

Secant Lines and Average Rates of Change

> If f is a function, then the average rate of change of $f(x)$ with respect to x as x changes from $x = a$ to $x = b$ is the slope of the secant line joining the points $(a, f(a))$ and $(b, f(b))$ on the graph of f.

EXAMPLE 4

Current data and projections from the Office of Management and Budget show that between 1990 and 2004, the national debt (the amount the U.S. government owes) can be approximated by

$$f(x) = 199x + 3760 \qquad (0 \le x \le 14),$$

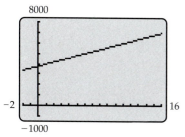

Figure 3–43

where $x = 0$ corresponds to 1990 and $f(x)$ is in billions of dollars. At what average rate is the national debt changing from year to year during this period?

SOLUTION   The graph of $f(x) = 199x + 3760$ is a straight line (Figure 3–43). The secant line joining any two points on the graph is just *the graph itself,* that is, the line $y = 199x + 3760$. As we saw in Section 1.3, the slope of this line is 199. Therefore, the average rate of change of the national debt function between any two values of x is 199. In other words, at any time during this period, the national debt is increasing at a constant rate of $199 billion per year. ■

The argument used in Example 4 works for any function whose graph is a straight line and leads to this conclusion.

Rates of Change of Linear Functions

> The average rate of change of a linear function $f(x) = mx + b$. as x changes from one value to another, is the slope m of the line.

✓ EXERCISES 3.6

1. A car moves along a straight test track. The distance traveled by the car at various times is shown in this table:

Time (seconds)	0	5	10	15	20	25	30
Distance (feet)	0	20	140	400	680	1400	1800

Find the average speed of the car over the interval from
(a) 0 to 10 seconds   (b) 10 to 20 seconds
(c) 20 to 30 seconds   (d) 15 to 30 seconds

2. Find the average rate of change of the volume of the balloon in Example 1 as the radius increases from
(a) 2 to 5 inches   (b) 4 to 8 inches

3. Use the function in Example 2 to find the average rate of change of Cleveland's population from
(a) 1920 to 1960   (b) 1990 to 2000.

4. The graph shows the total amount spent on advertising (in millions of dollars) in the United States, as estimated by a leading advertising publication.[*] Find the average rate of change in advertising expenditures over the following time periods:
(a) 1960–1970   (b) 1970–1980
(c) 1980–2001   (d) 1960–2001

(e) During which of these periods were expenditures increasing at the fastest rate?

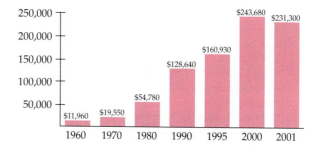

5. The table below shows the total projected elementary and secondary school enrollment (in thousands) for selected years.[†]

Find the average rate of change of enrollment from
(a) 1980 to 1985   (b) 1985 to 1995
(c) 1995 to 2005   (d) 2005 to 2010
(e) During which of these periods is enrollment increasing at the fastest rate? At the slowest rate?
(f) During which of these periods is enrollment decreasing at the fastest rate? At the slowest rate?

Year	1980	1985	1990	1995	2000	2005	2010
Enrollment	46,208	44,979	46,448	50,503	52,989	53,465	53,017

---

[*]*Advertising Age.*

[†]U.S. National Center for Education Statistics.

6. The graph shows the minimum wage (in constant 2000 dollars), with $x = 0$ corresponding to 1950.[*] Find the average rate of change in the minimum wage from
(a) 1950 to 1976 (b) 1976 to 1995
(c) 1995 to 2003 (d) 1976 to 2003

7. The table shows the total number of shares traded (in millions) on the New York Stock Exchange in selected years.[†]

Year	1993	1995	1997	1999	2001	2002
Volume	66.9	87.2	133.3	203.9	307.5	363.1

Find the average rate of change in share volume from
(a) 1993 to 1997 (b) 1995 to 1999
(c) 1999 to 2002 (d) 1993 to 2002
(e) During which of these periods did share volume increase at the fastest rate?

8. The graph at the top of the next column shows the approximate monthly supply of natural gas in the United States (in trillions of cubic feet), with $x = 0$ corresponding to March 1, 2001 and $x = 24$ to March 1, 2003.[‡] Find the average rate of change (in trillion cubic feet per month) over the interval:
(a) March 1, 2001, to December 1, 2001 ($x = 0$ to $x = 9$)
(b) March 1, 2001, to March 1, 2002
(c) December 1, 2001, to April 1, 2002
(d) April 1, 2002, to December 1, 2002
(e) December 1, 2002, to March 1, 2003
Find the average rate of change for each of the following periods and explain why this average is somewhat misleading.
(f) March 1, 2002, to June 1, 2002
(g) March 1, 2001, to March 1, 2003.

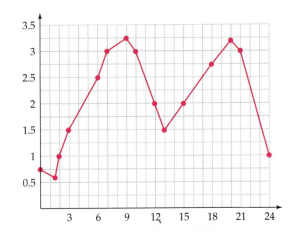

9. The Twells and Penrod Company has found that its sales are related to the amount of advertising it does in trade magazines. The graph in the figure shows the sales (in thousands of dollars) as a function of the amount of advertising (in number of magazine ad pages). Find the average rate of change of sales when the number of ad pages increases from
(a) 10 to 20 (b) 20 to 60
(c) 60 to 100 (d) 0 to 100
(e) Is it worthwhile to buy more than 70 pages of ads, if the cost of a one-page ad is $2000? If the cost is $5000? If the cost is $8000?

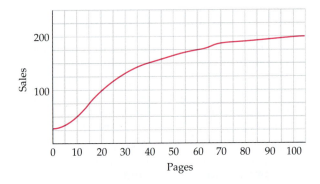

10. When blood flows through an artery (which can be thought of as a cylindrical tube) its velocity is greatest at the center of the artery. Because of friction along the walls of the tube, the blood's velocity decreases as the distance r from the center of the artery increases, finally becoming 0 at the wall of the artery. The velocity (in centimeters per second) is given by the function $v = 18{,}500(.000065 - r^2)$, where r is measured in centimeters. Find the average rate of change of the velocity as the distance from the center changes from
(a) $r = .001$ to $r = .002$ (b) $r = .002$ to $r = .003$
(c) $r = 0$ to $r = .005$

———————————
[*]U.S. Employment Standards Administration.
[†]*NYSE Factbook.*
[‡]Energy Information Administration.

11. A car is stopped at a traffic light and begins to move forward along a straight road when the light turns green. The distance (in feet) traveled by the car in t seconds is given by $s(t) = 2t^2$ $(0 \le t \le 30)$. What is the average speed of the car from

   (a) $t = 0$ to $t = 5$?   (b) $t = 5$ to $t = 10$?
   (c) $t = 10$ to $t = 30$?

In Exercises 12–20, find the average rate of change of the function f over the given interval.

12. $f(x) = 15x - 6$ from $x = 97$ to $x = 107$

13. $f(x) = -72x + 144$ from $x = 37$ to $x = 99$

14. $f(x) = 13x + 8$ from $x = -12$ to $x = 15$

15. $f(x) = 2 - x^2$ from $x = 0$ to $x = 2$

16. $f(x) = .25x^4 - x^2 - 2x + 4$ from $x = -1$ to $x = 4$

17. $f(x) = x^3 - 3x^2 - 2x + 6$ from $x = -1$ to $x = 3$

18. $f(x) = -\sqrt{x^4 - x^3 + 2x^2 - x + 4}$ from $x = 0$ to $x = 3$

19. $f(x) = \sqrt{x^3 + 2x^2 - 6x + 5}$ from $x = 1$ to $x = 2$

20. $f(x) = \dfrac{x^2 - 3}{2x - 4}$ from $x = 3$ to $x = 6$

21. Two cars race on a straight track, beginning from a dead stop. The distance (in feet) each car has covered at each time during the first 16 seconds is shown in the figure.

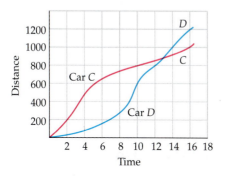

   (a) What is the average speed of each car during this 16-second interval?
   (b) Find an interval beginning at $t = 4$ during which the average speed of car D was approximately the same as the average speed of car C from $t = 2$ to $t = 10$.
   (c) Use secant lines and slopes to justify the statement "car D traveled at a higher average speed than car C from $t = 4$ to $t = 10$."

22. The figure shows the profits earned by a certain company during the last quarters of three consecutive years.

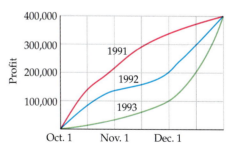

   (a) Explain why the average rate of change of profits from October 1 to December 31 was the same in all three years.
   (b) During what month in what year was the average rate of change of profits the greatest?

23. The graph in the figure shows the chipmunk population in a certain wilderness area. The population increases as the chipmunks reproduce but then decreases sharply as predators move into the area.

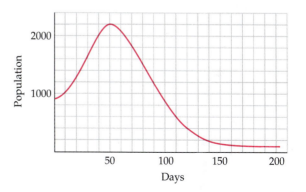

   (a) During what approximate time period (beginning on day 0) is the average growth rate of the chipmunk population positive?
   (b) During what approximate time period, beginning on day 0, is the average growth rate of the chipmunk population 0?
   (c) What is the average growth rate of the chipmunk population from day 50 to day 100? What does this number mean?
   (d) What is the average growth rate from day 45 to day 50? From day 50 to day 55? What is the approximate average growth rate from day 49 to day 51?

24. Lucy has a viral flu. How bad she feels depends primarily on how fast her temperature is rising at that time. The figure shows her temperature during the first day of the flu.

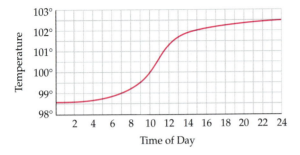

(a) At what average rate does her temperature rise during the entire day?

(b) During what two-hour period during the day does she feel worst?

(c) Find two time intervals, one in the morning and one in the afternoon, during which she feels about the same (that is, during which her temperature is rising at the same average rate).

25. The table shows the average weekly earnings (including overtime) of production workers and nonsupervisory employees in industry (excluding agriculture) in selected years.*

Year	1980	1985	1990	1995	2000	2001
Weekly Earnings	$235	$299	$345	$394	$474	$490

*U.S. Bureau of Labor Statistics.

(a) Use linear regression to find a function that models this data, with $x = 0$ corresponding to 1980.

(b) According to your function, what is the average rate of change in earnings over any time period between 1980 and 2001?

(c) Use the data in the table to find the average rate of change in earnings from 1980 to 1990 and from 1990 to 2001. How do these rates compare with the ones given by the model?

(d) If the model remains accurate, when will average weekly earnings reach $600?

26. The estimated number of 15- to 24-year-old people worldwide (in millions) who are or will be living with HIV/AIDS in selected years is given in the table.[†]

Year	2001	2003	2005	2007	2009
15- to 24-year-olds with HIV/AIDS	12	14.5	17	19	20.5

(a) Use linear regression to find a function that models this data, with $x = 0$ corresponding to 2000.

(b) According to your function, what is the average rate of change in this HIV/AIDS population over any time period between 2001 and 2009?

(c) Use the data in the table to find the average rate of change from 2001 to 2009. How does this rate compare with the one given by the model?

(d) If the model remains accurate, when will the number of people in this age group with HIV/AIDS reach 25 million?

[†]Kaiser Family Foundation, UNICEF, U.S. Census Bureau.

C 3.6.A *SPECIAL TOPICS* Instantaneous Rates of Change

OMIT

Recall that the distance (in feet) traveled by a falling rock in t seconds is given by $d(t) = 16t^2$. Consider the distances fallen by the rock over various time intervals:

Time Interval	Distance Traveled
$t = 0$ to $t = 1$	$d(1) - d(0) = 16 - 0 = 16$ feet
$t = 1$ to $t = 2$	$d(2) - d(1) = 64 - 16 = 48$ feet
$t = 2$ to $t = 3$	$d(3) - d(2) = 144 - 64 = 80$ feet

During each second, the rock falls farther than it did in the preceding second. So it must be traveling at faster and faster speeds as it falls. We know how to compute the *average* speed of the rock over any time interval, but what about its speed at a particular instant?

To make this discussion concrete, consider the exact speed of the rock at the instant when $t = 4$. We could approximate this speed by looking at a very small time interval near $t = 4$, say, from 4 to 4.01, or from 4 to 4.001. The average speed over such a short time interval cannot differ much from the exact speed at any time during the interval, so these average speeds should be reasonable approximations of the exact speed at $t = 4$. Note that

$$4.01 = 4 + .01 \qquad \text{and} \qquad 4.001 = 4 + .001.$$

Thus, we would do essentially the same thing in both cases: Compute the average speed (average rate of change of distance with respect to time) over an interval from 4 to $4 + h$, for some small nonzero quantity h.

EXAMPLE 1

Find the average speed of the falling rock from $t = 4$ to $t = 4 + h$, where h is a small nonzero quantity. Use the result to compute the average speed of the rock from 4 to 4.01 seconds and from 4 to 4.001 seconds.

SOLUTION   Using the average speed (rate of change) formula from Section 3.6, we have

$$\text{Average speed} = \frac{d(4 + h) - d(4)}{(4 + h) - 4} = \frac{16(4 + h)^2 - 16 \cdot 4^2}{h}$$

$$= \frac{16(16 + 8h + h^2) - 256}{h} = \frac{256 + 128h + 16h^2 - 256}{h}$$

$$= \frac{128h + 16h^2}{h} = \frac{h(128 + 16h)}{h} = 128 + 16h.$$

Therefore, from 4 to 4.01 seconds (that is, 4 to $4 + h$, with $h = .01$), the rock's average speed is

$$128 + 16h = 128 + 16(.01) = 128.16 \text{ feet per second.}$$

Similarly, its average speed from 4 to 4.001 seconds (here $h = .001$) is

$$128 + 16h = 128 + 16(.001) = 128.016 \text{ feet per second.}$$

These results suggest that the exact speed of the rock when $t = 4$ is about 128 feet per second.   ■

To approximate the speed of the rock at other times, we could perform a calculation similar to that in Example 1, with another number in place of 4. We can cover all the cases at once by considering a time interval from x to $x + h$.

EXAMPLE 2

Find the average speed of the falling rock from time x to time $x + h$, where x is a fixed number and h is a small nonzero quantity.

SOLUTION   The average rate of change formula shows that[*]

$$\text{Average speed} = \frac{d(x+h) - d(x)}{(x+h) - x} = \frac{16(x+h)^2 - 16x^2}{h}$$

$$= \frac{16(x^2 + 2xh + h^2) - 16x^2}{h} = \frac{16x^2 + 32xh + 16h^2 - 16x^2}{h}$$

$$= \frac{32xh + 16h^2}{h} = \frac{h(32x + 16h)}{h} = 32x + 16h. \quad \blacksquare$$

When $x = 4$, then the formula in Example 2 states that the average speed from 4 to $4 + h$ is $32(4) + 16h = 128 + 16h$, which is exactly what we found in Example 1. To find the average speed from 3 to 3.1 seconds, apply the formula

$$\text{Average speed} = 32x + 16h$$

with $x = 3$ and $h = .1$:

$$\text{Average speed} = 32 \cdot 3 + 16(.1) = 96 + 1.6 = 97.6 \text{ ft per second.}$$

More generally, we can compute the average rate of change of any function f over the interval from x to $x + h$ just as we did in Example 2: Apply the definition of average rate of change in the box on page 256 with x in place of a and $x + h$ in place of b:

$$\textbf{Average rate of change} = \frac{f(b) - f(a)}{b - a} = \frac{f(x+h) - f(x)}{(x+h) - x}$$

$$= \frac{f(x+h) - f(x)}{h}.$$

This last quantity is just the difference quotient of f (see page 205). Therefore,

Difference Quotients and Rates of Change

> If f is a function, then the average rate of change of f over the interval from x to $x + h$ is given by the difference quotient
>
> $$\frac{f(x+h) - f(x)}{h}.$$

EXAMPLE 3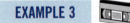

Find the difference quotient of $V(x) = x^3/55$, and use it to find the average rate of change of V as x changes from 8 to 8.01.

---

[*]Note that this calculation is the same as that in Example 1, except that 4 has been replaced by x.

SOLUTION   Use the definition of the difference quotient and algebra.

$$\frac{V(x+h)-V(x)}{h} = \frac{\overbrace{\frac{(x+h)^3}{55}}^{V(x+h)} - \overbrace{\frac{x^3}{55}}^{V(x)}}{h} = \frac{\frac{1}{55}[(x+h)^3 - x^3]}{h}$$

$$= \frac{1}{55} \cdot \frac{(x+h)^3 - x^3}{h} = \frac{1}{55} \cdot \frac{x^3 + 3x^2h + 3xh^2 + h^3 - x^3}{h}$$

$$= \frac{1}{55} \cdot \frac{3x^2h + 3xh^2 + h^3}{h} = \frac{1}{55} \cdot \frac{h(3x^2 + 3xh + h^2)}{h}$$

$$= \frac{3x^2 + 3xh + h^2}{55}.$$

When x changes from 8 to $8.01 = 8 + .01$, we have $x = 8$ and $h = .01$. So the average rate of change is

$$\frac{3x^2 + 3xh + h^2}{55} = \frac{3 \cdot 8^2 + 3 \cdot 8(.01) + (.01)^2}{55} \approx 3.495. \quad \blacksquare$$

▪▪ INSTANTANEOUS RATES OF CHANGE

Calculus is needed to determine exactly the instantaneous rate of change of a function (that is, its rate of change at a particular instant). As the preceding discussion suggests, however, accurate approximations of instantaneous rates of change may be found by evaluating the difference quotient for very small values of h.

EXAMPLE 4

A rock is dropped from a high place. What is its speed exactly 3 seconds after it is dropped?

SOLUTION   The distance the rock has fallen at time t is given by the function $d(t) = 16t^2$. The exact speed at $t = 3$ can be approximated by finding the average speed over very small time intervals, say, 3 to 3.01 or even shorter intervals. Over a very short time span, such as a hundredth of a second, the rock cannot change speed very much, so these average speeds should be a reasonable approximation of its speed at the instant $t = 3$. Example 2 shows that the average speed is given by the difference quotient $32x + 16h$. When $x = 3$, the difference quotient is $32 \cdot 3 + 16h = 96 + 16h$, and we have the following.

Change in Time 3 to 3 + h	h	Average Speed [Difference Quotient at $x = 3$] $96 + 16h$
3 to 3.1	.1	$96 + 16(.1) = 97.6$ ft per second
3 to 3.01	.01	$96 + 16(.01) = 96.16$ ft per second
3 to 3.005	.005	$96 + 16(.005) = 96.08$ ft per second
3 to 3.00001	.00001	$96 + 16(.00001) = 96.00016$ ft per second

The table suggests that the exact speed of the rock at the instant $t = 3$ seconds is very close to 96 feet per second. ■

EXAMPLE 5

A balloon is being filled with water in such a way that when its radius is x inches, then its volume is $V(x) = x^3/55$ gallons. What is the rate of change of the volume at the instant when the radius is 7 inches?

SOLUTION   The average rate of change when the radius goes from x to $x + h$ inches is given by the difference quotient of $V(x)$, which was found in Example 3:

$$\frac{V(x + h) - V(x)}{h} = \frac{3x^2 + 3xh + h^2}{55}.$$

Therefore, when $x = 7$, the difference quotient is

$$\frac{3 \cdot 7^2 + 3 \cdot 7 \cdot h + h^2}{55} = \frac{147 + 21h + h^2}{55},$$

and we have the following average rates of change over small intervals near 7.

Change in Radius 7 to 7 + h	h	Average Rate of Change of Volume [Difference Quotient at $x = 7$] $\dfrac{147 + 21h + h^2}{55}$
7 to 7.01	.01	2.6765 gallons per inch
7 to 7.001	.001	2.6731 gallons per inch
7 to 7.0001	.0001	2.6728 gallons per inch
7 to 7.00001	.00001	2.6727 gallons per inch

The chart suggests that at the instant the radius is 7 inches, the volume is changing at a rate of approximately 2.673 gallons per inch. ■

✓ EXERCISES 3.6.A

In Exercises 1–8, compute the difference quotient
$$\frac{f(x + h) - f(x)}{h}.$$

1. $f(x) = x + 5$

2. $f(x) = 7x + 2$

3. $f(x) = x^2 + 3$

4. $f(x) = x^2 + 3x - 1$

5. $f(t) = 160{,}000 - 8000t + t^2$

6. $V(x) = x^3$ **7.** $A(r) = \pi r^2$ **8.** $V(p) = 5/p$

9. Water is draining from a large tank. After t minutes, there are $160{,}000 - 8000t + t^2$ gallons of water in the tank.
   (a) Use the results of Exercise 5 to find the average rate at which the water runs out in the interval from 10 to 10.1 minutes.
   (b) Do the same for the interval from 10 to 10.01 minutes.
   (c) Estimate the rate at which the water runs out after exactly 10 minutes.

10. Use the results of Exercise 6 to find the average rate of change of the volume of a cube whose side has length x as x changes from
   (a) 4 to 4.1   (b) 4 to 4.01   (c) 4 to 4.001.
   (d) Estimate the rate of change of the volume at the instant when $x = 4$.

11. Use the results of Exercise 7 to find the average rate of change of the area of a circle of radius r as r changes from
   (a) 3 to 3.5   (b) 3 to 3.2   (c) 3 to 3.1.
   (d) Estimate the rate of change at the instant when $r = 3$.
   (e) How is your answer in part (d) related to the circumference of a circle of radius 3?

12. Under certain conditions, the volume V of a quantity of air is related to the pressure p (which is measured in kilopascals) by the equation $V = 5/p$. Use the results of Exercise 8 to estimate the rate at which the volume is changing at the instant when the pressure is 50 kilopascals.

Chapter 3 Review

IMPORTANT FACTS & FORMULAS

- The average rate of change of a function f as x changes from a to b is the number

$$\frac{f(b) - f(a)}{b - a}.$$

- The difference quotient of the function f is the quantity

$$\frac{f(x + h) - f(x)}{h},$$

- The average rate of change of a function f as x changes from a to b is the slope of the secant line joining the points $(a, f(a))$ and $(b, f(b))$.

IMPORTANT FACTS & FORMULAS

CATALOG OF BASIC FUNCTIONS—PART 1

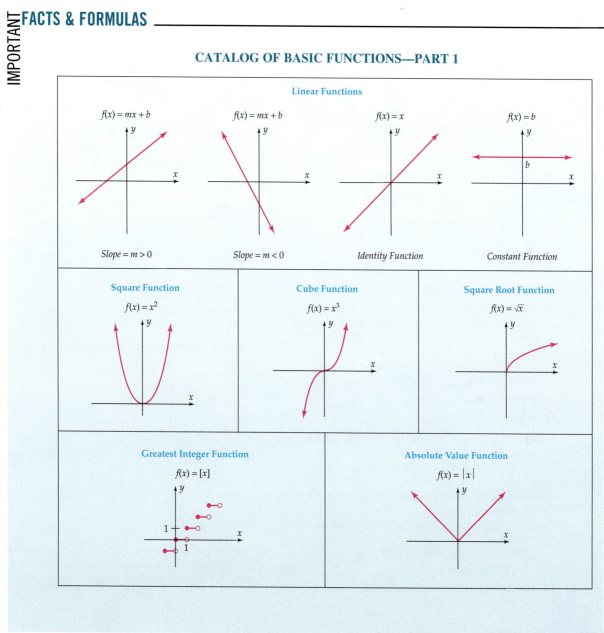

REVIEW QUESTIONS

1. Let $[x]$ denote the greatest integer function and evaluate

   (a) $[-5/2] = $ _____.   (b) $[1755] = $ _____.

   (c) $[18.7] + [-15.7] = $ _____.

   (d) $[-7] - [7] = $ _____.

2. If $f(x) = x + |x| + [x]$, then find $f(0), f(-1), f(1/2)$, and $f(-3/2)$.

3. Let f be the function given by the rule $f(x) = 7 - 2x$. Complete this table.

x	0	1	2	-4	t	k	$b-1$	$1-b$	$6-2u$
$f(x)$	7								

REVIEW QUESTIONS

4. What is the domain of the function g given by

$$g(t) = \frac{\sqrt{t - 2}}{t - 3}?$$

5. In each case, give a *specific* example of a function and numbers a, b to show that the given statement may be *false*.
(a) $f(a + b) = f(a) + f(b)$ (b) $f(ab) = f(a)f(b)$

6. If $f(x) = |3 - x| \sqrt{x - 3} + 7$, then

$$f(7) - f(4) = \underline{\quad\quad} .$$

7. What is the domain of the function given by

$$g(r) = \sqrt{r - 4} + \sqrt{r - 2}?$$

8. What is the domain of the function $f(x) = \sqrt{-x + 2}$?

9. If $h(x) = x^2 - 3x$, then $h(t + 2) = \underline{\quad}$.

10. Which of the following statements about the greatest integer function $f(x) = [x]$ is true for *every* real number x?
(a) $x - [x] = 0$ (b) $x - [x] \leq 0$
(c) $[x] + [-x] \leq 0$ (d) $[-x] \geq [x]$
(e) $3[x] = [3x]$

11. If $f(x) = 2x^3 + x + 1$, then $f(x/2) = \underline{\quad}$.

12. If $g(x) = x^2 - 1$, then $g(x - 1) - g(x + 1) = \underline{\quad}$.

13. The radius of an oil spill (in meters) is 50 times the square root of the time t (in hours).
(a) Write the rule of a function f that gives the radius of the spill at time t.
(b) Write the rule of a function g that gives the area of the spill at time t.
(c) What are the radius and area of the spill after 9 hours?
(d) When will the spill have an area of 100,000 square meters?

14. The cost of renting a limousine for 24 hours is given by

$$C(x) = \begin{cases} 150 & \text{if } 0 < x \leq 25 \\ 1.75x + 150 & \text{if } x > 25, \end{cases}$$

where x is the number of miles driven.
(a) What is the cost if the limo is driven 20 miles? 30 miles?
(b) If the cost is $218.25, how many miles were driven?

15. Sketch the graph of the function f given by

$$f(x) = \begin{cases} x^2 & \text{if } x \leq 0 \\ x + 1 & \text{if } 0 < x < 4 \\ \sqrt{x} & \text{if } x \geq 4. \end{cases}$$

16. U.S. Express Mail rates in 2003 are shown in the following table. Sketch the graph of the function e, whose rule is $e(x)$ = cost of sending a package weighing x pounds by Express Mail.

EXPRESS MAIL Letter Rate—Post Office to Addressee Service	
Up to 8 ounces	$13.65
Over 8 ounces to 2 pounds	$17.85
Up to 3 pounds	$21.05
Up to 4 pounds	$24.20
Up to 5 pounds	$27.30
Up to 6 pounds	$30.40
Up to 7 pounds	$33.45

17. Which of the following are graphs of functions of x?
(a)

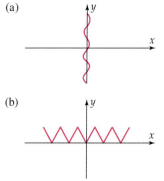

(b)

18. The function whose graph is shown gives the number of Americans who snowboard (in millions) in year x where $x = 0$ corresponds to 1988.[*]

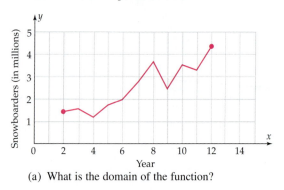

(a) What is the domain of the function?

*National Sporting Goods Association, TransWorld Snowboarding Business.

(b) Approximately, what is the range of the function?

(c) Find the average rate of change of the function from $x = 6$ to $x = 8$ and from $x = 9$ to $x = 10$.

In Questions 19–22, determine the local maxima and minima of the function, the intervals on which the function is increasing, and the intervals on which it is decreasing.

19. $g(x) = \sqrt{x^2 + x + 1}$

20. $f(x) = 2x^3 - 5x^2 + 4x - 3$

21. $g(x) = x^3 + 8x^2 + 4x - 3$

22. $f(x) = .5x^4 + 2x^3 - 6x^2 - 16x + 2$

In Questions 23 and 24, sketch the graph of the curve given by the parametric equations.

23. $x = t^2 - 4$ and $y = 2t + 1$ $(-3 \le t \le 3)$

24. $x = t^3 + 3t^2 - 1$ and $y = t^2 + 1$ $(-3 \le t \le 2)$

25. Sketch a graph that is symmetric with respect to both the x-axis and the y-axis. (There are many correct answers and your graph need not be the graph of an equation.)

26. Sketch the graph of a function that is symmetric with respect to the origin. (There are many correct answers and you don't have to state the rule of your function.)

In Questions 27 and 28, determine algebraically whether the graph of the given equation is symmetric with respect to the x-axis, the y-axis, or the origin.

27. $x^2 = y^2 + 2$

28. $5y = 7x^2 - 2x$

In Questions 29–31, determine whether the given function is even, odd, or neither.

29. $g(x) = 9 - x^2$

30. $f(x) = |x|x + 1$

31. $h(x) = 3x^5 - x(x^4 - x^2)$

32. (a) Draw some coordinate axes and plot the points $(-2, 1)$, $(-1, 3)$, $(0, 1)$, $(3, 2)$, $(4, 1)$.

(b) Suppose the points plotted in part (a) lie on the graph of an *even* function f. Plot these points: $(2, f(2))$, $(1, f(1))$, $(0, f(0))$, $(-3, f(-3))$, $(-4, f(-4))$.

33. Determine whether the circle with equation

$$x^2 + y^2 + 6y = -5$$

is symmetric with respect to the x-axis, the y-axis, or the origin.

34. Sketch the graph of a function f that satisfies all of these conditions:

(i) The domain of f consists of all x such that $-3 \le x \le 4$.

(ii) The range of f consists of all y such that $-2 \le y \le 5$.

(iii) $f(-2) = 0$

(iv) $f(1) > 2$

[*Note:* There are many possible correct answers and the function whose graph you sketch need *not* have an algebraic rule.]

35. Sketch the graph of $g(x) = 5 + \dfrac{4}{x - 5}$.

Use the graph of the function f in the figure to answer Questions 36–39.

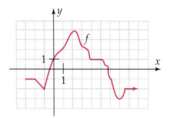

36. What is the domain of f?

37. What is the range of f?

38. Find all numbers x such that $f(x) = 1$.

39. Find a number x such that $f(x + 1) < f(x)$. (Many correct answers are possible.)

Use the graph of the function f in the figure to answer Questions 40–46.

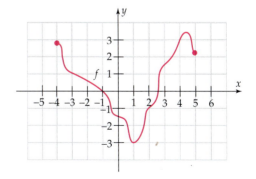

40. What is the domain of f?   **41.** $f(-3) = \underline{\hspace{1cm}}$.

42. $f(2 + 2) = \underline{\hspace{1cm}}$.

43. $f(-1) + f(1) = \underline{\hspace{1cm}}$.

44. True or false: $2f(2) = f(4)$.

45. True or false: $3f(2) = -f(4)$.

46. True or false: $f(x) = 3$ for exactly one number x.

Use the graphs of the functions f and g in the figure to answer Questions 47–52.

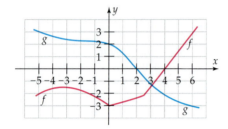

47. For which values of x is $f(x) = 0$?

48. True or false: If a and b are numbers such that $-5 \leq a < b \leq 6$, then $g(a) < g(b)$.

49. For which values of x is $g(x) \geq f(x)$?

50. Find $f(0) - g(0)$.

51. For which values of x is $f(x + 1) < 0$?

52. What is the distance from the point $(-5, g(-5))$ to the point $(6, g(6))$?

53. Fireball Bob and King Richard are two NASCAR racers. The graph shows their distance traveled in a recent race as a function of time.
 (a) Which car made the most pit stops?
 (b) Which car started out the fastest?
 (c) Which car won the race?

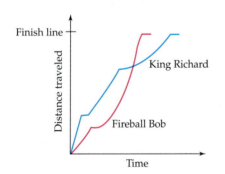

In Questions 54–57, list the transformations, in the order in which they should be performed on the graph of $g(x) = x^2$, so as to produce a complete graph of the function f.

54. $f(x) = (x - 2)^2$

55. $f(x) = .25x^2 + 2$

56. $f(x) = -(x + 4)^2 - 5$

57. $f(x) = -3(x - 7)^2 + 2$

58. The graph of a function f is shown in the figure. On the same coordinate plane, carefully draw the graphs of the functions g and h whose rules are

$$g(x) = -f(x) \quad \text{and} \quad h(x) = 1 - f(x).$$

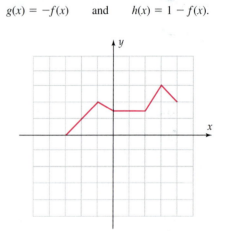

59. The figure shows the graph of a function f. If g is the function given by $g(x) = f(x + 2)$, then which of these statements about the graph of g is true?
 (a) It does not cross the x-axis.
 (b) It does not cross the y-axis.
 (c) It crosses the y-axis at $y = 4$.
 (d) It crosses the y-axis at the origin.
 (e) It crosses the x-axis at $x = -3$.

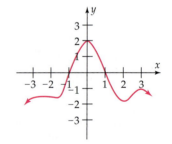

60. If $f(x) = 3x + 2$ and $g(x) = x^3 + 1$, find
 (a) $(f + g)(-1)$ (b) $(f - g)(2)$ (c) $(fg)(0)$.

61. If $f(x) = \dfrac{1}{x - 1}$ and $g(x) = \sqrt{x^2 + 5}$, find
 (a) $(f/g)(2)$ (b) $(g/f)(x)$
 (c) $(fg)(c + 1)$ $(c \neq 0)$.

62. Find two functions f and g such that neither is the identity function, and

$$(f \circ g)(x) = (2x + 1)^2.$$

QUESTIONS

REVIEW QUESTIONS

63. Use the graph of the function g in the figure to fill in the following table, in which h is the composite function $g \circ g$.

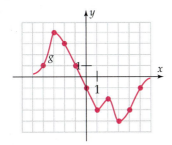

x	-4	-3	-2	-1	0	1	2	3	4
$g(x)$					-1				
$h(x) = g(g(x))$									

Questions 64–69 refer to the functions $f(x) = \dfrac{1}{x+1}$ *and* $g(t) = t^3 + 3.$

64. $(f \circ g)(1) = $ _____.   **65.** $(g \circ f)(2) = $ _____.

66. $g(f(-2)) = $ _____.   **67.** $(g \circ f)(x - 1) = $ _____.

68. $g(2 + f(0)) = $ _____.   **69.** $f(g(1) - 1) = $ _____.

70. Let f and g be the functions given by

$$f(x) = 4x + x^4 \quad \text{and} \quad g(x) = \sqrt{x^2 + 1}$$

(a) $(f \circ g)(x) = $ _____.   (b) $(g - f)(x) = $ _____.

71. If $f(x) = \dfrac{1}{x}$ and $g(x) = x^2 - 1$, then

$(f \circ g)(x) = $ _____   and $(g \circ f)(x) = $ _____.

72. Let $f(x) = x^2$. Give an example of a function g with domain all real numbers such that $g \circ f \neq f \circ g$.

73. If $f(x) = \dfrac{1}{1 - x}$ and $g(x) = \sqrt{x}$, then find the domain of the composite function $f \circ g$.

74. These tables show the values of the functions f and g at certain numbers:

x	-1	0	1	2	3
$f(x)$	1	0	1	3	5

and

t	0	1	2	3	4
$g(t)$	-1	0	1	2	5

Which of the following statements are *true*?
(a) $(g - f)(1) = 1$ (b) $(f \circ g)(2) = (f - g)(0)$
(c) $f(1) + f(2) = f(3)$ (d) $(g \circ f)(2) = 1$
(e) None of the above is true.

75. Find the average rate of change of the function

$$g(x) = \frac{x^3 - x + 1}{x + 2}$$

as x changes from
(a) -1 to 1   (b) 0 to 2.

76. Find the average rate of change of the function $f(x) = \sqrt{x^2 - x + 1}$ as x changes from
(a) -3 to 0   (b) -3 to 3.5   (c) -3 to 5.

77. If $f(x) = 2x + 1$ and $g(x) = 3x - 2$, find the average rate of change of the composite function $f \circ g$ as x changes from 3 to 5.

78. If $f(x) = x^2 + 1$ and $g(x) = x - 2$, find the average rate of change of the composite function $f \circ g$ as x changes from -1 to 1.

In Questions 79–82, find the difference quotient of the function.

79. $f(x) = 3x + 4$ **80.** $g(x) = \sqrt{x}$

81. $g(x) = x^2 - 1$ **82.** $f(x) = x^2 + x$

83. The profit (in hundreds of dollars) from selling x tons of Wonderchem is given by

$$P(x) = .2x^2 + .5x - 1.$$

What is the average rate of change of profit when the number of tons of Wonderchem sold increases from
(a) 4 to 8 tons?   (b) 4 to 5 tons?   (c) 4 to 4.1 tons?

84. On the planet Mars, the distance traveled by a falling rock (ignoring atmospheric resistance) in t seconds is $6.1t^2$ feet. How far must a rock fall in order to have an average speed of 25 feet per second over that time interval?

(*Exercises continue on next page.*)

REVIEW QUESTIONS

85. The graph in the figure shows the population of fruit flies during a 50-day experiment in a controlled atmosphere.

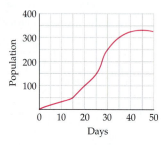

(a) During what 5-day period is the average rate of population growth the slowest?

(b) During what 10-day period is the average rate of population growth the fastest?

(c) Find an interval beginning at the 30th day during which the average rate of population growth is the same as the average rate from day 10 to day 20.

86. The graph of the function g in the figure consists of straight line segments. Find an interval over which the average rate of change of g is

(a) 0   (b) −3   (c) .5.

(d) Explain why the average rate of change of g is the same from −3 to −1 as it is from −2.5 to 0.

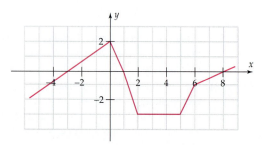

87. The table shows the median sale price for single-family homes in the western United States in selected years.*

Year	1990	1995	1998	2000	2002	2003
Median Price	$129,600	$141,000	$164,800	$183,000	$215,000	$217,200

(a) Find the average rate of change of the median price from 1990 to 1998, from 1998 to 2003, and from 1990 to 2003.

(b) If the average rate of change from 1998 to 2003 continues in years after 2003, what is the median price in 2007?

---

*National Association of Realtors; 2003 data is for the first quarter.

DISCOVERY PROJECT 3   Feedback—Good and Bad

Whether it's for a concert or a worship service, a sporting event or a graduation ceremony, a shopping mall or a lecture hall, audio engineers are concerned to avoid audio feedback. If the system is not set up correctly, then sound from the speakers reaches the microphone with enough clarity that it is fed back into the audio system and amplified. This feedback cycle repeats again and again. Each time the amplified sound follows very quickly behind the previous one so that the audience hears an unpleasant hum that rapidly becomes a loud screech. This is feedback we would prefer to avoid.

Other types of feedback are considerably more pleasing. Consider, for example, an investment of x dollars that earns 4% interest compounded annually. Then the amount in the account (principal plus interest) after one year is $1.04x$. Assume the initial deposit is left in the account without adding deposits or withdrawing funds. If $1000 is invested, then the balance (rounded to the nearest penny) is

Beginning	1000
End of year 1	$1.04 \cdot 1000 = 1040$
End of year 2	$1.04 \cdot 1081.60 = 1081.60$
End of year 3	$1.04 \cdot 1124.86 = 1124.86$

and so on. You can easily carry out this process on a calculator as follows: key in 1000 and press ENTER; then key in "× 1.04" and press ENTER repeatedly. Each time you press ENTER, the calculator computes

$$\text{(Answer for preceding calculation)} \times 1.04$$

as shown in Figure 1. The result is an ever-increasing bank balance.

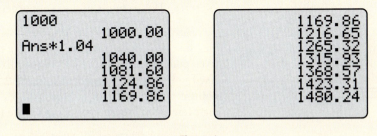

Figure 1

In each of the preceding examples, the output of the process was fed back as input and the process repeated to generate the result—unpleasant in the first case and quite pleasant in the second. Now we strip away the specifics and look at the process in more general terms. Suppose we have a function f and x is in its domain. We find $f(x)$ and feed that value back to the function to find

275

$f(f(x))$. Then feed that value back to find $f(f(f(x)))$. The sequence of values obtained by continuing this process

$$f(x), f(f(x)), f(f(f(x))), f(f(f(f(x)))), \ldots$$

is called the **orbit** of x and the feedback process is called **iteration.** The functions themselves $(f, f^2 = f \circ f, f^3 = f \circ f \circ f, \ldots)$ are called **iterates** of f.

Technology makes it easy to compute orbits. For example, let $f(x) = \sqrt{x}$ and enter f as Y_1 in your calculator. Key in 8 and press ENTER. Then key in "STO➡ X : Y_1" and press ENTER repeatedly.* Each time you press ENTER, the calculator stores the answer from the preceding calculation as X and evaluates Y_1 at X, as shown in Figure 2.

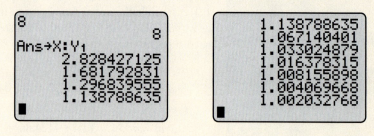

Figure 2

Thus, the (approximate) orbit of 8 is

$$\{8, 2.828, 1.682, 1.297, 1.139, 1.067, 1.033, \ldots\}.$$

Note that the terms of the orbit are getting closer and closer to 1.

1. Find the orbit of .2 and two other numbers of your choice (except 0 and 1). Do the terms of these orbits get closer and closer to 1?

For the function $f(x) = \sqrt{x}$, we can see that $f(1) = 1$, so iteration produces the orbit $\{1, 1, 1, \ldots\}$; we say that 1 is a **fixed point** for this function. Another fixed point is zero because $f(0) = 0$. A fixed point a is called an **attracting fixed point** if nearby numbers have orbits that approach a or a **repelling fixed point** if nearby numbers have orbits that move farther away from a. The following exercises illustrate these ideas for the function $f(x) = \sqrt{x}$.

2. Confirm numerically that 1 is an attracting fixed point by computing the orbits of .5, 2, and 11.†

3. Explain mathematically why 1 is an attracting fixed point. [*Hint:* You may assume that if $u < v$, then $\sqrt{u} < \sqrt{v}$. Is $\sqrt{x}$ larger or smaller than x when $x < 1$? When $x > 1$?]

---

*The colon (:) is on the TI and HP-39+ keyboards and in the Casio PRGM menu (press SHIFT VARS).

†Your calculator may tell you that the terms of the orbits eventually are equal to 1, but this is due to round-off error.

4. Determine whether zero is an attracting fixed point or a repelling fixed point.

5. Do you think this function has any other fixed points? Why or why not?

Generalizations are interesting in their own right but you may be wondering whether these ideas have any relevance to the real world. So let's look at another application. Ecologists want to model the growth of the population of some animal in an ecosystem. To simplify notation, let's let the variable x represent a fraction of the theoretical maximum population, so that $0 \le x \le 1$. [If the population were the maximum the ecosystem could sustain, then $x = 1$ and if there are none of these animals in the ecosystem, then $x = 0$. We're interested in the usual situation where x is somewhere in between.] With a growth rate r, the function $f(x) = rx$ seems to fill the bill.* But logistics, the limitation of resources (food, nesting locations, etc.), demands an adjustment. The **logistic function** $f(x) = rx(1 - x)$ gives a more realistic model of the growth, which is nearly exponential when x is small (and $1 - x$ is nearly 0). It also reflects the influence of logistics in that $(1 - x)$ approaches 0 as x approaches 1.

Do these exercises for the logistic function $f(x) = rx(1 - x)$ and the given value of r.

6. Use $r = 2$ and different initial values (such as .25, .8, and .1) to confirm that $f(x) = 2x(1 - x)$ has fixed point .5.

7. Find the fixed point (to the nearest hundredth) when $r =$

   (a) 2.5   (b) 2.8   (c) 3

8. Verify that for $r = 3.3$, the orbit of any point x (with $0 < x < 1$) seems to (eventually) jump back and forth between *two* values. Somewhere between 3 and 3.3 a **bifurcation** has occurred. Instead of a single fixed point we now have a periodic oscillation, with **period 2.** Find the two values.

9. Verify that periodic oscillation also occurs when $r = 3.4$. Is the period still 2?

10. Verify that $r = 3.5$ eventually oscillates among *four* values (that is, the period is $4 = 2^2$). Another bifurcation has occurred. Find the four values.

11. Verify that for $r = 3.54$ the period is still four but at $r = 3.56$ there are *eight* values (period is $8 = 2^3$). Find them.

12. Is it possible to determine exactly where bifurcations occur? When r is very large, do you think that the system breaks down to "chaos" or that there are values of r that have period 2^n for large values of n?

13. If you are able to consult a naturalist, determine whether some animal populations follow yearly patterns consistent with the logistic model.

*Notice the similarity to the 4% interest example, where the "balance function" was $f(x) = 1.04x$. Notice also, that if r is less than 1, the population declines to extinction.

Polynomial and Rational Functions

Rob Crandall/Stock, Boston Inc./PictureQuest

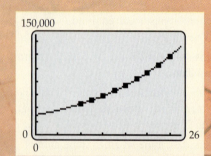

Can you afford to go to college?

As you (and your parents) know, the cost of a four-year college education (tuition, fees, room, board, books) has steadily increased. It is expected to continue to do so in the foreseeable future, according to projections by a large insurance company. This growth can be modeled by a fourth-degree polynomial function. See Exercise 38 on page 363.

Chapter Outline

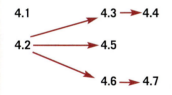

Polynomial functions arise naturally in many applications. Many complicated functions in applied mathematics can be approximated by polynomial functions or their quotients (rational functions).

4.1   Quadratic Functions and Models*

A **quadratic function** is a function whose rule can be written in the form

$$f(x) = ax^2 + bx + c$$

for some constants a, b, c, with $a \neq 0$. The graph of a quadratic function is called a **parabola.**

GRAPHING EXPLORATION

Using the standard viewing window, graph the following quadratic functions on the same screen.

$$f(x) = x^2, \qquad f(x) = 3x^2 + 30x + 77, \qquad f(x) = -x^2 + 4x,$$

$$f(x) = -.2x^2 + 1.5x - 5.$$

As the Exploration illustrates, all parabolas have the same basic "cup" shape, though the cup may be broad or narrow. The parabola opens upward when the coefficient of x^2 is positive and downward when this coefficient is negative.

---

*This section may be omitted or postponed if desired. Section 3.4 (Graphs and Transformations) is a prerequisite.

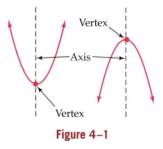

Figure 4–1

If a parabola opens upward, its **vertex** is the lowest point on the graph and if a parabola opens downward, its **vertex** is the highest point on the graph, as shown in Figure 4–1. Every parabola is symmetric with respect to the vertical line through its vertex; this line is called the **axis** of the parabola.

Parabolas are easily graphed on a calculator. The vertex can always be approximated by using the trace feature or a maximum/minimum finder. However, algebraic techniques can be used to find the vertex precisely and to graph the parabola.

EXAMPLE 1

Show that the function $g(x) = 2(x - 3)^2 + 1$ is quadratic, graph the function, and find its vertex.

SOLUTION   The function g is quadratic because its rule can be written in the form $g(x) = ax^2 + bx + c$:

$$g(x) = 2(x - 3)^2 + 1 = 2(x^2 - 6x + 9) + 1 = 2x^2 - 12x + 19.$$

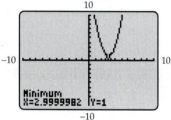

Figure 4–2

Graphing the function in the standard viewing window (Figure 4–2) and using the minimum finder, we see that the vertex is approximately $(2.999, 1)$. To find the vertex exactly, we use the techniques of Section 3.4. The graph of

$$g(x) = 2(x - 3)^2 + 1$$

can be obtained from the graph of $y = x^2$ as follows.

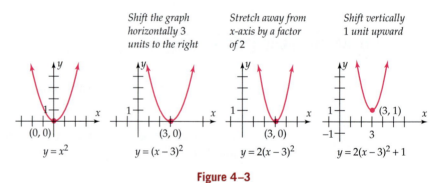

Figure 4–3

Figure 4–3 shows that when the vertex $(0, 0)$ of $y = x^2$ is shifted 3 units to the right and 1 unit upward, it moves to $(3, 1)$. Therefore, $(3, 1)$ is the vertex of $g(x) = 2(x - 3)^2 + 1$. Note how the coordinates of the vertex are related to the rule of the function g.

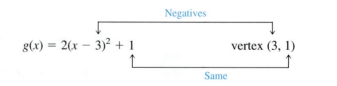

The vertex of the function g in Example 1 was easily determined because the rule of g had a special algebraic form. The vertex of the graph of any quadratic function can be determined in a similar fashion by first rewriting its rule.

EXAMPLE 2

Find the vertex of the graph of $g(x) = 3x^2 + 30x + 77$ algebraically.

SOLUTION   We rewrite the rule of g as follows, being careful at each step not to change its value.

$$g(x) = 3x^2 + 30x + 77$$

Factor out 3: $$= 3(x^2 + 10x) + 77$$

Add $25 - 25$ inside parentheses:* $$= 3(x^2 + 10x + 25 - 25) + 77$$

Use the distributive law: $$= 3(x^2 + 10x + 25) - 3 \cdot 25 + 77$$

Simplify: $$= 3(x^2 + 10x + 25) + 2$$

Factor expression in parentheses: $$= 3(x + 5)^2 + 2$$

As we saw in Section 3.4, the graph of $g(x)$ is the graph of $f(x) = x^2$ shifted horizontally 5 units to the left, stretched by a factor of 3, and shifted 2 units upward, as shown in Figure 4–4. In this process, the vertex $(0, 0)$ of f moves to $(-5, 2)$. Therefore, $(-5, 2)$ is the vertex of $g(x) = 3(x + 5)^2 + 2$. Once again, note how the coordinates of the vertex are related to the rule of the function.

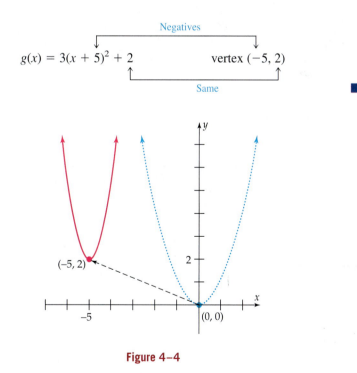

Negatives

$g(x) = 3(x + 5)^2 + 2$ vertex $(-5, 2)$

Same

Figure 4–4

---

*Since $25 - 25 = 0$, we haven't changed the value of the function. The number 25, which is the square of half the coefficient of x, is chosen because it will enable us to factor an expression below. This technique is called *completing the square*.

The technique used to find the vertex in Example 2 works for any quadratic function $f(x) = ax^2 + bx + c$. In what follows, don't let all the letters scare you. In Example 2 we replace 3 by a, 30 by b, and 77 by c, and then use exactly the same five-step computation to rewrite the rule of f.

$$f(x) = ax^2 + bx + c$$

Factor out a:
$$= a\left(x^2 + \frac{b}{a}x\right) + c$$

Add $\dfrac{b^2}{4a^2} - \dfrac{b^2}{4a^2}$ inside parentheses.[*]
$$= a\left(x^2 + \frac{b}{a}x + \frac{b^2}{4a^2} - \frac{b^2}{4a^2}\right) + c$$

Use the distributive law:
$$= a\left(x^2 + \frac{b}{a}x + \frac{b^2}{4a^2}\right) - a \cdot \frac{b^2}{4a^2} + c$$

Simplify and rearrange:
$$= a\left(x^2 + \frac{b}{a}x + \frac{b^2}{4a^2}\right) + \left(c - \frac{b^2}{4a}\right)$$

Factor first expression in parentheses:[†]
$$= a\left(x + \frac{b}{2a}\right)^2 + \left(c - \frac{b^2}{4a}\right).$$

As in the preceding examples, the graph of f is just the graph of x^2 shifted horizontally, stretched by a factor of a, and shifted vertically. As above, the vertex of this parabola can be read from the rule of the function:

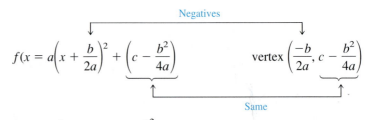

If we let $h = -\dfrac{b}{2a}$ and $k = c - \dfrac{b^2}{4a}$, then we have these useful facts.

Quadratic Functions

> The rule of the quadratic function $f(x) = ax^2 + bx + c$ can be rewritten in the form
>
> $$f(x) = a(x - h)^2 + k,$$
>
> where $h = -b/2a$. The graph of f is a parabola with vertex (h, k). It opens upward if $a > 0$ and downward if $a < 0$.

---

[*]Since $\dfrac{b^2}{4a^2} - \dfrac{b^2}{4a^2} = 0$, we haven't changed the value of the function. This is another example of "completing the square" (the number $b^2/4a^2$ is the square of half the coefficient of x). Note that when $a = 3$ and $b = 30$ (as in Example 2), then $\dfrac{b^2}{4a^2} = \dfrac{30^2}{4 \cdot 3^2} = \dfrac{900}{36} = 25$.

[†]Verify that this factorization is correct by multiplying out $\left(x + \dfrac{b}{2a}\right)^2$.

EXAMPLE 3

Describe the graph of

$$f(x) = -x^2 + 5x + 1$$

and find its vertex and x-intercepts exactly.

SOLUTION   The graph is a downward-opening parabola (because f is a quadratic function and the coefficient of x^2 is negative). According to the preceding box, the x-coordinate of its vertex is

$$-\frac{b}{2a} = -\frac{5}{2(-1)} = \frac{-5}{-2} = \frac{5}{2} = 2.5.$$

To find the y-coordinate of the vertex, we need only evaluate f at this number.

$$f(2.5) = -(2.5^2) + 5(2.5) + 1 = 7.25$$

Therefore, the vertex is (2.5, 7.25).

The x-intercepts are the points on the graph with y-coordinate 0, that is, the points where $f(x) = 0$. So we must solve the following equation.

$$-x^2 + 5x + 1 = 0$$

Multiply both sides by -1: $x^2 - 5x - 1 = 0$

Since the left-hand side doesn't readily factor, we use the quadratic formula.

$$x = \frac{-b \pm \sqrt{b^2 - 4ac}}{2a} = \frac{-(-5) \pm \sqrt{(-5)^2 - 4(1)(-1)}}{2 \cdot 1} = \frac{5 \pm \sqrt{29}}{2}$$

Therefore, the x intercepts are at $\dfrac{5 - \sqrt{29}}{2} \approx -.1926$ and $\dfrac{5 + \sqrt{29}}{2} \approx 5.1926$. ■

GRAPHING EXPLORATION

Graphically confirm the results of Example 3 by graphing $y = -x^2 + 5x + 1$ and using your maximum finder to approximate the vertex and your root finder to approximate the x-intercepts.

EXAMPLE 4

Find the rule of the quadratic function whose graph is a parabola with vertex (3, 4) that passes through the point $(-1, 36)$.

SOLUTION   The rule of a quadratic function can be written in the form $f(x) = a(x - h)^2 + k$ and its graph is a parabola with vertex (h, k). In this case, the vertex is (3, 4), so we have $h = 3$ and $k = 4$. Hence, the rule of f is $f(x) = a(x - 3)^2 + 4$. Since $(-1, 36)$ is on the graph, we have

$$f(-1) = 36$$

Substitute -1 for x in the rule of f: $a(-1 - 3)^2 + 4 = 36$

Simplify: $16a + 4 = 36$

Subtract 4 from both sides: $16a = 32$

Divide both sides by 16: $a = 2.$

Therefore, the rule of the function is $f(x) = 2(x - 3)^2 + 4.$ ■

▊▊ APPLICATIONS

Quadratic functions arise in the design of satellite dishes, microphones, and searchlights. They can also be used to track the path of a projectile, such as an artillery shell or a baseball.

EXAMPLE 5

Sammy Sosa hits a baseball. The height of the ball is given by

$$h(x) = -.001x^2 + .45x + 3,$$

where x is the distance on the ground from the batter to the point directly under the ball.

(a) How high does the ball go?

(b) How far from the batter does it land?

SOLUTION

(a) Imagine a coordinate plane with the batter at the origin, the y-axis perpendicular to the ground, and the ball traveling in the direction of the x-axis, as shown in Figure 4–5 (which is not to scale). The ball follows a parabolic path, namely, the graph of h. The highest point it reaches (the vertex of the parabola) has x-coordinate

$$-\frac{b}{2a} = -\frac{.45}{2(-.001)} = 225$$

and y-coordinate

$$f(225) = -.001(225^2) + .45(225) + 3 = 53.625.$$

Therefore, the maximum height of the ball is 53.625 feet.

(b) The ball hits ground when its height is 0, that is, when

$$h(x) = 0$$

$$-.001x^2 + .45x + 3 = 0.$$

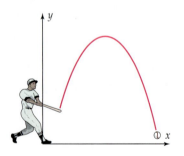

Figure 4–5

This equation can be solved by the quadratic formula:

$$x = \frac{-b \pm \sqrt{b^2 - 4ac}}{2a} = \frac{-.45 \pm \sqrt{.45^2 - 4(-.001)(3)}}{2(-.001)}$$

$$= \frac{-.45 \pm \sqrt{.2145}}{-.002} \approx \begin{cases} -6.57 \\ \text{or} \\ 456.57. \end{cases}$$

The negative solution does not apply here, and we see that the ball hits ground 456.57 feet from the batter. ■

EXPLORATION

Graphically confirm the results of Example 5 by graphing $y = -.001x^2 + .45x + 3$ and using your maximum finder to approximate the vertex and your root finder to approximate the appropriate x-intercept.

EXAMPLE 6

Find the area and dimensions of the largest rectangular field that can be enclosed with 3000 feet of fence.

SOLUTION　Let x denote the length and y the width of the field, as shown in Figure 4–6.

$$\text{Perimeter} = x + y + x + y$$

$$= 2x + 2y$$

$$\text{Area} = xy.$$

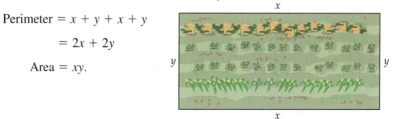

Figure 4–6

Since the perimeter is the length of the fence, $2x + 2y = 3000$. Hence,

$$2y = 3000 - 2x$$

and $y = 1500 - x$. Consequently, the area is

$$A = xy = x(1500 - x) = 1500x - x^2 = -x^2 + 1500x.$$

The largest possible area is just the maximum value of the quadratic function $A(x) = -x^2 + 1500x$. This maximum occurs at the vertex of the graph of $A(x)$ (which is a downward-opening parabola because the coefficient of x^2 is negative). The vertex may be found by using the fact in the box on page 282 (with $a = -1$ and $b = 1500$).

The x-coordinate of the vertex is $-\dfrac{1500}{2(-1)} = 750$ feet.

Hence, the y-coordinate of the vertex, the maximum value of $A(x)$, is

$$A(750) = -750^2 + 1500 \cdot 750 = 562,500 \text{ square feet.}$$

It occurs when the length is $x = 750$. In this case the width is

$$y = 1500 - x = 1500 - 750 = 750. \quad \blacksquare$$

EXAMPLE 7

Mary and Dirk Haraburd own a 20-unit apartment complex. They have found that each $50 increase in monthly rent results in a vacant apartment. All units are now rented at $400 per month. How many $50 increases in rent will produce the largest possible income? What is the maximum income?

SOLUTION   Let x represent the number of $50 increases. Then the monthly rent will be $400 + 50x$ dollars. Since one apartment goes vacant for each increase, the number of occupied apartments will be $20 - x$. Then the monthly income $R(x)$ is given by

$$R(x) = (\text{number of apartments rented}) \times (\text{rent per apartment})$$

$$R(x) = (20 - x)(400 + 50x)$$

$$R(x) = -50x^2 + 600x + 8000.$$

There are three ways to find the maximum possible income.

Algebraic Method: The graph of $R(x) = -50x^2 + 600x + 8000$ is a downward-opening parabola (why?). Maximum income occurs at the vertex of this parabola, that is, when

$$x = \frac{-b}{2a} = \frac{-600}{2(-50)} = \frac{-600}{-100} = 6.$$

Therefore, six increases of $50 will produce maximum income.

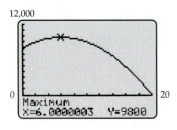

12,000

Maximum
X=6.0000003   Y=9800

0   20

Figure 4–7

Graphical Method: Graphing $R(x) = -50x^2 + 600x + 8000$ and using a maximum finder to determine the coordinates of the vertex, as in Figure 4–7, shows that maximum income of $9800 occurs when there are six rent increases.

Table Method: Make a table of values of $R(x)$ for $0 \le x \le 20$, as in Figure 4–8.[*] The table shows that the maximum income of $9800 occurs when $x = 6$. In this case, there will be $20 - 6 = 14$ apartments rented at a monthly rent of $400 + 6(50) = \$700.$ \quad \blacksquare

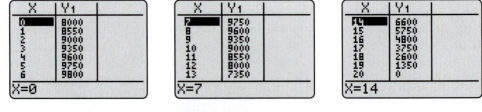

X	Y₁			X	Y₁			X	Y₁
0	8000			7	9750			14	6600
1	8550			8	9600			15	5750
2	9000			9	9350			16	4800
3	9350			10	9000			17	3750
4	9600			11	8550			18	2600
5	9750			12	8000			19	1350
6	9800			13	7350			20	0
X=0				X=7				X=14	

Figure 4–8

---

[*]This method is feasible here because there are only 20 apartments, but it cannot be used when the number of possibilities is very large or infinite.

EXAMPLE 8

The table shows the number of women in the full-time civilian labor force (in millions) in selected years.[*]

Year	1910	1930	1950	1970	1990	2010
Working Women	7.4	10.8	18.6	31.5	56.8	75.5

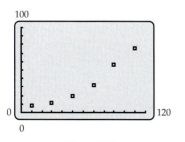

Figure 4–9

If you let $x = 0$ correspond to 1900 and plot the data points (10, 7.4), (30, 10.8), . . . , (110, 75.5), they appear to lie on a parabola, as shown in Figure 4–9.

(a) Find a quadratic function of the form $f(x) = a(x - h)^2 + k$ that models the data in the table.

(b) Use this function to estimate the number of working women in 2000 and 2005.

(c) According to this model, in what year will there be 80 million working women?

SOLUTION

(a) Figure 4–9 suggests that the vertex should be at $(h, k) = (10, 7.4)$, so the rule of the quadratic function is

$$f(x) = a(x - h)^2 + k$$
$$f(x) = a(x - 10)^2 + 7.4.$$

Next, choose one of the data points, say (110, 75.5). If this point is on the graph of f, then $f(110) = 75.5$. We use this fact to determine the value of a, as in Example 4.

$$f(110) = 75.5$$
$$a(110 - 10)^2 + 7.4 = 75.5$$
$$a(100^2) = 68.1$$
$$10{,}000a = 68.1$$
$$a = \frac{68.1}{10{,}000} = .00681$$

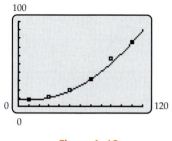

Figure 4–10

Therefore, the rule of a quadratic function that models this data is

$$f(x) = .00681(x - 10)^2 + 7.4[†]$$

The graph of f in Figure 4–10 suggests that f fits the data reasonably well.

(b) To find the number of working women in 2000 and 2005, we evaluate f at 100 and 105.

$$f(100) = .00681(100 - 10)^2 + 7.4 = 62.561 \text{ million}$$
$$f(105) = .00681(105 - 10)^2 + 7.4 = 68.86 \text{ million.}$$

[*]Data and projections from the U.S. Bureau of Labor Statistics.
[†]Other choices of a data point to determine a may produce a different function.

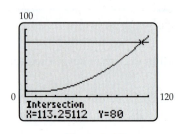

Figure 4–11

(c) There will be 80 million working women when $f(x) = 80$, so we must solve the following equation for x.

$$.00681(x - 10)^2 + 7.4 = 80.$$

This can be done algebraically with the quadratic formula, or technologically with an equation solver or graphical solver, as in Figure 4–11. The positive solution (the only one that applies here) is $x \approx 113.25112$. So the number of working women will reach 80 million in 2013. ∎

> NOTE: The model in Example 8 was constructed by choosing one data point as vertex and a second data point to lie on the parabola. If different choices had been made, a different quadratic model would have resulted.

EXERCISES 4.1

In Exercises 1–8, without graphing, determine the vertex of the parabola and state whether it opens upward or downward.

1. $f(x) = 3(x - 5)^2 + 2$ **2.** $g(x) = -6(x - 2)^2 - 5$

3. $y = -(x - 1)^2 + 2$ **4.** $h(x) = -x^2 + 1$

5. $f(x) = x^2 - 6x + 3$ **6.** $g(x) = x^2 + 8x - 1$

7. $h(x) = x^2 + 3x + 6$ **8.** $f(x) = x^2 - 5x - 7$

In Exercises 9–16, match the function with its graph, which is one of those shown here.

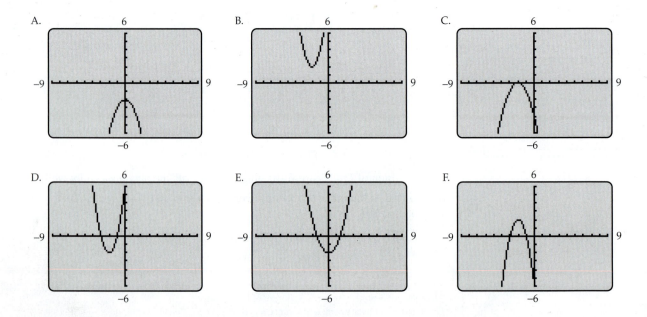

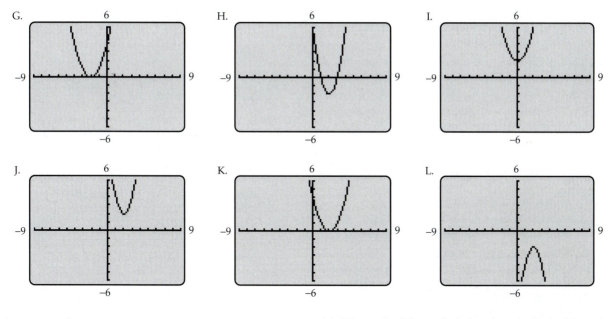

9. $f(x) = x^2 + 2$

10. $g(x) = x^2 - 2$

11. $g(x) = (x - 2)^2$

12. $f(x) = -(x + 2)^2$

13. $f(x) = 2(x - 2)^2 + 2$

14. $g(x) = -2(x - 2)^2 - 2$

15. $g(x) = -2(x + 2)^2 + 2$

16. $f(x) = 2(x + 2)^2 - 2$

In Exercises 7–24, find the rule of the quadratic function whose graph satisfies the given conditions.

17. Vertex at $(0, 0)$; passes through $(2, 12)$

18. Vertex at $(0, 1)$; passes through $(2, -7)$

19. Vertex at $(2, 5)$; passes through $(-3, 80)$

20. Vertex at $(4, 1)$; passes through $(2, -11)$

21. Vertex at $(-3, 4)$; passes through $(-2, 6)$

22. Vertex at $(-1, -2)$; passes through $(2, -29)$

23. Vertex at $(4, 3)$; passes through $(6, 5)$

24. Vertex at $(1/2, -3)$; passes through $(5, 3)$

In Exercises 25–32, sketch the graph of the parabola and determine its vertex and x-intercepts exactly.

25. $y = 2x^2 + 12x - 3$

26. $y = 3x^2 + 6x + 1$

27. $f(x) = -x^2 + 8x - 2$

28. $g(x) = -x^2 - 6x + 4$

29. $f(x) = -3x^2 + 4x + 5$

30. $g(x) = 2x^2 - x - 1$

31. $y = -x^2 + x$

32. $y = -2x^2 + 2x - 1$

33. The graph of the quadratic function g is obtained from the graph of $f(x) = x^2$ by vertically stretching it by a factor of 2 and then shifting vertically 5 units downward. What is the rule of the function g? What is the vertex of its graph?

34. The graph of the quadratic function g is obtained from the graph of $f(x) = x^2$ by shifting it horizontally 4 units to the left, then vertically stretching it by a factor of 3, and then shifting vertically 2 units upward. What is the rule of the function g? What is the vertex of its graph?

35. If the graph of the quadratic function h is shifted vertically 4 units downward, then shrunk by a factor of $1/2$, and then shifted horizontally 3 units to the left, the resulting graph is the parabola $f(x) = x^2$. What is the rule of the function h? What is the vertex of its graph?

36. If the graph of the quadratic function h is shifted vertically 3 units upward, then reflected in the x-axis, and then shifted horizontally 5 units to the right, the resulting graph is the parabola $f(x) = x^2$. What is the rule of the function h? What is the vertex of its graph?

C *In Exercises 37–40, find*

(a) The difference quotient of the functions;

(b) The vertex of the function's graph;

(c) The value of the difference quotient at the x-coordinate of the vertex.

37. $f(x) = -x^2 + x$

38. $g(x) = 2x^2 - x - 1$

39. $f(x) = -2x^2 + 2x - 1$

40. $g(x) = -3x^2 + 4x + 5$

In Exercises 41–44, use the formula for the height h of an object (that is traveling vertically subject only to gravity) at time t: $h = -16t^2 + v_0 t + h_0$, where h_0 is the initial height and v_0 the initial velocity.

41. A ball is thrown upward from the top of a 96-foot-high tower with an initial velocity of 80 feet per second. When does the ball reach its maximum height and how high is it at that time?

42. A rocket is fired upward from ground level with an initial velocity of 1600 feet per second. When does it attain its maximum height, and what is that height?

43. A ball is thrown upward from a height of 6 feet with an initial velocity of 32 feet per second. Find its maximum height.

44. A bullet is fired upward from ground level with an initial velocity of 1500 feet per second. How high does it go?

45. The Leslie Lahr Luggage Company has determined that its profit on its Luxury ensemble is given by $p(x) = 1600x - 4x^2 - 50{,}000$, where x is the number of units sold.
(a) What is the profit on 50 units? On 250 units?
(b) How many units should be sold to maximize profit? In that case, what will be the profit on each unit?

46. On the basis of data from past years, a consultant informs Bob's Bicycles that their profit from selling x bicycles is given by the function $p(x) = 250x - x^2/4 - 15{,}000$.
(a) How much profit do they make by selling 100 bicycles? By selling 400 bicycles?
(b) How many bicycles should be sold to maximize profit? In that case, what will be the profit per bicycle?

47. A projectile is fired at an angle of 45° upward. Exactly t seconds after firing, its vertical height above the ground is $500t - 16t^2$ feet.
(a) What is the greatest height the projectile reaches, and at what time does that occur?
(b) When does the projectile hit the ground? [*Hint:* It's on the ground when its height is 0.]

48. Jack throws a baseball. Its height above the ground is given by $h(x) = -.0013x^2 + .26x + 5$, where x is the distance from Jack to a point on the ground directly below the ball.
(a) How high is the ball at its highest point?
(b) At the ball's highest point, what is the distance from the ball to Jack?
(c) How far from Jack does the ball hit the ground?

49. During the Civil War, the standard heavy gun for coastal artillery was the 15-inch Rodman cannon, which fired a 330-pound shell. If one of these guns is fired from the top of a 50-foot-high shoreline embankment, then the height of the shell above the water (in feet) can be approximated by the function

$$p(x) = -.0000167x^2 + .23x + 50,$$

where x is the horizontal distance (in feet) from the foot of the embankment to a point directly under the shell. How high does the shell go, and how far away does it hit the water?

50. Travel spending by Americans (in billions of dollars) is approximated by

$$f(x) = 4.99x^2 - 23.83x + 489.5,$$

where $x = 0$ corresponds to the year 2000.* In what year was spending the lowest? When will spending reach $500 billion?

51. The braking distance (in meters) for a car with excellent brakes on a good road with an alert driver can be modeled by the quadratic function $B(s) = .01s^2 + .7s$, where s is the car's speed in kilometers per hour.
(a) What is the braking distance for a car traveling 30 km/h? For one traveling 100 km/h?
(b) If the car takes 60 meters to come to a complete stop, what was its speed?

52. The Golden Gate Bridge is supported by two huge cables strung between the towers at each end of the bridge. The function

$$f(x) = .0001193x^2 - .50106x + 526.113$$

gives the approximate height of the cables above the roadway at a point on the road x feet from one of the towers. The cables touch the road halfway between the two towers. How far apart are the towers?

53. Find the dimensions of the rectangle with perimeter 250 and largest possible area. [*Hint:* See Exercise 35 in Section 3.3.]

54. A gutter is to be made by bending up the edges of a 15-inch-wide piece of aluminum. What depth should the gutter be to have the maximum possible cross-sectional area?

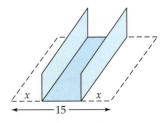

55. A gardener wants to use 130 feet of fencing to enclose a rectangular garden and divide it into two plots, as shown in the figure. What is the largest possible area for such a garden?

---

*Based on data from the Travel Industry Association of America.

56. A rectangular box (with top) has a square base. The sum of the lengths of its 12 edges is 8 feet. What dimensions should the box have so that its surface area is as large as possible?

57. A field bounded on one side by a river is to be fenced on three sides so as to form a rectangular enclosure. If 200 ft of fencing is to be used, what dimensions will yield an enclosure of the largest possible area?

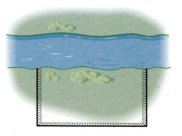

58. A rectangular garden next to a building is to be fenced on three sides. Fencing for the side parallel to the building costs $80 per foot and material for the other two sides costs $20 per foot. If $1800 is to be spent on fencing, what are the dimensions of the garden with the largest possible area?

59. At Middleton Place, a historic plantation near Charleston, South Carolina, there is a "joggling board" that was once used for courting. A young girl would sit at one end, her suitor at the other end, and her mother in the center. The mother would bounce on the board, thus causing the girl and her suitor to move closer together. A joggling board is 8 feet long, and an average mother sitting at its center causes the board to deflect 2 inches, as shown in the figure. The shape of the deflected board is parabolic.

(a) Find the equation of the parabola, assuming that the joggling board (with no one on it) lies on the x-axis with its center at the origin.

(b) How far from the center of the board is the deflection 1 inch?

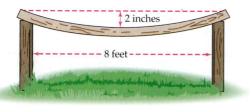

2 inches

8 feet

60. A salesperson finds that her sales average 40 cases per store when she visits 20 stores a week. Each time she visits an additional store per week, the average sales per store decreases by 1 case. How many stores should she visit each week if she wants to maximize her sales?

61. A potter can sell 120 bowls per week at $4 per bowl. For each 50¢ decrease in price, 20 more bowls are sold. What price should be charged to maximize sales income?

62. A vendor can sell 200 souvenirs per day at a price of $2 each. Each 10¢ price increase decreases the number of sales by 25 per day. Souvenirs cost the vendor $1.50 each. What price should be charged to maximize the profit?

63. When a basketball team charges $4 per ticket, average attendance is 500 people. For each 20¢ decrease in ticket price, average attendance increases by 30 people. What should the ticket price be to ensure maximum income?

64. Construct a different quadratic model in Example 8 by using the data point (90, 56.8) to determine the value of a. Use this function to answer parts (b) and (c) of Example 8.

65. Private health care expenditures were $1546 billion in 2002 and are projected to be $2816 billion in 2011.[*]
(a) Let $x = 0$ correspond to 2000 and construct a quadratic function that models this data. [*Hint:* Let (2, 1546) be the vertex; see Example 8.]
(b) What are these expenditures expected to be in 2007?
(c) When will expenditures reach $4000 billion?

66. Carbon dioxide concentrations (in parts per million) at Mauna Loa, Hawaii, have increased quadratically over the past half century, from 315 in 1958 to 371 in 2001.[†] Let $x = 0$ correspond to 1950 and construct a quadratic function that models this data. When will concentrations reach 400 parts per million?

67. National defense expenditures grew from $268.5 billion in 1998 to $348 billion in 2002.[‡] Let $x = 0$ correspond to 1990 and construct a quadratic function that models this data. If this model remains accurate, when do defense expenditures reach $600 billion?

*U.S. Centers for Medicare and Medicaid Services.
†C. D. Keeling and T. P. Whorf, Scripps Institute of Oceanography.
‡U.S. Office of Management and Budget.

Thinker

68. The *discriminant* of a quadratic function

$$f(x) = ax^2 + bx + c$$

is the number $b^2 - 4ac$. For each of the discriminants listed here, state which graphs could possibly be the graph of f.

(a) $b^2 - 4ac = 25$
(b) $b^2 - 4ac = 0$
(c) $b^2 - 4ac = -49$
(d) $b^2 - 4ac = 72$

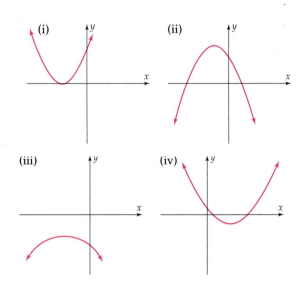

4.2   Polynomial Functions and Roots

A **polynomial function** is a function whose rule is given by a polynomial, such as

$$f(x) = x^4 - 2x^3 + 4x - 5 \qquad \text{or} \qquad g(x) = 2x^3 + x^2 - 3x + 1.$$

Long division of polynomials, which we now present, will play a key role in analyzing polynomial functions. The basic procedure, as illustrated in the following example, is quite similar to long division of numbers.

EXAMPLE 1

Divide $8x^3 + 2x^2 + 1$ by $2x^2 - x$.

SOLUTION   We set up the division in the same way that is used for numbers.

$$\text{Divisor} \rightarrow \quad 2x^2 - x \overline{\smash{\big)}\, 8x^3 + 2x^2 + 1} \quad \leftarrow \text{Dividend}$$

Begin by dividing the first term of the divisor ($2x^2$) into the first term of the dividend ($8x^3$) and putting the result $\left(\text{namely, } \dfrac{8x^3}{2x^2} = 4x\right)$ on the top line, as shown below. Then multiply $4x$ times the entire divisor, put the result on the third line, and subtract.

$$
\begin{array}{r}
4x \qquad\qquad \leftarrow \text{Partial quotient} \\
2x^2 - x \overline{\smash{\big)}\, 8x^3 + 2x^2 + 1} \\
\underline{8x^3 - 4x^2} \qquad \leftarrow 4x(2x^2 - x) \\
6x^2 + 1 \qquad \leftarrow \text{Subtraction}^*
\end{array}
$$

---

*If this subtraction is confusing, write it out horizontally and watch the signs:

$$(8x^3 + 2x^2 + 1) - (8x^3 - 4x^2) = 8x^3 + 2x^2 + 1 - 8x^3 + 4x^2 = 6x^2 + 1.$$

Solve
$x^3 + 5x^2 + x - 15 = 0$

Now divide the first term of the divisor $(2x^2)$ into $6x^2$ and put the result $\left(\dfrac{6x^2}{2x^2} = 3\right)$ on the top line, as shown below. Then multiply 3 times the entire divisor, put the result on the fifth line, and subtract.

$$
\begin{array}{r}
4x + 3 \qquad \leftarrow \text{Quotient}\\
2x^2 - x\,\overline{)\,8x^3 + 2x^2 \qquad\quad + 1}\\
\underline{8x^3 - 4x^2} \qquad\qquad \leftarrow 4x(2x^2 - x)\\
6x^2 \qquad + 1 \quad \leftarrow \text{Subtraction}\\
\underline{6x^2 - 3x} \qquad\quad \leftarrow 3(2x^2 - x)\\
3x + 1 \quad \leftarrow \text{Subtraction}
\end{array}
$$

The division process stops when the remainder is 0 or has smaller degree than the divisor, which is the case here. ■

Recall how you check a long division problem with numbers.

$$
\begin{array}{r}
145\\
31\,\overline{)\,4509}\\
\underline{31}\\
140\\
\underline{124}\\
169\\
\underline{155}\\
14
\end{array}
\qquad
\begin{array}{rl}
\text{Check:} & 145 \quad \leftarrow \text{Quotient}\\
& \underline{\times\, 31} \quad \leftarrow \text{Divisor}\\
& 145\\
& \underline{435}\\
& 4495\\
& \underline{+\,14} \quad \leftarrow \text{Remainder}\\
& 4509. \quad \leftarrow \text{Dividend}
\end{array}
$$

We can summarize this process in one line:

$$\text{Divisor} \cdot \text{Quotient} + \text{Remainder} = \text{Dividend}.$$

The same thing works for division of polynomials, as you can see by examining the division problem from Example 1.

$$
\overset{\text{Divisor} \cdot \ \text{Quotient} \ + \ \text{Remainder}}{(2x^2 - x) \cdot (4x + 3) + (3x + 1)} = (8x^3 + 2x^2 - 3x) + (3x + 1)
$$
$$
= 8x^3 + 2x^2 + 1
$$
$$
\text{Dividend}
$$

This fact is so important that it is given a special name and a formal statement.

The Division Algorithm

If a polynomial $f(x)$ is divided by a nonzero polynomial $h(x)$, then there is a quotient polynomial $q(x)$ and a remainder polynomial $r(x)$ such that

$$\text{Dividend} = \text{Divisor} \cdot \text{Quotient} + \text{Remainder}$$

$$f(x) = h(x)\,q(x) + r(x),$$

where either $r(x) = 0$ or $r(x)$ has degree less than the degree of the divisor $h(x)$.

∷ REMAINDERS

The remainder in long division of polynomials provides some interesting information, as we now see.

EXAMPLE 2

Show that $2x^2 + 1$ is a factor of $6x^3 - 4x^2 + 3x - 2$.

SOLUTION   We divide $6x^3 - 4x^2 + 3x - 2$ by $2x^2 + 1$ and find that the remainder is 0.

$$
\begin{array}{r}
3x \quad - 2 \\
2x^2 + 1 \overline{\smash{)}\ 6x^3 - 4x^2 + 3x - 2} \\
\underline{6x^3 \qquad\quad + 3x} \\
-4x^2 \qquad - 2 \\
\underline{-4x^2 \qquad - 2} \\
0.
\end{array}
$$

Since the remainder is 0, the Division Algorithm tells us that

$$\text{Dividend} = \text{Divisor} \cdot \text{Quotient} + \text{Remainder}$$
$$6x^3 - 4x^2 + 3x - 2 = (2x^2 + 1)(3x - 2) + 0$$
$$= (2x^2 + 1)(3x - 2).$$

Therefore, $2x^2 + 1$ is a factor of $6x^3 - 4x^2 + 3x - 2$, and the other factor is the quotient $3x - 2$. ∎

Example 2 illustrates this fact.

Remainders and Factors

> The remainder in polynomial division is 0 exactly when the divisor is a factor of the dividend. In this case, the quotient is the other factor.

When a polynomial $f(x)$ is divided by a first-degree polynomial, such as $x - 3$ or $x + 5$, the remainder is a constant (because constants are the only polynomials of degree less than 1). This remainder has an interesting connection with the values of the polynomial function $f(x)$.

EXAMPLE 3

Let $f(x) = x^3 - 2x^2 - 4x + 5$.

(a) Find the quotient and remainder when $f(x)$ is divided by $x - 3$.

(b) Find $f(3)$.

NOTE

When the divisor is a first-degree polynomial such as $x - 3$, there is a convenient shorthand method of division, called **synthetic division.** See Special Topics 4.2.A for details.

SOLUTION

(a) Using long division, we have

$$\begin{array}{r} x^2 + x - 1 \\ x - 3 \overline{\smash{\big)}\ x^3 - 2x^2 - 4x + 5} \\ \underline{x^3 - 3x^2} \\ x^2 - 4x + 5 \\ \underline{x^2 - 3x} \\ -x + 5 \\ \underline{-x + 3} \\ 2. \end{array}$$

Therefore, the quotient is $x^2 + x - 1$, and the remainder is 2.

(b) Using the Division Algorithm, we can write the dividend $f(x) = x^3 - 2x^2 - 4x + 5$ as

$$\text{Dividend} = \text{Divisor} \cdot \text{Quotient} + \text{Remainder}$$
$$f(x) = (x - 3)(x^2 + x - 1) + 2.$$

Hence,

$$f(3) = (3 - 3)(3^2 + 3 - 1) + 2 = 0 + 2 = 2.$$

Note that the number $f(3)$ is the same as the remainder when $f(x)$ is divided by $x - 3$. ■

The argument used in Example 3 to show that $f(3)$ is the remainder when $f(x)$ is divided by $x - 3$ also works in the general case and proves this fact.

Remainder Theorem

If a polynomial $f(x)$ is divided by $x - c$, then the remainder is the number $f(c)$.

EXAMPLE 4

To find the remainder when $f(x) = x^{79} + 3x^{24} + 5$ is divided by $x - 1$, we apply the Remainder Theorem with $c = 1$. The remainder is

$$f(1) = 1^{79} + 3 \cdot 1^{24} + 5 = 1 + 3 + 5 = 9. \quad ■$$

EXAMPLE 5

To find the remainder when $f(x) = 3x^4 - 8x^2 + 11x + 1$ is divided by $x + 2$, we must apply the Remainder Theorem *carefully*. The divisor in the theorem is $x - c$, not $x + c$. So we rewrite $x + 2$ as $x - (-2)$ and apply the theorem with $c = -2$. The remainder is

$$f(-2) = 3(-2)^4 - 8(-2)^2 + 11(-2) + 1 = 48 - 32 - 22 + 1 = -5. \quad ■$$

▪▪ REMAINDERS AND ROOTS

If $f(x)$ is a polynomial, then a solution of the equation $f(x) = 0$ is called a **root** or **zero** of $f(x)$. Thus, a number c is a root of $f(x)$ if $f(c) = 0$. A root that is a real number is called a **real root.** For example, 4 is a real root of the polynomial $f(x) = 3x - 12$ because $f(4) = 3 \cdot 4 - 12 = 0$. There is an interesting connection between the roots of a polynomial and its factors.

EXAMPLE 6

Let $f(x) = x^3 - 4x^2 + 2x + 3$.

(a) Show that 3 is a root of $f(x)$.

(b) Show that $x - 3$ is a factor of $f(x)$.

SOLUTION

(a) Evaluating $f(x)$ at 3 shows that

$$f(3) = 3^3 - 4(3^2) + 2(3) + 3 = 0.$$

Therefore, 3 is a root of $f(x)$.

(b) If $f(x)$ is divided by $x - 3$, then by the Division Algorithm, there is a quotient polynomial $q(x)$ such that

$$f(x) = (x - 3)q(x) + \text{Remainder}.$$

The Remainder Theorem shows that the remainder when $f(x)$ is divided by $x - 3$ is the number $f(3)$, which is 0, as we saw in part (a). Therefore,

$$f(x) = (x - 3)q(x) + 0 = (x - 3)q(x).$$

Thus, $x - 3$ is a factor of $f(x)$. (To determine the other factor, the quotient $q(x)$, you have to perform the division.) ■

Example 6 illustrates this fact, which can be proved by the same argument used in the example.

Factor Theorem

> The number c is a root of the polynomial $f(x)$ exactly when $x - c$ is a factor of $f(x)$.

EXAMPLE 7

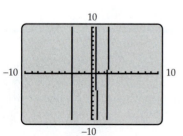

Figure 4–12

The graph of $f(x) = 15x^3 - x^2 - 114x + 72$ in the standard viewing window (Figure 4–12) is obviously not complete but suggests that -3 is an x-intercept and hence a root of $f(x)$. It is easy to verify that this is indeed the case.

$$f(-3) = 15(-3)^3 - (-3)^2 - 114(-3) + 72 = -405 - 9 + 342 + 72 = 0$$

Since -3 is a root, $x - (-3) = x + 3$ is a factor of $f(x)$. Use synthetic or long division to verify that the other factor is $15x^2 - 46x + 24$. By factoring this quadratic, we obtain a complete factorization of $f(x)$.

$$f(x) = (x + 3)(15x^2 - 46x + 24) = (x + 3)(3x - 2)(5x - 12) \quad ■$$

EXAMPLE 8

Find three polynomials of different degrees that have 1, 2, 3, and -5 as roots.

SOLUTION   A polynomial that has 1, 2, 3, and -5 as roots must have $x - 1$, $x - 2$, $x - 3$, and $x - (-5) = x + 5$ as factors. Many polynomials satisfy these conditions, such as

$$g(x) = (x - 1)(x - 2)(x - 3)(x + 5) = x^4 - x^3 - 19x^2 + 49x - 30$$

$$h(x) = 8(x - 1)(x - 2)(x - 3)^2(x + 5)$$

$$k(x) = 2(x + 4)^2(x - 1)(x - 2)(x - 3)(x + 5)(x^2 + x + 1).$$

Note that g has degree 4. When h is multiplied out, its leading term is $8x^5$, so h has degree 5. Similarly, k has degree 8, since its leading term is $2x^8$. ∎

If a polynomial $f(x)$ has four roots, say a, b, c, d, then by the same argument used in Example 8, it must have

$$(x - a)(x - b)(x - c)(x - d)$$

as a factor. Since $(x - a)(x - b)(x - c)(x - d)$ has degree 4 (multiply it out—its leading term is x^4), $f(x)$ must have degree at least 4. In particular, this means that no polynomial of degree 3 can have four or more roots. A similar argument works in the general case.

Number of Roots

> A polynomial of degree n has at most n distinct roots.

∷ RATIONAL ROOTS

Finding the real roots of polynomials is the same as solving polynomial equations. The root of a first-degree polynomial, such as $5x - 3$, can be found by solving the equation $5x - 3 = 0$. Similarly, the roots of any second-degree polynomial can be found by using the quadratic formula (Section 2.2). Although the roots of higher-degree polynomials can always be approximated graphically as in Section 2.3, it is better to find exact solutions, if possible. When a polynomial has integer coefficients, all of its **rational roots** (roots that are rational numbers) can be found exactly by using the following result.

The Rational Root Test

> If a rational number r/s (in lowest terms) is a root of the polynomial
>
> $$a_nx^n + \cdots + a_1x + a_0,$$
>
> where the coefficients $a_n, \ldots, a_1, a_0$ are integers with $a_n \neq 0$, $a_0 \neq 0$, then
>
> r is a factor of the constant term a_0 and
>
> s is a factor of the leading coefficient a_n.

The test states that every rational root must satisfy certain conditions.[*] By finding all the numbers that satisfy these conditions, we produce a list of *possible* rational roots. Then we must evaluate the polynomial at each number on the list to see whether the number actually is a root. This testing process can be considerably shortened by using a calculator, as in the next example.

EXAMPLE 9

Find the rational roots of

$$f(x) = 2x^4 + x^3 - 17x^2 - 4x + 6.$$

SOLUTION   If $f(x)$ has a rational root r/s, then by the Rational Root Test r must be a factor of the constant term 6. Therefore, r must be one of ± 1, ± 2, ± 3, or ± 6 (the only factors of 6). Similarly, s must be a factor of the leading coefficient 2, so s must be one of ± 1 or ± 2 (the only factors of 2). Consequently, the only *possibilities* for r/s are

$$\frac{\pm 1}{\pm 1}, \frac{\pm 2}{\pm 1}, \frac{\pm 3}{\pm 1}, \frac{\pm 6}{\pm 1}, \frac{\pm 1}{\pm 2}, \frac{\pm 2}{\pm 2}, \frac{\pm 3}{\pm 2}, \frac{\pm 6}{\pm 2}.$$

Eliminating duplications from this list, we see that the only *possible* rational roots are

$$1, -1, 2, -2, 3, -3, 6, -6, \frac{1}{2}, -\frac{1}{2}, \frac{3}{2}, -\frac{3}{2}.$$

Now graph $f(x)$ in a viewing window that includes all of these numbers on the x-axis, say $-7 \le x \le 7$ and $-5 \le y \le 5$ (Figure 4–13). A complete graph isn't necessary, since we are interested only in the x-intercepts.

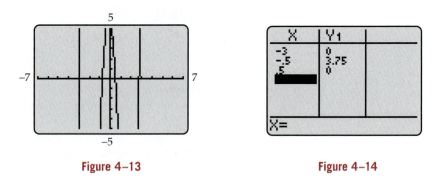

Figure 4–13   **Figure 4–14**

Figure 4–13 shows that the only numbers on our list that could possibly be roots (x-intercepts) are -3, $-1/2$, and $1/2$, so these are the only ones that need be tested. We use the table feature to evaluate $f(x)$ at these three numbers (Figure 4–14). The table shows that -3 and $1/2$ are the only rational roots of $f(x)$. Its other roots (x-intercepts) in Figure 4–13 must be irrational numbers. ∎

---

[*]Since the proof of the Rational Root Test sheds no light on how the test is actually used to solve equations, it will be omitted.

■■ ROOTS AND THE FACTOR THEOREM

Once some roots of a polynomial have been found, the Factor Theorem can be used to factor the polynomial, which may lead to additional roots.

EXAMPLE 10

Find all the roots of $f(x) = 2x^4 + x^3 - 17x^2 - 4x + 6$.

SOLUTION   In Example 9, we saw that -3 and $1/2$ are the rational roots of $f(x)$. By the Factor Theorem, $x - (-3) = x + 3$ and $x - 1/2$ are factors of $f(x)$. Using synthetic or long division twice, we have

$$2x^4 + x^3 - 17x^2 - 4x + 6 = (x + 3)(2x^3 - 5x^2 - 2x + 2)$$
$$= (x + 3)(x - .5)(2x^2 - 4x - 4).$$

The remaining roots of $f(x)$ are the roots of $2x^2 - 4x - 4$, that is, the solutions of

$$2x^2 - 4x - 4 = 0$$
$$x^2 - 2x - 2 = 0.$$

They are easily found by the quadratic formula.

$$x = \frac{-(-2) \pm \sqrt{(-2)^2 - 4 \cdot 1 \cdot (-2)}}{2 \cdot 1}$$
$$= \frac{2 \pm \sqrt{12}}{2} = \frac{2 \pm 2\sqrt{3}}{2} = 1 \pm \sqrt{3}.$$

Therefore, $f(x)$ has rational roots -3 and $1/2$, and irrational roots $1 + \sqrt{3}$ and $1 - \sqrt{3}$.   ■

EXAMPLE 11

Factor $f(x) = 2x^5 - 10x^4 + 7x^3 + 13x^2 + 3x + 9$ completely.

SOLUTION   We begin by finding as many roots of $f(x)$ as we can. By the Rational Root Test, every rational root is of the form r/s where $r = \pm 1, \pm 3$, or ± 9 and $s = \pm 1$ or ± 2. Thus, the possible rational roots are

$$\pm 1, \quad \pm 3, \quad \pm 9, \quad \pm \frac{1}{2}, \quad \pm \frac{3}{2}, \quad \pm \frac{9}{2}.$$

The partial graph of $f(x)$ in Figure 4–15 shows that the only possible roots (x-intercepts) are -1 and 3. You can easily verify that both -1 and 3 are roots of $f(x)$.

Since -1 and 3 are roots, $x - (-1) = x + 1$ and $x - 3$ are factors of $f(x)$ by the Factor Theorem. Division shows that

$$f(x) = 2x^5 - 10x^4 + 7x^3 + 13x^2 + 3x + 9$$
$$= (x + 1)(2x^4 - 12x^3 + 19x^2 - 6x + 9)$$
$$= (x + 1)(x - 3)(2x^3 - 6x^2 + x - 3).$$

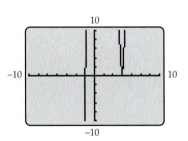

Figure 4–15

The other roots of $f(x)$ are the roots of $g(x) = 2x^3 - 6x^2 + x - 3$. We first check for rational roots of $g(x)$. Since every root of $g(x)$ is also a root of $f(x)$ (why?), the only possible rational roots of $g(x)$ are -1 and 3 (the rational roots of $f(x)$). We have

$$g(-1) = 2(-1)^3 - 6(-1)^2 + (-1) - 3 = -12;$$

$$g(3) = 2(3^3) - 6(3^2) + 3 - 3 = 0.$$

So -1 is not a root, but 3 is a root of $g(x)$. By the Factor Theorem, $x - 3$ is a factor of $g(x)$. Division shows that

$$f(x) = (x + 1)(x - 3)(2x^3 - 6x^2 + x - 3)$$

$$= (x + 1)(x - 3)(x - 3)(2x^2 + 1).$$

Since $2x^2 + 1$ has no real roots, it cannot be factored. So the factorization of $f(x)$ is complete. ■

SUMMARY

TIP

The polynomial solvers on TI-86/89, Casio FX 2.0, and HP-39+ can find or approximate all the roots of a polynomial simultaneously. The solver on Casio 9850 is limited to polynomials of degree 2 or 3. See the Calculator Investigation on page 147 for details.

The examples above illustrate the following guidelines for finding all the real roots of a polynomial $f(x)$.

1. Use the Rational Root Test to find all the rational roots of $f(x)$. [*Examples 9–11*]

2. Write $f(x)$ as the product of linear factors (one for each rational root) and another factor $g(x)$. [*Examples 10, 11*]

3. If $g(x)$ has degree 2, find its roots by factoring or the quadratic formula. [*Example 10*]

4. If $g(x)$ has degree 3 or more, approximate the remaining roots graphically. [*Section 2.3*]

✓ EXERCISES 4.2

In Exercises 1–6, state the quotient and remainder when the first polynomial is divided by the second. Check your division by calculating (divisor)(quotient) + remainder.

1. $3x^4 + 2x^2 - 6x + 1$; $x + 1$

2. $x^5 - x^3 + x - 5$; $x - 2$

3. $x^5 + 2x^4 - 6x^3 + x^2 - 5x + 1$; $x^3 + 1$

4. $3x^4 - 3x^3 - 11x^2 + 6x - 1$; $x^3 + x^2 - 2$

5. $5x^4 + 5x^2 + 5$; $x^2 - x + 1$

6. $x^5 - 1$; $x - 1$

In Exercises 7–10, determine whether the first polynomial is a factor of the second.

7. $x^2 + 3x - 1$; $x^3 + 2x^2 - 5x - 6$

8. $x^2 + 9$; $x^5 + x^4 - 81x - 81$

9. $x^2 + 3x - 1$; $x^4 + 3x^3 - 2x^2 - 3x + 1$

10. $x^2 - 5x + 7$; $x^3 - 3x^2 - 3x + 9$

In Exercises 11–14, determine which of the given numbers are roots of the given polynomial.

11. $2, 3, 0, -1$; $g(x) = x^4 + 6x^3 - x^2 - 30x$

12. $1, 1/2, 2, -1/2, 3$; $f(x) = 6x^2 + x - 1$

13. $2\sqrt{2}, \sqrt{2}, -\sqrt{2}, 1, -1$; $h(x) = x^3 + x^2 - 8x - 8$

14. $\sqrt{3}, -\sqrt{3}, 1, -1$; $k(x) = 8x^3 - 12x^2 - 6x + 9$

In Exercises 15–22, find the remainder when f(x) is divided by g(x), without using division.

15. $f(x) = x^{10} + x^8$; $g(x) = x - 1$

16. $f(x) = x^6 - 10$; $g(x) = x - 2$

17. $f(x) = 3x^4 - 6x^3 + 2x - 1$; $g(x) = x + 1$

18. $f(x) = x^5 - 3x^2 + 2x - 1$; $g(x) = x - 2$

19. $f(x) = x^3 - 2x^2 + 5x - 4$; $g(x) = x + 2$

20. $f(x) = 10x^{75} - 8x^{65} + 6x^{45} + 4x^{32} - 2x^{15} + 5$;
    $g(x) = x - 1$

21. $f(x) = 2x^5 - 3x^4 + x^3 - 2x^2 + x - 8$; $g(x) = x - 10$

22. $f(x) = x^3 + 8x^2 - 29x + 44$; $g(x) = x + 11$

In Exercises 23–26, use the Factor Theorem to determine whether or not h(x) is a factor of f(x).

23. $h(x) = x - 1$; $f(x) = x^5 + 1$

24. $h(x) = x + 2$; $f(x) = x^3 - 3x^2 - 4x - 12$

25. $h(x) = x + 1$; $f(x) = x^3 - 4x^2 + 3x + 8$

26. $h(x) = x - 1$; $f(x) = 14x^{99} - 65x^{56} + 51$

In Exercises 27–30, use the Factor Theorem and a calculator to factor the polynomial, as in Example 7.

27. $f(x) = 6x^3 - 7x^2 - 89x + 140$

28. $g(x) = x^3 - 5x^2 - 5x - 6$

29. $h(x) = 4x^4 + 4x^3 - 35x^2 - 36x - 9$

30. $f(x) = x^5 - 5x^4 - 5x^3 + 25x^2 + 6x - 30$

In Exercises 31–34, each graph is of a polynomial function f(x) of degree 5 whose leading coefficient is 1. The graph is not drawn to scale. Use the Factor Theorem to find the polynomial. [Hint: What are the roots of f(x)? What does the Factor Theorem tell you?]

31.

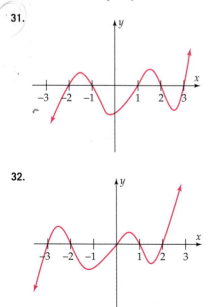

32.

33.

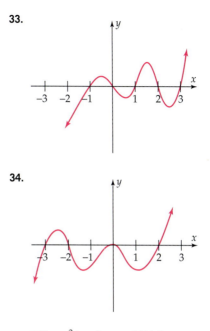

34.

[*Hint:* x^2 is a factor of $f(x)$.]

In Exercises 35–38, find a polynomial with the given degree n, the given roots, and no other roots.

35. $n = 3$; roots $1, 7, -4$

36. $n = 3$; roots $1, -1$

37. $n = 6$; roots $1, 2, \pi$

38. $n = 5$; root 2

39. Find a polynomial function f of degree 3 such that $f(10) = 17$ and the roots of $f(x)$ are 0, 5, and 8.

40. Find a polynomial function g of degree 4 such that the roots of g are $0, -1, 2, -3$ and $g(3) = 288$.

In Exercises 41–51, find all the rational roots of the polynomial.

41. $x^3 + 3x^2 - x - 3$ **42.** $x^3 - x^2 - 3x + 3$

43. $x^3 + 5x^2 - x - 5$ **44.** $3x^3 + 8x^2 - x - 20$

45. $f(x) = 2x^5 + 5x^4 - 11x^3 + 4x^2$ [*Hint:* The Rational Root Test can only be used on polynomials with nonzero constant terms. Factor $f(x)$ as a product of a power of x and a polynomial $g(x)$ with nonzero constant term. Then use the Rational Root Test on $g(x)$.]

46. $2x^6 - 3x^5 - 7x^4 - 6x^3$

47. $f(x) = \dfrac{1}{12}x^3 - \dfrac{1}{12}x^2 - \dfrac{2}{3}x + 1$ [*Hint:* The Rational Root Test can only be used on polynomials with integer coefficients. Note that $f(x)$ and $12f(x)$ have the same roots. (Why?)]

48. $\frac{2}{3}x^4 + \frac{1}{2}x^3 - \frac{5}{4}x^2 - x - \frac{1}{6}$

49. $\frac{1}{3}x^4 - x^3 - x^2 + \frac{13}{3}x - 2$

50. $\frac{1}{3}x^7 - \frac{1}{2}x^6 - \frac{1}{6}x^5 + \frac{1}{6}x^4$

51. $.1x^3 - 1.9x + 3$

52. The supply of natural gas in underground storage (in trillions of cubic feet), from July 2001 through January 2003 can be approximated by

$$f(x) = \frac{-1}{500}(x^4 - 36x^3 + 401x^2 - 1386x) + 2,$$

where x is measured in months and $x = 0$ corresponds to July 1, 2001.* Determine the months when the gas supply was 2 trillion cubic feet. [*Hint:* One solution of $f(x) = 2$ is $x = 18$; use the Rational Root Test to find the others.]

In Exercises 53–58, factor the polynomial as a product of linear factors and a factor $g(x)$ such that $g(x)$ is either a constant or a polynomial that has no rational roots.

53. $2x^3 - 4x^2 + x - 2$

54. $6x^3 - 5x^2 + 3x - 1$

55. $x^6 + 2x^5 + 3x^4 + 6x^3$

56. $x^5 - 2x^4 + 2x^3 - 3x + 2$

57. $x^5 - 4x^4 + 8x^3 - 14x^2 + 15x - 6$

58. $x^5 + 4x^3 + x^2 + 6x$

In Exercises 58–62, find all real roots of the polynomial.

59. $2x^3 - 5x^2 + x + 2$ **60.** $t^4 - t^3 + 2t^2 - 4t - 8$

61. $6x^3 - 11x^2 + 6x - 1$ **62.** $z^3 + z^2 + 2z + 2$

63. (a) Show that $\sqrt{2}$ is an irrational number. [*Hint:* $\sqrt{2}$ is a root of $x^2 - 2$, Does this polynomial have any rational roots?]
(b) Show that $\sqrt{3}$ is irrational.

64. Graph $f(x) = .001x^3 - .199x^2 - .23x + 6$ in the standard viewing window.
(a) How many roots does $f(x)$ appear to have? Without changing the viewing window, explain why $f(x)$ must have an additional root. [*Hint:* Each root corresponds to a factor of $f(x)$. What does the rest of the factorization consist of?]
(b) Find all the roots of $f(x)$.

65. According to data from the FBI, the number of people murdered each year per 100,000 population can be approximated by the polynomial function

$$f(x) = .00038x^4 + .0037x^3 - .1528x^2 + .3879x + 9.42$$
$$(0 \le x \le 11),$$

where $x = 0$ corresponds to 1990.
(a) What was the murder rate in 1995 and in 2000?
(b) In what year was the murder rate 7 people per 100,000?
(c) In what year between 1990 and 2001 was the murder rate the highest?

66. The percentage of new mothers who breast-feed their child in the hospital is approximated by

$$g(x) = -.001145x^4 + .0979x^3 - 2.9175x^2$$
$$+ 35.865x - 96.57 \qquad (7 \le x \le 32),$$

where $x = 7$ corresponds to 1977.† Use this model to find the year in which the percentage of mothers who breast-fed was
(a) 50   (b) 60   (c) 70.

67. During the first 150 hours of an experiment, the growth rate of a bacteria population at time t hours is $g(t) = -.0003t^3 + .04t^2 + .3t + .2$ bacteria per hour.
(a) What is the growth rate at 50 hours? At 100 hours?
(b) What is the growth rate at 145 hours? What does this mean?
(c) At what time is the growth rate 0?
(d) At what time is the growth rate -50 bacteria per hour?

68. In a sealed chamber where the temperature varies, the instantaneous rate of change of temperature with respect to time over an 11-day period is given by $F(t) = .0035t^4 - .4t^2 - .2t + 6$, where time is measured in days and temperature in degrees Fahrenheit (so the rate of change is in degrees per day).
(a) At what rate is the temperature changing at the beginning of the period ($t = 0$)? At the end of the period ($t = 11$)?
(b) When is the temperature increasing at a rate of 4°F per day?
(c) When is the temperature decreasing at a rate of 3°F per day?

Thinkers

69. For what value of k is the difference quotient of

$$g(x) = kx^2 + 2x + 1$$

equal to $7x + 2 + 3.5h$?

70. For what value of k is the difference quotient of

$$f(x) = x^2 + kx$$

equal to $2x + 5 + h$?

---

*Energy Information Administration.

†Based on data from Ross Products Division of Abbott Laboratories.

4.2.A   *SPECIAL TOPICS* Synthetic Division

Synthetic division is a fast method of doing polynomial division when the divisor is a first-degree polynomial of the form $x - c$ for some real number c. To see how it works, we first consider an example of ordinary long division.

$$
\begin{array}{r}
3x^3 + 6x^2 + 4x - 3 \quad \leftarrow \text{Quotient} \\
x - 2\;\overline{)\;3x^4 \qquad\quad - 8x^2 - 11x + 1}\quad \leftarrow \text{Dividend} \\
\underline{3x^4 - 6x^3} \\
6x^3 - 8x^2 \\
\underline{6x^3 - 12x^2} \\
4x^2 - 11x \\
\underline{4x^2 - 8x} \\
-3x + 1 \\
\underline{-3x + 6} \\
-5 \quad \leftarrow \text{Remainder}
\end{array}
$$

Divisor → ; Quotient ←; Dividend ←; Remainder ←

This calculation obviously involves a lot of repetitions. If we insert 0 coefficients for terms that don't appear above and keep the various coefficients in the proper columns, we can eliminate the repetition and all the x's.

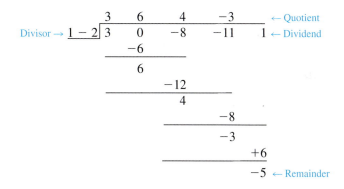

We can save space by moving the lower lines upward and writing 2 in the divisor position (since that's enough to remind us that the divisor is $x - 2$).

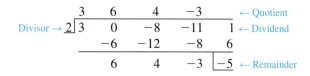

Since the last line contains most of the quotient line, we can save more space and still preserve the essential information by inserting a 3 in the last line and omitting the top line.

Synthetic division is a quick method for obtaining the last row of this array. Here is a step-by-step explanation of the division of $3x^4 - 8x^2 - 11x + 1$ by $x - 2$.

Step 1 In the first row, list the 2 from the divisor and the coefficients of the dividend in order of decreasing powers of x (insert 0 coefficients for missing powers of x).

$$2\rfloor \quad 3 \quad 0 \quad -8 \quad -11 \quad 1$$

Step 2 Bring down the first dividend coefficient (namely, 3) to the third row

$$2\rfloor \quad 3 \quad 0 \quad -8 \quad -11 \quad 1$$
$$ 3$$

Step 3 Multiply $2 \cdot 3$ and insert the answer 6 in the second row, in the position shown here.

$$2\rfloor \quad 3 \quad 0 \quad -8 \quad -11 \quad 1$$
$$ 6$$
$$ 3$$

Step 4 Add $0 + 6$ and write the answer 6 in the third row.

$$2\rfloor \quad 3 \quad 0 \quad -8 \quad -11 \quad 1$$
$$ 6$$
$$ 3 \quad 6$$

Step 5 Multiply $2 \cdot 6$ and insert the answer 12 in the second row.

$$2\rfloor \quad 3 \quad 0 \quad -8 \quad -11 \quad 1$$
$$ 6 \quad 12$$
$$ 3 \quad 6$$

Step 6 Add $-8 + 12$ and write the answer 4 in the third row.

$$2\rfloor \quad 3 \quad 0 \quad -8 \quad -11 \quad 1$$
$$ 6 \quad 12$$
$$ 3 \quad 6 \quad 4$$

Step 7 Multiply $2 \cdot 4$ and insert the answer 8 in the second row.

$$2\rfloor \quad 3 \quad 0 \quad -8 \quad -11 \quad 1$$
$$ 6 \quad 12 \quad 8$$
$$ 3 \quad 6 \quad 4$$

Step 8 Add $-11 + 8$ and write the answer -3 in the third row.

$$2\rfloor \quad 3 \quad 0 \quad -8 \quad -11 \quad 1$$
$$ 6 \quad 12 \quad 8$$
$$ 3 \quad 6 \quad 4 \quad -3$$

Step 9 Multiply $2 \cdot (-3)$ and insert the answer -6 in the second row.

$$2\rfloor \quad 3 \quad 0 \quad -8 \quad -11 \quad 1$$
$$ 6 \quad 12 \quad 8 \quad -6$$
$$ 3 \quad 6 \quad 4 \quad -3$$

Step 10 Add $1 + (-6)$ and write the answer -5 in the third row.

$$2\rfloor \quad 3 \quad 0 \quad -8 \quad -11 \quad 1$$
$$ 6 \quad 12 \quad 8 \quad -6$$
$$ 3 \quad 6 \quad 4 \quad -3 \quad \boxed{-5}$$

Except for the signs in the second row, this last array is the same as the array obtained from the long division process, and we can read off the quotient and remainder:

The last number in the third row is the remainder.

The other numbers in the third row are the coefficients of the quotient (arranged in order of decreasing powers of x).

Since we are dividing the *fourth*-degree polynomial $3x^4 - 8x^2 - 11x + 1$ by the *first*-degree polynomial $x - 2$, the quotient must be a polynomial of degree *three* with coefficients $3, 6, 4, -3$, namely, $3x^3 + 6x^2 + 4x - 3$. The remainder is -5.

> **CAUTION**
>
> Synthetic division can be used *only* when the divisor is a first-degree polynomial of the form $x - c$. In the example above, $c = 2$. If you want to use synthetic division with a divisor such as $x + 3$, you must write it as $x - (-3)$, which is of the form $x - c$ with $c = -3$.

EXAMPLE 1

TIP

Synthetic division programs are in the Program Appendix.

To divide $x^5 + 5x^4 + 6x^3 - x^2 + 4x + 29$ by $x + 3$, we write the divisor as $x - (-3)$ and proceed as above.

$$
\begin{array}{r|rrrrrr}
-3 & 1 & 5 & 6 & -1 & 4 & 29 \\
   &   & -3 & -6 & 0 & 3 & -21 \\
\hline
   & 1 & 2 & 0 & -1 & 7 & \;|\;\underline{8}
\end{array}
$$

The last row shows that the quotient is $x^4 + 2x^3 - x + 7$ and the remainder is 8. ∎

EXAMPLE 2

Show that $x - 7$ is a factor of $8x^5 - 52x^4 + 2x^3 - 198x^2 - 86x + 14$ and find the other factor.

SOLUTION $x - 7$ is a factor exactly when division by $x - 7$ leaves remainder 0, in which case the quotient is the other factor. Using synthetic division, we have

$$
\begin{array}{r|rrrrrr}
7 & 8 & -52 & 2 & -198 & -86 & 14 \\
  &   & 56 & 28 & 210 & 84 & -14 \\
\hline
  & 8 & 4 & 30 & 12 & -2 & \;|\;\underline{0}
\end{array}
$$

Since the remainder is 0, the divisor $x - 7$ and the quotient

$$8x^4 + 4x^3 + 30x^2 + 12x - 2$$

are factors.

$$8x^5 - 52x^4 + 2x^3 - 198x^2 - 86x + 14$$
$$= (x - 7)(8x^4 + 4x^3 + 30x^2 + 12x - 2).$$ ∎

✓ **EXERCISES** 4.2.A 📼

In Exercises 1–8, use synthetic division to find the quotient and remainder.

1. $(3x^4 - 8x^3 + 9x + 5) \div (x - 2)$

2. $(4x^3 - 3x^2 + x + 7) \div (x - 2)$

3. $(2x^4 + 5x^3 - 2x - 8) \div (x + 3)$

4. $(3x^3 - 2x^2 - 8) \div (x + 5)$

5. $(5x^4 - 3x^2 - 4x + 6) \div (x - 7)$

6. $(3x^4 - 2x^3 + 7x - 4) \div (x - 3)$

7. $(x^4 - 6x^3 + 4x^2 + 2x - 7) \div (x - 2)$

8. $(x^6 - x^5 + x^4 - x^3 + x^2 - x + 1) \div (x + 3)$

In Exercises 9–12, use synthetic division to find the quotient and the remainder. In each divisor $x - c$, the number c is not an integer, but the same technique will work.

9. $(3x^4 - 2x^2 + 2) \div \left(x - \dfrac{1}{4}\right)$

10. $(2x^4 - 3x^2 + 1) \div \left(x - \dfrac{1}{2}\right)$

11. $(2x^4 - 5x^3 - x^2 + 3x + 2) \div \left(x + \dfrac{1}{2}\right)$

12. $\left(10x^5 - 3x^4 + 14x^3 + 13x^2 - \dfrac{4}{3}x + \dfrac{7}{3}\right) \div \left(x + \dfrac{1}{5}\right)$

In Exercises 13–16, use synthetic division to show that the first polynomial is a factor of the second and find the other factor.

13. $x + 4;$ $3x^3 + 9x^2 - 11x + 4$

14. $x - 5;$ $x^5 - 8x^4 + 17x^2 + 293x - 15$

15. $x - 1/2;$ $2x^5 - 7x^4 + 15x^3 - 6x^2 - 10x + 5$

16. $x + 1/3;$ $3x^6 + x^5 - 6x^4 + 7x^3 + 3x^2 - 15x - 5$

In Exercises 17 and 18, use a calculator and synthetic division to find the quotient and remainder.

17. $(x^3 - 5.27x^2 + 10.708x - 10.23) \div (x - 3.12)$

18. $(2.79x^4 + 4.8325x^3 - 6.73865x^2 + .9255x - 8.125) \div (x - 1.35)$

Thinkers

19. When $x^3 + cx + 4$ is divided by $x + 2$, the remainder is 4. Find c.

20. If $x - d$ is a factor of $2x^3 - dx^2 + (1 - d^2)x + 5$, what is d?

4.3   Graphs of Polynomial Functions

The graphs of first- and second-degree polynomial functions are straight lines and parabolas respectively (Sections 1.3 and 4.1). The emphasis here will be on higher degree polynomial functions.

The polynomials with the simplest graphs are

$$f(x) = x^2, \qquad f(x) = x^3, \qquad f(x) = x^5, \qquad f(x) = x^6, \qquad \text{etc.}$$

Their graphs, which are shown on the left side of the chart on the next page, are part of the catalog of basic functions. You should memorize the shapes of these graphs. More generally, you should know the graphs of functions of the form $f(x) = ax^n$ (where a is a constant and n a positive integer). They are of four types, as shown in the following chart.

GRAPH OF $f(x) = ax^n$

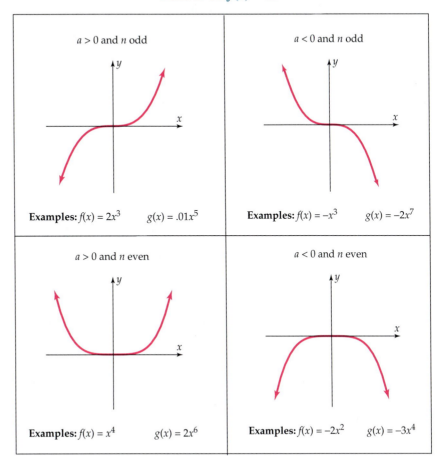

$a > 0$ and n odd	$a < 0$ and n odd
Examples: $f(x) = 2x^3$ $g(x) = .01x^5$	**Examples:** $f(x) = -x^3$ $g(x) = -2x^7$
$a > 0$ and n even	$a < 0$ and n even
Examples: $f(x) = x^4$ $g(x) = 2x^6$	**Examples:** $f(x) = -2x^2$ $g(x) = -3x^4$

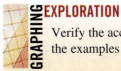

GRAPHING EXPLORATION

Verify the accuracy of the preceding summary by graphing each of the examples in the window with $-5 \le x \le 5$ and $-30 \le y \le 30$.

The graphs of more complicated polynomial functions can vary considerably in shape. Understanding the properties discussed below should assist you to interpret screen images correctly and to determine when a polynomial graph is complete.

▪▪ CONTINUITY

Every polynomial graph is **continuous,** meaning that it is an unbroken curve, with no jumps, gaps, or holes. Furthermore, polynomial graphs have no sharp corners. Thus, neither of the graphs in Figure 4–16 is the graph of a polynomial function. On a calculator screen, however, a polynomial graph may look like a series of juxtaposed line segments, rather than a smooth, continuous curve.

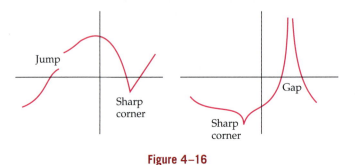

Figure 4–16

SHAPE OF THE GRAPH WHEN $|x|$ IS LARGE

The shape of a polynomial graph at the far left and far right of the coordinate plane is easily determined by using our knowledge of graphs of functions of the form $f(x) = ax^n$.

EXAMPLE 1

Consider the function $f(x) = 2x^3 + x^2 - 6x$ and the function determined by its leading term $g(x) = 2x^3$.

GRAPHING EXPLORATION

Using the standard viewing window, graph f and g on the same screen.

Do the graphs look different? Now graph f and g in the viewing window with $-20 \leq x \leq 20$ and $-10{,}000 \leq y \leq 10{,}000$. Do the graphs look almost the same?

Finally, graph f and g in the viewing window with

$$-100 \leq x \leq 100 \text{ and } -1{,}000{,}000 \leq y \leq 1{,}000{,}000.$$

Do the graphs look virtually identical?

The reason the answer to the last question is "yes" can be understood from this table.

x	-100	-50	70	100
$-6x$	600	300	-420	-600
x^2	10,000	2,500	4,900	10,000
$g(x) = 2x^3$	$-2{,}000{,}000$	$-250{,}000$	686,000	2,000,000
$f(x) = 2x^3 + x^2 - 6x$	$-1{,}989{,}400$	$-247{,}200$	690,480	2,009,400

It shows that when $|x|$ is large, the terms x^2 and $-6x$ are insignificant compared with $2x^3$ and play a very minor role in determining the value of $f(x)$. Hence, the values of $f(x)$ and $g(x)$ are relatively close. ■

Example 1 is typical of what happens in every case: When $|x|$ is very large, the highest power of x totally overwhelms all lower powers and plays the greatest role in determining the value of the function.

Behavior When
$|x|$ Is Large

> When $|x|$ is very large, the graph of a polynomial function closely resembles the graph of its highest-degree term.
>
> In particular, when the polynomial function has odd degree, one end of its graph shoots upward and the other end downward.
>
> When the polynomial function has even degree, both ends of its graph shoot upward or both ends shoot downward.

▪▪ *x*-INTERCEPTS

The x-intercepts of the graph of a polynomial function f are the solutions of the equation $f(x) = 0$, that is, the roots of the polynomial $f(x)$. As we saw in Section 4.2, a polynomial of degree n has at most n distinct roots. So we have the following fact.

x-Intercepts

> The graph of a polynomial function of degree n meets the x-axis at most n times.

There is another connection between roots and graphs. For example, it is easy to see that the roots of

$$f(x) = (x + 3)^2(x + 1)(x - 1)^3$$

are -3, -1, and 1. We say that

-3 is a root of multiplicity 2;

-1 is a root of multiplicity 1;

1 is a root of multiplicity 3.

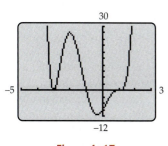

Figure 4–17

Observe that the graph of $f(x)$ in Figure 4–17 does not cross the x-axis at -3 (a root whose multiplicity is an *even* number) but does cross the x-axis at -1 and 1 (roots of *odd* multiplicity). More generally, a number c is a **root of multiplicity k** of a polynomial $f(x)$ if $(x - c)^k$ is a factor of $f(x)$ and no higher power of $(x - c)$ is a factor, and we have this fact.

Multiplicity
and Graphs

> Let c be a root of multiplicity k of a polynomial function f.
>
> If k is odd, the graph of f crosses the x-axis at c.
>
> If k is even, the graph of f touches, but does not cross, the x-axis at c.

▪▪ LOCAL EXTREMA

The term **local extremum** (plural, extrema) refers to either a local maximum or a local minimum, that is, a point where the graph is a peak or a valley.

GRAPHING EXPLORATION

Graph $f(x) = x^3 + 2x^2 - 4x - 3$ in the standard viewing window. What is the total number of peaks and valleys on the graph? What is the degree of $f(x)$?

Now graph $g(x) = x^4 - 3x^3 - 2x^2 + 4x + 5$ in the standard viewing window. What is the total number of peaks and valleys on the graph? What is the degree of $g(x)$? ▪

The two polynomials you have just graphed are illustrations of the following fact, which is proved in calculus.

Local Extrema

> A polynomial function of degree n has at most $n - 1$ local extrema. In other words, the total number of peaks and valleys on the graph is at most $n - 1$.

▪▪ COMPLETE GRAPHS OF POLYNOMIAL FUNCTIONS

By using the facts discussed earlier, you can often determine whether or not the graph of a polynomial function is complete (that is, shows all the important features).

EXAMPLE 2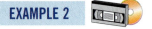

Find a complete graph of

$$f(x) = x^4 + 10x^3 + 21x^2 - 40x - 80.$$

SOLUTION   Since $f(0) = -80$, the standard viewing window probably won't show a complete graph, so we try the window with

$$-10 \le x \le 10 \qquad \text{and} \qquad -100 \le y \le 100$$

and obtain Figure 4–18. The three peaks and valleys shown here are the only ones because a fourth-degree polynomial graph has at most three local extrema. There cannot be more x-intercepts than the two shown here because if the graph turned toward the x-axis farther out, there would be an additional peak, which is impossible. Finally, the outer ends of the graph resemble the graph of x^4, the highest-degree term (see the chart on page 307). Hence, Figure 4–18 includes all the important features of the graph and is therefore complete. ▪

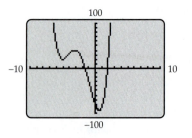

Figure 4–18

EXAMPLE 3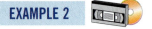

Find a complete graph of $f(x) = x^3 - 1.8x^2 + x + 2.$

SOLUTION   We first try the standard window (Figure 4–19). The graph is similar to the graph of the leading term $y = x^3$ but does not appear to have any local extrema. However, if you use the trace feature on the flat portion of the graph to the right of the x-axis, you see that the y-coordinates increase, then decrease, then increase (try it!). Zooming in on the portion of the graph between 0 and 1 (Figure 4–20), we see that the graph actually has a tiny peak and valley (the maximum possible number of local extrema for a cubic). So Figures 4–19 and 4–20 together provide a complete graph of f.   ■

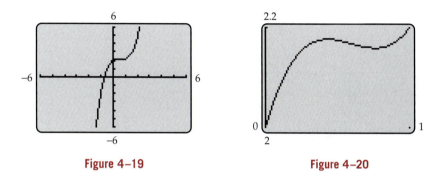

Figure 4–19   Figure 4–20

The following fact is proved in Exercise 53.

No polynomial graph contains any horizontal line segments.

However, a calculator may erroneously show horizontal segments, as in Figure 4–19. So always investigate such segments, by using trace or zoom-in, to determine any hidden behavior, such as that in Example 3.

EXAMPLE 4

The graph of $f(x) = .01x^5 + x^4 - x^3 - 6x^2 + 5x + 4$ in the standard window is shown in Figure 4–21. Explain why this is *not* a complete graph and find a complete graph of f.

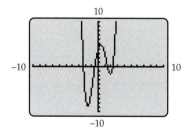

Figure 4–21

SOLUTION   When $|x|$ is large, the graph of f must resemble the graph of $y = .01x^5$, whose left end goes downward (see the chart on page 307). Since Figure 4–21 does not show this, it is not a complete graph. To have the same shape as the graph of $y = .01x^5$, the graph of f must turn downward and cross the x-axis somewhere to the left of the origin. Figure 4–21 shows three local extrema. Even without graphing, we can see that there must be one more peak (where the graph turns downward on the left), making a total of four local extrema (the most a fifth-degree polynomial can have), and another x-intercept, for a total of five. When these additional features are shown, we will have a complete graph.

EXPLORATION

GRAPHING   Find a viewing window that includes the local maximum and x-intercept not shown in Figure 4–21. When you do, the scale will be such that the local extrema and x-intercepts shown in Figure 4–21 will no longer be visible.

Consequently, a complete graph of $f(x)$ requires several viewing windows in order to see all the important features. ■

The graphs obtained in Examples 2–4 were known to be complete because in each case, they included the maximum possible number of local extrema. In many cases, however, a graph may not have the largest possible number of peaks and valleys. In such cases, use any available information and try several viewing windows to obtain the most likely complete graph.

✓ EXERCISES 4.3

In Exercises 1–6, decide whether the given graph could possibly be the graph of a polynomial function.

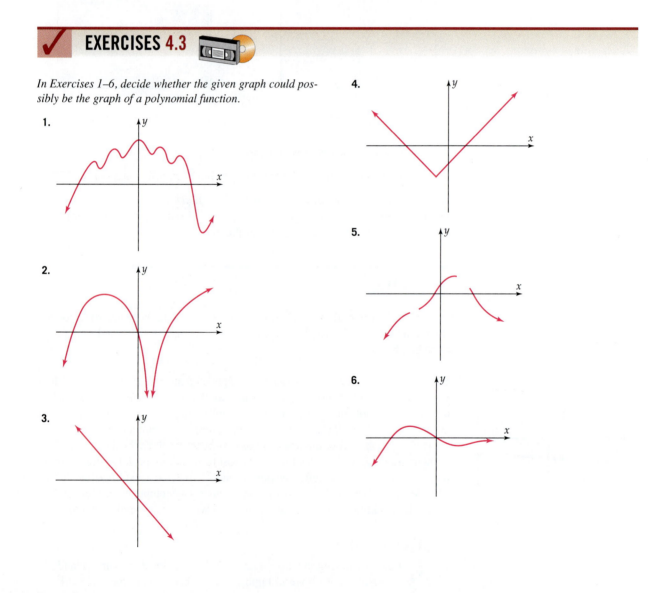

In Exercises 7–12, determine whether the given graph could possibly be the graph of a polynomial function of degree 3, of degree 4, or of degree 5.

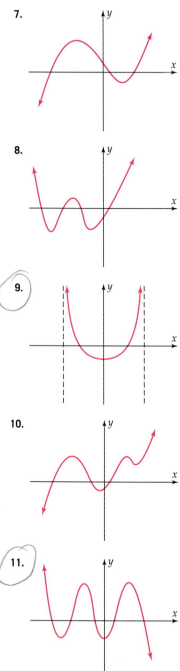

7.

8.

9.

10.

11.

12.

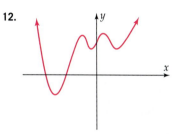

In Exercises 13 and 14, find a viewing window in which the graph of the given polynomial function f appears to have the same general shape as the graph of its leading term.

13. $f(x) = x^4 - 6x^3 + 9x^2 - 3$

14. $f(x) = x^3 - 5x^2 + 4x - 2$

In Exercises 15–18, the graph of a polynomial function is shown. List each root of the polynomial and state whether its multiplicity is even or odd.

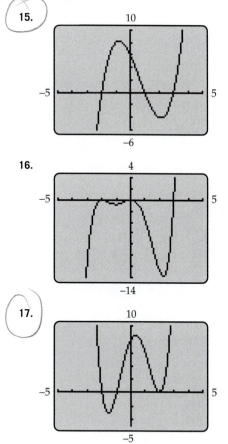

15.

16.

17.

18.

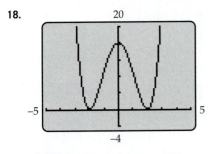

In Exercises 19–24, use your knowledge of polynomial graphs, not *a calculator, to match the given function with its graph, which is one of those shown here.*

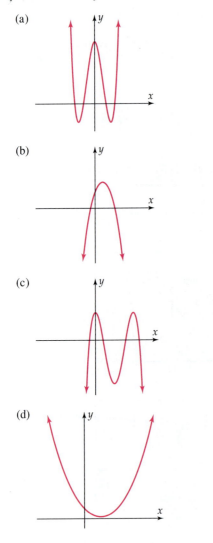

(a)

(b)

(c)

(d)

(e)

(f)

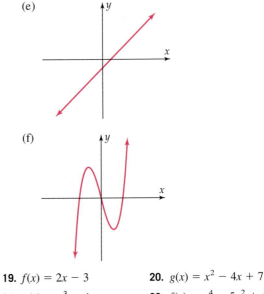

19. $f(x) = 2x - 3$ **20.** $g(x) = x^2 - 4x + 7$

21. $g(x) = x^3 - 4x$ **22.** $f(x) = x^4 - 5x^2 + 4$

23. $f(x) = -x^4 + 6x^3 - 9x^2 + 2$

24. $g(x) = -2x^2 + 3x + 1$

In Exercises 25–28, graph the function in the standard viewing window and explain why that graph cannot possibly be complete.

25. $f(x) = .01x^3 - .2x^2 - .4x + 7$

26. $g(x) = .01x^4 + .1x^3 - .8x^2 - .7x + 9$

27. $h(x) = .005x^4 - x^2 + 5$

28. $f(x) = .001x^5 - .01x^4 - .2x^3 + x^2 + x - 5$

In Exercises 29–34, find a single viewing window that shows a complete graph of the function.

29. $f(x) = x^3 + 8x^2 + 5x - 14$

30. $g(x) = x^3 - 3x^2 - 4x - 5$

31. $g(x) = -x^4 - 3x^3 + 24x^2 + 80x + 15$

32. $f(x) = x^4 - 10x^3 + 35x^2 - 50x + 24$

33. $f(x) = 2x^5 - 3.5x^4 - 10x^3 + 5x^2 + 12x + 6$

34. $g(x) = x^5 + 8x^4 + 20x^3 + 9x^2 - 27x - 7$

In Exercises 35–38, find a complete graph of the function and list the viewing window(s) that show this graph.

35. $f(x) = .1x^5 + 3x^4 - 4x^3 - 11x^2 + 3x + 2$

36. $g(x) = x^4 - 48x^3 - 101x^2 + 49x + 50$

37. $g(x) = .03x^3 - 1.5x^2 - 200x + 5$

38. $f(x) = .25x^6 + .25x^5 - 35x^4 - 7x^3 + 823x^2 + 25x - 2750$

39. (a) Explain why the graph of a cubic polynomial function has either two local extrema or none at all. [*Hint:* If it had only one, what would the graph look like when $|x|$ is very large?]

(b) Explain why the general shape of the graph of a cubic polynomial function must be one of the following:

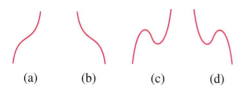

(a)   (b)   (c)   (d)

40. The figure shows an incomplete graph of a fourth-degree even polynomial function f. (Even functions were defined in Special Topics 3.4.A.)

(a) Find the roots of f.

(b) Explain why

$$f(x) = k(x - a)(x - b)(x - c)(x - d),$$

where a, b, c, d are the roots of f.

(c) Experiment with your calculator to find the value of k that produces the graph in the figure.

(d) Find all local extrema of f.

(e) List the approximate intervals on which f is increasing and those on which it is decreasing.

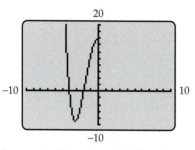

41. A complete graph of a polynomial function g is shown below.

(a) Is the degree of $g(x)$ even or odd?

(b) Is the leading coefficient of $g(x)$ positive or negative?

(c) What are the real roots of $g(x)$?

(d) What is the smallest possible degree of $g(x)$?

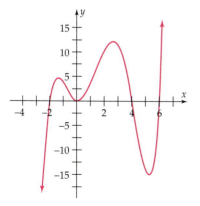

42. Do Exercise 41 for the polynomial function g whose complete graph is shown here.

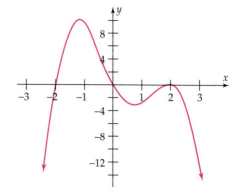

43. The figure is a partial view of the graph of a cubic polynomial whose leading coefficient is negative. Which of the patterns shown in Exercise 39 does this graph have?

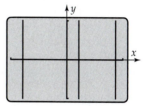

44. The figure is a partial view of the graph of a fourth-degree polynomial. Sketch the general shape of the graph and state whether the leading coefficient is positive or negative.

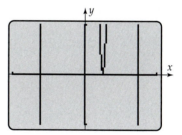

In Exercises 45–54, sketch a complete graph of the function. Label each x-intercept and the coordinates of each local extremum; find intercepts and coordinates exactly when possible and otherwise approximate them.

45. $f(x) = x^3 - 3x^2 + 4$ **46.** $g(x) = 4x - 4x^3/3$

47. $h(x) = .25x^4 - 2x^3 + 4x^2$

48. $f(x) = .25x^4 - 2x^3/3$

49. $g(x) = 3x^3 - 18.5x^2 - 4.5x - 45$

50. $h(x) = 2x^3 + x^2 - 4x - 2$

51. The crime rate (per 100,000 inhabitants) in the United States is approximated by

$$f(x) = .1414x^4 - 6.81x^3 + 98.68x^2 - 473.2x + 6001.6$$
$$(0 \le x \le 20),$$

where $x = 0$ corresponds to 1980.*

(a) Use your knowledge of polynomial functions (not graphing) to explain why this model predicts that the crime rate will eventually rise sharply and keep rising. [*Hint:* Consider the highest-degree term.]

(b) Confirm your answer in part (a) by graphing f in a viewing window with $0 \le x \le 30$.

(c) According to this model, in what year between 1980 and 2000 (inclusive) was the crime rate the highest?

52. The United Nations projects that the population of China (in millions) will be approximated by

$$g(x) = -.00096x^3 - .1x^2 + 11.3x + 1274$$
$$(0 \le x \le 50),$$

where $x = 0$ corresponds to 2000.

(a) Use your knowledge of polynomial functions (not graphing) to explain why this model is extremely unlikely to be accurate long into the future. [See the Hint for Exercise 51.]

(b) According to this model, when in the next half century will China's population be largest (round your answer to the nearest year).

(c) What will the population be in that year?

Thinkers

53. (a) Graph $g(x) = .01x^3 - .06x^2 + .12x + 3.92$ in the viewing window with $-3 \le x \le 3$ and $0 \le y \le 6$ and verify that the graph appears to coincide with the horizontal line $y = 4$ between $x = 1$ and $x = 3$. In other words, it appears that every x with $1 \le x \le 3$ is a solution of the equation

$$.01x^3 - .06x^2 + .12x + 3.92 = 4.$$

Explain why this is impossible. Conclude that the actual graph is not horizontal between $x = 1$ and $x = 3$.

(b) Use the trace feature to verify that the graph is actually rising from left to right between $x = 1$ and $x = 3$. Find a viewing window that shows this.

(c) Show that it is not possible for the graph of a polynomial $f(x)$ to contain a horizontal segment. [*Hint:* A horizontal line segment is part of the horizontal line $y = k$ for some constant k. Adapt the argument in part (a), which is the case $k = 4$.]

54. (a) Let $f(x)$ be a polynomial of odd degree. Explain why $f(x)$ must have at least one real root. [*Hint:* Why must the graph of f cross the x-axis, and what does this mean?]

(b) Let $g(x)$ be a polynomial of even degree, with a negative leading coefficient and a positive constant term. Explain why $g(x)$ must have at least one positive and at least one negative root.

55. For x-values in a particular interval, a nonpolynomial function $g(x)$ can often be approximated by a polynomial function, meaning that there is some polynomial $f(x)$ such that $g(x) \approx f(x)$ for every x in the interval.

(a) In the standard viewing window, graph both

$$g(x) = \sqrt{x}$$

(which is not a polynomial function) and the polynomial function

$$f(x) = .26705x^3 - .78875x^2 + 1.3021x + .22033.$$

Are the graphs similar?

(b) Graph $f(x)$ and $g(x)$ in the viewing window with $0.26 \le x \le 1$ and $0 \le y \le 1$. Does it now appear that $f(x)$ is a good approximation of $g(x)$ when

$$.26 \le x \le 1?$$

(c) For any particular value of x, the error in this approximation is the difference between $f(x)$ and $g(x)$. In other words, $h(x) = f(x) - g(x)$ measures the error in the approximation. Graph the function $h(x)$ in the viewing window with

$$.26 \le x \le 1 \quad \text{and} \quad -.001 \le y \le .001$$

and use the trace feature to determine the maximum error in the approximation.

56. (a) Graph $f(x) = x^3 - 4x$ in the viewing window with $-3 \le x \le 3$ and $-5 \le y \le 5$.

(b) Graph the difference quotient of $f(x)$ (with $h = .01$) on the same screen.

(c) Find the x-coordinates of the relative extrema of $f(x)$. How do these numbers compare with the x-intercepts of the difference quotient?

57. The graph of

$$f(x) = (x + 18)(x^2 - 20)(x - 2)^2(x - 10)$$

has x-intercepts at each of its roots, that is, at $x = -18$, $\pm\sqrt{20} \approx \pm4.472$, 2, and 10. It is also true that $f(x)$ has a relative minimum at $x = 2$.

(a) Draw the x-axis and mark the roots of $f(x)$. Then use the fact that $f(x)$ has degree 6 (why?) to sketch the general shape of the graph (as was done for cubics in Exercise 39).

(b) Now graph $f(x)$ in the standard viewing window. Does the graph resemble your sketch? Does it even show all the x-intercepts between -10 and 10?

(c) Graph $f(x)$ in the viewing window with $-19 \le x \le 11$ and $-10 \le y \le 10$. Does this window include all the x-intercepts as it should?

(d) List viewing windows that give a complete graph of $f(x)$.

---

*FBI Uniform Crime Reports.

OMIT

C 4.3.A *SPECIAL TOPICS* Optimization Applications

Many real-life situations require you to find the largest or smallest quantity satis-fying certain conditions. For instance, automotive engineers want to design engines with maximum fuel efficiency. A cereal manufacturer might want to know the dimensions of the box that requires the least amount of cardboard (and hence, is cheapest). Physicists need to find the maximum speed of an object and the minimum distance between two moving bodies.

Many such optimization problems can be modeled by polynomial functions and solved by finding the local extrema of these functions. Although calculus is usually required for exact solutions, graphing technology can provide accurate approximate solutions, as the examples below demonstrate. We begin with a quadratic model. [For other optimization situations that can be modeled by quad-ratic functions, see Examples 5–7 in Section 4.1.]

EXAMPLE 1

An auto parts manufacturer makes radiators that sell for $350 each. The cost of producing x radiators is approximated by the function

$$C(x) = .01x^2 + 25x + 600,000.$$

(a) Find the revenue and profit functions.

(b) What number of radiators will produce the largest profit? What is the maxi-mum profit?

SOLUTION

(a) Since each radiator sells for $350, the revenue from selling x radiators is given by $R(x) = 350x$. So the profit function P is given by

$$P(x) = \text{Revenue} - \text{Costs}$$

$$= R(x) - C(x)$$

$$= 350x - (.01x^2 + 25x + 600,000)$$

$$P(x) = -.01x^2 + 325x - 600,000.$$

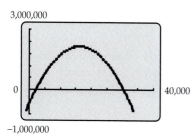

Figure 4–22

(b) The graph of $P(x)$ is a downward-opening parabola (why?), as shown in Fig-ure 4–22. Maximum profit occurs at the vertex of the parabola. The vertex can be found with a graphical maximum finder (Figure 4–23) or by using the formula on page 282.

$$x\text{-coordinate of vertex} = \frac{-b}{2a} = \frac{-325}{2(-.01)} = 16,250.$$

Therefore, maximum profit occurs when 16,250 radiators are sold. The profit is

$$P(16,250) = -.01(16,250)^2 + 325(16,250) - 600,000 = \$2,040,625.$$

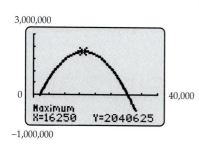

Figure 4–23

EXAMPLE 2

A box with no top is to be made from a 22×30 inch sheet of cardboard by cutting squares of equal size from each corner and bending up the flaps, as shown in Figure 2–24. To the nearest hundredth of an inch, what size square should be cut from each corner to obtain a box with the largest possible volume, and what is the volume of this box?

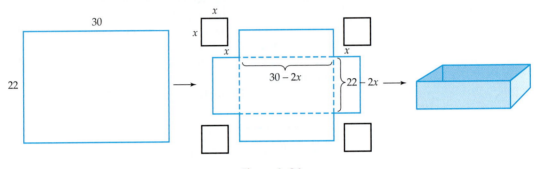

Figure 4–24

SOLUTION   Let x denote the length of the side of the square to be cut from each corner. Then,

$$\text{Volume of box} = \text{Length} \times \text{Width} \times \text{Height}$$

$$= (30 - 2x) \cdot (22 - 2x) \cdot x$$

$$= 4x^3 - 104x^2 + 660x.$$

Thus, the equation $y = 4x^3 - 104x^2 + 660x$ gives the volume y of the box that results from cutting a square of side x from each corner. Since the shortest side of the cardboard is 22 inches, the length x of the side of the cut-out square must be less than 11 (why?).

Each point on the graph of $y = 4x^3 - 104x^2 + 660x$ ($0 < x < 11$) in Figure 4–25 represents one of the possibilities:

The x-coordinate is the size of the square to be cut from each corner;

The y-coordinate is the volume of the resulting box.

The box with the largest volume corresponds to the point with the largest y-coordinate, that is, the highest point in the viewing window. A maximum finder (Figure 4–26) shows that this point is approximately (4.182, 1233.809). Therefore, a

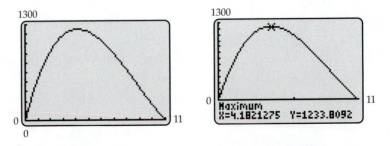

Figure 4–25   **Figure 4–26**

square measuring approximately 4.18 × 4.18 inches should be cut from each corner, producing a box of volume approximately 1233.81 cubic inches. ■

EXAMPLE 3

The graph of $y = 8 - x^2$ is a downward-opening parabola, as shown in Figure 4–27. If (x, y) is a point on the graph, with x positive, find the value of x for which the right triangle with vertices at $(0, 0)$, $(x, 0)$ and (x, y) has the largest possible area.

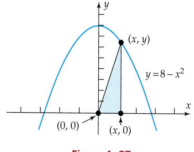

Figure 4–27

SOLUTION   Recall that the area of a triangle is given by

$$A = \frac{1}{2}bh,$$

Where b is the base and h the height. Figure 4–27 shows that the base b has length x and that the height h is the number $y = 8 - x^2$. So the area is given by

$$A(x) = \frac{1}{2}x(8 - x^2).$$

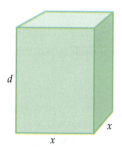

Figure 4–28

We must find the positive value of x that makes $A(x)$ as large as possible. So we graph the function and use a maximum finder (Figure 4–28) to determine that $x \approx 1.633$ produces the maximum area $A(x) \approx 4.355$. ■

EXAMPLE 4

A rectangular box with a square base (Figure 4–29) is to be mailed. The sum of the height of the box and the perimeter of the base is to be 84 inches, the maximum allowable under postal regulations. What are the dimensions of the box with largest possible volume that meets these conditions?

Figure 4–29

SOLUTION   If the length of one side of the base is x, then the perimeter of the base (the sum of the length of its four sides) is $4x$. If the height of the box is d, then $4x + d = 84$, so $d = 84 - 4x$, and hence, the volume is

$$V = x \cdot x \cdot d = x \cdot x \cdot (84 - 4x) = 84x^2 - 4x^3.$$

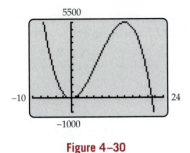

Figure 4–30

The graph of the polynomial function $V(x) = 84x^2 - 4x^3$ in Figure 4–30 is complete (why?). However, the only relevant part of the graph in this situation is the portion with x and $V(x)$ positive (because x is a length and $V(x)$ is a volume). The graph of $V(x)$ has a local maximum between 10 and 20, and this local maximum value is the largest possible volume for the box.

GRAPHING EXPLORATION

Use a maximum finder to find the x-value at which the local maximum occurs. State the dimensions of the box in this case.   ■

EXAMPLE 5

Nancy who is standing at point A on the bank of a 2.5-kilometer-wide river plans to row to point C on the opposite shore and then run to point B, as shown in Figure 4–31. Nancy can row at a rate of 4 km/h and can run at 8 km/h. How far from B should she land in order to make the trip in the shortest possible time?

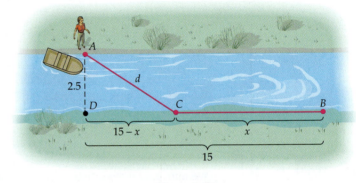

Figure 4–31

SOLUTION   The situation here is the same as in Example 6 of Section 2.4, where we showed that time for Nancy's trip is given by

$$T(x) = \frac{\sqrt{(x-15)^2 + 6.25}}{4} + \frac{x}{8},$$

where x is the distance from C to B (see Figure 4–31). To find the shortest possible time, we must find the value of x that makes $T(x)$ as small as possible.

GRAPHING EXPLORATION

Using the viewing window with $0 \le x \le 15$ and $0 \le y \le 4$, graph $T(x)$ and use a minimum finder to verify that the lowest point on the graph (that is, the point with the y-coordinate $T(x)$ as small as possible) is approximately $(13.56, 2.42)$.

Therefore, the shortest time for the trip will be 2.42 hours and will occur if Nancy rows to a point 13.56 kilometers from B.   ■

EXERCISES 4.3.A

1. Total commercial and industrial loans by large banks (in billions of dollars) are approximated by the polynomial function

$$f(x) = -.1341x^4 + 7.953x^3 - 167.75x^2 + 1495.01x - 4400$$
$$(8 \leq x < 14),$$

where $x = 8$ corresponds to 1988.*
   (a) In what year was the total loan volume the smallest? What might be the reason for this?
   (b) In what year was the total loan volume the largest?

2. The number of non-Hispanic black students (in thousands) enrolled at Ohio's public universities and colleges is approximated by

$$g(x) = -.0042x^3 + .1616x^2 - .9125x + 30.94$$
$$(0 \leq x \leq 22),$$

where $x = 0$ corresponds to 1980.† In what year between 1980 and 2002 was the number of such students the smallest?

3. The cost of manufacturing radiators in Example 1 has remained the same, but due to competitive pressure, the manufacturer must now sell each radiator for $300.
   (a) Find the profit function under these new circumstances.
   (b) What number of radiators will now produce the largest profit? What is the maximum profit?

4. A manufacturer sells loose-leaf notebooks for $3 each. It costs $.0002x^2 + 3000$ dollars to make x notebooks.
   (a) What is the weekly profit function?
   (b) Approximately how many notebooks should be made per week to make a profit of exactly $5000?
   (c) How many notebooks should be made each week to have the largest possible profit? What is that profit?

5. Name tags can be sold for $29 per thousand. The cost of manufacturing x thousand tags is $.001x^3 + .06x^2 - 1.5x$ dollars. Assume that all tags manufactured are sold.
   (a) Write the rule of the revenue function in this situation.
   (b) Write the rule of the profit function.
   (c) What number of tags should be made to guarantee a maximum profit? What will that profit be?

6. (a) A company makes novelty bookmarks that sell for $142 per hundred. The cost (in dollars) of making x hundred bookmarks is $x^3 - 8x^2 + 20x + 40$. Because of other projects, a maximum of 600 bookmarks per day can be manufactured. Assuming that the company can sell all the bookmarks it makes, how many should it make each day to maximize profits?
   (b) Owing to a change in other orders, as many as 1600 bookmarks can now be manufactured each day. How many should be made to maximize profits?

7. A 20-inch-square piece of metal is to be used to make an open-top box by cutting equal-sized squares from each corner and folding up the sides (as in Example 2). The length, width, and height of the box are each to be no more than 12 inches. What size squares should be cut out to produce a box with
   (a) volume 550 cubic inches?
   (b) largest possible volume?

8. A certain type of fencing comes in rigid 10-ft-long segments. Four uncut segments are to be used to fence in a garden on the side of a building, as shown in the figure. What value of x will result in a garden of the largest possible area and what is that area?

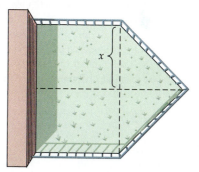

9. Graph $y = 24 - .5x^4$. Let (x, y) be a point on the graph, with $x > 0$.
   (a) Find a formula for the area of the right triangle with vertices $(0, 0)$, $(x, 0)$, and (x, y).
   (b) Find the value of x for which the triangle in part (a) has the largest area.

*Federal Reserve.
†Ohio Board of Regents.

10. Let (x, y) be a point in the first quadrant that is on the graph of $y = 24 - x^2$ and consider the rectangle with vertices $(-x, 0)$, $(-x, y)$, $(x, 0)$, and (x, y), as shown in the figure. For what value of x does the rectangle have maximal area?

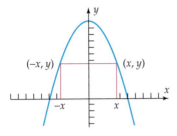

11. Find the dimensions of the rectangular box with a square base and no top that has volume 20,000 cubic centimeters and the smallest possible surface area. [*Hint:* See Example 3 in Section 2.4.]

12. An open-top box with a square base is to be constructed from 120 sq cm of material. What dimensions will produce a box
    (a) of volume 100 cubic centimeters?
    (b) with largest possible volume?

13. Anne is standing on a straight road and wants to reach her helicopter, which is located 2 miles down the road from her, a mile from the road in a field (see figure). She can run 5 miles per hour on the road and 3 miles per hour in the field. She plans to run down the road, then cut diagonally across the field to reach the helicopter. Where should she leave the road in order to reach the helicopter in the shortest possible time? [Compare with Exercise 50 in Section 2.4.]

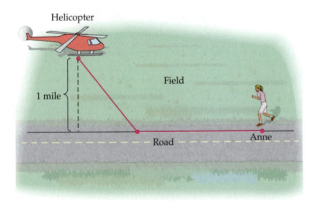

14. Tom, a lifeguard, is 20 feet from the edge of the ocean, and a swimmer is drowning 50 feet out to sea, as shown in the figure. The lifeguard can run at 15.7 feet per second and can swim at 5.9 feet per second. If the distance from A to B is 40 feet, at what point E should Tom enter the water so that he minimizes the time it takes to reach the drowning swimmer?

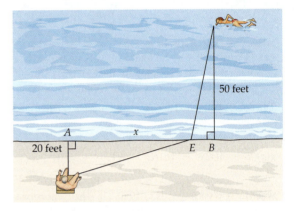

15. (a) Graph $f(x) = x^3 - 3x$
    (b) Find the distance from the point $(1, 1)$ to each of these points on the graph of f: $(-1, 2)$, $(0, 0)$, and $(2, 2)$.
    (c) For each point (x, y) on the graph of f, let

    $$g(x) = \text{distance from } (x, y) \text{ to } (1, 1)$$

    Write the rule of the function g.
    (d) Graph $g(x)$ in a viewing window with $-5 \leq x \leq 5$.
    (e) Find the point on the graph of f whose distance to $(1, 1)$ is minimal. [*Hint:* How is the answer to this question related to the graph of $g(x)$?]

16. Find the minimum distance from the graph of $f(x) = x^2$ to the point $(3, 0)$. [*Hint:* See Exercise 15.]

17. Find the minimum distance from the graph of $f(x) = x^3 + x^2 + 6$ to the point $(1, 2)$.

18. A cylindrical waste container with no top, a diameter of at least 2 feet, and a volume of 25 cubic feet is to be constructed. What should its radius be if
    (a) 65 sq ft of material are to be used to construct it?
    (b) the smallest possible amount of material is to be used to construct it? In this case, how much material is needed?

19. If $c(x)$ is the cost of producing x units, then $c(x)/x$ is the *average cost* per unit.[*] Suppose the cost of producing x units is given by $c(x) = .13x^3 - 70x^2 + 10,000x$ and that no more than 300 units can be produced per week.

(a) If the average cost is $1100 per unit, how many units are being produced?

(b) What production level should be used to minimize the average cost per unit? What is the minimum average cost?

---

*Depending on the situation, a unit of production might consist of a single item or several thousand items. Similarly, the cost of x units might be measured in thousands of dollars.

20. A mathematics book is to contain 36 square inches of printed matter per page, with margins of 1 inch along the sides and 1.5 inches at the top and bottom. Find the dimensions of the page that will lead to the minimum amount of paper being used for a page.

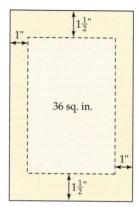

OMIT

4.4 Polynomial Models*

Linear regression was used in Section 1.4 to construct a linear function that modeled a set of data points. When the scatter plot of the data points looks more like a higher degree polynomial graph than a straight line, similar least squares regression procedures are available on most calculators for constructing quadratic, cubic, and quartic (fourth-degree) polynomial functions to model the data.

EXAMPLE 1

The table shows the estimated medical cost per claim under workers' compensation in selected years.[†]

Year	1992	1994	1996	1998	2000	2002
Cost per Claim	$9300	$10,700	$12,900	$18,600	$24,200	$34,300

(a) Use the data in the table to construct a cubic polynomial model.

(b) Use the model to estimate medical claim costs in 1999 and 2001.

---

*This section is optional; its prerequisite is Section 1.4.
†Workers' Compensation Insurance Rating Bureau.

SOLUTION

(a) Let $x = 2$ correspond to 1992. Then the data points are (2, 9300), (4, 10700), . . . , (12, 34300), as shown in Figure 4–32. The procedure for cubic regression is the same as for linear regression, with a few obvious changes; see the Technology Tips on pages 106 and 323. The resulting model is

$$f(x) = 8.912x^3 + 67.758x^2 + 7.011x + 8933.333.^*$$

The graph of f in Figure 4–33 appears to fit the data well.

(b) To estimate the costs in 1999 and 2001, we evaluate f at $x = 9$ and $x = 11$, as shown in Figure 4–34. An average claim cost \$20,982 in 1999 and \$29,071 in 2001. ■

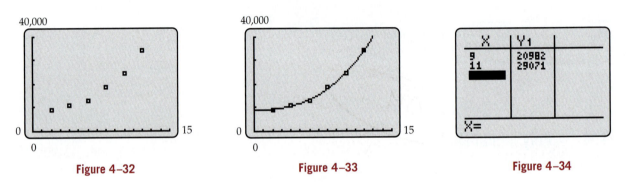

Figure 4–32   Figure 4–33   Figure 4–34

EXAMPLE 2

The table shows the population of San Francisco in selected years.[†]

Year	1950	1960	1970	1980	1990	2000
Population	775,357	740,316	715,674	678,974	723,959	776,733

(a) Find a polynomial model for this data.

(b) Use the model to estimate the population of San Francisco in 1996 and 2006.

SOLUTION

(a) Let $x = 0$ correspond to 1950 and plot the data points, as in Figure 4–35. We use quartic regression to obtain the following function:

$$f(x) = -.1072x^4 + 13.2x^3 - 387.24x^2 - 168.65x + 774,231.$$

The graph of f in Figure 4–36 appears to fit the data well.

---

*Here and later, coefficients are rounded for convenient reading, but the full coefficients are used to produce the graphs and estimates.
[†]U.S. Census Bureau.

(b) To estimate the population in 1996 and 2006, we evaluate f at $x = 46$ and $x = 56$, as shown in Figure 4–37. According to this model, the 1996 population was about 752,036 and the 2006 population is about 814,540.   ■

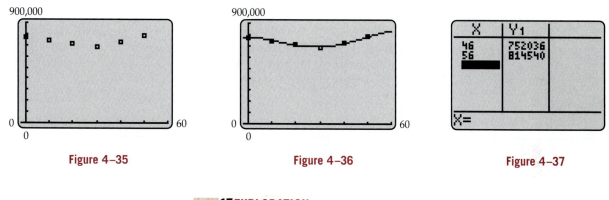

Figure 4–35   Figure 4–36   Figure 4–37

GRAPHING EXPLORATION

Use your minimum finder and the population function in Example 2 to estimate the year since 1950 when the population of San Francisco was smallest.

EXAMPLE 3

The table below, which is based on statistics from the Department of Health and Human Services, gives the cumulative number of reported cases of AIDS in the United States from 1982 through 2001. It shows, for example, that 41,662 cases were reported from 1982 through 1986.

Year	Cases	Year	Cases	Year	Cases
1982	1,563	1992	278,189	1997	652,439
1984	10,845	1993	380,601	1998	698,527
1986	41,662	1994	457,789	1999	743,418
1988	105,489	1995	528,421	2000	784,518
1990	188,872	1996	595,559	2001	816,149

Find a suitable polynomial model for this data.

SOLUTION   Let $x = 0$ correspond to 1980 and measure the cases in thousands, so that the data points are (2, 1.563), (4, 10.845), etc. The scatter plot of the data points in Figure 4–38 shows that the points do not lie on a straight line but could be part of a polynomial graph of higher degree.

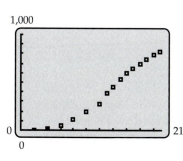

Figure 4–38

Since the data points suggest a curve that bends upward at the left and downward at the right, a third- or fourth-degree polynomial might provide a reasonable model. We use regression on a calculator to obtain the following models.

Cubic: $f(x) = -.2254x^3 + 9.22x^2 - 59.888x + 101.86$

Quartic: $g(x) = -.0135x^4 + .4014x^3 - .484x^2 - 3.636x + 9.55$

The graphs of both f and g are shown in Figure 4–39; as you can see, they are virtually identical in this viewing window. If you change the viewing window, however, you see that g is the better model in the 1980s (Figure 4–40), although there is little difference after 1990 (Figure 4–41).

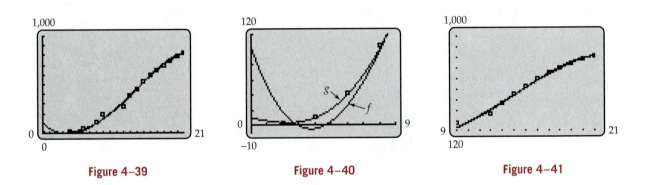

Figure 4–39   Figure 4–40   Figure 4–41

Although both functions provide reasonable models after 1990, our knowledge of polynomial graphs suggests that they may not be accurate in the future. As x gets larger, the graphs of f and g will resemble the graphs of $y = -.2254x^3$ and $y = -.0135x^4$, respectively (why?), both of which turn downward to the right (see the chart on page 307). However, the cumulative number of cases can't decrease. (Even when there are no new cases, the cumulative total stays the same.)

GRAPHING EXPLORATION

Graph f and g in a window with $10 \le x \le 28$. Which graph provides a plausible model for any years after 2000? How long will this model remain reasonable? ∎

NOTE: You must have at least three data points for quadratic regression, at least four for cubic regression, and at least five for quartic regression. If you have exactly the required minimum number data points, no two of them can have the same first coordinate. In this case, the polynomial regression function will pass through all of the data points (an exact fit). When you have more than the minimum number of data points required, the fit will generally be approximate rather than exact.

EXERCISES 4.4

In Exercises 1–4, a scatter plot of data is shown. State the type of polynomial model that seems most appropriate for the data (linear, quadratic, cubic, or quartic). If none of them is likely to provide a reasonable model, say so.

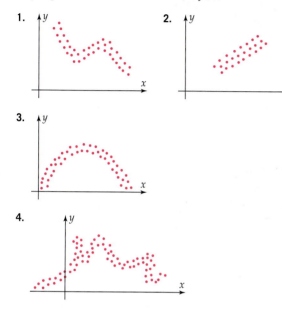

1.

2.

3.

4.

5. The table, which is based on the FBI Uniform Crime Reports, shows the rate of property crime per 100,000 population.

Year	Crimes	Year	Crimes
1984	4492.1	1996	4450.1
1986	4862.6	1997	4318.7
1988	5027.1	1998	4051.8
1990	5088.5	1999	3743.6
1992	4902.7	2000	3617.9
1994	4660.0	2001	3656.1

(a) Use cubic regression to find a polynomial function that models this data, with $x = 0$ corresponding to 1980.

(b) According to this model, what was the property crime rate in 1987 and 1993?

(c) The actual crime rate was about 4591 in 1995. What does the model predict?

(d) For how many years in the future is this model likely to be a reasonable one?

6. The table shows actual and projected enrollment (in millions) in public high schools in selected years.[*]

Year	Enrollment	Year	Enrollment
1975	14.3	1995	12.5
1980	13.2	2000	13.5
1985	12.4	2005	14.4
1990	11.3	2010	14.1

(a) Use quartic regression to find a polynomial function that models this data, with $x = 0$ corresponding to 1975.

(b) According to the model, what was enrollment in 1998 and in 1999?

(c) Estimate the year between 1975 and 2000 in which enrollment was the lowest. Does this estimate appear to be accurate?

7. The table shows the air temperature at various times during a spring day in Gainesville, Florida.

Time	Temp (F°)	Time	Temp (F°)
6 A.M.	52	1 P.M.	82
7 A.M.	56	2 P.M.	86
8 A.M.	61	3 P.M.	85
9 A.M.	67	4 P.M.	83
10 A.M.	72	5 P.M.	78
11 A.M.	77	6 P.M.	72
noon	80		

(a) Sketch a scatter plot of the data, with $x = 0$ corresponding to midnight.

(b) Find a quadratic polynomial model for the data.

(c) What is the predicted temperature for noon? For 9 A.M.? For 2 P.M.?

_____

*U.S. National Center for Educational Statistics.

8. The table shows median income of U.S. households in constant 2001 dollars.[*]

Year	Median Income	Year	Median Income
1989	$39,850	1997	$40,699
1991	$38,183	1999	$43,355
1993	$37,668	2000	$43,162
1995	$39,306	2001	$42,228

(a) Make a scatter plot of the data, with $x = 9$ corresponding to 1989 and income measured in thousands.
(b) Decide whether a quadratic or a quartic model seems more appropriate and state its rule.
(c) According to the model, what was the median income in 1998?
(d) Does this model seem likely to be accurate in 2002 and later? Why?

9. The average health care cost per employee paid by U.S. employers is shown in the table.[†]

Year	Cost	Year	Cost
1997	$3820	2001	$5162
1999	$4320	2002	$5758
2000	$4604	2003	$6535

(a) Make a scatter plot of the data, with $x = 7$ corresponding to 1997.
(b) Find a quadratic polynomial model for this data.
(c) Estimate the cost in 1998.
(d) If the model remains accurate, what will average cost per employee be in 2006?

10. The average time (in weeks) that a jobless worker remained unemployed is shown in the table.[‡]

Year	Weeks Jobless	Year	Weeks Jobless
1989	11	1997	15
1991	14	1999	13
1993	18	2001	12.5
1995	16	2003	19.6

(a) Make a scatter plot of the data, with $x = 9$ corresponding to 1989.
(b) Find a quartic model for the data.
(c) According to this model, how many weeks did an unemployed worker remain jobless in 2002?
(d) Explain why this model is not likely to be accurate much beyond 2003.

11. The table shows the total number of Black and Hispanic students enrolled at Ohio's public universities and colleges in various years.[§]

Year	Enrollment	Year	Enrollment
1980	33,318	1995	44,562
1985	32,487	2000	51,060
1990	38,442	2002	51,139

(a) Find a cubic polynomial model for this data, with $x = 0$ corresponding to 1980 and enrollment in thousands.
(b) Estimate Black/Hispanic enrollments in 1998 and 2001.
(c) If this model remains accurate, what can be said about Black/Hispanic enrollment from 2000 to 2005?

12. The U.S. public debt per person (in dollars) in selected years is shown in the table.

Year	Debt	Year	Debt
1991	$14,436	1999	$20,746
1993	$17,105	2001	$20,353
1995	$18,930	2002	$21,603
1997	$20,026		

(a) Find a cubic polynomial model for this data.
(b) Use the model to estimate the public debt per person in 2000. How does your estimate compare with the actual figure of $20,353?

13. The table shows total advertising expenditures (in billions of dollars) in selected years.[¶]

Year	Expenditures	Year	Expenditures
1994	152	2000	247
1996	175	2002	239
1998	206	2004	264

(a) Find both a cubic and quartic model for the data. Which seems to model the given data more accurately?
(b) Does either model seem likely to be accurate into the future? Explain your answer.

---

[*]U.S. Census Bureau.
[†]Mercer National Survey of Employer-Sponsored Health Plans.
[‡]Bureau of Labor Statistics, U.S. Department of Labor.

[§]Ohio Board of Regents.
[¶]Data and projections from Universal McCann.

14. The table shows the percentage of male high school students who currently smoke cigarettes in various years.[*]

Year	Percent	Year	Percent
1991	27.6	1997	37.7
1993	29.8	1999	34.7
1995	35.4	2001	29.2

*Youth at Risk Behavior Survey.

(a) Sketch a scatter plot of the data, with $x = 0$ corresponding to 1990.
(b) Find quadratic, cubic, and quartic polynomial models for the data.
(c) Which model seems to fit this data best? Which one seems most reasonable for future years.

4.5   Rational Functions

A **rational function** is a function whose rule is the quotient of two polynomials, such as

$$f(x) = \frac{1}{x}, \qquad t(x) = \frac{4x - 3}{2x + 1}, \qquad k(x) = \frac{2x^3 + 5x + 2}{x^2 - 7x + 6}.$$

A polynomial function is defined for every real number, but the rational function $f(x) = g(x)/h(x)$ is defined only when its denominator is nonzero. Hence,

Domain

> The domain of the rational function $f(x) = \dfrac{g(x)}{h(x)}$ is the set of all real numbers that are *not* roots of the denominator $h(x)$.

For instance, the domain of

$$f(x) = \frac{x^2 + 3x + 1}{x^2 - x - 6}$$

is the set of all real numbers except -2 and 3 [the roots of $x^2 - x - 6 = (x + 2)(x - 3)$]. The key to understanding the behavior of rational functions is this fact from arithmetic.

The Big-Little Principle

> If c is a number far from 0, then $1/c$ is a number close to 0. Conversely, if c is close to 0, then $1/c$ is far from 0. In less precise but more suggestive terms:
>
> $$\frac{1}{\text{big}} = \text{little} \qquad \text{and} \qquad \frac{1}{\text{little}} = \text{big.}$$

For example, 5000 is big (far from 0), and $1/5000$ is little (close to 0). Similarly, $-1/1000$ is very close to 0, but

$$\frac{1}{-1/1000} = -1000$$

is far from 0. To see the role played by the Big-Little Principle, we examine two rational functions that are part of the catalog of basic functions.

EXAMPLE 1   **Reciprocal Functions**

Graph

$$f(x) = \frac{1}{x} \quad \text{and} \quad g(x) = \frac{1}{x^2}.$$

SOLUTION   Note that f and g are not defined when $x = 0$ (root of the denominator). The graphs are easily obtained, either by hand or by calculator (see Figures 4–42 and 4–43), and you should know their shapes by heart. ■

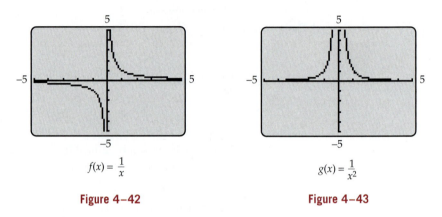

$$f(x) = \frac{1}{x}$$

Figure 4–42

$$g(x) = \frac{1}{x^2}$$

Figure 4–43

The Big-Little Principle explains why these graphs have the shapes they do. When x is very close to 0, both $1/x$ and $1/x^2$ are very far from 0. That's why the graphs "explode" near the undefined place at $x = 0$, shooting sharply upward or downward on either side of the y-axis. We say that the y-axis is the **vertical asymptote** of the graph: The graph gets closer and closer to this line, but never touches it. A vertical asymptote can occur only where the function is not defined.

Similarly, when x is very far from 0, then $1/x$ and $1/x^2$ are very small numbers. So the farther you go from the origin, the closer the graphs get to the line $y = 0$ (the x-axis), which is called the **horizontal asymptote** of the graph.

EXAMPLE 2

Without using technology, describe the graphs of

(a) $h(x) = \dfrac{1}{x + 3}$ \quad and \quad (b) $k(x) = \dfrac{1}{x^2 - 4x + 4}$.

SOLUTION

(a) Recall the graph of $f(x) = \dfrac{1}{x}$ in Figure 4–42, and note that

$$f(x + 3) = \frac{1}{x + 3} = h(x).$$

From Section 3.4, we know that the graph of h is the graph of f shifted horizontally 3 units to the left, as shown in Figure 4–44(a). The vertical asymptote of this graph is at $x = -3$ [the root of the denominator of $h(x)$].

(b) To understand the graph of k, recall the graph of $g(x) = \dfrac{1}{x^2}$ in Figure 4–43 and factor the denominator of $k(x)$:

$$k(x) = \frac{1}{x^2 - 4x + 4} = \frac{1}{(x-2)^2} = g(x - 2).$$

Thus, the graph of k is the graph of g shifted horizontally 2 units to the right, as shown in Figure 4–44(b). The vertical asymptote is at $x = 2$ [the root of the denominator of $k(x)$]. ∎

NOTE

The asymptotes are shown as dashed lines in Figure 4–44 and in other figures in this section. The asymptotes are included for easier visualization, but they are *not* part of the graph.

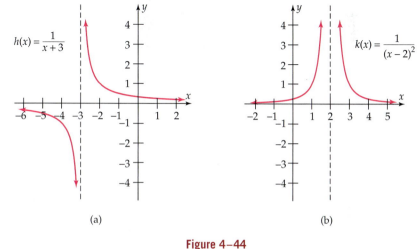

(a)   (b)

Figure 4–44

▚ LINEAR RATIONAL FUNCTIONS

Next we consider the graphs of rational functions, where both the numerator and denominator are either constants or first-degree polynomials.

EXAMPLE 3

Find the asymptotes of the graph of

$$f(x) = \frac{x + 1}{2x - 4}$$

and graph the function.

SOLUTION   The function is not defined when $x = 2$ because the denominator is 0 there. When x is a number very close to 2, then

The numerator $x + 1$ is very close to 3;

The denominator $2x - 4$ is very close to 0.

Therefore,

$$f(x) = \frac{x+1}{2x-4} \approx \frac{3}{\text{little}} = \text{BIG (far from 0).}$$

Consequently, the graph explodes near $x = 2$, as shown in the table and partial graph in Figure 4–45. The vertical line $x = 2$ is the vertical asymptote of the graph.

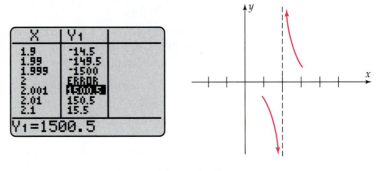

Figure 4–45

To determine the horizontal asymptote of the graph, we rewrite the rule of f like this:

$$f(x) = \frac{x+1}{2x-4} = \frac{\dfrac{x+1}{x}}{\dfrac{2x-4}{x}} = \frac{1 + \dfrac{1}{x}}{2 - \dfrac{4}{x}}.$$

As x gets larger in absolute value (far from 0), both $1/x$ and $4/x$ get very close to 0 by the Big-Little Principle. Consequently,

$$f(x) = \frac{1 + (1/x)}{2 - (4/x)}$$

gets very close to

$$\frac{1+0}{2-0} = \frac{1}{2}.$$

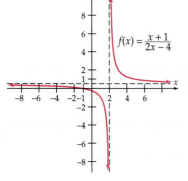

Figure 4–46

So when $|x|$ is large, the graph gets closer and closer to the horizontal line $y = 1/2$, but never touches it, as shown in Figure 4–46. Thus, the line $y = 1/2$ is the horizontal asymptote of the graph.

The preceding information, together with a few hand-plotted points, produces the graph in Figure 4–46. ∎

The analysis in Example 3 works for all linear rational functions.

Linear Rational Functions

The graph of $f(x) = \dfrac{ax+b}{cx+d}$ (with $c \neq 0$ and $ad \neq bc$) has two asymptotes:

The vertical asymptote occurs at the root of the denominator.

The horizontal asymptote is the line $y = a/c$.

Example 3 is the case where $a = 1$ and $c = 2$. Similarly, in Example 1, we saw that the horizontal asymptote of

$$f(x) = \frac{1}{x} = \frac{0x + 1}{1x + 0}$$

was the line

$$y = \frac{0}{1} = 0.$$

Figure 4–47 shows some additional examples.

$$f(x) = \frac{-5x + 12}{2x - 4} \qquad\qquad k(x) = \frac{3x + 6}{x} = \frac{3x + 6}{1x + 0}$$

Vertical asymptote $x = 2$ Vertical asymptote $x = 0$

Horizontal asymptote $y = -\frac{5}{2}$ Horizontal asymptote $y = \frac{3}{1} = 3$

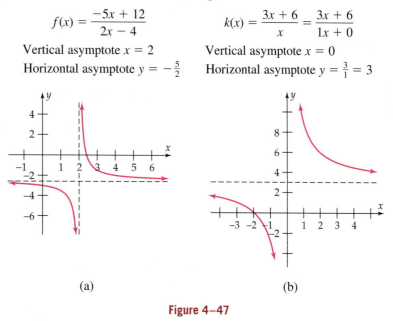

(a)   (b)

Figure 4–47

■■ RATIONAL FUNCTIONS AND TECHNOLOGY

Getting an accurate graph of a rational function on a calculator often depends on choosing an appropriate viewing window. For example, a TI-83+ produced the following graphs of

$$f(x) = \frac{x + 1}{2x - 4}$$

in Figure 4–48, two of which do not look like Figure 4–46 as they should.

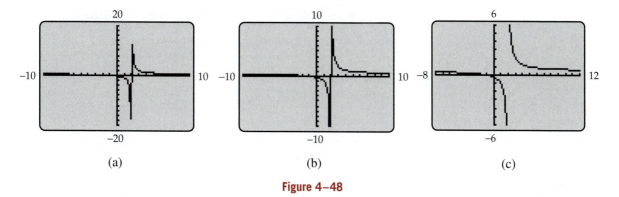

(a)   (b)   (c)

Figure 4–48

The vertical segments in graphs (a) and (b) are *not* representations of the vertical asymptote. They are a result of the calculator evaluating $f(x)$ just to the left of $x = 2$ and just to the right of $x = 2$ but not at $x = 2$ and then erroneously connecting these points with a near vertical segment that looks like an asymptote.

In the accurate graph (c), the calculator attempted to plot a point with $x = 2$ and, when it found that $f(2)$ was not defined, skipped a pixel and did not join the points on either side of the skipped one.

TECHNOLOGY TIP

To avoid erroneous vertical lines, use a window with the vertical asymptote in the center, as in Figure 4–48(c) (where the asymptote at $x = 2$ is half-way between -8 and 12). Also see the Technology Tip on page 337.

GRAPHING EXPLORATION

Find a viewing window that displays the graph of $f(x)$ in Figure 4–47(a), without any erroneous vertical line segments being shown. The Technology Tip in the margin may be helpful.

PROPERTIES OF RATIONAL GRAPHS

Here is a summary of the important characteristics of graphs of more complicated rational functions.

CONTINUITY

There will be breaks in the graph of a rational function wherever the function is not defined. Except for breaks at these undefined points, the graph is a continuous unbroken curve. In addition, the graph has no sharp corners.

LOCAL MAXIMA AND MINIMA

The graph may have some local extrema (peaks and valleys), and calculus is needed to determine their exact location. There are no simple rules for the possible number of peaks and valleys as there were with polynomial functions.

INTERCEPTS

As with any function, the y-intercept of the graph of a rational function f occurs at $f(0)$, provided that f is defined at $x = 0$. The x-intercepts of the graph of any function f occur at each number c for which $f(c) = 0$. Now a fraction is 0 only when its numerator is 0 and its denominator is nonzero (since division by 0 is not defined). Thus, we have the following.

Intercepts

> The x-intercepts of the graph of the rational function $f(x) = \dfrac{g(x)}{h(x)}$ occur at the numbers that are roots of the numerator $g(x)$ but *not* of the denominator $h(x)$.
>
> If f has a y-intercept, it occurs at $f(0)$.

For example, the graph of

$$f(x) = \frac{x^2 - x - 2}{x - 5}$$

has x-intercepts at $x = -1$ and $x = 2$ [which are the roots of $x^2 - x - 2 = (x + 1)(x - 2)$, but not of $x - 5$] and y-intercept at $y = 2/5$ (the value of f at $x = 0$).

VERTICAL ASYMPTOTES

In Example 3, we saw that the graph of

$$f(x) = \frac{x + 1}{2x - 4}$$

had a vertical asymptote at $x = 2$. Note that $x = 2$ is a root of the denominator $2x - 4$, but not of the numerator $x + 1$. The same thing occurs in the general case.

Vertical Asymptotes

> The function $f(x) = \dfrac{g(x)}{h(x)}$ has a vertical asymptote at every number that is a root of the denominator $h(x)$, but *not* of the numerator $g(x)$.

BEHAVIOR WHEN $|x|$ IS LARGE

The shape of a rational graph at the far left and far right (that is, when $|x|$ is large) can usually be found by algebraic analysis, as in the next example.

EXAMPLE 4

Determine the shape of the graph when $|x|$ is large for the following functions.

(a) $f(x) = \dfrac{7x^4 - 6x^3 + 4}{2x^4 + x^2}$

(b) $g(x) = \dfrac{x^2 - 2}{x^3 - 3x^2 + x - 3}$

SOLUTION

(a) When $|x|$ is very large, a polynomial function behaves in essentially the same way as its highest degree term, as we saw on page 309. Consequently, we have this approximation:

$$f(x) = \frac{7x^4 - 6x^3 + 4}{2x^4 + x^2} \approx \frac{7x^4}{2x^4} = \frac{7}{2} = 3.5.$$

Thus, when $|x|$ is large, the graph of $f(x)$ is very close to the horizontal line $y = 3.5$, which is a horizontal asymptote of the graph.

 GRAPHING EXPLORATION

Confirm the last statement by graphing $f(x)$ and $y = 3.5$ in the window with $-10 \le x \le 10$ and $-2 \le y \le 12$.

(b) When $|x|$ is large, the graph of g closely resembles the graph of

$$y = \frac{x^2}{x^3} = \frac{1}{x}.$$

By the Big-Little Principle, $1/x$ is very close to 0 when $|x|$ is large. So the line $y = 0$ (that is, the x-axis) is the horizontal asymptote. ∎

Arguments similar to those in the preceding example, using the highest-degree terms in the numerator and denominator, carry over to the general case and lead to this conclusion.

Horizontal
Asymptotes

> Let $f(x) = \dfrac{ax^n + \cdots}{cx^k + \cdots}$ be a rational function whose numerator has degree n and whose denominator has degree k.
>
> If $n = k$, then the line $y = a/c$ is a horizontal asymptote.
>
> If $n < k$, then the x-axis (the line $y = 0$) is a horizontal asymptote.

The asymptotes of rational functions in which the denominator has smaller degree than the numerator are discussed in Special Topics 4.5.A.

■■ GRAPHS OF RATIONAL FUNCTIONS

The procedure for finding accurate graphs of rational functions is summarized here.

Graphing
$$f(x) = \frac{g(x)}{h(x)} \text{ When}$$
Degree $g(x) \le$
Degree $h(x)$

> 1. Analyze the function algebraically to determine its vertical asymptotes and intercepts.
>
> 2. Determine the horizontal asymptote of the graph when $|x|$ is large by using the facts in the preceding box.
>
> 3. Use the preceding information to select an appropriate viewing window (or windows), to interpret the calculator's version of the graph (if necessary), and to sketch an accurate graph.

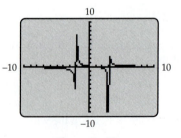

Figure 4–49

EXAMPLE 5

If you ignore the preceding advice and simply graph $f(x) = \dfrac{x - 1}{x^2 - x - 6}$ in the standard viewing window, you get garbage (Figure 4–49). So let's try analyzing the function. We begin by factoring.

$$f(x) = \frac{x - 1}{x^2 - x - 6} = \frac{x - 1}{(x + 2)(x - 3)}.$$

The factored form allows us to read off the necessary information:

Vertical Asymptotes: $x = -2$ and $x = 3$ (roots of the denominator but not of the numerator).

Horizontal Asymptote: x-axis (because denominator has larger degree than the numerator).

Intercepts: y-intercept at $f(0) = \dfrac{0 - 1}{0^2 - 0 - 6} = \dfrac{1}{6}$; x-intercept at $x = 1$ (root of the numerator but not of the denominator).

Interpreting Figure 4–49 in light of this information suggests that a complete graph of f looks something like Figure 4–50.

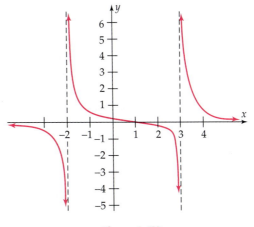

Figure 4–50

GRAPHING EXPLORATION

Find a viewing window in which the graph of f looks similar to Figure 4–50. The Technology Tip in the margin may be helpful. ■

NOTE: The graph of a rational function never touches a horizontal asymptote when *x* is large in absolute value. For smaller values of *x*, however, the graph may cross the asymptote, as in Example 5.

EXAMPLE 6

To graph

$$f(x) = \frac{2x^2}{x^2 + x - 2},$$

we factor and then read off the necessary information:

$$f(x) = \frac{2x^2}{x^2 + x - 2} = \frac{2x^2}{(x + 2)(x - 1)}.$$

Vertical Asymptotes: $x = -2$ and $x = 1$ (roots of denominator).

Horizontal Asymptote: $y = 2/1 = 2$ (because numerator and denominator have the same degree; see the first box on the preceding page).

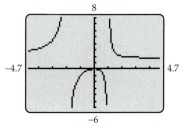

Figure 4–51

Intercepts: x-intercept at $x = 0$ (root of numerator); y-intercept at $f(0) = 0$.

Using this information and selecting a decimal viewing window that will accurately portray the graph near the vertical asymptotes, we obtain what seems to be a reasonably complete graph in Figure 4–51. The graph appears to be falling to the right of $x = 1$, but this is deceptive.

 EXPLORATION

Graph f in this same viewing window and use the trace feature, beginning at approximately $x = 1.1$ and moving to the right. For what values of x is the graph above the horizontal asymptote $y = 2$? For what values of x is the graph below the horizontal asymptote?

This use of the trace feature indicates that there is some *hidden behavior* of the graph that is not visible in Figure 4–51.

EXPLORATION

To see this hidden behavior, graph both f and the line $y = 2$ in the viewing window with $1 \le x \le 50$ and $1.7 \le y \le 2.1$.

This Exploration shows that the graph has a local minimum near $x = 4$ and then stays below the asymptote, moving closer and closer to it as x takes larger values. ■

▦ APPLICATIONS

If $C(x)$ is the cost function for producing x items, then the **average cost** per item $\overline{C}(x)$ is the cost of producing x items divided by the number of items; that is, $\overline{C}(x) = C(x)/x$.

EXAMPLE 7

The cost (in thousands of dollars) of producing x thousand mathematics texts is given by

$$C(x) = .000098x^4 - .022x^3 + 1.65x^2 + 10.2x + 600.$$

(a) Find the average cost per book when the number of copies produced is 2000, 25,000, and 60,000.

(b) Graph the average cost function.

(c) What production level leads to the lowest average cost per book?

SOLUTION

(a) The average cost function is

$$\overline{C}(x) = \frac{C(x)}{x} = \frac{.000098x^4 - .022x^3 + 1.65x^2 + 10.2x + 600}{x}.$$

Figure 4–52 shows the values of $\overline{C}(x)$ when $x = 2$, $x = 25$ and $x = 60$ (remember x is in thousands). Thus, the average cost is about \$313.41 per book when 2000 are produced, about \$63.23 when 25,000 are produced, and about \$61.17 when 60,000 are produced.

(b) The graph of $\overline{C}(x)$ is shown in Figure 4–53. It confirms what we found in part (a): The average cost per book decreases as more and more books are produced. Eventually, however, the average cost starts to increase when more than approximately 90,000 copies are made. (What might cause this?)

(c) The minimum finder shows that producing about 90,251 copies leads to the lowest average cost of about \$58.61 per book (Figure 4–54). ∎

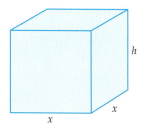

Figure 4–52

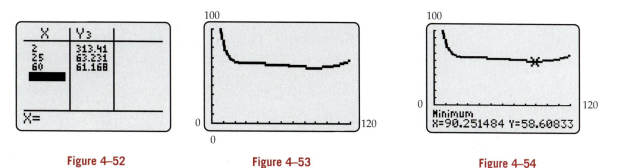

Figure 4–53

Figure 4–54

EXAMPLE 8

Figure 4–55

A cardboard box with a square base and a volume of 1000 cubic inches is to be constructed (Figure 4–55). The box must be at least 2 inches in height.

(a) What are the possible lengths for a side of the base if no more than 1100 square inches of cardboard can be used to construct the box?

(b) What is the least possible amount of cardboard that can be used?

(c) What are the dimensions of the box that uses the least possible amount of cardboard?

SOLUTION   The amount of cardboard needed is given by the surface area S of the box. From Figure 4–55, we have

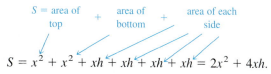

$$S = x^2 + x^2 + xh + xh + xh + xh = 2x^2 + 4xh.$$

Since the volume of the box is given by

$$\text{Length} \times \text{Width} \times \text{Height} = \text{Volume},$$

we have

$$x \cdot x \cdot h = 1000 \qquad \text{or, equivalently,} \qquad h = \frac{1000}{x^2}.$$

Substituting this into the surface area formula allows us to express the surface area as a function of x:

$$S(x) = 2x^2 + 4xh = 2x^2 + 4x\left(\frac{1000}{x^2}\right) = 2x^2 + \frac{4000}{x} = \frac{2x^3 + 4000}{x}.$$

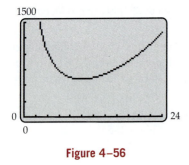

Figure 4–56

Although the rational function $S(x)$ is defined for all nonzero real numbers, x is a length here and must be positive. Furthermore, $x^2 \leq 500$ because if $x^2 > 500$, then $h = \frac{1000}{x^2}$ would be less than 2, contrary to specifications. Hence, the only values of x that make sense in this context are those with $0 < x \leq \sqrt{500}$. Since $\sqrt{500} \approx 22.4$, we choose the viewing window in Figure 4–56. For each point (x, y) on the graph, x is a possible side length for the base of the box, and y is the corresponding surface area.

(a) The points on the graph corresponding to the requirement that no more than 1100 square inches of cardboard be used are those whose y-coordinates are less than or equal to 1100. The x-coordinates of these points are the possible side lengths. The x-coordinates of the points where the graph of S meets the horizontal line $y = 1100$ are the smallest and largest possible values for x, as indicated schematically in Figure 4–57.

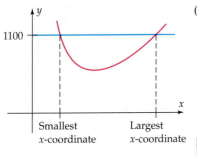

Figure 4–57

GRAPHING EXPLORATION

Graph $S(x)$ and $y = 1100$ on the same screen. Use an intersection finder to show that the possible side lengths that use no more than 1100 square inches of cardboard are those with $3.73 \leq x \leq 21.36$.

(b) The least possible amount of cardboard corresponds to the point on the graph of $S(x)$ with the smallest y-coordinate.

GRAPHING EXPLORATION

Show that the graph of S has a local minimum at the point $(10.00, 600.00)$. Consequently, the least possible amount of cardboard is 600 square inches and this occurs when $x = 10$.

(c) When $x = 10$, $h = 1000/10^2 = 10$. So the dimensions of the box using the least amount of cardboard are $10 \times 10 \times 10$. ■

✓ EXERCISES 4.5 📼

In Exercises 1–6, find the domain of the function. You may need to use some of the techniques of Section 4.2.

1. $f(x) = \dfrac{-3x}{2x + 5}$

2. $g(x) = \dfrac{x^3 + x + 1}{2x^2 - 5x - 3}$

3. $h(x) = \dfrac{6x - 5}{x^2 - 6x + 4}$

4. $g(x) = \dfrac{x^3 - x^2 - x - 1}{x^5 - 36x}$

5. $f(x) = \dfrac{x^5 - 2x^3 + 7}{x^3 - x^2 - 2x + 2}$

6. $h(x) = \dfrac{x^5 - 5}{x^4 + 12x^3 + 60x^2 + 50x - 125}$

In Exercises 7–10, use the graphs in Example 1 and the information in Section 3.4 to match the function with its graph, which is one of those shown here.

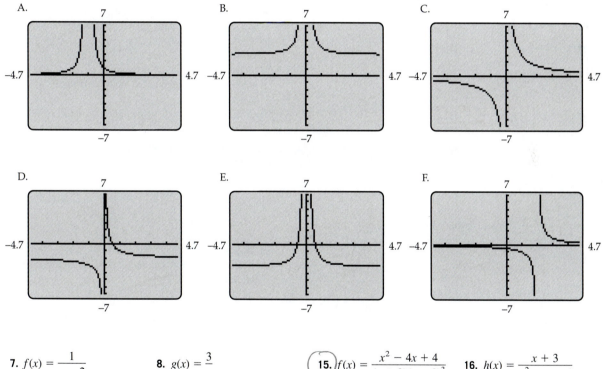

A.   7
−4.7   4.7
    −7

B.   7
−4.7   4.7
    −7

C.   7
−4.7   4.7
    −7

D.   7
−4.7   4.7
    −7

E.   7
−4.7   4.7
    −7

F.   7
−4.7   4.7
    −7

7. $f(x) = \dfrac{1}{x - 2}$

8. $g(x) = \dfrac{3}{x}$

9. $h(x) = \dfrac{1}{(x + 1)^2}$

10. $f(x) = \dfrac{1}{x^2} - 3$

In Exercises 11–16, use algebra to determine the location of the vertical asymptotes in the graph of the function.

11. $f(x) = \dfrac{x^2 + 4}{x^2 - 5x - 6}$

12. $g(x) = \dfrac{x - 5}{x^3 + 7x^2 + 2x}$

13. $f(x) = \dfrac{2}{x^3 + 2x^2 + x}$

14. $g(x) = \dfrac{1}{x^3 + 5x}$

15. $f(x) = \dfrac{x^2 - 4x + 4}{(x + 2)(x - 2)^3}$

16. $h(x) = \dfrac{x + 3}{x^2 - x - 6}$

In Exercises 17–22, find the horizontal asymptote of the graph of the function when $|x|$ is large and find a viewing window in which the ends of the graph are within .1 of this asymptote.

17. $f(x) = \dfrac{3x - 2}{x + 3}$

18. $g(x) = \dfrac{3x^2 + x}{2x^2 - 2x + 4}$

19. $h(x) = \dfrac{5 - x}{x - 2}$

20. $f(x) = \dfrac{4x^2 - 5}{2x^3 - 3x^2 + x}$

21. $g(x) = \dfrac{5x^3 - 8x^2 + 4}{2x^3 + 2x}$

22. $h(x) = \dfrac{8x^5 - 6x^3 + 2x - 1}{.5x^5 + x^4 + 3x^2 + x}$

In Exercises 23–40, analyze the function algebraically: List its vertical asymptotes and horizontal asymptote. Then sketch a complete graph of the function.

23. $f(x) = \dfrac{1}{x + 5}$

24. $q(x) = \dfrac{-7}{x - 6}$

25. $k(x) = \dfrac{-3}{2x + 5}$

26. $g(x) = \dfrac{-4}{2 - x}$

27. $f(x) = \dfrac{3x}{x - 1}$

28. $p(x) = \dfrac{x - 2}{x}$

29. $f(x) = \dfrac{2 - x}{x - 3}$

30. $g(x) = \dfrac{3x - 2}{x + 3}$

31. $f(x) = \dfrac{1}{x(x + 1)^2}$

32. $g(x) = \dfrac{x}{2x^2 - 5x - 3}$

33. $f(x) = \dfrac{x - 3}{x^2 + x - 2}$

34. $g(x) = \dfrac{x + 2}{x^2 - 1}$

35. $h(x) = \dfrac{(x^2 + 6x + 5)(x + 5)}{(x + 5)^3(x - 1)}$

36. $f(x) = \dfrac{x^2 - 1}{x^3 - 2x^2 + x}$

37. $f(x) = \dfrac{-4x^2 + 1}{x^2}$

38. $k(x) = \dfrac{x^2 + 1}{x^2 - 1}$

39. $q(x) = \dfrac{x^2 + 2x}{x^2 - 4x - 5}$

40. $F(x) = \dfrac{x^2 + x}{x^2 - 2x + 4}$

In Exercises 41–48, find a viewing window, or windows, that shows a complete graph of the function (if possible, with no erroneous vertical line segments). Be alert for hidden behavior, such as that in Example 6.

41. $f(x) = \dfrac{x^3 + 4x^2 - 5x}{(x^2 - 4)(x^2 - 9)}$

42. $g(x) = \dfrac{x^2 + x - 6}{x^3 - 19x + 30}$

43. $h(x) = \dfrac{2x^2 - x - 6}{x^3 + x^2 - 6x}$

44. $f(x) = \dfrac{x^3 - x + 1}{x^4 - 2x^3 - 2x^2 + x - 1}$

45. $h(x) = \dfrac{3x^2 + x - 4}{2x^2 - 5x}$

46. $f(x) = \dfrac{2x^2 - 1}{3x^3 + 2x + 1}$

47. $g(x) = \dfrac{x - 4}{2x^3 - 5x^2 - 4x + 12}$

48. $h(x) = \dfrac{x^2 - 9}{x^3 + 2x^2 - 23x - 60}$

C

In Exercises 49–52, find the difference quotient of the function. [See page 205.]

49. $f(x) = \dfrac{1}{x}$

50. $h(x) = \dfrac{1}{x + 5}$

51. $g(x) = \dfrac{1}{x^2}$

52. $g(x) = \dfrac{3}{2x^2}$

53. The graph of

$$f(x) = \dfrac{2x^3 - 2x^2 - x + 1}{3x^3 - 3x^2 + 2x - 1}$$

has a vertical asymptote. Find a viewing window that demonstrates this fact.

54. The percentage c of a drug in a person's bloodstream t hours after its injection is approximated by

$$c(t) = \dfrac{5t}{4t^2 + 5}.$$

(a) Approximately what percentage of the drug is in the person's bloodstream after four and a half hours?

(b) Graph the function c in an appropriate window for this situation.

(c) What is the horizontal asymptote of the graph? What does it tell you about the amount of the drug in the bloodstream?

(d) At what time is the percentage the highest? What is the percentage at that time?

55. The length of a small fish (in millimeters) is approximataed by

$$f(x) = \dfrac{30x}{x + 1.8},$$

where x is measured in weeks.

(a) When does the fish reach 25 mm?

(b) How long is the fish after one year?

(c) If this model is accurate, is there any limit on how large the fish can grow?

56. The **cost-benefit function**

$$g(x) = \dfrac{18x}{100 - x}$$

gives the cost (in thousands of dollars) to remove x percent of a pollutant from the emissions at a factory.

(a) What is the cost to remove 70% of the pollutant? To remove 80? To remove 90%? [Use $x = 70$, not .70, and similarly in the other cases.]

(b) The company wants to spend a maximum of $300,000 to remove the pollutant? How much can be removed?

(c) According to this model, is it possible to remove 100% of the pollutant? Why?

57. The temperature of a person (in degrees Fahrenheit) x hours after an infection begins is given by

$$T(x) = \frac{5.88x^2 + 2x + 98}{.06x^2 + 1}.$$

(a) What was the person's temperature when the infection began?
(b) When was her temperature the highest?
(c) When did her temperature return to normal (98.6°)?

58. According to an economic theory developed by Arthur Laffer, people will work harder and thus earn more if the tax rate is lower. In this case, the total revenue to the government may be more than would be the case at a higher tax rate (which, in theory, would discourage people from working harder). Suppose that the **Laffer curve** that models this situation is

$$L(x) = \frac{300x - 3x^2}{10x + 200},$$

where $L(x)$ is government revenue (in billions of dollars) from a tax rate of x percent.

(a) Find the revenue from the following tax rates: 15%, 25%, and 50%.
(b) What tax rate produces maximum revenue? What is this maximum revenue?

59. A specialty cycle maker determines that his monthly cost (in dollars) for producing x professional trail bikes is given by

$$C(x) = .04x^3 - 18x^2 + 2740x + 125,000.$$

(a) Find the total cost and the average cost per bike when the production level is: 100 bikes per month, 200 bikes per month, and 300 bikes per month.
(b) At what production level is the average cost per bike the lowest? What is that cost?
(c) If the average cost per bike is to be below $1600, how many bikes can be produced each month?

60. The Embree Company has fixed costs of $40,000 and variable costs of $2.60 per unit.

(a) Let x be the number of units produced. Find the rule of the average cost function.
(b) Graph the average cost function in a window with $0 \leq x \leq 100,000$ and $0 \leq y \leq 20$.
(c) Find the horizontal asymptote of the average cost function. Explain what the asymptote means in this situation. (How low can the average cost possibly be?)

61. It costs 2.5 cents per square inch to make the top and bottom of the box in Example 8. The sides cost 1.5 cents per square inch. What are the dimensions of the cheapest possible box?

62. A box with a square base and a volume of 1000 cubic inches is to be constructed. The material for the top and bottom of the box costs $3 per 100 square inches, and the material for the sides costs $1.25 per 100 square inches.

(a) If x is the length of a side of the base, express the cost of constructing the box as a function of x.
(b) If the side of the base must be at least 6 inches long, for what value of x will the cost of the box be $7.50?

63. A truck traveling at a constant speed on a reasonably straight, level road burns fuel at the rate of $g(x)$ gallons per mile, where x is the speed of the truck (in miles per hour) and $g(x)$ is given by

$$g(x) = \frac{800 + x^2}{200x}.$$

(a) If fuel costs $1.40 per gallon, find the rule of the cost function $c(x)$ that expresses the cost of fuel for a 500-mile trip as a function of the speed. [*Hint:* $500g(x)$ gallons of fuel are needed to go 500 miles. (Why?)]
(b) What driving speed will make the cost of fuel for the trip $250?
(c) What driving speed will minimize the cost of fuel for the trip?

64. Pure alcohol is being added to 50 gallons of a coolant mixture that is 40% alcohol.

(a) Find the rule of the concentration function $c(x)$ that expresses the percentage of alcohol in the resulting mixture as a function of the number x of gallons of pure alcohol that are added. [*Hint:* The final mixture contains $50 + x$ gallons (why?). So $c(x)$ is the amount of alcohol in the final mixture divided by the total amount $50 + x$. How much alcohol is in the original 50-gallon mixture? How much is in the final mixture?]
(b) How many gallons of pure alcohol should be added to produce a mixture that is at least 60% alcohol and no more than 80% alcohol?
(c) Determine algebraically the exact amount of pure alcohol that must be added to produce a mixture that is 70% alcohol.

65. A rectangular garden with an area of 250 square meters is to be located next to a building and fenced on three sides, with the building acting as a fence on the fourth side.

(a) If the side of the garden parallel to the building has length x meters, express the amount of fencing needed as a function of x.
(b) For what values of x will less than 60 meters of fencing be needed?
(c) What value of x will result in the least possible amount of fencing being used? What are the dimensions of the garden in this case?

66. The relationship between the fixed focal length F of a camera, the distance u from the object being photographed to the lens, and the distance v from the lens to the film is given by

$$\frac{1}{F} = \frac{1}{u} + \frac{1}{v}.$$

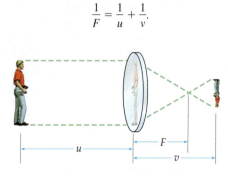

(a) If the focal length is 50 millimeters, express v as a function of u.

(b) What is the horizontal asymptote of the graph of the function in part (a)?

(c) Graph the function in part (a) when 50 millimeters $< u <$ 35,000 millimeters.

(d) When you focus the camera on an object, the distance between the lens and the film is changed. If the distance from the lens to the film changes by less than .1 millimeter, the object will remain in focus. Explain why you have more latitude in focusing on distant objects than on very close ones.

67. Radioactive waste is stored in a cylindrical tank, whose exterior has radius r and height h as shown in the figure. The sides, top, and bottom of the tank are 1 foot thick, and the tank has a volume of 150 cubic feet (including top, bottom, and walls).

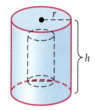

(a) Express the interior height h_1 (that is, the height of the storage area) as a function of h.

(b) Express the interior height as a function of r.

(c) Express the volume of the interior as a function of r.

(d) Explain why r must be greater than 1.

(e) What should the dimensions of the tank be in order for it to hold as much as possible?

68. The formula for the gravitational acceleration (in units of meters per second squared) of an object relative to the earth is

$$g(r) = \frac{3.987 \times 10^{14}}{(6.378 \times 10^6 + r)^2},$$

where r is the distance in meters above the earth's surface.

(a) What is the gravitational acceleration at the earth's surface?

(b) Graph the function $g(r)$ for $r \geq 0$.

(c) Can you ever escape the pull of gravity? [Does the graph have any r-intercepts?]

4.5.A SPECIAL TOPICS Other Rational Functions

We now examine the graphs of rational functions in which the degree of the denominator is smaller than the degree of the numerator. Such a graph has no horizontal asymptote. However, it does have some polynomial curve as an asymptote, which means that the graph will get very close to this curve when $|x|$ is very large.

EXAMPLE 1

To graph

$$f(x) = \frac{x^3 + 3x^2 + x + 1}{x^2 + 2x - 1},$$

we begin by finding the vertical asymptotes and the x- and y-intercepts. The quadratic formula can be used to find the roots of the denominator.

$$x = \frac{-2 \pm \sqrt{2^2 - 4 \cdot 1(-1)}}{2 \cdot 1} = \frac{-2 \pm \sqrt{8}}{2} = \frac{-2 \pm 2\sqrt{2}}{2} = -1 \pm \sqrt{2}.$$

It is easy to verify that neither of these numbers is a root of the numerator, so the graph has vertical asymptotes at $x = -1 - \sqrt{2}$ and $x = -1 + \sqrt{2}$. The y-intercept is $f(0) = -1$. The x-intercepts are the roots of the numerator.

GRAPHING EXPLORATION

Use a calculator to verify that $x^3 + 3x^2 + x + 1$ has exactly one real root, located between -3 and -2.

Therefore, the graph of $f(x)$ has one x-intercept. Using this information and the calculator graph in Figure 4–58 (which erroneously shows some vertical segments), we conclude that the graph looks approximately like Figure 4–59.

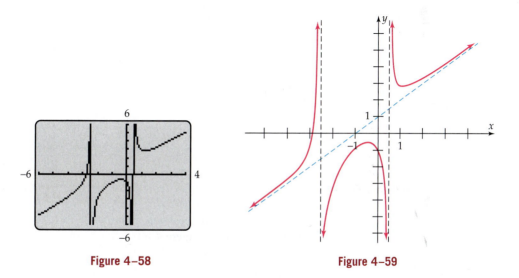

Figure 4–58   **Figure 4–59**

At the left and right ends, the graph moves away from the x-axis. To understand the behavior of the graph when $|x|$ is large, divide the numerator $f(x)$ by its denominator.

$$
\begin{array}{r}
x + 1 \\
x^2 + 2x - 1 \overline{\smash{\big)}\, x^3 + 3x^2 + x + 1} \\
\underline{x^3 + 2x^2 - x } \\
x^2 + 2x + 1 \\
\underline{x^2 + 2x - 1} \\
2.
\end{array}
$$

By the Division Algorithm,

$$x^3 + 3x^2 + x + 1 = (x^2 + 2x - 1)(x + 1) + 2.$$

Dividing both sides by $x^2 + 2x - 1$, we have

$$\frac{x^3 + 3x^2 + x + 1}{x^2 + 2x - 1} = \frac{(x^2 + 2x - 1)(x + 1) + 2}{x^2 + 2x - 1}$$

$$f(x) = (x + 1) + \frac{2}{x^2 + 2x - 1}.$$

Now when x is very large in absolute value, so is $x^2 + 2x - 1$. Hence, $2/(x^2 + 2x - 1)$ is very close to 0 by the Big-Little Principle, and $f(x)$ is very close to $(x + 1) + 0$. Therefore, as x gets larger in absolute value, the graph of $f(x)$ gets closer and closer to the line $y = x + 1$ (the dashed blue line in Figure 4–59) and this line is an asymptote of the graph.* Note that $x + 1$ is just the quotient obtained in the long division above.   ■

It is instructive to examine the graph in Example 1 further to see that the asymptote accurately indicates the behavior of the function when $|x|$ is large.

EXPLORATION

Using the viewing window with $-20 \le x \le 20$ and $-20 \le y \le 20$, graph both $f(x)$ and $y = x + 1$ on the same screen.

Except near the vertical asymptotes of $f(x)$, the two graphs are virtually identical.

EXAMPLE 2

To graph

$$g(x) = \frac{x^3 + 2x^2 - 7x + 5}{x - 1},$$

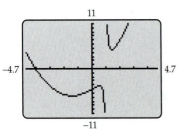

Figure 4–60

we first note that there is a vertical asymptote at $x = 1$ (root of the denominator, but not the numerator). The y-intercept is at $g(0) = -5$. By carefully choosing a viewing window that accurately portrays the behavior of $g(x)$ near its vertical asymptote, we obtain Figure 4–60.

EXPLORATION

Verify that the x-intercept near $x = -4$ is the only one by showing graphically that the numerator of $g(x)$ has exactly one real root.

To confirm that Figure 4–60 is a complete graph, we find its asymptote when $|x|$ is large. Divide the numerator by the denominator.

$$\begin{array}{r} x^2 + 3x - 4 \\ x - 1 \overline{\smash{\big)}\ x^3 + 2x^2 - 7x + 5} \\ \underline{x^3 - x^2} \\ 3x^2 - 7x + 5 \\ \underline{3x^2 - 3x} \\ -4x + 5 \\ \underline{-4x + 4} \\ 1 \end{array}$$

*An asymptote that is a nonvertical and nonhorizontal straight line is called an **oblique asymptote**.

Hence, by the Division Algorithm,

$$x^3 + 2x^2 - 7x + 5 = (x - 1)(x^2 + 3x - 4) + 1$$

$$\frac{x^3 + 2x^2 - 7x + 5}{x - 1} = \frac{(x - 1)(x^2 + 3x - 4) + 1}{x - 1}$$

$$g(x) = (x^2 + 3x - 4) + \frac{1}{x - 1}.$$

When $|x|$ is large, $1/(x - 1)$ is very close to 0 (why?), so $y = x^2 + 3x - 4$ is the asymptote. Once again, the asymptote is given by the quotient of the division.

GRAPHING EXPLORATION

Graph $g(x)$ and $y = x^2 + 3x - 4$ on the same screen to show that the graph of $g(x)$ does get very close to the asymptote when $|x|$ is large. Then find a large enough viewing window that the two graphs appear to be identical. ■

The procedures used in the preceding examples may be summarized as follows.

Graphing

$$f(x) = \frac{g(x)}{h(x)} \text{ When}$$

Degree $g(x) >$ Degree $h(x)$

1. Analyze the function algebraically to determine its vertical asymptotes and intercepts.

2. Divide the numerator $g(x)$ by the denominator $h(x)$. The quotient $q(x)$ is the nonvertical asymptote of the graph, which describes the behavior of the graph when $|x|$ is large.

3. Use the preceding information to select an appropriate viewing window (or windows), to interpret the calculator's version of the graph (if necessary), and to sketch an accurate graph.

✓ EXERCISES 4.5.A

In Exercises 1–4, find the nonvertical asymptote of the graph of the function when $|x|$ is large and find a viewing window in which the ends of the graph are within .1 of this asymptote.

1. $f(x) = \dfrac{x^3 - 1}{x^2 - 4}$

2. $g(x) = \dfrac{x^3 - 4x^2 + 6x + 5}{x - 2}$

3. $h(x) = \dfrac{x^3 + 3x^2 - 4x + 1}{x + 4}$

4. $f(x) = \dfrac{x^3 + 3x^2 - 4x + 1}{x^2 - x}$

In Exercises 5–12, analyze the function algebraically. List its vertical asymptotes and determine its nonvertical asymptote. Then sketch a complete graph of the function.

5. $f(x) = \dfrac{x^2 - x - 6}{x - 2}$

6. $k(x) = \dfrac{x^2 + x - 2}{x}$

7. $Q(x) = \dfrac{4x^2 + 4x - 3}{2x - 5}$

8. $K(x) = \dfrac{3x^2 - 12x + 15}{3x + 6}$

9. $f(x) = \dfrac{x^3 - 2}{x - 1}$

10. $p(x) = \dfrac{x^3 + 8}{x + 1}$

11. $q(x) = \dfrac{x^3 - 1}{x - 2}$

12. $f(x) = \dfrac{x^4 - 1}{x^2}$

In Exercises 13–18, find a viewing window (or windows) that shows a complete graph of the function (if possible, with no erroneous vertical line segments). Be alert for hidden behavior.

13. $f(x) = \dfrac{2x^2 + 5x + 2}{2x + 7}$

14. $g(x) = \dfrac{2x^3 + 1}{x^2 - 1}$

15. $h(x) = \dfrac{x^3 - 2x^2 + x - 2}{x^2 - 1}$

16. $f(x) = \dfrac{3x^3 - 11x - 1}{x^2 - 4}$

17. $g(x) = \dfrac{2x^4 + 7x^3 + 7x^2 + 2x}{x^3 - x + 50}$

18. $h(x) = \dfrac{2x^3 + 7x^2 - 4}{x^2 + 2x - 3}$

19. (a) Show that when $0 < x < 4$, the rational function

$$r(x) = \dfrac{4096x^3 + 34{,}560x^2 + 19{,}440x + 729}{18{,}432x^2 + 34{,}560x + 5832}$$

is a good approximation of the function $s(x) = \sqrt{x}$ by graphing both functions in the viewing window with $0 \le x \le 4$ and $0 \le y \le 2$.

(b) For what values of x is $r(x)$ within .01 of $s(x)$?

20. Find a rational function f that has these properties:

   (i) The curve $y = x^3 - 8$ is an asymptote of the graph of f.

   (ii) $f(2) = 1$.

   (iii) The line $x = 1$ is a vertical asymptote of the graph of f.

4.6   Complex Numbers

The equation $x^2 = 2$ has no solutions in the rational number system but has $\sqrt{2}$ and $-\sqrt{2}$ as solutions in the real number system. So the idea of enlarging a number system to solve an equation that can't be solved in the present system is a natural one.

Equations such as $x^2 = -1$ and $x^2 = -4$ have no solutions in the real number system because $\sqrt{-1}$ and $\sqrt{-4}$ are not real numbers. To solve such equations (or, equivalently, to find square roots of negative numbers), we must enlarge the number system again. We claim that there is a number system, called the **complex number system,** with the following properties.

Properties of the Complex
Number System

1. The complex number system contains all real numbers.

2. Addition, subtraction, multiplication, and division of complex numbers obey the same rules of arithmetic that hold in the real number system, with one exception: The exponent laws hold for *integer* exponents, but not necessarily for fractional ones.

3. The complex number system contains a number (usually denoted by i) such that $i^2 = -1$.

4. Every complex number can be written in the **standard form** $a + bi$, where a and b are real numbers.[*]

5. Two complex numbers $a + bi$ and $c + di$ are equal exactly when $a = c$ and $b = d$.

In view of our past experience with enlarging the number system, this claim *ought* to appear plausible. But the mathematicians who invented the complex numbers in the seventeenth century were very uneasy about a number i such that $i^2 = -1$ (that is, $i = \sqrt{-1}$). Consequently, they called numbers of the form bi (b any real number), such as $5i$ and $-\frac{1}{4}i$, **imaginary numbers.** The old familiar numbers (integers, rationals, irrationals) were called **real numbers.** Sums of real and imaginary numbers, numbers of the form $a + bi$, such as

$$5 + 2i, \qquad 7 - 4i, \qquad 18 + \frac{3}{2}i, \qquad \sqrt{3} - 12i$$

were called **complex numbers.**[†]

Every real number is a complex number; for instance, $7 = 7 + 0i$. Similarly, every imaginary number bi is a complex number, since $bi = 0 + bi$. Since the usual laws of arithmetic still hold, it's easy to add, subtract, and multiply complex numbers. As the following examples demonstrate, *all symbols can be treated as if they were real numbers, provided that i^2 is replaced by -1.* Unless directed otherwise, express your answers in the standard form $a + bi$.

EXAMPLE 1

(a) $(1 + i) + (3 - 7i) = 1 + i + 3 - 7i$
$$= (1 + 3) + (i - 7i) = 4 - 6i.$$

(b) $(4 + 3i) - (8 - 6i) = 4 + 3i - 8 - (-6i)$
$$= (4 - 8) + (3i + 6i) = -4 + 9i.$$

---

[*]Hereafter, whenever we write $a + bi$ or $c + di$, it is assumed that a, b, c, d are real numbers and $i^2 = -1$.

[†]This terminology is still used, even though there is nothing complicated, unreal, or imaginary about complex numbers—they are just as valid mathematically as are real numbers.

(c) $4i\left(2 + \dfrac{1}{2}i\right) = 4i \cdot 2 + 4i\left(\dfrac{1}{2}i\right) = 8i + 4 \cdot \dfrac{1}{2} \cdot i^2$

$$= 8i + 2i^2 = 8i + 2(-1) = -2 + 8i.$$

(d) $(2 + i)(3 - 4i) = 2 \cdot 3 + 2(-4i) + i \cdot 3 + i(-4i)$

$$= 6 - 8i + 3i - 4i^2 = 6 - 8i + 3i - 4(-1)$$

$$= (6 + 4) + (-8i + 3i) = 10 - 5i. \quad \blacksquare$$

The familiar multiplication patterns and exponent laws for integer exponents hold in the complex number system.

EXAMPLE 2

(a) $(3 + 2i)(3 - 2i) = 3^2 - (2i)^2$

$$= 9 - 4i^2 = 9 - 4(-1) = 9 + 4 = 13.$$

(b) $(4 + i)^2 = 4^2 + 2 \cdot 4 \cdot i + i^2 = 16 + 8i + (-1) = 15 + 8i.$

(c) To find i^{54}, we first note that $i^4 = i^2 i^2 = (-1)(-1) = 1$ and that

$$54 = 52 + 2 = 4 \cdot 13 + 2.$$

Consequently,

$$i^{54} = i^{52+2} = i^{52}i^2 = i^{4 \cdot 13}i^2 = (i^4)^{13}i^2 = 1^{13}(-1) = -1. \quad \blacksquare$$

TECHNOLOGY

TIP

To do complex arithmetic on TI-86 and HP-39+, enter $a + bi$ as (a, b). On other calculators, use the special i key whose location is

TI-83+/89: keyboard

Casio FX 2.0: keyboard

Casio: 9850: OPTN/CPLX

The **conjugate** of the complex number $a + bi$ is the number $a - bi$, and the conjugate of $a - bi$ is $a + bi$. For example, the conjugate of $3 + 4i$ is $3 - 4i$ and the conjugate of $-3i = 0 - 3i$ is $0 + 3i = 3i$. *Every real number is its own conjugate;* for instance, the conjugate of $17 = 17 + 0i$ is $17 - 0i = 17$.

For any complex number $a + bi$, we have

$$(a + bi)(a - bi) = a^2 - (bi)^2 = a^2 - b^2 i^2 = a^2 - b^2(-1) = a^2 + b^2.$$

Since a^2 and b^2 are nonnegative real numbers, so is $a^2 + b^2$. Therefore, *the product of a complex number and its conjugate is a nonnegative real number.* This fact enables us to express quotients of complex numbers in standard form.

EXAMPLE 3

To express $\dfrac{3 + 4i}{1 + 2i}$ in the form $a + bi$, *multiply both numerator and denominator by the conjugate of the denominator,* namely, $1 - 2i$:

$$\frac{3 + 4i}{1 + 2i} = \frac{3 + 4i}{1 + 2i} \cdot \frac{1 - 2i}{1 - 2i} = \frac{(3 + 4i)(1 - 2i)}{(1 + 2i)(1 - 2i)}$$

$$= \frac{3 + 4i - 6i - 8i^2}{1^2 - (2i)^2} = \frac{3 + 4i - 6i - 8(-1)}{1 - 4i^2} = \frac{11 - 2i}{1 - 4(-1)}$$

$$= \frac{11 - 2i}{5} = \frac{11}{5} - \frac{2}{5}i.$$

This is the form $a + bi$ with $a = 11/5$ and $b = -2/5$. $\quad \blacksquare$

EXAMPLE 4

To express $\dfrac{1}{1-i}$ in standard form, note that the conjugate of the denominator is $1 + i$, and therefore

$$\frac{1}{1-i} = \frac{1 \cdot (1+i)}{(1-i)(1+i)} = \frac{1+i}{1^2 - i^2} = \frac{1+i}{1-(-1)} = \frac{1+i}{2} = \frac{1}{2} + \frac{1}{2}i.$$

We can check this result by multiplying $\dfrac{1}{2} + \dfrac{1}{2}i$ by $1 - i$ to see whether the product is 1 $\left(\text{which it should be if } \dfrac{1}{2} + \dfrac{1}{2}i = \dfrac{1}{1-i}\right)$:

$$\left(\frac{1}{2} + \frac{1}{2}i\right)(1-i) = \frac{1}{2} \cdot 1 - \frac{1}{2}i + \frac{1}{2}i \cdot 1 - \frac{1}{2}i^2 = \frac{1}{2} - \frac{1}{2}(-1) = 1. \quad ■$$

Since $i^2 = -1$, we define $\sqrt{-1}$ to be the complex number i. Similarly, since $(5i)^2 = 5^2i^2 = 25(-1) = -25$, we define $\sqrt{-25}$ to be $5i$. In general,

Square Roots of Negative Numbers

> Let b be a positive real number.
>
> $$\sqrt{-b} \text{ is defined to be } \sqrt{b}\,i$$
>
> because $(\sqrt{b}\,i)^2 = (\sqrt{b})^2 i^2 = b(-1) = -b.$

CAUTION

$\sqrt{b}\,i$ is *not* the same as $\sqrt{bi}$. To avoid confusion, it may help to write $\sqrt{b}\,i$ as $i\sqrt{b}$.

EXAMPLE 5

(a) $\sqrt{-3} = \sqrt{3}\,i = i\sqrt{3}.$

(b) $\dfrac{1 - \sqrt{-7}}{3} = \dfrac{1 - \sqrt{7}\,i}{3} = \dfrac{1}{3} - \dfrac{\sqrt{7}}{3}i.$ ■

CAUTION

The property $\sqrt{cd} = \sqrt{c}\,\sqrt{d}$ (or equivalently in exponential notation, $(cd)^{1/2} = c^{1/2}d^{1/2}$), which is valid for positive real numbers, does *not* hold when both c and d are negative.

$$\sqrt{-20}\,\sqrt{-5} = \sqrt{20}\,i \cdot \sqrt{5}\,i = \sqrt{20}\,\sqrt{5} \cdot i^2 = \sqrt{20 \cdot 5}(-1)$$

$$= \sqrt{100}(-1) = -10.$$

But $\sqrt{(-20)(-5)} = \sqrt{100} = 10$, so

$$\sqrt{(-20)(-5)} \neq \sqrt{-20}\,\sqrt{-5}.$$

To avoid difficulty, *always write square roots of negative numbers in terms of i before doing any simplification.*

TECHNOLOGY TIP

Most calculators that do complex number arithmetic automatically return a complex number when asked for the square root of a negative number. On TI-83+/89, however, the MODE must be set to "rectangular" or "$a + bi$." On Casio FX 2.0, the COMPLEX MODE (in the SET UP menu) must be set to "$a + bi$."

EXAMPLE 6

$$(7 - \sqrt{-4})(5 + \sqrt{-9}) = (7 - \sqrt{4}i)(5 + \sqrt{9}i)$$
$$= (7 - 2i)(5 + 3i)$$
$$= 35 + 21i - 10i - 6i^2$$
$$= 35 + 11i - 6(-1) = 41 + 11i. \quad \blacksquare$$

Since every negative real number has a square root in the complex number system, we can now find complex solutions for equations that have no real solutions. For example, the solutions of $x^2 = -25$ are $x = \pm\sqrt{-25} = \pm 5i$. In fact,

Every quadratic equation with real coefficients has solutions in the complex number system.

EXAMPLE 7

To solve the equation $2x^2 + x + 3 = 0$, we apply the quadratic formula.

$$x = \frac{-1 \pm \sqrt{1^2 - 4 \cdot 2 \cdot 3}}{2 \cdot 2} = \frac{-1 \pm \sqrt{-23}}{4}.$$

Since $\sqrt{-23}$ is not a real number, this equation has no real number solutions. But $\sqrt{-23}$ *is* a complex number, namely, $\sqrt{-23} = \sqrt{23}i$. Thus, the equation does have solutions in the complex number system.

$$x = \frac{-1 \pm \sqrt{-23}}{4} = \frac{-1 \pm \sqrt{23}i}{4} = -\frac{1}{4} \pm \frac{\sqrt{23}}{4}i.$$

Note that the two solutions, $-\dfrac{1}{4} + \dfrac{\sqrt{23}}{4}i$ and $-\dfrac{1}{4} - \dfrac{\sqrt{23}}{4}i$, are conjugates of each other. $\quad \blacksquare$

TECHNOLOGY TIP

The polynomial solvers on TI-86, HP-39+ and Casio produce all real and complex solutions of any polynomial equation that they can solve. See Calculator Investigation 1 on page 147 for details.

On TI-89, use cSOLVE in the COMPLEX submenu of the ALGEBRA menu to find all solutions.

EXAMPLE 8

To find *all* solutions of $x^3 = 1$, we rewrite the equation and use the Difference of Cubes pattern (see page 38) to factor:

$$x^3 = 1$$
$$x^3 - 1 = 0$$
$$(x - 1)(x^2 + x + 1) = 0$$
$$x - 1 = 0 \quad \text{or} \quad x^2 + x + 1 = 0.$$

The solution of the first equation is $x = 1$. The solutions of the second can be obtained from the quadratic formula.

$$x = \frac{-1 \pm \sqrt{1^2 - 4 \cdot 1 \cdot 1}}{2 \cdot 1} = \frac{-1 \pm \sqrt{-3}}{2} = \frac{-1 \pm \sqrt{3}i}{2} = -\frac{1}{2} \pm \frac{\sqrt{3}}{2}i.$$

Therefore, the equation $x^3 = 1$ has one real solution ($x = 1$) and two nonreal complex solutions [$x = -1/2 + (\sqrt{3}/2)i$ and $x = -1/2 - (\sqrt{3}/2)i$]. Each of these solutions is said to be a **cube root of 1** or a **cube root of unity**. Observe that the two nonreal complex cube roots of unity are conjugates of each other. ∎

The preceding examples illustrate this useful fact (whose proof is discussed in Section 4.7).

Conjugate Solutions

If $a + bi$ is a solution of a polynomial equation with *real* coefficients, then its conjugate $a - bi$ is also a solution of this equation.

EXERCISES 4.6

In Exercises 1–54, perform the indicated operation and write the result in the form a + bi.

1. $(2 + 3i) + (6 - i)$ **2.** $(-5 + 7i) + (14 + 3i)$

3. $(2 - 8i) - (4 + 2i)$ **4.** $(3 + 5i) - (3 - 7i)$

5. $\dfrac{5}{4} - \left(\dfrac{7}{4} + 2i\right)$ **6.** $(\sqrt{3} + i) + (\sqrt{5} - 2i)$

7. $\left(\dfrac{\sqrt{2}}{2} + i\right) - \left(\dfrac{\sqrt{3}}{2} - i\right)$

8. $\left(\dfrac{1}{2} + \dfrac{\sqrt{3}i}{2}\right) + \left(\dfrac{3}{4} - \dfrac{5\sqrt{3}i}{2}\right)$

9. $(2 + i)(3 + 5i)$ **10.** $(2 - i)(5 + 2i)$

11. $(-3 + 2i)(4 - i)$ **12.** $(4 + 3i)(4 - 3i)$

13. $(2 - 5i)^2$ **14.** $(1 + i)(2 - i)i$

15. $(\sqrt{3} + i)(\sqrt{3} - i)$ **16.** $\left(\dfrac{1}{2} - i\right)\left(\dfrac{1}{4} + 2i\right)$

17. i^{15} **18.** i^{26} **19.** i^{33} **20.** $(-i)^{53}$

21. $(-i)^{107}$ **22.** $(-i)^{213}$ **23.** $\dfrac{1}{5 - 2i}$ **24.** $\dfrac{1}{i}$

25. $\dfrac{1}{3i}$ **26.** $\dfrac{i}{2 + i}$ **27.** $\dfrac{3}{4 + 5i}$ **28.** $\dfrac{2 + 3i}{i}$

29. $\dfrac{1}{i(4 + 5i)}$ **30.** $\dfrac{1}{(2 - i)(2 + i)}$ **31.** $\dfrac{2 + 3i}{i(4 + i)}$

32. $\dfrac{2}{(2 + 3i)(4 + i)}$ **33.** $\dfrac{2 + i}{1 - i} + \dfrac{1}{1 + 2i}$

34. $\dfrac{1}{2 - i} + \dfrac{3 + i}{2 + 3i}$ **35.** $\dfrac{i}{3 + i} - \dfrac{3 + i}{4 + i}$

36. $6 + \dfrac{2i}{3 + i}$ **37.** $\sqrt{-36}$ **38.** $\sqrt{-81}$

39. $\sqrt{-14}$ **40.** $\sqrt{-50}$ **41.** $-\sqrt{-16}$

42. $-\sqrt{-12}$ **43.** $\sqrt{-16} + \sqrt{-49}$

44. $\sqrt{-25} - \sqrt{-9}$ **45.** $\sqrt{-15} - \sqrt{-18}$

46. $\sqrt{-12}\,\sqrt{-3}$ **47.** $\sqrt{-16}/\sqrt{-36}$

48. $-\sqrt{-64}/\sqrt{-4}$ **49.** $(\sqrt{-25} + 2)(\sqrt{-49} - 3)$

50. $(5 - \sqrt{-3})(-1 + \sqrt{-9})$

51. $(2 + \sqrt{-5})(1 - \sqrt{-10})$

52. $\sqrt{-3}(3 - \sqrt{-27})$ **53.** $1/(1 + \sqrt{-2})$

54. $(1 + \sqrt{-4})(3 - \sqrt{-9})$

In Exercises 55–58, find x and y. Remember that

$$a + bi = c + di$$

exactly when a = c and b = d.

55. $3x - 4i = 6 + 2yi$ **56.** $8 - 2yi = 4x + 12i$

57. $3 + 4xi = 2y - 3i$ **58.** $8 - xi = \dfrac{1}{2}y + 2i$

In Exercises 59–70, solve the equation and express each solution in the form a + bi.

59. $3x^2 - 2x + 5 = 0$ **60.** $5x^2 + 2x + 1 = 0$

61. $x^2 + x + 2 = 0$ **62.** $5x^2 - 6x + 2 = 0$

63. $2x^2 - x = -4$ **64.** $x^2 + 1 = 4x$

65. $2x^2 + 3 = 6x$ **66.** $3x^2 + 4 = -5x$

67. $x^3 - 8 = 0$ **68.** $x^3 + 125 = 0$

69. $x^4 - 1 = 0$ **70.** $x^4 - 81 = 0$

71. Simplify: $i + i^2 + i^3 + \cdots + i^{15}$

72. Simplify: $i - i^2 + i^3 - i^4 + i^5 - \cdots + i^{15}$

Thinkers

73. If $z = a + bi$ is a complex number, then its conjugate is usually denoted $\bar{z}$, that is $\bar{z} = a - bi$. Verify that

$$\bar{\bar{z}} = z.$$

74. The **real part** of the complex number $a + bi$ is defined to be the real number a. The **imaginary part** of $a + bi$ is defined to the real number b (*not bi*).

ONIT

(a) Show that the real part of $z = a + bi$ is $\dfrac{z + \bar{z}}{2}$.

(b) Show that the imaginary part of $z = a + bi$ is $\dfrac{z - \bar{z}}{2i}$.

75. If $z = a + bi$ (with a, b real numbers, not both 0), express $1/z$ in standard form.

4.7   Theory of Equations*

The complex numbers were constructed to obtain a solution for the equation $x^2 = -1$, that is, a root of the polynomial $x^2 + 1$. In Section 4.6, we saw that *every* quadratic polynomial with real coefficients has roots in the complex number system. A natural question now arises:

> Do we have to enlarge the complex number system (perhaps many times) to find roots for higher degree polynomials?

In this section, we shall see that the somewhat surprising answer is no.

To give the full answer, we shall consider not just polynomials with real coefficients, but also those with complex coefficients, such as

$$x^3 - ix^2 + (4 - 3i)x + 1 \qquad \text{or} \qquad (-3 + 2i)x^6 - 3x + (5 - 4i).$$

The discussion of polynomial division in Section 4.2 can easily be extended to include polynomials with complex coefficients. In fact, *all of the results in Section 4.2 are valid for polynomials with complex coefficients.* For example, you can check that i is a root of $f(x) = x^2 + (i - 1)x + (2 + i)$ and that $x - i$ is a factor of $f(x)$:

$$f(x) = x^2 + (i - 1)x + (2 + i) = (x - i)[x - (1 - 2i)].$$

Since every real number is also a complex number, polynomials with real coefficients are just special cases of polynomials with complex coefficients. So in the rest of this section, "polynomial" means "polynomial with complex (possibly real) coefficients" unless specified otherwise. We can now answer the question posed in the first paragraph.

Fundamental Theorem of Algebra

> Every nonconstant polynomial has a root in the complex number system.

Although this is obviously a powerful result, neither the Fundamental Theorem nor its proof provides a practical method for *finding* a root of a given polynomial.[†] The proof of the Fundamental Theorem is beyond the scope of this

---

*Section 4.6 is a prerequisite for this section.

[†]It might seem strange that you can prove that a root exists without actually exhibiting one. But such "existence theorems" are quite common. A rough analogy is the situation that occurs when someone is killed by a sniper's bullet. The police know that there is a killer, but *finding* the killer may be impossible.

book, but we shall explore some of the useful implications of the theorem, such as this one.

Factorization over the Complex Numbers

Let $f(x)$ be a polynomial of degree $n > 0$ with leading coefficient d. Then there are (not necessarily distinct) complex numbers $c_1, c_2, \ldots, c_n$ such that

$$f(x) = d(x - c_1)(x - c_2)(x - c_3) \cdots (x - c_n).$$

Furthermore, $c_1, c_2, \ldots, c_n$ are the only roots of $f(x)$.

Proof By the Fundamental Theorem, $f(x)$ has a complex root c_1. The Factor Theorem shows that $x - c_1$ must be a factor of $f(x)$, say,

$$f(x) = (x - c_1)g(x),$$

where $g(x)$ has degree $n - 1$.[*] If $g(x)$ is nonconstant, then it has a complex root c_2 by the Fundamental Theorem. Hence, $x - c_2$ is a factor of $g(x)$, so

$$f(x) = (x - c_1)(x - c_2)h(x)$$

for some $h(x)$ of degree $n - 2$ [1 less than the degree of $g(x)$]. If $h(x)$ is nonconstant, then it has a complex root c_3, and the argument can be repeated. Continuing in this way, with the degree of the last factor going down by 1 at each step, we reach a factorization in which the last factor is a constant (degree 0 polynomial).

$(*)$ $$f(x) = (x - c_1)(x - c_2)(x - c_3) \cdots (x - c_n)d.$$

If the right-hand side were multiplied out, it would look like

$$dx^n + \text{lower degree terms}.$$

So the constant factor d is the leading coefficient of $f(x)$.

It is easy to see from the factored form $(*)$ that the numbers $c_1, c_2, \ldots, c_n$ are roots of $f(x)$. If k is *any* root of $f(x)$, then

$$0 = f(k) = d(k - c_1)(k - c_2)(k - c_3) \cdots (k - c_n).$$

The product on the right is 0 only when one of the factors is 0. Since the leading coefficient d is nonzero, we must have

$$k - c_1 = 0 \quad \text{or} \quad k - c_2 = 0 \quad \text{or} \quad \cdots \quad k - c_n = 0$$
$$k = c_1 \quad \text{or} \quad k = c_2 \quad \text{or} \quad \cdots \quad k = c_n.$$

Therefore, k is one of the c's, and $c_1, \ldots, c_n$ are the only roots of $f(x)$. This completes the proof. ■

Since the n roots $c_1, \ldots, c_n$ of $f(x)$ might not all be distinct, we have the following.

Number of Roots

Every polynomial of degree $n > 0$ has at most n different roots in the complex number system.

TECHNOLOGY TIP

To find all roots of a polynomial, see the second Technology Tip on page 352.

To factor a polynomial as a product of linear factors on TI-89, try

ALGEBRA/COMPLEX/cFACTOR.

In the Casio FX 2.0 main menu, try

CAS/TRANS/rFACTOR.

The Answer Type should be set to "complex" in the Casio SET-UP menu.

---

[*]The degree of $g(x)$ is 1 less than the degree n of $f(x)$ because $f(x)$ is the product of $g(x)$ and $x - c_1$ (which has degree 1).

Suppose $f(x)$ has repeated roots, meaning that some of the $c_1, c_2, \ldots, c_n$ are the same in factorization (∗). Recall that a root c is said to have multiplicity k if $(x - c)^k$ is a factor of $f(x)$ but no higher power of $(x - c)$ is a factor. Consequently, if every root is counted as many times as its multiplicity, then the statement in the preceding box implies that

A polynomial of degree n has exactly n roots.

EXAMPLE 1

Find a polynomial $f(x)$ of degree 5 such that 1, -2, and 5 are roots, 1 is a root of multiplicity 3, and $f(2) = -24$.

SOLUTION   Since 1 is a root of multiplicity 3, $(x - 1)^3$ must be a factor of $f(x)$. There are at least two other factors corresponding to the roots -2 and 5: $x - (-2) = x + 2$ and $x - 5$. The product of these factors $(x - 1)^3(x + 2)(x - 5)$ has degree 5, as does $f(x)$, so $f(x)$ must look like this:

$$f(x) = d(x - 1)^3(x + 2)(x - 5),$$

where d is the leading coefficient. Since $f(2) = -24$, we have

$$d(2 - 1)^3(2 + 2)(2 - 5) = f(2) = -24,$$

which reduces to $-12d = -24$. Therefore, $d = (-24)/(-12) = 2$, and

$$f(x) = 2(x - 1)^3(x + 2)(x - 5)$$
$$= 2x^5 - 12x^4 + 4x^3 + 40x^2 - 54x + 20. \quad \blacksquare$$

▪▪ POLYNOMIALS WITH REAL COEFFICIENTS

Recall that the **conjugate** of the complex number $a + bi$ is the number $a - bi$ (see page 350). We usually write a complex number as a single letter, say z, and indicate its conjugate by $\bar{z}$ (sometimes read "z bar"). For instance, if $z = 3 + 7i$, then $\bar{z} = 3 - 7i$. Conjugates play a role whenever a quadratic polynomial with real coefficients has complex roots.

EXAMPLE 2

The quadratic formula shows that $x^2 - 6x + 13$ has two complex roots.

$$\frac{-(-6) \pm \sqrt{(-6)^2 - 4 \cdot 1 \cdot 13}}{2 \cdot 1} = \frac{6 \pm \sqrt{-16}}{2} = \frac{6 \pm 4i}{2} = 3 \pm 2i.$$

The complex roots are $z = 3 + 2i$ and its conjugate $\bar{z} = 3 - 2i$. $\quad \blacksquare$

The preceding example is a special case of a more general theorem, whose proof is outlined in Exercises 59 and 60.

Conjugate Roots Theorem

Let $f(x)$ be a polynomial with *real* coefficients. If the complex number z is a root of $f(x)$, then its conjugate $\bar{z}$ is also a root of $f(x)$.

EXAMPLE 3

Find a polynomial with real coefficients whose roots include the numbers 2 and $3 + i$.

SOLUTION   Since $3 + i$ is a root, its conjugate $3 - i$ must also be a root. Consider the polynomial

$$f(x) = (x - 2)[(x - (3 + i)][x - (3 - i)].$$

Obviously, 2, $3 + i$. and $3 - i$ are roots of $f(x)$. Multiplying out this factored form shows that $f(x)$ *does* have real coefficients.

$$f(x) = (x - 2)[x^2 - (3 - i)x - (3 + i)x + (3 + i)(3 - i)]$$

$$= (x - 2)(x^2 - 3x + ix - 3x - ix + 9 - i^2)$$

$$= (x - 2)(x^2 - 6x + 10)$$

$$= x^3 - 8x^2 + 22x - 20.$$

The next-to-last line of this calculation also shows that $f(x)$ can be factored as a product of a linear and a quadratic polynomial, each with *real* coefficients.   ■

The technique in Example 3 works because the polynomial

$$[x - (3 + i)][x - (3 - i)]$$

turns out to have real coefficients. The proof of the following result shows why this must always be the case.

Factorization over the Real Numbers

> Every nonconstant polynomial with real coefficients can be factored as a product of linear and quadratic polynomials with real coefficients in such a way that the quadratic factors, if any, have no real roots.

Proof   The first box on page 355 shows that

$$f(x) = d(x - c_1)(x - c_2) \cdots (x - c_n),$$

where $c_1. \ldots, c_n$ are the roots of $f(x)$. If some c_i is a real number, then the factor $x - c_i$ is a linear polynomial with real coefficients.[*] If some c_j is a nonreal complex root, then its conjugate must also be a root. Thus some c_k is the conjugate of c_j, say, $c_j = a + bi$ (with a, b real) and $c_k = a - bi$.[†] In this case,

$$(x - c_j)(x - c_k) = [x - (a + bi)][x - (a - bi)]$$

$$= x^2 - (a - bi)x - (a + bi)x + (a + bi)(a - bi)$$

$$= x^2 - ax + bix - ax - bix + a^2 - (bi)^2$$

$$= x^2 - 2ax + (a^2 + b^2).$$

---

[*]In Example 3, for instance, 2 is a real root and $x - 2$ a linear factor.
[†]In Example 3, for instance, $c_j = 3 + i$ and $c_k = 3 - i$ are conjugate roots.

Therefore, the factor $(x - c_j)(x - c_k)$ of $f(x)$ is a quadratic with real coefficients (because a and b are real numbers). Its roots (c_j and c_k) are nonreal. By taking the real roots of $f(x)$ one at a time and the nonreal ones in conjugate pairs in this fashion, we obtain the desired factorization of $f(x)$. ■

EXAMPLE 4

Given that $1 + i$ is a root of $f(x) = x^4 - 2x^3 - x^2 + 6x - 6$, factor $f(x)$ completely over the real numbers.

SOLUTION   Since $1 + i$ is a root of $f(x)$, so is its conjugate $1 - i$, and hence, $f(x)$ has this quadratic factor:

$$[x - (1 + i)][x - (1 - i)] = x^2 - 2x + 2.$$

Dividing $f(x)$ by $x^2 - 2x + 2$ shows that the other factor is $x^2 - 3$, which factors as $(x + \sqrt{3})(x - \sqrt{3})$. Therefore,

$$f(x) = (x + \sqrt{3})(x - \sqrt{3})(x^2 - 2x + 2). \quad ■$$

✓ EXERCISES 4.7

In Exercises 1–6, find the remainder when $f(x)$ is divided by $g(x)$ without using synthetic or long division.

1. $f(x) = x^{10} + x^8;$ $g(x) = x - 1$

2. $f(x) = x^6 - 10;$ $g(x) = x - 2$

3. $f(x) = 3x^4 - 6x^3 + 2x - 1;$ $g(x) = x + 1$

4. $f(x) = x^5 - 3x^2 + 2x - 1;$ $g(x) = x - 2$

5. $f(x) = x^3 - 2x^2 + 5x - 4;$ $g(x) = x + 2$

6. $f(x) = 10x^{75} - 8x^{65} + 6x^{45} + 4x^{32} - 2x^{15} + 5;$
$g(x) = x - 1$

In Exercises 7–10, list the roots of the polynomial and state the multiplicity of each root.

7. $f(x) = x^{54}\left(x + \dfrac{4}{5}\right)$

8. $g(x) = 3\left(x + \dfrac{1}{6}\right)\left(x - \dfrac{1}{5}\right)\left(x + \dfrac{1}{4}\right)$

9. $h(x) = 2x^{15}(x - \pi)^{14}[x - (\pi + 1)]^{13}$

10. $k(x) = (x - \sqrt{7})^7(x - \sqrt{5})^5(2x - 1)$

In Exercises 11–22, find all the roots of $f(x)$ in the complex number system; then write $f(x)$ as a product of linear factors.

11. $f(x) = x^2 - 2x + 5$ **12.** $f(x) = x^2 - 4x + 13$

13. $f(x) = 3x^2 + 2x + 7$ **14.** $f(x) = 3x^2 - 5x + 2$

15. $f(x) = x^3 - 27$ [*Hint:* Factor first.]

16. $f(x) = x^3 + 125$ **17.** $f(x) = x^3 + 8$

18. $f(x) = x^6 - 64$ [*Hint:* Let $u = x^3$ and factor $u^2 - 64$ first.]

19. $f(x) = x^4 - 1$ **20.** $f(x) = x^4 - x^2 - 6$

21. $f(x) = x^4 - 3x^2 - 10$ **22.** $f(x) = 2x^4 - 7x^2 - 4$

In Exercises 23–44, find a polynomial $f(x)$ with real coefficients that satisfies the given conditions. Some of these problems have many correct answers.

23. Degree 3; only roots are 1, 7, −4.

24. Degree 3; only roots are 1 and −1.

25. Degree 6; only roots are 1, 2, π.

26. Degree 5; only root is 2.

27. Degree 3; roots $-3, 0, 4; f(5) = 80$.

28. Degree 3; roots $-1, 1/2, 2; f(0) = 2$.

29. Roots include $2 + i$ and $2 - i$.

30. Roots include $1 + 3i$ and $1 - 3i$.

31. Roots include 2 and $2 + i$.

32. Roots include 3 and $4i - 1$.

33. Roots include $-3, 1 - i, 1 + 2i$.

34. Roots include $1, 2 + i, 3i, -1$.

35. Degree 2; roots $1 + 2i$ and $1 - 2i$.

36. Degree 4; roots $3i$ and $-3i$, each of multiplicity 2.

37. Degree 4; only roots are $4, 3 + i$, and $3 - i$.

38. Degree 5; roots 2 (of multiplicity 3), i, and $-i$.

39. Degree 6; roots 0 (of multiplicity 3) and $3, 1 + i, 1 - i$, each of multiplicity 1.

40. Degree 6; roots include i (of multiplicity 2) and 3.

41. Degree 2; roots include $1 + i; f(0) = 6$.

42. Degree 2; roots include $3 + i; f(2) = 3$.

43. Degree 3; roots include i and $1; f(-1) = 8$.

44. Degree 3; roots include $2 + 3i$ and $-2; f(2) = -3$.

In Exercises 45–48, find a polynomial with complex coefficients that satisfies the given conditions.

45. Degree 2; roots i and $1 - 2i$.

46. Degree 2; roots $2i$ and $1 + i$.

47. Degree 3; roots $3, i$, and $2 - i$.

48. Degree 4; roots $\sqrt{2}, -\sqrt{2}, 1 + i$, and $1 - i$.

In Exercises 49–56, one root of the polynomial is given; find all the roots.

49. $x^3 - 2x^2 - 2x - 3$; root 3.

50. $x^3 + x^2 + x + 1$; root i.

51. $x^4 + 3x^3 + 3x^2 + 3x + 2$; root i.

52. $x^4 - x^3 - 5x^2 - x - 6$; root i.

53. $x^4 - 2x^3 + 5x^2 - 8x + 4$; root 1 of multiplicity 2.

54. $x^4 - 6x^3 + 29x^2 - 76x + 68$; root 2 of multiplicity 2.

55. $x^4 - 4x^3 + 6x^2 - 4x + 5$; root $2 - i$.

56. $x^4 - 5x^3 + 10x^2 - 20x + 24$; root $2i$.

57. Let $z = a + bi$ and $w = c + di$ be complex numbers (a, b, c, d are real numbers). Prove the given equality by computing each side and comparing the results.

   (a) $\overline{z + w} = \overline{z} + \overline{w}$ (The left side says: First find $z + w$ and then take the conjugate. The right side says: First take the conjugates of z and w and then add.)

   (b) $\overline{z \cdot w} = \overline{z} \cdot \overline{w}$

58. Let $g(x)$ and $h(x)$ be polynomials of degree n and assume that there are $n + 1$ numbers $c_1, c_2, \ldots, c_n, c_{n+1}$ such that

$$g(c_i) = h(c_i) \qquad \text{for every } i.$$

Prove that $g(x) = h(x)$. [*Hint:* Show that each c_i is a root of $f(x) = g(x) - h(x)$. If $f(x)$ is nonzero, what is its largest possible degree? To avoid a contradiction, conclude that $f(x) = 0$.]

59. Suppose $f(x) = ax^3 + bx^2 + cx + d$ has real coefficients and z is a complex root of $f(x)$.

   (a) Use Exercise 57 and the fact that $\overline{r} = r$, when r is a real number, to show that

$$\overline{f(x)} = \overline{az^3 + bz^2 + cz + d}$$

$$= a\overline{z}^3 + b\overline{z}^2 + c\overline{z} + d = f(\overline{z}).$$

   (b) Conclude that $\overline{z}$ is also a root of $f(x)$.
   [*Note:* $f(\overline{z}) = \overline{f(z)} = \overline{0} = 0$.]

60. Let $f(x)$ be a polynomial with real coefficients and z a complex root of $f(x)$. Prove that the conjugate $\overline{z}$ is also a root of $f(x)$. [*Hint:* Exercise 59 is the case when $f(x)$ has degree 3; the proof in the general case is similar.]

61. Use the statement in the box on page 357 to show that every polynomial with real coefficients and *odd* degree must have at least one real root.

62. Give an example of a polynomial $f(x)$ with complex, non-real coefficients and a complex number z such that z is a root of $f(x)$, but its conjugate is not. Hence, the conclusion of the Conjugate Roots Theorem (page 356) may be *false* if $f(x)$ doesn't have real coefficients.

Chapter 4 Review

FACTS & FORMULAS

- The graph of $f(x) = ax^2 + bx + c$ is a parabola whose vertex has x-coordinate $-b/2a$.

CATALOG OF BASIC FUNCTIONS—PART 2

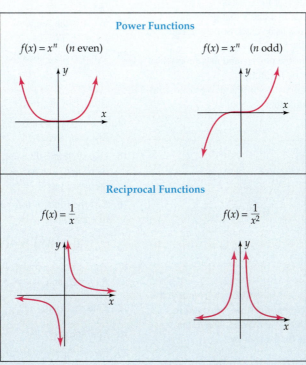

Power Functions

$f(x) = x^n$ (n even) $f(x) = x^n$ (n odd)

Reciprocal Functions

$f(x) = \dfrac{1}{x}$ $f(x) = \dfrac{1}{x^2}$

QUESTIONS

In Questions 1–5, find the vertex of the graph of the quadratic function.

1. $f(x) = (x - 2)^2 + 3$

2. $f(x) = 2(x + 1)^2 - 1$

3. $f(x) = x^2 - 8x + 12$

4. $f(x) = x^2 - 7x + 6$

5. $f(x) = 3x^2 - 9x + 1$

6. Which of the following statements about the functions

$$f(x) = 3x^2 + 2 \quad \text{and} \quad g(x) = -3x^2 + 2$$

is *false*?

(a) The graphs of f and g are parabolas.

(b) The graphs of f and g have the same vertex.

(c) The graphs of f and g open in opposite directions.

(d) The graph of f is the graph of $y = 3x^2$ shifted 2 units to the right.

7. A preschool wants to construct a fenced playground. The fence will be attached to the building at two corners, as shown in the figure. There is 400 feet of fencing available, all of it to be used.
   (a) Write an equation in x and y that gives the amount of fencing to be used. Solve the equation for y.
   (b) Write the area of the playground as a function of x. [Part (a) may be helpful.]
   (c) What are the dimensions of the playground with the largest possible area?

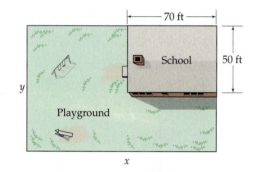

8. A model rocket is launched straight up from a platform at time $t = 0$ (where t is time measured in seconds). The altitude $h(t)$ of the rocket above the ground at given time (t) is given by

   $$h(t) = 10 + 112t - 16t^2$$

   (where $h(t)$ is measured in feet).
   (a) What is the altitude of the rocket the instant it is launched?
   (b) What is the altitude of the rocket 2 seconds after launching?
   (c) What is the maximum altitude attained by the rocket?
   (d) At what time does the rocket return to the altitude at which it was launched?

9. A rectangular garden next to a building is to be fenced with 120 feet of fencing. The side against the building will not be fenced. What should the lengths of the other three sides be to ensure the largest possible area?

10. A factory offers 100 calculators to a retailer at a price of $20 each. The price per calculator on the entire order will be reduced 5¢ for each additional calculator over 100. What number of calculators will produce the largest possible sales revenue for the factory?

11. Which of the following are polynomials?
   (a) $2^3 + x^2$
   (b) $x + \dfrac{1}{x}$

   (c) $x^3 - \dfrac{1}{\sqrt{2}}$
   (d) $\sqrt[3]{x^4}$
   (e) $\pi^3 - x$
   (f) $\sqrt{2} + 2x^2$
   (g) $\sqrt{x} + 2x^2$
   (h) $|x|$

12. What is the remainder when $x^4 + 3x^3 + 1$ is divided by $x^2 + 1$?

13. What is the remainder when

   $$x^{112} - 2x^8 + 9x^5 - 4x^4 + x - 5$$

   is divided by $x - 1$?

14. Is $x - 1$ a factor of $f(x) = 14x^{87} - 65x^{56} + 51$? Justify your answer.

15. Use synthetic division to show that $x - 2$ is a factor of $x^6 - 5x^5 + 8x^4 + x^3 - 17x^2 + 16x - 4$ and find the other factor.

16. List the roots of this polynomial and the multiplicity of each root:

   $$f(x) = 5(x - 4)^3(x - 2)(x + 17)^3(x^2 - 4).$$

17. Find a polynomial f of degree 3 such that $f(-1) = 0$, $f(1) = 0$, and $f(0) = 5$.

18. Find the root(s) of $2\left(\dfrac{x}{5} + 7\right) - 3x - \dfrac{x + 2}{5} + 4$.

19. Find the roots of $3x^2 - 2x - 5$.

20. Factor the polynomial $x^3 - 8x^2 + 9x + 6$. [*Hint:* 2 is a root.]

21. Find all real roots of $9x^3 - 6x^2 - 35x + 26$. [*Hint:* Try $x = -2$.]

22. Find the rational roots of $x^4 - 2x^3 - 4x^2 + 1$.

23. Consider the polynomial $2x^3 - 8x^2 + 5x + 3$.
   (a) List the only *possible* rational roots.
   (b) Find one rational root.
   (c) Find all the roots of the polynomial.

24. (a) Find all rational roots of $x^3 + 2x^2 - 2x - 2$.
   (b) Find two consecutive integers such that an irrational root of $x^3 + 2x^2 - 2x - 2$ lies between them.

25. The polynomial $x^3 - 2x + 1$ has
   (a) no real roots.   (d) only one rational root.
   (b) only one real root.   (e) none of the above.
   (c) three rational roots.

26. Without using technology, explain why every fifth-degree polynomial function has at least one real root. [*Hint:* What must its graph look like?]

C

In Questions 27 and 28, compute and simplify the difference quotient of the function.

27. $f(x) = x^2 + x$
28. $g(x) = x^3 - x + 1$

29. Draw the graph of a function that could not possibly be the graph of a polynomial function and explain why.

30. Draw a graph that could be the graph of a polynomial function of degree 5. You need not list a specific polynomial, nor do any computation.

31. Which of the statements below is *not* true about the polynomial function f whose graph is shown in the figure?

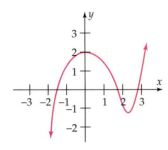

(a) f has three roots between -2 and 3.
(b) $f(x)$ could possibly be a fifth-degree polynomial.
(c) $(f \circ f)(0) > 0$.
(d) $f(2) - f(-1) < 3$.
(e) $f(x)$ is positive for all x with $-1 \le x \le 0$.

32. Which of the statements (i)–(v) about the polynomial function f whose graph is shown in the figure are *false*?

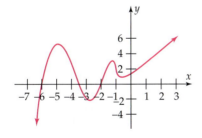

(i) f has 2 roots in the interval $-6 \le x < -3$.
(ii) $f(-3) - f(-6) < 0$.
(iii) $f(0) < f(1)$.
(iv) $f(2) - 2 = 0$.
(v) f has degree ≤ 4.

In Questions 33–36, find a viewing window (or windows) that shows a complete graph of the function. Be alert for hidden behavior.

33. $f(x) = .5x^3 - 4x^2 + x + 1$

34. $g(x) = .3x^5 - 4x^4 + x^3 - 4x^2 + 5x + 1$

35. $h(x) = 4x^3 - 100x^2 + 600x$

36. $f(x) = 32x^3 - 99x^2 + 100x + 2$

37. HomeArt makes plastic replicas of famous statues. Their total cost to produce copies of a particular statue are shown in the table below.
(a) Sketch a scatter plot of the data.
(b) Use cubic regression to find a function $C(x)$ that models the data [that is, $C(x)$ is the cost of making x statues]. Assume that C is reasonably accurate when $x \le 100$.
(c) Use C to estimate the cost of making the seventy-first statue.
(d) Use C to approximate the average cost per statue when 35 are made and when 75 are made.

Number of Statues	Total Cost
0	$2,000
10	2,519
20	2,745
30	2,938
40	3,021
50	3,117
60	3,269
70	3,425

38. The table gives the estimated cost of a college education at a public institution. Costs include tuition, fees, books, and room and board for four years. (*Source:* Teachers Insurance and Annuity Association College Retirement Equities Fund)

Enrollment Year	Costs	Enrollment Year	Costs
1998	$46,691	2008	$ 83,616
2000	52,462	2010	93,951
2002	58,946	2012	105,564
2004	66,232	2014	118,611
2006	74,418		

(a) Sketch a scatter plot of the data (with $x = 0$ corresponding to 1990).
(b) Use quartic regression to find a function C that models the data.
(c) Estimate the cost of a college education in 2007 and in 2015.

In Questions 39–46, sketch a complete graph of the function. Label the x-intercepts, all local extrema and asymptotes.

39. $f(x) = x^3 - 9x$

40. $g(x) = x^3 - 2x^2 + 3$

41. $h(x) = x^4 - x^3 - 4x^2 + 4x + 2$

42. $f(x) = x^4 - 3x - 2$

43. $g(x) = \dfrac{-2}{x + 4}$

44. $h(x) = \dfrac{3 - x}{x - 2}$

45. $k(x) = \dfrac{4x + 10}{3x - 9}$

46. $f(x) = \dfrac{x + 1}{x^2 - 1}$

In Questions 47 and 48, list all asymptotes of the graph of the function.

47. $f(x) = \dfrac{x^2 - 1}{x^3 - 2x^2 - 5x + 6}$

48. $g(x) = \dfrac{x^4 - 6x^3 + 2x^2 - 6x + 2}{x^2 - 3}$

In Questions 49–52, find a viewing window (or windows) that shows a complete graph of the function. Be alert for hidden behavior.

49. $f(x) = \dfrac{x - 3}{x^2 + x - 2}$

50. $g(x) = \dfrac{x^2 - x - 6}{x^3 - 3x^2 + 3x - 1}$

51. $h(x) = \dfrac{x^4 + 4}{x^4 - 99x^2 - 100}$

52. $k(x) = \dfrac{x^3 - 2x^2 - 4x + 8}{x - 10}$

53. It costs the Junkfood Company 50¢ to produce a bag of Munchies. There are fixed costs of $500 per day for building, equipment, etc. The company has found that if the price of a bag of Munchies is set at $1.95 - \dfrac{x}{2000}$ dollars, where x is the number of bags produced per day, then all the bags that are produced will be sold. What number of bags can be produced each day if all are to be sold and the company is to make a profit? What are the possible prices?

54. Highway engineers have found that a good model of the relationship between the density of automobile traffic and the speed at which it moves along a particular section of highway is given by the function

$$s = \frac{100}{1 + d^2} \qquad (0 \le d \le 3)$$

where the density of traffic d is measured in hundreds of cars per mile and the speed s at which traffic moves is measured in miles per hour. Then the traffic flow q is given by the product of the density and the speed, that is, $q = ds$, with q being measured in hundreds of cars per hour.

(a) Express traffic flow as a function of traffic density.

(b) For what densities will traffic flow be at least 3000 cars per hour?

(c) What traffic density will maximize traffic flow? What is the maximum flow?

55. Charlie lives 150 miles from the city. He drives 40 miles to the station (obeying the 55 mph speed limit) and catches a train to the city. The average speed of the train is 25 mph faster than the average speed of the car.

(a) Express the total time for the journey as a function of the speed of the car. What speeds make sense in this context?

(b) How fast should Charlie drive if the entire journey is to take no more than 2.5 hours?

56. The survival rate s of seedlings in the vicinity of a parent tree is given by

$$s = \frac{.5x}{1 + .4x^2},$$

where x is the distance from the seedling to the tree (in meters) and $0 < x \le 10$.

(a) For what distances is the survival rate at least .21?

(b) What distance produces the maximum survival rate?

C

In Questions 57 and 58, find the average rate of change of the function between x and x + h.

57. $f(x) = \dfrac{x}{x + 1}$

58. $g(x) = \dfrac{1}{x^2 + 1}$

REVIEW **QUESTIONS**

In Questions 59–62, perform the indicated operations and write the result in the form $a + bi$.

59. $(\sqrt{2} + i) + (\sqrt{3} - 2i)$ **60.** $(3 + i) + (4 + 2i)$

61. i^{27} **62.** $\dfrac{1}{3 - 4i}$

In Questions 63–74, solve the equation in the complex number system.

63. $x^2 + 3x + 10 = 0$ **64.** $x^2 + 2x + 5 = 0$

65. $5x^2 + 2 = 3x$

66. $-3x^2 + 4x - 5 = 0$

67. $3x^4 + x^2 - 2 = 0$

68. $8x^4 + 10x^2 + 3 = 0$

69. $x^3 + 8 = 0$ **70.** $x^3 - 27 = 0$

71. One root of $x^4 - x^3 - x^2 - x - 2$ is i. Find all the roots.

72. One root of $x^4 + x^3 - 5x^2 + x - 6$ is i. Find all the roots.

73. Give an example of a fourth-degree polynomial with real coefficients whose roots include 0 and $1 + i$.

74. Find a fourth-degree polynomial f whose only roots are $2 + i$ and $2 - i$, such that $f(-1) = 50$.

DISCOVERY PROJECT 4   Architectural Arches

You can see arches almost everywhere you look—in windows, entryways, tunnels, and bridges. Common arch shapes are semicircles, semiellipses, and parabolas. When constructed on a level base, arches are symmetric from left to right. This means that a mathematical function that describes an arch must be an even function.

A *semicircular arch* always has the property that its base is twice as wide as its height. This ratio can be modified by placing a rectangular area under the semicircle, giving a shape known as a *Norman arch*. This approach gives a tunnel or room a vaulted ceiling. *Parabolic arches* can also be created to give a more vaulted appearance.

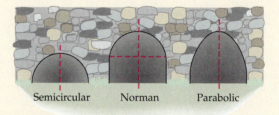

Semicircular   Norman   Parabolic

For the following exercises in arch modeling, you should always set the origin of your coordinate system to be the center of the base of the arch.

1. Show that the function that models a semicircular arch of radius r is
$$h(x) = \sqrt{r^2 - x^2}.$$

2. Write a function $h(x)$ that models a semicircular arch that is 15 feet tall. How wide is the arch?

3. Write a function $n(x)$ that models a Norman arch that is 15 feet tall and 16 feet wide at the base.

4. Parabolic arches are typically modeled by using the function
$$p(x) = H - ax^2,$$
where H is the height of the arch. Write a function $p(x)$ for an arch that is 15 feet tall and 16 feet wide at the base.

5. Would a truck that is 12 feet tall and 9 feet wide fit through all three arches? How could you fix any of the arches that is too small so that the truck would fit through?

Exponential and Logarithmic Functions

Walter Stuart/Index Stock Imagery/PictureQuest

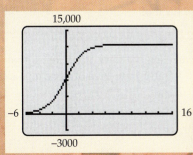

How old is that dinosaur?

Population growth (of humans, fish, bacteria, etc.), compound interest, radioactive decay, and a host of other phenomena can be mathematically described by exponential functions (see Exercises 50 on page 378, 21 on page 389, and 45 on page 390). Archeologists sometimes use carbon-14 dating to determine the approximate age of an artifact (such as a dinosaur skeleton, a mummy, or a wooden statue). This involves using logarithms to solve an appropriate exponential equation. See Exercise 53 on page 427.

367

Chapter Outline

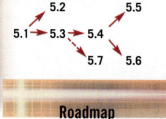
Exponential and logarithmic functions are essential for the mathematical description of a variety of phenomena in the physical sciences, economics, and engineering. Although a calculator is necessary to evaluate these functions at most numbers, you will not be able to use your calculator efficiently or interpret its answers unless you understand the properties of these functions. When calculations can readily be done by hand, you will be expected to do them without a calculator.

5.1   Exponential Functions

We know how to compute powers of a positive number when the exponent is an integer or a rational number (finite decimal), such as 10^{-4} or $10^{3.7}$ (see Sections 0.2 and 0.3). The next step is to define powers of a positive number when the exponent is an irrational real number. A mathematically rigorous definition is beyond the scope of this book, but we can illustrate the underlying idea.

To compute $10^{\sqrt{2}}$, for example, we use the infinite decimal expansion $\sqrt{2} \approx 1.414213562\cdots$ (see Special Topics 0.1.A). Each of

$$1.4, \ 1.41, \ 1.414, \ 1.4142, \ 1.41421, \ldots$$

is a rational number approximation of $\sqrt{2}$, and each is a more accurate approximation than the preceding one. We know how to raise 10 to each of these rational numbers:

$$10^{1.4} \approx 25.1189 \qquad 10^{1.4142} \approx 25.9537$$

$$10^{1.41} \approx 25.7040 \qquad 10^{1.41421} \approx 25.9543$$

$$10^{1.414} \approx 25.9418 \qquad 10^{1.414213} \approx 25.9545.$$

It appears that as the exponent r gets closer and closer to $\sqrt{2}$, 10^r gets closer and closer to a real number whose decimal expansion begins $25.954\cdots$. We define $10^{\sqrt{2}}$ to be this number.

Similarly, for any $a > 0$,

a^t is a well-defined *positive* number for each real exponent t.

We shall also assume this fact.

The exponent laws (page 27) are valid for *all* real exponents.

Consequently, for each positive real number a, the rule $f(x) = a^x$ defines a function whose domain is the set of all real numbers. The function $f(x) = a^x$ is called the **exponential function with base a.** For example, we have exponential functions:

$$f(x) = 10^x, \qquad g(x) = 2^x, \qquad h(x) = \left(\frac{1}{2}\right)^x, \qquad k(x) = \left(\frac{3}{2}\right)^x.$$

The graph of $f(x) = a^x$ is the next entry in the catalog of basic functions. To see how its graph depends on the size of the base a, do the following Graphing Exploration.

GRAPHING EXPLORATION

Using viewing window with $-3 \le x \le 7$ and $-2 \le y \le 18$, graph

$$f(x) = 1.3^x, \qquad g(x) = 2^x, \qquad \text{and} \qquad h(x) = 10^x$$

on the same screen. How is the steepness of the graph of $f(x) = a^x$ related to the size of a?

The Graphing Exploration illustrates these facts.

The Exponential Function
$$f(x) = a^x \,(a > 1)$$

When $a > 1$, the graph of $f(x) = a^x$ has the shape shown here and the properties listed below.

The graph is above the x-axis.

The y-intercept is 1.

$f(x)$ is an increasing function.

The negative x-axis is a horizontal asymptote.

The larger the base a, the more steeply the graph rises to the right.

The preceding figure doesn't really show how explosively an exponential graph rises.

EXAMPLE 1

Graph $f(x) = 2^x$ and estimate the height of the graph when $x = 50$.

SOLUTION   A small portion of the graph is shown in Figure 5–1. If the x-axis were to be extended with the same scale, $x = 50$ would be at the right edge of the page. At that point, the height of the graph is $f(50) = 2^{50}$. Now the y-axis scale in Figure 5–1 is approximately 12 units to the inch, which is equivalent to 760,320 units per mile, as you can readily verify. Therefore, the height of the graph at $x = 50$ is

$$\frac{2^{50}}{760,320} = 1,480,823,741 \text{ MILES},$$

which would put that part of the graph well beyond the planet Saturn! ■

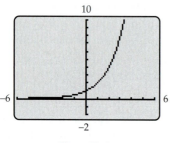

Figure 5–1

EXAMPLE 2

Use the graph of $f(x) = 2^x$ in Figure 5–1 to describe the graphs of the following functions without actually graphing them.

(a) $g(x) = 2^{x+3}$ (b) $h(x) = 2^{x-3} - 4$

SOLUTION

(a) Since $f(x) = 2^x$, we have $f(x + 3) = 2^{x+3} = g(x)$. So the graph of $g(x)$ is just the graph of $f(x) = 2^x$ shifted horizontally 3 units to the left. [See page 234.]

GRAPHING EXPLORATION

Verify the preceding statement graphically by graphing $f(x)$ and $g(x)$ on the same screen.

(b) In this case, $f(x - 3) - 4 = 2^{x-3} - 4 = h(x)$. So the graph of $h(x)$ is the graph of $f(x) = 2^x$ shifted horizontally 3 units to the right and vertically 4 units downward. [See pages 234 and 232.]

GRAPHING EXPLORATION

Verify the preceding statement graphically by graphing $f(x)$ and $h(x)$ on the same screen. ■

When the base a is between 0 and 1, then the graph of $f(x) = a^x$ has a different shape.

EXPLORATION

Using viewing window with $-4 \le x \le 4$ and $-1 \le y \le 4$, graph

$$f(x) = .2^x, \qquad g(x) = .4^x, \qquad h(x) = .6^x, \qquad \text{and} \qquad k(x) = .8^x$$

on the same screen. How is the steepness of the graph of $f(x) = a^x$ related to the size of a?

The exploration supports this conclusion.

The Exponential Function
$$f(x) = a^x \ (0 < a < 1)$$

When $0 < a < 1$, the graph of $f(x) = a^x$ has the shape shown here and the properties listed below.

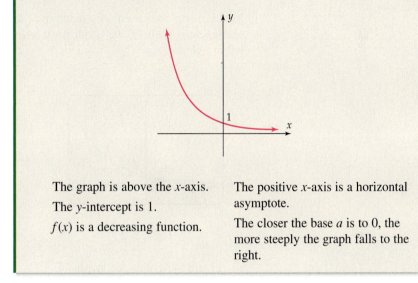

The graph is above the x-axis.

The y-intercept is 1.

$f(x)$ is a decreasing function.

The positive x-axis is a horizontal asymptote.

The closer the base a is to 0, the more steeply the graph falls to the right.

EXAMPLE 3

Without graphing, describe the graph of $g(x) = 3^{-x}$.

SOLUTION   Note that

$$g(x) = 3^{-x} = (3^{-1})^x = \left(\frac{1}{3}\right)^x.$$

So $g(x)$ is an exponential function with a positive base less than 1. Its graph has the shape shown in the preceding box: It falls quickly to the right and rises very steeply to the left of the y-axis.   ∎

EXPLORATION

Verify the analysis in Example 3 by graphing $g(x) = 3^{-x}$ in the viewing window with $-4 \le x \le 4$ and $-2 \le y \le 18$.

Since most quantities that grow exponentially don't increase or decrease as dramatically as the graphs of $f(x) = a^x$, exponential functions that model real-life situations generally have the form $f(x) = Pa^{kx}$, such as

$$f(x) = 5 \cdot 2^{.45x}, \qquad g(x) = 3.5(10^{-.03x}), \qquad h(x) = (-6)(1.076^{2x}).$$

Their graphs have the same basic shape as the graph of $f(x) = a^x$, but rise or fall at different rates, depending on the constants P, a, and k.

EXAMPLE 4

Figure 5–2 shows the graphs of

$$f(x) = 3^x, \qquad g(x) = 3^{.15x}, \qquad h(x) = 3^{.35x}, \qquad k(x) = 3^{-x}, \qquad p(x) = 3^{-.4x}.$$

Note how the coefficient of x determines the steepness of the graph. When this coefficient is positive, the graph rises; when it is negative, the graph falls from left to right.

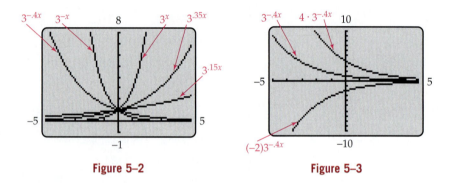

Figure 5–2   **Figure 5–3**

Figure 5–3 shows the graphs of

$$p(x) = 3^{-.4x}, \qquad q(x) = 4 \cdot 3^{-.4x}, \qquad r(x) = (-2)3^{-.4x}.$$

As we saw in Section 3.4, the graph of $q(x) = 4 \cdot 3^{-.4x}$ is the graph of $p(x) = 3^{-.4x}$ stretched away from the x-axis by a factor of 4. The graph of $r(x) = (-2)3^{-.4x}$ is the graph of $p(x) = 3^{-.4x}$ stretched away from the x-axis by a factor of 2 *and* reflected in the x-axis. ■

▌▌ EXPONENTIAL GROWTH AND DECAY

Exponential functions have a number of practical applications, as illustrated in the following examples. We shall learn how to construct functions such as the ones in these examples in Section 5.2, but for now we concentrate on using them.

EXAMPLE 5   Finance

Bill and Mary Ann Burger invest $5000 in a high-flying stock that is increasing in value at the rate of 12% per year. The value of their stock is given by the function $f(x) = 5000(1.12^x)$, where x is measured in years.

(a) Assuming that the value of the stock continues growing at this rate, how much will their investment be worth in four years?

(b) When will their investment be worth $14,000?

SOLUTION

(a) In four years ($x = 4$), their stock will be worth

$$f(4) = 5000(1.12^4) \approx \$7867.60.$$

(b) To determine when the stock will be worth $14,000, we must find the value of x for which $f(x) = 14,000$. In other words, we must solve the equation

$$5000(1.12^x) = 14,000.$$

The solution is the x-coordinate of the intersection point of the graphs of $f(x) = 5000(1.12^x)$ and $y = 14,000$. Figure 5–4 shows that this point is approximately $(9.085, 14,000)$. Therefore, the stock will be worth $14,000 in about 9.085 years. ∎

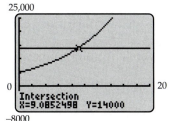

25,000

0

20

−8000

Intersection
X=9.0852498 Y=14000

Figure 5–4

EXAMPLE 6   Population Growth

Based on figures from the past 25 years, the world population (in billions) can be approximated by the function $g(x) = 4.5(1.015^x)$, where $x = 0$ corresponds to 1980.

(a) Estimate the world population in 2009.

(b) In what year will the world population be double what it is in 2009?

SOLUTION

(a) Since 2009 corresponds to $x = 29$, the population is

$$g(29) = 4.5(1.015^{29}) \approx 6.93 \text{ billion people.}$$

(b) Twice the population in 2009 is $2(6.93) = 13.86$ billion. We must find the number x such that $g(x) = 13.86$; that is, we must solve the equation

$$4.5(1.015^x) = 13.86.$$

This can be done with an equation solver or by graphical means, as in Figure 5–5, which shows the intersection point of $g(x) = 4.5(1.015^x)$ and $y = 13.86$. The solution is $x \approx 75.6$, which corresponds to the year 2055. According to this model, the world population will double in your lifetime. This is what is meant by the population explosion. ∎

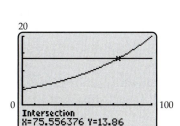

20

0

100

Intersection
X=75.556376 Y=13.86

Figure 5–5

EXAMPLE 7   Radioactive Decay

One kilogram of plutonium (^{239}Pu) is stored in a nuclear waste facility, where it slowly decays. The amount remaining after x years is approximated by

$$M(x) = .99997^x.$$

(a) Without graphing, describe the shape of the graph of M. Then confirm your analysis by graphing.

(b) How much plutonium will be left after 10,000 years?

SOLUTION

(a) Since M is an exponential function whose base is smaller than, but very close to, 1, its graph falls *very slowly* from left to right.

> **GRAPHING EXPLORATION**
> Confirm this statement by graphing $M(x)$ in the window with $0 \leq x \leq 5000$ and $0 \leq y \leq 2$.

(b) After 10,000 years, the amount of plutonium remaining is

$$M(10,000) = .99997^{10,000} \approx .74 \text{ kilogram.}$$

So almost three-fourths of the original amount remains. This is the reason that nuclear waste disposal is such a serious problem. ■

■■ THE NUMBER e AND THE NATURAL EXPONENTIAL FUNCTION

There is an irrational number, denoted e, that arises naturally in a variety of phenomena and plays a central role in the mathematical description of the physical universe. Its decimal expansion begins

$$e = 2.718281828459045 \cdots .$$

Your calculator has an e^x key that can be used to evaluate the **natural exponential function** $f(x) = e^x$. If you key in e^1, the calculator will display the first part of the decimal expansion of e.

The graph of $f(x) = e^x$ has the same shape as the graph of $g(x) = 2^x$ in Figure 5–1 but climbs more steeply.

> **TECHNOLOGY TIP**
> On most calculators, you use the e^x key, but not the x^y or $\wedge$ keys to enter the function $f(x) = e^x$.

> **GRAPHING EXPLORATION**
> Graph $f(x) = e^x$, $g(x) = 2^x$, and $h(x) = 3^x$ on the same screen in a window with $-5 \leq x \leq 5$. The Technology Tip in the margin may be helpful.

EXAMPLE 8   **Transistor Growth**

For the past 30 years, silicon chip technology has advanced dramatically. The approximate number of transistors that can be placed on a single chip in year x is given by

$$N(x) = 2418e^{.30895x},$$

where $x = 2$ corresponds to 1972.[*]

*Based on data from Intel.

(a) Estimate the number of transistors per chip in 1995 and 2003.

(b) Assuming that this model remains accurate for the next decade (which is not at all certain), when will it be possible to put 420 million transistors on a chip?

SOLUTION

(a) The year 1995 corresponds to $x = 25$, so the number of transistors was

$$N(25) = 2418e^{.30895(25)} \approx 5{,}468{,}123.$$

In 2003 ($x = 33$), the number was

$$N(33) = 2418e^{.30895(33)} \approx 64{,}750{,}115.$$

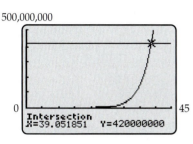

500,000,000

0   45

Intersection
X=39.051851   Y=420000000

Figure 5–6

(b) We must find the value of x for which $N(x) = 420$ million; that is, we must solve the equation

$$2418e^{.30895x} = 420{,}000{,}000.$$

This can be done with an equation solver or by graphical means. Figure 5–6 shows that the solution is $x \approx 39$, which corresponds to the year 2009. ■

▟ OTHER EXPONENTIAL FUNCTIONS

The population growth models in earlier examples do not take into account factors that may limit population growth in the future (wars, new diseases, etc.). Example 9 illustrates a function, called a **logistic model,** that is designed to model such situations more accurately.

EXAMPLE 9   Inhibited Population Growth

The world population model in Example 6 is based on data from 1980 to the present and shows a constantly increasing population. Because of decreasing fertility rates and aging populations, however, the population growth rate is expected to slow significantly during this century. The logistic function

$$f(x) = \frac{10}{1 + 232.97e^{-.02944x}}$$

models this situation.[*] It gives the world population (in billions) in year x, where $x = 0$ corresponds to 1800.

(a) According to this model, what will the world population be in 2009 and in 2055?

(b) If this model is accurate, is there an upper limit on the world's population?

SOLUTION

(a) The year 2009 corresponds to $x = 209$ and

$$f(209) = \frac{10}{1 + 232.97e^{-0.2944(209)}} \approx 6.69 \text{ billion.}$$

---

[*]Based on data from the United Nations Population Division and the U.S. Census Bureau.

This is quite close to the 6.93 billion given by the function in Example 6. For 2055 ($x = 255$), the model gives a population of

$$f(255) = \frac{10}{1 + 232.97e^{-.02944(255)}} \approx 8.87 \text{ billion,}$$

whereas Example 6 predicts close to 14 billion.

(b) The graph of $f(x)$ in Figure 5–7 suggests that the horizontal line $y = 10$ is a horizontal asymptote for the graph, which means that the world population will level off and never grow larger than 10 billion according to this model. You can confirm this algebraically by writing the rule of f in this form:

$$f(x) = \frac{10}{1 + 232.97e^{-.02944x}} = \frac{10}{1 + \dfrac{232.97}{e^{.02944x}}}.$$

When x gets larger, so does $.02944x$, so $e^{.02944x}$ eventually gets huge. Hence, by the Big-Little Principle (page 329), $\dfrac{232.97}{e^{.02944x}}$ is very close to 0, and $f(x)$ is very close to $\dfrac{10}{1 + 0} = 10$. Since $\dfrac{232.97}{e^{.02944x}}$ is positive, the denominator of $f(x)$ is slightly bigger than 1, so $f(x)$ is always less than 10. ■

When a cable, such as a power line, is suspended between towers of equal height, it forms a curve called a **catenary,** which is the graph of a function of the form

$$f(x) = A(e^{kx} + e^{-kx})$$

for suitable constants A and k. The Gateway Arch in St. Louis (Figure 5–8) has the shape of an inverted catenary, which was chosen because it evenly distributes the internal structural forces.

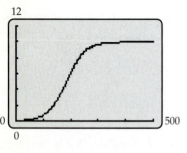

Figure 5–7

Figure 5–8

**EXPLORATION**

Graph each of the following functions in the window with
$-5 \le x \le 5$ and $-10 \le y \le 80$.

$$y_1 = 10(e^{.4x} + e^{-.4x}), \qquad y_2 = 10(e^{2x} + e^{-2x}),$$

$$y_3 = 10(e^{3x} + e^{-3x}).$$

How does the coefficient of x affect the shape of the graph?

Predict the shape of the graph of $y = -y_1 + 80$. Confirm your answer
by graphing.

✓ EXERCISES 5.1

In Exercises 1–6, sketch a complete graph of the function.

1. $f(x) = 4^{-x}$

2. $f(x) = (5/2)^{-x}$

3. $f(x) = 2^{3x}$

4. $g(x) = 3^{x/2}$

5. $f(x) = 2^{x^2}$

6. $g(x) = 2^{-x^2}$

In Exercises 7–12, list the transformations needed to transform the graph of $h(x) = 2^x$ into the graph of the given function. [Section 3.4 may be helpful.]

7. $f(x) = 2^x - 5$

8. $g(x) = -(2^x)$

9. $k(x) = 3(2^x)$

10. $g(x) = 2^{x-1}$

11. $f(x) = 2^{x+2} - 5$

12. $g(x) = -5(2^{x-1}) + 7$

In Exercises 13 and 14, match the functions to the graphs. Assume that $a > 1$ and $c > 1$.

13. $f(x) = a^x$
$g(x) = a^x + 3$
$h(x) = a^{x+5}$

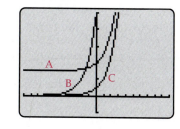

14. $f(x) = c^x$
$g(x) = -3c^x$
$h(x) = c^{x+5}$
$k(x) = -3c^x - 2$

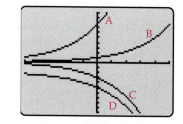

In Exercises 15–19, determine whether the function is even, odd, or neither (see Special Topics 3.4.A).

15. $f(x) = 10^x$

16. $g(x) = 2^x - x$

17. $f(x) = \dfrac{e^x + e^{-x}}{2}$

18. $f(x) = \dfrac{e^x - e^{-x}}{2}$

19. $f(x) = e^{-x^2}$

20. Use the Big-Little Principle to explain why $e^x + e^{-x}$ is approximately equal to e^x when x is large.

C *In Exercises 21–24, find the average rate of change of the function.**

21. $f(x) = x2^x$ as x goes from 1 to 3

22. $g(x) = 3^{x^2 - x}$ as x goes from -1 to 1

23. $h(x) = 5^{-x^2}$ as x goes from -1 to 0

24. $f(x) = e^x - e^{-x}$ as x goes from -3 to -1

C *In Exercises 25–28, find the difference quotient of the function.*

25. $f(x) = 10^x$

26. $g(x) = 5^{x^2}$

27. $f(x) = 2^x + 2^{-x}$

28. $f(x) = e^x - e^{-x}$

---

* **C** and ⬤ indicate examples, exercises, and sections that are
relevant to calculus.

In Exercises 29–36, find a viewing window (or windows) that shows a complete graph of the function.

29. $k(x) = e^{-x}$

30. $f(x) = e^{-x^2}$

31. $f(x) = \dfrac{e^x + e^{-x}}{2}$

32. $h(x) = \dfrac{e^x - e^{-x}}{2}$

33. $g(x) = 2^x - x$

34. $k(x) = \dfrac{2}{e^x + e^{-x}}$

35. $f(x) = \dfrac{5}{1 + e^{-x}}$

36. $g(x) = \dfrac{10}{1 + 9e^{-x/2}}$

In Exercises 37–42, list all asymptotes of the graph of the function and the approximate coordinates of each local extremum.

37. $f(x) = x2^x$

38. $g(x) = x2^{-x}$

39. $h(x) = e^{x^2/2}$

40. $k(x) = 2^{x^2 - 6x + 2}$

41. $f(x) = e^{-x^2}$

42. $g(x) = -xe^{x^2/20}$

43. (a) A genetic engineer is growing cells in a fermenter. The cells multiply by splitting in half every 15 minutes. The new cells have the same DNA as the original ones. Complete the following table.

Time (hours)	Number of Cells
0	1
.25	2
.5	4
.75	
1	

    (b) Write the rule of the function that gives the number of cells C at time t hours.

44. Do Exercise 43, using the following table.

Time (hours)	Number of Cells
0	300
.25	600
.5	1200
.75	
1	

45. The Gateway Arch (Figure 5–8) is 630 feet high and 630 feet wide at ground level. Suppose it were placed on a coordinate plane with the x-axis at ground level and the y-axis going through the center of the arch. Find a catenary function $g(x) = A(e^{kx} + e^{-kx})$ and a constant C such that the graph of the function $f(x) = g(x) + C$ provides a model of the arch. [*Hint*: Experiment with various values of A, k, C as in the Graphing Exploration on page 377. Many correct answers are possible.]

46. If you deposit $750 at 2.2% interest, compounded annually and paid from the day of deposit to the day of withdrawal, your balance at time t is given by $B(t) = 750(1.022)^t$. How much will you have after two years? After three years and nine months?

47. The number of digital devices (cell phones, PCs, digital cameras, MP3 players, etc.) used worldwide in year x is approximated by

$$g(x) = 1.244 \cdot 1.263^x,$$

where $x = 0$ corresponds to 2000 and $g(x)$ is in billions.[*]
    (a) Estimate the number of digital devices in 2002 and 2005.
    (b) When will the number of devices be twice the number in 2005? (Round your answer to the nearest year.)

48. The population of a colony of fruit flies t days from now is given by the function $p(t) = 100 \cdot 3^{t/10}$.
    (a) What will the population be in 15 days? In 25 days?
    (b) When will the population reach 2500?

49. If the population of the United States continues to grow as it has since 1980, then the approximate population (in millions) in year t is given by

$$P(t) = 227e^{.0093t},$$

where $t = 0$ corresponds to 1980.
    (a) Estimate the population in 2005 and 2010.
    (b) When will the population reach 400 million?

50. If current rates of deforestation and fossil fuel consumption continue, then the amount of atmospheric carbon dioxide in parts per million (ppm) will be given by $f(x) = 375e^{.00609x}$, where $x = 0$ corresponds to 2000.[†]
    (a) What is the amount of carbon dioxide in 2003? In 2022?
    (b) In what year will the amount of carbon dioxide reach 500 ppm?

---

[*]Based on data and projections from *IDC*.
[†]Based on projections from the International Panel on Climate Change.

51. The pressure of the atmosphere $p(x)$ (in pounds per square inch) is given by

$$p(x) = ke^{-.0000425x},$$

where x is the height above sea level (in feet) and k is a constant.

(a) Use the fact that the pressure at sea level is 15 pounds per square inch to find k.

(b) What is the pressure at 5000 feet?

(c) If you were in a spaceship at an altitude of 160,000 feet, what would the pressure be?

52. (a) The function $g(t) = 1 - e^{-.0479t}$ gives the percentage of the population (expressed as a decimal) that has seen a new TV show t weeks after it goes on the air. What percentage of people have seen the show after 24 weeks?

(b) Approximately when will 90% of the people have seen it?

53. According to data from the National Center for Health Statistics, the life expectancy at birth for a person born in a year x is approximated by the function

$$D(x) = \frac{79.257}{1 + 9.7135 \times 10^{24} \cdot e^{-.0304x}}$$

$$(1900 \le x \le 2050).$$

(a) What is the life expectancy of someone born in 1980? In 2000?

(b) In what year was life expectancy at birth 60 years?

54. (a) The beaver population near a certain lake in year t is approximately

$$p(t) = \frac{2000}{1 + 199e^{-.5544t}}.$$

What is the population now ($t = 0$) and what will it be in 5 years?

(b) Approximately when will there be 1000 beavers?

55. The number of subscribers to basic cable TV (in millions) can be approximated by

$$g(x) = \frac{76.7}{1 + 16(.8444^x)},$$

where $x = 0$ corresponds to 1970.*

(a) Estimate the number of subscribers in 1995 and 2005.

(b) When does the number of subscribers reach 70 million?

(c) According to this model, will the number of subscribers ever reach 90 million?

*Based on data from *The Cable TV Financial Datebook* and *The Pay TV Newsletter*.

56. An eccentric billionaire offers you a job for the month of September. She says that she will pay you 2¢ on the first day, 4¢ on the second day, 8¢ on the third day, and so on, doubling your pay on each successive day.

(a) Let $P(x)$ denote your salary in *dollars* on day x. Find the rule of the function P.

(b) Would you be better off financially if instead you were paid a flat rate of $10,000 per day? [*Hint:* Consider $P(30)$.]

57. Take an ordinary piece of printer paper and fold it in half; the folded sheet is twice as thick as the single sheet was. Fold it in half again so that it is twice as thick as before. Keep folding it in half as long as you can. Soon the folded paper will be so thick and small that you will be unable to continue, but suppose you could keep folding the paper as many times as you wanted. Assume that the paper is .002 inches thick.

(a) Make a table showing the thickness of the folded paper for the first four folds (with fold 0 being the thickness of the original unfolded paper).

(b) Find a function of the form $f(x) = Pa^x$ that describes the thickness of the folded paper after x folds.

(c) How thick would the paper be after 20 folds?

(d) How many folds would it take to reach the moon (which is 243,000 miles from the earth)? [*Hint:* One mile is 5280 feet.]

58. The figure is the graph of an exponential growth function $f(x) = Pa^x$.

(a) In this case, what is P? [*Hint:* What is $f(0)$?]

(b) Find the rule of the function f by finding a. [*Hint:* What is $f(2)$?]

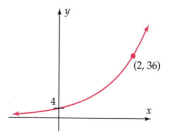

Thinkers

59. Find a function $f(x)$ with the property $f(r + s) = f(r)f(s)$ for all real numbers r and s. [*Hint:* Think exponential.]

60. Find a function $g(x)$ with the property $g(2x) = (g(x))^2$ for every real number x.

61. (a) Using the viewing window with $-4 \leq x \leq 4$ and $-1 \leq y \leq 8$, graph $f(x) = \left(\frac{1}{2}\right)^x$ and $g(x) = 2^x$ on the same screen. If you think of the y-axis as a mirror, how would you describe the relationship between the two graphs?

(b) Without graphing, explain how the graphs of $g(x) = 2^x$ and $k(x) = 2^{-x}$ are related.

62. Approximating exponential functions by polynomials. For each positive integer n, let f_n be the polynomial function whose rule is

$$f_n(x) = 1 + x + \frac{x^2}{2!} + \frac{x^3}{3!} + \frac{x^4}{4!} + \cdots + \frac{x^n}{n!},$$

where $k!$ is the product $1 \cdot 2 \cdot 3 \cdots k$.

(a) Using the viewing window with $-4 \leq x \leq 4$ and $-5 \leq y \leq 55$, graph $g(x) = e^x$ and $f_4(x)$ on the same screen. Do the graphs appear to coincide?

(b) Replace the graph of $f_4(x)$ by that of $f_5(x)$, then by $f_6(x)$, $f_7(x)$, and so on until you find a polynomial $f_n(x)$ whose graph appears to coincide with the graph of $g(x) = e^x$ in this viewing window. Use the trace feature to move from graph to graph at the same value of x to see how accurate this approximation is.

(c) Change the viewing window so that $-6 \leq x \leq 6$ and $-10 \leq y \leq 400$. Is the polynomial you found in part (b) a good approximation for $g(x)$ in this viewing window? What polynomial is?

5.2 Applications of Exponential Functions

In the previous section, we encountered several exponential functions that modeled growth and decay. In this section, we learn how to construct such exponential models in a variety of real-life situations.

■■ COMPOUND INTEREST

We begin with a subject that interests almost everyone: money and how it grows.

EXAMPLE 1

If you deposit $5000 in a savings account that pays 3% interest, compounded annually, how much money is in the account after nine years?

SOLUTION   After one year, the account balance is

$$5000 + 3\% \text{ of } 5000 = 5000 + (.03)5000 = 5000(1 + .03)$$

$$= 5000(1.03) = \$5150.$$

The initial balance has grown by a factor of 1.03. If you leave the $5150 in the account, then at the end of the second year, the balance is

$$5150 + 3\% \text{ of } 5150 = 5150 + (.03)5150 = 5150(1 + .03) = 5150(1.03).$$

Once again, the amount at the beginning of the year has grown by a factor of 1.03. The same thing happens every year. A balance of P dollars at the beginning of the year grows to $P(1.03)$. So the balance grows like this:

Year 1	Year 2	Year 3

$$5000 \to 5000(1.03) \to \underbrace{[5000(1.03)](1.03)}_{5000(1.03)^2} \to \underbrace{[5000(1.03)(1.03)](1.03)}_{5000(1.03)^3} \to \cdots.$$

Consequently, the balance at the end of year x is given by

$$f(x) = 5000 \cdot 1.03^x.$$

The balance at the end of nine years is $f(9) = 5000(1.03^9) = \$6523.87$ (rounded to the nearest penny). ■

The argument used in Example 1 applies in the general case.

Compound Interest Formula

> If P dollars is invested at interest rate r per time period (expressed as a decimal), then the amount A after t periods is
>
> $$A = P(1 + r)^t.$$

In Example 1, for instance, we had $P = 5000$ and $r = .03$ (so $1 + r = 1 + .03 = 1.03$); the number of periods (years) was $t = 9$.

EXAMPLE 2

Suppose you borrow \$50 from your friendly neighborhood loan shark, who charges 18% interest per week. How much do you owe after one year (assuming that he lets you wait that long to pay)?

SOLUTION   You use the compound interest formula with $P = 50$, $r = .18$, and $t = 52$ (because interest is compounded weekly and there are 52 weeks in a year). So you figure that you owe

$$A = P(1 + r)^t = 50(1 + .18)^{52} = 50 \cdot 1.18^{52} = \$273,422.58.^*$$

When you try to pay the loan shark this amount, however, he points out that a 365-day year has more than 52 weeks, namely, $\dfrac{365}{7} = 52\dfrac{1}{7}$ weeks. So you recalculate with $t = 365/7$ (and careful use of parentheses, as shown in Figure 5–9) and find that you actually owe

$$A = P(1 + r)^t = 50(1 + .18)^{365/7} = 50 \cdot 1.18^{365/7} = \$279,964.68.$$

Ouch! ■

```
50*1.18^(365/7)
       279964.6785
■
```

Figure 5–9

As Example 2 illustrates, the compound interest formula can be used even when the number of periods t is not an integer. You must also learn how to read "financial language" to apply the formula correctly, as shown in the following example.

---

*Here and below, all financial answers are rounded to the nearest penny.

EXAMPLE 3

Determine the amount a $3500 investment is worth after three and a half years at the following interest rates:

(a) 6.4% compounded annually;

(b) 6.4% compounded quarterly;

(c) 6.4% compounded monthly.

SOLUTION

(a) Using the compound interest formula with $P = 3500$, $r = .064$, and $t = 3.5$, we have

$$A = 3500(1 + .064)^{3.5} = \$4348.74.$$

(b) "6.4% interest, compounded quarterly" means that the interest period is one-fourth of a year and the interest rate per period is $.064/4 = .016$. Since there are four interest periods per year, the number of periods in 3.5 years is $4(3.5) = 14$, so

$$A = 3500\left(1 + \frac{.064}{4}\right)^{14} = 3500(1 + .016)^{14} = \$4370.99.$$

(c) Similarly, "6.4% compounded monthly" means that the interest period is one month (1/12 of a year) and the interest rate per period is $.064/12$. The number of periods (months) in 3.5 years is 42, so

$$A = 3500\left(1 + \frac{.064}{12}\right)^{42} = \$4376.14.$$

Note that the more often interest is compounded, the larger is the final amount. ∎

EXAMPLE 4

If $5000 is invested at 6.5% annual interest, compounded monthly, how long will it take for the investment to double?

SOLUTION

The compound interest formula (with $P = 5000$ and $r = .065/12$) shows that the amount in the account after t months is $5000\left(1 + \frac{.065}{12}\right)^{t}$. We must find the value of t such that

$$5000\left(1 + \frac{.065}{12}\right)^{t} = 10,000.$$

Algebraic methods for solving this equation will be considered in Section 5.5. For now, we use technology.

GRAPHING EXPLORATION

Solve the equation, either by using an equation solver or by graphical means as follows. Graph

$$y = 5000\left(1 + \frac{.065}{12}\right)^t - 10,000$$

in a viewing window with $0 \le t \le 240$ (that's 20 years) and find the t-intercept.

The exploration shows that it will take 128.3 months (approximately 10.7 years) for the investment to double. ∎

EXAMPLE 5

What interest rate, compounded annually, is needed for a $16,000 investment to grow to $50,000 in 18 years?

SOLUTION   In the compound interest formula, we have $A = 50,000$, $P = 16,000$ and $t = 18$. We must find r in the equation

$$16,000(1 + r)^{18} = 50,000.$$

The equation can be solved numerically with an equation solver or by one of the following methods.

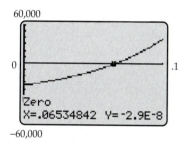

60,000

0   .1

Zero
X=.06534842 Y=-2.9E-8

−60,000

Figure 5–10

Graphical: Rewrite the equation as $16,000(1 + r)^{18} - 50,000 = 0$. Then the solution is the r-intercept of the graph of $y = 16,000(1 + r)^{18} - 50,000$, as shown in Figure 5–10.

Algebraic:

$$16,000(1 + r)^{18} = 50,000$$

Divide both sides by 16,000: $$(1 + r)^{18} = \frac{50,000}{16,000} = 3.125$$

Take 18th roots on both sides: $$\sqrt[18]{(1 + r)^{18}} = \sqrt[18]{3.125}$$

$$1 + r = \sqrt[18]{3.125}$$

$$r = \sqrt[18]{3.125} - 1 \approx .06535.$$

So the necessary interest rate is about 6.535%.   ∎

■■ CONTINUOUS COMPOUNDING AND THE NUMBER *e*

As a general rule, the more often interest is compounded, the better off you are, as we saw in Example 3. But there is, alas, a limit.

EXAMPLE 6   The Number *e*

You have $1 to invest for 1 year. The Exponential Bank offers to pay 100% annual interest, compounded n times per year and rounded to the nearest penny.

You may pick any value you want for n. Can you choose n so large that your \$1 will grow to some huge amount?

SOLUTION Since interest is compounded n times per year and the annual rate is 100% ($= 1.00$), the interest rate per period is $r = 1/n$ and the number of periods in 1 year is n. According to the formula, the amount at the end of the year will be $A = \left(1 + \dfrac{1}{n}\right)^n$. Here's what happens for various values of n:

Interest Is Compounded	$n =$	$\left(1 + \dfrac{1}{n}\right)^n =$
Annually	1	$\left(1 + \frac{1}{1}\right)^1 = 2$
Semiannually	2	$\left(1 + \frac{1}{2}\right)^2 = 2.25$
Quarterly	4	$\left(1 + \frac{1}{4}\right)^4 \approx 2.4414$
Monthly	12	$\left(1 + \frac{1}{12}\right)^{12} \approx 2.6130$
Daily	365	$\left(1 + \frac{1}{365}\right)^{365} \approx 2.71457$
Hourly	8760	$\left(1 + \frac{1}{8760}\right)^{8760} \approx 2.718127$
Every minute	525,600	$\left(1 + \frac{1}{525,600}\right)^{525,600} \approx 2.7182792$
Every second	31,536,000	$\left(1 + \frac{1}{31,536,000}\right)^{31,536,000} \approx 2.7182818$

Since interest is rounded to the nearest penny, your dollar will grow no larger than \$2.72, no matter how big n is. ∎

The last entry in the preceding table, 2.7182818, is the number e to seven decimal places. This is just one example of how e arises naturally in real-world situations. In calculus, it is provided that e is the *limit* of $\left(1 + \dfrac{1}{n}\right)^n$, meaning that as n gets larger and larger, $\left(1 + \dfrac{1}{n}\right)^n$ gets closer and closer to e.

GRAPHING EXPLORATION

Confirm this fact graphically by graphing the function

$$f(x) = \left(1 + \frac{1}{x}\right)^x$$

and the horizontal line $y = e$ in the viewing window with $0 \le x \le 5000$ and $-1 \le y \le 4$ and noting that the two graphs appear to be identical.

When interest is compounded n times per year for larger and larger values of n, as in Example 6, we say that the interest is **continuously compounded.** In this terminology, Example 6 says that \$1 will grow to \$2.72 in 1 year at an interest rate of 100% compounded continuously. A similar argument with more realistic interest rates (see Exercise 42) produces the following result (Example 6 is the case when $P = 1$, $r = 1$, and $t = 1$).

Continuous Compounding

> If P dollars is invested at interest rate r, compounded continuously, then the amount A after t years is
>
> $$A = Pe^{rt}.$$

EXAMPLE 7

If $3800 is invested at 9.2% interest, compounded continuously, find:

(a) The amount in the account after seven and a half years.

(b) The number of years for the account balance to reach $5000.

SOLUTION

(a) Apply the continuous compounding formula with $P = 3800$, $r = .092$, and $t = 7.5$.

$$A = 3800e^{(.092)7.5} = 3800e^{.69} \approx \$7576.12.$$

(b) We must solve the equation

$$3800e^{.092t} = 5000, \qquad \text{or, equivalently,} \qquad 3800e^{.092t} - 5000 = 0.$$

EXPLORATION

Solve the equation graphically and verify that it will take just a few days less than three years for the investment to be worth $5000.   ■

▦ EXPONENTIAL GROWTH

The rules of exponential growth functions are found in the same way that the formula for compound interest was developed.

EXAMPLE 8

The world population in 1980 was about 4.5 billion people and has been increasing at approximately 1.5% per year. Find the rule of the function that gives the world population in year x, where $x = 0$ corresponds to 1980.

SOLUTION

From one year to the next, the population increases by 1.5%. If P is the population at the start of a year, then the population at the end of that year is

$$P + 1.5\% \text{ of } P = P + .015P = P(1 + .015) = P(1.015).$$

So the population increases by a factor of 1.015 every year.

$$4.5 \underset{\text{Year 1}}{\rightarrow} 4.5 \cdot 1.015 \underset{\text{Year 2}}{\rightarrow} \underbrace{[4.5 \cdot 1.015]1.015}_{4.5 \,\cdot\, 1.015^2} \underset{\text{Year 3}}{\rightarrow} \underbrace{[4.5 \cdot 1.015 \cdot 1.015]1.015}_{4.5 \,\cdot\, 1.015^3} \rightarrow \cdots.$$

Consequently, the population (in billions) in year x is given by

$$g(x) = 4.5 \cdot 1.015^x.$$

This function was used in Example 6 of Section 5.1.   ■

Note the form of the function in Example 8: The constant 4.5 is the initial population (when $x = 0$), and the number 1.015 is the factor by which the population changes in one year when population grows at a rate of 1.5% per year. This function and the compound interest formula are special cases of the following.

Exponential Growth

Exponential growth can be described by a function of the form

$$f(x) = Pa^x,$$

where $f(x)$ is the quantity at time x, P is the initial quantity (when $x = 0$) and $a > 1$ is the factor by which the quantity changes when x increases by 1.

If the quantity is growing at the rate r per time period, then $a = 1 + r$, and

$$f(x) = Pa^x = P(1 + r)^x.$$

EXAMPLE 9

At the beginning of an experiment, a culture contains 1000 bacteria. Five hours later, there are 7600 bacteria. Assuming that the bacteria grow exponentially, how many will there be after 24 hours?

SOLUTION   The bacterial population is given by $f(x) = Pa^x$, where P is the initial population, a is the change factor, and x is the time in hours. We are given that $P = 1000$, so $f(x) = 1000a^x$. The next step is to determine a. Since there are 7600 bacteria when $x = 5$, we have

$$7600 = f(5) = 1000a^5,$$

so

$$1000\, a^5 = 7600$$

$$a^5 = 7.6$$

$$a = \sqrt[5]{7.6} = (7.6)^{1/5} = (7.6)^{.2}.$$

Therefore, the population function is $f(x) = 1000(7.6^{.2})^x = 1000 \cdot (7.6)^{.2x}$. After 24 hours, the bacteria population will be

$$f(24) = 1000(7.6)^{.2(24)} \approx 16{,}900{,}721.   ■$$

▪▪ EXPONENTIAL DECAY

In some situations, a quantity *decreases* by a fixed factor as time goes on.

EXAMPLE 10

When tap water is filtered through a layer of charcoal and other purifying agents, 30% of the chemical impurities in the water are removed, and 70% remain. If the water is filtered through a second purifying layer, then the amount of impurities remaining is 70% of 70%, that is, $(.7)(.7) = .7^2 = .49$ or 49%. A third layer results in $.7^3$ of the impurities remaining. Thus, the function

$$f(x) = .7^x$$

gives the percentage of impurities remaining in the water after it passes through x layers of purifying material. How many layers are needed to ensure that 95% of the impurities are removed from the water?

SOLUTION   If 95% of the impurities are removed, then 5% will remain. Hence, we must find x such that $f(x) = .05$, that is, we must solve the equation $.7^x = .05$. This can be done numerically or graphically. Figure 5–11 shows that the solution is $x \approx 8.4$, so 8.4 layers of material are needed.

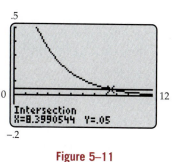

Figure 5–11

Intersection
X=8.3990544 Y=.05

GRAPHING EXPLORATION

How many layers are needed to ensure that 99% of the impurities are removed? ■

Example 10 illustrates **exponential decay.** Note that the impurities were removed at a rate of $30\% = .3$ and that the amount of impurities remaining in the water was changing by a factor of $1 - .30 = .7$. The same thing is true in the general case.

Exponential Decay

Exponential decay can be described by a function of the form

$$f(x) = Pa^x,$$

where $f(x)$ is the quantity at time x, P is the initial quantity (when $x = 0$) and $0 < a < 1$. Here, a is the factor by which the quantity changes when x increases by 1.

If the quantity is decaying at the rate r per time period, then $a = 1 - r$, and

$$f(x) = Pa^x = P(1 - r)^x.$$

One of the important uses of exponential functions is to describe radioactive decay. The **half-life** of a radioactive element is the time it takes a given quantity to decay to one-half of its original mass. The half-life depends only on the substance and not on the size of the sample. Exercise 61 proves the following result.

Radioactive Decay

The mass $M(x)$ of a radioactive element at time x is given by

$$M(x) = c(.5^{x/h}),$$

where c is the original mass and h is the half-life of the element.

EXAMPLE 11

When a living organism dies, its carbon-14 decays exponentially. An archeologist determines that the skeleton of a mastodon has lost 64% of its carbon-14.* Use the fact that the half-life of carbon-14 is 5730 years to estimate how long ago the mastodon died.

SOLUTION   The amount of carbon-14 in the skeleton at time x is given by $f(x) = P(.5^{x/5730})$, where $x = 0$ corresponds to the death of the mastodon. Since the skeleton has lost 64% of its carbon-14, we know that the amount in the skeleton now is 36% of P, that is, $.36P$. So we must find the value of x for which

$$f(x) = .36P$$

Use the rule of f: $$P(.5^{x/5730}) = .36P$$

Divide both sides by P: $$.5^{x/5730} = 36.$$

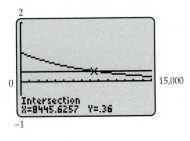

Figure 5–12

To solve this equation, we graph $y_1 = .5^{x/5730}$ and $y_2 = .36$ on the same screen and find the x-coordinate of the intersection point. Figure 5–12 shows that $x \approx 8445.63$. Therefore, the mastodon died about 8446 years ago. ∎

✓ EXERCISES 5.2

1. If $1,000 is invested at 8%, find the value of the investment after 5 years if interest is compounded
(a) annually.   (b) quarterly.   (c) monthly.
(d) weekly.

2. If $2500 is invested at 11.5%, what is the value of the investment after 10 years if interest is compounded
(a) annually?   (b) monthly?   (c) daily?

In Exercises 3–12, determine how much money will be in a savings account if the initial deposit was $500 and the interest rate is:

3. 2% compounded annually for 8 years.

4. 2% compounded annually for 10 years.

5. 2% compounded quarterly for 10 years.

6. 2.3% compounded monthly for 9 years.

7. 2.9% compounded daily for 8.5 years.

8. 3.5% compounded weekly for 7 years and 7 months.

9. 3% compounded continuously for 4 years.

10. 3.5% compounded continuously for 10 years.

11. 2.45% compounded continuously for 6.2 years.

12. 3.25% compounded continuously for 11.6 years.

*A sum of money P that can be deposited today to yield some larger amount A in the future is called the **present value** of A. In Exercises 13–18, find the present value of the given amount A. [Hint: Substitute A, the interest rate per period r, and the number t of periods in the compound interest formula and solve for P.]*

13. $5000 at 6% compounded annually for 7 years.

14. $3500 at 5.5% compounded annually for 4 years.

15. $4800 at 7.2% compounded quarterly for 5 years.

16. $7400 at 5.9% compounded quarterly for 8 years.

17. $8900 at 11.3% compounded monthly for 3 years.

18. $9500 at 9.4% compounded monthly for 6 years.

In Exercises 19–36, use the compound interest formula. In most cases, you will be given three of the quantities A, P, r, t and will have to solve an equation to find the remaining one.

19. A typical credit card company charges 18% annual interest, compounded monthly, on the unpaid balance. If your current balance is $520 and you don't make any payments for 6 months, how much will you owe (assuming they don't sue you in the meantime)?

---

*The technique involves measuring the ratio of the radioactive isotope of carbon, carbon-14, to ordinary nonradioactive carbon-12 in the skeleton.

20. When his first child was born a father put $3000 in a savings account that pays 4% annual interest, compounded quarterly. How much will be in the account on the child's 18th birthday?

21. Mary Alice has $10,000 to invest for two years. Fund *A* pays 13.2% interest, compounded annually. Fund *B* pays 12.7% interest, compounded quarterly. Fund *C* pays 12.6% interest, compounded monthly. Which fund will return the most money?

22. If you invest $7400 for five years, are you better off with an interest rate of 5% compounded quarterly or 4.8% compounded continuously?

23. If you borrow $1200 at 14% interest, compounded monthly, and pay off the loan (principal and interest) at the end of two years, how much interest will you have paid?

24. A developer borrows $150,000 at 6.5% interest, compounded quarterly, and agrees to pay off the loan in four years. How much interest will she owe?

25. A manufacturer has settled a lawsuit out of court by agreeing to pay $1.5 million four years from now. At this time, how much should the company put in an account paying 6.4% annual interest, compounded monthly, to have $1.5 million in four years? [*Hint:* See Exercises 13–18.]

26. Darlene Amidon-Brent wants to have $30,000 available in five years for a down payment on a house. She has inherited $25,000. How much of the inheritance should be invested at 5.7% annual interest, compounded quarterly, to accumulate the $30,000?

27. You win a contest and have a choice of prizes. You can take $3000 now, or you can receive $4000 in four years. If money can be invested at 6% interest, compounded annually, which prize is more valuable in the long run?

28. If money can be invested at 7% compounded quarterly, which is worth more: $9000 now or $12,500 in five years?

29. If an investment of $1000 grows to $1407.10 in seven years with interest compounded annually, what is the interest rate?

30. If an investment of $2000 grows to $2700 in three and a half years, with an annual interest rate that is compounded quarterly, what is the annual interest rate?

31. If you put $3000 in a savings account today, what interest rate (compounded annually) must you receive to have $4000 after five years?

32. If interest is compounded continuously, what annual rate must you receive if your investment of $1500 is to grow to $2100 in six years?

33. At an interest rate of 8% compounded annually, how long will it take to double an investment of
(a) $100   (b) $500   (c) $1200?
(d) What conclusion about doubling time do parts (a)–(c) suggest?

34. At an interest rate of 6% compounded annually, how long will it take to double an investment of *P* dollars?

35. How long will it take to double an investment of $500 at 7% annual interest, compounded continuously?

36. How long will it take to triple an investment of $5000 at 8% annual interest, compounded continuously?

Exercises 37–40 deal with zero-coupon (noninterest-bearing) bonds.

37. A zero-coupon bond is redeemable for $10,000 in five years. If an investor wants a return of 5% compounded yearly on her money, how much should she offer to pay for the bond? [*Hint:* In other words, what is the present value of the bond under these conditions? See Exercises 13–18.]

38. What price should the investor in Exercise 37 pay for the bond if she wants a return of 6% compounded monthly?

39. A father plans to purchase a $25,000 zero-coupon bond for his daughter as a college graduation present four years from now. (We should all have such nice fathers.) If he wants his investment to earn at least 4% compounded daily, what is the maximum price he should pay for the bond? [Assume a 365-day year.]

40. The father in Exercise 39 actually paid 21,500 for the bond. What was the return on his investment?

41. (a) Suppose *P* dollars is invested for one year at 12% interest compounded quarterly. What interest rate *r* would yield the same amount in one year with annual compounding? *r* is called the **effective rate of interest.** [*Hint:* Solve the equation $P(1 + .12/4)^4 = P(1 + r)$ for *r*. The left side of the equation is the yield after one year at 12% compounded quarterly and the right side is the yield after one year at *r*% compounded annually.]
(b) Fill the blanks in the following table.

12% Compounded	Effective Rate
Annually	12%
Quarterly	
Monthly	
Daily	

42. This exercise provides an illustration of why the continuous compounding formula (page 385) is valid, using a realistic interest rate. We shall determine the value of $4000 deposited for three years at 5% interest compounded n times per year for larger and larger values of n. In this case, the interest rate per period is $.05/n$ and the number of periods in three years is $3n$, so the amount in the account at the end of three years is

$$A = 4000\left(1 + \frac{.05}{n}\right)^{3n} = 4000\left[\left(1 + \frac{.05}{n}\right)^{n}\right]^{3}.$$

(a) Fill in the missing entries in the following table.

n	$\left(1 + \dfrac{.05}{n}\right)^{n}$
1,000	
10,000	
500,000	
1,000,000	
5,000,000	
10,000,000	

(b) Compare the entries in the second column of the table with the number $e^{.05}$ and fill the blank in the following sentence:

As n gets larger and larger, the value of $\left(1 + \dfrac{.05}{n}\right)^{n}$

gets closer and closer to the number _____.

(c) Use you answer to part (b) to fill the blank in the following sentence:

As n gets larger and larger, the value of

$$A = 4000\left[\left(1 + \frac{.05}{n}\right)^{n}\right]^{3}$$

gets closer and closer to _____.

(d) Compare your answer in part (c) to the value of the investment given by the continuous compounding formula.

43. A weekly census of the tree-frog population in Frog Hollow State Park produces the following results.

Week	1	2	3	4	5	6
Population	18	54	162	486	1458	4374

(a) Find a function of the form $f(x) = Pa^x$ that describes the frog population at time x weeks.
(b) What is the growth factor in this situation (that is, by what number must this week's population be multiplied to obtain next week's population)?

(c) Each tree frog requires 10 square feet of space and the park has an area of 6.2 square miles. Will the space required by the frog population exceed the size of the park in 12 weeks? In 14 weeks? [Remember: 1 square mile = 5280^2 square feet.]

44. The fruit fly population in a certain laboratory triples every day. Today there are 200 fruit flies.
(a) Make a table showing the number of fruit flies present for the first four days (today is day 0, tomorrow is day 1, etc.).
(b) Find a function of the form $f(x) = Pa^x$ that describes the fruit fly population at time x days.
(c) What is the growth factor here (that is, by what number must each day's population be multiplied to obtain the next day's population)?
(d) How many fruit flies will there be a week from now?

45. The population of Mexico was 100.4 million in 2000 and is expected to grow at the rate of 1.4% per year.
(a) Find the rule of the function f that gives Mexico's population (in millions) in year x, with $x = 0$ corresponding to 2000.
(b) Estimate Mexico's population in 2010.
(c) When will the population reach 125 million people?

46. The United States imported $15 billion in goods from China in 1990, and imports have been growing at the rate of 19.3% per year.
(a) Find the rule of a function that gives the value of U.S. imports from China (in billions of dollars) in year x, with $x = 0$ corresponding to 1990.
(b) Approximately how much did the United States import from China in 2002?
(c) If this model remains accurate, when will imports reach $360 billion?

47. Average annual expenditure per pupil in public elementary and secondary schools was $4966 in 1989–1990 and has been increasing at about 3.9% a year.[*]
(a) Write the rule of a function that gives the expenditure per pupil in year x, where $x = 0$ corresponds to the 1989–1990 school year.
(b) According to this model, what are the expenditures per pupil in 2003–2004?
(c) In what year do expenditures first exceed $9000 per pupil?

48. There were four bald eagle nests in Ohio in 1979. Because of a restoration program undertaken by the Ohio Department of Natural Resources, this number has been increasing about 13.7% per year.
(a) Find the rule of a function that gives the approximate number of eagle nests in Ohio in year x, where $x = 0$ corresponds to 1979.
(b) How many nests were there in 2003?
(c) If this growth continues, when will there be 150 nests?

[*]Based on data from the National Education Association.

49. The U.S. Census Bureau estimates that the Hispanic population in the United States will increase from 32.44 million in 2000 to 98.23 million in 2050.[*][*]

(a) Find an exponential function that gives the Hispanic population in year x, with $x = 0$ corresponding to 2000.

(b) What is the projected Hispanic population in 2010 and 2025?

(c) In what year will the Hispanic population reach 55 million?

50. The U.S. Department of Commerce estimated that there were 56.8 million Internet users in the United States in 1997 and 157.6 million in 2002.

(a) Find an exponential function that models the number of Internet users in year x, with $x = 0$ corresponding to 1997.

(b) For how long is this model likely to remain accurate? [*Hint:* The current U.S. population is about 294 million.]

51. At the beginning of an experiment, a culture contains 200 *H. pylori* bacteria. An hour later there are 205 bacteria. Assuming that the *H. pylori* bacteria grow exponentially, how many will there be after 10 hours? After 2 days?

52. The population of India was approximately 1030 million in 2001 and was 865 million a decade earlier. If the population continues to grow exponentially at the same rate, what will it be in 2006?

53. Kerosene is passed through a pipe filled with clay to remove various pollutants. Each foot of pipe removes 25% of the pollutants.

(a) Write the rule of a function that gives the percentage of pollutants remaining in the kerosene after it has passed through x feet of pipe. [See Example 10.]

(b) How many feet of pipe are needed to ensure that 90% of the pollutants have been removed from the kerosene?

54. If inflation runs at a steady 3% per year, then the amount a dollar is worth decreases by 3% each year.

(a) Write the rule of a function that gives the value of a dollar in year x.

(b) How much will the dollar be worth in 5 years? In 10 years?

(c) How many years will it take before today's dollar is worth only a dime?

55. Many people make minimum monthly payments on their credit card balances and wonder why they aren't getting ahead. If you owe $800 on your credit card and you make a minimal payment of 2% of the balance each month, how long will it take to reduce your balance to $100? Make the unlikely assumption that there is no interest—if there were, it would take much longer. [*Hint:* If your balance is P dollars, then after your payment, what is it?]

56. As the number of transistors per chip has increased (see Example 8 in Section 5.1), the average price of a transistor has dropped exponentially, from $1 in 1972 to a fraction of a cent in 2002, namely $.0000001. Assuming that this trend continues, estimate the average transistor price in 2006.

57. (a) The half-life of radium is 1620 years. Find the rule of the function that gives the amount remaining from an initial quantity of 100 milligrams of radium after x years.

(b) How much radium is left after 800 years? After 1600 years? After 3200 years?

58. (a) The half-life of polonium-210 is 140 days. Find the rule of the function that gives the amount of polonium-210 remaining from an initial 20 milligrams after t days.

(b) How much polonium-210 is left after 15 weeks? After 52 weeks?

(c) How long will it take for the 20 milligrams to decay to 4 milligrams?

59. How old is a piece of ivory that has lost 58% of its carbon-14? [See Example 11.]

60. How old is a mummy that has lost 49% of its carbon-14?

61. This exercise provides a justification for the claim that the function $M(x) = c(.5)^{x/h}$ gives the mass after x years of a radioactive element with half-life h years. Suppose we have c grams of an element that has a half-life of 50 years. Then after 50 years, we would have $c\left(\frac{1}{2}\right)$ grams. After another 50 years, we would have half of that, namely, $c\left(\frac{1}{2}\right)\left(\frac{1}{2}\right) = c\left(\frac{1}{2}\right)^2$.

(a) How much remains after a third 50-year period? After a fourth 50-year period?

(b) How much remains after t 50-year periods?

(c) If x is the number of years, then $x/50$ is the number of 50-year periods. By replacing the number of periods t in part (b) by $x/50$, you obtain the amount remaining after x years. This gives the function $M(x)$ when $h = 50$. The same argument works in the general case (just replace 50 by h).

---

[*] *Statistical Abstracts of the United States* 2002.

5.3 Common and Natural Logarithmic Functions

Roadmap

We begin with common and natural logarithms because they are the most widely used. Instructors who prefer to begin with logarithms to an arbitrary base b should cover Special Topics 5.4.A before covering this section.

Instructors who want to introduce logarithmic functions as the inverse functions of exponential functions should first cover Section 5.7.

From their invention in the 17th century until the development of computers and calculators, logarithms were the only effective tool for numerical computation in the natural sciences and engineering. Although they are no longer needed for computation, logarithmic functions still play an important role in science and engineering. In this section, we examine the two most important types of logarithms: those to base 10 and those to base e.

▪▪ COMMON LOGARITHMS

The basic idea of logarithms can be seen from an example. We'll use the number 704. The *logarithm* of 704 is denoted log 704 and is defined to be the solution of the equation $10^x = 704$. This number can be approximated in two ways.

> Solve the equation $10^x = 704$ graphically (that is, find the intersection of $y = 10^x$ and $y = 704$), as in Figure 5–13,

or

> Use the LOG key on your calculator, as in Figure 5–14.

Except for rounding, you get the same answer.

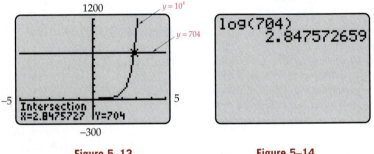

Figure 5–13   Figure 5–14

Figure 5–15

The fact that the number log 704 is the solution of $10^x = 704$ means that

> log 704 is *the exponent to which* 10 *must be raised to produce* 704.

Similarly, the solution of $10^x = 56$ is the number log 56. Figure 5–15 shows that log 56 ≈ 1.748188027 and that 10 raised to this exponent is indeed 56.

More generally, whenever c is a positive real number, the horizontal line $y = c$ lies above the x-axis and hence intersects the graph of $y = 10^x$, so that the equation $10^x = c$ has a solution. Consequently, we have the following definition.

Definition of Common Logarithms

If $c > 0$, then the *common logarithm* of c, denoted log c, is the solution of the equation $10^x = c$.* In other words,

> log c is the exponent to which 10 must be raised to produce c.

---

*The word "common" is usually omitted except when it is necessary to distinguish these logarithms from other types that are introduced below.

EXAMPLE 1

Without using a calculator, find:

(a) log 1000;   (b) log 1:   (c) log 10.

SOLUTION

(a) To find log 1000, ask yourself "what power of 10 equals 1000?" The answer is 3 because $10^3 = 1000$. Therefore, log 1000 = 3.

(b) To what power must 10 be raised to produce 1? Since $10^0 = 1$, we conclude that log 1 = 0.

(c) Log $\sqrt{10} = 1/2$ because $1/2$ is the exponent to which 10 must be raised to produce $\sqrt{10}$, that is, $10^{1/2} = \sqrt{10}$. ■

A calculator is necessary to find most logarithms, but even then you should proceed as in Example 1 to get a rough estimate. For instance, log 795 is the exponent to which 10 must be raised to produce 795. Since $10^2 = 100$ and $10^3 = 1000$, this exponent must be between 2 and 3, that is, $2 < \log 795 < 3$.

Since logarithms are exponents, every statement about logarithms is equivalent to a statement about exponents. For instance, "log $v = u$" means "u is the exponent to which 10 must be raised to produce v," or in symbols, "$10^u = v$." In other words,

Logarithmic and Exponential Equivalence

Let u and v be real numbers with $v > 0$. Then

$$\log v = u \quad \text{exactly when} \quad 10^u = v.$$

Note for those who have read Section 5.7.   If you apply the statement in the preceding box to the functions $g(x) = \log x$ and $f(x) = 10^x$, you obtain

$$\log v = u \quad \text{exactly when} \quad 10^u = v$$

$$g(v) = u \quad \text{exactly when} \quad f(u) = v.$$

Thus, the box says that the common logarithm function $g(x) = \log x$ is the *inverse function* of the exponential function $f(x) = 10^x$.

EXAMPLE 2

Translate each of the following logarithmic statements into an equivalent exponential statement.

$$\log 29 = 1.4624 \quad \log .47 = -.3279 \quad \log (k + t) = d.$$

SOLUTION   Using the preceding box, we have these translations.

Logarithmic Statement	Equivalent Exponential Statement
$\log 29 = 1.4624^*$	$10^{1.4624} = 29$
$\log .47 = -.3279$	$10^{-.3279} = .47$
$\log (k + t) = d$	$10^d = k + t$ ■

EXAMPLE 3

Translate each of the following exponential statements into an equivalent logarithmic statement:

$$10^{5.5} = 316{,}227.766 \qquad 10^{.66} = 4.5708819 \qquad 10^{rs} = t$$

SOLUTION   Translate as follows.

Exponential Statement	Equivalent Logarithmic Statement
$10^{5.5} = 316{,}277.766$	$\log 316{,}227.766 = 5.5$
$10^{.66} = 4.5708819$	$\log 4.5708819 = .66$
$10^{rs} = t$	$\log t = rs$ ■

EXAMPLE 4

Solve the equation $\log x = 4$.

SOLUTION $\log x = 4$ is equivalent to $10^4 = x$. So the solution is $x = 10{,}000$.
■

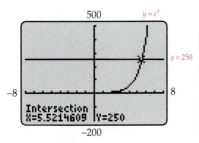

Figure 5–16

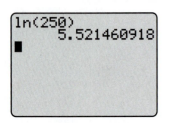

Figure 5–17

▪▪ NATURAL LOGARITHMS

Common logarithms are closely related to the exponential function $f(x) = 10^x$. With the advent of calculus, however, it became clear that the most useful exponential function in science and engineering is $g(x) = e^x$. Consequently, a new type of logarithm, based on the number e instead of 10, was developed. This development is essentially a carbon copy of what was done above, with some minor changes in notation.

For example, the *natural logarithm* of 250 is denoted ln 250 and is defined to be the solution of the equation $e^x = 250$. This number can be approximated in two ways.

Solve the equation $e^x = 250$ graphically (that is, find the intersection of $y = e^x$ and $y = 250$), as in Figure 5–16,

or

Use the LN key on your calculator, as in Figure 5–17.

Except for rounding, you get the same answer.

---

*Here and below, all logarithms are rounded to four decimal places and an equal sign is used rather than the more correct "approximately equal."

The fact that the number ln 250 is the solution of $e^x = 250$ means that

ln 250 is *the exponent to which e must be raised to produce* 250.

The same procedure works for any positive number.

EXPLORATION

Find ln 87. Then raise e to this exponent. Except possibly for rounding, the answer will be 87.

More generally, we have the following definition.

Definition of Natural Logarithms

> If $c > 0$, then the *natural logarithm* of c, denoted ln c, is the solution of the equation $e^x = c$. In other words,
>
> ln c is the exponent to which e must be raised to produce c.

Since natural logarithms are exponents, every statement about them is equivalent to a statement about exponents to the base e. For instance, "ln $v = u$" means "u is the exponent to which e must be raised to produce v," or in symbols, "$e^u = v$." In other words, we have the following.

Logarithmic and Exponential Equivalence

> Let u and v be real numbers, with $v > 0$. Then
>
> $$\ln v = u \qquad \text{exactly when} \qquad e^u = v.$$

Note for those who have read Section 5.7.   The situation here is the same as with common logarithms. If you apply the statement in the preceding box to the functions $g(x) = \ln x$ and $f(x) = e^x$, you obtain

$$\ln v = u \qquad \text{exactly when} \qquad e^u = v$$
$$g(v) = u \qquad \text{exactly when} \qquad f(u) = v.$$

Thus, the box says that the natural logarithm function $g(x) = \ln x$ is the *inverse function* of the exponential function $f(x) = e^x$.

EXAMPLE 5

Translate:

(a) ln 14 = 2.6391 into an equivalent exponential statement.

(b) $e^{5.0626} = 158$ into an equivalent logarithmic statement.

SOLUTION

(a) Using the preceding box, we see that $\ln 14 = 2.6391$ is equivalent to $e^{2.6391} = 14$.

(b) Similarly, $e^{5.0626} = 158$ is equivalent to $\ln 158 = 5.0626$.   ∎

▪▪ PROPERTIES OF LOGARITHMS

Since common and natural logarithms have almost identical definitions (just replace 10 by e), it is not surprising that they have the same essential properties. You don't need a calculator to understand these properties. You need only use the definition of logarithms or translate logarithmic statements into equivalent exponential ones (or vice versa).

EXAMPLE 6

What is $\log(-2)$?

Translation: To what power must 10 be raised to produce -2?

Answer: The graph of $f(x) = 10^x$ (see Figure 5–13) shows that every power of 10 is *positive*. So 10^x can *never* be -2 or any negative number or zero; hence, $\log(-2)$ is not defined. Similarly, $\ln(-2)$ is not defined because every power of e is positive. Therefore,

$$\textbf{ln } \textit{v} \textbf{ and log } \textit{v} \textbf{ are defined only when } \textit{v} > 0. \quad ∎$$

EXAMPLE 7

What is $\ln 1$?

Translation: To what power must e be raised to produce 1?

Answer: We know that $e^0 = 1$, which means that $\ln 1 = 0$. Combining this fact with Example 1(b), we have

$$\textbf{ln 1} = \textbf{0} \quad \textbf{and} \quad \textbf{log 1} = \textbf{0.} \quad ∎$$

EXAMPLE 8

What is $\ln e^9$?

Translation: To what power must e be raised to produce e^9?

Answer: Obviously, the answer is 9. So $\ln e^9 = 9$ and in general

$$\textbf{ln } \textit{e}^{\textit{k}} = \textit{k} \quad \textbf{for every real number } \textit{k.}$$

Similarly,

$$\textbf{log 10}^{\textit{k}} = \textit{k} \quad \textbf{for every real number } \textit{k}$$

because k is the exponent to which 10 must be raised to produce 10^k. In particular, when $k = 1$, we have

$$\ln e = 1 \quad \text{and} \quad \log 10 = 1. \qquad \blacksquare$$

EXAMPLE 9

Find $10^{\log 678}$ and $e^{\ln 678}$.

SOLUTION   By definition, log 678 is the exponent to which 10 must be raised to produce 678. So if you raise 10 to this exponent, the answer will be 678, that is, $10^{\log 678} = 678$. Similarly, ln 678 is the exponent to which e must be raised to produce 678, so that $e^{\ln 678} = 678$. The same argument works with any positive number v in place of 678:

$$e^{\ln v} = v \quad \text{and} \quad 10^{\log v} = v \quad \text{for every } v > 0. \qquad \blacksquare$$

The facts presented in the preceding examples may be summarized as follows.

Properties of Logarithms

Natural Logarithms	Common Logarithms
1. $\ln v$ is defined only when $v > 0$;	$\log v$ is defined only when $v > 0$.
2. $\ln 1 = 0$ and $\ln e = 1$;	$\log 1 = 0$ and $\log 10 = 1$.
3. $\ln e^k = k$ for every real number k;	$\log 10^k = k$ for every real number k.
4. $e^{\ln v} = v$ for every $v > 0$;	$10^{\log v} = v$ for every $v > 0$.

EXAMPLE 10

Applying Property 3 with $k = 2x^2 + 7x + 9$ shows that

$$\ln e^{2x^2+7x+9} = 2x^2 + 7x + 9. \qquad \blacksquare$$

EXAMPLE 11

Solve the equation $\ln (x + 1) = 2$.

SOLUTION   Since $\ln (x + 1) = 2$, we have

$$e^{\ln(x+1)} = e^2.$$

Applying Property 4 with $v = x + 1$ shows that

$$x + 1 = e^{\ln(x+1)} = e^2$$

$$x = e^2 - 1 \approx 6.3891. \qquad \blacksquare$$

Property 4 has another interesting consequence. If a is any positive number, then $e^{\ln a} = a$. Hence, the rule of the exponential function $f(x) = a^x$ can be written as

$$f(x) = a^x = (e^{\ln a})^x = e^{(\ln a)x}.$$

For example, $f(x) = 2^x = e^{(\ln 2)x} \approx e^{.6931x}$. Thus, we have this useful result.

Exponential Functions

Every exponential growth or decay function can be written in the form

$$f(x) = Pe^{kx},$$

where $f(x)$ is the amount at time x, P is the initial quantity, and k is positive for growth and negative for decay.

▓ GRAPHS OF LOGARITHMIC FUNCTIONS

Figure 5–18 shows the graphs of two more entries in the catalog of basic functions: $f(x) = \log x$ and $g(x) = \ln x$. Both are increasing functions with these four properties:

Domain: all positive real numbers;   ***x*-intercept:** 1;
Range: all real numbers;   **Vertical Asymptote:** *y*-axis.

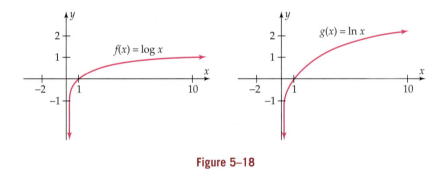

Figure 5–18

Calculators and computers do not accurately show that the *y*-axis is a vertical asymptote of these graphs. By evaluating the functions at very small numbers (such as $x = 1/10^{500}$), you can see that the graphs go lower and lower as x gets closer to 0. On a calculator, however, the graph will appear to end abruptly near the *y*-axis (try it!).

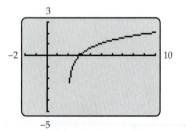

Figure 5–19

EXAMPLE 12

Sketch the graph of $f(x) = \ln (x - 2)$.

SOLUTION   Using a calculator to graph $f(x) = \ln (x - 2)$, we obtain Figure 5–19, in which the graph appears to end abruptly near $x = 2$. Fortunately, however, we have read Section 3.4, so we know that this is *not* how the graph looks. From Section 3.4, we know that the graph of $f(x) = \ln (x - 2)$ is the graph

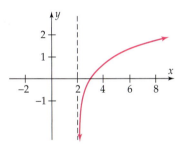

Figure 5–20

of $g(x) = \ln x$ shifted horizontally 2 units to the right, as shown in Figure 5–20. In particular, the graph of f has a vertical asymptote at $x = 2$ and drops sharply downward there. ∎

EXPLORATION

Graph $y_1 = \ln (x - 2)$ and $y_2 = -5$ in the same viewing window and verify that the graphs do not appear to intersect, as they should. Nevertheless, try to solve the equation $\ln (x - 2) = -5$ by finding the intersection point of y_1 and y_2. Some calculators will find the intersection point even though it does not show on the screen. Others produce an error message, in which case the SOLVER feature should be used instead of a graphical solution.

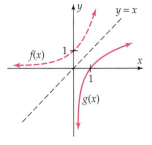

Figure 5–21

EXAMPLE 13

How is the graph of $g(x) = \log x$ related to the graph of $f(x) = 10^x$?

SOLUTION   We graph both functions on the same set of axes, together with the line $y = x$, and obtain Figure 5–21.[*] As the figure suggests, the graph of $g(x) = \log x$ is the mirror image of the graph of $f(x) = 10^x$, with the line $y = x$ being the mirror. In technical terms, the graph of g is the graph of f *reflected in the line $y = x$.*[†] The graphs of $k(x) = \ln x$ and $h(x) = e^x$ are related similarly: The graph of k is the graph of h reflected in the line $y = x$. ∎

Although logarithms are defined only for positive numbers, many logarithmic functions include negative numbers in their domains.

EXAMPLE 14

Find the domain of each of the following functions.

(a) $f(x) = \ln (x + 4)$ (b) $g(x) = \log x^2$

SOLUTION

(a) $f(x) = \ln (x + 4)$ is defined only when $x + 4 > 0$, that is, when $x > -4$. So the domain of f consists of all real numbers greater than -4.

(b) Since $x^2 > 0$ for all nonzero x, the domain of $g(x) = \log x^2$ consists of all real numbers except 0. ∎

EXPLORATION

Verify the conclusions of Example 14 by graphing each of the functions. What is the vertical asymptote of each graph?

---

[*]On a calculator screen, the parts of the graph near the axes may not be visible. Furthermore, to see the situation accurately, you must use the same scale on both axes.
[†]This should be no surprise to those who have read Section 5.7 because they know that $g(x) = \log x$ is the inverse function of $f(x) = 10^x$ and that all inverse functions have this reflection property (see page 446).

Logarithms have a variety of applications. For example, the Value Line Geometric Composite Index uses logarithms to compute the average return for a group of 1700 stocks.* Here is another financial application.

EXAMPLE 15

If you invest money at interest rate r (expressed as a decimal and compounded annually), then the time it takes to double your investment is given by

$$D(r) = \frac{\ln 2}{\ln (1 + r)}.$$

(a) How long will it take to double an investment of $2500 at 6.5%?

(b) What interest rate is needed for the investment in part (a) to double in six years?

SOLUTION

(a) The interest rate is $r = .065$, so the doubling time is

$$D(.065) = \frac{\ln 2}{\ln (1 + .065)} \approx 11 \text{ years.}$$

(b) For the investment to double in six years, we must have $D(r) = 6$. Thus, we must solve the equation

$$\frac{\ln 2}{\ln (1 + r)} = 6.$$

Algebraic solution techniques for such equations will be considered in Section 5.5. For now, we solve the equation graphically by finding the intersection of $y_1 = \dfrac{\ln 2}{\ln (1 + r)}$ and $y_2 = 6$. Figure 5–22 shows that $r \approx .1225$, so an interest rate of about 12.25% is needed to double the investment in six years. ∎

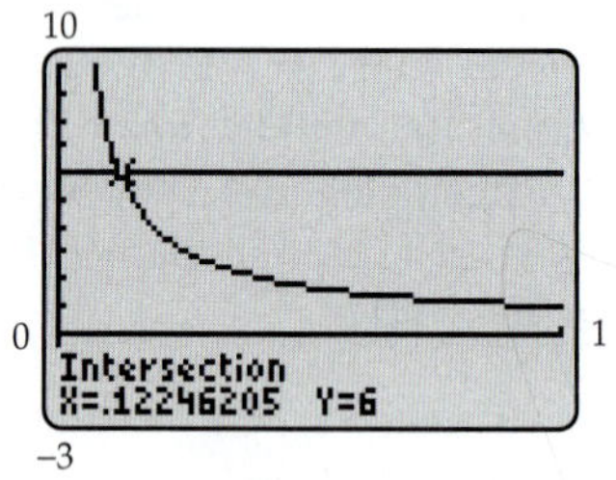

10

0   1

Intersection
X=.12246205 Y=6

−3

Figure 5–22

✓ **EXERCISES 5.3**

Unless stated otherwise, all letters represent positive numbers.

In Exercises 1–4, find the logarithm, without using a calculator.

1. log 10,000

2. log .001

3. $\log \dfrac{\sqrt{10}}{1000}$

4. $\log \sqrt[3]{.01}$

In Exercises 5–14, translate the given logarithmic statement into an equivalent exponential statement.

5. log 1000 = 3

6. log .001 = −3

7. log 750 = 2.88

8. log (.8) = −.097

9. ln 3 = 1.0986

10. ln 10 = 2.3026

11. ln .01 = −4.6052

12. ln $s = r$

13. ln $(x^2 + 2y) = z + w$

14. log $(a + c) = d$

In Exercises 15–24, translate the given exponential statement into an equivalent logarithmic one.

15. $10^{-2} = .01$

16. $10^3 = 1000$

17. $10^{.4771} = 3$

18. $10^{7k} = r$

19. $e^{3.25} = 25.79$

20. $e^{-4} = .0183$

21. $e^{12/7} = 5.5527$

22. $e^k = t$

23. $e^{2/r} = w$

24. $e^{4uv} = m$

---

*New York Times, August 31, 2003.

In Exercises 25–36, evaluate the given expression without using a calculator.

25. $\log 10^{\sqrt{43}}$ **26.** $\log 10^{\sqrt{x^2+y^2}}$ **27.** $\ln e^{15}$

28. $\ln e^{3.78}$ **29.** $\ln \sqrt{e}$ **30.** $\ln \sqrt[5]{e}$

31. $e^{\ln 931}$ **32.** $e^{\ln 34.17}$ **33.** $\ln e^{x+y}$

34. $\ln e^{x^2+2y}$ **35.** $e^{\ln x^2}$ **36.** $e^{\ln \sqrt{x+3}}$

In Exercises 37–40, write the rule of the function in the form $f(x) = Pe^{kx}$. (See the discussion and box after Example 11.)

37. $f(x) = 4(25^x)$ **38.** $g(x) = 3.9(1.03^x)$

39. $g(x) = -16(30.5^x)$ **40.** $f(x) = -2.2(.75^x)$

In Exercises 41–44, find the domain of the given function (that is, the largest set of real numbers for which the rule produces well-defined real numbers).

41. $f(x) = \ln (x + 1)$ **42.** $g(x) = \ln (x + 2)$

43. $h(x) = \log (-x)$ **44.** $k(x) = \log (2 - x)$

45. $f(x) = \log x^3$ **46.** $h(x) = \ln x^4$

47. (a) Graph $y = x$ and $y = e^{\ln x}$ in separate viewing windows [or use a split-screen if your calculator has that feature]. For what values of x are the graphs identical?
    (b) Use the properties of logarithms to explain your answer in part (a).

48. (a) Graph $y = x$ and $y = \ln (e^x)$ in separate viewing windows [or a split-screen if your calculator has that feature]. For what values of x are the graphs identical?
    (b) Use the properties of logarithms to explain your answer in part (a).

In Exercises 49–54, list the transformations that will change the graph of $g(x) = \ln x$ into the graph of the given function. [Section 3.4 may be helpful.]

49. $f(x) = 2 \cdot \ln x$ **50.** $f(x) = \ln x - 7$

51. $h(x) = \ln (x - 4)$ **52.** $k(x) = \ln (x + 2)$

53. $h(x) = \ln (x + 3) - 4$ **54.** $k(x) = \ln (x - 2) + 2$

In Exercises 55–58, sketch the graph of the function.

55. $f(x) = \log (x - 3)$ **56.** $g(x) = 2 \ln x + 3$

57. $h(x) = -2 \log x$ **58.** $f(x) = \ln (-x) - 3$

In Exercises 59–64, find a viewing window (or windows) that shows a complete graph of the function.

59. $f(x) = \dfrac{x}{\ln x}$ **60.** $g(x) = \dfrac{\ln x}{x}$

61. $h(x) = \dfrac{\ln x^2}{x}$ **62.** $k(x) = e^{2/\ln x}$

63. $f(x) = 10 \log x - x$ **64.** $f(x) = \dfrac{\log x}{x}$

C

In Exercises 65–68, find the average rate of change of the function.

65. $f(x) = \ln (x - 2)$, as x goes from 3 to 5.

66. $g(x) = x - \ln x$, as x goes from .5 to 1.

67. $g(x) = \log (x^2 + x + 1)$, as x goes from -5 to -3.

68. $f(x) = x \log |x|$, as x goes from 1 to 4.

69. (a) What is the average rate of change of $f(x) = \ln x$, as x goes from 3 to $3 + h$?
    (b) What is the value of h when the average rate of change of $f(x) = \ln x$, as x goes from 3 to $3 + h$, is .25?

70. (a) Find the average rate of change of $f(x) = \ln x^2$, as x goes from .5 to 2.
    (b) Find the average rate of change of $g(x) = \ln (x - 3)^2$, as x goes from 3.5 to 5.
    (c) What is the relationship between your answers in parts (a) and (b) and why is this so?

71. Use the doubling function D of Example 15 for this exercise.
    (a) Find the time it takes to double your money at each of these interest rates: 4%, 6%, 8%, 12%, 18%, 24%, 36%.
    (b) Round the answers in part (a) to the nearest year and compare them with these numbers:

$$72/4, 72/6, 72/8, 72/12, 72/18, 72/24, 72/36.$$

Use this evidence to state a "rule of thumb" for determining approximate doubling time, without using the function D. This rule of thumb, which has long been used by bankers, is called the **rule of 72.**

72. The concentration of hydrogen ions in a given solution is denoted $[H^+]$ and is measured in moles per liter. For example, $[H^+] = .00008$ for beer and $[H^+] = .0004$ for wine. Chemists define the pH of the solution to be the number pH $= -\log [H^+]$. The solution is said to be an *acid* if pH < 7 and a *base* if pH > 7.
    (a) Is beer an acid or a base? What about wine?
    (b) If a solution has a pH of 2, what is its $[H^+]$?
    (c) For hominy, $[H^+] = 5 \cdot 10^{-8}$. Is hominy a base?

73. The life expectancy of a woman born in year x is approximated by

$$f(x) = 17.7 + 30.855 \log x,$$

where $x = 10$ corresponds to 1910.[*]
    (a) What is the life expectancy of a woman born in 1988?
    (b) In what year will the life expectancy of a woman born in that year be 81 years?

---

[*]National Center for Health Statistics.

74. The number of multiple births (babies born as twins, triplets, etc.) each year is approximated by

$$g(x) = 93,201.973 + 10,467.499 \log x,$$

where $x = 1$ corresponds to 1989.
(a) Use this model to estimate the number of multiple births in 2005 and 2008.
(b) If this model remains accurate, in what year will there be 108,000 multiple births?

75. The percentage of first-year college students who are male is approximated by

$$M(x) = 53.145 - 6.185 \log x,$$

where $x = 5$ corresponds to 1985.[*]
(a) Estimate the percentage of male first-year students in 2006.
(b) If this model remains accurate, in what year will the percentage of male first-year students drop below 44%?

76. The height h above sea level (in miles) is related to the barometric pressure p (in inches of mercury) by the equation

$$h = -5 \ln\left(\frac{p}{29.92}\right).$$

If a weather balloon records the following pressures, how high is the balloon?
(a) 29.91 inches
(b) 20.04 inches
(c) 11.89 inches
(d) What is the pressure at a height of 4 miles?

77. Suppose $f(x) = A \ln x + B$, where A and B are constants. If $f(1) = 10$ and $f(e) = 1$, what are A and B?

78. (Section 5.7 is a prerequisite for this exercise.) Show that

$$g(x) = \ln\left(\frac{x}{1-x}\right) \text{ is the inverse function of}$$

$$f(x) = \frac{1}{1 + e^{-x}}.$$

79. The height h above sea level (in meters) is related to air temperature t (in degrees Celsius), the atmospheric pressure p (in centimeters of mercury at height h), and the atmospheric pressure c at sea level by

$$h = (30t + 8000) \ln (c/p).$$

If the pressure at the top of Mount Rainier is 44 centimeters on a day when sea level pressure is 75.126 centimeters and the temperature is 7°C, what is the height of Mount Rainier?

80. Mount Everest is 8850 meters high. What is the atmospheric pressure at the top of the mountain on a day when the temperature is −25°C and the atmospheric pressure at sea level is 75 centimeters? [See Exercise 79.]

81. Beef consumption in the United States (in billions of pounds) in year x can be approximated by the function

$$f(x) = -154.41 + 39.38 \ln x \qquad (x \geq 90).$$

where $x = 90$ corresponds to 1990.[†]
(a) How much beef was consumed in 1999 and in 2002?
(b) When will beef consumption reach 30 billion pounds per year?

82. Students in a college algebra class were given a final exam. Each month thereafter, they took an equivalent exam. The class average on the exam taken after t months is given by

$$F(t) = 82 - 8 \cdot \ln (t + 1).$$

(a) What was the class average after six months?
(b) After a year?
(c) When did the class average drop below 55?

83. One person with a flu virus visited the campus. The number T of days it took for the virus to infect x people was given by:

$$T = -.93 \ln\left[\frac{7000 - x}{6999x}\right].$$

(a) How many days did it take for 6000 people to become infected?
(b) After two weeks, how many people were infected?

84. The population of St. Petersburg, Florida (in thousands) can be approximated by the function

$$g(x) = -127.9 + 81.91 \ln x \qquad (x \geq 70),$$

where $x = 70$ corresponds to 1970.
(a) Estimate the population in 1995 and 2003.
(b) If this model remains accurate, when will the population be 260,000?

85. A bicycle store finds that the number N of bikes sold is related to the number d of dollars spent on advertising by $N = 51 + 100 \cdot \ln (d/100 + 2)$.
(a) How many bikes will be sold if nothing is spent on advertising? If $1000 is spent? If $10,000 is spent?
(b) If the average profit is $25 per bike, is it worthwhile to spend $1000 on advertising? What about $10,000?
(c) What are the answers in part (b) if the average profit per bike is $35?

---

[*]Based on data from the Higher Education Research Institute at UCLA.

[†]Based on data from the U.S. Department of Agriculture.

86. **Approximating Logarithmic Functions by Polynomials.** For each positive integer n, let f_n be the polynomial function whose rule is

$$f_n(x) = x - \frac{x^2}{2} + \frac{x^3}{3} - \frac{x^4}{4} + \frac{x^5}{5} - \cdots \pm \frac{x^n}{n},$$

where the sign of the last term is $+$ if n is odd and $-$ if n is even. In the viewing window with $-1 \leq x \leq 1$ and $-4 \leq y \leq 1$, graph $g(x) = \ln(1 + x)$ and $f_4(x)$ on the same screen. For what values of x does f_4 appear to be a good approximation of g?

87. Using the viewing window in Exercise 86, find a value of n for which the graph of the function f_n (as defined in Exercise 86) appears to coincide with the graph of $g(x) = \ln(1 + x)$. Use the trace feature to move from graph to graph to see how good this approximation actually is.

5.4 Properties of Logarithms

Logarithms have several important properties in addition to those presented in Section 5.3. These properties, which we shall call *logarithm laws,* arise from the fact that logarithms are exponents. Essentially, they are properties of exponents translated into logarithmic language.

The first law of exponents says that $b^m b^n = b^{m+n}$, or in words,

The exponent of a product is the sum of the exponents of the factors.

Since logarithms are just particular kinds of exponents, this statement translates as follows.

The logarithm of a product is the sum of the logarithms of the factors.

Here is the same statement in symbolic language.

Product Law for Logarithms

For all $v, w > 0$,

$$\ln(vw) = \ln v + \ln w,$$

and

$$\log(vw) = \log v + \log w.$$

Proof According to Property 4 of logarithms (in the box on page 397),

$$e^{\ln v} = v \quad \text{and} \quad e^{\ln w} = w.$$

Therefore, by the first law of exponents (with $m = \ln v$ and $n = \ln w$),

$$vw = e^{\ln v} e^{\ln w} = e^{\ln v + \ln w}.$$

So raising e to the exponent ($\ln v + \ln w$) produces vw. But the definition of logarithm says that $\ln vw$ is the exponent to which e must be raised to produce vw. Therefore, we must have $\ln vw = \ln v + \ln w$. A similar argument works for common logarithms (just replace e by 10 and "ln" by "log"). ■

EXAMPLE 1

A calculator shows that $\ln 7 = 1.9459$ and $\ln 9 = 2.1972$. Therefore,

$$\ln 63 = \ln (7 \cdot 9) = \ln 7 + \ln 9 = 1.9459 + 2.1972 = 4.1341. \quad \blacksquare$$

CALCULATOR EXPLORATION

We know that $5 \cdot 7 = 35$. Key in LOG(35) ENTER. Then key in LOG(5) + LOG(7) ENTER. The answers are the same by the Product Law. Do you get the same answer if you key in LOG(5) × LOG(7) ENTER?

EXAMPLE 2

Use the Product Law to write

(a) $\log (7xy)$ as a sum of three logarithms.

(b) $\log x^2 + \log y + 1$ as a single logarithm.

SOLUTION

(a) $\log (7xy) = \log 7x + \log y = \log 7 + \log x + \log y$

(b) Note that $\log 10 = 1$ (why?). Hence,

$$\log x^2 + \log y + 1 = \log x^2 + \log y + \log 10$$
$$= \log (x^2 y) + \log 10$$
$$= \log (10x^2 y). \quad \blacksquare$$

CAUTION

A common error in applying the Product Law for Logarithms is to write the *false* statement

$$\ln 7 + \ln 9 = \ln (7 + 9)$$
$$= \ln 16$$

instead of the correct statement

$$\ln 7 + \ln 9 = \ln (7 \cdot 9)$$
$$= \ln 63.$$

GRAPHING EXPLORATION

Illustrate the Caution in the margin graphically by graphing both

$$f(x) = \ln x + \ln 9 \qquad \text{and} \qquad g(x) = \ln (x + 9)$$

in the standard viewing window and verifying that the graphs are not the same. In particular, the functions have different values at $x = 7$.

The second law of exponents, namely, $b^m / b^n = b^{m-n}$, may be roughly stated in words as follows.

The exponent of the quotient is the difference of exponents.

When the exponents are logarithms, this says

The logarithm of a quotient is the difference of the logarithms.

In other words,

Quotient Law for Logarithms

For all $v, w > 0$,

$$\ln\left(\frac{v}{w}\right) = \ln v - \ln w$$

and

$$\log\left(\frac{v}{w}\right) = \log v - \log w.$$

The proof of the Quotient Law is very similar to the proof of the Product Law (see Exercise 34).

Figure 5–23

EXAMPLE 3

Figure 5–23 illustrates the Quotient Law by showing that

$$\log\left(\frac{297}{39}\right) = \log 297 - \log 39. \quad \blacksquare$$

EXAMPLE 4

For any $w > 0$,

$$\ln\left(\frac{1}{w}\right) = \ln 1 - \ln w = 0 - \ln w = -\ln w,$$

and

$$\log\left(\frac{1}{w}\right) = \log 1 - \log w = 0 - \log w = -\log w. \quad \blacksquare$$

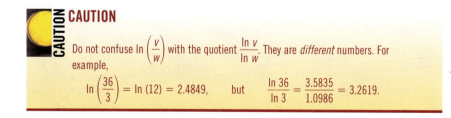

CAUTION

Do not confuse $\ln\left(\dfrac{v}{w}\right)$ with the quotient $\dfrac{\ln v}{\ln w}$. They are *different* numbers. For example,

$$\ln\left(\frac{36}{3}\right) = \ln(12) = 2.4849, \quad \text{but} \quad \frac{\ln 36}{\ln 3} = \frac{3.5835}{1.0986} = 3.2619.$$

GRAPHING EXPLORATION

Illustrate the preceding Caution graphically by graphing both $f(x) = \ln(x/3)$ and $g(x) = (\ln x)/(\ln 3)$ and verifying that the graphs are not the same at $x = 36$ (or anywhere else, for that matter).

The third law of exponents, namely, $(b^m)^k = b^{mk}$, can also be translated into logarithmic language.

Power Law for Logarithms

For all k and all $v > 0$,

$$\ln (v^k) = k(\ln v),$$

and

$$\log (v^k) = k(\log v).$$

Proof   Since $v = 10^{\log v}$ (why?), the third law of exponents (with $b = 10$ and $m = \log v$) shows that

$$v^k = (10^{\log v})^k = 10^{(\log v)k} = 10^{k(\log v)}.$$

So raising 10 to the exponent $k(\log v)$ produces v^k. But the exponent to which 10 must be raised to produce v^k is, by definition, $\log (v^k)$. Therefore, $\log (v^k) = k(\log v)$, and the proof is complete. A similar argument with e in place of 10 and "ln" in place of "log" works for natural logarithms. ■

EXAMPLE 5

Express $\ln \sqrt{19}$ without radicals or exponents.

SOLUTION   First write $\sqrt{19}$ in exponent notation, then use the Power Law:

$$\ln \sqrt{19} = \ln 19^{1/2}$$

$$= \frac{1}{2} \ln 19 \qquad \text{or} \qquad \frac{\ln 19}{2}. \qquad ■$$

EXAMPLE 6

Express as a single logarithm: $\dfrac{\log (x^2 + 1)}{3} - \log x.$

SOLUTION

$$\frac{\log (x^2 + 1)}{3} - \log x = \frac{1}{3} \log (x^2 + 1) - \log x$$

$$= \log (x^2 + 1)^{1/3} - \log x \qquad \text{[Power Law]}$$

$$= \log \sqrt[3]{x^2 + 1} - \log x$$

$$= \log \left(\frac{\sqrt[3]{x^2 + 1}}{x} \right) \qquad \text{[Quotient Law]} \qquad ■$$

EXAMPLE 7

Express as a single logarithm: $\ln 3x + 4 \ln x - \ln 3xy.$

SOLUTION

$$\ln 3x + 4 \cdot \ln x - \ln 3xy = \ln 3x + \ln x^4 - \ln 3xy \qquad \text{[Power Law]}$$

$$= \ln (3x \cdot x^4) - \ln 3xy \qquad \text{[Product Law]}$$

$$= \ln \frac{3x^5}{3xy} \qquad \text{[Quotient Law]}$$

$$= \ln \frac{x^4}{y} \qquad \text{[Cancel } 3x\text{]} \quad \blacksquare$$

EXAMPLE 8

Simplify: $\ln \left(\dfrac{\sqrt{x}}{x} \right) + \ln \sqrt[4]{ex^2}$.

SOLUTION   Begin by changing to exponential notation.

$$\ln \left(\frac{x^{1/2}}{x} \right) + \ln (ex^2)^{1/4} = \ln (x^{-1/2}) + \ln (ex^2)^{1/4}$$

$$= -\frac{1}{2} \cdot \ln x + \frac{1}{4} \cdot \ln ex^2 \qquad \text{[Power Law]}$$

$$= -\frac{1}{2} \cdot \ln x + \frac{1}{4}(\ln e + \ln x^2) \qquad \text{[Product Law]}$$

$$= -\frac{1}{2} \cdot \ln x + \frac{1}{4}(\ln e + 2 \cdot \ln x) \qquad \text{[Power Law]}$$

$$= -\frac{1}{2} \cdot \ln x + \frac{1}{4} \cdot \ln e + \frac{1}{2} \cdot \ln x$$

$$= \frac{1}{4} \cdot \ln e = \frac{1}{4} \qquad \text{[}\ln e = 1\text{]} \quad \blacksquare$$

▪▪ APPLICATIONS

Because logarithmic growth is slow, measurements on a logarithmic scale (that is, on a scale determined by a logarithmic function) can sometimes be deceptive.

EXAMPLE 9   **Earthquakes**

The magnitude $R(i)$ of an earthquake on the Richter scale is given by $R(i) = \log (i/i_0)$, where i is the amplitude of the ground motion of the earthquake and i_0 is the amplitude of the ground motion of the so-called zero earthquake.[*] A moderate earthquake might have 1000 times the ground motion of the zero earthquake (that is, $i = 1000i_0$). So its magnitude would be

$$\log (1000i_0/i_0) = \log 1000 = \log 10^3 = 3.$$

---

[*]The zero earthquake has ground motion amplitude of less than 1 micron on a standard seismograph 100 kilometers from the epicenter.

An earthquake with 10 times this ground motion (that is, $i = 10 \cdot 1000i_0 = 10,000i_0$) would have a magnitude of

$$\log (10,000i_0/i_0) = \log 10,000 = \log 10^4 = 4.$$

So a *tenfold* increase in ground motion produces only a one-point change on the Richter scale. In general,

Increasing the ground motion by a factor of 10^k increases the Richter magnitude by k units. [*]

For instance, the 1989 World Series earthquake in San Francisco measured 7.0 on the Richter scale, and the great earthquake of 1906 measured 8.3. The difference of 1.3 points means that the 1906 quake was $10^{1.3} \approx 20$ times more intense than the 1989 one in terms of ground motion.   ◼

✓ EXERCISES 5.4

In Exercises 1–18, write the given expression as a single logarithm.

1. $\ln x^2 + 3 \ln y$

2. $\ln 2x + 2(\ln x) - \ln 3y$

3. $\log (x^2 - 9) - \log (x + 3)$

4. $\log 3x - 2[\log x - \log (2 + y)]$

5. $2(\ln x) - 3(\ln x^2 + \ln x)$

6. $\ln (e/\sqrt{x}) - \ln \sqrt{ex}$

7. $3 \ln (e^2 - e) - 3$

8. $2 - 2 \log (20)$

9. $\log (10x) + \log (20y) - 1$

10. $\ln (e^2 x) + \ln (ey) - 3$

11. $\log (x - 3) + \log (x + 3)$

12. $\log (x - 3) - \log (x + 3)$

13. $2 \log x - 2[\log (x + 1) + \log (x - 2) + \log x]$

14. $\ln (x - 1) - 2[\ln (x + 5) + \ln x]$

15. $\dfrac{1}{2}[\ln x - \ln (3x + 5)]$

16. $\dfrac{1}{3} \ln x - \dfrac{1}{2}[\ln (x^2 + 1) + \ln x]$

17. $\log (\sqrt{x^2 + 2} + \sqrt{2}) + \log (\sqrt{x^2 + 2} - \sqrt{2}) - \log x$

18. $\ln (\sqrt{x + 1} - \sqrt{x}) + \ln (\sqrt{x + 1} + \sqrt{x})$

In Exercises 19–24, let $u = \ln x$ and $v = \ln y$. Write the given expression in terms of u and v. For example,

$$\ln x^3 y = \ln x^3 + \ln y = 3 \ln x + \ln y = 3u + v.$$

19. $\ln (x^2 y^5)$

20. $\ln (x^3 y^2)$

21. $\ln (\sqrt{x} \cdot y^2)$

22. $\ln \left(\dfrac{\sqrt{x}}{y}\right)$

23. $\ln (\sqrt[3]{x^2 \sqrt{y}})$

24. $\ln \left(\dfrac{\sqrt{x^2 y}}{\sqrt[3]{y}}\right)$

In Exercises 25–30, use graphical or algebraic means to determine whether the statement is true or false.

25. $\ln |x| = |\ln x|$?

26. $\ln \left(\dfrac{1}{x}\right) = \dfrac{1}{\ln x}$?

27. $\log x^5 = 5(\log x)$?

28. $e^{x \ln x} = x^x \quad (x > 0)$?

29. $\ln x^3 = (\ln x)^3$?

30. $\log \sqrt{x} = \sqrt{\log x}$?

In Exercises 31–32, find values of a and b for which the statement is false.

31. $\dfrac{\log a}{\log b} = \log \left(\dfrac{a}{b}\right)$

32. $\log (a + b) = \log a + \log b$

33. If $\ln b^7 = 7$, what is b?

---

[*]*Proof:* If one quake has ground motion amplitude i and the other $10^k i$, then

$$R(10^k i) = \log (10^k i/i_0) = \log 10^k + \log (i/i_0)$$

$$= k + \log (i/i_0) = k + R(i).$$

34. Prove the Quotient Law for Logarithms: For v, $w > 0$, $\ln\left(\dfrac{v}{w}\right) = \ln v - \ln w$. (Use properties of exponents and the fact that $v = e^{\ln v}$ and $w = e^{\ln w}$.)

35. According to the Power Law for Logarithms, $\log x^2 = 2 \log x$. Compare the graphs of $f(x) = \log x^2$ and $g(x) = 2 \log x$. Are they the same? Explain what's going on here.

36. Do Exercise 35 for the functions $f(x) = \log x^3$ and $g(x) = 3 \log x$.

In Exercises 37–40, use the logarithm laws to answer the question. Section 3.4 may be helpful.

37. Why can the graph of $f(x) = \log (5x)$ be obtained by shifting the graph of $g(x) = \log x$ upward? How far up must it be shifted?

38. Why can the graph of $f(x) = \log (x/5)$ be obtained by shifting the graph of $g(x) = \log x$ downward? How far down must it be shifted?

39. Why can the graph of $f(x) = \ln (x^5)$ be obtained by stretching the graph of $g(x) = \ln x$ away from the x-axis? What is the stretching factor?

40. How is the graph of $f(x) = \ln \sqrt{x}$ related to the graph of $g(x) = \ln x$?

41. Compute and simplify the difference quotient of $f(x) = \log x$. Your answer should be a single logarithm.

42. If $f(x) = \ln x$, show that $f\left(\dfrac{1}{x}\right) = -f(x)$.

43. The size of China's automobile market (in millions of vehicles sold per year) is approximated by

$$f(x) = 1.66 + 1.91 \ln x,$$

where $x = 1$ corresponds to 2001.*
(a) How many Chinese vehicles will be sold in 2007?
(b) When will sales of Chinese vehicles reach 6 million?

44. The size of Japan's automobile market (in millions of vehicles sold per year) is approximated by

$$g(x) = \frac{x}{45} + \frac{533}{90},$$

where $x = 1$ corresponds to 2001.* When will the Chinese automobile market be the same size as the Japanese market (see Exercise 43)?

In Exercises 45–48, state the magnitude on the Richter scale of an earthquake that satisfies the given condition.

45. 100 times stronger than the zero quake.

46. $10^{4.7}$ times stronger than the zero quake.

47. 350 times stronger than the zero quake.

48. 2500 times stronger than the zero quake.

Exercises 49–52 deal with the energy intensity i of a sound, which is related to the loudness of the sound by the function $L(i) = 10 \cdot \log (i/i_0)$, where i_0 is the minimum intensity detectable by the human ear and $L(i)$ is measured in decibels. Find the decibel measure of the sound.

49. Ticking watch (intensity is 100 times i_0).

50. Soft music (intensity is 10,000 times i_0).

51. Loud conversation (intensity is 4 million times i_0).

52. Victoria Falls in Africa (intensity is 10 billion times i_0).

53. How much louder is the sound in Exercise 50 than the sound in Exercise 49?

54. The perceived loudness L of a sound of intensity I is given by $L = k \cdot \ln I$, where k is a certain constant. By how much must the intensity be increased to double the loudness? (That is, what must be done to I to produce $2L$?)

Thinkers

55. Compute each of the following pairs of numbers.

(a) $\log 18$ and $\dfrac{\ln 18}{\ln 10}$

(b) $\log 456$ and $\dfrac{\ln 456}{\ln 10}$

(c) $\log 8950$ and $\dfrac{\ln 8950}{\ln 10}$

(d) What do these results suggest?

56. Prove that for any positive number c, $\log c = \dfrac{\ln c}{\ln 10}$. [*Hint:*

We know that $10^{\log c} = c$ (why?). Take natural logarithms on both sides and use a logarithm law to simplify and solve for $\log c$.]

57. Find each of the following logarithms.
(a) $\log 8.753$ (b) $\log 87.53$ (c) $\log 875.3$

(d) $\log 8753$ (e) $\log 87{,}530$

(f) How are the numbers 8.753, 87.53, . . . , 87,530 related to one another? How are their logarithms related? State a general conclusion that this evidence suggests.

58. Prove that for every positive number c, $\log c$ can be written in the form $k + \log b$, where k is an integer and $1 \le b < 10$. [*Hint:* Write c in scientific notation and use logarithm laws to express $\log c$ in the required form.]

---

*Based on forecasts by J. D. Power and Associates.

5.4.A   *SPECIAL TOPICS* Logarithmic Functions to Other Bases*

The same procedure used in Sections 5.3–5.4 can be carried out with any positive number b in place of 10 and e.

Throughout this section, b is a fixed positive number with $b > 1$.†

The basic idea of logarithms can be seen from an example. The *logarithm of 150 to base* 7 is defined to be the solution of the equation $7^x = 150$ and is denoted $\log_7 150$. The solution of $7^x = 150$ (that is, $\log_7 150$) can be approximated graphically by finding the intersection of $y = 7^x$ and $y = 150$, as in Figure 5–24. Subject to rounding, Figure 5–25 shows that 7 raised to this power is 150. In other words, $\log_7 150$ is the exponent to which 7 must be raised to produce 150.

$\log_7 150 \approx 2.574957$.

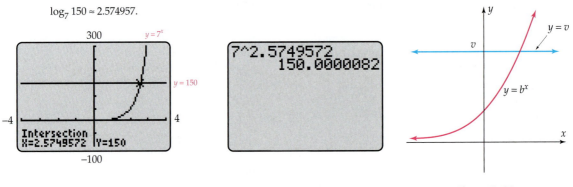

Figure 5–24   **Figure 5–25**   **Figure 5–26**

More generally, whenever b and v are positive numbers, then the horizontal line $y = v$ lies above the x-axis and hence intersects the graph of $y = b^x$ (Figure 5–26), so the equation $b^x = v$ has a solution. Consequently, we have this definition.

Definition of Logarithms to Base b

> If b and v are positive numbers, then the *logarithm of v to base b*, denoted $\log_b v$, is the solution of the equation $b^x = v$. In other words,
>
> $\log_b v$ is the exponent to which b must be raised to produce v.

EXAMPLE 1

To find $\log_2 16$, ask yourself, "What power of 2 equals 16?" Since $2^4 = 16$, we see that $\log_2 16 = 4$. Similarly, $\log_2 (1/8) = -3$ because $2^{-3} = 1/8$. ■

---

*This section replicates the discussion of Sections 5.3 and 5.4 in a more general context. It may be read before Section 5.3, if desired, and is not used in the sequel. Instructors who prefer to introduce logarithms as inverse functions of exponential functions should first cover Section 5.7.
†The discussion is also valid when $0 < b < 1$, but in that case the graphs have a different shape.

Since logarithms to base b are exponents, every statement about them is equivalent to a statement about exponents to base b. For instance, "$\log_b v = u$" means "u is the exponent to which b must be raised to produce v" or, in symbols, "$b^u = v$." In other words, we have the following.

Alternate Definition of Logarithms

Let u and v be numbers with $v > 0$. Then

$$\log_b v = u \qquad \text{exactly when} \qquad b^u = v.$$

Note for those who have read Section 5.7.   If you apply the statement in the preceding box to the functions $g(x) = \log_b x$ and $f(x) = b^x$, you obtain

$$\log_b v = u \qquad \text{exactly when} \qquad b^u = v$$
$$g(v) = u \qquad \text{exactly when} \qquad f(u) = v.$$

Thus, the box says that the logarithm function $g(x) = \log_b x$ is the *inverse function* of the exponential function $f(x) = b^x$.

EXAMPLE 2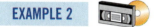

Since logarithms are just exponents, every logarithmic statement can be translated into exponential language.

Logarithmic Statement	Equivalent Exponential Statement
$\log_3 81 = 4$	$3^4 = 81$
$\log_4 64 = 3$	$4^3 = 64$
$\log_{125} 5 = \dfrac{1}{3}$	$125^{1/3} = 5^{*}$
$\log_8 \left(\dfrac{1}{4}\right) = -\dfrac{2}{3}$	$8^{-2/3} = \dfrac{1}{4}$ (verify!) ∎

EXAMPLE 3

Solve: $\log_5 x = 3$.

SOLUTION   The equation $\log_5 x = 3$ is equivalent to the exponential statement $5^3 = x$, so the solution is $x = 125$.   ∎

---

*Because $125^{1/3} = \sqrt[3]{125} = 5$.

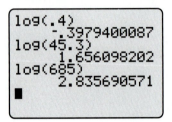

Figure 5–27

EXAMPLE 4

Logarithms to the base 10 are called **common logarithms.** It is customary to write log v instead of $\log_{10} v$. Then

$$\log 100 = 2 \qquad \text{because} \qquad 10^2 = 100;$$

$$\log .001 = -3 \qquad \text{because} \qquad 10^{-3} = \frac{1}{10^3} = \frac{1}{1000} = .001.$$

Calculators have a LOG key for evaluating common logarithms, as shown in Figure 5–27.[*]   ■

EXAMPLE 5

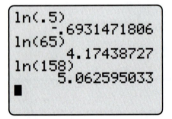

Figure 5–28

The most frequently used base for logarithms in modern applications is the number e ($\approx 2.71828 \ldots$). Logarithms to the base e are called **natural logarithms** and use a different notation: We write ln v instead of $\log_e v$. Calculators also have an LN key for evaluating natural logarithms. For example, see Figure 5–28.   ■

You *don't* need a calculator to understand the essential properties of logarithms. You need only translate logarithmic statements into exponential ones (or vice versa).

EXAMPLE 6

What is log (-25)?

Translation: To what power must 10 be raised to produce -25? The graph of $f(x) = 10^x$ lies entirely above the x-axis (use your calculator), which means that *every* power of 10 is *positive*. So 10^x can *never* be -25, or any negative number, or zero. The same argument works for any base b.

$$\log_b v \text{ is defined only when } v > 0. \qquad ■$$

EXAMPLE 7

What is $\log_5 1$?

Translation: To what power must 5 be raised to produce 1? The answer, of course, is $5^0 = 1$. So $\log_5 1 = 0$. Similarly, $\log_5 5 = 1$ because 1 is the answer to "what power of 5 equals 5?" In general,

$$\log_b 1 = 0 \qquad \text{and} \qquad \log_b b = 1. \qquad ■$$

---

[*]For convenient reading, logarithms printed in the text are rounded to four decimal places. For example, log .4 = $-.3979$. It is customary to use an equal sign in such cases rather than an "approximately equal" sign ($\approx$).

EXAMPLE 8

What is $\log_2 2^9$?

Translation: To what power must 2 be raised to produce 2^9? Obviously, the answer is 9. So $\log_2 2^9 = 9$, and, in general,

$$\log_b b^k = k \qquad \textbf{for every real number } k.$$

This property holds even when k is a complicated expression. For instance, if x and y are positive, then

$$\log_6 6^{\sqrt{3x+y}} = \sqrt{3x + y} \qquad \text{(here } k = \sqrt{3x + y}\text{).} \qquad \blacksquare$$

EXAMPLE 9

What is $10^{\log 439}$? Well, $\log 439$ is the power to which 10 must be raised to produce 439, that is $10^{\log 439} = 439$. Similarly,

$$b^{\log_b v} = v \qquad \textbf{for every } v > 0. \qquad \blacksquare$$

Here is a summary of the facts illustrated in the preceding examples.

*Properties
of Logarithms*

> 1. $\log_b v$ is defined only when $v > 0$.
>
> 2. $\log_b 1 = 0$ and $\log_b b = 1$.
>
> 3. $\log_b (b^k) = k$ for every real number k.
>
> 4. $b^{\log_b v} = v$ for every $v > 0$.

▉ LOGARITHM LAWS

The first law of exponents states that $b^m b^n = b^{m+n}$, or, in words,

> The exponent of a product is the sum of the exponents of the factors.

Since logarithms are just particular kinds of exponents, this statement translates as follows.

> **The logarithm of a product is the sum of the logarithms of the factors.**

The second and third laws of exponents, namely, $b^m / b^n = b^{m-n}$ and $(b^m)^k = b^{mk}$, can also be translated into logarithmic language.

*Logarithm
Laws*

> Let b, v, w, k be real numbers, with b, v, w positive and $b \neq 1$.
>
> *Product Law:* $\log_b (vw) = \log_b v + \log_b w$.
>
> *Quotient Law:* $\log_b\left(\dfrac{v}{w}\right) = \log_b v - \log_b w$.
>
> *Power Law:* $\log_b (v^k) = k(\log_b v)$.

Proof of the Quotient Law   According to Property 4 in the first box on page 413,

$$b^{\log_b v} = v \quad \text{and} \quad b^{\log_b w} = w.$$

Therefore, by the second law of exponents (with $m = \log_b v$ and $n = \log_b w$), we have

$$\frac{v}{w} = \frac{b^{\log_b v}}{b^{\log_b w}} = b^{\log_b v - \log_b w}.$$

Since $\log_b (v/w)$ is the exponent to which b must be raised to produce v/w, we must have $\log_b (v/w) = \log_b v - \log_b w$. This proves the Quotient Law. The Product and Power Laws are proved in a similar fashion.   ■

EXAMPLE 10

Simplify and write as a single logarithm.

(a) $\log_3 (x + 2) + \log_3 y - \log_3 (x^2 - 4)$

(b) $3 - \log_5 (125x)$

SOLUTION

(a) $\log_3 (x + 2) + \log_3 y - \log_3 (x^2 - 4)$

$$= \log_3 [(x + 2)y] - \log_3 (x^2 - 4) \qquad \text{[Product Law]}$$

$$= \log_3 \left(\frac{(x + 2)y}{x^2 - 4} \right) \qquad \text{[Quotient Law]}$$

$$= \log_3 \left(\frac{(x + 2)y}{(x + 2)(x - 2)} \right) \qquad \text{[Factor denominator]}$$

$$= \log_3 \left(\frac{y}{x - 2} \right) \qquad \text{[Cancel common factor]}$$

(b) $3 - \log_5 (125x)$

$$= 3 - (\log_5 125 + \log_5 x) \qquad \text{[Product Law]}$$

$$= 3 - \log_5 125 - \log_5 x$$

$$= 3 - 3 - \log_5 x \qquad \text{[$\log_5 125 = 3$ because $5^3 = 125$]}$$

$$= -\log_5 x$$

$$= \log_5 x^{-1} = \log_5 \left(\frac{1}{x} \right) \qquad \text{[Power Law]} \qquad ■$$

CAUTION

1. A common error in using the Product Law is to write something like $\log 6 + \log 7 = \log (6 + 7) = \log 13$ instead of the correct statement $\log 6 + \log 7 = \log (6 \cdot 7) = \log 42$.

2. Do not confuse $\log_b \left(\frac{v}{w} \right)$ with the quotient $\frac{\log_b v}{\log_b w}$. They are *different* numbers. For example, when $b = 10$,

$$\log \left(\frac{48}{4} \right) = \log 12 = 1.0792 \qquad \text{but} \qquad \frac{\log 48}{\log 4} = \frac{1.6812}{0.6021} = 2.7922.$$

For graphic illustrations of the errors mentioned in the Caution, see Exercises 86 and 87.

EXAMPLE 11

Given that

$$\log_7 2 = .3562, \qquad \log_7 3 = .5646, \qquad \text{and} \qquad \log_7 5 = .8271,$$

find:

(a) $\log_7 10$;   (b) $\log_7 2.5$;   (c) $\log_7 48$.

SOLUTION

(a) By the Product Law,

$$\log_7 10 = \log_7 (2 \cdot 5) = \log_7 2 + \log_7 5 = .3562 + .8271 = 1.1833.$$

(b) By the Quotient Law,

$$\log_7 2.5 = \log_7 \left(\frac{5}{2}\right) = \log_7 5 - \log_7 2 = .8271 - .3562 = .4709.$$

(c) By the Product and Power Laws,

$$\log_7 48 = \log_7 (3 \cdot 16) = \log_7 3 + \log_7 16 = \log_7 3 + \log_7 2^4$$

$$= \log_7 3 + 4 \cdot \log_7 2 = .5646 + 4(.3562)$$

$$= 1.9894. \quad \blacksquare$$

Example 11 worked because we were *given* several logarithms to base 7. But there's no $\log_7$ key on the calculator, so how do you find logarithms to base 7 or to any base other than e or 10? *Answer:* Use the LN key on the calculator and the following formula.

Change of Base Formula

For any positive numbers b and v,

$$\log_b v = \frac{\ln v}{\ln b}.$$

Proof   By Property 4 in the box on page 413, $b^{\log_b v} = v$. Take the natural logarithm of each side of this equation.

$$\ln (b^{\log_b v}) = \ln v$$

Apply the Power Law for natural logarithms on the left side.

$$(\log_b v)(\ln b) = \ln v$$

Dividing both sides by $\ln b$ finishes the proof.

$$\log_b v = \frac{\ln v}{\ln b} \quad \blacksquare$$

To find $\log_7 3$, apply the Change of Base Formula with $b = 7$.

$$\log_7 3 = \frac{\ln 3}{\ln 7} = \frac{1.0986}{1.9459} = .5646 \qquad \blacksquare$$

EXAMPLE 13

Environmental scientists study the diversity of species in an ecological community. If there are k different species, with n_1 individuals of species 1, n_2 individuals of species 2, and so on, then the *Shannon index of diversity H* is given by

$$H = \frac{N \log_2 N - [n_1 \log_2 n_1 + n_2 \log_2 n_2 + \cdots + n_k \log_2 n_k]}{N},$$

where $N = n_1 + n_2 + \cdots + n_k$. A study of the prey species of barn owls (that is, the creatures the typical barn owl ate) obtained the following data.

Species	Number
Rats	143
Mice	1405
Birds	452

Find the index of diversity for this prey community.

SOLUTION   In this case, $n_1 = 143$, $n_2 = 1405$, $n_3 = 452$, and

$$N = n_1 + n_2 + n_3 = 143 + 1405 + 452 = 2000.$$

So the index of diversity is

$$N = \frac{N \log_2 N - [n_1 \log_2 n_1 + n_2 \log_2 n_2 + n_3 \log_2 n_3]}{N}$$

$$= \frac{2000 \log_2 2000 - [143 \log_2 143 + 1405 \log_2 1405 + 452 \log_2 452]}{2000}.$$

To compute H, we use the Change of Base Formula.

$$H = \frac{2000 \dfrac{\ln 2000}{\ln 2} - \left[143 \dfrac{\ln 143}{\ln 2} + 1405 \dfrac{\ln 1405}{\ln 2} + 452 \dfrac{\ln 452}{\ln 2} \right]}{2000}$$

$$\approx 1.1149. \qquad \blacksquare$$

✓ **EXERCISES 5.4.A**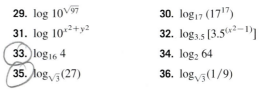

Note: *Unless stated otherwise, all letters represent positive numbers and $b \neq 1$.*

In Exercises 1–8, fill in the missing entries in each table.

1.

x	0	1	2	4
$f(x) = \log_4 x$				

2.

x	1/25	5	25	$\sqrt{5}$
$g(x) = \log_5 x$				

3.

x		1/6	1	216
$h(x) = \log_6 x$	-2			

4.

x	10/3	4	6	12
$k(x) = \log_3 (x - 3)$				

5.

x	0	1/7	$\sqrt{7}$	49
$f(x) = 2 \log_7 x$				

6.

x			100	1000
$g(x) = 3 \log x$	6	3		

7.

x	-2.75	-1	1	29
$h(x) = 3 \log_2 (x + 3)$				

8.

x	1/e	1	e	e^2
$k(x) = 2 \ln x$				

In Exercises 9–18, translate the given exponential statement into an equivalent logarithmic one.

9. $10^{-2} = .01$ **10.** $10^3 = 1000$

11. $\sqrt[3]{10} = 10^{1/3}$ **12.** $10^{.4771} \approx 3$

13. $10^{7k} = r$ **14.** $10^{(a+b)} = c$

15. $7^8 = 5{,}764{,}801$ **16.** $2^{-3} = 1/8$

17. $3^{-2} = 1/9$ **18.** $b^{14} = 3379$

In Exercises 19–28, translate the given logarithmic statement into an equivalent exponential one.

19. $\log 10{,}000 = 4$ **20.** $\log .001 = -3$

21. $\log 750 \approx 2.88$ **22.** $\log (.8) = -.097$

23. $\log_5 125 = 3$ **24.** $\log_8 (1/4) = -2/3$

25. $\log_2 (1/4) = -2$ **26.** $\log_2 \sqrt{2} = 1/2$

27. $\log (x^2 + 2y) = z + w$ **28.** $\log (a + c) = d$

In Exercises 29–36, evaluate the given expression without using a calculator.

29. $\log 10^{\sqrt{97}}$ **30.** $\log_{17} (17^{17})$

31. $\log 10^{x^2 + y^2}$ **32.** $\log_{3.5} [3.5^{(x^2-1)}]$

33. $\log_{16} 4$ **34.** $\log_2 64$

35. $\log_{\sqrt{3}} (27)$ **36.** $\log_{\sqrt{3}} (1/9)$

In Exercises 37–40, a graph or a table of values for the function $f(x) = \log_b x$ is given. Find b.

37.

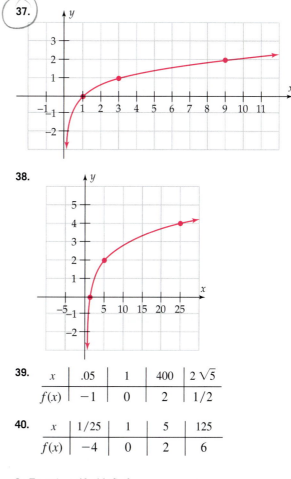

38.

39.

x	.05	1	400	$2\sqrt{5}$
$f(x)$	-1	0	2	1/2

40.

x	1/25	1	5	125
$f(x)$	-4	0	2	6

In Exercises 41–46, find x.

41. $\log_3 243 = x$ **42.** $\log_{81} 27 = x$

43. $\log_{27} x = 1/3$ **44.** $\log_5 x = -4$

45. $\log_x 64 = 3$ **46.** $\log_x (1/9) = -2/3$

In Exercises 47–60, write the given expression as the logarithm of a single quantity, as in Example 10.

47. $2 \log x + 3 \log y - 6 \log z$

48. $5 \log_8 x - 3 \log_8 y + 2 \log_8 z$

49. $\log x - \log (x + 3) + \log (x^2 - 9)$

50. $\log_3 (y + 2) + \log_3 (y - 3) - \log_3 y$

51. $\dfrac{1}{2} \log_2 (25c^2)$ **52.** $\dfrac{1}{3} \log_2 (27b^6)$

53. $-2 \log_4 (7c)$ **54.** $\dfrac{1}{3} \log_5 (x + 1)$

55. $2 \ln (x + 1) - \ln (x + 2)$

56. $\ln (z - 3) + 2 \ln (z + 3)$

57. $\log_2 (2x) - 1$

58. $2 - \log_5 (25z)$

59. $2 \ln (e^2 - e) - 2$

60. $4 - 4 \log_5 (20)$

In Exercises 61–68, use a calculator and the change of base formula to find the logarithm.

61. $\log_2 10$ **62.** $\log_2 22$ **63.** $\log_7 5$

64. $\log_5 7$ **65.** $\log_{500} 1000$ **66.** $\log_{500} 250$

67. $\log_{12} 56$ **68.** $\log_{12} 725$

Exercises 69–72 deal with the Shannon index of diversity (Example 13). Note that in two communities with the same number of species, a larger index indicates greater diversity.

69. A study of barn owl prey in another area produced the following data.

Species	Number
Rats	662
Mice	907
Birds	531

Find the index of diversity of this community. Is this community more or less diverse than the one in Example 13?

70. An eastern forest is composed of the following trees.

Species	Number
Beech	2754
Birch	689
Hemlock	4428
Maple	629

What is the index of diversity of this community?

71. A community has high divesity when all species have approximately the same number of individuals. It has low diversity when a few species account for most of the total population. Illustrate this fact for the following two communities.

Community 1	Number
Species A	1000
Species B	1000
Species C	1000

Community 2	Number
Species A	2500
Species B	200
Species C	300

72. In a community of k species, the maximum possible index is the number $\log_2 k$ (as is shown in Exercise 90). Use properties of logarithms (*not* a calculator) to show that Community 1 in Exercise 71 has the largest possible index.

In Exercises 73–78, answer true or false and give reasons for your answer.

73. $\log_b (r/5) = \log_b r - \log_b 5$

74. $\dfrac{\log_b a}{\log_b c} = \log_b \left(\dfrac{a}{c} \right)$

75. $(\log_b r)/t = \log_b (r^{1/t})$

76. $\log_b (cd) = \log_b c + \log_b d$

77. $\log_5 (5x) = 5(\log_5 x)$

78. $\log_b (ab)^t = t(\log_b a) + t$

Thinkers

79. Which is larger: 397^{398} or 398^{397}? [*Hint:* $\log 397 \approx 2.5988$ and $\log 398 \approx 2.5999$ and $f(x) = 10^x$ is an increasing function.]

80. If $\log_b 9.21 = 7.4$ and $\log_b 359.62 = 19.61$, then what is $\log_b 359.62/\log_b 9.21$?

In Exercises 81–84, assume that a and b are positive, with $a \neq 1$ and $b \neq 1$.

81. Express $\log_b u$ in terms of logarithms to the base a.

82. Show that $\log_b a = 1/\log_a b$.

83. How are $\log_{10} u$ and $\log_{100} u$ related?

84. Show that $a^{\log b} = b^{\log a}$.

85. If $\log_b x = \dfrac{1}{2}\log_b v + 3$, show that $x = (b^3)\sqrt{v}$.

86. Graph the functions $f(x) = \log x + \log 7$ and $g(x) = \log(x + 7)$ on the same screen. For what values of x is it true that $f(x) = g(x)$? What do you conclude about the statement

$$\log 6 + \log 7 = \log(6 + 7)?$$

87. Graph the functions $f(x) = \log(x/4)$ and $g(x) = (\log x)/(\log 4)$. Are they the same? What does this say about a statement such as

$$\log\left(\frac{48}{4}\right) = \frac{\log 48}{\log 4}?$$

In Exercises 88–89, sketch a complete graph of the function, labeling any holes, asymptotes, or local extrema.

88. $f(x) = \log_5 x + 2$ **89.** $h(x) = x \log x^2$

90. Suppose an ecological community with a total of N individuals has maximum diversity: It has k species, each of the same size N/k. Use the properties of logarithms to show that the index of diversity of this community is $\log_2 k$.

5.5   Algebraic Solutions of Exponential and Logarithmic Equations

Most of the exponential and logarithmic equations solved by graphical means earlier in this chapter could also have been solved algebraically. The algebraic techniques for solving such equations are based on the properties of logarithms.

■■ EXPONENTIAL EQUATIONS

The easiest exponential equations to solve are those in which both sides are powers of the same base.

EXAMPLE 1

Solve $8^x = 2^{x+1}$.

SOLUTION   Using the fact that $8 = 2^3$, we rewrite the equation as follows.

$$8^x = 2^{x+1}$$
$$(2^3)^x = 2^{x+1}$$
$$2^{3x} = 2^{x+1}$$

Since the powers of 2 are equal, the exponents must be the same, that is,

$$3x = x + 1$$
$$2x = 1$$
$$x = \frac{1}{2}. \quad ■$$

When different bases are involved in an exponential equation, a different solution technique is needed.

EXAMPLE 2

To solve $5^x = 2$,

Take logarithms on each side:[*] $\ln 5^x = \ln 2$

Use the Power Law: $x(\ln 5) = \ln 2$

Divide both sides by ln 5: $x = \dfrac{\ln 2}{\ln 5} \approx \dfrac{.6931}{1.6094} \approx .4307.$

Remember: $\dfrac{\ln 2}{\ln 5}$ is *not* $\ln \dfrac{2}{5}$ or $\ln 2 - \ln 5$. ■

EXAMPLE 3

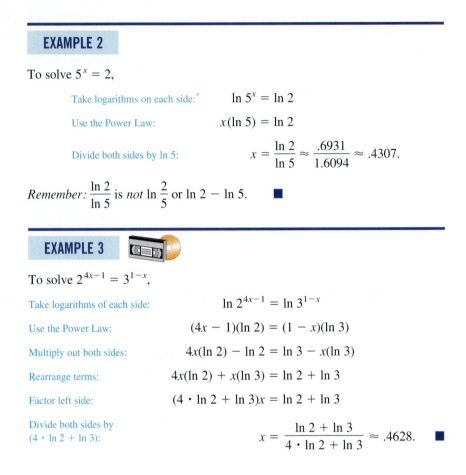

To solve $2^{4x-1} = 3^{1-x}$,

Take logarithms of each side: $\ln 2^{4x-1} = \ln 3^{1-x}$

Use the Power Law: $(4x - 1)(\ln 2) = (1 - x)(\ln 3)$

Multiply out both sides: $4x(\ln 2) - \ln 2 = \ln 3 - x(\ln 3)$

Rearrange terms: $4x(\ln 2) + x(\ln 3) = \ln 2 + \ln 3$

Factor left side: $(4 \cdot \ln 2 + \ln 3)x = \ln 2 + \ln 3$

Divide both sides by
$(4 \cdot \ln 2 + \ln 3)$: $x = \dfrac{\ln 2 + \ln 3}{4 \cdot \ln 2 + \ln 3} \approx .4628.$ ■

■■ APPLICATIONS OF EXPONENTIAL EQUATIONS

As we saw in Section 5.2, the mass of a radioactive element at time x is given by

$$M(x) = c(.5^{x/h}),$$

where c is the initial mass and h is the half-life of the element.

EXAMPLE 4

After 43 years, a 20-milligram sample of strontium-90 (^{90}Sr) decays to 6.071 mg. What is the half-life of strontium-90?

SOLUTION   The mass of the sample at time x is given by

$$f(x) = 20(.5^{x/h}),$$

where h is the half-life of strontium-90. We know that $f(x) = 6.071$ when $x = 43$,

---

*We shall use natural logarithms, but using logarithms to other bases would produce the same answer (Exercise 24).

that is, $6.071 = 20(.5^{43/h})$. We must solve this equation for h.

Divide both sides by 20: $$\dfrac{6.071}{20} = .5^{43/h}$$

Take logarithms on both sides: $$\ln \dfrac{6.071}{20} = \ln .5^{43/h}$$

Use the Power Law: $$\ln \dfrac{6.071}{20} = \dfrac{43}{h} \ln .5$$

Multiply both sides by h: $$h \ln \dfrac{6.071}{20} = 43 \ln .5$$

Divide both sides by $\ln \frac{6.071}{20}$: $$h = \dfrac{43 \ln .5}{\ln (6.071/20)} \approx 25$$

Therefore, strontium-90 has a half-life of 25 years. ■

EXAMPLE 5

A certain bacteria is known to grow exponentially, with the population at time t given by a function of the form $g(t) = Pe^{kt}$, where P is the original population and k is the continuous growth rate. A culture shows 1000 bacteria present. Seven hours later, there are 5000.

(a) Find the continuous growth rate k.

(b) Determine when the population will reach one billion.

SOLUTION

(a) The original population is $P = 1000$, so the growth function is $g(t) = 1000e^{kt}$. We know that $g(7) = 5000$, that is,

$$1000e^{k \cdot 7} = 5000.$$

To determine the growth rate, we solve this equation for k.

Divide both sides by 1000: $$e^{7k} = 5$$

Take logarithms of both sides: $$\ln e^{7k} = \ln 5$$

Use the Power Law: $$7k \ln e = \ln 5$$

Since $\ln e = 1$ (why?), this equation becomes

$$7k = \ln 5$$

Divide both sides by 7: $$k = \dfrac{\ln 5}{7} \approx .22992$$

Therefore, the growth function is $g(t) \approx 1000e^{.22992t}$.

(b) The population will reach one billion when $g(t) = 1{,}000{,}000{,}000$, that is, when

$$1000e^{.22992t} = 1{,}000{,}000{,}000.$$

So we solve this equation for *t*:

Divide both sides by 1000:	$e^{.22992t} = 1,000,000$
Take logarithms on both sides:	$\ln e^{.22992t} = \ln 1,000,000$
Use the Power Law:	$.22992t \ln e = \ln 1,000,000$
Remember $\ln e = 1$:	$.22992t = \ln 1,000,000$
Divide both sides by .22992:	$t = \dfrac{\ln 1,000,000}{.22992} \approx 60.09.$

Therefore, it will take a bit more than 60 hours for the culture to grow to one billion.   ■

EXAMPLE 6   **Inhibited Population Growth**

The population of fish in a lake at time *t* months is given by the function

$$p(t) = \frac{20,000}{1 + 24e^{-t/4}}.$$

How long will it take for the population to reach 15,000?

SOLUTION   We must solve this equation for *t*.

$$15,000 = \frac{20,000}{1 + 24e^{-t/4}}$$

$$15,000(1 + 24e^{-t/4}) = 20,000$$

$$1 + 24e^{-t/4} = \frac{20,000}{15,000} = \frac{4}{3}$$

$$24e^{-t/4} = \frac{1}{3}$$

$$e^{-t/4} = \frac{1}{3} \cdot \frac{1}{24} = \frac{1}{72}$$

$$\ln e^{-t/4} = \ln\left(\frac{1}{72}\right)$$

$$\left(-\frac{t}{4}\right)(\ln e) = \ln 1 - \ln 72$$

$$-\frac{t}{4} = -\ln 72 \qquad \text{[ln } e = 1 \text{ and ln } 1 = 0]$$

$$t = 4(\ln 72) \approx 17.1067$$

So the population reaches 15,000 in a little over 17 months.   ■

▰▰ LOGARITHMIC EQUATIONS

Equations that involve only logarithmic terms may be solved by using this fact, which is proved in Exercise 23 (and is valid with log replaced by ln).

If log u = log v, then $u = v$.

EXAMPLE 7

Solve: $\log (3x + 2) + \log (x + 2) = \log (7x + 6)$.

SOLUTION   First we write the left side as a single logarithm.

$$\log (3x + 2) + \log (x + 2) = \log (7x + 6)$$

Use the Product Law: $$\log[(3x + 2)(x + 2)] = \log (7x + 6)$$

Multiply out left side: $$\log (3x^2 + 8x + 4) = \log (7x + 6).$$

Since the logarithms are equal, we must have

$$3x^2 + 8x + 4 = 7x + 6$$

Subtract $7x + 6$ from both sides: $$3x^2 + x - 2 = 0$$

Factor: $$(3x - 2)(x + 1) = 0$$

$$3x - 2 = 0 \quad \text{or} \quad x + 1 = 0$$

$$3x = 2 \qquad\qquad x = -1$$

$$x = \frac{2}{3}.$$

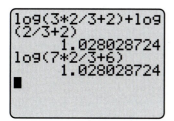

Figure 5–29

Thus, $x = 2/3$ and $x = -1$ are the *possible* solutions and must be checked in the *original* equation. When $x = 2/3$, both sides of the original equation have the same value, as shown in Figure 5–29. So 2/3 is a solution. When $x = -1$, however, the right side of the equation is

$$\log (7x + 6) = \log [7(-1) + 6] = \log (-1),$$

which is not defined. So -1 is not a solution.   ■

EXAMPLE 8

Solve: $\ln (x + 4) - \ln (x + 2) = \ln x$.

SOLUTION   First, write the left side as a single logarithm.

Use the Quotient Law: $$\ln \left(\frac{x + 4}{x + 2} \right) = \ln x$$

Therefore,

$$\frac{x + 4}{x + 2} = x$$

Multiply both sides by $x + 2$: $$x + 4 = x(x + 2)$$

Multiply out right side: $$x + 4 = x^2 + 2x$$

Rearrange terms: $$x^2 + x - 4 = 0$$

This equation does not readily factor, so we use the quadratic formula.

$$x = \frac{-1 \pm \sqrt{1^2 - 4 \cdot 1 \cdot (-4)}}{2 \cdot 1} = \frac{-1 \pm \sqrt{17}}{2}.$$

```
(-1+J(17))/2→A
        1.561552813
ln(A+4)-ln(A+2)
          .445680719
ln(A)
          .445680719
```

Figure 5–30

An easy way to verify that $\dfrac{-1 + \sqrt{17}}{2}$ is a solution is to store this number as A and then evaluate both sides of the original equation at $x = A$, as shown in Figure 5–30. The second possibility, however, is not a solution (because $x = \dfrac{-1 - \sqrt{17}}{2}$ is negative, so $\ln x$ is not defined). ■

Equations that involve both logarithmic and constant terms may be solved by using the basic property of logarithms (see page 397).

$$(*) \qquad 10^{\log v} = v \quad \text{and} \quad e^{\ln v} = v.$$

EXAMPLE 9

Solve $7 + 2 \log 5x = 11$.

SOLUTION   We start by getting all the logarithmic terms on one side and the constant on the other.

| Subtract 7 from both sides: | $2 \log 5x = 4$ |
| Divide both sides by 2: | $\log 5x = 2$ |

We know that if two quantities are equal, say $a = b$, then $10^a = 10^b$. We use this fact here, with the two sides of the preceding equation as a and b.

Exponentiate both sides:	$10^{\log 5x} = 10^2$
Use the basic logarithm property ($*$):	$5x = 100$
Divide both sides by 5:	$x = 20$

Verify that 20 is actually a solution of the original equation. ■

EXAMPLE 10

Solve $\ln (x - 3) = 5 - \ln (x - 3)$.

SOLUTION   We proceed as in Example 9, but since the base for these logarithms is e, we use e rather than 10 when we exponentiate.

$$\ln (x - 3) = 5 - \ln (x - 3)$$

Add $\ln (x - 3)$ to both sides:	$2 \ln (x - 3) = 5$
Divide both sides by 2:	$\ln (x - 3) = \dfrac{5}{2}$
Exponentiate both sides:	$e^{\ln(x-3)} = e^{5/2}$
Use the basic property of logarithms ($*$):	$x - 3 = e^{5/2}$
Add 3 to both sides:	$x = e^{5/2} + 3 \approx 15.1825$

This is the only possibility for a solution. A calculator shows that it actually is a solution of the original equation.   ■

EXAMPLE 11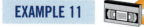

Solve $\log (x - 16) = 2 - \log (x - 1)$.

SOLUTION

$$\log (x - 16) = 2 - \log (x - 1)$$

Add $\log (x - 1)$ to both sides:	$\log (x - 16) + \log (x - 1) = 2$
Use the Product Law:	$\log [(x - 16)(x - 1)] = 2$
Multiply out left side:	$\log (x^2 - 17x + 16) = 2$
Exponentiate both sides:	$10^{\log (x^2 - 17x + 16)} = 10^2$
Use the basic logarithm property ($*$):	$x^2 - 17x + 16 = 100$
Subtract 100 from both sides:	$x^2 - 17x - 84 = 0$
Factor:	$(x + 4)(x - 21) = 0$

$$x + 4 = 0 \quad \text{or} \quad x - 21 = 0$$
$$x = -4 \quad \text{or} \quad x = 21$$

You can easily verify that 21 is a solution of the original equation, but -4 is not [when $x = -4$, then $\log (x - 16) = \log (-20)$, which is not defined].   ■

▪▪ APPLICATIONS OF LOGARITHMIC EQUATIONS

Many phenomena can be modeled by logarithmic functions. In such cases, logarithmic equations are used to answer various questions.

EXAMPLE 12

The number of pounds of fish (in billions) used for human consumption in the United States in year x is approximated by the function

$$f(x) = 10.57 + 1.75 \ln x,$$

where $x = 5$ corresponds to 1995.[*]

(a) How many pounds of fish were used in 2004?

(b) When will fish consumption reach 16 billion pounds?

SOLUTION

(a) Since 2004 corresponds to $x = 14$, we evaluate $f(x)$ at 14.

$$f(14) = 10.57 + 1.75 \ln 14 \approx 15.19 \text{ billion pounds.}$$

---

[*]Based on data from the U.S. National Oceanic and Atmospheric Administration and the National Marine Fisheries Service.

(b) Fish consumption is 16 billion pounds when $f(x) = 16$, so we must solve the equation

$$10.57 + 1.75 \ln x = 16$$

Subtract 10.57 from both sides: $\qquad 1.75 \ln x = 5.43$

Divide both sides by 1.75: $\qquad \ln x = \dfrac{5.43}{1.75}$

Exponentiate both sides: $\qquad e^{\ln x} = e^{5.43/1.75}$

Use the basic property of logarithms (∗): $\qquad x \approx 22.26.$

Since $x = 22$ corresponds to 2012, fish consumption will reach 16 billion pounds in 2012. ∎

✓ EXERCISES 5.5

In Exercises 1–8, solve the equation without using logarithms.

1. $3^x = 81$ $\qquad$ **2.** $3^x + 3 = 30$ $\qquad$ **3.** $3^{x+1} = 9^{5x}$

4. $4^{5x} = 16^{2x-1}$ $\qquad$ **5.** $3^{5x}9^{x^2} = 27$

6. $2^{x^2+5x} = 1/16$ $\qquad$ **7.** $9^{x^2} = 3^{-5x-2}$

8. $4^{x^2-1} = 8^x$

In Exercises 9–22, solve the equation. First express your answer in terms of natural logarithms (for instance, $x = (2 + \ln 5)/(\ln 3)$). Then use a calculator to find an approximation for the answer.

9. $3^x = 5$ $\qquad$ **10.** $5^x = 4$ $\qquad$ **11.** $2^x = 3^{x-1}$

12. $4^{x+2} = 2^{x-1}$ $\qquad$ **13.** $3^{1-2x} = 5^{x+5}$

14. $4^{3x-1} = 3^{x-2}$ $\qquad$ **15.** $2^{1-3x} = 3^{x+1}$

16. $3^{z+3} = 2^z$ $\qquad$ **17.** $e^{2x} = 5$

18. $e^{-3x} = 2$ $\qquad$ **19.** $6e^{-1.4x} = 21$

20. $3.4e^{-x/3} = 5.6$ $\qquad$ **21.** $2.1e^{(x/2)\ln 3} = 5$

22. $7.8e^{(x/3)\ln 5} = 14$

23. Prove that if $\ln u = \ln v$, then $u = v$. [*Hint:* Property (∗) on page 424.]

24. (a) Solve $7^x = 3$, using natural logarithms. Leave your answer in logarithmic form; don't approximate with a calculator.
   (b) Solve $7^x = 3$, using common (base 10) logarithms. Leave your answer in logarithmic form.
   (c) Use the Change of Base Formula in Special Topics 5.4.A to show that your answers in parts (a) and (b) are the same.

In Exercises 25–34, solve the equation as in Examples 7 and 8.

25. $\ln (3x - 5) = \ln 11 + \ln 2$

26. $\log (4x - 1) = \log (x + 1) + \log 2$

27. $\log (3x - 1) + \log 2 = \log 4 + \log (x + 2)$

28. $\ln (x + 6) - \ln 10 = \ln (x - 1) - \ln 2$

29. $2 \ln x = \ln 36$

30. $2 \log x = 3 \log 4$

31. $\ln x + \ln (x + 1) = \ln 3 + \ln 4$

32. $\ln (6x - 1) + \ln x = \frac{1}{2} \ln 4$

33. $\ln x = \ln 3 - \ln (x + 5)$

34. $\ln (2x + 3) + \ln x = \ln e$

In Exercises 35–42, solve the equation, as in Examples 9–11.

35. $\ln (x + 9) - \ln x = 1$

36. $\ln (2x + 1) - 1 = \ln (x - 2)$

37. $\log x + \log (x - 3) = 1$

38. $\log (x - 1) + \log (x + 2) = 1$

39. $\log \sqrt{x^2 - 1} = 2$

40. $\log \sqrt[3]{x^2 + 21x} = 2/3$

41. $\ln (x^2 + 1) - \ln (x - 1) = 1 + \ln (x + 1)$

42. $\dfrac{\ln (x + 1)}{\ln (x - 1)} = 2$

43. Book sales on the Internet (in billions of dollars) in year x are approximated by

$$f(x) = 1.84 + 2.1 \ln x,$$

where $x = 2$ corresponds to 2002.*
(a) How much was spent on Internet book sales in 2004?
(b) When will book sales reach \$6.5 billion?

44. According to data from the U.S. Bureau of Labor Statistics, the consumer price index (CPI) for food is approximated in the year x by

$$g(x) = 10.13 + 52.1 \ln x,$$

where $x = 7$ corresponds to 1987.
(a) What was the CPI for food in 2003?
(b) If this model remains accurate, when will the CPI for food reach 185?

45. U.S. Census Bureau data shows that the number of families in the United States (in millions) in year x is given by

$$h(x) = 51.42 + 15.473 \log x,$$

where $x = 5$ is 1985.
(a) How many families were there in 2002?
(b) When will there be 74 million families?

46. The number of automobiles sold in the United States (in millions of vehicles) in year x is approximated by

$$g(x) = 15.27 + 2.282 \log x,$$

where $x = 3$ corresponds to 2003.†
(a) How many cars were sold in 2004?
(b) If this model remains accurate, when will sales reach 18 million cars?

Exercises 47–56, deal with radioactive decay and the function $M(x) = c(.5^{x/h})$; see Example 4.

47. A sample of 300 grams of uranium decays to 200 grams in .26 billion years. Find the half-life of uranium.

48. It takes 1000 years for a sample of 100 mg of radium-226 to decay to 65 mg. Find the half-life of radium-226.

49. A 3-gram sample of an isotope of sodium decays to 1 gram in 23.7 days. Find the half-life of the isotope of sodium.

50. The half-life of cobalt-60 is 4.945 years. How long will it take for 25 grams to decay to 15 grams?

51. After six days a sample of radon-222 decayed to 33.6% of its original mass. Find the half-life of radon-222. [*Hint:* When $x = 6$, then $M(x) = .336P$.]

52. Krypton-85 loses 6.44% of its mass each year. What is its half-life?

53. A Native American mummy was found recently. If it has lost 26.4% of its carbon-14, approximately how long ago did the Native American die? [*Hint:* Remember that the half-life of carbon-14 is 5730 years and see Example 11 in Section 5.2.]

54. How old is a wooden statue that has only one-third of its original carbon-14?

55. How old is a piece of ivory that has lost 36% of its carbon-14?

56. How old is a mummy that has lost 49% of its carbon-14?

Exercises 57–62, deal with the compound interest formula $A = P(1 + r)^t$, which was discussed in Section 5.2.

57. At what annual rate of interest should \$1000 be invested so that it will double in 10 years if interest is compounded quarterly?

58. How long does it take \$500 to triple if it is invested at 6% compounded: (a) annually, (b) quarterly, (c) daily?

59. (a) How long will it take to triple your money if you invest \$500 at a rate of 5% per year compounded annually?
(b) How long will it take at 5% compounded quarterly?

60. At what rate of interest (compounded annually) should you invest \$500 if you want to have \$1500 in 12 years?

61. How much money should be invested at 5% interest, compounded quarterly, so that 9 years later the investment will be worth \$5000? This amount is called the **present value** of \$5000 at 5% interest.

62. Find a formula that gives the time needed for an investment of P dollars to double, if the interest rate is r% compounded annually. [*Hint:* Solve the compound interest formula for t, when $A = 2P$.]

Exercises 63–70 deal with functions of the form $f(x) = Pe^{kx}$, where k is the continuous exponential growth rate (see Example 5).

63. The present concentration of carbon dioxide in the atmosphere is 364 parts per million (ppm) and is increasing exponentially at a continuous yearly rate of .4% (that is, $k = .004$). How many years will it take for the concentration to reach 500 ppm?

64. The amount P of ozone in the atmosphere is currently decaying exponentially each year at a continuous rate of $\frac{1}{4}$% (that is, $k = -.0025$). How long will it take for half the ozone to disappear (that is, when will the amount be $P/2$)? [Your answer is the half-life of ozone.]

---

*Based on forecasts by Forrester Research.
†Based on data from J. D. Power and Associates.

65. The population of Brazil increased from 151 million in 1990 to 173 million in 2000.[*]
   (a) At what continuous rate was the population growing during this period?
   (b) Assuming that Brazil's population continues to increase at this rate, when will it reach 250 million?

66. Outstanding consumer debt increased exponentially from $781.5 billion in 1990 to $1765.5 billion in 2002.[†]
   (a) At what continuous growth rate is consumer debt growing?
   (b) Assuming that this rate continues, when will consumer debt reach $2500 billion?

67. The probability P percent of having an accident while driving a car is related to the alcohol level of the driver's blood by the formula $P = e^{kt}$, where k is a constant. Accident statistics show that the probability of an accident is 25% when the blood alcohol level is $t = .15$.
   (a) Find k. [Use $P = 25$, not .25.]
   (b) At what blood alcohol level is the probability of having an accident 50%?

68. Under normal conditions, the atmospheric pressure (in millibars) at height h feet above sea level is given by $P(h) = 1015e^{-kh}$, where k is a positive constant.
   (a) If the pressure at 18,000 feet is half the pressure at sea level, find k.
   (b) Using the information from part (a), find the atmospheric pressure at 1000 feet, 5000 feet, and 15,000 feet.

69. One hour after an experiment begins, the number of bacteria in a culture is 100. An hour later, there are 500.
   (a) Find the number of bacteria at the beginning of the experiment and the number three hours later.
   (b) How long does it take the number of bacteria at any given time to double?

70. If the population at time t is given by $S(t) = ce^{kt}$, find a formula that gives the time it takes for the population to double.

---

*International Data Base, Bureau of the Census, U.S. Department of Commerce.
†*Federal Reserve Bulletin.*

71. The spread of a flu virus in a community of 45,000 people is given by the function

$$f(t) = \frac{45,000}{1 + 224e^{-.899t}},$$

where $f(t)$ is the number of people infected in week t.
   (a) How many people had the flu at the outbreak of the epidemic? After three weeks?
   (b) When will half the town be infected?

72. The beaver population near a certain lake in year t is approximately

$$p(t) = \frac{2000}{1 + 199e^{-.5544t}}.$$

   (a) When will the beaver population reach 1000?
   (b) Will the population ever reach 2000? Why?

Thinkers

73. According to one theory of learning, the number of words per minute N that a person can type after t weeks of practice is given by $N = c(1 - e^{-kt})$, where c is an upper limit that N cannot exceed and k is a constant that must be determined experimentally for each person.
   (a) If a person can type 50 wpm (words per minute) after four weeks of practice and 70 wpm after eight weeks, find the values of k and c for this person. According to the theory, this person will never type faster than c wpm.
   (b) Another person can type 50 wpm after four weeks of practice and 90 wpm after eight weeks. How many weeks must this person practice to be able to type 125 wpm?

74. Hilary Hungerford has been offered two jobs, each with the same starting salary of $32,000 and identical benefits. Assuming satisfactory performance, she will receive a $1600 raise each year at the Great Gizmo Company, whereas the Wonder Widget Company will give her a 4% raise each year.
   (a) In what year (after the first year) would her salary be the same at either company? Until then, which company pays better? After that, which company pays better?
   (b) Answer the questions in part (a) assuming that the annual raise at Great Gizmo is $2000.

OMIT

5.6 Exponential, Logarithmic, and Other Models*

Many data sets can be modeled by suitable exponential, logarithmic, and related functions. Most calculators have regression procedures for constructing the following models.

Model	Equation	Examples	
Power	$y = ax^r$	$y = 5x^{2.7}$	$y = 3.5x^{-.045}$
Exponential	$y = ab^x$ or $y = ae^{kx}$	$y = 2(1.64)^x$	$y = 2 \cdot e^{.4947x}$
Logistic	$y = \dfrac{a}{1 + be^{-kx}}$	$y = \dfrac{20{,}000}{1 + 24e^{-.25x}}$	$y = \dfrac{650}{1 + 6e^{.3x}}$
Logarithmic	$y = a + b \ln x$	$y = 5 + 4.2 \ln x$	$y = 2 - 3 \ln x$

We begin by examining exponential models, such as $y = 3 \cdot 2^x$. A table of values for this model is shown below. Look carefully at the ratio of successive entries (that is, each entry divided by its predecessor).

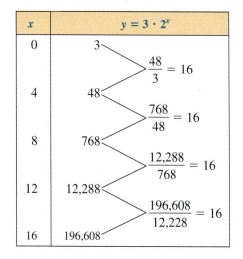

It should not be a surprise that the ratio of successive entries is constant. For at each step, x changes from x to $x + 4$ (from 0 to 4, from 4 to 8, and so on), and y changes from $3 \cdot 2^x$ to $3 \cdot 2^{x+4}$. Hence, the ratio of successive terms is always

$$\frac{3 \cdot 2^{x+4}}{3 \cdot 2^x} = \frac{3 \cdot 2^x \cdot 2^4}{3 \cdot 2^x} = 2^4 = 16.$$

A similar argument applies to any exponential model $y = ab^x$ and shows that if x changes by a fixed amount k, then the ratio of the corresponding y values is the constant b^k (in our example b was 2 and k was 4). This suggests that

> **when the ratio of successive entries in a table of data is approximately constant, an exponential model is appropriate.**

---

*This section is optional; its prerequisites are Sections 1.4 and 4.4. It will be used in clearly identifiable exercises but not elsewhere in the text.

EXAMPLE 1

In the years before the Civil War, the population of the United States grew rapidly, as shown in the following table from the U.S. Bureau of the Census. Find a model for this growth.

Year	Population in Millions	Year	Population in Millions
1790	3.93	1830	12.86
1800	5.31	1840	17.07
1810	7.24	1850	23.19
1820	9.64	1860	31.44

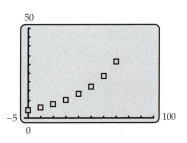

Figure 5–31

SOLUTION   The data points (with $x = 0$ corresponding to 1790) are shown in Figure 5–31. Their shape suggests either a polynomial graph of even degree or an exponential graph. Since populations generally grow exponentially, an exponential model is likely to be a good choice. We can confirm this by looking at the ratios of successive entries in the table.

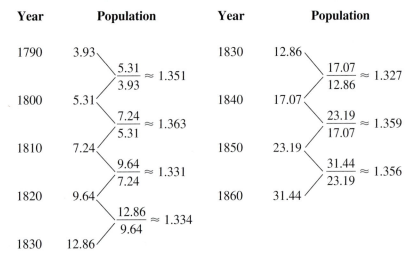

The ratios are almost constant, as they would be in an exponential model. So we use regression to find such an exponential model. The procedure is the same as that for linear and polynomial regression (see the Tips on pages 106 and 323). It produces this model:[*]

$$y = 3.9572 \, (1.0299^x).$$

The graph in Figure 5–32 appears to fit the data quite well. In fact, you can readily verify that the model has an error of less than 1% for each of the data points. Furthermore, as was discussed before the example, when x changes by 10, the value of y changes by approximately $1.0299^{10} \approx 1.343$, which is very close to the successive ratios of the data that were computed above.   ■

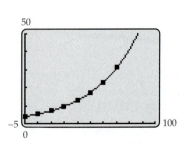

Figure 5–32

---

[*]Throughout this section, coefficients are rounded for convenient reading, but the full expansion is used for calculations and graphs.

EXAMPLE 2

After the Civil War, the U.S. population continued to increase, as shown below.

Year	Population in Millions	Year	Population in Millions	Year	Population in Millions
1870	38.56	1920	106.02	1960	179.32
1880	50.19	1930	123.20	1970	202.30
1890	62.98	1940	132.16	1980	226.54
1900	76.21	1950	151.33	1990	248.72
1910	92.23			2000	281.42

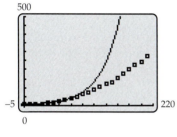

Figure 5–33

However, the model from Example 1 does not remain valid, as can be seen in Figure 5–33, which shows its graph together with all the data points from 1790 through 2000 ($x = 0$ corresponds to 1790).

The problem is that the *rate* of growth has steadily decreased since the Civil War. For instance, the ratio of the first two entries in the preceding table is

$$\frac{50.19}{38.56} \approx 1.302,$$

and the ratio of the last two is

$$\frac{281.42}{248.72} \approx 1.13.$$

So an exponential model might not be the best choice now. Other possibilities are polynomial models (which grow at a slower rate) or logistic models (in which the growth rate decreases with time). Figure 5–34 shows three possible models, each obtained by using the appropriate regression program on a calculator, with all the data points from 1790 through 2000.

Exponential Model

$$y = 6.06616 \cdot 1.02039^x$$

Polynominal Model

$$y = (7.94 \times 10^{-8})x^4 - (2.76 \times 10^{-5})x^3$$
$$+ .0093x^2 - .1621x + 5.462$$

Logistic Model*

$$y = \frac{442.1}{1 + 56.33e^{-.0216x}}$$

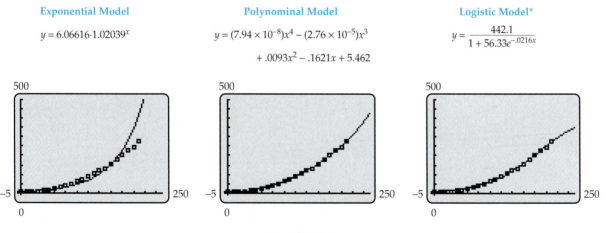

Figure 5–34

As expected, the polynomial and logistic models fit the data better than does the exponential model. The main difference between them is that the polynomial

*This model was obtained on a TI-83+. Other calculators may produce a slightly different model or an error message.

model indicates unlimited future growth, whereas the logistic model has the population growing more slowly in the future (and eventually leveling off—see Exercise 13). ∎

In Example 1, we used the ratios of successive entries of the data table to determine that an exponential model was appropriate. Here is another way to make that determination. Consider the exponential function $y = ab^x$. Taking natural logarithms of both sides and using the logarithm laws on the right-hand side shows that

$$\ln y = \ln (ab^x) = \ln a + \ln b^x = \ln a + x \ln b.$$

Now $\ln a$ and $\ln b$ are constants, say, $k = \ln a$ and $m = \ln b$, so

$$\ln y = mx + k.$$

Thus, the points $(x, \ln y)$ lie on the straight line with slope m and y-intercept k. Consequently, we have this guideline.

> **If (x, y) are data points and if the points $(x, \ln y)$ are approximately linear, then an exponential model may be appropriate for the data.**

Similarly, if $y = ax^r$ is a power function, then

$$\ln y = \ln (ax^r) = \ln a + r \ln x.$$

Since $\ln a$ is a constant, say, $k = \ln a$, we have

$$\ln y = r \ln x + k,$$

which means that the points $(\ln x, \ln y)$ lie on a straight line with slope r and y-intercept k. Consequently, we have this guideline.

> **If (x, y) are data points and if the points $(\ln x, \ln y)$ are approximately linear, then a power model may be appropriate for the data.**

EXAMPLE 3

The length of time that a planet takes to make one complete rotation around the sun is its year. The table shows the length (in earth years) of each planet's year and the distance of that planet from the sun (in millions of miles).[*] Find a model for this data in which x is the length of the year and y the distance from the sun.

Planet	Year	Distance	Planet	Year	Distance
Mercury	.24	36.0	Saturn	29.46	886.7
Venus	.62	67.2	Uranus	84.01	1783.0
Earth	1	92.9	Neptune	164.79	2794.0
Mars	1.88	141.6	Pluto	247.69	3674.5
Jupiter	11.86	483.6			

---

[*]Since the orbit of a planet around the sun is not circular, its distance from the sun varies through the year. Each given number is the average of its maximum and minimum distances from the sun.

SOLUTION   Figure 5–35 shows the data points for the five planets with the shortest years. Figure 5–36 shows all the data points, but on this scale, the first four points look like a single large one near the origin.

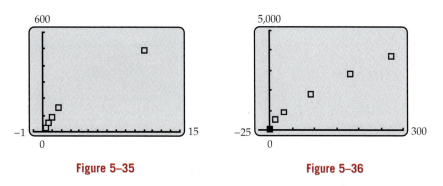

Figure 5–35   Figure 5–36

TECHNOLOGY TIP

Suppose the *x*- and *y*-coordinates of the data points are stored in lists L_1 and L_2, respectively. On calculators other than TI-89, keying in

$$\ln L_2 \text{ STO} \rightarrow L_4$$

produces the list L_4, whose entries are the natural logarithms of the numbers in list L_2, and stores it in the statistics editor. You can then use lists L_1 and L_4 to plot the points $(x, \ln y)$. For TI-89, check your instruction manual.

Plotting the point $(x, \ln y)$ for each data point (x, y) (see the Technology Tip in the margin) produces Figure 5–37. Its points do not form a linear pattern (four of them are almost vertical near the *y*-axis and the other five almost horizontal), so an exponential function is not an appropriate model. On the other hand, the points $(\ln x, \ln y)$ in Figure 5–38 do form a linear pattern, which suggests that a power model will work.

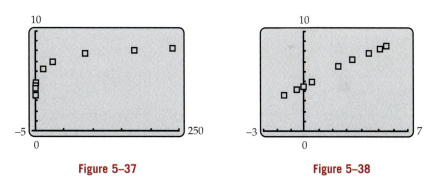

Figure 5–37   Figure 5–38

A calculator's power regression feature produces the following model.

$$y = 92.8935 \, x^{.6669}$$

Its graph in Figure 5–39 shows that it fits the original data points quite well.   ■

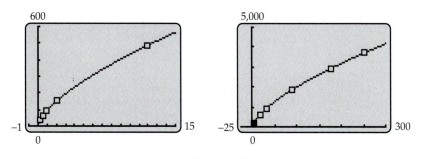

Figure 5–39

If $y = a + b \ln x$ is a logarithmic model, then the points $(\ln x, y)$ lie on the straight line with slope b and y-intercept a (why?). Thus, we have this guideline.

If (x, y) are data points and if the points $(\ln x, y)$ are approximately linear, then a logarithmic model may be appropriate for the data.

EXAMPLE 4

Find a model for population growth in El Paso, Texas, given the following data.[*]

Year	1950	1970	1980	1990	2000
Population	130,485	322,261	425,259	515,342	563,662

SOLUTION   The scatter plot of the data points (with $x = 50$ corresponding to 1950) in Figure 5–40 bends slightly, suggesting a logarithmic curve. So we plot the points $(\ln x, y)$, that is,

$$(\ln 50, 130485), (\ln 70, 322261), \ldots, (\ln 100, 563662),$$

in Figure 5–41. Since these points lie approximately on a straight line, a logarithmic model is appropriate. Using logarithmic regression on a calculator, we obtain the following model.

$$y = -2{,}382{,}368.345 + 640{,}666.815 \ln x$$

Its graph in Figure 5–42 is a good fit for the data. ■

> **CAUTION**
>
> When using logarithmic models, you must have data points with positive first coordinates (since logarithms of negative numbers and 0 are not defined).

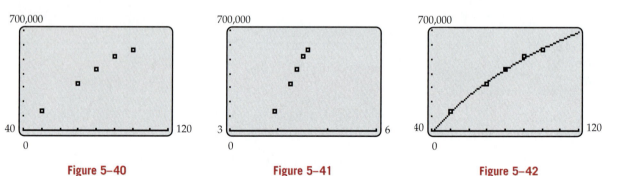

Figure 5–40   **Figure 5–41**   **Figure 5–42**

---

[*]Bureau of the Census, U.S. Department of Commerce.

✓ EXERCISES 5.6

In Exercises 1–10, state which of the following models might be appropriate for the given scatter plot of data (more than one model may be appropriate).

Model	Corresponding Function
A. Linear	$y = ax + b$
B. Quadratic	$y = ax^2 + bx + c$
C. Power	$y = ax^r$
D. Cubic	$y = ax^3 + bx^2 + cx + d$
E. Exponential	$y = ab^x$
F. Logarithmic	$y = a + b \ln x$
G. Logistic	$y = \dfrac{a}{1 + be^{-kx}}$

1.

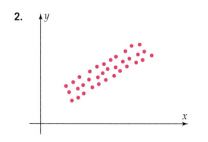

2.

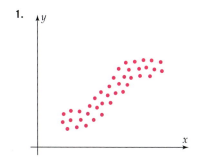

3.

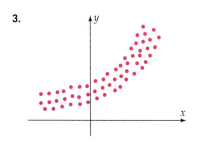

4.

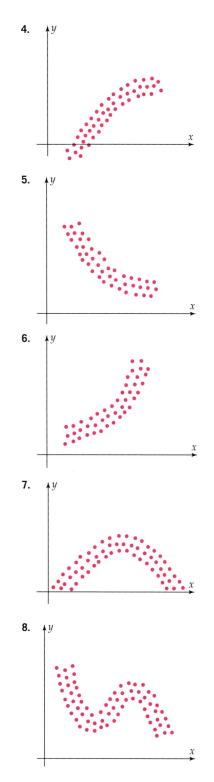

5.

6.

7.

8.

9.

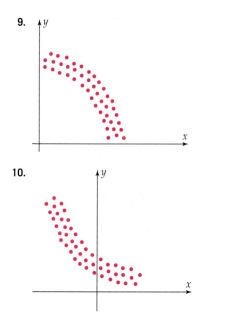

10.

In Exercises 11 and 12, compute the ratios of successive entries in the table to determine whether or not an exponential model is appropriate for the data.

11.

x	0	2	4	6	8	10
y	3	15.2	76.9	389.2	1975.5	9975.8

12.

x	1	3	5	7	9	11
y	3	21	55	105	171	253

13. (a) Show algebraically that in the logistic model for the U.S. population in Example 2, the population can never exceed 442.1 million people.
   (b) Confirm your answer in part (a) by graphing the logistic model in a window that includes the next three centuries.

14. According to estimates by the U.S. Bureau of the Census, the U.S. population was 287.7 million in 2002. On the basis of this information, which of the models in Example 2 appears to be the most accurate predictor?

15. Graph each of the following power functions in a window with $0 \le x \le 20$.
   (a) $f(x) = x^{-1.5}$ (b) $g(x) = x^{.75}$ (c) $h(x) = x^{2.4}$

16. On the basis of your graphs in Exercise 15, describe the general shape of the graph of $y = ax^r$ when $a > 0$ and
   (a) $r < 0$ (b) $0 < r < 1$ (c) $r > 1$

In Exercises 17–20, determine whether an exponential, power, or logarithmic model (or none or several of these) is appropriate for the data by determining which (if any) of the following sets of points are approximately linear:

$$\{(x, \ln y)\}, \qquad \{(\ln x, \ln y)\}, \qquad \{(\ln x, y)\}.$$

where the given data set consists of the points $\{(x, y)\}$.

17.

x	1	3	5	7	9	11
y	2	25	81	175	310	497

18.

x	3	6	9	12	15	18
y	385	74	14	2.75	.5	.1

19.

x	5	10	15	20	25	30
y	17	27	35	40	43	48

20.

x	5	10	15	20	25	30
y	2	110	460	1200	2500	4525

21. The table shows the number of babies born as twins, triplets, quadruplets, etc., in recent years.

Year	Multiple Births
1989	92,916
1990	96,893
1991	98,125
1992	99,255
1993	100,613
1994	101,658
1995	101,709

   (a) Sketch a scatter plot of the data, with $x = 1$ corresponding to 1989.
   (b) Plot each of the following models on the same screen as the scatter plot.

$$f(x) = 93{,}201.973 + 4{,}545.977 \ln x$$
$$g(x) = \frac{102{,}519.98}{1 + .1536e^{-.4263x}}.$$

   (c) Use the table feature to estimate the number of multiple births in 2000 and 2005.
   (d) Over the long run, which model do you think is the better predictor?

22. The graph shows the U.S. Census Bureau estimates of future U.S. population.

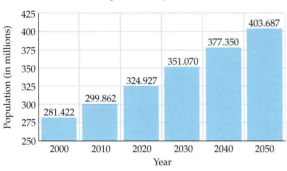

U.S. Population Projections: 2000–2050

(a) How well do the projections in the graph compare with those given by the logistic model in Example 2?

(b) Find a logistic model of the U.S. population, using the data given in Example 2 for the years from 1900 to 2000.

(c) How well do the projections in the graph compare with those given by the model in part (b)?

23. Infant mortality rates in the United States are shown in the table.

Year	Infant Mortality Rate[*]	Year	Infant Mortality Rate[*]
1920	76.7	1980	12.6
1930	60.4	1985	10.6
1940	47.0	1990	9.2
1950	29.2	1995	7.6
1960	26.0	2000	6.9
1970	20.0		

(a) Sketch a scatter plot of the data, with $x = 0$ corresponding to 1900.

(b) Verify that the set of points $(x, \ln y)$, where (x, y) are the original data points, is approximately linear.

(c) On the basis of part (b), what type of model would be appropriate for this data? Find such a model.

24. The number of children who were home schooled in the United States in selected years is shown in the table at the top of the next column.[†]

(a) Sketch a scatter plot of the data, with $x = 0$ corresponding to 1980.

(b) Find a quadratic model for the data.

(c) Find a logistic model for the data.

(d) What is the number of home-schooled children predicted by each model for the year 2003?

(e) What are the limitations of each model?

Fall of School Year	Number of Children (in thousands)
1985	183
1988	225
1990	301
1992	470
1993	588
1994	735
1995	800
1996	920
1997	1100
1999	1400
2000	1700

25. The average number of students per computer in the U.S. public schools (elementary through high school) is shown in the table.

(a) Sketch a scatter plot of the data, with $x = 7$ corresponding to 1987.

(b) Find an exponential model for the data.

(c) Use the model to estimate the number of students per computer in 2003.

(d) In what year, according to this model, will each student have his or her own computer in school?

(e) What are the limitations of this model?

Fall of School Year	Students/Computer
1987	32
1988	25
1989	22
1990	20
1991	18
1992	16
1993	14
1994	10.5
1995	10
1996	7.8
1997	6.1
1998	5.7
1999	5.4

[*]Rates are infant (under 1 year) deaths per 1000 live births.
[†]National Home Education Research Institute.

26. Projected Medicare spending on home health care is shown in the table.*

Year	Home Health Care Spending (in billions of dollars)
2000	8.3
2002	10
2004	11
2006	14
2008	18.3
2010	23.3

(a) Find cubic and exponential models for this data, with $x = 0$ corresponding to 2000.
(b) Use each model to estimate home health care spending in 2005 and 2007.
(c) As a taxpayer for the next 25 years, which model would you prefer to have happen?
(d) The actual projected spending for 2012 is about $32.7 billion. In light of this, which model appears to be the best fit for the next decade?

27. The number of U.S. households (in millions) from 1995 to 2000 are shown in the first two columns of the following table.†

Year	Households	Predicted Number of Households	Ratio
1995	99.0		
1996	99.6		
1997	101.0		
1998	102.5		
1999	103.6		
2000	104.8		

(a) Sketch a scatter plot of the data with $x = 0$ corresponding to 1995.
(b) Fill in column 4 by dividing each entry in column 2 by the preceding one; round your answers to two decimal places. In view of column 4, what type of model is appropriate here?
(c) Find an appropriate model for the data.
(d) Use the model to complete column 3 in the table. How well does the model fit the data?
(e) If the model remained accurate, how many U.S. households were there in 2003?

28. In the past two decades, more women than men have been entering college. The table shows the percentage of male first-year college students in selected years.‡

Year	1985	1990	1995	1997	1998	1999	2000
Percent	48.9	46.9	45.6	45.5	45.5	45.3	45.2

(a) Find three models for this data: exponential, logarithmic, and power, with $x = 5$ corresponding to 1985.
(b) For the years 1985–2000, is there any significant difference among the models?
(c) Assume that the models remain accurate. What year does each predict as the first year in which fewer than 43% of first-year college students will be male?

29. The table gives the life expectancy (at birth) of a woman born in the given year.§

Year	Life Expectancy (in years)
1910	51.8
1920	54.6
1930	61.6
1940	65.2
1950	71.1
1960	73.1
1970	74.7
1980	77.5
1990	78.8
2000	79.4

(a) Find a logarithmic model for the data, with $x = 10$ corresponding to 1910.
(b) Use the model to find the life expectancy of a woman born in 1986. [For comparison, the actual expectancy is 78.3 years.]
(c) Assume that the model remains accurate. In what year will the life expectancy of a woman born in that year be at least 81 years?

30. The table gives the death rate in motor vehicle accidents (per 100,000 population) in selected years.

Year	1970	1980	1985	1990	1995	2000
Death Rate	26.8	23.4	19.3	18.8	16.5	15.6

(a) Find an exponential model for the data, with $x = 0$ corresponding to 1970.
(b) What was the death rate in 1998 and in 2002?
(c) Assume that the model remains accurate, when will the death rate drop to 13 per 100,000?

*Congressional Budget Office.
†U.S. Department of Commerce.
‡Higher Education Research Institute at UCLA.
§National Center for Health Statistics.

5.7 Inverse Functions

Consider the functions *f* and *h* given by these tables.

f-input	−2	−1	0	1	2
f-output	−3	−2	1	4	5

h-input	1	2	3	4	5
h-output	−1	3	0	3	2

Roadmap

Readers who cover Section 5.7 before Section 5.3 should omit Examples 5, 8, and 10.

With the function *h*, two different inputs (2 and 4) produce the same output 3. With the function *f*, however, different inputs always produce different outputs. Functions with this property have a special name. A function *f* is said to be **one-to-one** if distinct inputs always produce distinct outputs, that is,

$$\text{if } a \neq b, \text{ then } f(a) \neq f(b)$$

In graphical terms, this means that two points on the graph, $(a, f(a))$ and $(b, f(b))$, that have different *x*-coordinates $[a \neq b]$ must also have different *y*-coordinates $[f(a) \neq f(b)]$. Consequently, these points cannot lie on the same horizontal line because all points on a horizontal line have the same *y*-coordinate. Therefore, we have this geometric test to determine whether a function is one-to-one.

The Horizontal Line Test

If a function *f* is one-to-one, then it has the following property.

No horizontal line intersects the graph of *f* more than once.

Conversely, if the graph of a function has this property, then the function is one-to-one.

EXAMPLE 1

Which of the following functions are one-to-one?

(a) $f(x) = 7x^5 + 3x^4 - 2x^3 + 2x + 1$

(b) $g(x) = x^3 - 3x - 1$

(c) $h(x) = 1 - .2x^3$

SOLUTION   Complete graphs of each function are shown in Figure 5–43.

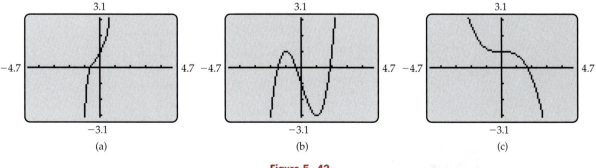

(a)   (b)   (c)

Figure 5–43

(a) The graph of f in Figure 5–43(a) passes the horizontal line test, since no horizontal line intersects the graph more than once. Hence, f is one-to-one.

(b) The graph of g in Figure 5–43(b) obviously fails the horizontal line test, since many horizontal lines (including the x-axis) intersect the graph more than once. Therefore, g is not one-to-one.

(c) The graph of h in Figure 5–43(c) appears to contain a horizontal line segment. So h appears to fail the Horizontal Line Test, since the horizontal line through (0, 1) seems to intersect the graph infinitely many times. But appearances are deceiving.

TECHNOLOGY TIP

Although a horizontal segment might appear on a calculator screen when the graph is actually rising or falling, there is another possibility. The graph may have a tiny wiggle (less than the height of a pixel) and thus fail the horizontal line test:

You can usually detect such a wiggle by zooming in to magnify that portion of the graph or by using the trace feature to see whether the y-coordinates increase and then decrease (or vice versa) along the "horizontal" segment.

GRAPHING EXPLORATION

Graph $h(x) = 1 - .2x^3$ and use the trace feature to move from left to right along the "horizontal" segment. Do the y-coordinates stay the same, or do they decrease?

The Graphing Exploration shows that the graph is actually falling from left to right, so each horizontal line intersects it only once. (It appears to have a horizontal segment because the amount the graph falls there is less than the height of a pixel on the screen.) Therefore, h is a one-to-one function. ■

The function f in Example 1 is an **increasing function** (its graph is always rising from left to right), and the function h is a **decreasing function** (its graph is always falling from left to right). Every increasing or decreasing function is necessarily one-to-one because its graph can never touch the same horizontal line twice (it would have to change from rising to falling, or vice versa, to do so).

■■ INVERSE FUNCTIONS

We begin with a simple example that illustrates the basic idea of an inverse function. Consider the one-to-one function f introduced at the beginning of this section.

f-input	−2	−1	0	1	2
f-output	−3	−2	1	4	5

Now define a new function g by the following table (which simply *reverses* the rows in the f table).

g-input	−3	−2	1	4	5
g-output	−2	−1	0	1	2

Note that the inputs of f are the outputs of g and the outputs of f are the inputs of g. In other words,

Domain of f = Range of g and   Range of f = Domain of g.

The rule of g *reverses* the action of f by taking each output of f back to the input it came from. For instance,

$$g(4) = 1 \quad \text{and} \quad f(1) = 4$$
$$g(-3) = -2 \quad \text{and} \quad f(-2) = -3,$$

and in general,

$$g(y) = x \quad \text{exactly when} \quad f(x) = y.$$

We say that g is the *inverse function* of f.

The preceding construction works for any one-to-one function f. Each output of f comes from exactly one input (because different inputs produce different outputs). Consequently, we can define a new function g that reverses the action of f by sending each output back to the unique input it came from. For instance, if $f(7) = 11$, then $g(11) = 7$. Thus, the outputs of f become the inputs of g, and we have the following definition.

Inverse Functions

Let f be a one-to-one function. Then the **inverse function** of f is the function g whose rule is

$$g(y) = x \quad \text{exactly when} \quad f(x) = y.$$

The domain of g is the range of f and the range of g is the domain of f.

EXAMPLE 2

The graph of $f(x) = 3x - 2$ is a straight line that certainly passes the Horizontal Line Test, so f is one-to-one and has an inverse function g. From the definition of g, we know that

$$g(y) = x \quad \text{exactly when} \quad f(x) = y,$$

that is,

$$g(y) = x \quad \text{exactly when} \quad 3x - 2 = y.$$

To find the rule of g, we need only solve this last equation for x.

$$3x - 2 = y$$

Add 2 to both sides: $\qquad 3x = y + 2$

Divide both sides by 3: $\qquad x = \dfrac{y + 2}{3}.$

Since $g(y) = x$, we see that the rule of g is $g(y) = \dfrac{y + 2}{3}.$ ■

Recall that the letter used for the variable of a function doesn't matter. For instance, $h(x) = x^2$ and $h(t) = t^2$ and $h(u) = u^2$ all describe the same function, whose rule is "square the input." When dealing with inverse functions, it is customary to use the same variable for both f and its inverse g. Consequently, the inverse function in Example 2 would normally be written as

$$g(x) = \frac{x + 2}{3}.$$

We can summarize this procedure as follows.

Finding Inverse Functions Algebraically

To find the inverse function of a one-to-one function f:

1. Solve the equation $f(x) = y$ for x.

2. The solution is an expression in y, which is the rule of the inverse function g, that is, $x = g(y)$.

3. Rewrite the rule of $x = g(y)$ by interchanging x and y.

EXAMPLE 3

Use your calculator to verify that the function $f(x) = x^3 + 5$ passes the Horizontal Line Test and hence is one-to-one. Its inverse can be found by solving for x in the equation $x^3 + 5 = y$.

Subtract 5 from both sides: $x^3 = y - 5$

Take cube roots on both sides: $x = \sqrt[3]{y - 5}.$

Therefore, $g(y) = \sqrt[3]{y - 5}$ is the inverse function of f. Interchanging x and y, we write this rule as $g(x) = \sqrt[3]{x - 5}.$ ∎

EXAMPLE 4

The function $f(x) = \sqrt{x - 3}$ is one-to-one, as you can verify with your calculator. To find its inverse, we solve the equation

$$y = \sqrt{x - 3}$$

Square both sides: $y^2 = x - 3$

Add 3 to both sides: $x = y^2 + 3.$

Although this last equation is defined for all real numbers y, the original equation $y = \sqrt{x - 3}$ has $y \geq 0$ (since square roots are nonnegative). In other words, the range of the function f (the possible values of y) consists of all nonnegative real numbers. Consequently, the domain of the inverse function g is the set of all nonnegative real numbers, and its rule is

$$g(y) = y^2 + 3 \qquad (y \geq 0).$$

Once again, it's customary to use the same variable to describe both f and its inverse function, so we write the rule of g as $g(x) = x^2 + 3 \ (x \geq 0)$. ∎

EXAMPLE 5*

Show that the inverse function of the exponential function $f(x) = 10^x$ is the logarithmic function $g(x) = \log x$.

---

*Skip this example if you have not read Section 5.3.

SOLUTION   The definition of common logarithms shows that

$$\log y = x \qquad \text{exactly when} \qquad 10^x = y$$

(see the box on page 393). If $f(x) = 10^x$ and $g(x) = \log x$, then this statement says that

$$g(y) = x \qquad \text{exactly when} \qquad f(x) = y,$$

which says that g is the inverse function of f (see the box preceding Example 2).

An alternative approach is to note that $f(x) = 10^x$ is an increasing function (its graph is always rising from left to right). So f is one-to-one and has an inverse function g. Its rule can be found by using the techniques of Section 5.5 to solve the equation $f(x) = y$ for x, as in the preceding examples.

$$f(x) = y$$
$$10^x = y$$

Take logarithms of both sides: $\log 10^x = \log y$

Use the Power Law for Logarithms: $x \log 10 = \log y$

Remember that $\log 10 = 1$: $x = \log y$

Therefore, the rule of the inverse function $g(y) = x$ is $g(y) = \log y$. Changing variables, as above, we write the inverse function as $g(x) = \log x$.

Note that the arguments used above can also be applied with e in place of 10 and "ln" in place of "log" to show that $g(x) = \ln x$ is the inverse function of $f(x) = e^x$. This will be shown by another method in Example 8.   ■

▪▪ THE ROUND-TRIP PROPERTIES

The inverse function g of a function f was designed to send each output of f back to the input it came from. Consequently, if you first apply f and then apply g to the result, you obtain the number you started with, as is illustrated in the next example.

EXAMPLE 6

As we saw in Example 2, the inverse function of $f(x) = 3x - 2$ is

$$g(x) = \frac{x + 2}{3}.$$

If we start with a number c and apply f, we obtain $f(c) = 3c - 2$. If we now apply g to this result, we obtain

$$g(f(c)) = g(3c - 2) = \frac{(3c - 2) + 2}{3} = c.$$

So we are back where we started. Similarly, if we first apply g and then apply f to a number, we end up where we started.

$$f(g(c)) = f\left(\frac{c + 2}{3}\right) = 3\left(\frac{c + 2}{3}\right) - 2 = c.$$

The function $f(x) = x^3 + 5$ of Example 3 and its inverse function

$$g(x) = \sqrt[3]{x - 5}$$

also have these "round-trip" properties. If you apply one function and then the other, you wind up at the number you started with.

$$g(f(x)) = g(x^3 + 5) = \sqrt[3]{(x^3 + 5) - 5} = \sqrt[3]{x^3} = x,$$

and

$$f(g(x)) = f(\sqrt[3]{x - 5}) = (\sqrt[3]{x - 5})^3 + 5 = (x - 5) + 5 = x. \quad \blacksquare$$

Not only do a function and its inverse have the round-trip properties illustrated in Example 6, but somewhat more is true.

Round-Trip Theorem

A one-to-one function f and its inverse function g have the following properties.

$$g(f(x)) = x \quad \text{for every } x \text{ in the domain of } f$$

$$f(g(x)) = x \quad \text{for every } x \text{ in the domain of } g$$

Conversely, if f and g are functions having these properties, then f is one-to-one, and its inverse is g.

Proof   By the definition of inverse function,

$$g(d) = c \qquad \text{exactly when} \qquad f(c) = d.$$

Consequently, for any c in the domain of f.

$$g(f(c)) = g(d) \quad \text{(because } f(c) = d\text{)}$$

$$= c \quad \text{(because } g(d) = c\text{)}.$$

A similar argument shows that $f(g(d)) = d$ for any d in the domain of g. The last statement in the Round-Trip Theorem is proved in Exercise 53. $\quad \blacksquare$

EXAMPLE 7

If

$$f(x) = \frac{5}{2x - 4} \qquad \text{and} \qquad g(x) = \frac{4x + 5}{2x},$$

then for every x in the domain of f (that is, all $x \neq 2$),

$$g(f(x)) = g\left(\frac{5}{2x - 4}\right) = \frac{4\left(\dfrac{5}{2x - 4}\right) + 5}{2\left(\dfrac{5}{2x - 4}\right)} = \frac{\dfrac{20 + 5(2x - 4)}{2x - 4}}{\dfrac{10}{2x - 4}}$$

$$= \frac{20 + 5(2x - 4)}{10} = \frac{20 + 10x - 20}{10} = \frac{10x}{10} = x,$$

and for every x in the domain of g (all $x \neq 0$),

$$f(g(x)) = f\left(\frac{4x+5}{2x}\right) = \frac{5}{2\left(\dfrac{4x+5}{2x}\right) - 4} = \frac{5}{\dfrac{4x+5}{x} - 4}$$

$$= \frac{5}{\dfrac{4x+5-4x}{x}} = \frac{5}{\dfrac{5}{x}} = x.$$

By the Round-Trip Theorem, f is a one-to-one function with inverse g. ■

EXAMPLE 8*

Use the Round-Trip Theorem to show that $g(x) = \ln x$ is the inverse function of $f(x) = e^x$.

SOLUTION   For every real number x (that is, every number in the domain of f),

$$g(f(x)) = g(e^x) \qquad \text{[Definition of } f]$$

$$= \ln e^x \qquad \text{[Definition of } g]$$

$$= x. \qquad \text{[Property 3 of logarithms (page 397)]}$$

Hence, $g(f(x)) = x$. Reversing the order of f and g, we have

$$f(g(x)) = f(\ln x) \qquad \text{[Definition of } g]$$

$$= e^{\ln x} \qquad \text{[Definition of } f]$$

$$= x. \qquad \text{Property 4 of logarithms (page 397)]}$$

Thus, $f(g(x)) = x$ for every $x > 0$ (every number in the domain of g). Therefore, by the Round-Trip Theorem, $g(x) = \ln x$ is the inverse function of $f(x) = e^x$. ■

▄▄ GRAPHS OF INVERSE FUNCTIONS

Finding the rule of the inverse function g of a one-to-one function f by solving the equation $y = f(x)$ for x, as in the preceding examples, is not always possible (some equations are hard to solve). But even if you don't know the rule of g, you can always find its graph, as shown below.

Suppose f is a one-to-one function and g is its inverse function. Then by the definition of inverse function,

$$f(a) = b \qquad \text{exactly when} \qquad g(b) = a.$$

But $f(a) = b$ means that (a, b) is on the graph of f, and $g(b) = a$ means that (b, a) is on the graph of g. Therefore, we have the following.

---

*Skip this example if you have not read Section 5.3.

Inverse Function Graphs

> If f is a one-to-one function and g is its inverse function, then
>
> (a, b) is on the graph of f exactly when (b, a) is on the graph of g.

Therefore, the graph of the inverse function g can be obtained by reversing the coordinates of each point on the graph of f. There are two practical ways of doing this, each of which is illustrated below.

EXAMPLE 9

Verify that $f(x) = .7x^5 + .3x^4 - .2x^3 + 2x + .5$ has an inverse function g. Use parametric graphing to graph both f and g.*

SOLUTION   First, we graph f in parametric mode by letting

$$x = t   \text{and}   y = f(t) = .7t^5 + .3t^4 - .2t^3 + 2t + .5.$$

The complete graph in Figure 5–44 shows that f is one-to-one (why?) and hence has an inverse function g. According to the preceding box, the graph of g can be obtained by taking each point on the graph of f and reversing its coordinates. Thus, g can be graphed parametrically by letting

$$x = f(t) = .7t^5 + .3t^4 - .2t^3 + 2t + .5   \text{and}   y = t.$$

Figure 5–45 shows the graphs of g and f on the same screen.   ■

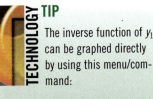

TECHNOLOGY TIP

The inverse function of y_1 can be graphed directly by using this menu/command:

TI: DRAW/DRAWINV y_1

Casio: SKETCH/INV

[DRAW is a submenu of the GRAPH menu on TI-86/89.]

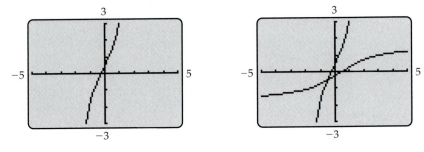

Figure 5–44

Figure 5–45

The second method of graphing inverse functions by reversing coordinates depends on this geometric fact, which is proved in Exercise 51:

> The line $y = x$ is the perpendicular bisector of
> the line segment from (a, b) to (b, a),

as shown in Figure 5–46 when $a = 7$, $b = 2$. Thus (a, b) and (b, a) lie on opposite sides of $y = x$, the same distance from it: They are mirror images of each other, with the line $y = x$ being the mirror.[†] Consequently, the graph of the inverse function g is the mirror image of the graph of f. In formal terms, we have the following.

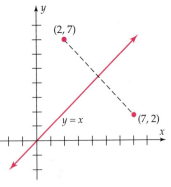

Figure 5–46

Inverse Function Graphs

> If g is the inverse function of f, then the graph of g is the reflection of the graph of f in the line $y = x$.

---

*Parametric graphing is explained in Special Topics 3.3.A.

[†]In technical terms, (a, b) and (b, a) are **symmetric with respect to the line $y = x$.**

EXPLORATION

Illustrate this fact by graphing the line $y = x$, the function

$$f(x) = x^3 + 5$$

of Example 3, and its inverse $g(x) = \sqrt[3]{x - 5}$ on the same screen. (Use a square viewing window so that the mirror effect won't be distorted.)

EXAMPLE 10*

We know that $g(x) = \ln x$ is the inverse function of $f(x) = e^x$. Figure 5–47 shows that the graph of g is the reflection (mirror image) of the graph of f in the line (mirror) $y = x$. ∎

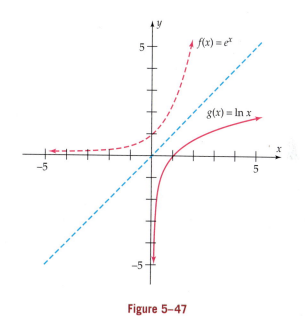

Figure 5–47

NOTE: In many texts, the inverse function of a function f is denoted f^{-1}. In this notation, for instance, the inverse of the function $f(x) = x^3 + 5$ in Example 3 would be written as $f^{-1}(x) = \sqrt[3]{x - 5}$. Similarly, the reversal properties of inverse functions become

$$f^{-1}(f(x)) = x \text{ for every } x \text{ in the domain of } f \text{ and}$$
$$f(f^{-1}(x)) = x \text{ for every } x \text{ in the domain of } f^{-1}.$$

In this context, f^{-1} does *not* mean $1/f$ (see Exercise 47).

---

*Skip this example if you have not read Section 5.3.

✓ EXERCISES 5.7

In Exercises 1–8, use a calculator and the Horizontal Line Test to determine whether or not the function f is one-to-one.

1. $f(x) = x^4 - 4x^2 + 3$

2. $f(x) = x^4 - 4x + 3$

3. $f(x) = x^3 + x - 5$

4. $f(x) = \begin{cases} x - 3 & \text{for } x \le 3 \\ 2x - 6 & \text{for } x > 3 \end{cases}$

5. $f(x) = x^5 + 2x^4 - x^2 + 4x - 5$

6. $f(x) = x^3 - 4x^2 + x - 10$

7. $f(x) = .1x^3 - .1x^2 - .005x + 1$

8. $f(x) = .1x^3 + .005x + 1$

In Exercises 9–22, use algebra to find the inverse of the given one-to-one function.

9. $f(x) = -x$

10. $f(x) = -x + 1$

11. $f(x) = 5x - 4$

12. $f(x) = -3x + 5$

13. $f(x) = 5 - 2x^3$

14. $f(x) = (x^5 + 1)^3$

15. $f(x) = \sqrt{4x - 7}$

16. $f(x) = 5 + \sqrt{3x - 2}$

17. $f(x) = 1/x$

18. $f(x) = 1/\sqrt{x}$

19. $f(x) = \dfrac{1}{2x + 1}$

20. $f(x) = \dfrac{x}{x + 1}$

21. $f(x) = \dfrac{x^3 - 1}{x^3 + 5}$

22. $f(x) = \sqrt[5]{\dfrac{3x - 1}{x - 2}}$

In Exercises 23–28, use the Round-Trip Theorem on page 444 to show that g is the inverse of f.

23. $f(x) = x + 1$, $g(x) = x - 1$

24. $f(x) = 2x - 6$, $g(x) = \dfrac{x}{2} + 3$

25. $f(x) = \dfrac{1}{x + 1}$, $g(x) = \dfrac{1 - x}{x}$

26. $f(x) = \dfrac{-3}{2x + 5}$, $g(x) = \dfrac{-3 - 5x}{2x}$

27. $f(x) = x^5$, $g(x) = \sqrt[5]{x}$

28. $f(x) = 5^x$, $g(x) = \log_5 x$

29. Show that the inverse function of the function f whose rule is $f(x) = \dfrac{2x + 1}{3x - 2}$ is f itself.

30. List three different functions (other than the one in Exercise 29), each of which is its own inverse. [Many correct answers are possible.]

In Exercises 31 and 32, the graph of a function f is given. Sketch the graph of the inverse function of f. [Reflect carefully.]

31.

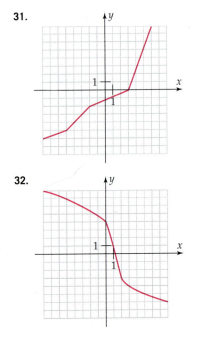

32.

In Exercises 33–38, each given function has an inverse function. Sketch the graph of the inverse function.

33. $f(x) = \sqrt{x + 3}$

34. $f(x) = \sqrt{3x - 2}$

35. $f(x) = .3x^5 + 2$

36. $f(x) = \sqrt[3]{x + 3}$

37. $f(x) = \sqrt[5]{x^3 + x - 2}$

38. $f(x) = \begin{cases} x^2 - 1 & \text{for } x \le 0 \\ -.5x - 1 & \text{for } x > 0 \end{cases}$

In Exercises 39–46, none of the functions has an inverse. State at least one way of restricting the domain of the function (that is, find a function with the same rule and a smaller domain) so that the restricted function has an inverse. Then find the rule of the inverse function.

Example: $f(x) = x^2$ has no inverse. But the function h with domain all $x \ge 0$ and rule $h(x) = x^2$ is increasing (its graph is the right half of the graph of f—see Figure 1–17 on page 73)—and therefore has an inverse.

39. $f(x) = |x|$

40. $f(x) = |x - 3|$

41. $f(x) = -x^2$

42. $f(x) = x^2 + 4$

43. $f(x) = \dfrac{x^2 + 6}{2}$

44. $f(x) = \sqrt{4 - x^2}$

45. $f(x) = \dfrac{1}{x^2 + 1}$

46. $f(x) = 3(x + 5)^2 + 2$

47. (a) Using the f^{-1} notation for inverse functions, find $f^{-1}(x)$ when $f(x) = 3x + 2$.

    (b) Find $f^{-1}(1)$ and $1/f(1)$. Conclude that f^{-1} is not the same function as $1/f$.

48. Let C be the temperature in degrees Celsius. Then the temperature in degrees Fahrenheit is given by $f(C) = \frac{9}{5}C + 32$. Let g be the function that converts degrees Fahrenheit to degrees Celsius. Show that g is the inverse function of f and find the rule of g.

Thinkers

49. Let m and b be constants with $m \neq 0$. Show that the function $f(x) = mx + b$ has an inverse function g and find the rule of g.

50. Prove that the function $h(x) = 1 - .2x^3$ of Example 1(c) is one-to-one by showing that it satisfies the definition:

$$\text{If } a \neq b, \text{ then } h(a) \neq h(b).$$

[*Hint:* Use the rule of h to show that when $h(a) = h(b)$, then $a = b$. If this is the case, then it is impossible to have $h(a) = h(b)$ when $a \neq b$.]

51. Show that the points $P = (a, b)$ and $Q = (b, a)$ are symmetric with respect to the line $y = x$ as follows.

    (a) Find the slope of the line through P and Q.

    (b) Use slopes to show that the line through P and Q is perpendicular to $y = x$.

    (c) Let R be the point where the line $y = x$ intersects line segment PQ. Since R is on $y = x$, it has coordinates (c, c) for some number c, as shown in the figure. Use the distance formula to show that segment PR has the same length as segment RQ. Conclude that the line $y = x$ is the perpendicular bisector of segment PQ. Therefore, P and Q are symmetric with respect to the line $y = x$.

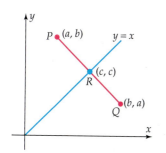

52. (a) Experiment with your calculator or use some of the preceding exercises to find four different increasing functions. For each function, sketch the graph of the function and the graphs of its inverse on the same set of axes.

    (b) On the basis of the evidence in part (a), do you think the following statement true or false: The inverse function of every increasing function is also an increasing function.

    (c) Do parts (a) and (b) with "increasing" replaced by "decreasing."

53. Suppose that functions f and g have these round-trip properties:

    (1) $g(f(x)) = x$ for every x in the domain of f.

    (2) $f(g(y)) = y$ for every y in the domain of g.

To complete the proof of the Round-Trip Theorem, we must show that g is the inverse function of f. Do this as follows.

    (a) Prove that f is one-to-one by showing that

$$\text{if } a \neq b, \text{ then } f(a) \neq f(b).$$

    [*Hint:* If $f(a) = f(b)$, apply g to both sides and use (1) to show that $a = b$. Consequently, if $a \neq b$, it is impossible to have $f(a) = f(b)$.]

    (b) If $g(y) = x$, show that $f(x) = y$. [*Hint:* Use (2).]

    (c) If $f(x) = y$, show that $g(y) = x$. [*Hint:* Use (1).]

Parts (b) and (c) prove that

$$g(y) = x \qquad \text{exactly when} \qquad f(x) = y.$$

Hence, g is the inverse function of f (see page 441).

54. Prove that every function f that has an inverse function g is one-to-one. [*Hint:* The proof of the Round-Trip Theorem on page 444 shows that f and g have the round-trip properties; use Exercise 53(a).]

Chapter 5 Review

IMPORTANT FACTS & FORMULAS

■ *Laws of Exponents:*

$$c^r c^s = c^{r+s} \qquad (cd)^r = c^r d^r$$

$$\frac{c^r}{c^s} = c^{r-s} \qquad \left(\frac{c}{d}\right)^r = \frac{c^r}{d^r}$$

$$(c^r)^s = c^{rs} \qquad c^{-r} = \frac{1}{c^r}$$

■ *Exponential Growth Functions:*

$$f(x) = P(1 + r)^x \qquad (r > 0),$$

$$f(x) = Pa^x \qquad (a > 1),$$

$$f(x) = Pe^{kx} \qquad (k > 0)$$

■ *Exponential Decay Functions:*

$$f(x) = P(1 - r)^x \qquad (0 < r < 1),$$

$$f(x) = Pa^x \qquad (0 < a < 1),$$

$$f(x) = Pe^{kx} \qquad (k < 0)$$

■ *Compound Interest Formula:* $A = P(1 + r)^t$

■ *Logarithm Laws:* For all v, $w > 0$ and any k:

$$\ln (vw) = \ln v + \ln w \qquad \log_b (vw) = \log_b v + \log_b w$$

$$\ln \left(\frac{v}{w}\right) = \ln v - \ln w \qquad \log_b \left(\frac{v}{w}\right) = \log_b v - \log_b w$$

$$\ln (v^k) = k(\ln v) \qquad \log_b (v^k) = k(\log_b v)$$

■ *Change of Base Formula:* $\log_b v = \dfrac{\ln v}{\ln b}$

■ $g(x) = \log x$ is the inverse function of $f(x) = 10^x$:

$$10^{\log v} = v \text{ for all } v > 0 \qquad \text{and} \qquad \log 10^u = u \text{ for all } u.$$

■ $g(x) = \ln x$ is the inverse function of $f(x) = e^x$:

$$e^{\ln v} = v \text{ for all } v > 0 \qquad \text{and} \qquad \ln (e^u) = u \text{ for all } u.$$

■ $h(x) = \log_b x$ is the inverse function of $k(x) = b^x$:

$$b^{\log_b v} = v \text{ for all } v > 0 \qquad \text{and} \qquad \log_b (b^u) = u \text{ for all } u.$$

CATALOG OF BASIC FUNCTIONS—PART 3

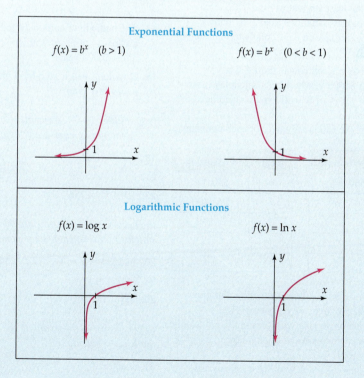

Exponential Functions

$f(x) = b^x \quad (b > 1)$ $f(x) = b^x \quad (0 < b < 1)$

Logarithmic Functions

$f(x) = \log x$ $f(x) = \ln x$

In Questions 1–8, find a viewing window (or windows) that shows a complete graph of the function.

1. $f(x) = 3.7^x$ **2.** $g(x) = 2.5^{-x}$

3. $g(x) = 2^x - 1$ **4.** $f(x) = 2^{x-1}$

5. $f(x) = 2^{x^3 - x - 2}$ **6.** $g(x) = \dfrac{850}{1 + 5e^{-.4x}}$

7. $h(x) = \ln(x + 4) - 2$ **8.** $k(x) = \ln\left(\dfrac{x}{x-2}\right)$

9. A computer software company claims the following function models the "learning curve" for their mathematical software.

$$P(t) = \frac{100}{1 + 48.2e^{-.52t}},$$

where t is measured in months and $P(t)$ is the average percent of the software program's capabilities mastered after t months.

(a) Initially, what percent of the program is mastered?

(b) After six months, what percent of the program is mastered?

(c) Roughly, when can a person expect to "learn the most in the least amount of time"?

(d) If the company's claim is true, how many months will it take to have completely mastered the program?

10. Compunote has offered you a starting salary of $60,000 with $1000 yearly raises. Calcuplay offers you an initial salary of $30,000 and a guaranteed 6% raise each year.

(a) Complete the following table for each company.

Year	Compunote
1	$60,000
2	$61,000
3	
4	
5	

Year	Calcuplay
1	$30,000
2	$31,800
3	
4	
5	

(b) For each company, write a function that gives your salary in terms of years employed.

(c) If you plan on staying with the company for only five years, which job should you take to earn the most money?

(d) If you plan on staying with the company for 20 years, which is your better choice?

(e) In what year does the salary at Calcuplay exceed the salary at Compunote?

11. Phil borrows $800 at 9% interest, compounded annually.

(a) How much does he owe after six years?

(b) If he pays off the loan at the end of six years, how much interest will he owe?

12. If you invest $5000 for five years at 9% interest, how much more will you make if interest is compounded continuously than if it is compounded quarterly?

13. Mary Karen invests $2000 at 5.5% interest, compounded monthly.

(a) How much is her investment worth in three years?

(b) When will her investment be worth $12,000?

14. If a $2000 investment grows to $5000 in 14 years, with interest compounded annually, what is the interest rate?

15. Company sales are increasing at 6.5% per year. If sales this year are $56,000, write the rule of a function that gives the sales in year x ($x = 0$ being the present year).

16. The population of Potterville is decreasing at an annual rate of 1.5%. If the population is now 38,500, what will the population be x years from now?

17. The half-life of carbon-14 is 5730 years. How much carbon-14 remains from an original 16-gram sample after 12,000 years?

18. How long will it take for 4 grams of carbon-14 to decay to 1 gram?

In Questions 19–24, translate the given exponential statement into an equivalent logarithmic one.

19. $e^{6.628} = 756$ **20.** $e^{5.8972} = 364$

21. $e^{r^2 - 1} = u + v$ **22.** $e^{a-b} = c$

23. $10^{2.8785} = 756$ **24.** $10^{c+d} = t$

In Questions 25–30, translate the given logarithmic statement into an equivalent exponential one.

25. $\ln 1234 = 7.118$ **26.** $\ln(ax + b) = y$

27. $\ln(rs) = t$ **28.** $\log 1234 = 3.0913$

29. $\log_5(cd - k) = u$ **30.** $\log_d(uv) = w$

In Questions 31–34, evaluate the given expression without using a calculator.

31. $\ln e^3$ **32.** $\ln \sqrt[3]{e}$

33. $e^{\ln 3/4}$ **34.** $e^{\ln(x+2y)}$

REVIEW QUESTIONS

35. Simplify: $3 \ln \sqrt{x} + (1/2)\ln x$
36. Simplify: $\ln (e^{4e})^{-1} + 4e$

In Questions 37–39, write the given expression as a single logarithm.

37. $\ln 3x - 3 \ln x + \ln 3y$
38. $\log_7 7x + \log_7 y - 1$
39. $4 \ln x - 2(\ln x^3 + 4 \ln x)$
40. $\log (-.01) = ?$
41. $\log_{20} 400 = ?$

42. You are conducting an experiment about memory. The people who participate agree to take a test at the end of your course and every month thereafter for a period of two years. The average score for the group is given by the model

$$M(t) = 91 - 14 \ln (t + 1) \qquad (0 \le t \le 24)$$

where t is time in months after the first test.
(a) What is the average score on the initial exam?
(b) What is the average score after three months?
(c) When will the average drop below 50%?
(d) Is the magnitude of the rate of memory loss greater in the first month after the course (from $t = 0$ to $t = 1$) or after the first year (from $t = 12$ to $t = 13$)?
(e) Hypothetically, if the model could be extended past $t = 24$ months, would it be possible for the average score to be 0%?

43. Which of the following statements are *true*?
(a) $\ln 10 = (\ln 2)(\ln 5)$ (b) $\ln (e/6) = \ln e + \ln 6$
(c) $\ln (1/7) + \ln 7 = 0$ (d) $\ln (-e) = -1$
(e) None of the above is true.

44. Which of the following statements are *false*?
(a) $10 (\log 5) = \log 50$ (b) $\log 100 + 3 = \log 10^5$
(c) $\log 1 = \ln 1$ (d) $\log 6/\log 3 = \log 2$
(e) All of the above are false.

Use the following six graphs for Questions 45 and 46.

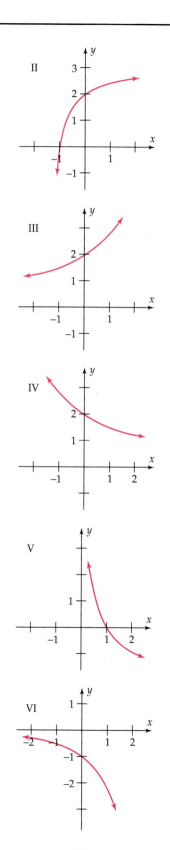

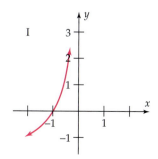

QUESTIONS

45. If $b > 1$, then the graph of $f(x) = -\log_b x$ could possibly be:

(a) I   (b) IV

(c) V   (d) VI

(e) none of these

46. If $0 < b < 1$, then the graph of $g(x) = b^x + 1$ could possibly be:

(a) II   (b) III

(c) IV   (d) VI

(e) none of these

47. If $\log_3 9^{x^2} = 4$, what is x?

48. What is the domain of the function $f(x) = \ln\left(\dfrac{x}{x-1}\right)$?

In Questions 49–57, solve the equation for x.

49. $8^x = 4^{x^2 - 3}$ **50.** $e^{3x} = 4$

51. $2 \cdot 4^x - 5 = -4$ **52.** $725e^{-4x} = 1500$

53. $u = c + d \ln x$ **54.** $2^x = 3^{x+3}$

55. $\ln x + \ln (3x - 5) = \ln 2$

56. $\ln (x + 8) - \ln x = 1$

57. $\log (x^2 - 1) = 2 + \log (x + 1)$

58. At a small community college the spread of a rumor through the population of 500 faculty and students can be modeled by the equation.

$$\ln (n) - \ln (1000 - 2n) = .65t - \ln 998,$$

where n is the number of people who have heard the rumor after t days.

(a) How many people know the rumor initially (at $t = 0$)?

(b) How many people have heard the rumor after four days?

(c) Roughly, in how many weeks will the entire population have heard the rumor?

(d) Use the properties of logarithms to write n as a function of t; in other words solve the model above for n in terms of t.

(e) Enter the function you found in part (d) into your calculator and use the table feature to check your answers to parts (a), (b), and (c). Do they agree?

(f) Now graph the function. Roughly over what time interval does the rumor seem to spread the fastest?

59. The half-life of polonium (^{210}Po) is 140 days. If you start with 10 milligrams, how much will be left at the end of a year?

60. An insect colony grows exponentially from 200 to 2000 in three months' time. How long will it take for the insect population to reach 50,000?

61. Hydrogen-3 decays at a rate of 5.59% per year. Find its half-life.

62. The half-life of radium-88 is 1590 years. How long will it take for 10 grams to decay to 1 gram?

63. How much money should be invested at 8% per year, compounded quarterly, in order to have $1000 in 10 years?

64. At what annual interest rate should you invest your money if you want to double it in six years?

65. One earthquake measures 4.6 on the Richter scale. A second earthquake is 1000 times more intense than the first. What does it measure on the Richter scale?

66. The table gives the population of Austin, Texas.[*]

Year	Population
1950	132,459
1970	253,539
1980	345,890
1990	465,622
2000	656,562

(a) Sketch a scatter plot of the data, with $x = 0$ corresponding to 1950.

(b) Find an exponential model for the data.

(c) Use the model to estimate the population of Austin in 1960 and 2005.

---

*Bureau of the Census, U.S. Department of Commerce.

REVIEW QUESTIONS

67. The wind-chill factor is the temperature that would produce the same cooling effect on a person's skin if there were no wind. The table shows the wind-chill factors for various wind speeds when the temperature is 25°F.[*]

Wind Speed (in mph)	Wind Chill Temperature (in °F)
0	25
5	19
10	15
15	13
20	11
25	9
30	8
35	7
40	6
45	5

(a) What does a 20-mph wind make 25°F feel like?

(b) Sketch a scatter plot of the data.

(c) Explain why an exponential model would be appropriate.

(d) Find an exponential model for the data.

(e) According to the model, what is the wind-chill factor for a 23-mph wind?

68. Cigarette consumption in the United States has been decreasing for some time, as is shown in the table (in which the number of cigarettes each year is in billions).[†]

Year	Cigarettes
1980	631.5
1990	525
1997	480
1998	470
1999	435
2000	430

(a) Let $x = 10$ correspond to 1980 and find exponential, logarithmic, and power models for the data.

(b) Which of the three models do you think is most appropriate? Justify your answer

(c) If you were a doctor, which model would you prefer for the future? Which one would you prefer if you manufactured cigarettes?

69. Find the inverse of the function $f(x) = 2x + 1$.

70. Find the inverse of the function $f(x) = \sqrt{5 - x} + 7$.

71. Find the inverse of the function $f(x) = \sqrt[5]{x^3 + 1}$.

72. What is the inverse function of $f(x) = 7^x$?

73. The graph of a function f is shown in the figure. Sketch the graph of the inverse function of f.

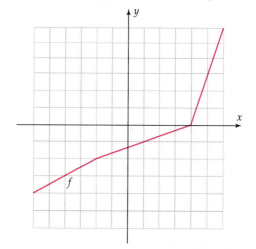

74. Which of the following functions have inverse functions (give reasons for your answers):

(a) $f(x) = x^3$ (b) $f(x) = 1 - x^2$, $x \le 0$

(c) $f(x) = |x|$

In Exercises 75–77, determine whether or not the given function has an inverse function [give reasons for your answer]. If it does, find the graph of the inverse function.

75. $f(x) = 1/x$

76. $f(x) = .02x^3 - .04x^2 + .6x - 4$

77. $f(x) = .2x^3 - 4x^2 + 6x - 15$

[*]National Weather Service.
[†]Centers for Disease Control and Prevention; Economic Research Services, U.S. Department of Agriculture.

Exponential and Logistic Modeling
of Diseases

Diseases that are contagious and are transmitted homogeneously through a population often appear to be spreading exponentially. That is, the rate of spread is proportional to the number of people in the population who are already infected. This is a reasonable model as long as the number of infected people is relatively small in comparison with the number of people in the population who can be infected. The standard exponential model looks like this:

$$f(t) = Y_0 e^{rt}.$$

Y_0 is the initial number of infected people (the number on the arbitrarily decided day 0), and r is the rate by which the disease spreads through the population. If the time t is measured in days, then r is the ratio of new infections to current infections each day.

Suppose that in Big City, population 3855, there is an outbreak of dingbat disease. On the first Monday after the outbreak was discovered (day 0), 72 people have dingbat disease. On the following Monday (day 7), 193 people have dingbat disease.

1. Using the exponential model, Y_0 is clearly 72. Calculate the value of r.

2. Using your values of Y_0 and r, predict the number of cases of dingbat disease that will be reported on day 14.

It turns out that eventually, the spread of disease must slow as the number of infected people approaches the number of susceptible people. What happens is that some of the people to whom the disease would spread are already infected. As time goes on, the spread of the disease becomes proportional to the number of susceptible and uninfected people. The disease then follows the logistic model

$$g(t) = \frac{rY_0}{aY_0 + (r - aY_0)e^{-rt}}.$$

Y_0 is still the initial value, and r serves the same function as before, at least at the initial time. The extra parameter a is not so obvious, but it is inversely related to the number of people susceptible to the disease. Unfortunately, the algebra to solve for a is quite complicated. It is much easier to approximate a using the same r from the exponential model.

3. On day 14 in Big City, 481 people have dingbat disease. Using the values of Y_0 and r from Exercise 1 in the rule of the function g, determine the value of a. [Hint: $g(14) = 481$.] Does g overestimate or underestimate the number of people with dingbat disease on day 7?

4. Use the function g from Exercise 3 to approximate the number of people in Big City who are susceptible to the disease. Does this model make sense? [Remember, as time goes on, the number of people infected approaches the number of people susceptible.]

5. In the logistic model, the rate at which the disease spreads tends to fall over time. This means that the value of r you calculated in Exercise 1 is a little low. Raise the value of r and find the new value of a as in Exercise 3. Experiment until you find a value of r for which $g(7) = 193$ (meaning that the model g matches the data on day 7).

6. Using the function g from Exercise 5, repeat Exercise 4.

Systems of Equations

S. Stammer/Photo Researchers, Inc.

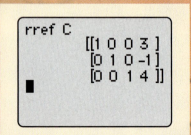

Is this a diamond in the rough?

The structure of certain crystals can be described by a large system of linear equations (more than 100 equations and variables). A variety of resource allocation problems involving many variables can be handled by solving an appropriate system of equations. The fastest solution methods involve matrices and are easily implemented on a computer or calculator. See Exercise 51 on page 499.

Chapter Outline

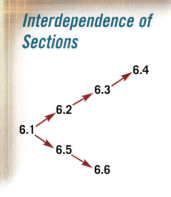
This chapter deals primarily with *systems* of equations and inequalities, such as

$$\begin{aligned} 2x - 5y + 3z &= 1 \\ x + 2y - z &= 2 \\ 3x + y + 2z &= 11 \end{aligned} \qquad \begin{aligned} 2x + 5y + z + w &= 0 \\ 2y - 4z + 41w &= 5 \\ 3x + 7y + 5z - 8w &= -6 \end{aligned} \qquad \begin{aligned} 2x + y &> 0 \\ x - y &< -3 \end{aligned}$$

Three equations in three variables   Three equations in four variables   Two inequalities in two variables

A *solution of a system* is a solution that satisfies all the equations or inequalities in the system. For instance, in the first system of equations above, $x = 1$, $y = 2$, $z = 3$ is a solution of all three equations (check it!) and hence is a solution for the system. On the other hand, $x = 0$, $y = 7$, $z = 12$ is a solution of the first two equations but not the third (verify!); hence, it is not a solution of the system.

We also consider matrices, which are used for solving systems of equations and other purposes, and linear programming, which is a widely used method for dealing with optimization problems involving several variables.

6.1 Systems of Linear Equations in Two Variables

Systems of linear equations in two variables may be solved graphically or algebraically. The graphical method is similar to what we have done previously.

EXAMPLE 1

Solve this system graphically.

$$2x - y = 1$$

$$3x + 2y = 4.$$

SOLUTION   First, we solve each equation for y.

$$2x - y = 1 \qquad\qquad 3x + 2y = 4$$
$$-y = -2x + 1 \qquad\qquad 2y = -3x + 4$$
$$y = 2x - 1 \qquad\qquad y = \frac{-3x + 4}{2}.$$

Next, we graph both equations on the same screen (Figure 6–1). As we saw in Section 1.3, each graph is a straight line, and every point on the graph represents a solution of the equation. Therefore, the solution of the system is given by the coordinates of the point that lies on both lines. An intersection finder (Figure 6–2) shows that the approximate coordinates of this point are

$$x \approx .85714286 \qquad \text{and} \qquad y \approx .71428571. \qquad \blacksquare$$

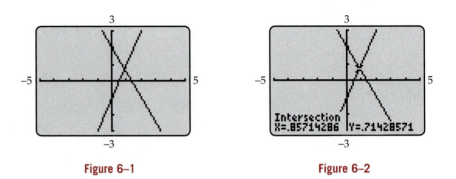

Figure 6–1   Figure 6–2

We can summarize the solution method used in Example 1 as follows.

Graphical Method

1. Solve each equation for y.

2. Graph both equations on the same screen.

3. Use an intersection finder to determine the point of intersection. Its coordinates are the solution of the system.

As is shown in Example 1, the solutions of a system of linear equations are determined by the points where their graphs intersect. There are exactly three geometric possibilities for two lines in the plane: They are parallel, they intersect at a single point, or they coincide, as illustrated in Figure 6–3 on the next page. Each of these possibilities leads to a different number of solutions for the system.

Number of Solutions of a System

A system of two linear equations in two variables must have

No solutions *or*

Exactly one solution *or*

An infinite number of solutions.

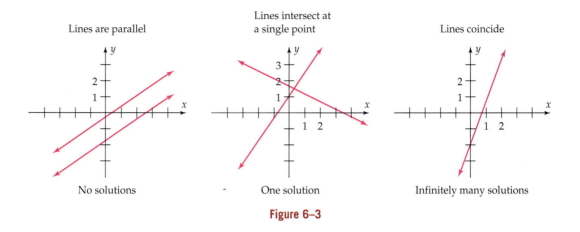

Figure 6–3

▪▪ THE SUBSTITUTION METHOD

When you use a calculator to solve systems graphically, you might have to settle for an approximate solution, as we did in Example 1. Algebraic methods, however, produce exact solutions. Furthermore, algebraic methods are often as easy to implement as geometric ones, so we shall use them, whenever practical, to obtain exact solutions. One algebraic method is **substitution,** which is explained in the next example.

EXAMPLE 2

Use substitution to find the exact solution of the system from Example 1.

$$2x - y = 1$$
$$3x + 2y = 4.$$

SOLUTION   Any solution of this system must satisfy the first equation, $2x - y = 1$. Solving this equation for y, as in Example 1, shows that

$$y = 2x - 1.$$

Substituting this expression for y in the second equation, we have

$$3x + 2y = 4$$
$$3x + 2(2x - 1) = 4$$
$$3x + 4x - 2 = 4$$
$$7x = 6$$
$$x = 6/7.$$

Therefore, every solution of the original system must have $x = 6/7$. But when $x = 6/7$, we see from the first equation that

$$2x - y = 1$$

$$2\left(\frac{6}{7}\right) - y = 1$$

$$\frac{12}{7} - y = 1$$

$$-y = -\frac{12}{7} + 1$$

$$y = \frac{12}{7} - 1 = \frac{5}{7}.$$

(We would also have found that $y = 5/7$ if we had substituted $x = 6/7$ in the second equation.) Consequently, the exact solution of the original system is $x = 6/7$, $y = 5/7$.   ■

The method used in Example 2 may be summarized as follows.

Substitution Method

> 1. Solve one of the equations for one of the variables.
>
> 2. Substitute the result of step 1 in the other equation.
>
> 3. Solve the resulting one-variable equation.
>
> 4. Substitute the number found in step 3 into one of the original equations to find the value of the remaining variable.

In Example 2, we solved for y in the first equation because that avoided the fractional expression that would have occurred if we had solved for x in the first equation or had solved the second equation for x or y.

▪▪ THE ELIMINATION METHOD

The **elimination method** of solving systems of linear equations is often more convenient than substitution. It depends on this fact.

Multiplying both sides of an equation by a nonzero constant does not change the solutions of the equation.

For example, the equation $x + 3 = 5$ has the same solution as $2x + 6 = 10$ (the first equation multiplied by 2). The elimination method also uses the following fact from basic algebra.

If $A = B$ and $C = D$, then $A + C = B + D$ and $A - C = B - D$.

EXAMPLE 3

To solve the system

$$x - 3y = 4$$
$$2x + y = 1,$$

we replace the first equation by an equivalent one (that is, one with the same solutions).

$$-2x + 6y = -8 \qquad \text{[First equation multiplied by } -2]$$
$$2x + y = 1.$$

The multiplier -2 was chosen so that the coefficients of x in the two equations would be negatives of each other. Any solution of this last system must also be a solution of the sum of the two equations.

$$-2x + 6y = -8$$
$$\underline{2x + y = 1}$$
$$7y = -7. \qquad \text{[The first variable has been eliminated]}$$

Solving this last equation, we see that $y = -1$. Substituting this value in the first of the original equations shows that

$$x - 3(-1) = 4$$
$$x = 1.$$

Therefore, $x = 1$, $y = -1$ is the solution of the original system. ∎

EXAMPLE 4

Any solution of the system

$$5x - 3y = 3$$
$$3x - 2y = 1$$

must also be a solution of the system

$$10x - 6y = 6 \qquad \text{[First equation multiplied by 2]}$$
$$-9x + 6y = -3. \qquad \text{[Second equation multiplied by } -3]$$

The multipliers 2 and -3 were chosen so that the coefficients of y in the new equations would be negatives of each other. Any solution of this last system must also be a solution of the equation obtained by adding these two equations.

$$10x - 6y = 6$$
$$\underline{-9x + 6y = -3}$$
$$x = 3. \qquad \text{[The second variable has been eliminated]}$$

Substituting $x = 3$ in the first of the original equations shows that

$$5(3) - 3y = 3$$
$$-3y = -12$$
$$y = 4.$$

Therefore, the solution of the original system is $x = 3$, $y = 4$.   ■

Here is a summary of the solution method used in Examples 3 and 4.

Elimination Method

1. Multiply one or both of the equations by suitable constants so that the coefficient of one variable in one of the equations is the negative of its coefficient in the other equation.

2. Add the two resulting equations to eliminate this variable.

3. Solve the equation in step 2 for the remaining variable.

4. Substitute the number found in step 3 into one of the original equations to find the value of the other variable.

When a system has no solution or has an infinite number of solutions (which was not the case in Examples 1–4), the elimination method gives a clear signal, as is shown in the next two examples.

EXAMPLE 5

To solve the system

$$2x - 3y = 5$$
$$4x - 6y = 1,$$

we multiply the first equation by -2 and add.

$$-4x + 6y = -10$$
$$\underline{4x - 6y = \quad 1}$$
$$0 \qquad = \ -9.$$

Since $0 = -9$ is always false, the original system cannot possibly have any solutions. A system with no solution is said to be **inconsistent.**   ■

GRAPHING EXPLORATION

Confirm the result of Example 5 geometrically by graphing the two equations in the system. Do these lines intersect, or are they parallel?

EXAMPLE 6

To solve the system

$$3x - y = 2$$
$$6x - 2y = 4,$$

we multiply the first equation by -2 to obtain the following system.

$$-6x + 2y = -4$$
$$\underline{6x - 2y = \quad 4}$$

Add the equations: $\qquad 0 = \quad 0$

The statement "$0 = 0$" is obviously true, but what does it indicate? You can see what's going on by looking at the original system more closely.

$$3x - y = 2$$
$$6x - 2y = 4$$

[The second equation is 2 times the first equation!]

So any solution of the first equation is necessarily a solution of the second equation, and hence, a solution of the system. In other words, the solutions of the system are the same as the solutions of the single equation $3x - y = 2$, which can be rewritten as

$$y = 3x - 2.$$

Thus, the system has an infinite number of solutions. They may be described as follows: Choose any real number for x, say, $x = b$. Then $y = 3x - 2 = 3b - 2$. So the solutions of the system are all pairs of numbers of the form

$$x = b \quad \text{and} \quad y = 3b - 2, \quad \text{where } b \text{ is any real number.}$$

For example, two of the solutions are

$$x = 2 \quad \text{and} \quad y = 3(2) - 2 = 4$$
$$x = -5 \quad \text{and} \quad y = 3(-5) - 2 = -17.$$

A system such as this is said to be **dependent.** ■

The arguments used in Examples 5 and 6 lead to this conclusion.

Inconsistent or Dependent Systems

If the elimination method produces a false statement (such as $0 = -9$), the system is inconsistent and has no solutions.

If the elimination method produces the true statement $0 = 0$, the system is dependent and has an infinite number of solutions.

Some nonlinear systems can be solved by replacing them with equivalent linear systems.

EXAMPLE 7

To solve the system

$$\frac{1}{x} + \frac{3}{y} = -1$$

$$\frac{2}{x} - \frac{1}{y} = 5,$$

we let $u = 1/x$ and $v = 1/y$ so that the system becomes

$$u + 3v = -1$$

$$2u - v = 5.$$

We can solve this system by multiplying the first equation by -2 and adding it to the second equation.

$$-2u - 6v = 2$$

$$\underline{2u - v = 5}$$

$$-7v = 7$$

$$v = -1.$$

Substituting $v = -1$ in the equation $u + 3v = -1$, we see that

$$u = -3(-1) - 1 = 2.$$

Consequently, the possible solution of the original system is

$$x = \frac{1}{u} = \frac{1}{2} \quad \text{and} \quad y = \frac{1}{v} = \frac{1}{(-1)} = -1.$$

You should substitute this possible solution in both equations of the original system to check that it is actually a solution of the system.   ■

⊞ APPLICATIONS

Producers are happy to supply items at a high price, but if they do, the consumers' demand for the items might be low. At lower prices, more items will be demanded, but if the price is too low, producers might not want to supply the items. The **equilibrium price** is the price at which the number of items demanded by consumers is the same as the number supplied by producers. The number of items demanded and supplied at the equilibrium price is the **equilibrium quantity.**

EXAMPLE 8

The consumer demand for a certain type of floor tile is related to its price by the equation $p = 60 - .75x$, where p is the price (in dollars) at which x thousand boxes of tile will be demanded. The supply of these tiles is related to their price by the equation $p = .8x + 5.75$, where p is the price at which x thousand boxes will be supplied by the producer. Find the equilibrium quantity and the equilibrium price.

SOLUTION   We must find the values of x and p that satisfy both the supply and demand equations. In other words, we must solve this system:

$$p = 60 - .75x$$

$$p = .8x + 5.75.$$

Algebraic Method: Since both equations are already solved for p, we use substitution. Substituting the value of p given by the first equation into the second, we obtain

$$60 - .75x = .8x + 5.75$$

Subtract $.8x$ from both sides: $60 - 1.55x = 5.75$

Subtract 60 from both sides: $-1.55x = -54.25$

Divide both sides by -1.55: $x = \dfrac{-54.25}{-1.55} = 35.$

We can determine p by substituting $x = 35$ in either equation, say, the first one.

$$p = 60 - .75x$$

$$p = 60 - .75(35) = 33.75.$$

Therefore, the equilibrium quantity is 35,000 boxes (x is measured in thousands), and the equilibrium price is \$33.75 per box.

Graphical Method: We graph the two equations in the form $y = 60 - .75x$ and $y = .8x + 5.75$ on the same screen and find their intersection point (Figure 6–4). The intersection point (35, 33.75) is called the **equilibrium point.** Its first coordinate is the equilibrium quantity, and its second coordinate is the equilibrium price.   ■

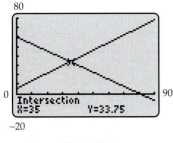

Figure 6–4

EXAMPLE 9

575 people attend a ball game, and total ticket sales are \$2575. If adult tickets cost \$5 and children's tickets cost \$3, how many adults attended the game? How many children?

SOLUTION   Let x be the number of adults, and let y be the number of children. Then,

Number of adults + Number of children = Total attendance

$$x + y = 575.$$

We can obtain a second equation by using the information about ticket sales.

Adult ticket sales   +   Children ticket sales   = Total ticket sales

$$\left(\begin{matrix}\text{Price}\\\text{per}\\\text{ticket}\end{matrix}\right) \times \left(\begin{matrix}\text{Number}\\\text{of}\\\text{adults}\end{matrix}\right) + \left(\begin{matrix}\text{Price}\\\text{per}\\\text{ticket}\end{matrix}\right) \times \left(\begin{matrix}\text{Number}\\\text{of}\\\text{children}\end{matrix}\right) = 2575$$

$$5x \quad + \quad 3y \qquad\qquad\qquad = 2575.$$

To find x and y, we need only solve the following system of equations.

$$x + y = 575$$
$$5x + 3y = 2575.$$

Multiplying the first equation by -3 and adding, we have

$$-3x - 3y = -1725$$
$$\underline{5x + 3y = 2575}$$
$$2x = 850$$
$$x = 425.$$

So 425 adults attended the game. The number of children was

$$y = 575 - x = 575 - 425 = 150. \qquad \blacksquare$$

EXAMPLE 10

A plane flies 3000 miles from San Francisco to Boston in five hours, with a tailwind all the way. The return trip on the same route, now with a headwind, takes six hours. Assuming that both remain constant, find the speed of the plane and the speed of the wind.

SOLUTION   Let x be the plane's speed, and let y be the wind speed (both in miles per hour). Then on the trip to Boston with a tailwind,

$x + y =$ actual speed of the plane (wind and plane go in same direction).

On the return trip against a headwind,

$x - y =$ actual speed of plane (wind and plane go in opposite directions).

Using the basic rate/distance equation, we have

Trip to Boston	**Return Trip**
time $\times$ rate = distance	time $\times$ rate = distance
$5(x + y) = 3000$	$6(x - y) = 3000$
$5x + 5y = 3000$	$6x - 6y = 3000$
$x + y = 600$	$x - y = 500$

Thus, we need only solve the following system of equations.

$$x + y = 600$$

$$x - y = 500.$$

Adding the two equations shows that

$$2x = 1100$$

$$x = 550.$$

Substituting this result in the first equation, we have

$$550 + y = 600$$

$$y = 50.$$

Thus, the plane's speed is 550 mph, and the wind speed is 50 mph. ■

EXAMPLE 11

How many pounds of tin and how many pounds of copper should be added to 1000 pounds of an alloy that is 10% tin and 30% copper to produce a new alloy that is 27.5% tin and 35% copper?

SOLUTION   Let x be the number of pounds of tin and y the number of pounds of copper to be added to the 1000 pounds of the old alloy. Then there will be $1000 + x + y$ pounds of the new alloy. We first find the *amounts* of tin and copper in the new alloy.

	Pounds in old alloy	**+ Pounds added**	**= Pounds in new alloy**
Tin	10% of 1000 +	x	= $100 + x$
Copper	30% of 1000 +	y	= $300 + y$

Now consider the *percentages* of tin and copper in the new alloy.

	Percentage in new alloy	**× Total weight of new alloy**	**= Pounds in new alloy**
Tin	27.5%	of $1000 + x + y$	= $.275(1000 + x + y)$
Copper	35%	of $1000 + x + y$	= $.35(1000 + x + y)$

The two ways of computing the weight of each metal in the alloy must produce the same result, that is,

$$100 + x = .275(1000 + x + y) \qquad \text{[Pounds of tin]}$$

$$300 + y = .35(1000 + x + y). \qquad \text{[Pounds of copper]}$$

Multiplying out the right-hand sides and rearranging terms produces this system of equations:

$$.725x - .275y = 175$$

$$-.35x + .65y = 50.$$

Multiplying the first equation by .65 and the second by .275 and adding the results, we have

$$.47125x - .17875y = 113.75$$

$$\underline{-.09625x + .17875y = 13.75}$$

$$.37500x \qquad\quad = 127.50$$

$$x = 340.$$

Substituting this in the first equation and solving for y shows that $y = 260$. Therefore, 340 pounds of tin and 260 pounds of copper should be added. ■

✓ EXERCISES 6.1

In Exercises 1–6, determine whether the given values of x, y, and z are a solution of the system of equations.

1. $x = -1, y = 3$

$$2x + y = 1$$
$$-3x + 2y = 9$$

2. $x = 3, y = 4$

$$2x + 6y = 30$$
$$x + 2y = 11$$

3. $x = 2, y = -1$

$$\frac{1}{3}x + \frac{1}{2}y = \frac{1}{6}$$
$$\frac{1}{2}x + \frac{1}{3}y = \frac{2}{3}$$

4. $x = .4, \quad y = .7$

$$3.1x - 2y = -.16$$
$$5x - 3.5y = -.48$$

5. $x = \frac{1}{2}, y = 3, z = -1$

$$2x - y + 4z = -6$$
$$3y + 3z = 6$$
$$2z = 2$$

6. $x = 2, y = \frac{3}{2}, z = -\frac{1}{2}$

$$3x + 4y - 2z = 13$$
$$\frac{1}{2}x + 8z = -3$$
$$x - 3y + 5z = -5$$

In Exercises 7–14, use substitution to solve the system.

7. $x - 2y = 5$

$$2x + y = 3$$

8. $3x - y = 1$

$$-x + 2y = 4$$

9. $3x - 2y = 4$

$$2x + y = -1$$

10. $5x - 3y = -2$

$$-x + 2y = 3$$

11. $r + s = 0$

$$r - s = 5$$

12. $t = 3u + 5$

$$t = u + 5$$

13. $x + y = c + d$ (where c, d are constants)

$$x - y = 2c - d$$

14. $x + 3y = c - d$ (where c, d are constants)

$$2x - y = c + d$$

In Exercises 15–34, use the elimination method to solve the system.

15. $2x - 2y = 12$

$$-2x + 3y = 10$$

16. $3x + 2y = -4$

$$4x - 2y = -10$$

17. $x + 3y = -1$

$$2x - y = 5$$

18. $4x - 3y = -1$

$$x + 2y = 19$$

19. $2x + 3y = 15$

$$8x + 12y = 40$$

20. $2x + 5y = 8$

$$6x + 15y = 18$$

21. $3x - 2y = 4$

$$6x - 4y = 8$$

22. $2x - 8y = 2$

$$3x - 12y = 3$$

23. $12x - 16y = 8$

$$42x - 56y = 28$$

24. $\frac{1}{3}x + \frac{2}{5}y = \frac{1}{6}$

$$20x + 24y = 10$$

[*Hint:* First, eliminate fractions by multiplying both sides of the first equation by 30.]

25. $9x - 3y = 1$

$$6x - 2y = -5$$

26. $8x + 4y = 3$

$$10x + 5y = 1$$

27. $\frac{x}{3} - \frac{y}{2} = -3$

$$\frac{2x}{5} + \frac{y}{5} = -2$$

28. $\frac{x}{3} + \frac{3y}{5} = 4$

$$\frac{x}{6} - \frac{y}{2} = -3$$

29. $\dfrac{x+y}{4} - \dfrac{x-y}{3} = 1$

$\dfrac{x+y}{4} + \dfrac{x-y}{2} = 9$

30. $\dfrac{x-y}{4} + \dfrac{x+y}{3} = 1$

$\dfrac{x+2y}{3} + \dfrac{3x-y}{2} = -2$

31. $\dfrac{1}{x} - \dfrac{3}{y} = 2$

$\dfrac{2}{x} + \dfrac{1}{y} = 3$

32. $\dfrac{5}{x} + \dfrac{2}{y} = 0$

$\dfrac{6}{x} + \dfrac{4}{y} = 3$

33. $\dfrac{2}{x} + \dfrac{3}{y} = 8$

$\dfrac{3}{x} - \dfrac{1}{y} = 1$

34. $\dfrac{3}{x+1} - \dfrac{4}{y-2} = 2$

$\dfrac{1}{x+1} + \dfrac{4}{y-2} = 5$

$\left[\textit{Hint: } \text{Let } u = \dfrac{1}{x+1} \text{ and} \right.$

$\left. v = \dfrac{1}{y-2}. \right]$

35. The population y in year x of Philadelphia and Houston is approximated by the equations

$$\textit{Philadelphia:} \quad 12.04x + y = 2099.64$$
$$\textit{Houston:} \quad -26.47x + y = 660.64,$$

where $x = 0$ corresponds to 1950 and y is in thousands.[*]
(a) How can you tell from the equation whether a city's population was increasing or decreasing since 1950?
(b) In what year do the two cities have the same population?

36. The population y in year x of Long Beach and New Orleans is approximated by the equations

$$\textit{Long Beach:} \quad -6.1x + 2y = 660$$
$$\textit{New Orleans:} \quad 3.82x + y = 628.4,$$

where $x = 0$ corresponds to 1960 and y is in thousands.[*] In what year do the two cities have the same population?

37. On the basis of data from 1995–2000, the median income y in year x for men and women is approximated by the equations

$$\textit{Men:} \quad -2106x + 3y = 74{,}148$$
$$\textit{Women:} \quad -1190x + 2y = 26{,}366,$$

where $x = 0$ corresponds to 1995 and y is in constant 1999 dollars.[*] If the equations remain valid in the future, when will the median income of men and women be the same?

38. The death rate per 100,000 population y in year x for heart disease and cancer is approximated by the equations

$$\textit{Heart disease:} \quad 6.9x + 2y = 728.4$$
$$\textit{Cancer:} \quad -1.3x + y = 167.5,$$

where $x = 0$ corresponds to 1970.[†] If the equations remain accurate, when will the death rates for heart disease and cancer be the same?

In Exercises 39 and 40, solve the system of equations.

39. $ax + by = r$ (where a, b, c, d, r, s are
$cx + dy = s$ constants and $ad - bc \neq 0$)

40. $ax + by = ab$ (where a, b are nonzero
$bx - ay = ab$ constants)

41. Let c be any real number. Show that this system has exactly one solution.

$$x + 2y = c$$
$$6x - 3y = 4.$$

42. (a) Find the values of c for which this system has an infinite number of solutions.

$$2x - 4y = 6$$
$$-3x + 6y = c.$$

(b) Find the values of c for which the system in part (a) has no solutions.

In Exercises 43 and 44, find the values of c and d for which both given points lie on the given straight line.

43. $cx + dy = 2$; (0, 4) and (2, 16)

44. $cx + dy = -6$; (1, 3) and (−2, 12)

In Exercises 45–48, find the equilibrium quantity and the equilibrium price. In the supply and demand equations, p is price (in dollars) and x is quantity (in thousands).

45. Supply: $p = .75x$
Demand: $p = 16 - 1.25x$

46. Supply: $p = 1.4x - .6$
Demand: $p = -2x + 3.2$

47. Supply: $p = 300 - 30x$
Demand: $p = 80 + 25x$

48. Supply: $p = 181 - .01x$
Demand: $p = 52 + .02x$

---
[*]Bureau of the Census, U.S. Department of Commerce.

[†]U.S. Department of Health and Human Services.

*In Exercises 49–50, R is the revenue (in dollars) from the sale of x units and C is the cost (in dollars) of producing x units. Find the **break-even point,** that is, the point where revenue equals cost.*

49. $R = 35x$ and $C = 28x + 840$

50. $R = 200x$ and $C = 150x + 18,000$

51. The cost C of making x hedge trimmers is given by $C = 45x + 6000$. Each hedge trimmer can be sold for $60.
  (a) Find an equation that expresses the revenue R from selling x hedge trimmers.
  (b) How many hedge trimmers must be sold for the company to break even?

52. The cost C of making x cases of pet food is given by $C = 100x + 2600$. Each case of 100 boxes sells for $120.
  (a) Find an equation that expresses the revenue from selling x cases.
  (b) How many cases must be sold for the company to break even?

In Exercises 53–56, you are the manager of a firm that must determine what it will take for a new product to break even and decide whether or not to produce the product.

53. The marketing department estimates that no more than 450 units can be sold. Costs are expected to be given by $C = 80x + 7500$ and revenues by $R = 95x$.

54. Costs are given by $C = 1750x + 96,175$ and revenue by $R = 1975x$; no more than 800 units can be sold.

55. Revenue is given by $R = 155x$ and costs by $C = 125x + 42,000$; no more than 2000 units can be sold.

56. Costs are given by $C = 140x + 3000$ and revenue by $R = 150x$; no more than 250 units can be sold.

57. A 200-seat theater charges $3 for adults and $1.50 for children. If all seats were filled and the total ticket income was $510, how many adults and how many children were in the audience?

58. A theater charges $4 for main floor seats and $2.50 for balcony seats. If all seats are sold, the ticket income is $2100. At one show, 25% of the main floor seats and 40% of the balcony seats were sold, and ticket income was $600. How many seats are on the main floor and how many are in the balcony?

59. A boat made a 4-mile trip upstream against a constant current in 15 minutes. The return trip at the same constant speed with the same current took 12 minutes. What is the speed of the boat and what is the speed of the current?

60. A plane flying into a headwind travels 2000 miles in 4 hours and 24 minutes. The return flight along the same route with a tailwind takes 4 hours. Find the wind speed and the plane's speed (assuming that both are constant).

61. At a certain store, cashews cost $4.40/pound and peanuts cost $1.20/pound. If you want to buy exactly 3 pounds of nuts for $6.00, how many pounds of each kind of nuts should you buy? [*Hint:* If you buy x pounds of cashews and y pounds of peanuts, then $x + y = 3$. Find a second equation by considering cost, and solve the resulting system.]

62. How many cubic centimeters (cm^3) of a solution that is 20% acid and of another solution that is 45% acid should be mixed to produce 100 cm^3 of a solution that is 30% acid?

63. How many grams of a 50%-silver alloy should be mixed with a 75%-silver alloy to obtain 40 grams of a 60%-silver alloy?

64. A winemaker has two large casks of wine. One wine is 8% alcohol, and the other is 18% alcohol. How many liters of each wine should be mixed to produce 30 liters of wine that is 12% alcohol?

65. Karin Sandberg plans to install a heating system for her swimming pool. Since gas is not available, she has a choice of electric or solar heat. She has gathered the following cost information.

System	Installation Costs	Monthly Operational Cost
Electric	$2,000	$80
Solar	$14,000	$9.50

  (a) Ignoring changes in fuel prices, write a linear equation for each heating system that expresses its total cost y in terms of the number of *years x* of operation.
  (b) What is the five-year total cost of electric heat? Of solar heat?
  (c) In what year will the total cost of the two heating systems be the same? Which is the cheaper system before that time? After that time?

66. A toy company makes Boomie Babies, as well as collector cases for each Boomie Baby. To make x cases costs the company $5000 in fixed overhead, plus $7.50 per case. An outside supplier has offered to produce any desired volume of cases for $8.20 per case.
  (a) Write an equation that expresses the company's cost to make x cases itself.
  (b) Write an equation that expresses the cost of buying x cases from the outside supplier.
  (c) Graph both equations on the same axes and determine when the two costs are the same.
  (d) When should the company make the cases themselves and when should they buy them from the outside supplier?

67. When Neil Simon planned to open his play *London Suite,* his producer, Emanuel Azenberg, made the following cost and revenue estimates for opening on Broadway or off Broadway.[*]

	On Broadway	Off Broadway
Initial cost to open[†]	$1,295,000	$440,000
Weekly costs[†]	$206,500	$82,000
Weekly revenue	$250,500	$109,000

For a production on Broadway that runs x weeks, find a linear equation that gives the
(a) total revenue R;
(b) total cost C;
(c) total profit P.
(d) Do parts (a)–(c) for an off-Broadway production.
(e) After how many weeks will the on-Broadway profit equal the off-Broadway profit? What is the profit then?
(f) When would it be better to open off Broadway rather than on Broadway?

68. A store sells deluxe tape recorders for $150. The regular model costs $120. The total tape recorder inventory would sell for $43,800. But during a recent month, the store actually sold half of its deluxe models and two-thirds of the regular models and took in a total of $26,700. How many of each kind of recorder did the store have at the beginning of the month?

69. Ann Melville has part of her money in an account that pays 9% annual interest and the rest in an account that pays 11% annual interest. If she has $8000 less in the higher-paying account than in the lower-paying one and her total annual interest income is $2010, how much does she have invested in each account?

70. Joyce Pluth has money in two investment funds. Last year, the first fund paid a dividend of 8%, the second paid a dividend of 2%, and Joyce received a total of $780. This year, the first fund paid a 10% dividend, the second paid only 1%, and Joyce received $810. How much money does she have invested in each fund?

71. The table shows the percentages of men and women in the labor force who are college graduates in selected years.[‡]

Year	1992	1995	1997	1998	1999	2000
Men	27.5	29.7	29.2	29.6	30.5	30.9
Women	25.0	26.6	28.0	28.6	29.5	29.8

(a) Use linear regression (with $x = 0$ corresponding to 1990) to find an equation that gives the percentage of men in the labor force in year x who are college graduates.
(b) Do part (a) for women.
(c) If these models remain accurate, when will the percentage of women equal that of men?

72. The table shows the number of passenger cars (in thousands) imported into the United States from Japan and Canada in selected years.[§]

Year	1995	1996	1997	1998	1999	2000
Japan	1114.4	1190.9	1387.8	1456.1	1707.3	1839.1
Canada	1552.7	1690.7	1731.2	1837.6	2170.4	2138.8

(a) Use linear regression to find an equation that approximates the number y of cars imported from Japan in year x, with $x = 5$ corresponding to 1995.
(b) Do part (a) for Canada.
(c) If these models remain accurate, in what year will the imports from Japan and Canada be the same? Approximately how many cars will be imported from those two countries that year?

73. A machine in a pottery factory takes three minutes to form a bowl and two minutes to form a plate. The material for a bowl costs $.25, and the material for a plate costs $.20. If the machine runs for eight hours straight and exactly $44 is spent for material, how many bowls and plates can be produced?

74. Because Chevrolet and Saturn produce cars in the same price range, Chevrolet's sales are a function not only of Chevy prices (x), but of Saturn prices (y) as well. Saturn prices are related similarly to both Saturn and Chevy prices. Suppose General Motors forecasts the demand z_1 for Chevrolets and the demand z_2 for Saturns to be given by

$$z_1 = 68{,}000 - 6x + 4y \quad \text{and} \quad z_2 = 42{,}000 + 3x - 3y.$$

Solve this system of equations and express
(a) the price x of Chevrolets as a function of z_1 and z_2;
(b) the price y of Saturns as a function of z_1 and z_2.

---

[*]From Albert Goetz, "Basic Economics: Calculating Against Theatrical Disaster," Media Clips, *Mathematics Teacher* 89, no. 1 (January 1996) © 1996. Reprinted with permission of the National Councils of Teachers of Mathematics. All rights reserved.
[†]Initial cost includes sets, costumes, rehearsals, etc. Weekly expenses include theater rent, salaries, advertising, etc.

[‡]Bureau of Labor Statistic, U.S. Department of Commerce.
[§]Foreign Trade Division, Bureau of the Census, U.S. Department of Commerce.

6.1.A   *SPECIAL TOPICS* Systems of Nonlinear Equations

Some systems that include nonlinear equations can be solved algebraically.

EXAMPLE 1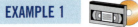

Solve the system

$$-2x + y = -1$$
$$xy = 3.$$

SOLUTION   Solve the first equation for y,

$$y = 2x - 1$$

and substitute this into the second equation.

$$xy = 3$$
$$x(2x - 1) = 3$$
$$2x^2 - x = 3$$
$$2x^2 - x - 3 = 0$$
$$(2x - 3)(x + 1) = 0$$
$$2x - 3 = 0 \quad \text{or} \quad x + 1 = 0$$
$$x = 3/2 \qquad\qquad x = -1$$

Using the equation $y = 2x - 1$ to find the corresponding values of y, we see that

$$\text{If } x = \frac{3}{2}, \quad \text{then} \quad y = 2\left(\frac{3}{2}\right) - 1 = 2.$$

$$\text{If } x = -1, \quad \text{then} \quad y = 2(-1) - 1 = -3.$$

Therefore, the solutions of the system are $x = 3/2$, $y = 2$, and $x = -1$, $y = -3$. ∎

EXAMPLE 2

Solve the system

$$x^2 + y^2 = 8$$
$$x^2 - y = 6.$$

SOLUTION   Solve the second equation for y, obtaining $y = x^2 - 6$, and substitute this into the first equation.

$$x^2 + y^2 = 8$$
$$x^2 + (x^2 - 6)^2 = 8$$
$$x^2 + x^4 - 12x^2 + 36 = 8$$
$$x^4 - 11x^2 + 28 = 0$$
$$(x^2 - 4)(x^2 - 7) = 0$$
$$x^2 - 4 = 0 \quad \text{or} \quad x^2 - 7 = 0$$
$$x^2 = 4 \qquad\qquad x^2 = 7$$
$$x = \pm 2 \qquad\qquad x = \pm\sqrt{7}.$$

Using the equation $y = x^2 - 6$ to find the corresponding values of y, we find that the solutions of the system are

$$x = 2, y = -2; \quad x = -2, y = -2; \quad x = \sqrt{7}, y = 1;$$
$$x = -\sqrt{7}, y = 1. \quad \blacksquare$$

Algebraic techniques were successful in Examples 1 and 2 because substitution led to equations whose solutions could be found exactly. When this is not the case, graphical methods are needed. The solutions of a system of equations are the points that are on the graphs of all the equations in the system. They can be approximated with a graphical intersection finder.

EXAMPLE 3

Solve the system

$$y = x^4 - 4x^3 + 9x - 1$$
$$y = 3x^2 - 3x - 7.$$

SOLUTION   If you try substitution on this system, say, by substituting the expression for y from the first equation into the second, you obtain

$$x^4 - 4x^3 + 9x - 1 = 3x^2 - 3x - 7$$
$$x^4 - 4x^3 - 3x^2 + 12x + 6 = 0.$$

This fourth-degree equation cannot be readily solved algebraically, so a graphical approach is appropriate.

Graph both equations of the original system on the same screen. In the viewing window of Figure 6–5, the graphs intersect at three points. However, the graphs seem to be getting closer together as they run off the screen at the top right, which suggests that there may be another intersection point.

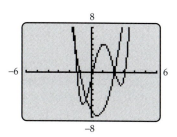

Figure 6–5

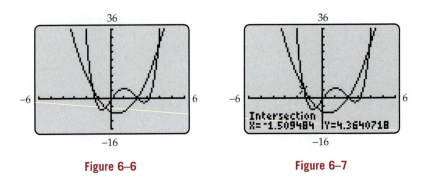

Figure 6–6   **Figure 6–7**

The larger window in Figure 6–6 shows four intersection points. There cannot be any more because, as we saw in the previous paragraph, the intersection points (solutions of the system) correspond to the solutions of a fourth-degree polynomial equation, which has a maximum of four solutions. An intersection finder (Figure 6–7) shows that one of the approximate solutions of the system is

$$x = -1.509484, \qquad y = 4.3640718.$$

GRAPHING EXPLORATION

Graph the two equations in the viewing window of Figure 6–6 and use your intersection finder to approximate the other three solutions of this system. ■

▪▪ SYSTEMS WITH SECOND-DEGREE EQUATIONS

Solving systems of equations graphically depends on our ability to graph each equation in the system. With some equations of higher degree, this may require special techniques.

EXAMPLE 4

Solve this system graphically.

$$x^2 - 4x - y + 1 = 0$$
$$10x^2 + 25y^2 = 100.$$

SOLUTION   It's easy to graph the first equation, since it can be rewritten as $y = x^2 - 4x + 1$. To graph the second equation, we must first solve for y.

$$10x^2 + 25y^2 = 100$$
$$25y^2 = 100 - 10x^2$$
$$y^2 = \frac{100 - 10x^2}{25}$$
$$y = \sqrt{\frac{100 - 10x^2}{25}} \quad \text{or} \quad y = -\sqrt{\frac{100 - 10x^2}{25}}.$$

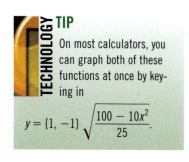

TIP

On most calculators, you can graph both of these functions at once by keying in

$$y = \{1, -1\} \sqrt{\dfrac{100 - 10x^2}{25}}.$$

By graphing both these functions on the same screen (see the Technology Tip), we obtain the complete graph of the equation $10x^2 + 25y^2 = 100$, as shown in Figure 6–8. The graphs of both equations in the system are shown in Figure 6–9.

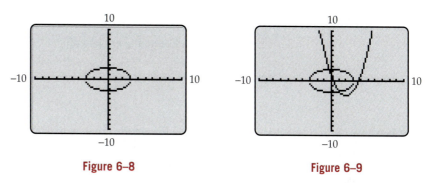

Figure 6–8   **Figure 6–9**

The two intersection points (solutions of the system) in Figure 6–9 can now be determined by an intersection finder.

$$x = -.2348, \quad y = 1.9945 \qquad \text{and} \qquad x = .9544, \quad y = -1.9067. \quad \blacksquare$$

The Global Positioning System (GPS) uses 24 satellites, each orbiting the earth every 12 hours at a height of approximately 12,700 miles. At virtually any time, a hand-held GPS device can receive signals from at least six satellites. By timing how long it takes a radio signal to travel from the satellite, the distance from the device to the satellite can be determined. This information is then used to determine position, as in the following simplified example.

EXAMPLE 5

Suppose the earth is a flat, circular disc of radius 4000 miles, centered at the origin of a two-dimensional coordinate system in which 1 unit represents 1000 miles. Figure 6–10 (which is not to scale) shows a GPS device at point (x, y) and two satellites at $(8, 15.6)$ and $(16.2, 4)$. Radio signals show that the distances of the satellites to (x, y) are 14 thousand and 13.1 thousand miles respectively, as shown in the figure. What are the coordinates of the device?

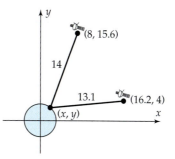

Figure 6–10

SOLUTION   The distance from (x, y) to $(8, 15.6)$ is 14. By the distance formula,

$$\sqrt{(x-8)^2 + (y-15.6)^2} = 14$$

Square both sides: $$(x-8)^2 + (y-15.6)^2 = 14^2.$$

Thus, (x, y) lies on the circle with center $(8, 15.6)$ and radius 14. Similarly, since the distance from (x, y) to $(16.2, 4)$ is 13.1,

$$\sqrt{(x-16.2)^2 + (y-4)^2} = 13.1$$

$$(x-16.2)^2 + (y-4)^2 = 13.1^2.$$

So (x, y) also lies on the circle with center $(16.2, 4)$ and radius 13.1.

Therefore, (x, y) is one of the intersection points of these two circles, as shown in Figure 6–11.* We must find this intersection point, that is, solve the system

$$(x-8)^2 + (y-15.6)^2 = 14^2$$

$$(x-16.2)^2 + (y-4)^2 = 13.1^2.$$

To solve this system graphically, we first solve each equation for y.

$$(x-8)^2 + (y-15.6)^2 = 14^2$$

$$(y-15.6)^2 = 14^2 - (x-8)^2$$

$$y - 15.6 = \pm\sqrt{14^2 - (x-8)^2}$$

$$y = \sqrt{14^2 - (x-8)^2} + 15.6 \quad \text{or} \quad y = -\sqrt{14^2 - (x-8)^2} + 15.6.$$

A similar computation with the equation $(x-16.2)^2 + (y-4)^2 = 13.1^2$ shows that

$$y = \sqrt{13.1^2 - (x-16.2)^2} + 4 \quad \text{or} \quad y = -\sqrt{13.1^2 - (x-16.2)^2} + 4.$$

Graphing the four preceding equations should produce something similar to Figure 6–11. Because of limited resolution, however, a graphing calculator will show only parts of the circles (try it!). Fortunately, we are interested only in the intersection point closest to the origin. Using an intersection finder in the partial graph in Figure 6–12, shows that this point is

$$(x, y) \approx (3.1918, 2.4516). \quad \blacksquare$$

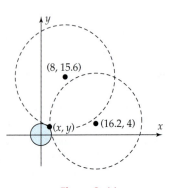

Figure 6–11

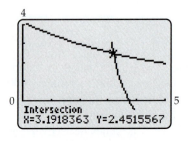

Figure 6–12

EXAMPLE 6

A 52-foot-long piece of wire is to be cut into three pieces, two of which are the same length. The two equal pieces are to be bent into squares, and the third piece is to be bent into a circle. What should the length of each piece be if the total area enclosed by the two squares and the circle is 100 square feet?

---

*In our actual three-dimensional world, the GPS device would be an intersection point of three or more spheres, that is, a solution of a system of three or more second-degree equations in three variables.

SOLUTION   Let x be the length of each piece of wire that is to be bent into a square, and let y be the length of the piece that is to be bent into a circle. Since the original wire is 52 feet long,

$$x + x + y = 52 \qquad \text{or, equivalently,} \qquad y = 52 - 2x.$$

If a piece of wire of length x is bent into a square, the side of the square will have length $x/4$; hence, the area of the square will be $(x/4)^2 = x^2/16$. The remaining piece of wire will be made into a circle of circumference (length) y. Since the circumference is 2π times the radius (that is, $y = 2\pi r$), the circle has radius $r = y/2\pi$. Therefore, the area of the circle is

$$\pi r^2 = \pi \left(\frac{y}{2\pi}\right)^2 = \frac{\pi y^2}{4\pi^2} = \frac{y^2}{4\pi}.$$

The sum of the areas of the two squares and the circle is 100, that is,

$$\frac{x^2}{16} + \frac{x^2}{16} + \frac{y^2}{4\pi} = 100$$

$$\frac{y^2}{4\pi} = 100 - \frac{2x^2}{16}$$

$$y^2 = 4\pi\left(100 - \frac{x^2}{8}\right).$$

Therefore, the lengths x and y are solutions of the following system of equations.

$$y = 52 - 2x$$

$$y^2 = 4\pi\left(100 - \frac{x^2}{8}\right)$$

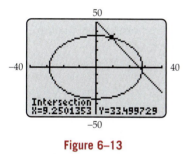

Figure 6–13

The system may be solved either algebraically or graphically, using Figure 6–13. Since x and y are lengths, both must be positive. Consequently, we need only consider the intersection point in the first quadrant. An intersection finder shows that its coordinates are $x \approx 9.25$, $y \approx 33.50$. Therefore, the wire should be cut into two 9.25-foot pieces and one 33.5-foot piece. ■

EXERCISES 6.1.A

In Exercises 1–12, solve the system algebraically.

1. $x^2 - y = 0$
  $-2x + y = 3$

2. $x^2 - y = 0$
  $-3x + y = -2$

3. $x^2 - y = 0$
  $x + 3y = 6$

4. $x^2 - y = 0$
  $x + 4y = 4$

5. $x + y = 10$
  $xy = 21$

6. $2x + y = 4$
  $xy = 2$

7. $xy + 2y^2 = 8$
  $x - 2y = 4$

8. $xy + 4x^2 = 3$
  $3x + y = 2$

9. $x^2 + y^2 - 4x - 4y = -4$
  $x - y = 2$

10. $x^2 + y^2 - 4x - 2y = -1$
   $x + 2y = 2$

11. $x^2 + y^2 = 25$
   $x^2 + y = 19$

12. $x^2 + y^2 = 1$
   $x^2 - y = 5$

In Exercises 13–20, solve the system by any means.

13. $y = x^3 - 3x^2 + 4$
   $y = -.5x^2 + 3x - 2$

14. $y = -x^3 + 3x^2 + x - 3$
   $y = -2x^2 + 5$

15. $y = x^3 - 3x + 2$

$y = \dfrac{3}{x^2 + 3}$

16. $y = .25x^4 - 2x^2 + 4$

$y = x^3 - x^2 - 2x + 1$

17. $25x^2 - 16y^2 = 400$

$-9x^2 + 4y^2 = -36$

18. $9x^2 + 16y^2 = 140$

$-x^2 + 4y^2 = -4$

19. $5x^2 + 3y^2 - 20x + 6y = -8$

$x - y = 2$

20. $4x^2 + 9y^2 = 36$

$2x - y = -1$

21. Suppose that the first satellite in Example 5 is located at (2.7, 15.9) and is 12.7 thousand miles from the GPS device. If the second satellite is at (13.9, 9.9) and is 13.3 thousand miles from the device, what are the coordinates of the device?

22. Internet sales of apparel (in billions of dollars) are projected to be given by

$$f(x) = .38x^2 - .35x + 4.85,$$

where $x = 2$ corresponds to 2002.[*] Similarly, sales of computer-related items are projected to be given by

$$g(x) = -.17x^2 + 2.73x + 2.71.$$

In what year were Internet sales of apparel and computer-related items at the same level?

23. What would be the answer in Example 6 if the original piece of wire were 70 feet long? Explain.

_____

[*]Based on data from Forrester Research.

24. A 52-foot-long piece of wire is to be cut into three pieces, two of which are the same length. The two equal pieces are to be bent into circles and the third piece into a square. What should the length of each piece be if the total area enclosed by the two circles and the square is 100 square feet?

25. A rectangular box (including top) with square ends and a volume of 16 cubic meters is to be constructed from 40 square meters of cardboard. What should its dimensions be?

26. A rectangular sheet of metal is to be rolled into a circular tube. If the tube is to have a surface area (excluding ends) of 210 square inches and a volume of 252 cubic inches, what size sheet of metal should be used?

27. Find two real numbers whose sum is -16 and whose product is 48.

28. Find two real numbers whose sum is 34.5 and whose product is 297.

29. Find two real numbers whose sum is 3 such that the sum of their squares is 369.

30. Find two real numbers whose sum is 2 such that the difference of their squares is 60.

31. Find the dimensions of a rectangular room whose perimeter is 58 feet and whose area is 204 square feet.

32. Find the dimensions of a rectangular room whose perimeter is 53 feet and whose area is 165 square feet.

33. A rectangle has area 120 square inches and a diagonal of length 17 inches. What are its dimensions?

34. A right triangle has area 225 square centimeters and a hypotenuse of length 35 centimeters. To the nearest tenth of a centimeter, how long are the legs of the triangle?

35. Find the equation of the straight line that intersects the parabola $y = x^2$ *only* at the point (3, 9). [*Hint:* What condition on the discriminant guarantees that a quadratic equation has exactly one real solution?]

6.2   Large Systems of Linear Equations

Roadmap

Instructors who want to use technology as soon as possible for solving large systems should cover Section 6.2 only through the box after Example 2 and then proceed to Section 6.3.

Systems of linear equations in three variables can be interpreted geometrically as the intersection of planes. However, algebraic methods are the only practical means to solve such systems or ones with more variables. Large systems can be solved by **Gaussian elimination,**[*] which is an extension of the elimination method used in Section 6.1.

Two systems of equations are said to be **equivalent** if they have the same solutions. The basic idea of Gaussian elimination is to transform a given system

_____

[*]Named after the German mathematician K. F. Gauss (1777–1855).

into an equivalent system that can easily be solved. There are several operations on a system of equations that leave the solutions to the system unchanged and, hence, produce an equivalent system. The first one is to

Interchange any two equations in the system,

which obviously won't affect the solutions of the system. The second is to

Multiply an equation in the system by a nonzero constant.

This does not change the solutions of the equation and therefore does not change the solutions of the system. To understand how the next operation works, we shall examine an earlier example from a different viewpoint.

EXAMPLE 1

In Example 3 of Section 6.1, we solved the system

$$x - 3y = 4$$
$$2x + y = 1$$

by multiplying the first equation by -2 and adding it to the second to eliminate the variable x.

$$
\begin{array}{ll}
-2x + 6y = -8 & \text{[}-2 \text{ times first equation]} \\
\underline{2x + y = 1} & \text{[Second equation]} \\
7y = -7 & \text{[Sum of second equation and } -2 \text{ times first equation]}
\end{array}
$$

We then solved this last equation for y and substituted the answer, $y = -1$, in the original first equation to find that $x = 1$. What we did, in effect, was

> **Replace the original system by the following system, in which x has been eliminated from the second equation; then solve this new system.**

$$
(*) \quad
\begin{array}{ll}
x - 3y = 4 & \text{[First equation]} \\
7y = -7. & \text{[Sum of second equation and } -2 \text{ times first equation]}
\end{array}
$$

The solution of system $(*)$ is easily seen to be $y = -1$, $x = 1$, and you can readily verify that this is the solution of the original system. So the two systems are equivalent. [*Note:* We are not claiming that the second equations in the two systems have the same solutions—they don't—but only that the two *systems* have the same solutions.] ■

Example 1 is an illustration of the third of the following operations.

Elementary Operations

Performing any of the following operations on a system of equations produces an equivalent system.

1. Interchange any two equations in the system.

2. Replace an equation in the system by a nonzero constant multiple of itself.

3. Replace an equation in the system by the sum of itself and a constant multiple of another equation in the system.

The next example shows how elementary operations can be used to transform a system into an equivalent system that can be solved.

EXAMPLE 2

Solve the system

$$x + 4y - 3z = 1 \qquad \text{[Equation A]}$$
$$-3x - 6y + z = 3 \qquad \text{[Equation B]}$$
$$2x + 11y - 5z = 0. \qquad \text{[Equation C]}$$

SOLUTION   We first use elementary operations to produce an equivalent system in which the variable x has been eliminated from the second and third equations.

To eliminate x from equation B, replace equation B by the sum of itself and 3 times equation A.

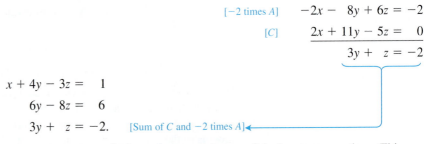

$$\text{[3 times } A\text{]} \qquad 3x + 12y - 9z = 3$$
$$\text{[}B\text{]} \qquad -3x - 6y + z = 3$$
$$6y - 8z = 6$$

$$x + 4y - 3z = 1 \qquad [A]$$
$$6y - 8z = 6 \qquad \text{[Sum of } B \text{ and 3 times } A\text{]}$$
$$2x + 11y - 5z = 0. \qquad [C]$$

To eliminate x from equation C, we replace equation C by the sum of itself and -2 times equation A.

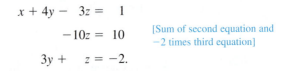

$$\text{[}-2 \text{ times } A\text{]} \qquad -2x - 8y + 6z = -2$$
$$\text{[}C\text{]} \qquad 2x + 11y - 5z = 0$$
$$3y + z = -2$$

$$x + 4y - 3z = 1$$
$$6y - 8z = 6$$
$$3y + z = -2. \qquad \text{[Sum of } C \text{ and } -2 \text{ times } A\text{]}$$

The next step is to eliminate the y term in one of the last two equations. This can be done by replacing the second equation by the sum of itself and -2 times the third equation.

$$x + 4y - 3z = 1$$
$$-10z = 10 \qquad \text{[Sum of second equation and } -2 \text{ times third equation]}$$
$$3y + z = -2.$$

Finally, interchange the last two equations.

$$(*) \qquad \begin{aligned} x + 4y - 3z &= 1 \\ 3y + z &= -2 \\ -10z &= 10. \end{aligned}$$

This last system, which is equivalent to the original one, is easily solved. The last equation shows that

$$-10z = 10 \quad \text{or, equivalently,} \quad z = -1.$$

Substituting $z = -1$ in the second equation shows that

$$3y \quad + z = -2$$
$$3y + (-1) = -2$$
$$3y = -1$$
$$y = -\frac{1}{3}.$$

Substituting $y = -1/3$ and $z = -1$ in the first equation yields

$$x \quad + 4y \quad -3z = 1$$
$$x + 4\left(-\frac{1}{3}\right) - 3(-1) = 1$$
$$x = 1 + \frac{4}{3} - 3 = -\frac{2}{3}.$$

Therefore, the original system has just one solution: $x = -2/3$, $y = -1/3$, $z = -1$. ■

The process used to solve the final system (∗) in Example 2 is called **back substitution** because you begin with the last equation and work back to the first. It works because system (∗) is in **triangular form:** The first variable in the first equation, x, does not appear in any subsequent equation; the first variable in the second equation, y, does not appear in any subsequent equation, and so on. It can be shown that the procedure in Example 2 works in every case.

Gaussian Elimination

> Any system of linear equations can be transformed into an equivalent system in triangular form by using a finite number of elementary operations. If the system has solutions, they can then be found by back substitution in the triangular form system.

In the next example of Gaussian elimination, we won't show the calculation of the sum of one equation and a constant multiple of another, as we did in Example 2. If you can't do them mentally by looking at the equations involved, write them out on scratch paper.

EXAMPLE 3

Solve the system

$$x + 2y + 4z = 6$$
$$x + 3y + 3z = 6$$
$$-2x + 6y - 12z = -24$$
$$2x + 5y + 7z = 12.$$

SOLUTION We begin by eliminating the x terms from the last three equations.

$$x + 2y + 4z = 6$$
$$y - z = 0 \quad \text{[Sum of second equation and } -1 \text{ times first equation]}$$
$$-2x + 6y - 12z = -24$$
$$2x + 5y + 7z = 12$$

$$x + 2y + 4z = 6$$
$$y - z = 0$$
$$10y - 4z = -12 \quad \text{[Sum of third equation and 2 times first equation]}$$
$$2x + 5y + 7z = 12$$

$$x + 2y + 4z = 6$$
$$y - z = 0$$
$$10y - 4z = -12$$
$$y - z = 0. \quad \text{[Sum of fourth equation and } -2 \text{ times first equation]}$$

Next, we use the second equation to eliminate the y term from the third and fourth equations.

$$x + 2y + 4z = 6$$
$$y - z = 0$$
$$6z = -12 \quad \text{[Sum of third equation and } -10 \text{ times second equation]}$$
$$y - z = 0$$

$$x + 2y + 4z = 6$$
$$y - z = 0$$
$$6z = -12$$
$$0 = 0. \quad \text{[Sum of fourth equation and } -1 \text{ times second equation]}$$

We now have a triangular form system that is easily solved. Use back substitution to verify that the solution is

$$x = 18, \quad y = -2, \quad z = -2. \quad \blacksquare$$

If your calculator has a solver for systems of equations, check your instruction manual to see how to use it. However, you should be aware of possible limitations (see the Technology Tip in the margin). Technological methods that can be used with all systems are introduced in Section 6.3.

TECHNOLOGY TIP

To solve a system of linear equations on a calculator, use SIMULT(ANEOUS) in this menu:

TI-86: Keyboard

Casio: EQUA (main menu)

TI-83+ users may download the PolySmlt solver from TI.

The TI-86 and Casio solvers work only on systems with the same number of equations as variables.

If "error" or "no solution found" appears with any of these solvers, no conclusion can be drawn. The system may have a solution (or an infinite number of them), or it may have no solutions.

EXAMPLE 4

John-Paul wants to invest $30,000 in corporate bonds. The bonds are rated AAA, AA, A, B, C, and those with lower ratings (more risk) pay a higher rate of interest. Currently, the average yield is 5% on AAA bonds, 6% on A bonds, and 10% on B bonds. Being conservative, he wants to have twice as much in AAA bonds as in B bonds. How much should he invest in each type of bond to have an interest income of $2000?

SOLUTION   Let x be the amount invested in AAA bonds, y the amount in A bonds, and z the amount in B bonds. Since the total amount to be invested is $30,000,

$$x + y + z = 30,000.$$

Since x must be twice as large as z, we have

$$x = 2z \qquad \text{or, equivalently,} \qquad x - 2z = 0.$$

Finally, the income from the bonds is

$$\underset{\text{AAA bonds}}{\text{Interest on}} + \underset{\text{A bonds}}{\text{Interest on}} + \underset{\text{B bonds}}{\text{Interest on}} = \text{Total interest}$$

$$(5\% \text{ of } x) + (6\% \text{ of } y) + (10\% \text{ of } z) = 2000.$$

$$.05x + .06y + .10z = 2000.$$

Therefore, we must solve the following system of equations.

$$
\begin{aligned}
x + \quad y + \quad z &= 30,000 \\
x \qquad\quad - \quad 2z &= 0 \\
.05x + .06y + .10z &= 2000
\end{aligned}
$$

$$
\begin{aligned}
x + \quad y + \quad z &= \quad 30,000 \\
-y - \quad 3z &= -30,000 \qquad &&\text{[Sum of second equation and} \\
.05x + .06y + .10z &= \quad 2000 \qquad &&-1 \text{ times first equation]}
\end{aligned}
$$

$$
\begin{aligned}
x + \quad y + \quad z &= \quad 30,000 \\
-y - \quad 3z &= -30,000 \\
.01y + .05z &= \quad 500 \qquad &&\text{[Sum of third equation and} \\
 & &&-.05 \text{ times first equation]}
\end{aligned}
$$

$$
\begin{aligned}
x + \quad y + \quad z &= \quad 30,000 \\
-y - \quad 3z &= -30,000 \\
.02z &= \quad 200 \qquad &&\text{[Sum of third equation and} \\
 & &&.01 \text{ times second equation]}
\end{aligned}
$$

Back substitution now shows that

$$z = \frac{200}{.02} = 10,000, \qquad y = 0, \qquad x = 20,000.$$

Therefore, $20,000 should be invested in AAA bonds, $10,000 should be invested in B bonds, and nothing should be invested in A bonds. ■

✓ EXERCISES 6.2

In Exercises 1–4, solve the system by back substitution.

1. $x + 3y - 4z + 2w = 1$
  $y + z - w = 4$
  $2z + 2w = -6$
  $3w = 9$

2. $x + 5z + 6w = 10$
  $y + 3z - 2w = 4$
  $z - 4w = -6$
  $2w = 4$

3. $2x + 2y - 4z + w = -5$
  $3y + 4z - w = 0$
  $2z - 7w = -6$
  $5w = 15$

4. $3x - 2y - 4z + 2w = 6$
  $2y + 5z - 3w = 7$
  $3z + 4w = 0$
  $3w = 15$

In Exercises 5–12, obtain an equivalent system by performing the stated elementary operation on the system.

5. Interchange equations 1 and 2.
  $2x - 4y + 5z = 1$
  $x - 3z = 2$
  $5x - 8y + 7z = 6$
  $3x - 4y + 2z = 3$

6. Interchange equations 1 and 3.
  $2x - 2y + z = -6$
  $3x + y + 2z = 2$
  $x + y - 2z = 0$

7. Multiply the second equation by -1.
  $3x + z + 2w + 18v = 0$
  $4x - y + w + 24v = 0$
  $7x - y + z + 3w + 42v = 0$
  $4x + z + 2w + 24v = 0$

8. Multiply the third equation by $1/2$.
  $x + 2y + 4z = 3$
  $x + 2z = 0$
  $2x + 4y + z = 3$

9. Replace the second equation by the sum of itself and -2 times the first equation.
  $x + y + 2z + 3w = 1$
  $2x + y + 3z + 4w = 1$
  $3x + y + 4z + 5w = 2$

10. Replace the third equation by the sum of itself and -1 times the first equation.
  $x + 2y + 4z = 6$
  $y + z = 1$
  $x + 3y + 5z = 10$

11. Replace the third equation by the sum of itself and -2 times the second equation.
  $x + 12y - 3z + 4w = 10$
  $2y + 3z + w = 4$
  $4y + 5z + 2w = 1$
  $6y - 2z - 3w = 0$

12. Replace the third equation by the sum of itself and 3 times the second equation.
  $2x + 2y - 4z + w = -5$
  $2y + 4z - w = 2$
  $-6y - 4z + 2w = 6$
  $2y + 5z - 3w = 7$

In Exercises 13–20, use Gaussian elimination to solve the system.

13. $-x + 3y + 2z = 0$
  $2x - y - z = 3$
  $x + 2y + 3z = 0$

14. $3x + 7y + 9z = 0$
  $x + 2y + 3z = 2$
  $x + 4y + z = 2$

15. $x - 2y + 4z = 6$
  $x + y + 13z = 6$
  $-2x + 6y - z = -10$

16. $x - y + 5z = -6$
$3x + 3y - z = 10$
$x + 3y + 2z = 5$

17. $x + y + z = 200$
$x - 2y = 0$
$2x + 3y + 5z = 600$
$2x - y + z = 200$

18. $2x - y + 2z = 3$
$-x + 2y - z = 0$
$3y - 2z = 1$
$x + y - z = 1$

19. $x + y = 3$
$5x - y = 3$
$9x - 4y = 1$

20. $3x - y + z = 6$
$x + 2y - z = 0$
$2x - 2y + z = -2$
$3x - 7y + 4z = 4$

In Exercises 21–24, solve the system graphically by finding the point that lies on the graphs of all equations in the system, if there is one. Since your intersection finder only works for two equations at a time, you may have to use it several times to verify that there is a solution.

21. $x + 2y = 6$
$2x - y = 3$
$15x + 5y = 27$

22. $x - y = 2$
$2x - y = 3$
$2x + y = 1$

23. $6x + 2y = 8$
$x + y = 2$
$10x - 5y = 6$

24. $2x - y = 1$
$x + 2y = 4$
$20x - 5y = 16$

25. An investor want to invest \$35,000 in the AAA, A, and B bonds of Example 4. If he wants to have twice as much in AAA bonds as in B bonds and wants to get annual interest of \$2200, how much should he invest in each type of bond?

26. An investor wants to invest \$30,000 in bonds. She wants to have twice as much in AAA bonds as in B bonds and wants to receive \$2000 in annual interest. If AAA bonds yield an average of 4%, A bonds an average of 7%, and B bonds an average of 9%, how much should she invest in each type of bond?

27. The Snack Company wants to make a 100-pound mixture of corn chips, nuts, and pretzels that will cost \$4 per pound. Corn chips cost \$2 per pound, nuts cost \$6 per pound, and pretzels cost \$3 per pound. If the mixture is to have three times as many corn chips as pretzels (by weight), how many pounds of each ingredient should be used?

28. An animal feed is to be made from alfalfa, cornmeal, and meat by-products. Each unit of alfalfa provides 10 units of fiber, 30 units of fat, and 20 units of protein. Each unit of cornmeal provides 20 units of fiber, 20 units of fat, and 40 units of protein. Each unit of by-products provides 30 units of fiber, 40 units of fat, and 25 units of protein. The animal feed must provide 1800 units of fiber, 2800 units of fat, and 2200 units of protein. How many units of alfalfa, cornmeal, and by-products should the feed contain?

29. A minor league baseball park has 7000 seats. Box seats cost \$6, grandstand seats cost \$4, and bleacher seats cost \$2. When all seats are sold, the revenue is \$26,400. If the number of box seats is one-third the number of bleacher seats, how many of each type of seat are there?

30. Shipping charges at the online bookstore Hercules.com are \$4 for one book, \$6 for two books, and \$7 for three to five books. Last week, there were 6400 orders of five or fewer books, and total shipping charges for these orders were \$33,600. The number of shipments with \$7 charges was 1000 less than the number with \$6 charges. How many shipments were made in each category (one book, two books, three to five books)?

6.3   Matrix Solution Methods

We now develop two versions of Gaussian elimination that can be implemented with technology. When solving systems by hand, a lot of time is wasted copying the x's, y's, z's, and so on. This fact suggests a shorthand system for representing a system of equations. For example, the system

$$x + 2y + 3z = -2$$

(∗)

$$2x + 6y + z = 2$$

$$3x + 3y + 10z = -2$$

can be represented by the following rectangular array of numbers, consisting of the coefficients of the variables and the constants on the right of the equal sign, arranged in the same order in which they appear in the system.

$$\begin{pmatrix} 1 & 2 & 3 & \vdots & -2 \\ 2 & 6 & 1 & \vdots & 2 \\ 3 & 3 & 10 & \vdots & -2 \end{pmatrix}$$

This array is called the **augmented matrix** of the system.[*] It has three horizontal **rows** and four vertical **columns.**

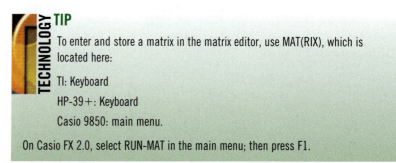

TECHNOLOGY

TIP

To enter and store a matrix in the matrix editor, use MAT(RIX), which is located here:

TI: Keyboard

HP-39+: Keyboard

Casio 9850: main menu.

On Casio FX 2.0, select RUN-MAT in the main menu; then press F1.

EXAMPLE 1

Use the matrix form of the preceding system (∗) to solve the system.

SOLUTION   To solve the system in its original equation form, we would use elementary operations to eliminate the x terms from the last two equations, and then eliminate the y term from the last equation. With matrices, we do essentially the same thing, with the elementary operations on equations being replaced by the corresponding **row operations** on the augmented matrix in order to make certain entries in the first and second columns 0. Here is a side-by-side development of the two solution methods.

---

[*]The plural of *matrix* is *matrices*.

Equation Method	**Matrix Method**

Replace the second equation by the sum of itself and -2 times the first equation:

$$x + 2y + 3z = -2$$
$$2y - 5z = 6$$
$$3x + 3y + 10z = -2$$

Replace the second row by the sum of itself and -2 times the first row:

$$\begin{pmatrix} 1 & 2 & 3 & | & -2 \\ 0 & 2 & -5 & | & 6 \\ 3 & 3 & 10 & | & -2 \end{pmatrix}$$

Replace the third equation by the sum of itself and -3 times the first equation:

$$x + 2y + 3z = -2$$
$$2y - 5z = 6$$
$$-3y + z = 4$$

Replace the third row by the sum of itself and -3 times the first row:

$$\begin{pmatrix} 1 & 2 & 3 & | & -2 \\ 0 & 2 & -5 & | & 6 \\ 0 & -3 & 1 & | & 4 \end{pmatrix}$$

Multiply the second equation by $1/2$ (so that y has coefficient 1):

$$x + 2y + 3z = -2$$
$$y - \frac{5}{2}z = 3$$
$$-3y + z = 4$$

Multiply the second row by $1/2$:

$$\begin{pmatrix} 1 & 2 & 3 & | & -2 \\ 0 & 1 & -\frac{5}{2} & | & 3 \\ 0 & -3 & 1 & | & 4 \end{pmatrix}$$

Replace the third equation by the sum of itself and 3 times the second equation:

$$x + 2y + 3z = -2$$
$$y - \frac{5}{2}z = 3$$
$$-\frac{13}{2}z = 13$$

Replace the third row by the sum of itself and 3 times the second row:

$$\begin{pmatrix} 1 & 2 & 3 & | & -2 \\ 0 & 1 & -\frac{5}{2} & | & 3 \\ 0 & 0 & -\frac{13}{2} & | & 13 \end{pmatrix}$$

Finally, multiply the last equation by $-2/13$:[*]

$$x + 2y + 3z = -2$$
$$(**) \qquad y - \frac{5}{2}z = 3$$
$$z = -2$$

Finally, multiply the last row by $-2/13$:

$$\begin{pmatrix} 1 & 2 & 3 & | & -2 \\ 0 & 1 & -\frac{5}{2} & | & 3 \\ 0 & 0 & 1 & | & -2 \end{pmatrix}$$

---

[*]This step isn't necessary, but it is often convenient to have 1 as the coefficient of the first variable in each equation.

System (∗∗) is easily solved. The third equation shows that $z = -2$, and substituting this in the second equation shows that

$$y - \frac{5}{2}(-2) = 3$$

$$y = 3 - 5 = -2.$$

Substituting $y = -2$ and $z = -2$ in the first equation yields

$$x + 2(-2) + 3(-2) = -2$$

$$x = -2 + 4 + 6 = 8.$$

Therefore, the only solution of the original system is $x = 8$, $y = -2$, $z = -2$. ■

When using matrix notation, row operations replace elementary operations on equations, as shown in Example 1. The solution process ends when you reach a matrix, (and corresponding system), such as the final matrix in Example 1:

(∗∗)
$$\begin{pmatrix} 1 & 2 & 3 & -2 \\ 0 & 1 & -\dfrac{5}{2} & 3 \\ 0 & 0 & 1 & -2 \end{pmatrix} \qquad \begin{aligned} x + 2y + 3z &= -2 \\ y - \frac{5}{2}z &= 3 \\ z &= -2. \end{aligned}$$

The final matrix should satisfy these conditions:

All rows consisting entirely of zeros (if any) are at the bottom.

The first nonzero entry in each nonzero row is a 1 (called a **leading 1**).

Each leading 1 appears to the right of leading 1's in any preceding rows.

Such a matrix is said to be in **row echelon form.**

TECHNOLOGY TIP

To put a matrix in row echelon form, use REF in this menu/submenu:

TI-83+: MATRIX/MATH

TI-86: MATRIX/OPS

TI-89: MATH/MATRIX

Casio and HP-39+ users should consult the Technology Tip after Example 3.

Most calculators have a key that uses row operations to put a given matrix into row echelon form (see the Technology Tip in the margin). For example, using the TI-83+ REF key on the first matrix in Example 1 produced this row echelon matrix and corresponding system of equations.

$$\begin{pmatrix} 1 & 1 & \dfrac{10}{3} & -\dfrac{2}{3} \\ 0 & 1 & -\dfrac{17}{12} & \dfrac{5}{6} \\ 0 & 0 & 1 & -2 \end{pmatrix} \qquad \begin{aligned} x + y + \frac{10}{3}z &= -\frac{2}{3} \\ y - \frac{17}{12}z &= \frac{5}{6} \\ z &= -2 \end{aligned}$$

Because the calculator used a different sequence of row operations than was used in Example 1, it produced a row echelon matrix (and corresponding system) that differs from matrix (∗∗) above. You can easily verify, however, that the preceding system has the same solutions as system (∗∗), namely,

$$x = 8, \qquad y = -2, \qquad z = -2.$$

EXAMPLE 2

Solve the system

$$x + y + 2z = 1$$

$$2x + 4y + 5z = 2$$

$$3x + 5y + 7z = 4.$$

SOLUTION   We form the augmented matrix and reduce it to row echelon form either by hand or by using the REF key. A TI-86 produced the row echelon matrix in Figure 6–14.

```
ref A▸Frac
     [[1 5/3 7/3 4/3]
      [0 1   1/2 1/2]
      [0 0   0   1  ]]
```

Figure 6–14

Look at the last row of the matrix in Figure 6–14; it represents the equation

$$0x + 0y + 0z = 1.$$

Since this equation has no solutions (the left side is always 0 and the right side is always 1), neither does the original system. Such a system is said to be **inconsistent.** ■

TECHNOLOGY TIP

On TI calculators, using the FRAC key in conjunction with the REF key, as in Figure 6–14, usually eliminates long decimals and makes the matrix easier to read.

▪▪ THE GAUSS-JORDAN METHOD

Gaussian elimination on a calculator is an efficient method of solving systems of equations but may involve some messy calculations when you solve the final triangular form system by hand. Most hand computations can be eliminated by using a slight variation, known as the **Gauss-Jordan method,**[*] which is illustrated in the next example.

EXAMPLE 3

Use the Gauss-Jordan method to solve the following system.

$$x - y + 5z = -6$$

$$3x + 3y - z = 10$$

$$x - 5y + 8z = -17$$

$$x + 3y + 2z = 5$$

---

[*]This method was developed by the German engineer Wilhelm Jordan (1842–1899).

SOLUTION   The augmented matrix of the system is shown in Figure 6–15, and an equivalent row echelon matrix (obtained by using the REF and FRAC keys) is shown in Figure 6–16.

$$
\begin{pmatrix}
1 & -1 & 5 & -6 \\
3 & 3 & -1 & 10 \\
1 & -5 & 8 & -17 \\
1 & 3 & 2 & 5
\end{pmatrix}
$$

Figure 6–15   **Figure 6–16**

At this point in Gaussian elimination, we would use back substitution to solve the triangular form system represented by the last matrix in Figure 6–16. In the Gauss-Jordan method, however, additional elimination of variables replaces back substitution. Look at the leading 1 in the third row (shown in color).

$$
\begin{pmatrix}
1 & 1 & -\dfrac{1}{3} & \dfrac{10}{3} \\
0 & 1 & -\dfrac{25}{18} & \dfrac{61}{18} \\
0 & 0 & 1 & -1 \\
0 & 0 & 0 & 0
\end{pmatrix}
$$

In Gauss-Jordan elimination, we make the entries above this leading 1 into 0's.

Replace the second row by the sum of itself and 25/18 times the third row:

$$
\begin{pmatrix}
1 & 1 & -\dfrac{1}{3} & \dfrac{10}{3} \\
0 & 1 & 0 & 2 \\
0 & 0 & 1 & -1 \\
0 & 0 & 0 & 0
\end{pmatrix}
$$

Replace the first row by the sum of itself and 1/3 times the third row:

$$
\begin{pmatrix}
1 & 1 & 0 & 3 \\
0 & 1 & 0 & 2 \\
0 & 0 & 1 & -1 \\
0 & 0 & 0 & 0
\end{pmatrix}
$$

Now consider the leading 1 in the second row (shown in color), and make the entry above it 0.

Replace the first row by the sum of itself and −1 times the second row:

$$
\begin{pmatrix}
1 & 0 & 0 & 1 \\
0 & 1 & 0 & 2 \\
0 & 0 & 1 & -1 \\
0 & 0 & 0 & 0
\end{pmatrix}
$$

This last matrix represents the following system, whose solution is obvious:

$$
x \quad\;\; = \quad 1
$$
$$
y \;\; = \quad 2
$$
$$
z = -1. \quad\blacksquare
$$

A row echelon form matrix, such as the last one in Example 3, in which any column containing a leading 1 has 0's in all other positions, is said to be in **reduced row echelon form.** The goal in the Gauss-Jordan method is to use row operations to put a given augmented matrix into reduced row echelon form (from which the solutions can be read immediately, as in Example 3).

As a general rule, Gaussian elimination (matrix version) is the method of choice when working by hand, (the additional row operations needed to put a matrix in reduced row echelon form are usually more time-consuming—and error-prone—than back substitution). With a calculator or computer, however, it's better to find a reduced row echelon matrix for the system. You can do this in one step on a calculator by using the RREF key (see the Technology Tip in the margin).

TIP

To put a matrix in reduced row echelon form, use RREF in this menu/submenu:

TI-83+: MATRIX/MATH

TI-86: MATRIX/OPS

TI-89: MATH/MATRIX

HP-39+: MATH/MATRIX

RREF programs for Casio and TI-82 are in the Program Appendix.

EXAMPLE 4

Solve this system.

$$2x + 5y + z + 3w = 0$$
$$2y - 4z + 6w = 0$$
$$2x + 17y - 23z + 40w = 0.$$

SOLUTION   A system such as this, in which all the constants on the right side are zero, is called a **homogeneous system.** Every homogeneous system has at least one solution, namely, $x = 0$, $y = 0$, $z = 0$, $w = 0$, which is called the **trivial solution.** The issue with homogeneous systems is whether or not they have any nonzero solutions. The augmented matrix of the system is shown in Figure 6–17, and an equivalent reduced row echelon form matrix is shown in Figure 6–18.*

```
B
        [[2  5   1   3 0]
         [0  2  -4   6 0]
         [2 17 -23 40 0]]
■
```

Figure 6–17

```
rref B▸Frac
        [[1 0 11/2 0 0]
         [0 1  -2  0 0]
         [0 0   0  1 0]]
■
```

Figure 6–18

The system corresponding to the reduced echelon form matrix in Figure 6–18 is

$$x + \frac{11}{2}z = 0$$
$$y - 2z = 0$$
$$w = 0.$$

*In dealing with homogeneous systems, it's not really necessary to include the last column of zeros, as is done here, because row operations do not change this column.

The second equation shows that

$$y = 2z.$$

This equation has an infinite number of solutions, for instance,

$$z = 1, \quad y = 2 \qquad \text{or} \qquad z = 3, \quad y = 6 \qquad \text{or} \qquad z = -2.5, \quad y = -5.$$

In fact, for each real number t, there is a solution: $z = t$, $y = 2t$. Substituting $z = t$ into the first equation shows that

$$x + \frac{11}{2}t = 0$$

$$x = -\frac{11}{2}t = -5.5t.$$

Therefore, this system, and hence the original one, has an infinite number of solutions, one for each real number t.

$$x = -5.5t, \qquad y = 2t, \qquad z = t, \qquad w = 0.$$

A system with infinitely many solutions, such as this one, is said to be **dependent**. ■

> **NOTE**
>
> Every system that has more variables than equations (as in Example 4) is dependent, but other systems may be dependent as well.

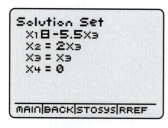

Figure 6–19

TECHNOLOGY TIP

The linear equation system solvers on TI-86 and Casio work only for systems that have the same number of equations as variables *and* have a unique solution. So they cannot handle the systems in Examples 2–4. Students who use these calculators are advised to use the matrix solution method and the RREF key, as in the preceding examples, rather than the solver.

TI-83+ users can download the PolySmlt solver from TI. It can handle virtually any system, provided that you know how to use it properly. For instance, Figure 6–19 shows how it displays the solutions of the system in Example 4. If you get the message "no solution found," as in Figure 6–20, choose RREF at the bottom of the screen. Then the reduced row echelon matrix of the system is displayed. You can use it to read off the solutions (as in Figure 6–21) or to determine that there are no solutions (as in Figure 6–22).

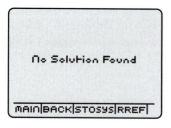

Figure 6–20

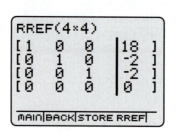

Figure 6–21   Figure 6–22

▪▪ APPLICATIONS

In calculus, it is sometimes necessary to write a complicated rational expression as the sum of simpler ones. One technique for doing this involves systems of equations.

* **C** ───────────────────────────────────

EXAMPLE 5

Find constants A, B, and C such that

$$\frac{2x^2 + 15x + 10}{(x - 1)(x + 2)^2} = \frac{A}{x - 1} + \frac{B}{x + 2} + \frac{C}{(x + 2)^2}.$$

SOLUTION Multiply both sides of the equation by the common denominator $(x - 1)(x + 2)^2$ and collect like terms on the right-hand side.

$$2x^2 + 15x + 10 = A(x + 2)^2 + B(x - 1)(x + 2) + C(x - 1)$$

$$= A(x^2 + 4x + 4) + B(x^2 + x - 2) + C(x - 1)$$

$$= Ax^2 + 4Ax + 4A + Bx^2 + Bx - 2B + Cx - C$$

$$= (A + B)x^2 + (4A + B + C)x + (4A - 2B - C)$$

Since the polynomials on the left- and right-hand sides of the last equation are equal, their coefficients must be equal term by term, that is,

$$A + B \quad\;\;\;\; = 2 \qquad \text{[Coefficients of } x^2\text{]}$$

$$4A + B + C = 15 \qquad \text{[Coefficients of } x\text{]}$$

$$4A - 2B - C = 10. \qquad \text{[Constant terms]}$$

We can consider this as a system of equations with unknowns A, B, C. The augmented matrix of the system is shown in Figure 6–23, and an equivalent reduced row echelon form matrix is shown in Figure 6–24.

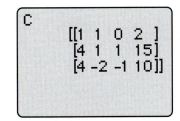

Figure 6–23

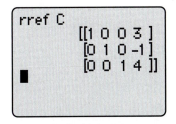

Figure 6–24

The solutions of the system can be read from the reduced row echelon form matrix in Figure 6–24:

$$A = 3, \qquad B = -1, \qquad C = 4.$$

Therefore,

$$\frac{2x^2 + 15x + 10}{(x - 1)(x + 2)^2} = \frac{3}{x - 1} + \frac{-1}{x + 2} + \frac{4}{(x + 2)^2}.$$

───────────────────────────────────

* **C** and ⬤ indicate examples, exercises, and sections that are relevant to calculus.

The right side of this equation is called the **partial fraction decomposition** of the fraction on the left side.   ■

EXAMPLE 6

Ellen Ryan is starting a small business and borrows $10,000 on three different credit cards, with annual interest rates of 18%, 15%, and 9%, respectively. She borrows three times as much on the 15% card as on the 18% card, and her total annual interest on all three cards is $1244.25. How much did she borrow on each credit card?

SOLUTION   Let x be the amount on the 18% card, let y be the amount on the 15% card, and let z be the amount on the 9% card. Then $x + y + z = 10,000$. Furthermore,

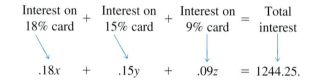

$$.18x \quad + \quad .15y \quad + \quad .09z \quad = 1244.25.$$

Finally, we have

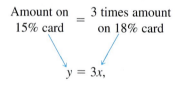

$$y = 3x,$$

which is equivalent to $3x - y = 0$. Therefore, we must solve this system of equations.

$$x + \quad y + \quad z = 10,000$$
$$.18x + .15y + .09z = \quad 1,244.25$$
$$3x - \quad y \quad \quad = \quad \quad 0,$$

whose augmented matrix is

$$\begin{pmatrix} 1 & 1 & 1 & 10,000 \\ .18 & .15 & .09 & 1,244.25 \\ 3 & -1 & 0 & 0 \end{pmatrix}.$$

EXPLORATION

Enter this matrix in your calculator. Use row operations or the RREF key to put it in reduced row echelon form. Read the solutions of the system from this last matrix.

The Calculator Exploration shows that Ellen borrowed $1275 on the 18% card, $3825 on the 15% card, and $4900 on the 9% card.   ■

The preceding examples illustrate the following fact, whose proof is omitted.

Number of Solutions of a System

Any system of linear equations must have

No solutions (an inconsistent system) *or*

Exactly one solution *or*

An infinite number of solutions (a dependent system).

EXERCISES 6.3

In Exercises 1–4, write the augmented matrix of the system.

1. $2x - 3y + 4z = 1$
$x + 2y - 6z = 0$
$3x - 7y + 4z = -3$

2. $x + 2y - 3w + 7z = -5$
$2x - y + 2z = 4$
$3x + 7w - 6z = 0$

3. $x - \frac{1}{2}y + \frac{7}{4}z = 0$
$2x - \frac{3}{2}y + 5z = 0$
$-2y + \frac{1}{3}z = 0$

4. $2x - \frac{1}{2}y + \frac{7}{2}w - 6z = 1$
$\frac{1}{4}x - 6y + 2w - z = 2$
$4y - \frac{1}{2}w + z = 3$
$2x + 3y + \frac{1}{2}z = 4$

In Exercises 5–8, the augmented matrix of a system of equations is given. Express the system in equation notation.

5. $\begin{pmatrix} 2 & -3 & 1 \\ 4 & 7 & 2 \end{pmatrix}$

6. $\begin{pmatrix} 2 & 3 & 5 & 2 \\ 1 & 6 & 9 & 0 \end{pmatrix}$

7. $\begin{pmatrix} 1 & 0 & 1 & 0 & 1 \\ 1 & -1 & 4 & -2 & 3 \\ 4 & 2 & 5 & 0 & 2 \end{pmatrix}$

8. $\begin{pmatrix} 1 & 7 & 0 & 4 \\ 2 & 3 & 1 & 6 \\ -1 & 0 & 2 & 3 \end{pmatrix}$

In Exercises 9–12, the reduced row echelon form of the augmented matrix of a system of equations is given. Find the solutions of the system.

9. $\begin{pmatrix} 1 & 0 & 0 & 0 & 3/2 \\ 0 & 1 & 0 & 0 & 5 \\ 0 & 0 & 1 & 0 & -2 \\ 0 & 0 & 0 & 1 & 0 \end{pmatrix}$

10. $\begin{pmatrix} 1 & 0 & 0 & 0 & 0 & 5 \\ 0 & 1 & 0 & 0 & 0 & 4 \\ 0 & 0 & 1 & 0 & 0 & 3 \\ 0 & 0 & 0 & 0 & 1 & 2 \\ 0 & 0 & 0 & 0 & 0 & 1 \end{pmatrix}$

11. $\begin{pmatrix} 1 & 0 & 0 & 1 & 2 \\ 0 & 1 & 0 & 2 & -3 \\ 0 & 0 & 1 & 0 & 4 \\ 0 & 0 & 0 & 0 & 0 \end{pmatrix}$

12. $\begin{pmatrix} 1 & 0 & 0 & 0 & 7 \\ 0 & 1 & 0 & 0 & 1 \\ 0 & 0 & 1 & 0 & -5 \\ 0 & 0 & 0 & 1 & 4 \\ 0 & 0 & 0 & 0 & 0 \\ 0 & 0 & 0 & 0 & 0 \end{pmatrix}$

In Exercises 13–16, use Gaussian elimination to solve the system.

13. $-x + 3y + 2z = 0$
$2x - y - z = 3$
$x + 2y + 3z = 0$

14. $3x + 7y + 9z = 0$
$x + 2y + 3z = 2$
$x + 4y + z = 2$

15. $x + y + z = 1$
$x - 2y + 2z = 4$
$2x - y + 3z = 5$

16. $2x - y + z = 1$
$3x + y + z = 0$
$7x - y + 3z = 2$

In Exercises 17–20, use the Gauss-Jordan method to solve the system.

17. $x - 2y + 4z = 6$
$x + y + 13z = 6$
$-2x + 6y - z = -10$

18.
$$x - y + 5z = -6$$
$$3x + 3y - z = 10$$
$$x + 3y + 2z = 5$$

19.
$$x + y + z = 200$$
$$x - 2y = 0$$
$$2x + 3y + 5z = 600$$
$$2x - y + z = 200$$

20.
$$3x - y + z = 6$$
$$x + 2y - z = 0$$

In Exercises 21–36, solve the system.

21.
$$11x + 10y + 9z = 5$$
$$x + 2y + 3z = 1$$
$$3x + 2y + z = 1$$

22.
$$-x + 2y - 3z + 4w = 8$$
$$2x - 4y + z + 2w = -3$$
$$5x - 4y + z + 2w = -3$$

23.
$$x + y = 3$$
$$5x - y = 3$$
$$9x - 4y = 1$$

24.
$$2x - y + 2z = 3$$
$$-x + 2y - z = 0$$
$$x + y - z = 1$$

25.
$$x - 4y - 13z = 4$$
$$x - 2y - 3z = 2$$
$$-3x + 5y + 4z = 2$$

26.
$$2x - 4y + z = 3$$
$$x + 3y - 7z = 1$$
$$-2x + 4y - z = 10$$

27.
$$4x + y + 3z = 7$$
$$x - y + 2z = 3$$
$$3x + 2y + z = 4$$

28.
$$x + 4y + z = 3$$
$$-x + 2y + 2z = 0$$
$$2x + 2y - z = 3$$

29.
$$x + y + z = 0$$
$$3x - y + z = 0$$
$$-5x - y + z = 0$$

30.
$$x + y + z = 0$$
$$x - y - z = 0$$
$$x - y + z = 0$$

31.
$$2x + y + 3z - 2w = -6$$
$$4x + 3y + z - w = -2$$
$$x + y + z + w = -5$$
$$-2x - 2y + 2z + 2w = -10$$

32.
$$x + y + z + w = -1$$
$$-x + 4y + z - w = 0$$
$$x - 2y + z - 2w = 11$$
$$-x - 2y + z + 2w = -3$$

33.
$$x - 2y - z - 3w = -3$$
$$-x + y + z = 0$$
$$4y + 3z - 2w = -1$$
$$2x - 2y + w = 1$$

34.
$$3x - y + 2z = 0$$
$$-x + 3y + 2z + 5w = 0$$
$$x + 2y + 5z - 4w = 0$$
$$2x - y + 3w = 0$$

35.
$$\frac{3}{x} - \frac{1}{y} + \frac{4}{z} = -13$$
$$\frac{1}{x} + \frac{2}{y} - \frac{1}{z} = 12$$
$$\frac{4}{x} - \frac{1}{y} + \frac{3}{z} = -7$$

[*Hint:* Let $u = 1/x$, $v = 1/y$, $w = 1/z$.]

36.
$$\frac{1}{x+1} - \frac{2}{y-3} + \frac{3}{z-2} = 4$$
$$\frac{5}{y-3} - \frac{10}{z-2} = -5$$
$$\frac{-3}{x+1} + \frac{4}{y-3} - \frac{1}{z-2} = -2$$

[*Hint:* Let $u = 1/(x+1)$, $v = 1/(y-3)$, $w = 1/(z-2)$.]

Exercises 37–40, solve the system. [Note: *The REF and RREF keys on some calculators produce an error message when there are more rows than columns in a matrix, in which case you will have to solve the system by some other means.*]

37.
$$2x - y = 1$$
$$3x + y = 2$$
$$4x - 2y = 2$$
$$5x + 5y = 4$$

38.
$$x + y = 3$$
$$-x + 2y = 3$$
$$5x - y = 3$$
$$-7x + 5y = 3$$

39.
$$x + 2y = 3$$
$$2x + 3y = 4$$
$$3x + 4y = 5$$
$$4x + 5y = 6$$

40.
$$x - y = 2$$
$$x + y = 4$$
$$2x + 3y = 9$$
$$3x - 2y = 6$$

C

In Exercises 41–46, find the constants A, B, and C.

41. $\dfrac{x}{(x+1)(x+2)} = \dfrac{A}{x+1} + \dfrac{B}{x+2}$

42. $\dfrac{1}{(x+1)(x-1)} = \dfrac{A}{x+1} + \dfrac{B}{x-1}$

43. $\dfrac{2x+1}{(x+2)(x-3)^2} = \dfrac{A}{x+2} + \dfrac{B}{x-3} + \dfrac{C}{(x-3)^2}$

44. $\dfrac{x^2 - x - 21}{(2x-1)(x^2+4)} = \dfrac{A}{2x-1} + \dfrac{Bx+C}{x^2+4}$

45. $\dfrac{5x^2+1}{(x+1)(x^2-x+1)} = \dfrac{A}{x+1} + \dfrac{Bx+C}{x^2-x+1}$

46. $\dfrac{x-2}{(x+4)(x^2+2x+2)} = \dfrac{A}{x+4} + \dfrac{Bx+C}{x^2+2x+2}$

47. Lillian borrows $10,000. She borrows some from her friend at 8% annual interest, twice as much as that from her bank at 9%, and the remainder from her insurance company at 5%. She pays a total of $830 in interest for the first year. How much did she borrow from each source?

48. An investor puts a total of $25,000 into three very speculative stocks. She invests some of it in Crystalcomp and $2000 more than one-half that amount in Flyboys. The remainder is invested in Zumcorp. Crystalcomp rises 16% in value, Flyboys rises 20%, and Zumcorp rises 18%. Her investment in the three stocks is now worth $29,440. How much was originally invested in each stock?

49. An investor has $70,000 invested in a mutual fund, bonds, and a fast food franchise. She has twice as much invested in bonds as in the mutual fund. Last year the mutual fund paid a 2% dividend, the bonds paid 10%, and the fast food franchise paid 6%; her dividend income was $4800. How much is invested in each of the three investments?

50. Tickets to a band concert cost $2 for children, $3 for teenagers, and $5 for adults. 570 people attended the concert and total ticket receipts were $1950. Three-fourths as many teenagers as children attended. How many children, adults, and teenagers attended?

51. The table shows the calories, sodium, and protein in one cup of various kinds of soup.

	Progresso Roasted Chicken Rotini	Healthy Choice Hearty Chicken	Campbell's Chunky Chicken Noodle
Calories	100	130	130
Sodium (mg)	970	480	880
Protein (g)	6	8	8

How many cups of each kind of soup should be mixed together to produce 10 servings of soup, each of which provides 203 calories, 1190 milligrams of sodium, and 12.4 grams of protein? What is the serving size (in cups)? (*Hint:* In 10 servings, there must be 2030 calories, 11,900 milligrams of sodium, and 124 grams of protein.)

52. The table shows the calories, sodium, and fat in one ounce of various snack foods (all produced by Planters).

	Cashews	Dry Roasted Honey Peanuts	Cajun Crunch Trail Mix
Calories	170	150	130
Sodium (mg)	120	95	10
Fat (g)	6	7	3

How many ounces of each kind of snack should be combined to produce 10 servings, each of which provides 288 calories, 115 milligrams of sodium, and 9.6 grams of fat? What is the serving size?

53. Comfort Systems, Inc., sells three models of humidifiers. The bedroom model weighs 10 pounds and comes in an 8-cubic-foot box; the living room model weighs 20 pounds and comes in an 8-cubic-foot box; the whole-house model weighs 60 pounds and comes in a 28-cubic-foot box. Each of the company's delivery vans has 248 cubic feet of space and can hold a maximum of 440 pounds. For a van to be as fully loaded as possible, how many of each model should it carry?

54. Peanuts cost $3 per pound, almonds cost $4 per pound, and cashews costs $8 per pound. How many pounds of each should be used to produce 140 pounds of a mixture costing $6 per pound in which there are twice as many peanuts as almonds?

Exercises 55 and 56 deal with computer-aided tomography (CAT) scanners that take X-rays of body parts from different directions to create a picture of a cross section of the body.[*] The amount by which the X-ray energy decreases (measured in linear-attenuation units) indicates whether the X-ray has passed through healthy tissue, tumorous tissue, or bone, according to the following table.

Tissue Type	Linear-Attenuation Units
Healthy	.1625–.2977
Tumorous	.2679–.3930
Bone	.3857–.5108

---

*Exercises 55 and 56 are based on D. Jabon, G. Nord, B. W. Wilson, and P. Coffman, "Medical Applications of Linear Equations," *Mathematics Teacher 89,* no. 5 (May 1996).

The body part being scanned is divided into cells. The total linear-attenuation value is the sum of the values for each cell the X-ray passes through. In the figure for Exercise 55, for example, let a, b, and c be the values for cells A, B, and C, respectively; then the attenuation value for X-ray 3 is b + c.

55. (a) In the figure, find the linear-attenuation values for X-rays 1 and 2.
   (b) If the total linear-attenuation values for X-rays 1, 2, and 3 are .75, .60, and .54, respectively, set up and solve a system of three equations in a, b, c.
   (c) What kind of tissue are cells A, B, and C?

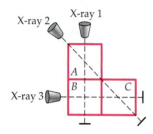

56. Four X-ray beams are aimed at four cells, as shown in the figure.

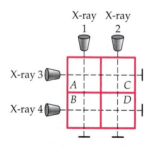

   (a) If the total linear-attenuation values for X-rays 1, 2, 3, and 4 are .60, .75, .65, and .70, respectively, is there enough information to determine the values of a, b, c, and d? Explain.
   (b) If an additional X-ray beam is added, with a linear-attenuation value of .85, as shown in the figure below, can the values of a, b, c, and d be determined? If so, what are they? What can be said about cells A, B, C, and D?

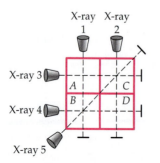

57. A furniture manufacturer has 1950 machine hours available each week in the cutting department, 1490 hours in the assembly department, and 2160 in the finishing department. Manufacturing a chair requires .2 hours of cutting, .3 hours of assembly, and .1 hours of finishing. A chest requires .5 hours of cutting, .4 hours of assembly, and .6 hours of finishing. A table requires .3 hours of cutting, .1 hours of assembly, and .4 hours of finishing. How many chairs, chests, and tables should be produced to use all the available production capacity?

58. A stereo equipment manufacturer produces three models of speakers, R, S, and T, and has three kinds of delivery vehicles: trucks, vans, and station wagons. A truck holds two boxes of model R, one of model S, and three of model T. A van holds one box of model R, three of model S, and two of model T. A station wagon holds one box of model R, three of model S, and one of model T. If 15 boxes of model R, 20 boxes of model S, and 22 boxes of model T are to be delivered, how many vehicles of each type should be used so that all operate at full capacity?

59. The diagram shows the traffic flow at four intersections during a typical one-hour period. The streets are all one-way, as indicated by the arrows. To adjust the traffic lights to avoid congestion, engineers must determine the possible values of x, y, z, and t.
   (a) Write a system of linear equations that describes congestion-free traffic flow. [*Hint:* 600 cars per hour come down Euclid to intersection A, and 400 come down 4th Avenue to intersection A. Also, x cars leave intersection A on Euclid, and t cars leave on 4th Avenue. To avoid congestion, the number of cars leaving the intersection must be the same as the number entering, that is, $x + t = 600 + 400$. Use intersections B, C, and D to find three more equations.]
   (b) Solve the system in part (a), which is dependent. Express your answers in terms of the variable t.
   (c) Find the largest and smallest number of cars that can leave the given intersection on the given street: A on 4th Avenue, A on Euclid, C on 5th Avenue, and C on Chester.

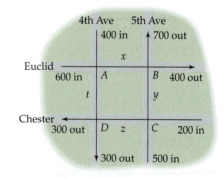

60. The diagram shows the traffic flow at four intersections during rush hour, as in Exercise 59.
  (a) What are the possible values of x, y, z, and t in order to avoid any congestion? [Express your answers in terms of t.]
  (b) What are the possible values of t?

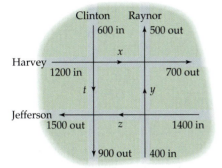

61. If Tom, Dick, and Harry work together, they can paint a large room in 4 hours. When only Dick and Harry work together, it takes 8 hours to paint the room. Tom and Dick, working together, take 6 hours to paint the room. How long would it take each of them to paint the room alone? [*Hint:* If x is the amount of the room painted in 1 hour by Tom, y the amount painted by Dick, and z the amount painted by Harry, then $x + y + z = 1/4$.]

62. Pipes R, S, and T are connected to the same tank. When all three pipes are running, they can fill the tank in 2 hours. When only pipes S and T are running, they can fill the tank in 4 hours. When only R and T are running, they can fill the tank in 2.4 hours. How long would it take each pipe running alone to fill the tank?

6.4 Matrix Methods for Square Systems

Matrices were used in Section 6.3 as a convenient shorthand for solving systems of linear equations. We now consider matrices in a more general setting and show how the algebra of matrices provides an alternative method for solving systems of equations that are not dependent and have the same number of equations as variables.

Let m and n be positive integers. An **$m \times n$ matrix** (read "m by n matrix") is a rectangular array of numbers, with m horizontal rows and n vertical columns. For example,

$$
\begin{pmatrix} 3 & 2 & -5 \\ 6 & 1 & 7 \\ -2 & 5 & 0 \end{pmatrix}
\qquad
\begin{pmatrix} -3 & 4 \\ 2 & 0 \\ 0 & 1 \\ 7 & 3 \\ 1 & -6 \end{pmatrix}
\qquad
\begin{pmatrix} 3 & 0 & 1 & 0 \\ \sqrt{2} & -\tfrac{1}{2} & 4 & \tfrac{8}{3} \\ 10 & 2 & -\tfrac{3}{4} & 12 \end{pmatrix}
\qquad
\begin{pmatrix} \sqrt{3} \\ 2 \\ 0 \\ 11 \end{pmatrix}
$$

3×3 matrix	5×2 matrix	3×4 matrix	4×1 matrix
3 rows	5 rows	3 rows	4 rows
3 columns	2 columns	4 columns	1 column

In a matrix, the rows are horizontal and are numbered from top to bottom. The columns are vertical and are numbered from left to right. For example,

$$
\begin{array}{l}
\text{Row 1} \rightarrow \\
\text{Row 2} \rightarrow \\
\text{Row 3} \rightarrow
\end{array}
\begin{pmatrix} 11 & 3 & 14 \\ -2 & 0 & -5 \\ \tfrac{1}{3} & 6 & 7 \end{pmatrix}
$$

$$
\begin{array}{ccc}
\uparrow & \uparrow & \uparrow \\
\text{Column 1} & \text{Column 2} & \text{Column 3}
\end{array}
$$

Each entry in a matrix can be located by stating the row and column in which it appears. For instance, in the preceding 3×3 matrix, 14 is the entry in row 1, column 3, and 0 is the entry in row 2, column 2. When you enter a matrix on a calculator, the words row and column won't be displayed, but the row numbers will always be listed before the column number. Thus, a display such as "$A[3, 2]$," or simply "3, 2," indicates the entry in row 3, column 2.

Two matrices are said to be **equal** if they have the same size (same number of rows and columns) and the corresponding entries are equal. For example,

$$\begin{pmatrix} 3 & (-1)^2 \\ 6 & 12 \end{pmatrix} = \begin{pmatrix} 3 & 1 \\ \sqrt{36} & 12 \end{pmatrix}, \quad \text{but} \quad \begin{pmatrix} 6 & 4 \\ 5 & 1 \end{pmatrix} \neq \begin{pmatrix} 6 & 5 \\ 4 & 1 \end{pmatrix}.$$

■■ MATRIX MULTIPLICATION

Although there is an extensive arithmetic of matrices, we shall need only matrix multiplication. The simplest case is the product of a matrix with a single row and a matrix with a single column, where the row and column have the same number of entries. This is done by multiplying corresponding entries (first by first, second by second, and so on) and then adding the results. An example is shown in Figure 6–25.

$$(3 \quad 1 \quad 2) \begin{pmatrix} 2 \\ 0 \\ 1 \end{pmatrix} = 3 \cdot 2 + 1 \cdot 0 + 2 \cdot 1 = 8$$

First Terms   Second Terms   Third Terms

Figure 6–25

Note that the product of a row and a column is a single number.

Now let A be an $m \times n$ matrix, and let B be an $n \times p$ matrix, so that the number of columns of A is the same as the number of rows of B (namely, n). The product matrix AB is defined to be an $m \times p$ matrix (same number of rows as A and same number of columns as B). The product AB is defined as follows.

Matrix Multiplication

If A is an $m \times n$ and B is an $n \times p$ matrix, then AB is the $m \times p$ matrix whose entry in the ith row and jth column is

the product of the ith row of A and the jth column of B.

EXAMPLE 1

If it is defined, find the product AB, where

$$A = \begin{pmatrix} 3 & 1 & 2 \\ -1 & 0 & 4 \end{pmatrix} \quad \text{and} \quad B = \begin{pmatrix} 2 & -3 & 0 & 1 \\ 0 & 5 & 2 & 7 \\ 1 & 8 & -4 & 1 \end{pmatrix}.$$

SOLUTION *A* has 3 columns, and *B* has 3 rows. So the product matrix *AB* is defined. *AB* has 2 rows (same as *A*) and 4 columns (same as *B*). Its entries are calculated as follows. The entry in row 1, column 1 of *AB* is the product of row 1 of *A* and column 1 of *B*, which is the number 8, as shown in Figure 6–25 and indicated at the right below.

row 1, column 1 $\begin{pmatrix} 3 & 1 & 2 \\ -1 & 0 & 4 \end{pmatrix} \begin{pmatrix} 2 & -3 & 0 & 1 \\ 0 & 5 & 2 & 7 \\ 1 & 8 & -4 & 1 \end{pmatrix} = \begin{pmatrix} 8 \\ \end{pmatrix}$ $3 \cdot 2 + 1 \cdot 0 + 2 \cdot 1 = 8$

The other entries in *AB* are obtained similarly.

row 1, column 2 $\begin{pmatrix} 3 & 1 & 2 \\ -1 & 0 & 4 \end{pmatrix} \begin{pmatrix} 2 & -3 & 0 & 1 \\ 0 & 5 & 2 & 7 \\ 1 & 8 & -4 & 1 \end{pmatrix} = \begin{pmatrix} 8 & 12 \\ \end{pmatrix}$ $3(-3) + 1 \cdot 5 + 2 \cdot 8 = 12$

row 1, column 3 $\begin{pmatrix} 3 & 1 & 2 \\ -1 & 0 & 4 \end{pmatrix} \begin{pmatrix} 2 & -3 & 0 & 1 \\ 0 & 5 & 2 & 7 \\ 1 & 8 & -4 & 1 \end{pmatrix} = \begin{pmatrix} 8 & 12 & -6 \\ \end{pmatrix}$ $3 \cdot 0 + 1 \cdot 2 + 2(-4) = -6$

row 1, column 4 $\begin{pmatrix} 3 & 1 & 2 \\ -1 & 0 & 4 \end{pmatrix} \begin{pmatrix} 2 & -3 & 0 & 1 \\ 0 & 5 & 2 & 7 \\ 1 & 8 & -4 & 1 \end{pmatrix} = \begin{pmatrix} 8 & 12 & -6 & 12 \\ \end{pmatrix}$ $3 \cdot 1 + 1 \cdot 7 + 2 \cdot 1 = 12$

row 2, column 1 $\begin{pmatrix} 3 & 1 & 2 \\ -1 & 0 & 4 \end{pmatrix} \begin{pmatrix} 2 & -3 & 0 & 1 \\ 0 & 5 & 2 & 7 \\ 1 & 8 & -4 & 1 \end{pmatrix} = \begin{pmatrix} 8 & 12 & -6 & 12 \\ 2 \end{pmatrix}$ $(-1)2 + 0 \cdot 0 + 4 \cdot 1 = 2$

row 2, column 2 $\begin{pmatrix} 3 & 1 & 2 \\ -1 & 0 & 4 \end{pmatrix} \begin{pmatrix} 2 & -3 & 0 & 1 \\ 0 & 5 & 2 & 7 \\ 1 & 8 & -4 & 1 \end{pmatrix} = \begin{pmatrix} 8 & 12 & -6 & 12 \\ 2 & 35 \end{pmatrix}$ $(-1)(-3) + 0 \cdot 5 + 4 \cdot 8 = 35$

row 2, column 3 $\begin{pmatrix} 3 & 1 & 2 \\ -1 & 0 & 4 \end{pmatrix} \begin{pmatrix} 2 & -3 & 0 & 1 \\ 0 & 5 & 2 & 7 \\ 1 & 8 & -4 & 1 \end{pmatrix} = \begin{pmatrix} 8 & 12 & -6 & 12 \\ 2 & 35 & -16 \end{pmatrix}$ $(-1)0 + 0 \cdot 2 + 4(-4) = -16$

row 2, column 4 $\begin{pmatrix} 3 & 1 & 2 \\ -1 & 0 & 4 \end{pmatrix} \begin{pmatrix} 2 & -3 & 0 & 1 \\ 0 & 5 & 2 & 7 \\ 1 & 8 & -4 & 1 \end{pmatrix} = \begin{pmatrix} 8 & 12 & -6 & 12 \\ 2 & 35 & -16 & 3 \end{pmatrix}$ $(-1)1 + 0 \cdot 7 + 4 \cdot 1 = 3$

The last matrix on the right is the product *AB*. ∎

EXAMPLE 2

Let *A*, *B*, *C*, and *D* be the following matrices.

$$A = \begin{pmatrix} 3 & 2 & 1 \\ -2 & 0 & 4 \\ 1 & -2 & 5 \end{pmatrix}, \quad B = \begin{pmatrix} 5 & -2 \\ 1 & -1 \\ 4 & 2 \end{pmatrix}, \quad C = \begin{pmatrix} 4 & -2 & 7 \\ 6 & 3 & 1 \\ 2 & 1 & 4 \end{pmatrix},$$

$$D = \begin{pmatrix} 1 & 0 & -5 \\ 2 & 3 & -4 \\ 3 & 7 & 2 \end{pmatrix}.$$

Find each of the following matrices, if possible.

(a) AB (b) BC (c) CD and DC

SOLUTION

(a) Following the same procedure as in Example 1, we have

$$AB = \begin{pmatrix} 3 & 2 & 1 \\ -2 & 0 & 4 \\ 1 & -2 & 5 \end{pmatrix} \begin{pmatrix} 5 & -2 \\ 1 & -1 \\ 4 & 2 \end{pmatrix}$$

$$= \begin{pmatrix} 3 \cdot 5 + 2 \cdot 1 + 1 \cdot 4 & 3(-2) + 2(-1) + 1 \cdot 2 \\ (-2) \cdot 5 + 0 \cdot 1 + 4 \cdot 4 & (-2)(-2) + 0(-1) + 4 \cdot 2 \\ 1 \cdot 5 + (-2) \cdot 1 + 5 \cdot 4 & 1(-2) + (-2)(-1) + 5 \cdot 2 \end{pmatrix}$$

$$= \begin{pmatrix} 21 & -6 \\ 6 & 12 \\ 23 & 10 \end{pmatrix}.$$

(b) Matrix B is 3×2, and C is 3×3. Since the number of columns in B is different from the number of rows in C, the product is not defined.

(c) We use the matrix editor of a calculator to enter the matrices (Figure 6–26).

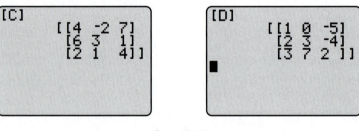

Figure 6–26

We then use the calculator to compute both products (Figure 6–27).

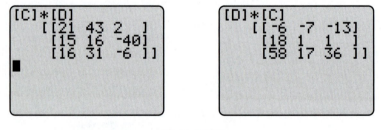

Figure 6–27

Note that DC is not equal to CD.   ■

    It can be shown that matrix multiplication is associative, meaning that $A(BC) = (AB)C$ for all matrices A, B, and C for which the products are defined. As we saw in Example 2, however, matrix multiplication is *not* commutative, that is, AB might not be equal to BA, even when both products are defined.

▉▉ IDENTITY MATRICES AND INVERSES

The $n \times n$ **identity matrix** I_n is the matrix with 1's on the diagonal from upper left to lower right and 0's everywhere else; for example,

$$I_2 = \begin{pmatrix} 1 & 0 \\ 0 & 1 \end{pmatrix}, \qquad I_3 = \begin{pmatrix} 1 & 0 & 0 \\ 0 & 1 & 0 \\ 0 & 0 & 1 \end{pmatrix}, \qquad I_4 = \begin{pmatrix} 1 & 0 & 0 & 0 \\ 0 & 1 & 0 & 0 \\ 0 & 0 & 1 & 0 \\ 0 & 0 & 0 & 1 \end{pmatrix}.$$

The number 1 is the multiplicative identity of the number system because $a \cdot 1 = a = 1 \cdot a$ for every number a. The identity matrix I_n is the multiplicative identity for $n \times n$ matrices.

Identity Matrix

For any $n \times n$ matrix A,

$$AI_n = A = I_nA.$$

For example, in the 2×2 case,

$$\begin{pmatrix} a & b \\ c & d \end{pmatrix}\begin{pmatrix} 1 & 0 \\ 0 & 1 \end{pmatrix} = \begin{pmatrix} a \cdot 1 + b \cdot 0 & a \cdot 0 + b \cdot 1 \\ c \cdot 1 + d \cdot 0 & c \cdot 0 + d \cdot 1 \end{pmatrix} = \begin{pmatrix} a & b \\ c & d \end{pmatrix}.$$

Verify that the same answer results if you reverse the order of multiplication.

Every nonzero number c has a multiplicative inverse $c^{-1} = 1/c$ with the property that $cc^{-1} = 1$. The analogous statement for matrix multiplication does not always hold, and special terminology is used when it does. An $n \times n$ matrix A is said to be **invertible** (or **nonsingular**) if there is an $n \times n$ matrix B such that $AB = I_n$. In this case, it can be proved that $BA = I_n$ also. The matrix B is called the **inverse** of A and is sometimes denoted A^{-1}.

EXAMPLE 3

You can readily verify that

$$\begin{pmatrix} 2 & 1 \\ 3 & 1 \end{pmatrix}\begin{pmatrix} -1 & 1 \\ 3 & -2 \end{pmatrix} = \begin{pmatrix} 1 & 0 \\ 0 & 1 \end{pmatrix} = \begin{pmatrix} -1 & 1 \\ 3 & -2 \end{pmatrix}\begin{pmatrix} 2 & 1 \\ 3 & 1 \end{pmatrix}.$$

Therefore, $A = \begin{pmatrix} 2 & 1 \\ 3 & 1 \end{pmatrix}$ is an invertible matrix with inverse

$$A^{-1} = \begin{pmatrix} -1 & 1 \\ 3 & -2 \end{pmatrix}. \qquad ■$$

EXAMPLE 4

Find the inverse of the matrix $\begin{pmatrix} 2 & 6 \\ 1 & 4 \end{pmatrix}$.

SOLUTION We must find numbers x, y, u, v such that

$$\begin{pmatrix} 2 & 6 \\ 1 & 4 \end{pmatrix} \begin{pmatrix} x & u \\ y & v \end{pmatrix} = \begin{pmatrix} 1 & 0 \\ 0 & 1 \end{pmatrix},$$

which is the same as

$$\begin{pmatrix} 2x + 6y & 2u + 6v \\ x + 4y & u + 4v \end{pmatrix} = \begin{pmatrix} 1 & 0 \\ 0 & 1 \end{pmatrix}.$$

Since corresponding entries in these last two matrices are equal, finding x, y, u, v amounts to solving these systems of equations:

$$\begin{aligned} 2x + 6y &= 1 \\ x + 4y &= 0 \end{aligned} \qquad \text{and} \qquad \begin{aligned} 2u + 6v &= 0 \\ u + 4v &= 1. \end{aligned}$$

We shall solve the systems by the Gauss-Jordan method of Section 6.3. The augmented matrices for the two systems are

$$A = \begin{pmatrix} 2 & 6 & \vdots & 1 \\ 1 & 4 & \vdots & 0 \end{pmatrix} \qquad \text{and} \qquad B = \begin{pmatrix} 2 & 6 & \vdots & 0 \\ 1 & 4 & \vdots & 1 \end{pmatrix}.$$

Note that the row operations that are needed for both A and B will be the same (because the first two columns are the same in both A and B). Consequently, we can save space and time by combining both of these matrices into the single matrix

$$\begin{pmatrix} 2 & 6 & \vdots & 1 & 0 \\ 1 & 4 & \vdots & 0 & 1 \end{pmatrix}.$$

The first three columns of the last matrix are matrix A, and the first two and last columns are matrix B. Performing row operations on this matrix amounts to doing the operations simultaneously on both A and B.

Multiply row 1 by $1/2$:

$$\begin{pmatrix} 1 & 3 & \vdots & \frac{1}{2} & 0 \\ 1 & 4 & \vdots & 0 & 1 \end{pmatrix}$$

Replace row 2 by the sum of itself and -1 times row 1:

$$\begin{pmatrix} 1 & 3 & \vdots & \frac{1}{2} & 0 \\ 0 & 1 & \vdots & -\frac{1}{2} & 1 \end{pmatrix}$$

Replace row 1 by the sum of itself and -3 times row 2:

$$\begin{pmatrix} 1 & 0 & \vdots & 2 & -3 \\ 0 & 1 & \vdots & -\frac{1}{2} & 1 \end{pmatrix}$$

The first three columns of the last matrix show that $x = 2$ and $y = -1/2$. Similarly, the first two and last columns show that $u = -3$ and $v = 1$. Therefore,

$$A^{-1} = \begin{pmatrix} 2 & -3 \\ -\frac{1}{2} & 1 \end{pmatrix}.$$

Observe that A^{-1} is just the right half of the final form of the preceding augmented matrix and that the left half is the identity matrix I_2. ∎

Although the technique in Example 4 can be used to find the inverse of any matrix that has one, it's quicker to use a calculator. Any calculator with matrix capabilities can find the inverse of an invertible matrix (see the Technology Tip in the margin).

TECHNOLOGY TIP

On calculators other than TI-89, you can find the inverse A^{-1} of matrix A by keying in A (or Mat A) and using the x^{-1} key. Using the $\wedge$ key and -1 produces A^{-1} on TI-89 and HP-39+ but leads to an error message on other calculators.

> **CAUTION**
>
> A calculator should produce an error message when asked for the inverse of a matrix A that does not have one. However, because of round-off errors, it may sometimes display a matrix that it says is A^{-1}. As an accuracy check when finding inverses, multiply A by A^{-1} to see whether the product is the identity matrix. If it isn't, A does not have an inverse.

▪▪ INVERSE MATRICES AND SYSTEMS OF EQUATIONS

Any system of linear equations can be expressed in matrix form, as is shown in the next example.

EXAMPLE 5

Use matrix multiplication to express the following system of equations in matrix form.

$$x + \ y + z = \ \ 2$$
$$2x + 3y \ \ \ \ \ \ = \ \ 5$$
$$x + 2y + z = -1$$

SOLUTION   Let A be the 3×3 matrix of coefficients on the left-hand side of the equations, let B be the column matrix of constants on the right-hand side, and let X be the column matrix of unknowns.

$$A = \begin{pmatrix} 1 & 1 & 1 \\ 2 & 3 & 0 \\ 1 & 2 & 1 \end{pmatrix}, \qquad X = \begin{pmatrix} x \\ y \\ z \end{pmatrix}, \qquad B = \begin{pmatrix} 2 \\ 5 \\ -1 \end{pmatrix}.$$

Then AX is a matrix with three rows and one column, as is B.

$$AX = \begin{pmatrix} 1 & 1 & 1 \\ 2 & 3 & 0 \\ 1 & 2 & 1 \end{pmatrix} \begin{pmatrix} x \\ y \\ z \end{pmatrix} = \begin{pmatrix} x + \ y + \ z \\ 2x + \ 3y + \ 0z \\ x + \ 2y + \ z \end{pmatrix} \quad \text{and} \quad B = \begin{pmatrix} 2 \\ 5 \\ -1 \end{pmatrix}.$$

The entries in AX are just the left-hand sides of the equations of the system, and the entries in B are the constants on the right-hand sides. Therefore, the system can be expressed as the matrix equation $AX = B$. ▪

Suppose a system of equations is written in matrix form $AX = B$ and that the matrix A has an inverse. Then we can solve $AX = B$ by multiplying both sides by A^{-1}.

$$A^{-1}(AX) = A^{-1}B$$
$$(A^{-1}A)X = A^{-1}B \qquad \text{[Matrix multiplication is associative]}$$
$$I_n X = A^{-1}B \qquad \text{[$A^{-1}A$ is the identity matrix]}$$
$$X = A^{-1}B \qquad \text{[Product of identity matrix and X is X]}$$

The next example shows how this works in practice.

EXAMPLE 6

Solve the system

$$x + y + z = 2$$
$$2x + 3y = 5$$
$$x + 2y + z = -1.$$

SOLUTION As we saw in Example 5, this system is equivalent to the matrix equation

$$AX = B$$

$$\begin{pmatrix} 1 & 1 & 1 \\ 2 & 3 & 0 \\ 1 & 2 & 1 \end{pmatrix} \begin{pmatrix} x \\ y \\ z \end{pmatrix} = \begin{pmatrix} 2 \\ 5 \\ -1 \end{pmatrix}.$$

Use a calculator to find the inverse of the coefficient matrix A and multiply both sides of the equation by A^{-1}.

$$A^{-1}AX = A^{-1}B$$

$$\begin{pmatrix} 1.5 & .5 & -1.5 \\ -1 & 0 & 1 \\ .5 & -.5 & .5 \end{pmatrix} \begin{pmatrix} 1 & 1 & 1 \\ 2 & 3 & 0 \\ 1 & 2 & 1 \end{pmatrix} \begin{pmatrix} x \\ y \\ z \end{pmatrix} = \begin{pmatrix} 1.5 & .5 & -1.5 \\ -1 & 0 & 1 \\ .5 & -.5 & .5 \end{pmatrix} \begin{pmatrix} 2 \\ 5 \\ -1 \end{pmatrix}$$

$$\begin{pmatrix} 1 & 0 & 0 \\ 0 & 1 & 0 \\ 0 & 0 & 1 \end{pmatrix} \begin{pmatrix} x \\ y \\ z \end{pmatrix} = \begin{pmatrix} 1.5 & .5 & -1.5 \\ -1 & 0 & 1 \\ .5 & -.5 & .5 \end{pmatrix} \begin{pmatrix} 2 \\ 5 \\ -1 \end{pmatrix} \quad \text{[Since } A^{-1}A = I_3]$$

$$\begin{pmatrix} x \\ y \\ z \end{pmatrix} = \begin{pmatrix} 7 \\ -3 \\ -2 \end{pmatrix} \quad \text{[Since } I_3X = X]$$

Therefore, the solution of the original system is $x = 7$, $y = -3$, $z = -2$. ■

Only a matrix with the same number of rows as columns can possibly have an inverse. Consequently, the method of Example 6 can be tried only when the system has the same number of equations as unknowns. In this case, you should use your calculator to verify that the coefficient matrix actually has an inverse (see the Caution on page 507). If it does not, other methods must be used. Here is a summary of the possibilities.

Matrix Solution of a System of Equations

Suppose a system with the same number of equations as unknowns is written in matrix form as $AX = B$.

If the matrix A has an inverse, then the unique solution of the system is

$$X = A^{-1}B.$$

If A does not have an inverse, then the system either has no solutions or has infinitely many solutions. Its solutions (if any) may be found by using Gauss-Jordan elimination (Section 6.3).

EXAMPLE 7

Solve the system

$$2x + y - z = 2$$
$$x + 3y + 2z = 1$$
$$x + y + z = 2.$$

SOLUTION   Since there are the same number of equations as unknowns, we can try the method of Example 6. In this case, we have

$$A = \begin{pmatrix} 2 & 1 & -1 \\ 1 & 3 & 2 \\ 1 & 1 & 1 \end{pmatrix} \quad \text{and} \quad B = \begin{pmatrix} 2 \\ 1 \\ 2 \end{pmatrix}.$$

CALCULATOR EXPLORATION

Verify that the matrix A does have an inverse. Show that the solutions of the system are $x = 2$, $y = -1$, $z = 1$ by computing $A^{-1}B$. ∎

■■ APPLICATIONS

Just as two points determine a unique line, three points (that aren't on the same line) determine a unique parabola, as the next example demonstrates.

EXAMPLE 8

Find the equation of the parabola that passes through the points $(-1, 6)$, $(3, -2)$, and $(4, 1)$.

SOLUTION   As we saw in Section 4.1, a parabola is the graph of an equation of the form

$$y = ax^2 + bx + c$$

for some constants a, b, and c. Since $(-1, 6)$ is on the graph, we know that when $x = -1$, then $y = 6$.

$$ax^2 + bx + c = y$$

Let $x = -1$ and $y = 6$: $a(-1)^2 + b(-1) + c = 6$

$$a - b + c = 6. \quad \text{[Equation 1]}$$

Similarly, since $(3, -2)$ is on the graph, we have

$$ax^2 + bx + c = y$$

Let $x = 3$ and $y = -2$: $a(3^2) + b(3) + c = -2$

$$9a + 3b + c = -2. \quad \text{[Equation 2]}$$

Finally, since $(4, 1)$ is on the graph,

$$ax^2 + bx + c = y$$

Let $x = 4$ and $y = 1$: $a(4^2) + b(4) + c = 1$

$$16a + 4b + c = 1. \qquad \text{[Equation 3]}$$

We can determine a, b, and c by solving the system determined by Equations 1–3.

$$a - b + c = 6$$
$$9a + 3b + c = -2$$
$$16a + 4b + c = 1$$

or, in matrix form,

$$\begin{pmatrix} 1 & -1 & 1 \\ 9 & 3 & 2 \\ 16 & 4 & 2 \end{pmatrix} \begin{pmatrix} a \\ b \\ c \end{pmatrix} = \begin{pmatrix} 6 \\ -2 \\ 1 \end{pmatrix}.$$

A calculator shows that the solution is

$$\begin{pmatrix} a \\ b \\ c \end{pmatrix} = \begin{pmatrix} 1 & -1 & 1 \\ 9 & 3 & 2 \\ 16 & 4 & 2 \end{pmatrix}^{-1} \begin{pmatrix} 6 \\ -2 \\ 1 \end{pmatrix} = \begin{pmatrix} 1 \\ -4 \\ 1 \end{pmatrix}.$$

Using these values for a, b, and c, we obtain the equation of the parabola.

$$y = ax^2 + bx + c$$

Let $a = 1$, $b = -4$, and $c = 1$: $y = x^2 - 4x + 1.$ ∎

EXAMPLE 9

A batter hits a baseball, and special measuring devices locate its position at various times during its flight. If the path of the ball is drawn on a coordinate plane, with the batter at the origin, it looks like Figure 6–28. According to the measuring devices, the ball passes through the points (7, 9), (47, 38), and (136, 70).

(a) What is the equation of the path of the ball?

(b) How far from the batter does the ball hit the ground?

SOLUTION

(a) The path of the ball appears to be part of a parabola (a fact that will not be proved here) and hence has an equation of the form

$$y = ax^2 + bx + c.$$

As in Example 8, each of the points (7, 9), (47, 38) and (136, 70) determines an equation.

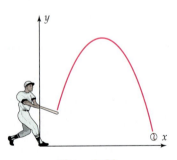

Figure 6–28

(7, 9) $a(7^2) \quad + b(7) \quad + c = 9$

(47, 38): $a(47^2) \quad + b(47) \quad + c = 38$

(136, 70): $a(136^2) + b(136) + c = 70.$

We must solve the resulting system.

$$49a + 7b + c = 9$$
$$2209a + 47b + c = 38$$
$$18,496a + 136b + c = 70$$

whose matrix form is

$$\begin{pmatrix} 49 & 7 & 1 \\ 2209 & 47 & 1 \\ 18,496 & 136 & 1 \end{pmatrix} \begin{pmatrix} a \\ b \\ c \end{pmatrix} = \begin{pmatrix} 9 \\ 38 \\ 70 \end{pmatrix}.$$

A calculator shows that the solution is

$$\begin{pmatrix} a \\ b \\ c \end{pmatrix} = \begin{pmatrix} 49 & 7 & 1 \\ 2209 & 47 & 1 \\ 18,496 & 136 & 1 \end{pmatrix}^{-1} \begin{pmatrix} 9 \\ 38 \\ 70 \end{pmatrix} \approx \begin{pmatrix} -.00283 \\ .87798 \\ 2.99296 \end{pmatrix}.$$

Therefore, the approximate equation for the ball's path is

$$y = -.00283x^2 + .87798x + 2.99296.$$

(b) The ball hits the ground at a point where its height y is 0, that is, when

$$-.00283x^2 + .87798x + 2.99296 = 0.$$

Use the quadratic formula or an equation solver to verify that the solutions of this equation are $x \approx -3.37$ (which is not applicable here) and $x \approx 313.61$. Therefore, the ball hits the ground approximately 314 feet from the batter. ∎

Given any three points not on a straight line, the method in Examples 8 and 9 can be used to find the unique parabola that passes through these points. This parabola can also be found by quadratic regression.[*]

CALCULATOR EXPLORATION

Use quadratic regression on the three points in Example 9 and verify that the equation obtained is the same as the one in Example 9.

To see why regression produces the parabola that actually passes through the three points, recall that the error in a quadratic model is measured by taking the difference between the y-coordinate of each data point and the y-coordinate of the corresponding point on the model and summing the squares of these errors. If the data points actually lie on a parabola (which is always the case with three points that are not on a line), then the error for that parabola will be 0 (the smallest possible error). Hence, it will be the parabola produced by the least squares quadratic regression procedure.

---

*If you have not read Sections 1.4 and 4.4, you may skip this discussion.

EXERCISES 6.4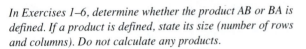

In Exercises 1–6, determine whether the product AB or BA is defined. If a product is defined, state its size (number of rows and columns). Do not calculate any products.

1. $A = \begin{pmatrix} 3 & 6 & 7 \\ 8 & 0 & 1 \end{pmatrix}, \qquad B = \begin{pmatrix} 2 & 5 & 9 & 1 \\ 7 & 0 & 0 & 6 \\ -1 & 3 & 8 & 7 \end{pmatrix}$

2. $A = \begin{pmatrix} -1 & -2 & -5 \\ 9 & 2 & -1 \\ 10 & 34 & 5 \end{pmatrix}, \qquad B = \begin{pmatrix} 17 & -9 \\ -6 & 12 \\ 3 & 5 \end{pmatrix}$

3. $A = \begin{pmatrix} 1 & 0 \\ 1 & 1 \\ 0 & 1 \end{pmatrix}, \qquad B = \begin{pmatrix} 5 & 6 & 11 \\ 7 & 8 & 15 \end{pmatrix}$

4. $A = \begin{pmatrix} 1 & -5 & 7 \\ 2 & 4 & 8 \\ 1 & -1 & 2 \end{pmatrix}, \qquad B = \begin{pmatrix} -2 & 4 & 9 \\ 13 & -2 & 1 \\ 5 & 25 & 0 \end{pmatrix}$

5. $A = \begin{pmatrix} -4 & 15 \\ 3 & -7 \\ 2 & 10 \end{pmatrix}, \qquad B = \begin{pmatrix} 1 & 2 \\ 3 & 4 \end{pmatrix}$

6. $A = \begin{pmatrix} 10 & 12 \\ -6 & 0 \\ 1 & 23 \\ -4 & 0 \end{pmatrix}, \qquad B = \begin{pmatrix} 1 & 2 & 3 \\ 3 & 2 & 1 \end{pmatrix}$

In Exercises 7–12, find AB.

7. $A = \begin{pmatrix} 3 & 2 \\ 2 & 4 \end{pmatrix}, \qquad B = \begin{pmatrix} 1 & -2 & 3 \\ 0 & 3 & 1 \end{pmatrix}$

8. $A = \begin{pmatrix} -1 & 2 & 3 \\ 0 & -1 & 2 \\ 1 & 2 & 0 \end{pmatrix}, \qquad B = \begin{pmatrix} 3 & -2 & -1 \\ 1 & 0 & 5 \\ 1 & -1 & -1 \end{pmatrix}$

9. $A = \begin{pmatrix} 1 & 0 & -4 \\ 0 & 2 & -1 \\ 2 & 3 & 4 \end{pmatrix}, \qquad B = \begin{pmatrix} 1 & 1 \\ 1 & 0 \\ 0 & 1 \end{pmatrix}$

10. $A = \begin{pmatrix} 1 & -2 \\ 3 & 0 \\ 0 & -1 \\ 2 & 1 \end{pmatrix}, \qquad B = \begin{pmatrix} -1 & 3 & -2 & 0 \\ 6 & 1 & 0 & -2 \end{pmatrix}$

11. $A = \begin{pmatrix} 2 & 0 & -1 \\ 1 & 1 & 2 \\ 0 & 2 & -3 \\ 2 & 3 & 0 \end{pmatrix}, \qquad B = \begin{pmatrix} 1 & 0 & 1 & 1 \\ 1 & 1 & 0 & 1 \\ 1 & 1 & 1 & 0 \end{pmatrix}$

12. $A = \begin{pmatrix} 10 & 0 & 1 & 0 \\ -1 & 1 & 0 & 1 \end{pmatrix}, \qquad B = \begin{pmatrix} 2 & -1 & 0 & 1 \\ -2 & 3 & 1 & -4 \\ 3 & 5 & 2 & -5 \end{pmatrix}$

In Exercises 13–16, show that AB is not equal to BA by computing both products.

13. $A = \begin{pmatrix} 3 & 2 \\ 5 & 1 \end{pmatrix}, \qquad B = \begin{pmatrix} 7 & -5 \\ -2 & 6 \end{pmatrix}$

14. $A = \begin{pmatrix} 3/2 & 2 \\ 4 & 7/2 \end{pmatrix}, \qquad B = \begin{pmatrix} 1 & -3/2 \\ 5/2 & 1 \end{pmatrix}$

15. $A = \begin{pmatrix} 4 & 2 & -1 \\ 0 & 1 & 2 \\ -3 & 0 & 1 \end{pmatrix}, \qquad B = \begin{pmatrix} 1 & 7 & -5 \\ 2 & -2 & 6 \\ 0 & 0 & 0 \end{pmatrix}$

16. $A = \begin{pmatrix} 1 & 1 & -1 & 1 \\ 2 & 0 & 3 & 2 \\ -3 & 0 & 0 & 1 \\ 1 & -1 & 1 & 2 \end{pmatrix}, \qquad B = \begin{pmatrix} 0 & 1 & 7 & 7 \\ 2 & 3 & -2 & 1 \\ 5 & 0 & 1 & 0 \\ -1 & 0 & 1 & 0 \end{pmatrix}$

In Exercises 17–24, find the inverse of the matrix, if it exists.

17. $\begin{pmatrix} 1 & 2 \\ 3 & 4 \end{pmatrix}$

18. $\begin{pmatrix} 3 & 5 \\ 1 & 4 \end{pmatrix}$

19. $\begin{pmatrix} 3 & -1 \\ -6 & 2 \end{pmatrix}$

20. $\begin{pmatrix} 1 & -1 & 0 \\ 1 & 0 & -1 \\ 6 & -2 & -3 \end{pmatrix}$

21. $\begin{pmatrix} 1 & 2 & 0 \\ 3 & -1 & 2 \\ -2 & 3 & -2 \end{pmatrix}$

22. $\begin{pmatrix} 1 & -3 & 4 \\ 2 & -5 & 7 \\ 0 & -1 & 1 \end{pmatrix}$

23. $\begin{pmatrix} 5 & 0 & 2 \\ 2 & 2 & 1 \\ -3 & 1 & -1 \end{pmatrix}$

24. $\begin{pmatrix} -1 & 3 & 1 \\ 2 & 5 & 0 \\ 3 & 1 & -2 \end{pmatrix}$

In Exercises 25–28, solve the system of equations by using the method of Example 6.

25.
$\begin{aligned} -x + y &= 1 \\ -x + z &= -2 \\ 6x - 2y - 3z &= 3 \end{aligned}$

26.
$\begin{aligned} x + 2y + 3z &= 1 \\ 2x + 5y + 3z &= 0 \\ x + 8z &= -1 \end{aligned}$

27.
$\begin{aligned} 2x + y &= 0 \\ -4x - y - 3z &= 1 \\ 3x + y + 2z &= 2 \end{aligned}$

28.
$\begin{aligned} -3x - 3y - 4z &= 2 \\ y + z &= 1 \\ 4x + 3y + 4z &= 3 \end{aligned}$

In Exercises 29–39, solve the system by any method.

29. $x + y + 2w = 3$
$2x - y + z - w = 5$
$3x + 3y + 2z - 2w = 0$
$x + 2y + z = 2$

30. $x - 2y + 3z = 1$
$y - z + w = -2$
$-2x + 2y - 2z + 4w = 5$
$2y - 3z + w = 8$

31. $x + y + 6z + 2v = 1.5$
$x + 5z + 2v - 3w = 2$
$3x + 2y + 17z + 6v - 4w = 2.5$
$4x + 3y + 21z + 7v - 2w = 3$
$-6x - 5y - 36z - 12v + 3w = 3.5$

32. $x - 1.5y + 1.5v - 4w = 3$
$-1.5y + .5w = 0$
$-x + 2y + .5z - 2v + 4.5w = 2$
$-.5y - 2.5z + .5v - .75w = 8$
$y - .5z + .5w = 4$

33. $x + 2y + 2z - 2w = -23$
$4x + 4y - z + 5w = 7$
$-2x + 5y + 6z + 4w = 0$
$5x + 13y + 7z + 12w = -7$

34. $x + 4y + 5z + 2w = 0$
$2x + y + 4z - 2w = 0$
$-x + 7y + 10z + 5w = 0$
$-4x + 2y + z + 5w = 0$

35. $x + 2y + 5z - 2v + 4w = 0$
$2x - 4y + 6z + v + 4w = 0$
$5x + 2y - 3z + 2v + 3w = 0$
$6x - 5y - 2z + 5v + 3w = 0$
$x + 2y - z - 2v + 4w = 0$

36. $4x + 2y + 3z + 3v + 2w = 1$
$2x + 2z + 2u + v - w = 2$
$10x + 2y + 10z + 10u + 3v + 4w = 5$
$16x + 4y + 16z + 18u + 7v - 2w = -3$
$x + 2y + 4z - 6u + 2v + w = -2$
$6x + 2y + 6z + 6u + 3v = 1$

37. $x + 2y + 3z = 1$
$3x + 2y + 4z = -1$
$2x + 6y + 8z + w = 3$
$2x + 2z - 2w = 3$

38. $x + 2y + 4z = 6$
$y + z = 1$
$x + 3y + 5z = 10$

39. $x + 3w = -2$
$x - 4y - z + 3w = -7$
$4y + z = 5$
$-x + 12y + 3z - 3w = 17$

In Exercises 40–42, find constants a, b, c such that the three given points lie on the parabola $y = ax^2 + bx + c$. [See Example 8.]

40. $(-3, 2), (1, 1), (2, -1)$

41. $(-3, 15), (1, -7), (5, 111)$

42. $(1, -2), (3, 1), (4, -1)$

43. Concentrations of the greenhouse gas carbon dioxide (CO_2) have increased quadratically over the past half-century.[*] So the concentration y of CO_2 (in parts per million) in year x is given by an equation of the form

$$y = ax^2 + bx + c.$$

(a) Let $x = 0$ correspond to 1958 and use the following data to find a, b, and c.

Year	1958	1979	2001
CO_2 Concentration	315	337	371

(b) Use this equation to estimate the CO_2 concentration in the years 1983, 1993, and 2003. [For comparison purposes, the actual concentrations in 1983 and 1993 were 343 ppm and 357 ppm, respectively.]

44. The table shows the per capita consumption of chicken (in pounds) in selected years.[†]

Year	1980	1990	1999
Chicken Consumption	32.7	42.4	54.2

(a) Let $x = 0$ correspond to 1980. Find a quadratic equation that models this data.
(b) Use the equation to estimate the consumption of chicken in 1985, 1995, and 2005. [The actual consumption in 1985 and 1995 was 36.4 and 49.0 pounds, respectively.]

---

[*]C. D. Keeling and T. P. Whorf, Scripps Institution of Oceanography.
[†]U.S. Department of Agriculture.

45. The table shows the per capita sales (in dollars) of electronics and appliances in selected years.[*]

Year	1992	1996	2000
Sales	$169	$260	$320

(a) Find a quadratic equation that models this data, with $x = 2$ corresponding to 1992.

(b) Use the equation to estimate sales of electronics and appliances in 1993, 1998, and 2002.

46. The table shows the percentage y of total consumer spending on sound recordings that is spent on pop music in year x.[†]

Year	1990	1994	1996	1998	2000
% Pop	13.7	10.3	9.3	10	11

(a) Assume that the data can be modeled by an equation of the form

$$y = ax^4 + bx^3 + cx^2 + dx + k.$$

Use the same method used in Exercises 43–45 (with five equations instead of three) to find the constants a, b, c, d, and k.

(b) Use the equation in part (a) to estimate the percentage of consumer spending spent on pop music in 1993 and 1999.

47. Find constants a, b, c such that the points $(0, -2)$, $(\ln 2, 1)$, and $(\ln 4, 4)$ lie on the graph of $f(x) = ae^x + be^{-x} + c$. [*Hint:* Proceed as in Example 8.]

48. Find constants a, b, c such that the points $(0, -1)$, $(\ln 2, 4)$, and $(\ln 3, 7)$ lie on the graph of $f(x) = ae^x + be^{-x} + c$.

49. New Army Stores, a national chain of casual clothing stores, recently sent shipments of jeans, jackets, sweaters, and shirts to its stores in various cities. The number of items shipped to each city and their total wholesale cost are shown in the table below. Find the wholesale price of each of the following: one pair of jeans, one jacket, one sweater, and one shirt.

City	Jeans	Jackets	Sweaters	Shirts	Total Cost
Cleveland	3000	3000	2200	4200	$507,650
St. Louis	2700	2500	2100	4300	459,075
Seattle	5000	2000	1400	7500	541,225
Phoenix	7000	1800	600	8000	571,500

50. A 100-bed nursing home provides two levels of long-term care: regular and maximum. Patients at each level have a choice of a private room or a less expensive semiprivate room. The table below shows the number of patients in each category at various times last year and the total daily cost for these patients. Find the daily cost of each of the following: a private room (regular care), a private room (maximum care), a semiprivate room (regular care), and a semiprivate room (maximum care).

Month	Regular Care Patients Semi-private	Regular Care Patients Private	Maximum Care Patients Semi-private	Maximum Care Patients Private	Daily Cost
January	22	8	60	10	$18,824
April	26	8	54	12	18,738
July	24	14	56	6	18,606
October	20	10	62	8	18,824

51. A candy company produces three types of gift boxes: A, B, and C. A box of variety A contains .6 pound of chocolates and .4 pound of mints. A box of variety B contains .3 pound of chocolates, .4 pound of mints, and .3 pound of caramels. A box of variety C contains .5 pound of chocolates, .3 pound of mints, and .2 pound of caramels. The company has 41,400 pounds of chocolates, 29,400 pounds of mints, and 16,200 pounds of caramels in stock. How many boxes of each variety should be made to use up all the stock?

52. Certain circus animals are fed the same three food mixes: R, S, and T. Lions receive 1.1 units of R, 2.4 units of S, and 3.7 units of T each day. Horses receive 8.1 units of R, 2.9 units of S, and 5.1 units of T each day. Bears receive 1.3 units of R, 1.3 units of S, and 2.3 units of T each day. If 16,000 units of R, 28,000 units of S, and 44,000 units of T are available each day, how many of each type of animal can be supported?

---

[*]Bureau of the Census, U.S. Department of Commerce.
[†]Recording Industry Association of America, Inc.

6.4.A SPECIAL TOPICS Matrix Algebra

Matrices and matrix multiplication were introduced in Section 6.4. In this section, we consider other aspects of matrix algebra. Recall that an *m* × *n* **matrix** is a rectangular array of numbers, with *m* horizontal **rows** (numbered from top to bottom) and *n* vertical **columns** (numbered from left to right), such as

$$\begin{pmatrix} 3 & 2 & -5 \\ 6 & 1 & 7 \\ -2 & 0 & 5 \end{pmatrix} \qquad \begin{pmatrix} \frac{3}{2} \\ -5 \\ 0 \\ 12 \end{pmatrix}$$

3 × 3 matrix   4 × 1 matrix

In discussing matrices in a general context, a typical matrix may be denoted by a capital letter, such as *A*, or by an array like

$$\begin{pmatrix} a_{11} & a_{12} & a_{13} & \cdots & a_{1n} \\ a_{21} & a_{22} & a_{23} & \cdots & a_{2n} \\ a_{31} & a_{32} & a_{33} & \cdots & a_{3n} \\ \vdots & \vdots & \vdots & & \vdots \\ a_{m1} & a_{m2} & a_{m3} & \cdots & a_{mn} \end{pmatrix},$$

in which a_{ij} denotes the entry in row i and column j. This array notation is often abbreviated as (a_{ij}).

We have already seen how matrices can be used in the solution of systems of linear equations. They also provide a convenient way to display data.

EXAMPLE 1

At the beginning of a laboratory experiment, five baby rats measured 5.5, 6.1, 6.8, 7.4, and 6.2 centimeters in length and weighed 139, 141, 148, 153, and 145 grams, respectively. We can summarize this data with the 2 × 5 matrix

$$\begin{pmatrix} 5.5 & 6.1 & 6.8 & 7.4 & 6.2 \\ 139 & 141 & 148 & 153 & 145 \end{pmatrix},$$

in which row 1 represents length and row 2 represents weight, with each column corresponding to one of the rats. ■

Recall that two matrices are said to be **equal** if they have the same size and the corresponding entries are equal. For example,

$$\begin{pmatrix} -3 & (-1)^2 & \frac{6}{8} \\ 6 & 12 & 0 \end{pmatrix} = \begin{pmatrix} -3 & 1 & \frac{3}{4} \\ \sqrt{36} & 3 \cdot 4 & 0 \end{pmatrix},$$

but

$$\begin{pmatrix} 6 & 4 \\ 5 & 1 \end{pmatrix} \neq \begin{pmatrix} 6 & 5 \\ 4 & 1 \end{pmatrix}$$

because the corresponding entries in row 1, column 2, aren't equal $(4 \neq 5)$, and similarly in row 2, column 1. More formally, we have the following.

Equality of Matrices

> If $A = (a_{ij})$ and $B = (b_{ij})$ are $m \times n$ matrices,
>
> then $A = B$ means $a_{ij} = b_{ij}$ for every i and j.

■■ MATRIX ADDITION AND SUBTRACTION

Matrix addition and subtraction are defined only when the two matrices have the same size. In that case, the rule is: Add (or subtract) the corresponding entries. In formal terms, we have the following.

Matrix Addition

> The sum of the $m \times n$ matrices $A = (a_{ij})$ and $B = (b_{ij})$ is the $m \times n$ matrix $A + B$ whose entry in row i, column j, is
>
> $$a_{ij} + b_{ij}.$$

EXAMPLE 2

Let A, B, C, and D be the following matrices.

$$A = \begin{pmatrix} 1 & 2 \\ 3 & 0 \\ -7 & 4 \end{pmatrix}, \qquad B = \begin{pmatrix} 5 & 8 \\ -3 & 7 \\ 4 & 5 \end{pmatrix}, \qquad C = \begin{pmatrix} 6 & -3 \\ -2 & 8 \end{pmatrix}, \qquad D = \begin{pmatrix} 1 & 4 \\ 5 & -4 \end{pmatrix}$$

Find each of these:

(a) $A + B$ (b) $B + A$ (c) $B + C$ (d) $C - D$.

SOLUTION

(a) $A + B = \begin{pmatrix} 1 & 2 \\ 3 & 0 \\ -7 & 4 \end{pmatrix} + \begin{pmatrix} 5 & 8 \\ -3 & 7 \\ 4 & 5 \end{pmatrix} = \begin{pmatrix} 1+5 & 2+8 \\ 3+(-3) & 0+7 \\ -7+4 & 4+5 \end{pmatrix} = \begin{pmatrix} 6 & 10 \\ 0 & 7 \\ -3 & 9 \end{pmatrix}.$

(b) $B + A = \begin{pmatrix} 5 & 8 \\ -3 & 7 \\ 4 & 5 \end{pmatrix} + \begin{pmatrix} 1 & 2 \\ 3 & 0 \\ -7 & 4 \end{pmatrix} = \begin{pmatrix} 5+1 & 8+2 \\ -3+3 & 7+0 \\ 4+(-7) & 5+4 \end{pmatrix} = \begin{pmatrix} 6 & 10 \\ 0 & 7 \\ -3 & 9 \end{pmatrix}.$

Note that $A + B = B + A$.

(c) The sum $B + C$ is not defined because B and C are different sizes.

(d) $C - D = \begin{pmatrix} 6 & -3 \\ -2 & 8 \end{pmatrix} - \begin{pmatrix} 1 & 4 \\ 5 & -4 \end{pmatrix} = \begin{pmatrix} 6-1 & -3-4 \\ -2-5 & 8-(-4) \end{pmatrix} = \begin{pmatrix} 5 & -7 \\ -7 & 12 \end{pmatrix}.$

The sums and differences in this example can also be found on a calculator (Figure 6–29).   ■

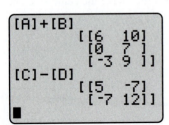

Figure 6–29

EXAMPLE 3

Back to the laboratory. As we saw in Example 1, the matrix

$$A = \begin{pmatrix} 5.5 & 6.1 & 6.8 & 7.4 & 6.2 \\ 139 & 141 & 148 & 153 & 145 \end{pmatrix}$$

summarizes the length (in centimeters) and weight (in grams) of five rats. Two weeks later, their lengths and weights are given by the matrix

$$B = \begin{pmatrix} 10.1 & 11.3 & 11.3 & 12.5 & 10.9 \\ 195 & 197 & 220 & 252 & 234 \end{pmatrix}.$$

The amount that the length and weight of each rat has changed during this two-week period can be represented by the matrix

$$B - A = \begin{pmatrix} 10.1 & 11.3 & 11.3 & 12.5 & 10.9 \\ 195 & 197 & 220 & 252 & 234 \end{pmatrix} - \begin{pmatrix} 5.5 & 6.1 & 6.8 & 7.4 & 6.2 \\ 139 & 141 & 148 & 153 & 145 \end{pmatrix}$$

$$= \begin{pmatrix} 4.6 & 5.2 & 4.5 & 5.1 & 4.7 \\ 56 & 56 & 72 & 99 & 89 \end{pmatrix}.$$

The matrix shows, for example that the second rat had the largest gain in length (5.2 cm) and the fourth rat had the largest weight gain (99 g). ■

▪▪ SCALAR MULTIPLICATION

Matrix arithmetic also has an operation that has no analogue in the arithmetic of real numbers. The product of a single number c and a matrix is defined by this rule: Multiply every entry in the matrix by c. For example,

$$2\begin{pmatrix} 3 & -4 & 0 \\ 1 & 5 & 2 \end{pmatrix} = \begin{pmatrix} 2 \cdot 3 & 2(-4) & 2 \cdot 0 \\ 2 \cdot 1 & 2 \cdot 5 & 2 \cdot 2 \end{pmatrix} = \begin{pmatrix} 6 & -8 & 0 \\ 2 & 10 & 4 \end{pmatrix}.$$

The process of multiplying a number by a matrix is called **scalar multiplication,** and the number is sometimes called a **scalar.** Here is the formal definition:

Scalar Multiplication

The product of an $m \times n$ matrix $A = (a_{ij})$ and a scalar c is the $m \times n$ matrix cA whose entry in row i, column j, is ca_{ij}.

EXAMPLE 4

Show that $3(A + B) = 3A + 3B$, where

$$A = \begin{pmatrix} 2 & -1 \\ 0 & 4 \end{pmatrix} \quad \text{and} \quad B = \begin{pmatrix} -3 & 2 \\ 5 & 7 \end{pmatrix}.$$

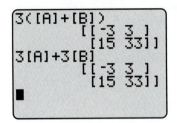

Figure 6–30

SOLUTION   Figure 6–30 shows that the matrix $3(A + B)$ is the same as the matrix $3A + 3B$; hence, $3(A + B) = 3A + 3B$. Here is how the same calculation is done by hand.

$$A + B = \begin{pmatrix} 2 & -1 \\ 0 & 4 \end{pmatrix} + \begin{pmatrix} -3 & 2 \\ 5 & 7 \end{pmatrix} = \begin{pmatrix} 2 + (-3) & -1 + 2 \\ 0 + 5 & 4 + 7 \end{pmatrix} = \begin{pmatrix} -1 & 1 \\ 5 & 11 \end{pmatrix},$$

so,

$$3(A + B) = 3\begin{pmatrix} -1 & 1 \\ 5 & 11 \end{pmatrix} = \begin{pmatrix} 3(-1) & 3 \cdot 1 \\ 3 \cdot 5 & 3 \cdot 11 \end{pmatrix} = \begin{pmatrix} -3 & 3 \\ 15 & 33 \end{pmatrix}.$$

On the other hand,

$$3A = 3\begin{pmatrix} 2 & -1 \\ 0 & 4 \end{pmatrix} = \begin{pmatrix} 3 \cdot 2 & 3(-1) \\ 3 \cdot 0 & 3 \cdot 4 \end{pmatrix} = \begin{pmatrix} 6 & -3 \\ 0 & 12 \end{pmatrix}$$

and

$$3B = 3\begin{pmatrix} -3 & 2 \\ 5 & 7 \end{pmatrix} = \begin{pmatrix} 3(-3) & 3 \cdot 2 \\ 3 \cdot 5 & 3 \cdot 7 \end{pmatrix} = \begin{pmatrix} -9 & 6 \\ 15 & 21 \end{pmatrix},$$

so

$$3A + 3B = \begin{pmatrix} 6 & -3 \\ 0 & 12 \end{pmatrix} + \begin{pmatrix} -9 & 6 \\ 15 & 21 \end{pmatrix} = \begin{pmatrix} -3 & 3 \\ 15 & 33 \end{pmatrix}.$$

Therefore, $3(A + B) = 3A + 3B$.   ■

Here is a summary of the properties of matrix addition and subtraction and scalar multiplication. Property 1 was illustrated in Example 2, and Property 3 was illustrated in Example 4. The others are illustrated in Exercises 23–27.

Properties of Matrix Addition and Scalar Multiplication

For any $m \times n$ matrices A, B, C and numbers c, d:

1. $A + B = B + A$.

2. $A + (B + C) = (A + B) + C$.

3. $c(A + B) = cA + cB$.

4. $(cd)A = c(dA)$.

5. $(c + d)A = cA + dA$.

▪▪ MATRIX MULTIPLICATION

In Example 2(c) of Section 6.4, we saw that matrix multiplication, unlike multiplication of numbers, is not commutative, that is, AB might not be equal to BA. Furthermore, the product of two nonzero matrices may be the zero matrix, unlike the situation with numbers (in which a product is zero only if one of the factors

is zero). For instance, if

$$A = \begin{pmatrix} 4 & 6 \\ 2 & 3 \end{pmatrix} \quad \text{and} \quad B = \begin{pmatrix} -3 & -9 \\ 2 & 6 \end{pmatrix},$$

then

$$AB = \begin{pmatrix} 4(-3) + 6 \cdot 2 & 4(-9) + 6 \cdot 6 \\ 2(-3) + 3 \cdot 2 & 2(-9) + 3 \cdot 6 \end{pmatrix} = \begin{pmatrix} 0 & 0 \\ 0 & 0 \end{pmatrix}.$$

However, matrix multiplication does have some other familiar properties in common with real-number arithmetic.

Properties of Matrix Multiplication

For any matrices A, B, C for which each of the following sums and products are defined:

1. $A(BC) = (AB)C.$

2. $A(B + C) = AB + AC.$

3. $(B + C)A = BA + CA.$

CALCULATOR EXPLORATION

Enter the following matrices in your calculator.

$$A = \begin{pmatrix} 2 & 0 & -2 \\ 4 & 1 & -5 \\ 7 & 3 & -6 \end{pmatrix}, \quad B = \begin{pmatrix} -2 & 1.5 & 4 \\ 1 & 2 & 0 \\ 0 & 3.2 & -1 \end{pmatrix}, \quad C = \begin{pmatrix} 9 & 8 & 7 \\ 4 & 5 & 6 \\ 3 & 2 & 1 \end{pmatrix}.$$

Verify Property 2 in the preceding box by showing that $A(B + C)$ and $AB + AC$ are the same matrix.

■■ APPLICATIONS

Matrix arithmetic can sometimes be used with data that is presented in matrix form.

EXAMPLE 5

Jerry and Jeanette's Chicken Palace has three locations in Roanoke, Virginia, each of which sells chicken nuggets at $2.35, Buffalo wings at $2.95, and filet of chicken sandwiches at $1.95. The number of each item sold daily at each location is given in the following matrix:

	Nuggets	Wings	Sandwiches
Store 1	220	400	280
Store 2	360	300	225
Store 3	180	380	300

Use matrix multiplication to find the daily revenue at each store.

SOLUTION   The daily revenue per item can be represented by the 3×1 matrix

$$
\begin{array}{r}
 \\
\text{Nuggets} \\
\text{Wings} \\
\text{Sandwiches}
\end{array}
\begin{array}{c}
\text{Revenue} \\
\begin{pmatrix} 2.35 \\ 2.95 \\ 1.95 \end{pmatrix}.
\end{array}
$$

The daily revenue per store is the product:

$$(\text{Store} \times \text{Item}) \cdot (\text{Item} \times \text{Revenue}) = \quad \text{Store} \times \text{Revenue}$$

$$
\begin{pmatrix} 220 & 400 & 280 \\ 360 & 300 & 225 \\ 180 & 380 & 300 \end{pmatrix} \cdot \begin{pmatrix} 2.35 \\ 2.95 \\ 1.95 \end{pmatrix} = \begin{pmatrix} 220(2.35) + 400(2.95) + 280(1.95) \\ 360(2.35) + 300(2.95) + 225(1.95) \\ 180(235) + 380(2.95) + 300(1.95) \end{pmatrix}
$$

$$
= \begin{pmatrix} \$2243 \\ \$2169.75 \\ \$2129 \end{pmatrix}.
$$

Therefore, the daily revenue is $2243 at Store 1, $2169.75 at Store 2, and $2129 at Store 3. ∎

Matrices also have applications in cryptography (the study of secret codes).

EXAMPLE 6

Use the matrix

$$
A = \begin{pmatrix} 3 & -1 & 2 & 6 \\ 1 & 3 & -1 & 0 \\ 5 & -2 & 1 & -3 \\ 3 & -7 & 4 & 4 \end{pmatrix}
$$

to encode the message "Attack at dawn."

SOLUTION   The first step is to translate the message into numerical form by assigning a number to each letter. For simplicity, we use the following.

0 = space	7 = G	14 = N	21 = U
1 = A	8 = H	15 = O	22 = V
2 = B	9 = I	16 = P	23 = W
3 = C	10 = J	17 = Q	24 = X
4 = D	11 = K	18 = R	25 = Y
5 = E	12 = L	19 = S	26 = Z
6 = F	13 = M	20 = T	

Since the encoding matrix A is 4×4, we break our message into blocks of four letters (using # for space) and write them in numerical form.

A	T	T	A		C	K	#	A		T	#	D	A		W	N	#	#
(1	20	20	1)		(3	11	0	1)		(20	0	4	1)		(23	1	0	0)

Each message block is a 4×1 matrix; to encode it, we multiply by the matrix A:

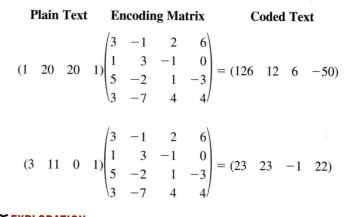

$$ (1 \quad 20 \quad 20 \quad 1) \begin{pmatrix} 3 & -1 & 2 & 6 \\ 1 & 3 & -1 & 0 \\ 5 & -2 & 1 & -3 \\ 3 & -7 & 4 & 4 \end{pmatrix} = (126 \quad 12 \quad 6 \quad -50) $$

$$ (3 \quad 11 \quad 0 \quad 1) \begin{pmatrix} 3 & -1 & 2 & 6 \\ 1 & 3 & -1 & 0 \\ 5 & -2 & 1 & -3 \\ 3 & -7 & 4 & 4 \end{pmatrix} = (23 \quad 23 \quad -1 \quad 22) $$

CALCULATOR EXPLORATION

Enter the matrix A in your calculator and encode the rest of the message by computing $(20 \quad 0 \quad 4 \quad 1)A$ and $(23 \quad 1 \quad 0 \quad 0)A$.

Using the results of the Calculator Exploration, we send the encoded message as a single list of numbers.

126, 12, 6, −50, 23, 23, −1, 22, 83, −35, 48, 112, 70, −20, 45, 138.

The story continues in the next example. ■

EXAMPLE 7

After the message in Example 6 was sent, the following reply was received in the same code.

$$ 122, -126, 75, 32, 68, 38, 14, 94, 94, 17, 33, 129, $$

$$ 180, -151, 99, 59, 129, -17, 43, 84, 174, -97, 84, 95. $$

Decode this message.

SOLUTION   We first break the message into blocks of four numbers each, which we think of as 1×4 matrices.

(122	−126	75	32),		(68	38	14	94),		(94	17	33	129),
(180	−151	99	59),		(129	−17	43	84),		(174	−97	84	95).

Recall how this message was encoded. Each four-letter block is written numerically as a 4×1 matrix $E,$ and the coded text is the product EA. To obtain the original matrix E from the coded matrix EA, we need only multiply by A^{-1} and use the fact that AA^{-1} is the identity matrix I_4.

$$(EA)A^{-1} = E(AA^{-1}) = EI_4 = E.$$

Using a calculator (Figures 6–31 and 6–32), we see that the decoded form of the first four blocks of text is

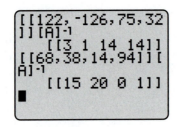

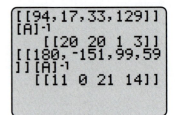

Figure 6–31   **Figure 6–32**

3	1	14	14		15	20	0	1		20	20	1	3		11	0	21	14
C	A	N	N		O	T	#	A		T	T	A	C		K	#	U	N

So the message begins "Cannot attack un . . . ."

EXPLORATION

Use your calculator to decode the last two blocks of the message by computing $(129 \quad -17 \quad 43 \quad 84)A^{-1}$ and $(174 \quad -97 \quad 84 \quad 95)A^{-1}$. What is the entire message?   ■

EXERCISES 6.4.A

In Exercises 1–8, compute $A + B$, AB, BA, and $2A - 3B$.

1. $A = \begin{pmatrix} 3 & 2 \\ 5 & 1 \end{pmatrix}$, $B = \begin{pmatrix} 7 & -5 \\ -2 & 6 \end{pmatrix}$

2. $A = \begin{pmatrix} -6 & 2 \\ 7 & -1 \end{pmatrix}$, $B = \begin{pmatrix} -8 & 4 \\ 2 & 7 \end{pmatrix}$

3. $A = \begin{pmatrix} \frac{3}{2} & 2 \\ 4 & \frac{7}{2} \end{pmatrix}$, $B = \begin{pmatrix} \frac{1}{2} & -\frac{3}{2} \\ \frac{5}{2} & 1 \end{pmatrix}$

4. $A = \begin{pmatrix} \frac{3}{4} & 7 \\ 6 & -\frac{5}{4} \end{pmatrix}$, $B = \begin{pmatrix} \frac{1}{2} & 3 \\ -5 & \frac{3}{2} \end{pmatrix}$

5. $A = \begin{pmatrix} 0 & 1 \\ 1 & 0 \end{pmatrix}$. $B = \begin{pmatrix} 3 & 5 \\ 7 & 9 \end{pmatrix}$

6. $A = \begin{pmatrix} 1 & -2 \\ -3 & 5 \end{pmatrix}$, $B = \begin{pmatrix} -1 & 2 \\ 3 & -5 \end{pmatrix}$

7. $A = \begin{pmatrix} 5 & 2 \\ 3 & -1 \end{pmatrix}$, $B = \begin{pmatrix} -1 & 2 \\ 3 & -4 \end{pmatrix}$

8. $A = \begin{pmatrix} 1 & -1 \\ 2 & -2 \end{pmatrix}$, $B = \begin{pmatrix} 3 & -1 \\ 3 & -1 \end{pmatrix}$

In Exercises 9–18, perform the indicated operations, where

$$A = \begin{pmatrix} 1 & -2 \\ 3 & 4 \end{pmatrix}, \quad B = \begin{pmatrix} -1 & 0 \\ 1 & 2 \end{pmatrix}, \quad C = \begin{pmatrix} 1 & 0 \\ 0 & 1 \end{pmatrix},$$

$$D = \begin{pmatrix} 1 & -2 & 3 \\ 0 & 4 & 1 \end{pmatrix}, \quad E = \begin{pmatrix} 1 & 2 \\ -2 & 1 \\ 0 & 5 \end{pmatrix},$$

$$F = \begin{pmatrix} 0 & 0 \\ 0 & 0 \\ 0 & 0 \end{pmatrix}, \quad G = \begin{pmatrix} 2 & -4 \\ 3 & 0 \\ -1 & 5 \end{pmatrix}.$$

9. $B - C$

10. $2A + C$

11. $-3B + 4C$

12. $A - 2B$

13. $AD + BD$

14. $EA + EB$

15. $2E + 3G$

16. $B^2 - AB$

17. $DG - B$

18. $DE + 2C$

In Exercises 19–22, find a matrix X satisfying the given equation, where

$$A = \begin{pmatrix} 1 & -2 \\ 4 & 3 \end{pmatrix} \quad \text{and} \quad B = \begin{pmatrix} 2 & -1 \\ 0 & 5 \end{pmatrix}.$$

19. $2X = 2A + 3B$

20. $3X = A - 3B$

21. $2X + 3A = 4B$

22. $3X + 2B = 3A$

In Exercises 23–30, verify that the given statement is true when c = 2, d = 3, and

$$A = \begin{pmatrix} 1 & 2 \\ 3 & 0 \end{pmatrix}, \quad B = \begin{pmatrix} -1 & 2 \\ 3 & 4 \end{pmatrix}, \quad C = \begin{pmatrix} 2 & 3 \\ 1 & 2 \end{pmatrix}.$$

23. $A + B = B + A$

24. $A + (B + C) = (A + B) + C$

25. $c(A + B) = cA + cB$

26. $(c + d)A = cA + dA$

27. $(cd)A = c(dA)$

28. $A(BC) = (AB)C$

29. $A(B + C) = AB + AC$

30. $(B + C)A = BA + CA$

31. If A is an $n \times k$ matrix and B is an $r \times t$ matrix, what conditions must n, k, r, and t satisfy so that both AB and BA are defined?

In Exercises 32–36, verify that the statement is false *for the given matrices.*

32. $AB = BA;$ $\quad A = \begin{pmatrix} 1 & 2 \\ 3 & 4 \end{pmatrix} \quad \text{and} \quad B = \begin{pmatrix} -1 & 4 \\ 5 & 2 \end{pmatrix}$

33. If $AB = AC$, then $B = C;$ $\quad A = \begin{pmatrix} 1 & 2 \\ 2 & 4 \end{pmatrix},$

$B = \begin{pmatrix} 3 & 6 \\ -\frac{3}{2} & -3 \end{pmatrix}, \quad C = \begin{pmatrix} 0 & 0 \\ 0 & 0 \end{pmatrix}$

34. $A^2 = \begin{pmatrix} 0 & 0 \\ 0 & 0 \end{pmatrix}$ only if $A = \begin{pmatrix} 0 & 0 \\ 0 & 0 \end{pmatrix};$ $\quad A = \begin{pmatrix} 0 & 2 \\ 0 & 0 \end{pmatrix}$

35. $(A + B)(A - B) = A^2 - B^2;$

$A = \begin{pmatrix} 3 & 1 \\ 2 & -4 \end{pmatrix}$ and $B = \begin{pmatrix} 2 & -1 \\ 5 & 3 \end{pmatrix}$

36. $(A + B)(A + B) = A^2 + 2AB + B^2;$

$A = \begin{pmatrix} 2 & -1 \\ 3 & 5 \end{pmatrix}$ and $B = \begin{pmatrix} 1 & 2 \\ -3 & 4 \end{pmatrix}$

In Exercises 37–40, the transpose of the matrix A is denoted by A^t and is defined by this rule: Row 1 of A is column 1 of A^t; row 2 of A is column 2 of A^t, etc.

37. Find A^t and B^t, when

$$A = \begin{pmatrix} a & b \\ c & d \end{pmatrix} \quad \text{and} \quad B = \begin{pmatrix} r & s \\ u & v \end{pmatrix}.$$

In Exercises 38–40, assume that A and B are as in Exercise 37 and show that:

38. $(A + B)^t = A^t + B^t$

39. $(A^t)^t = A$

40. $(AB)^t = B^t A^t$ (note the order)

41. There are long waiting lists of people in the United States who need organ transplants. In 2000, 4164 people were waiting for a heart transplant, 3732 for a lung transplant, 17,132 for a liver transplant, and 50,426 for a kidney transplant. Corresponding figures for 2001 were 4148, 3821, 17,546, and 52,216. In late 2003, the figures were 3568, 3927, 17,567, and 59,333.[*] Express this information as a matrix. Label each row and column.

42. A survey of credit card use by undergraduates in 2001 found that 54% of first-year students, 92% of sophomores, 87% of juniors, and 96% of seniors (including fifth-year seniors) have at least one credit card. The average number of cards per student for first-year students was 2.5, and their average credit card debt was $1533. The average number of cards per person for sophomores, juniors, and seniors was 3.67, 4.5, and 6.13, respectively; their average debt was $1825, $2705, and $3262, respectively. Finally, 40% of seniors had a balance exceeding $3000; the corresponding percentages for first year through junior year were 12%, 22%, and 31%.[†] Write this information as a 4 × 4 matrix with the columns corresponding to first, sophomore, junior, and senior years.

43. The Chicken Palace has raised its prices. Chicken nuggets now cost $2.75, Buffalo wings cost $3.25, and filet of chicken sandwiches cost $2.20. Assuming the same number of sales at each location as in Example 5, find a matrix that represents the daily revenue per store.

---

[*]United Network for Organ Sharing and National Organ Procurement and Transplantation Network.
[†]Nellie Mae.

44. The following table shows the population (in millions) of four countries in various years.[*]

	2001	2010
United States	278.1	300.1
Canada	31.6	34.3
Argentina	37.4	41.1
Japan	126.8	127.3

The next table gives the average birth and death rates for these countries.[*]

	Birth Rate	Death Rate
United States	.01425	.00865
Canada	.01145	.00775
Argentina	.0175	.00755
Japan	.00945	.00925

(a) Write the information in the first table as a 2 × 4 matrix A.

(b) Write the information in the second table as a 4 × 2 matrix B.

(c) Find the product AB.

(d) Explain what AB represents.

(e) According to matrix AB, what is the total number of people born in these four countries in 2001?

45. A drug company is testing 200 patients to determine whether Lopress (a new blood-pressure medicine) is effective. Half the patients take Lopress for six months, and the other half take a placebo. The results are summarized in this matrix:

$$
\begin{array}{cc}
 & \text{Effective?} \\
 & \text{Yes \quad No}
\end{array}
$$
$$
\begin{array}{c}
\text{Patient took Lopress} \\
\text{Patient took placebo}
\end{array}
\begin{pmatrix} 88 & 12 \\ 32 & 68 \end{pmatrix}
$$

The test was repeated for three more groups of 200, with the results given by

$$\begin{pmatrix} 92 & 8 \\ 12 & 88 \end{pmatrix}, \quad \begin{pmatrix} 84 & 16 \\ 24 & 76 \end{pmatrix}, \quad \begin{pmatrix} 76 & 24 \\ 40 & 60 \end{pmatrix}.$$

Use matrix addition to construct a matrix that gives total results of the test for all 800 patients.

*Figures and projections by the Bureau of the Census, U.S. Department of Commerce.

46. A new 300-home subdivision is to have two styles of homes: traditional and contemporary. Each style comes in three different models: 2 bedrooms, 2 baths, or 3 bedrooms, 3 baths, or 4 bedrooms, 3.5 baths (the half-bath is a powder room). The numbers of each model are given by this matrix:

$$
\begin{array}{cc}
\text{Traditional} & \text{Contemporary}
\end{array}
$$
$$
\begin{array}{c} 2/2 \\ 3/3 \\ 4/3.5 \end{array}
\begin{pmatrix} 0 & 90 \\ 30 & 60 \\ 60 & 60 \end{pmatrix} = A.
$$

The materials needed for the exterior of each style of house are given by

$$
\begin{array}{cccc}
\text{Concrete} & \text{Lumber} & \text{Bricks} & \text{Shingles}
\end{array}
$$
$$
\begin{array}{c} \text{Traditional} \\ \text{Contemporary} \end{array}
\begin{pmatrix} 30 & 6 & 0 & 6 \\ 150 & 3 & 60 & 6 \end{pmatrix} = B,
$$

where concrete is measured in cubic yards, lumber in thousands of board feet, bricks in thousands, and shingles in units of 100 square feet. The cost per unit of each kind of material is given by

$$
\begin{array}{c}
\text{Cost}
\end{array}
$$
$$
\begin{array}{c} \text{Concrete} \\ \text{Lumber} \\ \text{Bricks} \\ \text{Shingles} \end{array}
\begin{pmatrix} 25 \\ 200 \\ 70 \\ 30 \end{pmatrix} = C.
$$

(a) Explain why the matrix AB shows the amount of each material needed for each type of house. Find AB.

(b) Find $(AB)C$ and explain why this matrix gives the total cost for each model.

(c) Use matrix AB to find a 1 × 4 matrix D that represents the total amount of each type of material needed for the entire subdivision. [*Hint:* What do the columns of matrix AB represent?]

(d) What matrix product represents the total cost of materials for the entire subdivision? Find this product.

In Exercises 47 and 48, use matrix A in Example 6 to encode the given message.

47. Shipment arrives Monday.

48. Follow the fourth car.

In Exercises 49 and 50, decode the given message (which was encoded by using matrix A in Example 6).

49. 140, −1, 38, 81, 120, −43, 34, −9, 88, −64, 36, −16, 164, −148, 116, 168, 6, −2, 4, 12.

50. 102, −22, 22, −16, 104, −11, 10, −57, 116, −47, 40, 15, 67, −35, 45, 108, 123, −101, 92, 170, 128, −110, 73, 50, 175, −170, 128, 175, 76, 19, 11, 33

In Exercises 51 and 52, use this matrix:

$$\begin{pmatrix} 1 & 2 & 3 & 4 & 5 \\ 9 & 8 & 7 & 6 & -5 \\ 2 & 0 & -3 & 0 & 4 \\ 1 & -2 & 5 & -6 & 0 \\ 6 & 0 & -3 & 0 & 2 \end{pmatrix}.$$

51. Encode this message: "The trumpet sounds loudly."

52. Decode this message: 100, 18, 11, 10, 150, 89, 16, 22, 4, 146, 56, 0, 125, −38, 131, 250, 214, 241, 166, −58, 322, 146, 102, 124, 147, 76, −32, 57, −106, 105, 81, 20, 156, −50, 37.

53. (a) Encode the message "Rats!" using the matrix

$$\begin{pmatrix} 1 & 3 & 5 & 3 \\ 2 & 4 & -1 & 0 \\ 4 & -2 & 1 & 2 \\ -1 & 9 & 3 & 1 \end{pmatrix}.$$

(b) Now decode the message you sent in part (a). What goes wrong? What condition must a matrix satisfy if it is to be used for encoding and decoding messages?

54. Tom uses matrix A to encode a message and sends it to Anne. Anne, who does not know what matrix A is, uses matrix B to encode the message she receives from Tom and sends it on to Laura. Laura has both matrices A and B. What matrix should she use to decode the message?

6.5 Systems of Linear Inequalities

The graph of a linear equation, such as $y = 2x + 1$, consists of all points (c, d) such that $x = c, y = d$ is a solution of the equation. Similarly, the **graph of a linear inequality,** such as

$$y \le 2x + 1 \quad \text{or} \quad y > 2x + 1 \quad \text{or} \quad x - y < 3 \quad \text{or} \quad 3x + 2y \ge 6,$$

consists of all points (c, d) such that $x = c, y = d$ is a solution of the inequality.

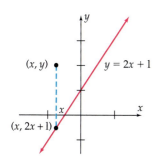

Figure 6–33

EXAMPLE 1

Graph each of these inequalities

(a) $y > 2x + 1$ (b) $y \le 2x + 1$.

SOLUTION First, graph the line $y = 2x + 1$. Any point (x, y) above the line $y = 2x + 1$ lies directly *above* the point $(x, 2x + 1)$ and hence has a *larger* y-coordinate, as in Figure 6–33. Therefore,

$(*)$ **(x, y) is above the line**   **exactly when**   **$y > 2x + 1$.**

Similarly, any point (x, y) below the line lies directly *below* the point $(x, 2x + 1)$ and hence has a *smaller* y-coordinate, as in Figure 6–34. Hence,

$(**)$ **(x, y) is below the line**   **exactly when**   **$y < 2x + 1$.**

Now the graphs are easily determined.

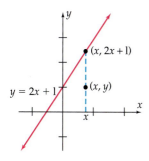

Figure 6–34

(a) Statement (∗) shows that the graph of $y > 2x + 1$ is the half-plane consisting of every point *above* the line $y = 2x + 1$, as shown by the shading in Figure 6–35. The line $y = 2x + 1$ is dashed to indicate that it is *not* part of the graph of $y > 2x + 1$.

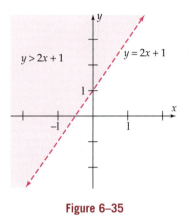

Figure 6–35

(b) Clearly, every point on the line $y = 2x + 1$ is a solution of $y \le 2x + 1$. Combining this fact with statement (∗∗), we see that the graph of $y \le 2x + 1$ is the half-plane consisting of every point *on or below* the line $y = 2x + 1$, as shown by the shading in Figure 6–36. The line $y = 2x + 1$ is solid to indicate that it *is* part of the graph of $y \le 2x + 1$. ■

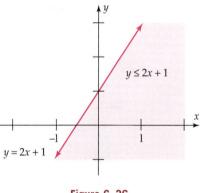

Figure 6–36

The argument used in Example 1 can be summarized as follows.

Solving a Linear Inequality

If an inequality is in this form:	then its solution consists of all points that are
$y < mx + b$	**Below** the line $y = mx + b$
$y \le mx + b$	**On or below** the line $y = mx + b$
$y > mx + b$	**Above** the line $y = mx + b$
$y \ge mx + b$	**On or above** the line $y = mx + b$

EXAMPLE 2

Graph $3x + 2y \geq 6$.

SOLUTION First, solve the inequality for y.

$$2y \geq -3x + 6$$

$$y \geq -\frac{3}{2}x + 3$$

$$y \geq -1.5x + 3$$

The $\geq$ symbol tells us that the solutions are all points whose y-coordinates lie *on or above* the line $y = -1.5x + 3$. The graph of the line can be sketched by hand and the area above it shaded (Figure 6–37), or the entire graph can be obtained on a calculator. See Figure 6–38 and the Technology Tip in the margin. ■

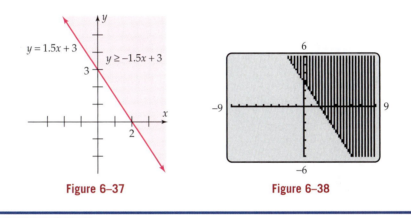

Figure 6–37   **Figure 6–38**

EXAMPLE 3

Graph each of these inequalities:

(a) $y \leq 4$;   (b) $x > 3$.

SOLUTION

(a) The inequality $y \leq 4$, which is understood to have two variables in this context, can be written as $y \leq 0x + 4$. So its graph consists of all points on or below the horizontal line $y = 4$, as shown in Figure 6–39.

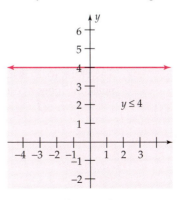

Figure 6–39

(b) This inequality does not fit the pattern of the preceding examples, but it can be solved by a similar technique. The graph of $x > 3$ consists of all points (x, y) with first coordinate greater than 3. So the graph is the half-plane to the right of the vertical line $x = 3$, as shown in Figure 6–40. The vertical line $x = 3$ is dashed because it is not part of the graph.   ■

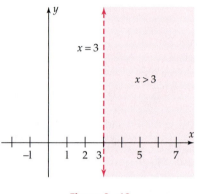

Figure 6–40

Although the solution method illustrated above is usually more efficient (particularly when using technology), the following alternative method is sometimes useful.

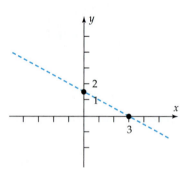

Figure 6–41

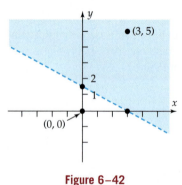

Figure 6–42

EXAMPLE 4

Graph $2x + 4y > 6$.

SOLUTION   The graph consists of a half-plane bounded by the line $2x + 4y = 6$. This line can be graphed by finding its intercepts.

x-intercept: Let $y = 0$. y-intercept: Let $x = 0$.

$$2x + 4(0) = 6 \qquad\qquad 2(0) + 4y = 6$$

$$x = 3 \qquad\qquad\qquad y = \frac{6}{4} = 1.5.$$

The line is graphed in Figure 6–41. To determine whether the graph of the inequality is the half-plane above or below the line, choose a *test point*—any point not on the line $2x + 4y = 6$—say, $(0, 0)$. Letting $x = 0$ and $y = 0$ in the inequality produces

$$2(0) + 4(0) > 6, \qquad \text{a } \textit{false} \text{ statement.}$$

Therefore, $(0, 0)$ is not in the solution of the inequality. So the solution is the half-plane that does not include $(0, 0)$, that is, the half-plane above the line, as shown in Figure 6–42.

If a different test point is used, the result is the same. If you choose (3, 5), for example, then substituting $x = 3$ and $y = 5$ in the inequality produces

$$2(3) + 4(5) > 6, \qquad \text{a } true \text{ statement.}$$

Therefore, the solution of the inequality is the half-plane containing (3, 5), that is, the half-plane above the line, as shown in Figure 6–42. ■

■■ SYSTEMS OF INEQUALITIES

The graph of a system of inequalities consists of all points that are solutions of *every* inequality in the system.

EXAMPLE 5

Sketch the graph of the system

$$2x + y > 0$$

$$x - y > -3.$$

SOLUTION   By rearranging each inequality the system becomes

$$y > -2x$$

$$y < x + 3.$$

The solutions of the first inequality are the points *above* the line $y = -2x$ (Figure 6–43). The solutions of the second are the points *below* the line $y = x + 3$ (Figure 6–44). So the solutions of the *system* are the points that satisfy both of these conditions (Figure 6–45). ■

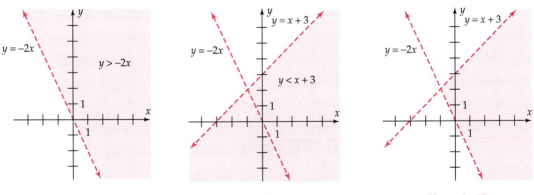

Figure 6–43   **Figure 6–44**   **Figure 6–45**

EXAMPLE 6

Graph the system

$$x + y > 11$$
$$2x - y < 10$$
$$x - 2y > -16.$$

SOLUTION Begin by rewriting the system as

$$y > -x + 11$$
$$y > 2x - 10$$
$$y < \frac{1}{2}x + 8.$$

The solutions of the system consist of all points that lie *above both* of the lines $y = -x + 11$ and $y = 2x - 10$, and also *below* the line $y = \frac{1}{2}x + 8$, as shown in Figure 6–46. ■

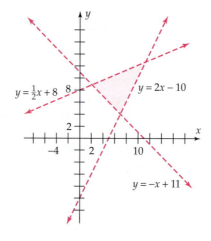

Figure 6–46

EXAMPLE 7

Graph the system

$$6x + 5y \leq 100$$
$$15x + 6y \leq 168$$
$$x \geq 0, \qquad y \geq 0.$$

SOLUTION We solve each inequality for y and rewrite the system as

$$y \le -\frac{6}{5}x + \frac{100}{5} \qquad\qquad y \le -1.2x + 20$$

$$y \le -\frac{15}{6}x + \frac{168}{6} \quad \text{or, equivalently,} \quad y \le -2.5x + 28$$

$$x \ge 0, \quad y \ge 0 \qquad\qquad x \ge 0, \quad y \ge 0.$$

A point (x, y) that is a solution of $x \ge 0$ and $y \ge 0$ must have both coordinates nonnegative; that is, it must lie in the first quadrant. Therefore, the solutions of the system are all points in the first quadrant that lie on or below both of the lines

$$y = -1.2x + 20 \quad \text{and} \quad y = -2.5x + 28,$$

as shown in Figure 6–47. ■

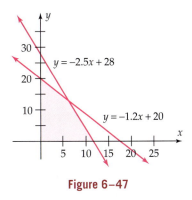

Figure 6–47

EXAMPLE 8

Rose and Joe's Candy Emporium sells two kinds of 4-pound gift boxes. The Regular Mix contains 2 pounds of assorted chocolates and 2 pounds of assorted caramels. The Chewy Mix has 1 pound of chocolates and 3 pounds of caramels. If 100 pounds of chocolates and 180 pounds of caramels are available, what possible numbers of both mixes can be made up?

SOLUTION Let x be the number of Regular Mixes to be made and y the number of Chewy Mixes. We know that

$$\left(\begin{array}{c} \text{Pounds of chocolates} \\ \text{in } x \text{ boxes of Regular} \\ \text{(2 pounds per box)} \end{array} \right) + \left(\begin{array}{c} \text{Pounds of chocolates} \\ \text{in } y \text{ boxes of Chewy} \\ \text{(1 pound per box)} \end{array} \right) \le 100$$

$$2x \quad + \quad 1y \qquad\qquad\qquad \le 100.$$

Similarly,

$$\left(\begin{array}{c}\text{Pounds of caramels}\\\text{in } x \text{ boxes of Regular}\\\text{(2 pounds per box)}\end{array}\right) + \left(\begin{array}{c}\text{Pounds of caramels}\\\text{in } y \text{ boxes of Chewy}\\\text{(3 pounds per box)}\end{array}\right) \le 180$$

$$2x \quad + \quad 3y \qquad\qquad \le 180.$$

Furthermore, both x and y must be nonnegative (you can't have a negative number of boxes). So the possible numbers of the two mixes must be solutions of this system of inequalities

$$2x + y \le 100$$
$$2x + 3y \le 180$$
$$x \ge 0$$
$$y \ge 0.$$

We can picture the possibilities by graphing this system. To do this, we rewrite it as

$$y \le -2x + 100$$
$$y \le -\frac{2}{3}x + 60$$
$$x \ge 0, \qquad y \ge 0.$$

So the graph (Figure 6–48) consists of the points that satisfy *all* of these conditions:

On or below the lines $y = -2x + 100$ and $y = -\frac{2}{3}x + 60$;

On or above the line $y = 0$ (the x-axis);

On or to the right of the vertical line $x = 0$ (the y-axis).

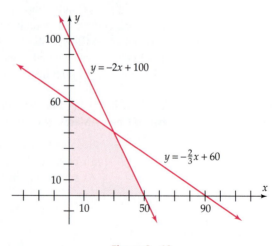

Figure 6–48

Since we can't have fractional boxes, the points on the graph with integer coordinates represent the possible numbers of Regular and Chewy Mixes that can be made from the available supplies. ∎

✓ EXERCISES 6.5 🎞️

In Exercises 1–14, sketch the graph of the inequality.

1. $y < x + 3$

2. $y > 5 - 2x$

3. $y \geq -1.5x - 2$

4. $y \leq 1.5x + 2$

5. $y \geq -3$

6. $x \leq -1$

7. $x > 2$

8. $y < 3$

9. $2x - y \leq 4$

10. $4x - 3y \leq 24$

11. $3x - 2y \geq 18$

12. $2x + 5y \geq 10$

13. $3x + 4y > 12$

14. $4x - 3y > 9$

In Exercises 15–20, match the inequality with its graph, which is one of those shown here.

15. $y \geq -x - 2$ **16.** $y \leq 2x - 2$ **17.** $y \leq x + 2$

18. $y \geq x + 1$ **19.** $6x + 4y \geq -12$ **20.** $3x - 2y \geq -4$

In Exercises 21–42, sketch the graph of the system of inequalities.

21. $y \geq 3x - 6$
$\quad\ y \geq -x + 1$

22. $x + y \leq 4$
$\quad\ x - y \geq 2$

23. $2x + \ y \leq 5$
$\quad\ x + 2y \leq 5$

24. $4x + \ y \geq 9$
$\quad 2x + 3y \leq 7$

25. $2x + y > 8$
$\quad 4x - y < 3$

26. $x + \ y > 5$
$\quad\ x - 2y < 2$

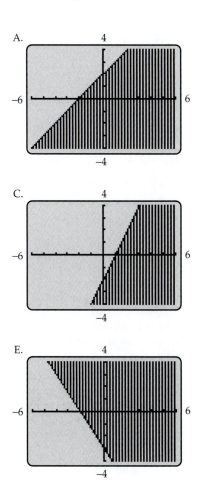

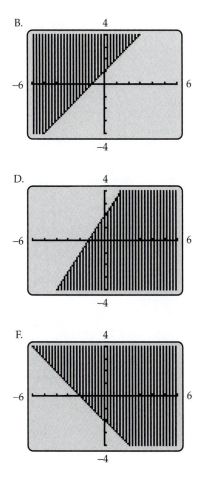

27. $3x + y \geq 6$
$x + 2y \geq 7$
$x \geq 0, \quad y \geq 0$

28. $2x + 3y \geq 12$
$x + y \geq 4$
$x \geq 0, \quad y \geq 0$

29. $3x + 2y < 18$
$x + 2y < 10$
$x \geq 0, \quad y \geq 0$

30. $x + 2y > 5$
$3x + 4y > 11$
$x \geq 0, \quad y \geq 0$

31. $2x + y \geq 8$
$2x - y \geq 0$
$x \leq 10, \quad y \geq 2$

32. $3x + y \leq -4$
$2x - y \leq 6$
$x \geq -5, \quad y \geq -8$

33. $3x + 2y \geq 12$
$3x + 2y \leq 18$
$3x - y \geq 0$

34. $4x + 5y \leq 20$
$x + 4y \geq 8$
$6x + y \geq 6$

35. $3x - y \leq 4$
$y - 3 \leq 0$
$x + y \geq 0$

36. $5x + 10y \leq 600$
$3x + 4y \leq 240$
$x \geq 0, \quad y \geq 0$

37. $x + y \geq 2$
$2x - y \leq 4$
$x \geq 0, \quad y \geq 0$

38. $x + y \geq 3$
$2x + 4y \geq 4$
$x \geq 0, \quad y \geq 0$

39. $x + y \leq 25$
$x \quad \geq 15$
$0 \leq y \leq 5$

40. $x + y \leq 1200$
$2y \leq x$
$x - 3y \leq 600$
$x \geq 0, \quad y \geq 0$

41. $x + 2y \geq 6$
$3x + y \geq 8$
$2y \leq x + 8$
$x \geq 0, \quad y \geq 0$

42. $2x + 5y \leq 80$
$x + y \leq 25$
$x - y \leq -11$
$x \geq 0, \quad y \geq 0$

In Exercises 43–46, find a system of inequalities that has the given graph.

43.

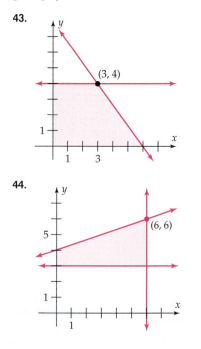

44.

In Exercises 47–50, find a system of inequalities whose graph is the interior of the given figure.

47. Rectangle with vertices $(2, 3)$, $(2, -1)$, $(7, 3)$, $(7, -1)$.

48. Parallelogram with vertices $(2, 4)$, $(2, -2)$, $(5, 6)$, $(5, 0)$.

49. Triangle with vertices $(2, 4)$, $(-4, 0)$, $(2, -1)$.

50. Quadrilateral with vertices $(-2, 3)$, $(2, 4)$, $(5, 0)$, $(0, -3)$.

In Exercises 51–56, find a system of inequalities that describes all the possible answers and sketch its graph.

51. For an oil company to make 1 storage tank of regular gasoline, it takes 2 hours of processing and 3 hours of refining. One tank of premium gas takes 3 hours of processing and 3.5 hours of refining. The processing plant can be operated for at most 12 hours a day, and the refinery can be operated for at most 16 hours a day. What are the possible outputs of regular and premium gas?

52. A can opener manufacturer makes both manual and electric models. The demand for manual can openers is never more than half of that for electric ones. The factory's production cannot exceed 1200 can openers per month. What are the possibilities for making x manual and y electric can openers each month?

53. Jack takes 4 hours to assemble a toy chest, and Jill takes 3 hours to decorate it. To make a silverware case takes Jack 2 hours for assembly and Jill 4 hours for decorating. Neither Jack nor Jill wants to work more than 20 hours a week. What are the possible production levels for toy and silverware chests?

45.

46.

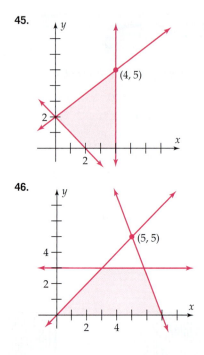

54. A builder plans to construct a building containing both one-bedroom apartments, each with 1000 square feet of space, and two-bedroom apartments, each with 1500 square feet. The maximum building size is 45,000 square feet. If there can be at most 40 apartments, what are the possibilities for the number of one- and two-bedroom units?

55. A pet food is to be made of grain and meat by-products. One serving must provide at most 12 units of fat, at least 2 units of carbohydrates, and at least 1 unit of protein. Each gram of grain contains 2 units of fat, 2 units of carbohydrates, and no units of protein. A gram of meat by-products has 3 units of fat, 1 unit of carbohydrate, and 1 unit of protein. What are the possible ways of combining x grams of grain and y grams of meat by-products in one serving?

56. An airline dietician is planning a snack package of candy and nuts. Each ounce of candy will supply 1 unit of protein, 2 units of carbohydrates, and 1 unit of fat. Each ounce of nuts will supply 1 unit of protein, 1 unit of carbohydrates, and 1 unit of fat. Every package must provide at least 7 units of protein, at least 10 units of carbohydrates, and no more than 9 units of fat. What are the possible ways of mixing x ounces of candy and y ounces of nuts in each package?

6.6 Introduction to Linear Programming

Many problems in business, economics, and the sciences are concerned with finding the optimal value of a function (for instance, the maximum value of a profit function or the minimum value of a cost function), subject to various constraints (transportation costs, environmental protection laws, interest rates, availability of parts, etc.). The function to be optimized is usually a function of several variables and the constraints are expressed as linear inequalities.

EXAMPLE 1

Find the maximum and minimum values of $F = 2x + 5y$, subject to these constraints:

$$x + y \geq 1$$
$$-2x + 4y \leq 8$$
$$3x + 2y \leq 6$$
$$x \geq 0, \qquad y \geq 0.$$

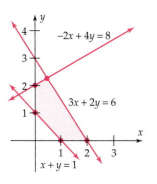

Figure 6–49

SOLUTION   The possible solutions of this problem are the points (x, y) that satisfy all of the constraint inequalities. We call them the *feasible solutions*. By graphing the system of constraint inequalities, we see that the feasible solutions are the points in the shaded area of Figure 6–49. We must find the feasible solution that makes the value of $F = 2x + 5y$ smallest and the one that makes it largest.

Consider different possible values of the function F, for instance,

$F = 0$	$F = 5$	$F = 10$	$F = 15$
$2x + 5y = 0$	$2x + 5y = 5$	$2x + 5y = 10$	$2x + 5y = 15.$

Each value of F leads to an equation, such as $2x + 5y = 10$, whose graph is a straight line. The points where this line intersects the shaded region are the feasible solutions for which $F = 10$ (Figure 6–50). Similarly, the points on the line $2x + 5y = 5$ that lie in the shaded region are the feasible solutions for which $F = 5$. The lines $2x + 5y = 0$ and $2x + 5y = 15$ do not intersect the shaded region, so there are no feasible solutions for which $F = 0$ or $F = 15$.

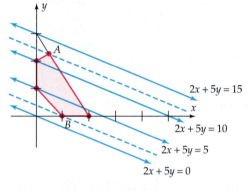

Figure 6–50

Look at the pattern in Figure 6–50. The lines

$$2x + 5y = 0, \qquad 2x + 5y = 5, \qquad 2x + 5y = 10, \qquad \text{etc.}$$

are all parallel (because they all have the same slope, $-2/5$) and the constant on the right side of the equal sign determines their positions: As this number (the value of F) gets larger, the lines move upward; as it gets smaller, the lines move downward.

The minimum value of F corresponds to the lowest line that contains feasible solutions. As Figure 6–50 shows, this is the dashed line through the vertex B. All lines below this one (corresponding to smaller values of F) do not contain any feasible solutions. Therefore, the coordinates of B will produce the minimum value of F. As Figure 6–49 shows, B has coordinates $(1, 0)$. Therefore, the minimum value of F occurs when $x = 1$ and $y = 0$. This mininum value is

$$F = 2x + 5y = 2(1) + 5(0) = 2.$$

Similarly, the maximum value of F corresponds to the highest line that contains feasible solutions, namely, the dashed line through vertex A. So the coordinates of A will produce the maximum value of F. Since A is the intersection of the lines

$$-2x + 4y = 8 \qquad \text{and} \qquad 3x + 2y = 6,$$

its coordinates can be found

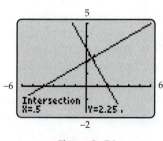

Figure 6–51

Algebraically	or	*Graphically*
Solve the system		Solve each equation for y:
$-2x + 4y = 8$		$y = .5x + 2$
$3x + 2y = 6$		$y = -1.5x + 3.$
as in Section 6.1 and find		Graph them and use an
that $x = .5$ and $y = 2.25$.		intersection finder, as in
		Figure 6–51.

Thus, the maximum value of F occurs when $x = .5$ and $y = 2.25$. This maximum value is

$$F = 2x + 5y = 2(.5) + 5(2.25) = 12.25. \quad \blacksquare$$

Example 1 is a **linear programming problem** in two variables. Such problems consist of a linear function of the form $F = ax + by$ (called the **objective function**), subject to certain **constraints** that are expressed as a system of linear inequalities. Any pair (x, y) that satisfies all the constraints is a **feasible solution.** The goal is to find an **optimal solution,** that is, a feasible solution that produces the maximum or minimum value of the objective function.

As Example 1 illustrates, the set of feasible solutions (the graph of the system of constraint inequalities) is a region of the plane (called the **feasible region**) whose edges are straight-line segments. The feasible region may be either **bounded** as in Example 1, or **unbounded,** as in Figure 6–52 on the next page. The points where the edges of the feasible region meet are the **vertices.** The analysis in Example 1 showed that the minimum F occurred at the vertex $(1, 0)$ of the feasible region and that the maximum value of F occurred at the vertex $(.5, 2.25)$. The same kind of analysis works in the general case and proves the following fact.

Linear Programming Theorem

If the feasible region is bounded, then the objective function has both a maximum and a minimum value, and each occurs at a vertex.

If the feasible region is an unbounded region in the first quadrant and both coefficients of the objective function are positive,[*] then the objective function has no maximum value but does have a minimum value, and this occurs at a vertex.

NOTE

In some cases, the same optimal solution may occur at more than one vertex. If it occurs at two adjacent vertices, it also occurs on the line segment joining them.

EXAMPLE 2

Find the maximum and minimum values of $F = 3x + 2y$, subject to the constraints

$$2x + y \geq 9$$

$$x + y \geq 5$$

$$y \geq 0$$

$$x \geq 1.5.$$

---

[*]This is the only case of an unbounded feasible region that occurs in most applications.

SOLUTION   By solving the first two constraints for y, we can easily find the graph of the system of constraint inequalities (the feasible region), as shown in Figure 6–52. It is an unbounded region that continues forever as you move to the right or upward.

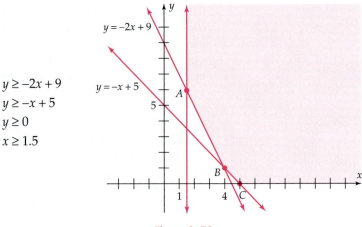

$y \geq -2x + 9$
$y \geq -x + 5$
$y \geq 0$
$x \geq 1.5$

Figure 6–52

Each vertex is the intersection of two edges. Its coordinates can be found by solving the system of equations given by the two edges (either algebraically, which is easiest here, or graphically).

Vertex:	A	B	C
Edges:	$y = -2x + 9$	$y = -2x + 9$	$y = -x + 5$
	$x = 1.5$	$y = -x + 5$	$y = 0$
Coordinates:	$(1.5, 6)$	$(4, 1)$	$(5, 0)$

Since the feasible region is unbounded, F has no maximum value. Its minimum value must occur at one of the vertices, so we test each one of them.

Vertex	$F = 3x + 2y$
$(1.5, 6)$	16.5
$(4, 1)$	14
$(5, 0)$	15

Therefore, the minimum value of F is 14, and it occurs when $x = 4$ and $y = 1$. ■

▦ APPLICATIONS

EXAMPLE 3

The Candy Emporium sells two kinds of 4-pound gift boxes: Regular and Chewy. A box of Regular has 2 pounds of chocolates and 2 pounds of caramels. A box of

Chewy has 1 pound of chocolates and 3 pounds of caramels. The store has 100 pounds of chocolates and 180 pounds of caramels available. If the profit is $3 on each box of Regular and $4 on each box of Chewy, how many boxes of each kind should be made to maximize the profit?

SOLUTION   First, we find the rule of the profit function. Let x be the number of boxes of Regular and y the number of boxes of Chewy. Then the profit P is given by

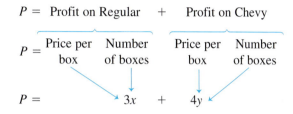

So the profit function is $P = 3x + 4y$.

In Example 8 on pages 531–532, we saw that the available amounts of chocolates and caramels led to the following system of constraints.

$$2x + y \leq 100 \qquad\qquad y \leq -2x + 100$$

$$2x + 3y \leq 180 \quad \text{or, equivalently,} \quad y \leq -\frac{2}{3}x + 60$$

$$x \geq 0, \qquad y \geq 0 \qquad\qquad x \geq 0, \qquad y \geq 0.$$

We must find the maximum value of $P = 3x + 4y$, subject to these constraints. The graph of the feasible region is shown in Figure 6–53.

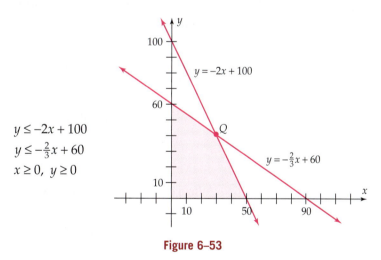

Figure 6–53

We need only evaluate $P = 3x + 4y$ at the four vertices of the graph, namely, $(0, 0)$, $(0, 60)$, $(50, 0)$ and Q. To find the coordinates of Q, we solve the system of equations

$$y = -2x + 100$$

$$y = -\frac{2}{3}x + 60$$

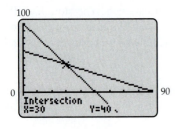

Figure 6–54

or use an intersection finder (Figure 6–54) to find that Q is (30, 40). So we have the following table.

Vertex	Profit = 3x + 4y
(0, 0)	$P = 3 \cdot 0 + 4 \cdot 0 = 0$
(0, 60)	$P = 3 \cdot 0 + 4 \cdot 60 = 240$
(50, 0)	$P = 3 \cdot 50 + 4 \cdot 0 = 150$
(30, 40)	$P = 3 \cdot 30 + 4 \cdot 40 = 250$

Therefore, making 30 boxes of Regular and 40 of Chewy will produce the maximum profit. ■

EXAMPLE 4

An animal food is to be made from grain and silage. Each pound of grain provides 2400 calories and 20 grams of protein and costs 18¢. Each pound of silage provides 600 calories and 2 grams of protein and costs 3¢. An animal must have at least 18,000 calories and 120 grams of protein each day. It is unhealthy for an animal to have more than 12 pounds of grain or 20 pounds of silage per day. What combination of grain and silage should be used to keep costs as low as possible?

SOLUTION   Let x be the number of pounds of grain an animal eats per day, and y the number of pounds of silage. Then the cost function to be minimized is given by $C = .18x + .03y$. The health limits on the amount of food imply these constraints:

$$0 \leq x \leq 12 \quad \text{and} \quad 0 \leq y \leq 20.$$

The minimal calorie requirement also leads to a constraint.

$$\left(\begin{matrix} \text{Calories in } x \\ \text{pounds of grain} \end{matrix} \right) + \left(\begin{matrix} \text{Calories in } y \\ \text{pounds of silage} \end{matrix} \right) \geq 18{,}000$$

$$2400x \quad + \quad 600y \quad \geq 18{,}000$$

A final constraint comes from the minimal protein requirement.

$$\left(\begin{matrix} \text{Protein in } x \\ \text{pounds of grain} \end{matrix} \right) + \left(\begin{matrix} \text{Protein in } y \\ \text{pounds of silage} \end{matrix} \right) \geq 120$$

$$20x \quad + \quad 2y \quad \geq 120$$

So the entire system of constraints is given by

$2400x + 600y \geq 18{,}000$	$4x + y \geq 30$
$20x + \quad 2y \geq 120$	$10x + y \geq 60$
$x \leq 12, \quad y \leq 20$	$x \leq 12, \quad y \leq 20$
$x \geq 0, \quad y \geq 0,$	$x \geq 0, \quad y \geq 0.$

which is equivalent to

The graph of the feasible region (Figure 6–55) and its vertices are found as above.

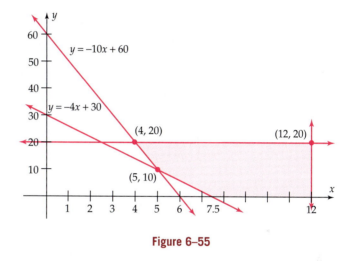

Figure 6–55

Since the set of feasible solutions is bounded, the cost function has both a maximum and a minimum value, each of which occurs at a vertex. The following table shows that the minimum is $1.20 and occurs when 5 pounds of grain and 10 pounds of silage are used.

Vertex	$C = .18x + .03y$
(4, 20)	$C = .18(4) + .03(20) \ = \$1.32$
(5, 10)	$C = .18(5) + .03(10) \ = \$1.20$
(7.5, 0)	$C = .18(7.5) + .03(0) \ = \$1.35$
(12, 0)	$C = .18(12) + .03(0) \ = \$2.16$
(12, 20)	$C = .18(12) + .03(20) = \$2.76$

EXERCISES 6.6

In Exercises 1–4, use the graph of the feasible region in the figure. Find the minimum and maximum values of the objective function F.

1. $F = 5x + 2y$

2. $F = x + 4y$

3. $F = 3x + y$

4. $F = 8x + 12y$

7. $F = 3x + 4y$

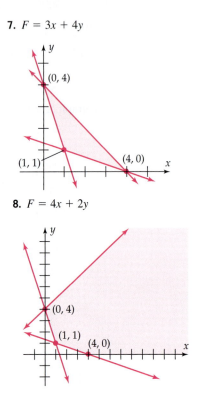

8. $F = 4x + 2y$

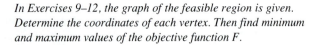

In Exercises 5–8, the graph of the feasible region is given. Find the minimum and maximum values of the objective function F.

5. $F = 3x + 2y$

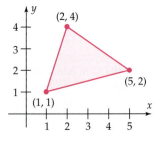

6. $F = 2x - y$

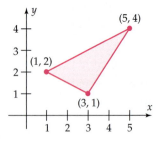

In Exercises 9–12, the graph of the feasible region is given. Determine the coordinates of each vertex. Then find minimum and maximum values of the objective function F.

9. $F = 12x + 8y$

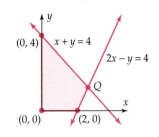

10. $F = 1.5x + 2.5y$

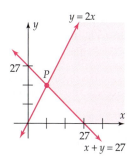

11. $F = 3.5x + 2y$

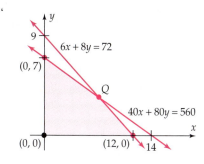

12. $F = 4x + 2.2y$

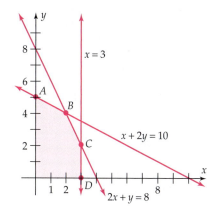

In Exercises 13–24, find the minimum and maximum values of the function F, subject to the given constraints.

13. $F = 4x + 3y$; constraints:
$$3x + y \leq 6$$
$$2x + y \leq 5$$
$$x \geq 0, \quad y \geq 0$$

14. $F = 3x + 7y$; constraints:
$$6x + 5y \leq 100$$
$$8x + 5y \leq 120$$
$$x \geq 0, \quad y \geq 0$$

15. $F = 3x + 2y$; constraints:
$$11x + 5y \geq 75$$
$$4x + 15y \geq 80$$
$$x \geq 0, \quad y \geq 0$$

16. $F = 6x + 3y$; constraints:
$$-2x + 3y \leq 9$$
$$-x + 3y \leq 12$$
$$x \geq 0, \quad y \geq 0$$

17. $F = 3x + y$; constraints:
$$2x + y \leq 20$$
$$10x + y \geq 36$$
$$2x + 5y \geq 36$$

18. $F = 2x + 6y$; constraints:
$$x + y \geq 6$$
$$-x + y \leq 2$$
$$2x - y \leq 8$$

19. $F = x + 3y$; constraints:
$$2x + 3y \leq 30$$
$$-x + y \leq 5$$
$$x + y \geq 5$$
$$0 \leq x \leq 10$$
$$y \geq 0$$

20. $F = 1100 - 2x - y$; constraints:
$$x + y \leq 80$$
$$x + y \geq 30$$
$$0 \leq x \leq 40$$
$$0 \leq y \leq 70$$

21. $F = 3x + 4y$; constraints:
$$x + y \leq 6$$
$$-x + y \leq 2$$
$$2x - y \leq 8$$
$$x \geq 0, \quad y \geq 0$$

22. $F = 4x + 3y$; constraints:
$$x + 2y \geq 10$$
$$2x + y \geq 12$$
$$x - y \leq 8$$
$$x \geq 0, \quad y \geq 0$$

23. $F = 2x + 3y$; constraints:
$$2x + 5y \leq 22$$
$$4x + 3y \geq 28$$
$$x - y \geq 20$$
$$x \geq 0, \quad y \geq 0$$

24. $F = 1.5x + 4.5y$; constraints:
$$10x + 7y \leq 252$$
$$4x + 10y \geq 85$$
$$-x + 4y \leq 48$$
$$x \geq 0, \quad y \geq 0$$

25. Wonderboat Company makes four-person and two-person inflatable rafts. A four-person raft requires 6 hours of fabrication and 1 hour of finishing. A two-person raft requires 4 hours of fabrication and 1 hour of finishing. Each week 108 hours of fabrication time and 24 hours of finishing time are available. If $40 is the profit on a four-person raft and $30 on a two-person one, how many of each kind should be made to maximize weekly profit?

26. The Coop sells summer-weight and winter-weight sleeping bags. A summer bag requires .9 hour of labor for cutting and .8 hour for assembly. A winter bag requires 1.8 hours for cutting and 1.2 hours for assembly. There are at most 864 hours of labor available in the cutting department each month and at most 672 hours in the assembly department. If the profit is $25 on a summer bag and $40 on a winter bag, how many of each kind should be made to maximize profit?

27. Sixty pounds of chocolates and 100 pounds of mints are available to make up 5-pound boxes of candy. A box of Choco-Mix has 4 pounds of chocolates and 1 pound of mints and sells for $10. A box of Minto-Mix has 2 pounds of chocolates and 3 pounds of mints and sells for $16. How many boxes of each kind should be made to maximize revenue?

28. CalenCo makes large wall calendars and small desk calendars. Printing a wall calendar takes 2 minutes and binding it 1 minute. A desk calendar requires 1 minute for printing and 3 minutes for binding. The printing machine is available for 3 hours a day and the binding machine for 5 hours. The profit is $1.00 on a wall calendar and $1.20 on a desk calendar. How many of each type should be made to maximize daily profit?

29. A person who is suffering from nutritional deficiencies is told to take at least 2400 mg of iron, 2100 mg of vitamin B_1, and 1500 mg of vitamin B_2. One Supervites pill contains 40 mg of iron, 10 mg of B_1, and 5 mg of B_2, and costs 6¢. One Vitahealth pill provides 10 mg of iron, 15 mg of B_1, and 15 mg of B_2, and costs 8¢. What combination of Supervites and Vitahealth pills will meet the requirements at lowest cost?

30. A fertilizer is to contain two major ingredients. Each pound of ingredient I contains 4 oz of nitrogen and 2 oz of phosphates and costs $1.00. Each pound of ingredient II contains 3 oz of nitrogen and 5 oz of phosphates and costs $2.50. A bag of fertilizer must contain at least 70 oz of nitrogen and 110 oz of phosphates. How much of each ingredient should be used to produce a bag of fertilizer at minimum cost? How many pounds of each ingredient does it contain?

31. A pound of chicken feed A contains 3000 units of Nutrigood and 1000 units of Fasgrow and costs 20¢. A pound of chicken feed B contains 4000 units of Nutrigood and 4000 units of Fasgrow and costs 40¢. The minimum daily requirement for one chicken farm is 36,000 units of Nutrigood and 20,000 units of Fasgrow. How many pounds of each chicken feed should be used each day to minimize food costs while meeting or exceeding the daily nutritional requirements?

32. A company has warehouses in Titusville and Rockland. It has 80 stereo systems stored in Titusville and 70 in Rockland. Superstore orders 35 systems, and Giantval orders 60. It costs $8 to ship a system from Titusville to Superstore and $12 to ship one to Giantval. It costs $10 to ship a system from Rockland to Superstore and $13 to ship one to Giantval. How should the orders be filled to keep shipping costs as low as possible?

33. A cereal company plans to combine bran and oats to manufacture at least 28 tons of a new cereal. The cereal must contain at least twice as much oats as bran. If bran costs $350 per ton and oats cost $225 per ton, how much of each should be used to minimize costs?

34. Newsmag publishes a U.S. and a Canadian edition each month. There are 30,000 subscribers in the United States and 20,000 in Canada. Other copies are sold at newsstands. Shipping costs average $80 per thousand copies for U.S. newsstands and $60 per thousand copies for Canadian newsstands. Surveys show that no more than 120,000 copies of each issue can be sold (including subscriptions) and that the number of copies of the Canadian edition should not exceed twice the number of copies of the U.S. edition. The publisher can spend at most $8400 a month on shipping costs to newsstands. If the profit is $200 for each thousand copies of the U.S. edition and $150 for each thousand copies of the Canadian edition, how many copies of each version should be printed to earn as large a profit as possible?

35. Students at Zonker College are required to take at least four humanities and four science courses. The maximum allowable number of science courses is 12. Each humanities course carries 4 credits, and each science course carries 5 credits. The total number of credits in humanities and science cannot exceed 92. Quality points for each course are assigned in the usual way: the number of credit hours times 4 for an A grade, times 3 for a B grade, and times 2 for a C grade. Terri expects to get B's in all her science courses. She expects to get C's in half her humanities courses, B's in one-fourth of them, and A's in the rest. Under these assumptions, how many courses of each kind should she take to earn the maximum possible number of quality points?

36. An investor has $20,000 to invest in bonds. He expects an annual yield of 10% for grade B bonds and 6% for grade AAA bonds. He decides that he shouldn't invest more than half his money in B bonds and should invest at least one-fourth in AAA bonds. He wants the amount invested in B bonds to be at least two-thirds of the amount invested in AAA bonds. How much should he invest in each type of bond to maximize his annual yield?

Thinkers

37. Explain why it is impossible to maximize the function $P = 3x + 4y$, subject to the constraints $x + y \geq 8$; $2x + y \leq 10$; $x + 2y \leq 8$; $x \geq 0$; $y \geq 0$.

38. The Linear Programming Theorem on page 537 guarantees that there is at least one vertex at which the optimal solution occurs (when it exists). This allows the possibility that the optimal solution might also occur at other points.

(a) Show that this is indeed the case here: Maximize $P = 3x + 2y$, subject to the constraints:

$$3x + 2y \leq 22; \; x \leq 6; \; y \leq 5; \; x \geq 0; \; y \geq 0.$$

(b) Show that the optimal value of P in part (a) also occurs at every point on the line segment joining the two vertices where the optimal value occurs. [*Hint:* The maximum value of P is 22, so any optimal solution satisfies $3x + 2y = 22$. Where does the line $3x + 2y = 22$ intersect the set of feasible solutions?]

39. In Exercise 35, find the student's grade point average (total number of quality points divided by total number of credit hours) at each vertex of the set of feasible solutions. Does the distribution of courses that produces the highest number of quality points also yield the highest grade point average? Is this a contradiction?

Chapter 6 Review

REVIEW QUESTIONS

In Questions 1–4, solve the system of linear equations by any means you want.

1. $-5x + 3y = 4$
$2x - y = -3$

2. $3x - y = 6$
$2x + 3y = 7$

3. $3x - 5y = 10$
$4x - 3y = 6$

4. $\dfrac{1}{4}x - \dfrac{1}{3}y = -\dfrac{1}{4}$
$\dfrac{1}{10}x + \dfrac{2}{5}y = \dfrac{2}{5}$

5. The number of days y in year x in which acceptable air quality standards were not met in two metropolitan areas are approximated by these equations (in which $x = 4$ corresponds to 1994).[*]

Riverside–San Bernadino, CA: $10.89x + y = 184.92$

Atlanta, GA: $16.06x - 2y = 32.04$

In what year is the number of unacceptable days the same in both areas? How many such days are there in that year?

---

*U.S. Environmental Protection Agency.

REVIEW QUESTIONS

6. The sum of one number and three times a second number is -20. The sum of the second number and two times the first number is 55. Find the two numbers.

7. An alloy containing 40% gold and an alloy containing 70% gold are to be mixed to produce 50 pounds of an alloy containing 60% gold. How much of each alloy is needed?

8. The table shows the population (in thousands) of the cities of Miami and Cleveland in selected years.[*]

Year	1950	1960	1970	1980	1990	2000
Miami	249.3	291.7	335.0	347.0	358.5	362.5
Cleveland	914.8	876.1	751.0	574.0	505.6	478.4

(a) Use linear regression to find an equation that gives the population y of Miami in year x, with $x = 0$ corresponding to 1950.
(b) Do part (a) for Cleveland.
(c) If these models remain accurate, when do the two cities have the same population and what is that population?

In Questions 9–12, solve the nonlinear system.

9. $x^2 - y = 0$
$\quad y - 2x = 3$

10. $x^2 + y^2 = 25$
$\quad\quad x^2 + y = 19$

11. $x^2 + y^2 = 16$
$\quad x + y = 2$

12. $6x^2 + 4xy + 3y^2 = 36$
$\quad\quad x^2 - xy + y^2 = 9$

13. Write the augmented matrix of the system

$$x - 2y + 3z = 4$$
$$2x + y - 4z = 3$$
$$-3x + 4y - z = -2.$$

14. Use matrix methods to solve the system in Question 13.

15. Write the coefficient matrix of the system

$$2x - y - 2z + 2u = 0$$
$$x + 3y - 2z + u = 0$$
$$-x + 4y + 2z - 3u = 0.$$

16. Use matrix methods to solve the system in Question 15.

In Questions 17–22, solve the system.

17. $3x + y - z = 13$
$\quad x + 2z = 9$
$\quad -3x - y + 2z = 9$

18. $x + 2y + 3z = 1$
$\quad 4x + 4y + 4z = 2$
$\quad 10x + 8y + 6z = 4$

19. $4x + 3y - 3z = 2$
$\quad 5x - 3y + 2z = 10$
$\quad 2x - 2y + 3z = 14$

20. $x + y - 4z = 0$
$\quad 2x + y - 3z = 2$
$\quad -3x - y + 2z = -4$

21. $x - 2y - 3z = 1$
$\quad 5y + 10z = 0$
$\quad 8x - 6y - 4z = 8$

22. $4x - y - 2z = 4$
$\quad x - y - \frac{1}{2}z = 1$
$\quad 2x - y - z = 8$

23. Let L be the line with equation $4x - 2y = 6$, and let M be the line with equation $-10x + 5y = -15$. Which of the following statements is true?
(a) L amd M do not intersect.
(b) L and M intersect at a single point.
(c) L and M are the same line.
(d) All of the above are true.
(e) None of the above are true.

24. Which of the following statements about this system of equations are *false*?

$$x + z = 2$$
$$6x + 4y + 14z = 24$$
$$2x + y + 4z = 7$$

(a) $x = 2, y = 3, z = 0$ is a solution.
(b) $x = 1, y = 1, z = 1$ is a solution.
(c) $x = 1, y = -3, z = 3$ is a solution.
(d) The system has an infinite number of solutions.
(e) $x = 2, y = 5, z = -1$ is not a solution.

In Questions 25 and 26, find the constants A, B, C that make the statement true.

25. $\dfrac{4x - 7}{x^2 - x - 6} = \dfrac{A}{x - 3} + \dfrac{B}{x + 2}$

26. $\dfrac{6x^2 + 6x - 6}{(x^2 - 1)(x + 2)} = \dfrac{A}{x + 1} + \dfrac{B}{x - 1} + \dfrac{C}{x + 2}$

In Questions 27–30, perform the indicated matrix multiplication or state that the product is not defined. Use these matrices.

$$A = \begin{pmatrix} -1 & 0 \\ 0 & -1 \end{pmatrix}, \quad B = \begin{pmatrix} 2 & -3 \\ 4 & 1 \end{pmatrix}, \quad C = \begin{pmatrix} 3 & 2 \\ 2 & 4 \end{pmatrix},$$

$$D = \begin{pmatrix} -3 & 1 & 2 \\ 1 & 0 & 4 \end{pmatrix}, \quad E = \begin{pmatrix} 1 & 2 \\ -3 & 4 \\ 0 & 5 \end{pmatrix}, \quad F = \begin{pmatrix} 2 & 3 \\ 6 & 3 \\ 6 & 1 \end{pmatrix}$$

---

[*]Bureau of the Census, U.S. Department of Commerce.

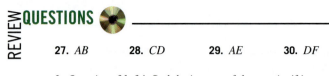

27. AB **28.** CD **29.** AE **30.** DF

In Questions 31–34, find the inverse of the matrix, if it exists.

31. $\begin{pmatrix} 3 & -7 \\ 4 & -9 \end{pmatrix}$ **32.** $\begin{pmatrix} 2 & 6 \\ 1 & 3 \end{pmatrix}$

33. $\begin{pmatrix} 3 & 2 & 6 \\ 1 & 1 & 2 \\ 2 & 2 & 5 \end{pmatrix}$ **34.** $\begin{pmatrix} 1 & -1 & 1 \\ 2 & -3 & 2 \\ -4 & 6 & 1 \end{pmatrix}$

In Questions 35 and 36, use matrix inverses to solve the system.

35.
$$\begin{aligned} x + \quad\; 2z + 6w &= \;\;\; 2 \\ 3x + 4y - 2z - \;\; w &= \;\;\; 0 \\ 5x + \quad\; 2z - 5w &= -4 \\ 4x - 4y + 2z + 3w &= \;\;\; 1 \end{aligned}$$

36.
$$\begin{aligned} 2x + \; y + 2z \quad\quad\; + \; u &= \;\;\; 2 \\ x + 3y - 4z - 2u + 2v &= -2 \\ 2x + 3y + 5z - 4u + \; v &= \;\;\; 1 \\ x \quad\quad - 2z \quad\quad + 4v &= \;\;\; 4 \\ 2x \quad\;\; + 6z \quad\quad - 5v &= \;\;\; 0 \end{aligned}$$

In Questions 37 and 38, find the equation of the parabola passing through the given points.

37. $(-3, 52), (2, 17), (8, 305)$

38. $(-2, -18), (2, 6), (4, -12)$

39. The table shows the number of hours spent per person per year on home video games.[*]

Year	1996	2000	2004
Hours	25	76	161

(a) Find a quadratic equation that models this data, with $x = 6$ corresponding to 1996.
(b) Use the equation to estimate the number of hours spent in 1998, 2002, and 2006.

40. The table shows the per capita consumption of eggs in selected years.[†]

Year	1980	1990	1999
Egg Consumption	271	234	255

(a) Find a quadratic equation that models this data, with $x = 0$ corresponding to 1980.
(b) Use the equation to estimate egg consumption in 1985, 1995, and 2005.

---

[*]Data and projections from *Statistical Abstract of the United States: 2001*.

[†]U.S. Department of Agriculture.

41. Tickets to a lecture cost $1 for students, $1.50 for faculty, and $2 for others. Total attendance at the lecture was 460, and the total income from tickets was $570. Three times as many students as faculty attended. How many faculty members attended the lecture?

42. If Andrew, Laura, and Ryan work together, using three lawnmowers, they can mow the lawn in 12 minutes. When only Laura and Ryan work together, it takes 36 minutes to mow the lawn. Andrew and Laura, working together, take 13.5 minutes to mow the lawn. How long would it take each of them to mow the lawn alone?

43. An animal feed is to be made from corn, soybeans, and meat by-products. One bag is to supply 1800 units of fiber, 2800 units of fat, and 2200 units of protein. Each pound of corn has 10 units of fiber, 30 units of fat, and 20 units of protein. Each pound of soybeans has 20 units of fiber, 20 units of fat, and 40 units of protein. Each pound of by-products has 30 units of fiber, 40 units of fat, and 25 units of protein. How many pounds of corn, soybeans, and by-products should each bag contain?

44. A company produces three camera models: *A, B,* and *C.* Each model *A* requires 3 hours of lens polishing, 2 hours of assembly time, and 2 hours of finishing time. Each model *B* requires 2 hours of lens polishing, 2 hours of assembly time, and 1 hour of finishing time. Each model *C* requires 1, 3, and 1 hours of lens polishing, assembly, and finishing time, respectively. There are 100 hours available for lens polishing, 100 hours for assembly, and 65 hours for finishing each week. How many of each model should be produced if all available time is to be used?

In Questions 45–50, use the matrices A, B, C, D, E, and F given before Question 27. Perform the indicated matrix operations (if they are defined). or state which operation is not defined.

45. $3F - 2E$ **46.** $3A + B - C$

47. $AB + AC$ **48.** $DA - A$

49. $DF + FE$ **50.** $DF - DE$

51. A recent survey on campaign reform legislation produced the following results. Among men, 227 favored the legislation, 124 opposed it, and 49 had no opinion. Among women, 266 favored the legislation, 40 opposed it, and 94 had no opinion. Write a 2×3 matrix that expresses this information.

REVIEW QUESTIONS

52. Use the matrix A to encode this message: "Fire when ready."

$$A = \begin{pmatrix} 2 & -5 & 4 \\ 1 & 6 & -2 \\ 3 & 4 & 5 \end{pmatrix}$$

In Questions 53–55, sketch the graph of the system.

53. $x + y \geq 0$
   $-x + y \leq 6$
   $x \leq 5$

54. $2x + y \geq 10$
   $-x + y \leq 1$
   $x \geq 0, \quad y \geq 0$

55. $x + y \leq 4$
   $x - y \geq 1$
   $y \leq 0$

In Questions 56–58, find the minimum and maximum values (if any) of F subject to the given constraints. State the values of x and y that produce each optimal value of F.

56. $F = 5x + 7y$;
   constraints:
   $10x + 20y \leq 480$
   $10x + 30y \leq 570$
   $x \geq 0, \qquad y \geq 0$

57. $F = .08x + .05y$;
   constraints:
   $3x + y \geq 20$
   $3x + 6y \geq 60$
   $x \geq 0, \quad y \geq 0$

58. $F = 2x - 3y$; constraints:
   $2x + 3y \geq 12$
   $3x + 2y \leq 18$
   $3x - y \leq 0$
   $x \geq 0, \quad y \geq 0$

59. A private trash collection firm charges $260 for each container it removes. A container weighs 20 pounds and is 5 cubic feet in volume. The firm charges $200 for each barrel it removes. A barrel weighs 20 pounds and is 3 cubic feet in volume. If the truck can carry at most 1000 pounds and 180 cubic feet of material, what combination of containers and barrels will produce the largest possible revenue? Will there be any unused space in the truck?

60. A grass seed mixture contains bluegrass seeds costing 30¢ an ounce and rye seeds costing 15¢ an ounce. What is the cheapest way to obtain a mixture of 500 pounds that is at least 60% bluegrass?

DISCOVERY PROJECT 6   Input-Output Analysis

Wassily Leontief won the Nobel Prize in Economics in 1973 for his method of input-output analysis of the economies of industrialized nations. This method has become a permanent part of production planning and forecasting by both national governments and private corporations. During the Arab oil boycott in 1973, for example, General Electric used input-output analysis on 184 sectors of the economy (such as energy, agriculture, and transportation) to predict the effect of the energy crisis on public demand for its products. The key to Leontief's method is knowing how much each sector of the economy needs from other sectors to do its job.

A simple economic model will illustrate the basic ideas behind input-output analysis. Suppose the country Hypothetica has just two sectors in its economy: agriculture and manufacturing. The production of a ton of agricultural products requires the use of .1 ton of agricultural products and .1 ton of manufactured products. Similarly, the production of a ton of manufactured products consumes .1 ton of agricultural products and .3 ton of manufactured products. The key economic question is: How many tons of each sector must Hypothetica produce to have enough surplus to export 10,000 tons of agricultural products and 10,000 tons of manufactured goods?

Steve Cole/Getty Images

Let A be the total amount of agricultural goods, and let M be the total amount of manufactured goods produced. The total agricultural production A is the sum of the amount needed for producing the agricultural products plus the amount needed for producing manufactured products plus the amount targeted for export. In other words,

Agricultural needs	+	Manufacturing needs	+	Export needs	=	Total agricultural production
$.1A$	+	$.1M$	+	$100,000$	=	$A.$

1. Write a similar equation for manufactured goods.

2. Solve the system of equations given by the agricultural equation above and the manufacturing equation of Problem 1. What level of production of agricultural and manufactured goods should Hypothetica work toward to reach its export goals?

3. The same kind of analysis can be used to determine the surplus available for exports in an economy. Suppose that Hypothetica can produce a total of 120,000 tons of agricultural products and 28,000 tons of manufactured goods. How much of each is available for export?

4. Input-output analysis can also be used for allocating resources in smaller-scale situations. Suppose a horse outfitter is hired by the State of Washington Department of Fish and Wildlife to haul 20 salt blocks into a game area for winter feeding. Each horse can carry a 200-pound load. A single person can handle three horses. Each person requires one horse to ride and one-half horse to carry his or her personal gear. A single horse can carry the feed and equipment for five horses. How many people and horses must go on the trip? [*Note:* Each horse load of salt must consist of an even number of blocks. Each salt block weighs 50 pounds. Also, fractional numbers of horses and people are not allowed. Make sure to adjust your answer to integer values, and be sure that the answer still provides sufficient carrying capacity for the salt.]

CHAPTER 7

Discrete Algebra

Bob Daemmrich/Stock, Boston Inc./PictureQuest

$$\sum_{m=1}^{k} a_m = \frac{k}{2}(a_1 + a_k)$$

$$\sum_{m=1}^{k} a_m = a_1\left(\frac{1-r^k}{1-r}\right)$$

What's next?

CD and CD-ROM players, fax machines, cameras, and other devices incorporate digital technology, which uses sequences of 0's and 1's to send signals. Determining the monthly payment on a car loan involves the sum of a geometric sequence. Problems involving the action of bouncing balls, vacuum pumps, and other devices can sometimes be solved by using sequences. See Exercises 53, 57, and 58 on page 578.

Chapter Outline

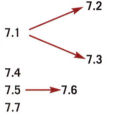
This chapter deals with a variety of subjects involving counting processes and the nonnegative integers 0, 1, 2, 3, ...

7.1 Sequences and Sums

A **sequence** is an ordered list of numbers, such as

$$2, 4, 6, 8, 10, 12, \ldots$$

$$1, -3, 5, -7, 9, -11, 13, \ldots$$

$$1, 0, 1, 0, 1, 0, 1, 0, \ldots$$

$$2, 1, \frac{2}{3}, \frac{3}{4}, \frac{4}{5}, \frac{5}{6}, \frac{6}{7}, \ldots,$$

where the dots indicate that the same pattern continues forever.[*] Each number on the list is called a **term** of the sequence.

$$
\begin{array}{cccccc}
2, & 1, & 0, & 1, & 2, & 3, \quad 2, 1, 0, 1, 2, 3, 2, \ldots . \\
\uparrow & \uparrow & \uparrow & \uparrow & \uparrow & \uparrow \\
\text{1st} & \text{2nd} & \text{3rd} & \text{4th} & \text{5th} & \text{6th} \\
\text{term} & \text{term} & \text{term} & \text{term} & \text{term} & \text{term}
\end{array}
$$

When the pattern isn't obvious, as in the preceding examples, sequences are usually described in terms of a formula.

---

[*]Such a list defines a function f whose domain is the set of positive integers. The rule is $f(1) =$ first number on the list, $f(2) =$ second number on the list, and so on. Conversely, any function g whose domain is the set of positive integers leads to an ordered list of numbers, namely, $g(1)$, $g(2)$, $g(3)$, ... So a sequence is formally defined to be a function whose domain is the set of positive integers.

EXAMPLE 1

Consider the sequence $a_1, a_2, a_3, \ldots, a_n, \ldots$, where a_n is given by the formula

$$a_n = \frac{n^2 - 3n + 1}{2n + 5}.$$

To find a_1, we substitute $n = 1$ in the formula for a_n; to find a_2, we substitute $n = 2$ in the formula; and so on.

$$a_1 = \frac{1^2 - 3 \cdot 1 + 1}{2 \cdot 1 + 5} = -\frac{1}{7},$$

$$a_2 = \frac{2^2 - 3 \cdot 2 + 1}{2 \cdot 2 + 5} = -\frac{1}{9},$$

$$a_3 = \frac{3^2 - 3 \cdot 3 + 1}{2 \cdot 3 + 5} = \frac{1}{11}.$$

Thus, the sequence begins $-1/7, -1/9, 1/11, \ldots$. The 39th term is

$$a_{39} = \frac{39^2 - 3 \cdot 39 + 1}{2 \cdot 39 + 5} = \frac{1405}{83}. \qquad \blacksquare$$

The subscript notation for sequences is sometimes abbreviated by writing $\{a_n\}$ in place of $a_1, a_2, a_3, \ldots$.

EXAMPLE 2

Find the first three terms, the 41st term, and the 206th term of the sequence

$$\left\{\frac{(-1)^n}{n + 2}\right\}.$$

SOLUTION   The formula is

$$a_n = \frac{(-1)^n}{n + 2}.$$

Substituting $n = 1$, $n = 2$, and $n = 3$ shows that

$$a_1 = \frac{(-1)^1}{1 + 2} = -\frac{1}{3}, \qquad a_2 = \frac{(-1)^2}{2 + 2} = \frac{1}{4}, \qquad a_3 = \frac{(-1)^3}{3 + 2} = -\frac{1}{5}.$$

Similarly,

$$a_{41} = \frac{(-1)^{41}}{41 + 2} = -\frac{1}{43} \qquad \text{and} \qquad a_{206} = \frac{(-1)^{206}}{206 + 2} = \frac{1}{208}. \qquad \blacksquare$$

EXAMPLE 3

Here are some other sequences whose nth term can be described by a formula.

Sequence	nth Term	First 5 Terms
$a_1, a_2, a_3, \ldots$	$a_n = n^2 + 1$	2, 5, 10, 17, 26
$b_1, b_2, b_3, \ldots$	$b_n = \dfrac{1}{n}$	$1, \dfrac{1}{2}, \dfrac{1}{3}, \dfrac{1}{4}, \dfrac{1}{5}$
$c_1, c_2, c_3, \ldots$	$c_n = \dfrac{(-1)^{n+1}2n}{(n+1)(n+2)}$	$\dfrac{1}{3}, -\dfrac{1}{3}, \dfrac{3}{10}, -\dfrac{4}{15}, \dfrac{5}{21}$
$x_1, x_2, x_3, \ldots$	$x_n = 3 + \dfrac{1}{10^n}$	3.1, 3.01, 3.001, 3.0001, 3.00001
$a_1, a_2, a_3, \ldots$	$a_n = 7$	7, 7, 7, 7, 7 ■

A sequence in which every term is the same, such as the last one in Example 3, is called a **constant sequence.** A calculator is often useful for computing and displaying the terms of more complicated sequences.

EXAMPLE 4

(a) Display the first five terms of the sequence $\{a_n = n^2 - n - 3\}$ on your calculator screen.

(b) Display the first, fifth, ninth, and thirteenth terms of this sequence.

SOLUTION

Method 1: Enter the sequence in the function memory as $y_1 = x^2 - x - 3$. In the table set-up screen, begin the table at $x = 1$, set the increment at 1, and display a table of values (Figure 7–1). To display the first, fifth, ninth, and thirteenth terms, set the table increment at 4 (Figure 7–2).

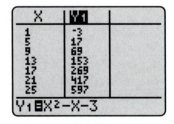

Figure 7–1   Figure 7–2

Method 2: Using the Technology Tip in the margin, enter the following, which produces Figure 7–3 on the next page.

$$\text{SEQ}(x^2 - x - 3, x, 1, 5, 1).$$

To display the first, fifth, ninth, and thirteenth terms, enter

$$\text{SEQ}(x^2 - x - 3, x, 1, 13, 4),$$

which tells the calculator to look at every fourth term from 1 to 13 and produces Figure 7–4.

Figure 7–3

Figure 7–4

Method 3: If possible, put your calculator in sequence graphing mode (see the Technology Tip in the margin). Enter the formula in the equation memory (Figure 7–5). Now you can either construct a table (as in *Method 1*) or graph the sequence and use the trace feature to determine its terms, as in Figure 7–6.* ■

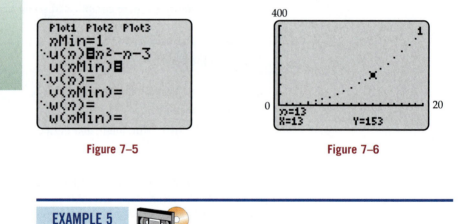

Figure 7–5

Figure 7–6

EXAMPLE 5

Display the first 10 terms of the sequence $\left\{\dfrac{n}{n+1}\right\}$ on your calculator screen in fractional form, if possible.

SOLUTION Creating a table always produces decimal approximations, as you can easily verify. The same is usually true of the SEQ key, unless you take special steps. On HP-39+, change the number format mode to "fraction" (MODE menu). On TI calculators (other than TI-89), either use the FRAC key after obtaining the sequence (Figure 7–7) or use parentheses and the FRAC key as part of the function. Entering

$$\text{SEQ}(x/(x+1) \blacktriangleright \text{FRAC}, x, 1, 10, 1)$$

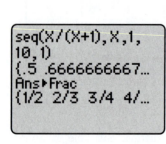

Figure 7–7

*Set TI calculators for DOT instead of CONNECTED graphing in the MODE menu. This is not necessary on Casio and not available on HP-39+ when it is in sequence mode.

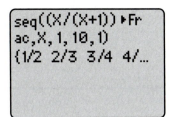

Figure 7–8

produces Figure 7–8. In each figure, you must use the arrow key to scroll to the right to see all the terms.   ∎

A sequence is said to be defined **recursively** (or **inductively**) if the first term is given (or the first several terms) and there is a method of determining the nth term by using the terms that precede it.

EXAMPLE 6

Consider the sequence whose first two terms are

$$a_1 = 1 \qquad \text{and} \qquad a_2 = 1$$

and whose nth term (for $n \geq 3$) is the sum of the two preceding terms.

$$a_3 = a_2 + a_1 = 1 + 1 = 2$$

$$a_4 = a_3 + a_2 = 2 + 1 = 3$$

$$a_5 = a_4 + a_3 = 3 + 2 = 5$$

For each integer n, the two preceding integers are $n - 1$ and $n - 2$. So

$$a_n = a_{n-1} + a_{n-2} \qquad (n \geq 3).$$

This sequence 1, 1, 2, 3, 5, 8, 13, . . . is called the **Fibonacci sequence,** and the numbers that appear in it are called **Fibonacci numbers.** Fibonacci numbers have many surprising and interesting properties. See Exercises 76–82 for details.   ∎

EXAMPLE 7

The sequence given by

$$a_1 = -7 \qquad \text{and} \qquad a_n = a_{n-1} + 3 \qquad \text{for } n \geq 2$$

is defined recursively. Its first three terms are

$$a_1 = -7$$

$$a_2 = a_1 + 3 = -7 + 3 = -4$$

$$a_3 = a_2 + 3 = -4 + 3 = -1 \qquad ∎$$

Entering a recursively defined sequence in a calculator (in sequence mode) might require the use of special keys; check your instruction manual. Figure 7–9 shows the Fibonacci sequence in the function memory of a TI-83+; the entry "u(nMin) = {1, 1}" indicates that the first two terms of the sequence are 1, 1. On other calculators, these terms are entered directly as a_1 and a_2, either in the function memory (HP-39+), or in the RANG menu (Casio).

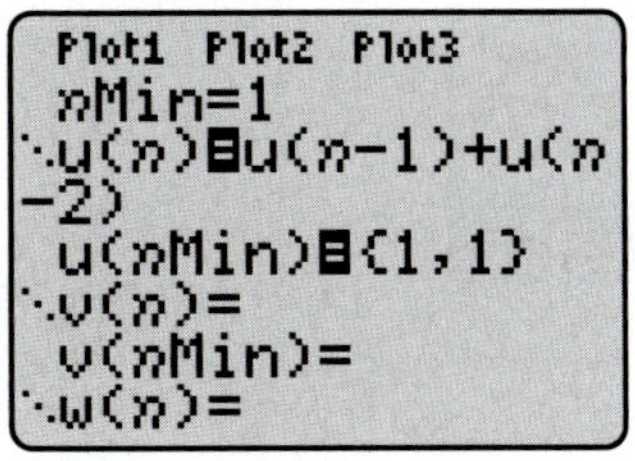

Figure 7–9

Sometimes, it is convenient or more natural to begin numbering the terms of a sequence with a number other than 1. So we may consider sequences such as

$$b_4, b_5, b_6, \ldots \quad \text{or} \quad c_0, c_1, c_2, \ldots.$$

EXAMPLE 8

The sequence $4, 5, 6, 7, \ldots$ can be conveniently described by saying $b_n = n$, with $n \geq 4$. In the brackets notation, we write $\{n\}_{n \geq 4}$. Similarly, the sequence

$$2^0, 2^1, 2^2, 2^3, \ldots$$

may be described as $\{2^n\}_{n \geq 0}$ or by saying $c_n = 2^n$, with $n \geq 0$. ■

EXAMPLE 9

To buy a car, Leslie Lahr borrows \$14,000 at 7% annual interest. Her monthly payment is \$277.22 for 60 months.

(a) Find a formula for a recursively defined sequence $\{u_n\}$ such that u_n is the balance due on the loan after the nth payment.

(b) Find the balance after 30 months.

SOLUTION   With car loans and home mortgages, interest is computed monthly on the unpaid balance. The monthly interest rate is understood to be one-twelfth of the annual rate, that is, $.07/12$.

(a) Let $u_0 = 14,000$, the balance after 0 payments. When the first payment is due, the loan balance is

$$\$14,000 + \text{one month's interest} = \$14,000 + \frac{.07}{12}(14,000).$$

Subtracting the first loan payment, gives the balance after one month (rounded to the nearest penny).

$$u_1 = \$14,000 + \frac{.07}{12}(14,000) - 277.22 = \$13,804.45.$$

Similarly,

$$u_n = u_{n-1} + \frac{.07}{12}u_{n-1} - 277.22,$$

Balance after n payments   Balance after $n-1$ payments   Interest on u_{n-1} nth payment

which can be written as

$$u_n = \left(1 + \frac{.07}{12}\right)u_{n-1} - 277.22.$$

(b) Using the table feature of a calculator in sequence mode, we find that $u_{30} = \$7609.07$, as shown in Figure 7–10. ■

n	$u(n)$
24	8978
25	8753.1
26	8527
27	8299.5
28	8070.7
29	7840.6
30	7609.1

$u(n)=7609.070348$

Figure 7–10

▉▉ SUMMATION NOTATION

It is sometimes necessary to find the sum of various terms in a sequence. For instance, we might want to find the sum of the first nine terms of the sequence $\{a_n\}$. Mathematicians often use the Greek letter sigma (Σ) to abbreviate such a sum:[*]

$$\sum_{k=1}^{9} a_k = a_1 + a_2 + a_3 + a_4 + a_5 + a_6 + a_7 + a_8 + a_9.$$

Similarly, for any positive integer m and numbers $c_1, c_2, \ldots, c_m$, we have the following.

Summation Notation

$$\sum_{k=1}^{m} c_k \quad \text{means} \quad c_1 + c_2 + c_3 + \cdots + c_m.$$

EXAMPLE 10

Compute each of these sums.

(a) $\displaystyle\sum_{k=1}^{5} k^2$ (b) $\displaystyle\sum_{k=1}^{4} k^2(k-2)$ (c) $\displaystyle\sum_{k=1}^{6} (-1)^k k$

SOLUTION

(a) We successively substitute 1, 2, 3, 4, 5 for k in the expression k^2 and add up the results.

$$\sum_{k=1}^{5} k^2 = 1^2 + 2^2 + 3^2 + 4^2 + 5^2 = 55$$

(b) Successively substituting 1, 2, 3, 4 for k in $k^2(k-2)$ and adding the results, we have

$$\sum_{k=1}^{4} k^2(k-2) = 1^2(1-2) + 2^2(2-2) + 3^2(3-2) + 4^2(4-2)$$
$$= 1(-1) + 4(0) + 9(1) + 16(2) = 40.$$

(c) $\displaystyle\sum_{k=1}^{6} (-1)^k k =$

$$(-1)^1 \cdot 1 + (-1)^2 \cdot 2 + (-1)^3 \cdot 3 + (-1)^4 \cdot 4 + (-1)^5 \cdot 5 + (-1)^6 \cdot 6$$
$$= -1 + 2 - 3 + 4 - 5 + 6 = 3 \quad ■$$

In sums such as $\displaystyle\sum_{k=1}^{5} k^2$ and $\displaystyle\sum_{k=1}^{6} (-1)^k k$, the letter k is called the **summation index.** Any letter may be used for the summation index, just as the rule of a function f may be denoted by $f(x)$ or $f(t)$ or $f(k)$. For example, $\displaystyle\sum_{n=1}^{5} n^2$ means "Take the sum of the terms n^2 as n takes values from 1 to 5." In other words,

$$\sum_{n=1}^{5} n^2 = \sum_{k=1}^{5} k^2.$$

---

[*]Σ is the letter S in the Greek alphabet, the first letter in *Sum*.

Similarly,

$$\sum_{k=1}^{4} k^2(k-2) = \sum_{j=1}^{4} j^2(j-2) = \sum_{n=1}^{4} n^2(n-2).$$

The Σ notation for sums can also be used for sums that don't begin with $k=1$. For instance,

$$\sum_{k=4}^{10} k^2 = 4^2 + 5^2 + 6^2 + 7^2 + 8^2 + 9^2 + 10^2 = 371$$

$$\sum_{j=0}^{3} j^2(2j+5) = 0^2(2\cdot0+5) + 1^2(2\cdot1+5) + 2^2(2\cdot2+5) + 3^2(2\cdot3+5) = 142.$$

EXAMPLE 11

Use a calculator to compute these sums.

(a) $\displaystyle\sum_{k=1}^{50} k^2$

(b) $\displaystyle\sum_{k=38}^{75} k^2$

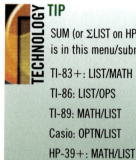

TIP

SUM (or ΣLIST on HP-39+) is in this menu/submenu:

TI-83+: LIST/MATH

TI-86: LIST/OPS

TI-89: MATH/LIST

Casio: OPTN/LIST

HP-39+: MATH/LIST

SOLUTION   In each case, use SUM together with SEQ (or ΣLIST and MAKELIST on HP-39+), with the same syntax for SEQ as in Example 4:

$$\sum_{k=1}^{50} k^2 = \text{SUM SEQ}(x^2, x, 1, 50, 1) = 42{,}925$$

and

$$\sum_{k=38}^{75} k^2 = \text{SUM SEQ}(x^2, x, 38, 75, 1) = 125{,}875,$$

as shown in Figure 7–11.   ■

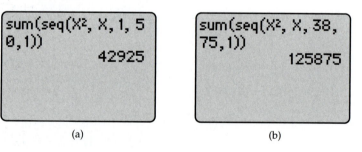

| (a) | (b) |

Figure 7–11

■■ PARTIAL SUMS

Suppose $\{a_n\}$ is a sequence and k is a positive integer. The sum of the first k terms of the sequence is called the **kth partial sum** of the sequence. Thus, we have the following.

Partial Sums

> The **kth partial sum** of $\{a_n\}$ is $\displaystyle\sum_{n=1}^{k} a_n = a_1 + a_2 + a_3 + \cdots + a_k.$

EXAMPLE 12

Here are some partial sums of the sequence $\{n^3\}$:

First partial sum: $\quad \displaystyle\sum_{n=1}^{1} n^3 = 1^3 = 1$

Second partial sum: $\quad \displaystyle\sum_{n=1}^{2} n^3 = 1^3 + 2^3 = 9$

Sixth partial sum: $\quad \displaystyle\sum_{n=1}^{6} n^3 = 1^3 + 2^3 + 3^3 + 4^3 + 5^3 + 6^3 = 441$ ■

EXAMPLE 13

The sequence $\{2^n\}_{n\geq 0}$ begins with the 0th term, so the fourth partial sum (the sum of the first four terms) is

$$2^0 + 2^1 + 2^2 + 2^3 = \sum_{n=0}^{3} 2^n.$$

Similarly, the fifth partial sum of the sequence $\left\{\dfrac{1}{n(n-2)}\right\}_{n\geq 3}$ is the sum of the first five terms.

$$\frac{1}{3(3-2)} + \frac{1}{4(4-2)} + \frac{1}{5(5-2)} + \frac{1}{6(6-2)} + \frac{1}{7(7-2)} = \sum_{n=3}^{7} \frac{1}{n(n-2)}.$$ ■

Certain calculations can be written very compactly in summation notation. For example, the distributive law shows that

$$ca_1 + ca_2 + ca_3 + \cdots + ca_r = c(a_1 + a_2 + a_3 + \cdots + a_r).$$

In summation notation, this becomes

$$\sum_{n=1}^{r} ca_n = c\left(\sum_{n=1}^{r} a_n\right).$$

This proves the first of the following statements.

Properties of Sums

1. $\displaystyle\sum_{n=1}^{r} ca_n = c\left(\sum_{n=1}^{r} a_n\right)$ for any number c.

2. $\displaystyle\sum_{n=1}^{r} (a_n + b_n) = \sum_{n=1}^{r} a_n + \sum_{n=1}^{r} b_n.$

3. $\displaystyle\sum_{n=1}^{r} (a_n - b_n) = \sum_{n=1}^{r} a_n - \sum_{n=1}^{r} b_n.$

To prove statement 2, use the commutative and associative laws repeatedly to show that

$$(a_1 + b_1) + (a_2 + b_2) + (a_3 + b_3) + \cdots + (a_r + b_r)$$
$$= (a_1 + a_2 + a_3 + \cdots + a_r) + (b_1 + b_2 + b_3 + \cdots + b_r),$$

which can be written in summation notation as

$$\sum_{n=1}^{r} (a_n + b_n) = \sum_{n=1}^{r} a_n + \sum_{n=1}^{r} b_n.$$

The last statement is proved similarly.

EXERCISES 7.1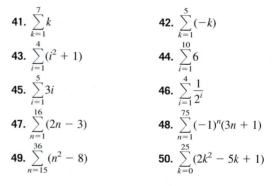

In Exercises 1–14, find the first five terms of the sequence $\{a_n\}$.

1. $a_n = 2n + 6$ **2.** $a_n = 2^n - 7$

3. $a_n = \dfrac{1}{n^3}$ **4.** $a_n = \dfrac{1}{(n + 3)(n + 1)}$

5. $a_n = \dfrac{n}{2^n}$ **6.** $a_n = \sqrt{n^2 + 1}$

7. $a_n = (-1)^n \sqrt{n + 2}$ **8.** $a_n = (-1)^{n+1} n(n - 1)$

9. $a_n = 4 + (-.1)^n$ **10.** $a_n = 5 - (.1)^n$

11. $a_n = (-1)^n + 3n$ **12.** $a_n = (-1)^{n+2} - (n + 1)$

13. a_n is the nth digit in the decimal expansion of π.

14. a_n is the nth digit in the decimal expansion of $1/13$.

In Exercises 15–24, find a formula for the nth term of the sequence whose first few terms are given.

15. $-1, 1, -1, 1, -1, 1, \ldots$

16. $2, -2, 2, -2, 2, -2, \ldots$

17. $\dfrac{1}{2}, \dfrac{2}{3}, \dfrac{3}{4}, \dfrac{4}{5}, \dfrac{5}{6}, \ldots$

18. $\dfrac{1}{2 \cdot 3}, \dfrac{1}{3 \cdot 4}, \dfrac{1}{4 \cdot 5}, \dfrac{1}{5 \cdot 6}, \dfrac{1}{6 \cdot 7}, \ldots$

19. $2, 7, 12, 17, 22, 27, \ldots$

20. $8, 5, 2, -1, -4, \ldots$

21. $3, 6, 12, 24, 48, \ldots$

22. $-\dfrac{1}{8}, -\dfrac{1}{2}, -2, -8, -32, \ldots$

23. $4, \sqrt{32}, \sqrt{48}, 8, \sqrt{80}, \ldots$

24. $8, -5, 2, -11, -4, -17, -10, \ldots$

In Exercises 25–34, find the first five terms of the recursively defined sequence.

25. $a_1 = 4$ and $a_n = 2a_{n-1} + 3$ for $n \geq 2$

26. $a_1 = 0$ and $a_n = 3a_{n-1} - 2$ for $n \geq 2$

27. $a_1 = -16$ and $a_n = \dfrac{a_{n-1}}{2}$ for $n \geq 2$

28. $a_1 = 3$ and $a_n = n + 2a_{n-1}$ for $n \geq 2$

29. $a_1 = 2$ and $a_n = n - a_{n-1}$ for $n \geq 2$

30. $a_1 = -3$ and $a_n = (-1)^n 4 a_{n-1} - 5$ for $n \geq 2$

31. $a_1 = 1, a_2 = -2, a_3 = 3$,   and
$a_n = a_{n-1} + a_{n-2} + a_{n-3}$ for $n \geq 4$

32. $a_1 = 1, a_2 = 3$, and $a_n = 2a_{n-1} + 3a_{n-2}$ for $n \geq 3$

33. $a_0 = 2, a_1 = 3$, and $a_n = (a_{n-1})\left(\dfrac{1}{2} a_{n-2}\right)$ for $n \geq 2$

34. $a_0 = 1, a_1 = 1$, and $a_n = na_{n-1}$ for $n \geq 2$

In Exercises 35–40, express the sum in Σ notation.

35. $1 + 2 + 3 + 4 + 5 + 6 + 7 + 8 + 9 + 10 + 11$

36. $1^1 + 2^2 + 3^3 + 4^4 + 5^5$

37. $\dfrac{1}{2^7} + \dfrac{1}{2^8} + \dfrac{1}{2^9} + \dfrac{1}{2^{10}} + \dfrac{1}{2^{11}} + \dfrac{1}{2^{12}} + \dfrac{1}{2^{13}}$

38. $(-6)^{11} + (-6)^{12} + (-6)^{13} + (-6)^{14} + (-6)^{15}$

39. $2 + \dfrac{2^2}{2} + \dfrac{2^3}{3} + \dfrac{2^4}{4} + \dfrac{2^5}{5} + \dfrac{2^6}{6} + \dfrac{2^7}{7}$

40. $2 + \dfrac{3}{2} + \dfrac{4}{3} + \dfrac{5}{4} + \dfrac{6}{5} + \dfrac{7}{6} + \dfrac{8}{7} + \dfrac{9}{8} + \dfrac{10}{9}$

In Exercises 41–50, find the sum.

41. $\displaystyle\sum_{k=1}^{7} k$ **42.** $\displaystyle\sum_{k=1}^{5} (-k)$

43. $\displaystyle\sum_{i=1}^{4} (i^2 + 1)$ **44.** $\displaystyle\sum_{i=1}^{10} 6$

45. $\displaystyle\sum_{i=1}^{5} 3i$ **46.** $\displaystyle\sum_{i=1}^{4} \dfrac{1}{2^i}$

47. $\displaystyle\sum_{n=1}^{16} (2n - 3)$ **48.** $\displaystyle\sum_{n=1}^{75} (-1)^n (3n + 1)$

49. $\displaystyle\sum_{n=15}^{36} (n^2 - 8)$ **50.** $\displaystyle\sum_{k=0}^{25} (2k^2 - 5k + 1)$

In Exercises 51–54, find the third and the sixth partial sums of the sequence.

51. $\{n^2 - 5n + 2\}$

52. $\{(2n - 3n^2)^2\}$

53. $\{(-1)^{n+1}5\}$

54. $\{2^n(2 - n^2)\}_{n \geq 0}$

In Exercises 55–58, express the given sum in Σ notation and find the sum.

55. $\dfrac{1}{3} + \dfrac{1}{5} + \dfrac{1}{7} + \dfrac{1}{9} + \dfrac{1}{11} + \dfrac{1}{13}$

56. $2 + 1 + \dfrac{4}{5} + \dfrac{5}{7} + \dfrac{2}{3} + \dfrac{7}{11} + \dfrac{8}{13}$

57. $\dfrac{1}{8} - \dfrac{2}{9} + \dfrac{3}{10} - \dfrac{4}{11} + \dfrac{5}{12}$

58. $\dfrac{2}{3 \cdot 5} + \dfrac{4}{5 \cdot 7} + \dfrac{8}{7 \cdot 9} + \dfrac{16}{9 \cdot 11}$

$+ \dfrac{32}{11 \cdot 13} + \dfrac{64}{13 \cdot 15} + \dfrac{128}{15 \cdot 17}$

In Exercises 59–64, use a calculator to approximate the required term or sum.

59. a_{12} where $a_n = \left(1 + \dfrac{1}{n}\right)^n$

60. a_{50} where $a_n = \dfrac{\ln n}{n^2}$

61. a_{102} where $a_n = \dfrac{n^3 - n^2 + 5n}{3n^2 + 2n - 1}$

62. a_{125} where $a_n = \sqrt[n]{n}$

63. $\displaystyle\sum_{k=1}^{14} \dfrac{1}{k^2}$

64. $\displaystyle\sum_{n=8}^{22} \dfrac{1}{n}$

In Exercises 65–66, use a recursively defined sequence, as in Example 9.

65. Suppose you take out a car loan for $9500 at 5% annual interest for 48 months. Your monthly payment is $218.78. How much do you owe after one year (12 payments)? After three years?

66. Lisa Chow is buying a condo. She takes out an $80,000 mortgage for 30 years at 6% annual interest. Her monthly payment is $479.64.
(a) How much does she owe after one year (12 payments)?
(b) How much interest does Lisa pay during the first five years? [*Hint:* The interest paid is the difference between her total payments and the amount of the loan paid off after five years

($80,000 − remaining balance).]

(c) After 15 years, Lisa sells the condo and pays off the remaining mortgage balance. How much does she pay?

67. The number of hours that the average person spends listening to the radio in year n is approximated by the sequence $\{a_n\}$, where $n = 1$ corresponds to 1998 and $a_n = 937.84 + 37.55 \ln n.$[*]
(a) For how many hours did the average person listen to the radio in 2000? In 2003?
(b) How much total time did the average person spend listening from 1998 to 2003 (inclusive)?

68. The budget of the National Institutes of Health (in billions of dollars) in year n is approximated by the sequence $\{b_n\}$, where $n = 0$ corresponds to 1998 and $b_n = 2.72n + 12.96.$[†]
(a) What was the NIH budget in 2002?
(b) Assuming that NIH kept to its budget and spent the entire amount each year, how much did NIH spend from 1998 to 2002?

69. Consumer spending per person (in dollars) on home video games each year is approximated by the sequence $\{c_n\}$, where $n = 0$ is 1995 and $c_n = .42n^2 + 1.528n + 10.396.$[*]
(a) What is the per person spending in 2002? In 2004?
(b) What was the total amount spent per person on home video games from 1995 to 2003?

70. Book sales (in billions of dollars) in the United States are approximated by the sequence $\{d_n\}$, where $n = 0$ corresponds to 1990 and $d_n = 1.26n + 18.83.$[‡]
(a) What were the sales in 2000?
(b) What was the total spent on books from 1990 to 2000?

Thinkers

*Exercises 71–75, deal with prime numbers. A positive integer greater than 1 is **prime** if its only positive integer factors are itself and 1. For example, 7 is prime because its only factors are 7 and 1, but 15 is not prime because it has factors other than 15 and 1 (namely, 3 and 5).*

71. (a) Let $\{a_n\}$ be the sequence of prime integers in their usual ordering. Verify that the first ten terms are 2, 3, 5, 7, 11, 13, 17, 19, 23, 29.
(b) Find $a_{17}, a_{18}, a_{19}, a_{20}$.

In Exercises 72–75, find the first five terms of the sequence.

72. a_n is the nth prime integer larger than 10. [*Hint:* $a_1 = 11$.]

73. a_n is the square of the nth prime integer.

74. a_n is the number of prime integers less than n.

75. a_n is the largest prime integer less than $5n$.

_____

*Based on data and projections from the *Statistical Abstract of the United States: 2001*.

†Federal budget.

‡Book Industry Study Group, Inc.

Exercises 76–82, deal with the Fibonacci sequence $\{a_n\}$ that was discussed in Example 6.

76. Leonardo Fibonacci discovered the sequence in the thirteenth century in connection with this problem: A rabbit colony begins with one pair of adult rabbits (one male, one female). Each adult pair produces one pair of babies (one male, one female) every month. Each pair of baby rabbits becomes adult and produces the first offspring at age two months. Assuming that no rabbits die, how many adult pairs of rabbits are in the colony at the end of n months ($n = 1, 2, 3, \ldots$)? [*Hint:* It may be helpful to make up a chart listing for each month the number of adult pairs, the number of one-month-old pairs, and the number of baby pairs.]

77. (a) List the first 10 terms of the Fibonacci sequence.
(b) List the first 10 partial sums of the sequence.
(c) Do the partial sums follow an identifiable pattern?

78. Verify that every positive integer less than or equal to 15 can be written as a sum of Fibonacci numbers, with none used more than once.

79. Verify that $5(a_n)^2 + 4(-1)^n$ is always a perfect square for $n = 1, 2, \ldots, 10$.

80. Verify that $(a_n)^2 = a_{n+1}a_{n-1} + (-1)^{n-1}$ for $n = 2, \ldots, 10$.

81. Show that $\sum\limits_{n=1}^{k} a_n = a_{k+2} - 1$. [*Hint:* $a_1 = a_3 - a_2$; $a_2 = a_4 - a_3$; etc.]

82. Show that $\sum\limits_{n=1}^{k} a_{2n-1} = a_{2k}$, that is, the sum of the first k odd-numbered terms is the kth even-numbered term. [*Hint:* $a_3 = a_4 - a_2$; $a_5 = a_6 - a_4$; etc.]

7.2   Arithmetic Sequences

An **arithmetic sequence** (sometimes called an **arithmetic progression**) is a sequence in which the difference between each term and the preceding one is always the same constant.

EXAMPLE 1

In the sequence 3, 8, 13, 18, 23, 28, $\ldots$, the difference between each term and the preceding one is always 5. So this is an arithmetic sequence.   ■

EXAMPLE 2

Which of the following sequences are arithmetic?

(a) $14, 10, 6, 2, -2, -6, -10, -14, \ldots$
(b) $7, 7^2, 7^3, 7^4, 7^5, \ldots$
(c) $\log 5, \log 5^2, \log 5^3, \log 5^4, \log 5^5, \ldots$

SOLUTION

(a) The difference between each term and the preceding one is -4.

$$10 - 14 = -4 \qquad -2 - 2 = -4,$$
$$6 - 10 = -4 \qquad -6 - (-2) = -4,$$
$$2 - 6 = -4 \qquad -10 - (-6) = -4,$$

and so on. Hence, the sequence is arithmetic.

(b) Note that

$$7^2 - 7 = 49 - 7 = 42, \qquad \text{but} \qquad 7^3 - 7^2 = 343 - 49 = 294.$$

Since the difference between consecutive terms is not always the same, this sequence is not arithmetic.

(c) Using the Power Law for logarithms (Section 5.4), we see that

$$\log 5^2 - \log 5 = 2 \log 5 - \log 5 = \log 5$$
$$\log 5^3 - \log 5^2 = 3 \log 5 - 2 \log 5 = \log 5$$
$$\log 5^4 - \log 5^3 = 4 \log 5 - 3 \log 5 = \log 5,$$

and so on. The difference between each term and the preceding one is always $\log 5 \approx .69897$, so the sequence is arithmetic.   ■

If $\{a_n\}$ is an arithmetic sequence, then for each $n \geq 2$, the term preceding a_n is a_{n-1}, and the difference $a_n - a_{n-1}$ is some constant—call it d. Therefore, $a_n - a_{n-1} = d$ or, equivalently,

Arithmetic Sequences

> In an arithmetic sequence $\{a_n\}$,
>
> $$a_n = a_{n-1} + d$$
>
> for some constant d and all $n \geq 2$.

The number d is called the **common difference** of the arithmetic sequence.

EXAMPLE 3

If $\{a_n\}$ is an arithmetic sequence with $a_1 = 3$ and $a_2 = 4.5$, find the common difference and list the first eight terms.

SOLUTION   The common difference is

$$d = a_2 - a_1 = 4.5 - 3 = 1.5.$$

Therefore, the sequence begins 3, 4.5, 6, 7.5, 9, 10.5, 12, 13.5, ...   ■

EXAMPLE 4

Show that the sequence $\{-7 + 4n\}$ is arithmetic and find the common difference.

SOLUTION   The sequence is arithmetic because for each $n \geq 2$,

$$a_n - a_{n-1} = (-7 + 4n) - [-7 + 4(n - 1)]$$
$$= (-7 + 4n) - (-7 + 4n - 4)$$
$$= -7 + 4n + 7 - 4n + 4 = 4.$$

Therefore, the common difference is $d = 4$.   ■

If $\{a_n\}$ is an arithmetic sequence with common difference d, then for each $n \geq 2$, we know that $a_n = a_{n-1} + d$. Applying this fact repeatedly shows that

$$a_2 = a_1 + d$$
$$a_3 = a_2 + d = (a_1 + d) + d = a_1 + 2d$$
$$a_4 = a_3 + d = (a_1 + 2d) + d = a_1 + 3d$$
$$a_5 = a_4 + d = (a_1 + 3d) + d = a_1 + 4d,$$

and in general, we have the following.

nth Term of an Arithmetic Sequence

In an arithmetic sequence $\{a_n\}$ with common difference d,

$$a_n = a_1 + (n - 1)d \qquad \text{for every } n \geq 1.$$

EXAMPLE 5

Find the nth term of the arithmetic sequence with first term -5 and common difference 3.

SOLUTION   Since $a_1 = -5$ and $d = 3$, the formula in the preceding box shows that

$$a_n = a_1 + (n - 1)d = -5 + (n - 1)3 = -5 + 3n - 3 = 3n - 8. \qquad \blacksquare$$

EXAMPLE 6

What is the 45th term of the arithmetic sequence whose first three terms are 5, 9, and 13?

SOLUTION   The first three terms show that $a_1 = 5$ and that the common difference d is 4. Applying the formula in the box with $n = 45$, we have

$$a_{45} = a_1 + (45 - 1)d = 5 + (44)4 = 181. \qquad \blacksquare$$

EXAMPLE 7

If $\{a_n\}$ is an arithmetic sequence with $a_6 = 57$ and $a_{10} = 93$, find a_1 and a formula for a_n.

SOLUTION   Apply the formula $a_n = a_1 + (n - 1)d$ with $n = 6$ and $n = 10$.

$$a_6 = a_1 + (6 - 1)d \qquad \text{and} \qquad a_{10} = a_1 + (10 - 1)d$$
$$57 = a_1 + 5d \qquad\qquad\qquad 93 = a_1 + 9d$$
$$a_1 = 57 - 5d \qquad\qquad\qquad a_1 = 93 - 9d$$

These two expressions for a_1 must be equal.

$$57 - 5d = 93 - 9d$$

$$4d = 36$$

$$d = 9$$

Substituting $d = 9$ in either of the equations above shows that $a_1 = 12$.

$$a_1 = 57 - 5(9) = 12 \qquad \text{or} \qquad a_1 = 93 - 9(9) = 12$$

So the formula for a_n is

$$a_n = a_1 + (n - 1)d = 12 + (n - 1)9 = 12 + 9n - 9 = 9n + 3. \qquad \blacksquare$$

▦ PARTIAL SUMS

It's easy to compute partial sums of arithmetic sequences by using the following formulas.

Partial Sums of an Arithmetic Sequence

If $\{a_n\}$ is an arithmetic sequence with common difference d, then for each positive integer k, the kth partial sum can be found by using *either* of these formulas.

1. $\displaystyle\sum_{n=1}^{k} a_n = \frac{k}{2}(a_1 + a_k)$ or

2. $\displaystyle\sum_{n=1}^{k} a_n = ka_1 + \frac{k(k - 1)}{2}d$

Proof   Let S denote the kth partial sum $a_1 + a_2 + \cdots + a_k$. For reasons that will become apparent later, we shall calculate the number $2S$.

$$2S = S + S = (a_1 + a_2 + \cdots + a_k) + (a_1 + a_2 + \cdots + a_k).$$

Now we rearrange the terms on the right by grouping the first and last terms together, then the first and last of the remaining terms, and so on.

$$2S = (a_1 + a_k) + (a_2 + a_{k-1}) + (a_3 + 2_{k-2}) + \cdots + (a_k + a_1).$$

Since adjacent terms of the sequence differ by d, we have

$$a_2 + a_{k-1} = (a_1 + d) + (a_k - d) = a_1 + a_k.$$

Using this fact, we have

$$a_3 + a_{k-2} = (a_2 + d) + (a_{k-1} - d) = a_2 + a_{k-1} = a_1 + a_k.$$

Continuing in this manner, we see that every pair in the sum for $2S$ is equal to $a_1 + a_k$. Therefore,

$$2S = (a_1 + a_k) + (a_2 + a_{k-1}) + (a_3 + a_{k-2}) + \cdots + (a_k + a_1)$$

$$= (a_1 + a_k) + (a_1 + a_k) + (a_1 + a_k) + \cdots + (a_1 + a_k) \qquad (k \text{ terms})$$

$$= k(a_1 + a_k).$$

Dividing both sides of this last equation by 2 shows that $S = \dfrac{k}{2}(a_1 + a_k)$. This proves the first formula. To obtain the second one, note that

$$a_1 + a_k = a_1 + [a_1 + (k-1)d] = 2a_1 + (k-1)d.$$

Substituting the right side of this equation in the first formula for S shows that

$$S = \frac{k}{2}(a_1 + a_k) = \frac{k}{2}[2a_1 + (k-1)d] = ka_1 + \frac{k(k-1)}{2}d.$$

This proves the second formula.   ■

EXAMPLE 8

Find the 12th partial sum of the arithmetic sequence that begins $-8, -3, 2, 7, \ldots$.

SOLUTION   We first note that the common difference d is 5. Since $a_1 = -8$ and $d = 5$, the second formula in the box with $k = 12$ shows that

$$\sum_{n=1}^{12} a_n = 12(-8) + \frac{12(11)}{2}5 = -96 + 330 = 234.   ■$$

EXAMPLE 9

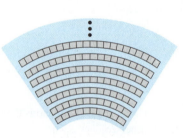

Figure 7–12

A corner section of a stadium has 28 rows of seats. There are 10 seats in the first row, 12 in the second row, 14 in the third row, and so on, as indicated in Figure 7–12.

(a) How many seats are in the 20th row?

(b) How many seats are in the entire section?

SOLUTION

(a) Let a_n be the number of seats in row n. Then $a_1 = 10$, $a_2 = 12$, $a_3 = 14$, and so on. Thus, we have an arithmetic sequence, with common difference $d = 2$. According to the box before Example 5, the nth term of the sequence is

$$a_n = a_1 + (n-1)d$$

$$a_n = 10 + (n-1)2 = 8 + 2n.$$

The number of seats in the 20th row is

$$a_{20} = 8 + 2(20) = 48.$$

(b) The total number of seats in all 28 rows is the 28th the partial sum of the sequence $\{a_n\}$, which is given by the second formula in the box on page 567.

$$\sum_{n=1}^{k} a_n = ka_1 + \frac{k(k-1)}{2}d.$$

Here $k = 28$, $a_1 = 10$, and $d = 2$, so

$$\sum_{n=1}^{28} a_n = 28(10) + \frac{28(28 - 1)}{2} \cdot 2$$

$$= 280 + 28(27) = 1036 \text{ seats.} \quad \blacksquare$$

EXAMPLE 10

Find the sum of all multiples of 3 from 3 to 333.

SOLUTION   The answer can be found in two ways.

Algebraic Method: This sum is a partial sum of the arithmetic sequence 3, 6, 9, 12, ... Since this sequence can be written in the form

$$3 \cdot 1, 3 \cdot 2, 3 \cdot 3, 3 \cdot 4, 3 \cdot 5, 3 \cdot 6, \ldots,$$

we see that $333 = 3 \cdot 111$ is the 111th term. The 111th partial sum of this sequence can be found by using the first formula in the box on page 567 with $k = 111$, $a_1 = 3$, and $a_{111} = 333$.

$$\sum_{n=1}^{111} a_n = \frac{111}{2}(3 + 333) = \frac{111}{2}(336) = 18{,}648.$$

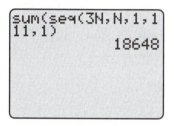

Figure 7–13

Calculator Method: Note that the nth term of the sequence 3, 6, 9, . . . is $3n$. Consequently, we can compute the 111th partial sum as in Figure 7–13. $\blacksquare$

EXAMPLE 11

If the starting salary for a job is $20,000 and you get a $2000 raise at the beginning of each subsequent year, what will your salary be during the tenth year? How much will you earn during the first 10 years?

SOLUTION   Your yearly salary rates form a sequence: 20,000, 22,000, 24,000, 26,000, and so on. It is an arithmetic sequence with $a_1 = 20{,}000$ and $d = 2000$. Your tenth-year salary is

$$a_{10} = a_1 + (10 - 1)d = 20{,}000 + 9 \cdot 2000 = \$38{,}000.$$

Your 10-year total earnings are the tenth partial sum of the sequence.

$$\frac{10}{2}(a_1 + a_{10}) + \frac{10}{2}(20{,}000 + 38{,}000) = 5(58{,}000) = \$290{,}000. \quad \blacksquare$$

✓ EXERCISES 7.2

In Exercises 1–8, determine whether the sequence is arithmetic or not. If it is, find the common difference.

1. $1, 3, 5, 7, 9, \ldots$

2. $\dfrac{1}{3}, \dfrac{2}{3}, \dfrac{3}{3}, \dfrac{4}{3}, \dfrac{5}{3}, \ldots$

3. $1, 2, 4, 8, 16, \ldots$

4. $-9, -6, -3, 0, \ldots$

5. $\log 1, \log 2, \log 4, \log 8, \log 16, \ldots$

6. $\log 3, \log 6, \log 9, \log 12, \log, 15, \ldots$

7. $\dfrac{1}{3}, -\dfrac{4}{6}, -\dfrac{15}{9}, -\dfrac{32}{12}, -\dfrac{55}{15}, -\dfrac{84}{18}, \ldots$

8. $\dfrac{1}{2}, \dfrac{3}{4}, 1, \dfrac{3}{4}, \dfrac{1}{2}, \dfrac{1}{4}, \ldots$

In Exercises 9–16, write the first five terms of the sequence whose nth term is given. Use them to decide whether the sequence is arithmetic. If it is, list the common difference.

9. $a_n = 5 + 4n$

10. $b_n = n - \dfrac{5}{4}$

11. $c_n = (-1)^n$

12. $d_n = 2n + \dfrac{1}{n}$

13. $a_n = 2 + (-1)^n n$

14. $c_n = 1 + \dfrac{n}{3}$

15. $a_n = e^{\ln n}$

16. $b_n = 3\sqrt{7n^2}$

In Exercises 17–24, show that the sequence is arithmetic and find its common difference.

17. $\{3 - 2n\}$

18. $\{1.5 + 1.5n\}$

19. $\left\{4 + \dfrac{n}{3}\right\}$

20. $\left\{-3 - \dfrac{n}{2}\right\}$

21. $\left\{\dfrac{5 + 3n}{2}\right\}$

22. $\left\{\dfrac{\pi - n}{2}\right\}$

23. $\{c + 2n\}$ (*c* constant)

24. $\{2b + 3nc\}$ (*b, c* constants)

In Exercises 25–32, the first term a_1 and the common difference d of an arithmetic sequence are given. Find the fifth term and the formula for the nth term.

25. $a_1 = 5, d = 2$

26. $a_1 = -4, d = 5$

27. $a_1 = 4, d = \dfrac{1}{4}$

28. $a_1 = -6, d = \dfrac{2}{3}$

29. $a_1 = 10, d = -\dfrac{1}{2}$

30. $a_1 = \pi, d = \dfrac{1}{5}$

31. $a_1 = 8, d = .1$

32. $a_1 = -.1, d = -8$

In Exercises 33–40, use the given information about the arithmetic sequence with common difference d to find a_1 and a formula for a_n.

33. $a_4 = 12, d = 2$

34. $a_7 = -8, d = 3$

35. $a_3 = 3, d = 5$

36. $a_4 = -5, d = -5$

37. $a_2 = 4, a_6 = 32$

38. $a_7 = 6, a_{12} = -4$

39. $a_5 = 0, a_9 = 6$

40. $a_5 = -3, a_9 = -18$

In Exercises 41–48, find the kth partial sum of the arithmetic sequence $\{a_n\}$ with common difference d.

41. $k = 6, a_1 = 2, d = 5$

42. $k = 8, a_1 = \dfrac{2}{3}, d = -\dfrac{4}{3}$

43. $k = 7, a_1 = \dfrac{3}{4}, d = -\dfrac{1}{2}$

44. $k = 9, a_1 = -4, d = \dfrac{1}{2}$

45. $k = 6, a_1 = -4, a_6 = 14$

46. $k = 10, a_1 = 0, a_{10} = 30$

47. $k = 9, a_1 = 6, a_9 = -24$

48. $k = 8, a_1 = -6, a_8 = 13$

In Exercises 49–54, find the sum.

49. $\displaystyle\sum_{n=1}^{20} (3n + 4)$

50. $\displaystyle\sum_{n=1}^{25} \left(\dfrac{n}{4} + 5\right)$

51. $\displaystyle\sum_{n=1}^{30} \left(3 - \dfrac{n}{2}\right)$

52. $\displaystyle\sum_{n=1}^{35} \left(\dfrac{2n + 4}{8}\right)$

53. $\displaystyle\sum_{n=1}^{40} \dfrac{n + 3}{6}$

54. $\displaystyle\sum_{n=1}^{30} \dfrac{4 - 6n}{3}$

55. The sequence in which

a_n = per capita amount spent on health care in year n,

with $n = 1$ corresponding to 1981, is approximately arithmetic.[*]

(a) If the per capita amount was \$1223.50 in 1981 and \$2786.80 in 1990, find a formula for a_n.

(b) Use the sequence to estimate the per capita amount in 2003 and 2005.

56. The sequence in which

b_n = remaining life expectancy of a man at age n

is approximately arithmetic.[*]

(a) Use the fact that a man's remaining life expectancy is 59.3 years at age 15 and 55 years at age 20 to find a formula for b_n.

(b) Determine the remaining life expectancy of a man at these ages: 18, 22, 25, and 30.

---

[*]Based on data from the U.S. Department of Health and Human Services.

57. A lecture hall has six seats in the first row, eight in the second, ten in the third, and so on, through row 12. Rows 12 through 20 (the last row) all have the same number of seats. Find the number of seats in the lecture hall.

58. A monument is constructed by first laying a row of 60 bricks at ground level. A second row, with two fewer bricks, is centered on that; a third row, with two fewer bricks, is centered on the second; and so on. The top row contains 10 bricks. How many bricks are there in the monument?

59. A ladder with nine rungs is to be built, with the bottom rung 24 inches wide and the top rung 18 inches wide. If the lengths of the rungs decrease uniformly from bottom to top, how long should each of the seven intermediate rungs be?

60. Find the first eight numbers in an arithmetic sequence in which the sum of the first and seventh term is 40 and the product of the first and fourth terms is 160.

61. Find the sum of all the even integers from 2 to 100.

62. Find the sum of all the integer multiples of 7 from 7 to 700.

63. Find the sum of the first 200 positive integers.

64. Find the sum of the positive integers from 101 to 200 (inclusive). [*Hint:* What's the sum from 1 to 100? Use it and Exercise 63.]

65. A business makes a $10,000 profit during its first year. If the yearly profit increases by $7500 in each subsequent year, what will the profit be in the tenth year and what will the total profit for the first 10 years be?

66. If a man's starting salary is $15,000 and he receives a $1000 increase every six months, what will his salary be during the last six months of the sixth year? How much will he earn during the first six years?

67. The sequence with

$$a_n = \begin{array}{l} \text{holiday retail sales in year } n \\ \text{(in billions of dollars)}, \end{array}$$

with $n = 1$ corresponding to 1994, is approximately arithmetic.[*]
(a) If sales were $136.4 billion in 1994 and $217.4 billion in 2003, find the formula for a_n.
(b) Estimate holiday sales in 2005.
(c) Find the holiday sales from 2000 to 2005.

68. Let $\{b_n\}$ be the sequence that gives U.S. personal income in year n (in trillions of dollars), where $n = 1$ corresponds to 1990. This sequence is approximately arithmetic.[†]
(a) If personal income was $4.9 trillion in 1990 and $8.3 trillion in 2000, find a formula for b_n.
(b) Estimate personal income in 2002.
(c) Find the total personal income from 1998 to 2004.

69. Let $n = 1$ correspond to 1981. The arithmetic sequence in which

$$c_n = \begin{array}{l} \text{average yearly earnings of a production} \\ \text{worker in manufacturing in year } n \end{array}$$

has $c_1 = \$15{,}828.80$ and $c_{19} = \$28{,}932.80$.[‡]
(a) Find a formula for c_n.
(b) Find the total earnings of an average production worker from 1981 to 2005.

70. Let $n = 1$ correspond to 1997. The arithmetic sequence in which

$$d_n = \text{hours per person spent on the Internet in year } n$$

has $d_1 = 39.9$ and $d_6 = 181.9$.[§]
(a) Find a formula for d_n.
(b) How many hours per person were spent on the Internet in 2001?
(c) Find the total number of hours per person on the Internet from 1997 to 2003.

---

[*]Based on data from the U.S. Department of Commerce.
[†]Based on data from the U.S. Bureau of Economic Analysis.
[‡]Based on data from the U.S. Bureau of Labor Statistics.
[§]Based on data and projections in *Statistical Abstract of the United States: 2002.*

7.3   Geometric Sequences

A **geometric sequence** (sometimes called a **geometric progression**) is a sequence in which the quotient of each term and the preceding one is the same constant r. This constant r is called the **common ratio** of the geometric sequence.

EXAMPLE 1

The sequence $3, 9, 27, \ldots 3^n, \ldots$ is geometric with common ratio 3. For instance, $a_2/a_1 = 9/3 = 3$ and $a_3/a_2 = 27/9 = 3$. If 3^n is any term ($n \geq 2$), then the preceding term is 3^{n-1}, and

$$\frac{3^n}{3^{n-1}} = \frac{3 \cdot 3^{n-1}}{3^{n-1}} = 3. \quad \blacksquare$$

EXAMPLE 2

Which of the following sequences are geometric?

(a) $\dfrac{1}{2}, \dfrac{2}{3}, \dfrac{3}{4}, \dfrac{4}{5}, \dfrac{5}{6}, \ldots$

(b) $e, -e^3, e^5, -e^7, e^9, \ldots$

(c) $\left\{ \dfrac{5}{2^n} \right\}$

SOLUTION

(a) Note that

$$\frac{a_2}{a_1} = \frac{2/3}{1/2} = \frac{2}{3} \cdot \frac{2}{1} = \frac{4}{3}, \qquad \text{but} \qquad \frac{a_3}{a_2} = \frac{3/4}{2/3} = \frac{3}{4} \cdot \frac{3}{2} = \frac{9}{8}.$$

Since the quotient of consecutive terms is not always the same, this sequence is not geometric.

(b) Compute the quotient of each term and the preceding one.

$$\frac{a_2}{a_1} = \frac{-e^3}{e} = -e^2, \qquad \frac{a_3}{a_2} = \frac{e^5}{-e^3} = -e^2,$$

$$\frac{a_4}{a_3} = \frac{-e^7}{e^5} = -e^2, \qquad \frac{a_5}{a_4} = \frac{e^9}{-e^7} = -e^2,$$

and so on. The consecutive quotients are always the number $-e^2$, so the sequence is geometric with common ratio $-e^2$.

(c) For each $n \geq 2$, the term preceding a_n is a_{n-1}. Here,

$$a_n = \frac{5}{2^n} \qquad \text{and} \qquad a_{n-1} = \frac{5}{2^{n-1}},$$

so

$$\frac{a_n}{a_{n-1}} = \frac{5/2^n}{5/2^{n-1}} = \frac{5}{2^n} \cdot \frac{2^{n-1}}{5} = \frac{2^{n-1}}{2^n} = \frac{1}{2}.$$

Since the quotient of each term and the preceding one is always $1/2$, the sequence is geometric with common ratio $1/2$. ■

Suppose $\{a_n\}$ is a geometric sequence with common ratio r. For each $n \geq 2$, the quotient of the term a_n and the preceding one is $\dfrac{a_n}{a_{n-1}} = r$. Multiplying both sides of the equation by a_{n-1} produces this fact.

Geometric Sequences

> In a geometric sequence $\{a_n\}$ with common ratio r,
>
> $$a_n = ra_{n-1} \qquad \text{for each } n \geq 2.$$

Applying this last formula for $n = 2, 3, 4, \ldots$, we have

$$a_2 = ra_1$$
$$a_3 = ra_2 = r(ra_1) = r^2 a_1$$
$$a_4 = ra_3 = r(r^2 a_1) = r^3 a_1$$
$$a_5 = ra_4 = r(r^3 a_1) = r^4 a_1,$$

and in general we have the following.

nth Term of a Geometric Sequence

> If $\{a_n\}$ is a geometric sequence with common ratio r, then for all $n \geq 1$,
>
> $$a_n = r^{n-1} a_1.$$

EXAMPLE 3

Find a formula for the nth term of the geometric sequence $\{a_n\}$ that satisfies the given conditions.

(a) $a_1 = 7$ and $r = 2$ \qquad (b) $a_3 = 8$ and $r = 1/2$

SOLUTION

(a) The equation in the preceding box shows that

$$a_n = r^{n-1} a_1 = 2^{n-1} \cdot 7.$$

(b) We must first find a_1.

$$a_n = r^{n-1} a_1$$

Let $n = 3$: $\qquad a_3 = r^{3-1} a_1$

Substitute $a_3 = 8$ and $r = 1/2$: $\qquad 8 = \left(\dfrac{1}{2}\right)^2 a_1$

$$8 = \frac{1}{4} a_1$$

Multiply both sides by 4: $\qquad 32 = a_1.$

Therefore,

$$a_n = r^{n-1}a_1 = \left(\frac{1}{2}\right)^{n-1} \cdot 32. \qquad \blacksquare$$

EXAMPLE 4

Find a formula for the nth term of the geometric sequence whose first two terms are 2 and $-2/5$.

SOLUTION   The common ratio is

$$r = \frac{a_2}{a_1} = \frac{-2/5}{2} = \frac{-2}{5} \cdot \frac{1}{2} = -\frac{1}{5}.$$

Using the equation in the box, we now see that the formula for the nth term is

$$a_n = r^{n-1}a_1 = \left(-\frac{1}{5}\right)^{n-1}(2) = \frac{(1)^{n-1}}{(-5)^{n-1}}(2) = \frac{2}{(-5)^{n-1}}.$$

So the sequence begins $2, -2/5, 2/5^2, -2/5^3, 2/5^4, \dots$. $\blacksquare$

EXAMPLE 5

Find a formula for the nth term of the geometric sequence $\{a_n\}$ in which $a_2 = 20/9$ and $a_5 = 160/243$.

SOLUTION   By the equation in the box above, we have

$$\frac{160/243}{20/9} = \frac{a_5}{a_2} = \frac{r^4 a_1}{r a_1} = r^3.$$

Consequently,

$$r = \sqrt[3]{\frac{160/243}{20/9}} = \sqrt[3]{\frac{160}{243} \cdot \frac{9}{20}} = \sqrt[3]{\frac{8 \cdot 9}{243}} = \sqrt[3]{\frac{8}{27}} = \frac{2}{3}.$$

Since $a_2 = r a_1$, we see that

$$a_1 = \frac{a_2}{r} = \frac{20/9}{2/3} = \frac{20}{9} \cdot \frac{3}{2} = \frac{10}{3}.$$

Therefore,

$$a_n = r^{n-1}a_1 = \left(\frac{2}{3}\right)^{n-1} \cdot \frac{10}{3} = \frac{2^{n-1} \cdot 2 \cdot 5}{3^{n-1} \cdot 3} = \frac{2^n \cdot 5}{3^n} = 5\left(\frac{2}{3}\right)^n. \qquad \blacksquare$$

▪▪ PARTIAL SUMS

If the common ratio r of a geometric sequence is the number 1, then we have

$$a_n = 1^{n-1}a_1 \quad \text{for every } n \geq 1.$$

Therefore, the sequence is just the constant sequence $a_1, a_1, a_1, \ldots$. For any positive integer k, the kth partial sum of this constant sequence is

$$\underbrace{a_1 + a_1 + \cdots + a_1}_{k \text{ terms}} = ka_1.$$

In other words, the kth partial sum of a constant sequence is just k times the constant. If a geometric sequence is not constant (that is, $r \neq 1$), then its partial sums are given by the following formula.

Partial Sums of a Geometric Sequence

The kth partial sum of the geometric sequence $\{a_n\}$ with common ratio $r \neq 1$ is

$$\sum_{n=1}^{k} a_n = a_1 \left(\frac{1 - r^k}{1 - r} \right).$$

Proof If S denotes the kth partial sum, then the formula for the nth term of a geometric sequence shows that

$$S = a_1 + a_2 + \cdots + a_k = a_1 + a_1 r + a_1 r^2 + a_1 r^3 + \cdots + a_1 r^{k-1}.$$

Use this equation to compute $S - rS$.

$$
\begin{array}{rl}
S = & a_1 + a_1 r + a_1 r^2 + a_1 r^3 + \cdots + a_1 r^{k-1} \\
rS = & a_1 r + a_1 r^2 + a_1 r^3 + \cdots + a_1 r^{k-1} + a_1 r^k \\
\hline
S - rS = & a_1 \phantom{+ a_1 r + a_1 r^2 + a_1 r^3 + \cdots + a_1 r^{k-1}} - a_1 r^k \\
\end{array}
$$

$$(1 - r)S = a_1(1 - r^k)$$

Since $r \neq 1$, we can divide both sides of this last equation by $1 - r$ to complete the proof.

$$S = \frac{a_1(1 - r^k)}{1 - r} = a_1 \left(\frac{1 - r^k}{1 - r} \right). \qquad \blacksquare$$

EXAMPLE 6

Find the sum

$$-\frac{3}{2} + \frac{3}{4} - \frac{3}{8} + \frac{3}{16} - \frac{3}{32} + \frac{3}{64} - \frac{3}{128} + \frac{3}{256} - \frac{3}{512}.$$

SOLUTION Note that this is the ninth partial sum of the geometric sequence $\left\{ 3 \left(\frac{-1}{2} \right)^n \right\}$. The common ratio is $r = -1/2$. The formula in the box shows that

$$\sum_{n=1}^{9} 3 \left(\frac{-1}{2} \right)^n = a_1 \left(\frac{1 - r^9}{1 - r} \right) = \left(\frac{-3}{2} \right) \left[\frac{1 - (-1/2)^9}{1 - (-1/2)} \right]$$

$$= \left(\frac{-3}{2} \right) \left(\frac{1 + 1/2^9}{3/2} \right) = \left(\frac{-3}{2} \right) \left(\frac{2}{3} \right) \left(1 + \frac{1}{2^9} \right)$$

$$= -1 - \frac{1}{2^9} = -1 - \frac{1}{512} = -\frac{513}{512}.$$

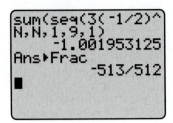

Figure 7–14

The sum can also be found with a calculator (Figure 7–14) and expressed in fractional form.* ■

EXAMPLE 7

A superball is dropped from a height of 9 feet. It hits the ground and bounces to a height of 6 feet. It continues to bounce up and down. On each bounce, it rises to 2/3 of the height of the previous bounce. How far has the ball traveled (both up and down) when it hits the ground for the seventh time?

SOLUTION   We first consider how far the ball travels on each bounce. On the first bounce, it rises 6 feet and falls 6 feet for a total of 12 feet. On the second bounce, it rises and falls 2/3 of the previous height and hence travels 2/3 of 12 feet. If a_n denotes the distance traveled on the nth bounce, then

$$a_1 = 12, \qquad a_2 = \left(\frac{2}{3}\right)a_1, \qquad a_3 = \left(\frac{2}{3}\right)a_2 = \left(\frac{2}{3}\right)^2 a_1,$$

and in general

$$a_n = \left(\frac{2}{3}\right)a_{n-1} = \left(\frac{2}{3}\right)^{n-1} a_1.$$

So $\{a_n\}$ is a geometric sequence with common ratio $r = 2/3$. When the ball hits the ground for the seventh time, it has completed six bounces. Therefore, the total distance it has traveled is the distance it was originally dropped (9 feet) plus the distance traveled in six bounces, namely,

$$9 + a_1 + a_2 + a_3 + a_4 + a_5 + a_6 = 9 + \sum_{n=1}^{6} a_n = 9 + a_1\left(\frac{1 - r^6}{1 - r}\right)$$

$$= 9 + 12\left[\frac{1 - (2/3)^6}{1 - (2/3)}\right] \approx 41.84 \text{ feet.} \quad ■$$

✓ EXERCISES 7.3

In Exercises 1–12, determine whether the sequence is arithmetic, geometric, or neither.

1. 2, 7, 12, 17, 22, . . .   **2.** 2, 6, 18, 54, 162, . . .

3. 13, 13/2, 13/4, 13/8, . . .

4. $-1, -\dfrac{1}{2}, 0, \dfrac{1}{2}, \ldots$ **5.** 50, 48, 46, 44, . . .

6. 2, −3, 9/2, −27/4, −81/8, . . .

7. 3, −3/2, 3/4, −3/8, 3/16, . . .

8. −6, −3.7, −1.4, .9, 3.2, . . .

9. 3, $3\sqrt{2}$, 6, $6\sqrt{2}$, 12, $12\sqrt{2}$, . . .

10. $\ln e$, $\ln e^2$, $\ln e^3$, $\ln e^4$, $\ln e^5$, . . .

11. 6, 6, 6, 6, 6, . . .

12. 1, $\sqrt[3]{3}$, $\sqrt[3]{9}$, 3, $3\sqrt[3]{3}$, $3\sqrt[3]{9}$, 9, . . .

---

*For fractional form on TI, use the FRAC key, as in Figure 7–14. On HP-39+, change the number format in the MODE menu to "fraction." On Casio 9850, use the FRAC program in the Program Appendix; on Casio FX 2.0, use the *a b/c* key.

In Exercises 13–22, one term and the common ratio r of a geometric sequence are given. Find the sixth term and a formula for the nth term.

13. $a_1 = 5, r = 2$

14. $a_1 = 1, r = -2$

15. $a_1 = 4, r = \dfrac{1}{4}$

16. $a_1 = -6, r = \dfrac{2}{3}$

17. $a_1 = 10, r = -\dfrac{1}{2}$

18. $a_1 = \pi, r = \dfrac{1}{5}$

19. $a_2 = 12, r = 1/3$

20. $a_3 = 1/2, r = 3$

21. $a_4 = -4/5, r = 2/5$

22. $a_5 = 2/3, r = -1/3$

In Exercises 23–28, show that the given sequence is geometric and find the common ratio.

23. $\left\{\left(-\dfrac{1}{2}\right)^n\right\}$

24. $\{2^{3n}\}$

25. $\{5^{n+2}\}$

26. $\{3^{n/2}\}$

27. $\{(\sqrt{5})^n\}$

28. $\{4^{n-4}\}$

In Exercises 29–36, use the given information about the geometric sequence $\{a_n\}$ to find a_5 and a formula for a_n.

29. $a_1 = 256, a_2 = -64$

30. $a_1 = 1/6, a_2 = -1/18$

31. $a_1 = 1/2, a_2 = 5$

32. $a_1 = \sqrt{7}, a_2 = \sqrt{42}$

33. $a_3 = 4, a_6 = 1/16$

34. $a_3 = 4, a_6 = -32$

35. $a_1 = 5, a_7 = 20$ (assume that $r > 0$)

36. $a_2 = 6, a_7 = 192$

In Exercises 37–40, find the kth partial sum of the geometric sequence $\{a_n\}$ with common ratio r.

37. $k = 6, a_1 = 5, r = \dfrac{1}{2}$

38. $k = 8, a_1 = 9, r = \dfrac{1}{3}$

39. $k = 7, a_2 = 6, r = 2$

40. $k = 9, a_2 = 6, r = \dfrac{1}{4}$

In Exercises 41–46, find the sum.

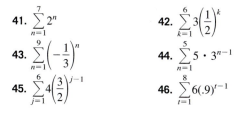

41. $\displaystyle\sum_{n=1}^{7} 2^n$

42. $\displaystyle\sum_{k=1}^{6} 3\left(\dfrac{1}{2}\right)^k$

43. $\displaystyle\sum_{n=1}^{9} \left(-\dfrac{1}{3}\right)^n$

44. $\displaystyle\sum_{n=1}^{5} 5 \cdot 3^{n-1}$

45. $\displaystyle\sum_{j=1}^{6} 4\left(\dfrac{3}{2}\right)^{j-1}$

46. $\displaystyle\sum_{t=1}^{8} 6(.9)^{t-1}$

In Exercises 47–52, you are asked to find a geometric sequence. In each case, round the common ratio r to four decimal places.

47. The number of students per computer in U.S. schools in year n (with $n = 1$ corresponding to 1985) can be approximated by a geometric sequence whose first two terms are $a_1 = 44.4313$ and $a_2 = 38.0154$.[*]
 (a) Find a formula for a_n.
 (b) What is the number of students per computer in 2003?
 (c) In what year will there be a computer for each student?

48. According to data from the U.S. Census Bureau, the population of the United States (in millions) in year n can be approximated by a geometric sequence $\{b_n\}$, where $n = 1$ corresponds to 1971.
 (a) If $b_1 = 207.5489$ and $b_3 = 211.7207$, find a formula for b_n.
 (b) Estimate the U.S. population in 2005 and 2010.

49. The value of all group life insurance (in billions of dollars) in year n can be approximated by a geometric sequence $\{c_n\}$, where $n = 1$ corresponds to 1991.[†]
 (a) If there was $3.9631 billion in effect in 1991 and $4.1672 billion in 1992, find a formula for c_n.
 (b) How much group life insurance is in effect in 1998? In 2002? In 2004?

50. Data from the U.S. Centers for Disease Control and Prevention indicates that the number of new cases of AIDS each year can be approximated by a geometric sequence $\{a_n\}$, where $n = 1$ corresponds in 1994.
 (a) If there were 80,691 new cases in 1994 and 71,605 in 1995, find a formula for a_n.
 (b) About how many new cases were reported in 2000?
 (c) Find the total number of new AIDS cases reported from 1994 to 2000.

51. The amount spent per person per year on cable and satellite TV can be approximated by a geometric sequence $\{b_n\}$, where $n = 1$ corresponds to 1995.[‡]
 (a) If $130.76 was spent in 1995 and $151.39 was spent in 1997, find a formula for b_n.
 (b) Find the total spent per person from 1995 to 2003.

52. According to data from the U.S. National Center for Education Statistics, the number of bachelor's degrees earned by women can be approximated by a geometric sequence $\{c_n\}$, where $n = 1$ corresponds to 1970.
 (a) If 365,916 degrees were earned in 1970 and 373,454 in 1971, find a formula for c_n.
 (b) How many degrees were earned in 1997? In 2000? In 2003?
 (c) Find the total number of degrees earned from 1995 to 2003.

[*]Based on data from Quality Education Data, Inc.
[†]Based on data from the American Council of Life Insurance.
[‡]Based on data in the *Statistical Abstract of the United States: 2001.*

53. A ball is dropped from a height of 8 feet. On each bounce, it rises to half its previous height. When the ball hits the ground for the seventh time, how far has it traveled?

54. A ball is dropped from a height of 10 feet. On each bounce, it rises to 45% of its previous height. When it hits the ground for the tenth time, how far has it traveled?

55. If you are paid a salary of 1¢ on the first day of March and 2¢ on the second day, and your salary continues to double each day, how much will you earn in the month of March?

56. Starting with your parents, how many ancestors do you have for the preceding ten generations?

57. A car that sold for $8000 depreciates in value 25% each year. What is it worth after five years?

58. A vacuum pump removes 60% of the air in a container at each stroke. What percentage of the original amount of air remains after six strokes?

Thinkers

59. Suppose $\{a_n\}$ is a geometric sequence with common ratio $r > 0$ and each $a_n > 0$. Show that the sequence $\{\log a_n\}$ is an arithmetic sequence with common difference $\log r$.

60. Suppose $\{a_n\}$ is an arithmetic sequence with common difference d. Let C be any positive number. Show that the sequence $\{C^{a_n}\}$ is a geometric sequence with common ratio C^d.

61. In the geometric sequence 1, 2, 4, 8, 16, . . . , show that each term is 1 plus the sum of all preceding terms.

62. In the geometric sequence 2, 6, 18, 54, . . . , show that each term is twice the sum of 1 and all preceding terms.

63. The minimum monthly payment for a certain bank credit card is the larger of $1/25$ of the outstanding balance or $5. If the balance is less than $5, the entire balance is due. If you make only the minimum payment each month, how long will it take to pay off a balance of $200 (excluding any interest that might be due)?

7.3.A   *SPECIAL TOPICS* Infinite Series

An **infinite series** (or simply **series**) is an expression of the form

$$a_1 + a_2 + a_3 + a_4 + a_5 + \cdots,$$

where each a_n is a real number. This notation suggests the "sum" of the infinite sequence $a_1, a_2, a_3, a_4, a_5, \ldots$. However, ordinary addition is defined only for a finite list of numbers. So we must *define* what is meant by the sum of an infinite series. Consider, for example, the infinite series

$$\frac{3}{10} + \frac{3}{10^2} + \frac{3}{10^3} + \frac{3}{10^4} + \cdots.$$

We begin with the *sequence* $3/10, 3/10^2, 3/10^3, 3/10^4, \ldots$ and compute its partial sums.

$$S_1 = \frac{3}{10}$$

$$S_2 = \frac{3}{10} + \frac{3}{10^2} = \frac{33}{100}$$

$$S_3 = \frac{3}{10} + \frac{3}{10^2} + \frac{3}{10^3} = \frac{333}{1000}$$

$$S_4 = \frac{3}{10} + \frac{3}{10^2} + \frac{3}{10^3} + \frac{3}{10^4} = \frac{3333}{10,000}$$

These partial sums $S_1, S_2, S_3, S_4, \ldots$ themselves form a sequence.

$$\frac{3}{10}, \frac{33}{100}, \frac{333}{1000}, \frac{3333}{10,000}, \ldots$$

The terms in the sequence of partial sums appear to be getting closer and closer to $1/3$. In other words, as k gets larger and larger, the corresponding partial sum S_k gets closer and closer to $1/3$. So we say that $1/3$ is the *sum* of the infinite series

$$\frac{3}{10} + \frac{3}{10^2} + \frac{3}{10^3} + \frac{3}{10^4} + \cdots$$

or that this series *converges* to $1/3$, and we write

$$\frac{3}{10} + \frac{3}{10^2} + \frac{3}{10^3} + \frac{3}{10^4} + \cdots = \frac{1}{3}.$$

The general case is handled similarly. If $a_1 + a_2 + a_3 + a_4 + \cdots$ is an infinite series, then its **partial sums** are the numbers

$$S_1 = a_1$$

$$S_2 = a_1 + a_2$$

$$S_3 = a_1 + a_2 + a_3,$$

and in general, for any $k \geq 1$,

$$S_k = a_1 + a_2 + a_3 + a_4 + \cdots + a_k.$$

If it happens that the terms $S_1, S_2, S_3, S_4, \ldots$ of the *sequence* of partial sums get closer and closer to a particular real number S in such a way that the partial sum S_k is arbitrarily close to S when k is large enough, then we say that the series **converges** and that S is the **sum of the convergent series.**

Not every series has a sum. For instance, the partial sums of the series

$$1 + 2 + 3 + 4 + \cdots$$

get larger and larger and never get closer and closer to a single real number. So this series is not convergent and does not have a sum.

EXAMPLE 1

Although no proof will be given here, it is intuitively clear that every infinite decimal may be thought of as the sum of a convergent series. For instance,

$$\pi = 3.1415926 \cdots = 3 + .1 + .04 + .001 + .0005 + .00009 + \cdots.$$

Note that the third partial sum is $3 + .1 + .04 = 3.14$, which is π to two decimal places. Similarly, the kth partial sum of this series is just π to $k - 1$ decimal places. ■

INFINITE GEOMETRIC SERIES

If $\{a_n\}$ is a geometric sequence with common ratio r, then the corresponding infinite series

$$a_1 + a_2 + a_3 + a_4 + a_5 + \cdots$$

is called an **infinite geometric series.** By using the formula for the nth term of a geometric sequence, we can also express the corresponding geometric series in the form

$$a_1 + ra_1 + r^2a_1 + r^3a_1 + r^4a_1 + \cdots.$$

Under certain circumstances, an infinite geometric series is convergent and has a sum.

Sum of an Infinite Geometric Series

If $|r| < 1$, then the infinite geometric series

$$a_1 + ra_1 + r^2a_1 + r^3a_1 + r^4a_1 + \cdots$$

converges, and its sum is

$$\frac{a_1}{1 - r}.$$

Although we cannot prove this fact rigorously here, we can make it highly plausible both geometrically and algebraically.

EXAMPLE 2

$\dfrac{8}{5} + \dfrac{8}{5^2} + \dfrac{8}{5^3} + \cdots$ is an infinite geometric series with $a_1 = 8/5$ and $r = 1/5$. The kth partial sum of this series is the same as the kth partial sum of the sequence $\{8/5^n\}$, and hence, from the box on page 575, we know that

$$S_k = a_1\left(\frac{1 - r^k}{1 - r}\right) = \frac{8}{5}\left[\frac{1 - \left(\dfrac{1}{5}\right)^k}{1 - \dfrac{1}{5}}\right]$$

$$= \frac{8}{5}\left(\frac{1 - \dfrac{1}{5^k}}{\dfrac{4}{5}}\right) = \frac{8}{5} \cdot \frac{5}{4}\left(1 - \frac{1}{5^k}\right)$$

$$= 2\left(1 - \frac{1}{5^k}\right) = 2 - \frac{2}{5^k}.$$

The function $f(x) = 2 - 2/5^x$ is defined for all real numbers. When $x = k$ is a positive integer, then $f(k)$ is S_k, the kth partial sum of the series. Using a calculator, we obtain the graph of $f(x)$ in Figure 7–15.

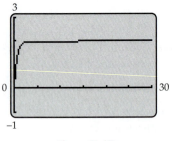

Figure 7–15

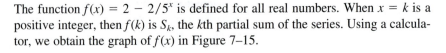

GRAPHING EXPLORATION

Graph $f(x)$ in the same viewing window as in Figure 7–15. Use the trace feature to move the cursor along the graph. As x gets larger, what is the apparent value of $f(x)$ (that is, the value of the partial sum)?

Your calculator will probably tell you that every partial sum is 2, once you move beyond approximately $x = 15$. Actually, the partial sums are slightly smaller than 2 but are rounded to 2 by the calculator. In any case, the horizontal line through 2 is a horizontal asymptote of the graph (meaning that the graph gets very close to the line as x gets larger), so it is very plausible that the sequence converges to the number 2. But 2 is exactly what the preceding box says the sum should be.

$$\frac{a_1}{1-r} = \frac{\frac{8}{5}}{1-\frac{1}{5}} = \frac{\frac{8}{5}}{\frac{4}{5}} = \frac{8}{4} = 2. \quad \blacksquare$$

Example 2 is typical of the general case, as can be seen algebraically. Consider the geometric series $a_1 + a_2 + a_3 + \cdots$ with common ratio r such that $|r| < 1$. The kth partial sum S_k is the same as the kth partial sum of the geometric sequence $\{a_n\}$, and hence,

$$S_k = a_1\left(\frac{1-r^k}{1-r}\right).$$

As k gets larger and larger, the number r^k gets very close to 0 because $|r| < 1$ (for instance, $(-.6)^{20} \approx .0000366$ and $.2^9 \approx .000000512$). Consequently, when k is very large, $1 - r^k$ is very close to $1 - 0$ so that

$$S_k = a_1\left(\frac{1-r^k}{1-r}\right) \quad \text{is very close to} \quad a_1\left(\frac{1-0}{1-r}\right) = \frac{a_1}{1-r}.$$

EXAMPLE 3

$-\dfrac{1}{2} + \dfrac{1}{4} - \dfrac{1}{8} + \dfrac{1}{16} + \cdots$ is an infinite geometric series with $a_1 = -1/2$ and $r = -1/2$. Since $|r| < 1$, this series converges, and its sum is

$$\frac{a_1}{1-r} = \frac{-\frac{1}{2}}{1-\left(-\frac{1}{2}\right)} = \frac{-\frac{1}{2}}{\frac{3}{2}} = -\frac{1}{3}. \quad \blacksquare$$

Infinite geometric series provide another way of writing an infinite repeating decimal as a rational number.

EXAMPLE 4

To express $6.8573573573 \cdots$ as a rational number, we first write it as $6.8 + .0573573573 \cdots$. Consider $.0573573573 \cdots$ as an infinite series.

$$.0573 + .0000573 + .0000000573 + .0000000000573 + \cdots,$$

which is the same as

$$.0573 + (.001)(.0573) + (.001)^2(.0573) + (.001)^3(.0573) + \cdots.$$

This is a convergent geometric series with $a_1 = .0573$ and $r = .001$. Its sum is

$$\frac{a_1}{1-r} = \frac{.0573}{1-.001} = \frac{.0573}{.999} = \frac{573}{9990}.$$

Therefore,

$$6.8573573573 \cdots = 6.8 + [.0573 + .0000573 + \cdots]$$

$$= 6.8 + \frac{573}{9990}$$

$$= \frac{68}{10} + \frac{573}{9990}$$

$$= \frac{68,505}{9990} = \frac{4567}{666}. \quad \blacksquare$$

We close this section by mentioning an alternative notation for infinite series. The series

$$a_1 + a_2 + a_3 + a_4 + \cdots \qquad \text{is denoted by} \qquad \sum_{n=1}^{\infty} a_n.$$

For example, $\displaystyle\sum_{n=1}^{\infty} 2(.6^n)$ denotes the series

$$2(.6) + 2(.6^2) + 2(.6^3) + 2(.6^4) + \cdots,$$

and $\displaystyle\sum_{n=1}^{\infty} \left(\frac{-1}{2}\right)^n$ denotes

$$-\frac{1}{2} + \left(\frac{-1}{2}\right)^2 + \left(\frac{-1}{2}\right)^3 + \left(\frac{-1}{2}\right)^4 + \cdots = -\frac{1}{2} + \frac{1}{4} - \frac{1}{8} + \frac{1}{16} - \frac{1}{32} + \cdots.$$

✓ EXERCISES 7.3.A 📼

In Exercises 1–8, find the sum of the infinite series, if it has one.

1. $\displaystyle\sum_{n=1}^{\infty} \frac{1}{2^n}$ **2.** $\displaystyle\sum_{n=1}^{\infty} \left(-\frac{3}{4}\right)^n$ **3.** $\displaystyle\sum_{n=1}^{\infty} (.06)^n$

4. $1 - .5 + .25 - .125 + .0625 - \cdots$

5. $500 + 200 + 80 + 32 + \cdots$

6. $9 - 3\sqrt{3} + 3 - \sqrt{3} + 1 - \dfrac{1}{\sqrt{3}} + \cdots$

7. $2 + \sqrt{2} + 1 + \dfrac{1}{\sqrt{2}} + \dfrac{1}{2} + \cdots$

8. $\displaystyle\sum_{n=1}^{\infty} \left(\frac{1}{2^n} - \frac{1}{3^n}\right)$

In Exercises 9–15, express the repeating decimal as a rational number.

9. $.22222 \cdots$ **10.** $.37373737 \cdots$

11. $5.4272727 \cdots$ **12.** $85.131313 \cdots$

13. $2.1425425425 \cdots$ **14.** $3.7165165165 \cdots$

15. $1.74241241241 \cdots$

16. If $\{a_n\}$ is an arithmetic sequence with common difference $d > 0$ and each $a_i > 0$, explain why the infinite series $a_1 + a_2 + a_3 + a_4 + \cdots$ is not convergent.

17. (a) Verify that $\sum_{n=1}^{\infty} 2(1.5)^n$ is a geometric series with $a_1 = 3$ and $r = 1.5$.

(b) Find the kth partial sum of the series and use this expression to define a function f, as in Example 2.

(c) Graph the function f in a viewing window with $0 \leq x \leq 30$. As x gets very large, what happens to the corresponding value of $f(x)$? Does the graph get closer and closer to some horizontal line, as in Example 2? What does this say about the convergence of the series?

18. Use the graphical approach illustrated in Example 2 to find the sum of the series in Example 3. Does the graph get very close to the horizontal line through $-1/3$? What's going on?

In Exercises 19 and 20, use your calculator to compute partial sums and estimate the sum of the convergent infinite series.

19. $\dfrac{\pi}{2} - \dfrac{(\pi/2)^3}{3!} + \dfrac{(\pi/2)^5}{5!} - \dfrac{(\pi/2)^7}{7!} + \cdots$, where $n!$ is the product $1 \cdot 2 \cdot 3 \cdot 4 \cdots \cdot n$.*

20. $1 + 1 + \dfrac{1}{2!} + \dfrac{1}{3!} + \dfrac{1}{4!} + \cdots$ [*Hint:* What is the decimal expansion of e?]

---

*To compute $n!$ quickly, look for ! in the PROB submenu of the MATH or OPTN menu.

7.4 The Binomial Theorem

The Binomial Theorem provides a formula for calculating the product $(x + y)^n$ for any positive integer n. Before we state the theorem, some preliminaries are needed.

Let n be a positive integer. The symbol $n!$ (read **n factorial**) denotes the product of all the integers from 1 to n. For example,

$$2! = 1 \cdot 2 = 2, \qquad 3! = 1 \cdot 2 \cdot 3 = 6, \qquad 4! = 1 \cdot 2 \cdot 3 \cdot 4 = 24,$$

$$5! = 1 \cdot 2 \cdot 3 \cdot 4 \cdot 5 = 120,$$

$$10! = 1 \cdot 2 \cdot 3 \cdot 4 \cdot 5 \cdot 6 \cdot 7 \cdot 8 \cdot 9 \cdot 10 = 3,628,800.$$

So, we have this definition.

n Factorial

Let n be a positive integer. Then

$$n! = 1 \cdot 2 \cdot 3 \cdot 4 \cdots (n - 2)(n - 1)n.$$

$0!$ is defined to be the number 1.

Learn to use your calculator to compute factorials. You will find ! in the PROB (or PRB) submenu of the MATH or OPTN menu.

CALCULATOR EXPLORATION

15! is such a large number your calculator will switch to scientific notation to approximate it. What is this approximation? Many calculators cannot compute factorials larger than 69! If yours does compute larger ones, how large a one can you compute without getting an error message [or on HP-39+, getting the number 9.9999 ... E499]?

If r and n are integers with $0 \le r \le n$, then we have the following.

Binomial
Coefficients

> Either of the symbols $\binom{n}{r}$ or $_nC_r$ denotes the number $\dfrac{n!}{r!(n-r)!}$.
> $\binom{n}{r}$ is called a **binomial coefficient.**

For example,

$$_5C_3 = \binom{5}{3} = \frac{5!}{3!(5-3)!} = \frac{5}{3!2!} = \frac{1 \cdot 2 \cdot 3 \cdot 4 \cdot 5}{(1 \cdot 2 \cdot 3)(1 \cdot 2)} = \frac{4 \cdot 5}{2} = 10$$

$$_4C_2 = \binom{4}{2} = \frac{4!}{2!(4-2)!} = \frac{4!}{2!2!} = \frac{1 \cdot 2 \cdot 3 \cdot 4}{(1 \cdot 2)(1 \cdot 2)} = \frac{3 \cdot 4}{2} = 6.$$

Binomial coefficients can be computed on a calculator by using $_nC_r$ or Comb in the PROB (or PRB) submenu of the MATH or OPTN menu.

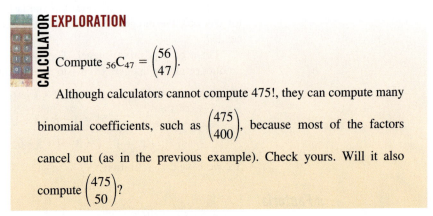

CALCULATOR EXPLORATION

Compute $_{56}C_{47} = \binom{56}{47}$.

Although calculators cannot compute 475!, they can compute many binomial coefficients, such as $\binom{475}{400}$, because most of the factors cancel out (as in the previous example). Check yours. Will it also compute $\binom{475}{50}$?

The preceding examples illustrate a fact whose proof will be omitted: *Every binomial coefficient is an integer.* Furthermore, for every nonnegative integer n,

$$\binom{n}{0} = 1 \qquad \text{and} \qquad \binom{n}{n} = 1$$

because

$$\binom{n}{0} = \frac{n!}{0!(n-0)!} + \frac{n!}{0!n!} = \frac{n!}{n!} = 1 \qquad \text{and}$$

$$\binom{n}{n} = \frac{n!}{n!(n-n)!} = \frac{n!}{n!0!} = \frac{n!}{n!} = 1.$$

If we list the binomial coefficients for each value of *n* in this manner:

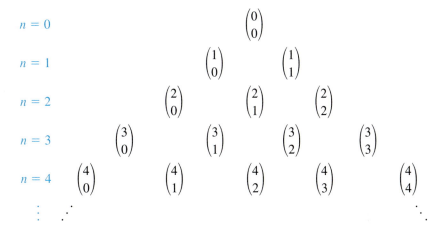

$n = 0$

$n = 1$

$n = 2$

$n = 3$

$n = 4$

and then calculate each of them, we obtain the following array of numbers.

row 0   1

row 1   1   1

row 2   1   2   1

row 3 1   3   ③   1

row 4 1   4   ⑥   4   1

This array is called **Pascal's triangle.** It has an interesting property:

> *Each interior entry of Pascal's triangle is the sum of the two closest entries in the row above it.*

Two examples of this are shown above by the colored circles and arrows. Using this fact we can easily construct row 5 of the triangle from row 4.

row 4:   1   4   6   4   1

row 5:   1   5   10   10   5   1

This property of Pacal's triangle is proved in Exercise 67.

■■ THE BINOMIAL THEOREM

To develop a formula for calculating $(x + y)^n$, we first calculate these products for small values of *n* to see whether we can find some kind of pattern.

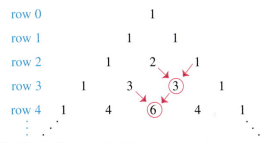

$n = 0 \quad (x + y)^0 = \qquad\qquad 1$

$n = 1 \quad (x + y)^1 = \qquad\qquad 1x + 1y$

(∗) $\quad n = 2 \quad (x + y)^2 = \qquad 1x^2 + 2xy + 1y^2$

$n = 3 \quad (x + y)^3 = \quad 1x^3 + 3x^2y + 3xy^2 + 1y^3$

$n = 4 \quad (x + y)^4 = 1x^4 + 4x^3y + 6x^2y^2 + 4xy^3 + 1y^4$

One pattern is immediately obvious: The coefficients shown in color on the preceding page are the top part of Pascal's triangle! In the case $n = 4$, for example, this means that the coefficients are the numbers

$$\begin{matrix} 1 & 4 & 6 & 4 & 1 \\ \binom{4}{0}, & \binom{4}{1}, & \binom{4}{2}, & \binom{4}{3}, & \binom{4}{4}. \end{matrix}$$

If this pattern holds for larger n, then the coefficients in the expansion of $(x + y)^n$ are

$$\binom{n}{0}, \binom{n}{1}, \binom{n}{2}, \binom{n}{3}, \ldots, \binom{n}{n-1}, \binom{n}{n}.$$

As for the xy-terms associated with each of these coefficients, look at the pattern in (∗) on the preceding page: The exponent of x goes down by 1 and the exponent of y goes up by 1 as you go from term to term, which suggests that the terms of the expansion of $(x + y)^n$ (without the coefficients) are

$$x^n, \quad x^{n-1}y, \quad x^{n-2}y^2, \quad x^{n-3}y^3, \quad \ldots, \quad xy^{n-1}, \quad y^n.$$

Combining the patterns of coefficients and xy-terms suggests that the following result is true about the expansion of $(x + y)^n$.

The Binomial Theorem

For each positive integer n,

$$(x + y)^n = x^n + \binom{n}{1}x^{n-1}y + \binom{n}{2}x^{n-2}y^2 +$$

$$\binom{n}{3}x^{n-3}y^3 + \cdots + \binom{n}{n-1}xy^{n-1} + y^n.$$

Using summation notation and the fact that

$$\binom{n}{0} = 1 = \binom{n}{n},$$

we can write the Binomial Theorem compactly as

$$(x + y)^n = \sum_{j=0}^{n} \binom{n}{j} x^{n-j}y^j.$$

The Binomial Theorem will be proved in Section 7.7 by means of mathematical induction. We shall assume its truth for now and illustrate some of its uses.

EXAMPLE 1

Expand $(x + y)^8$.

SOLUTION   We apply the Binomial Theorem in the case $n = 8$.

$$(x + y)^8 = x^8 + \binom{8}{1}x^7y + \binom{8}{2}x^6y^2 + \binom{8}{3}x^5y^3$$

$$+ \binom{8}{4}x^4y^4 + \binom{8}{5}x^3y^5 + \binom{8}{6}x^2y^6 + \binom{8}{7}xy^7 + y^8.$$

The coefficients can be computed individually by hand or by using $_nC_r$ (or COMB) on a calculator; for instance,

$$_8C_2 = \binom{8}{2} = \frac{8!}{2!6!} = 28 \quad \text{and} \quad _8C_3 = \binom{8}{3} = \frac{8!}{3!5!} = 56.$$

Alternatively, you can display all the coefficients at once by making a table of values for the function $f(x) = {_8C_x}$, as shown in Figure 7–16.[*]

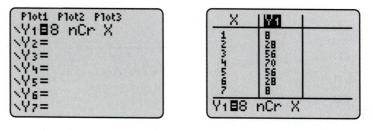

Figure 7–16

Substituting these values in the preceding expansion, we have

$$(x + y)^8 = x^8 + 8x^7y + 28x^6y^2$$
$$+ 56x^5y^3 + 70x^4y^4 + 56x^3y^5 + 28x^2y^6 + 8xy^7 + y^8. \qquad ■$$

EXAMPLE 2

Expand $(1 - z)^6$.

SOLUTION   Note that $1 - z = 1 + (-z)$ and apply the Binomial Theorem with $x = 1$, $y = -z$, and $n = 6$.

$$(1 - z)^6 = 1^6 + \binom{6}{1}1^5(-z) + \binom{6}{2}1^4(-z)^2 + \binom{6}{3}1^3(-z)^3$$

$$+ \binom{6}{4}1^2(-z)^4 + \binom{6}{5}1(-z)^5 + (-z)^6$$

$$= 1 - \binom{6}{1}z + \binom{6}{2}z^2 - \binom{6}{3}z^3 + \binom{6}{4}z^4 - \binom{6}{5}z^5 + z^6$$

$$= 1 - 6z + 15z^2 - 20z^3 + 15z^4 - 6z^5 + z^6. \qquad ■$$

EXAMPLE 3

If $f(x) = x^4 + 2x^3 + 5$ and $g(x) = f(x + 1)$, expand and simplify the rule of g.

SOLUTION   The rule of g is found by replacing x by $x + 1$ in the rule of f.

$$f(x) = x^4 + 2x^3 + 5$$

$$g(x) = f(x + 1) = (x + 1)^4 + 2(x + 1)^3 + 5.$$

---

[*]Thanks to Nick Goodbody for suggesting this.

Now we apply the Binomial Theorem to the terms $(x + 1)^4$ and $(x + 1)^3$.

$$g(x) = (x + 1)^4 + 2(x + 1)^3 + 5$$
$$= (x^4 + 4x^3 + 6x^2 + 4x + 1) + 2(x^3 + 3x^2 + 3x + 1) + 5$$
$$= x^4 + 4x^3 + 6x^2 + 4x + 1 + 2x^3 + 6x^2 + 6x + 2 + 5$$
$$= x^4 + 6x^3 + 12x^2 + 10x + 8. \quad \blacksquare$$

EXAMPLE 4

TIP

Binomial expansions, such as those in Examples 1–4 can be done on some calculators. Use EXPAND in this menu/submenu:

TI-89:   ALGEBRA

Casio FX 2.0:   CAS/TRNS

[CAS is in the main menu.]

Expand $(x^2 + x^{-1})^4$.

SOLUTION   Use the Binomial Theorem with x^2 in place of x and x^{-1} in place of y.

$$(x^2 + x^{-1})^4 = (x^2)^4 + \binom{4}{1}(x^2)^3(x^{-1}) + \binom{4}{2}(x^2)^2(x^{-1})^2 + \binom{4}{3}(x^2)(x^{-1})^3 + (x^{-1})^4$$
$$= x^8 + 4x^6x^{-1} + 6x^4x^{-2} + 4x^2x^{-3} + x^{-4}$$
$$= x^8 + 4x^5 + 6x^2 + 4x^{-1} + x^{-4}. \quad \blacksquare$$

EXAMPLE 5

Show that $(1.001)^{1000} > 2$ without using a calculator.

SOLUTION   We write 1.001 as $1 + .001$ and apply the Binomial Theorem with $x = 1$, $y = .001$, and $n = 1000$.

$$(1.001)^{1000} = (1 + .001)^{1000}$$
$$= 1^{1000} + \binom{1000}{1}1^{999}(.001) + \text{other positive terms}$$
$$= 1 + \binom{1000}{1}(.001) + \text{other positive terms}.$$

But

$$\binom{1000}{1} = \frac{1000!}{1!999!} = \frac{1000 \cdot 999!}{999!} = 1000.$$

Therefore,

$$\binom{1000}{1}(.001) = 1,000(.001) = 1$$

and

$$(1.001)^{1000} = 1 + 1 + \text{other positive terms} = 2 + \text{other positive terms}.$$

Hence, $(1.001)^{1000} > 2$. $\quad \blacksquare$

Sometimes we need to know only one term in the expansion of $(x + y)^n$. If you examine the expansion given by the Binomial Theorem, you will see that in the second term y has exponent 1, in the third term y has exponent 2, and so on. Thus, we have the following.

Properties of the Binomial Expansion

> In the binomial expansion of $(x + y)^n$,
>
> The exponent of y is always one less than the number of the term.
>
> Furthermore, in each of the middle terms of the expansion,
>
> The coefficient of the term containing y^r is $\binom{n}{r}$.
>
> The sum of the x exponent and the y exponent is n.

For instance, in the *ninth* term of the expansion of $(x + y)^{13}$, y has exponent 8, the coefficient is $\binom{13}{8}$, and x must have exponent 5 (since $8 + 5 = 13$). Thus, the ninth term is $\binom{13}{8} x^5 y^8$.

EXAMPLE 6

Find the ninth term of the expansion of $\left(2x^2 + \dfrac{\sqrt[4]{y}}{\sqrt{6}}\right)^{13}$.

SOLUTION   We use the Binomial Theorem with $n = 13$ and with $2x^2$ in place of x and $\sqrt[4]{y}/\sqrt{6}$ in place of y. The remarks above show that the ninth term is

$$\binom{13}{8}(2x^2)^5\left(\frac{\sqrt[4]{y}}{\sqrt{6}}\right)^8.$$

Since $\sqrt[4]{y} = y^{1/4}$ and $\sqrt{6} = \sqrt{3}\,\sqrt{2} = 3^{1/2}2^{1/2}$, we can simplify as follows.

$$\binom{13}{8}(2x^2)^5\left(\frac{\sqrt[4]{y}}{\sqrt{6}}\right)^8 = \binom{13}{8}2^5(x^2)^5\frac{(y^{1/4})^8}{(3^{1/2})^8(2^{1/2})^8} = \binom{13}{8}2^5 x^{10}\frac{y^2}{3^4 \cdot 2^4}$$

$$= \binom{13}{8}\frac{2}{3^4}x^{10}y^2 = \frac{13 \cdot 12 \cdot 11 \cdot 10 \cdot 9}{5 \cdot 4 \cdot 3 \cdot 2} \cdot \frac{2}{3^4}x^{10}y^2$$

$$= \frac{286}{9}x^{10}y^2. \qquad \blacksquare$$

✓ EXERCISES 7.4 📼

In Exercises 1–18, evaluate the expression.

1. $6!$ **2.** $10!$ **3.** $\dfrac{8!}{6!}$

4. $\dfrac{11!}{8!}$ **5.** $\dfrac{12!}{9!3!}$ **6.** $\dfrac{9! - 8!}{7!}$

7. $\dbinom{6}{2}$ **8.** $\dbinom{8}{4}$ **9.** $\dbinom{100}{99}$

10. $\dbinom{200}{198}$ **11.** $\dbinom{4}{1}\dbinom{5}{3}$ **12.** $\dbinom{5}{3}\dbinom{4}{2}$

13. $\dbinom{5}{3} + \dbinom{5}{2} - \dbinom{6}{3}$ **14.** $\dbinom{12}{11} - \dbinom{11}{10} + \dbinom{7}{0}$

15. $\dbinom{6}{0} + \dbinom{6}{1} + \dbinom{6}{2} + \dbinom{6}{3} + \dbinom{6}{4} + \dbinom{6}{5} + \dbinom{6}{6}$

16. $\dbinom{6}{0} - \dbinom{6}{1} + \dbinom{6}{2} - \dbinom{6}{3} + \dbinom{6}{4} - \dbinom{6}{5} + \dbinom{6}{6}$

17. $\dbinom{100}{96}$ **18.** $\dbinom{75}{72}$

In Exercises 19–38, expand and (where possible) simplify the expression.

19. $(x + 1)^4$ **20.** $(x - 1)^5$ **21.** $(x + 2)^4$

22. $(x - 2)^6$ **23.** $(x + y)^5$ **24.** $(a + b)^7$

25. $(a - b)^5$ **26.** $(c - d)^8$ **27.** $(2x + y^2)^5$

28. $(3u - v^3)^6$ **29.** $(\sqrt{x} + 1)^6$ **30.** $(2 - \sqrt{y})^5$

31. $(1 - c)^{10}$ **32.** $\left(\sqrt{c} + \dfrac{1}{\sqrt{c}}\right)^7$

33. $(x^{-3} + x)^4$ **34.** $(3x^{-2} - x^2)^6$

35. $(1 + \sqrt{3})^4 + (1 - \sqrt{3})^4$

36. $(\sqrt{3} + 1)^6 - (\sqrt{3} - 1)^6$

37. $(1 + i)^6$, where $i^2 = -1$

38. $(\sqrt{2} - i)^4$, where $i^2 = -1$

In Exercises 39–42, expand and simplify the rule of the function g.

39. $f(x) = x^3 - 2x^2 + x - 4$ and $g(x) = f(x + 1)$

40. $f(x) = 2x^3 + 3x^2 + x$ and $g(x) = f(x - 1)$

41. $f(x) = x^4 + 3x + 5$ and $g(x) = f(x - 2)$

42. $f(x) = 2x^3 - x + 4$ and $g(x) = f(x + 3)$

In Exercises 43–48, find the indicated term of the expansion of the given expression.

43. third, $(x + y)^5$ **44.** fourth, $(a + b)^6$

45. fifth, $(c - d)^7$ **46.** third, $(a + 2)^8$

47. fourth, $\left(u^{-2} + \dfrac{u}{2}\right)^7$ **48.** fifth, $(\sqrt{x} - \sqrt{2})^7$

49. Find the coefficient of x^5y^8 in the expansion of $(2x - y^2)^9$.

50. Find the coefficient of $x^{12}y^6$ in the expansion of $(x^3 - 3y)^{10}$.

51. Find the coefficient of $1/x^3$ in the expansion of $\left(2x + \dfrac{1}{x^2}\right)^6$.

52. Find the constant term in the expansion of $\left(y - \dfrac{1}{2y}\right)^{10}$.

In Exercises 53–56, use the Binomial Theorem to factor the expression.

53. $x^5 + 5x^4 + 10x^3 + 10x^2 + 5x + 1$

54. $x^4 - 4x^3 + 6x^2 - 4x + 1$

55. $16z^4 + 32z^3 + 24z^2 + 8z + 1$

56. $27y^3 - 27y^2 + 9y - 1$

57. (a) Verify that $\dbinom{9}{1} = 9$ and $\dbinom{9}{8} = 9$.

(b) Prove that for each positive integer n, $\dbinom{n}{1} = n$ and $\dbinom{n}{n-1} = n$. [*Note:* Part (a) is the case when $n = 9$ and $n - 1 = 8$.]

58. (a) Verify that $\dbinom{7}{2} = \dbinom{7}{5}$.

(b) Let r and n be integers with $0 \le r \le n$. Prove that $\dbinom{n}{r} = \dbinom{n}{n-r}$. [*Note:* Part (a) is the case when $n = 7$ and $r = 2$.]

59. Prove that for any positive integer n,
$$2^n = \dbinom{n}{0} + \dbinom{n}{1} + \dbinom{n}{2} + \cdots + \dbinom{n}{n}.$$

[*Hint:* $2 = 1 + 1$.]

60. Prove that for any positive integer n,
$$\dbinom{n}{0} - \dbinom{n}{1} + \dbinom{n}{2} - \dbinom{n}{3} + \dbinom{n}{4} - \cdots$$
$$+ (-1)^k\dbinom{n}{k} + \cdots + (-1)^n\dbinom{n}{n} = 0.$$

61. (a) Let f be the function given by $f(x) = x^5$. Let h be a nonzero number and compute $f(x + h) - f(x)$ (but leave all binomial coefficients in the form $\dbinom{5}{r}$ here and below).

(b) Use part (a) to show that h is a factor of $f(x + h) - f(x)$ and find $\dfrac{f(x + h) - f(x)}{h}$.

(c) If h is *very* close to 0, find a simple approximation of the quantity $\dfrac{f(x + h) - f(x)}{h}$. [See part (b).]

---

* 🟢 indicates exercises that are relevant to calculus.

62. Do Exercise 61 with $f(x) = x^8$ in place of $f(x) = x^5$.

63. Do Exercise 61 with $f(x) = x^{12}$ in place of $f(x) = x^5$.

64. Let n be a fixed positive integer. Do Exercise 61 with $f(x) = x^n$ in place of $f(x) = x^5$.

65. Total credit card debt (in billions of dollars) in the United States is approximated by

$$f(x) = .0021x^3 + .89x^2 + 9.1x + 32.1 \qquad (0 \le x \le 24).$$

where $x = 0$ corresponds to 1980.* Suppose we want to describe this situation by a function g, for which $x = 0$ corresponds to 1990 instead of 1980.
(a) Explain why the function $g(x) = f(x + 10)$ will work.
(b) How is the graph of $g(x)$ related to the graph of $f(x)$? [*Hint:* See Section 3.4.]
(c) Compute and simplify the rule of $g(x)$.

66. According to data and projections in an article in *USA Today*, the number of Internet phone calls (in millions) is approximated by

$$f(x) = .075x^3 - .86x^2 + 3.66x - 4 \qquad (3 \le x \le 7),$$

where $x = 3$ corresponds to 2003.† Write the rule of a function $g(x)$, which provides the same information as f, but has $x = 0$ corresponding to 2003. [*Hint:* See Exercise 65.]

---

*Federal Reserve.
†November 28, 2003.

Thinkers

67. Let r and n be integers such that $0 \le r \le n$.
(a) Verify that $(n - r)! = (n - r)[n - (r + 1)]!$
(b) Verify that $(n - r)! = [(n + 1) - (r + 1)]!$
(c) Prove that $\binom{n}{r + 1} + \binom{n}{r} = \binom{n + 1}{r + 1}$ for any $r \le n - 1$.
[*Hint:* Write out the terms on the left side and use parts (a) and (b) to express each of them as a fraction with denominator $(r + 1)!(n - r)!$. Then add these two fractions, simplify the numerator, and compare the result with $\binom{n + 1}{r + 1}$.]
(d) Use part (c) to explain why each entry in Pascal's triangle (except the 1's at the beginning or end of a row) is the sum of the two closest entries in the row above it.

68. (a) Find these numbers and write them one *below* the next: 11^0, 11^1, 11^2, 11^3, 11^4.
(b) Compare the list in part (a) with rows 0 to 4 of Pascal's triangle. What's the explanation?
(c) What can be said about 11^5 and row 5 of Pascal's triangle?
(d) Calculate all integer powers of 101 from 101^0 to 101^8, list the results one under the other, and compare the list with rows 0 to 8 of Pascal's triangle. What's the explanation? What happens with 101^9?

7.5 Permutations and Combinations

How many different choices are there in a lottery in which you select six numbers from 1 to 44? How many seven-digit phone numbers can there be in any one area code? The answers to these and many other questions require systematic counting techniques that are introduced here.

EXAMPLE 1

Anne, Bill, Charlie, Dana, Elsie, and Fred enter a short-story competition in which three prizes will be awarded (and there will be no ties). If they are the only contestants, in how many possible ways can the prizes be awarded?

SOLUTION For convenience, we use initials. There are six possibilities for first place: *A, B, C, D, E, F*. For each possible first-place winner, there are five possible second-place winners.

1. 2.	1. 2.	1. 2.	1. 2.	1. 2.	1. 2.
A B	B A	C A	D A	E A	F A
A C	B C	C B	D B	E B	F B
A D	B D	C D	D C	E C	F C
A E	B E	C E	D E	E D	F D
A F	B F	C F	D F	E F	F E

So the total number of ways in which first and second prize can be awarded is $6 \cdot 5 = 30$. If, for instance, A and B take first and second, then there are four possibilities for third.

$$A, B, C; \quad A, B, D; \quad A, B, E; \quad A, B, F$$

Similarly, for *each* of the 30 ways in which first and second prizes can be awarded, there are four ways to award third prize. So there are $30 \cdot 4 = 120$ possible ways to award all three prizes.　■

The argument in Example 1 may be summarized like this.

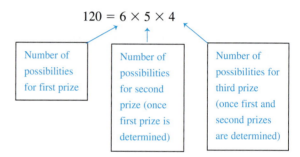

$$120 = 6 \times 5 \times 4$$

| Number of possibilities for first prize | Number of possibilities for second prize (once first prize is determined) | Number of possibilities for third prize (once first and second prizes are determined) |

When thought of this way, the example is an illustration of the following principle.

Fundamental Counting Principle

Consider a list of k events. Suppose the first event can occur in n_1 ways; the second event can occur in n_2 ways (once the first event has occurred); the third event can occur in n_3 ways (once the first two events have occurred); and so on. Then the total number of ways in which all k events can occur is the product $n_1 n_2 n_3 \cdots n_k$.

EXAMPLE 2

How many license plates can be made consisting of three letters followed by 3 one-digit numbers, subject to these conditions: No plate may begin with I or O, and the first number may not be a zero?

SOLUTION   Consider the filling of each of the six positions on a license plate as an event. Since O and I are excluded, there are 24 possible letters for the first position ($n_1 = 24$). There are 26 possible letters for each of the second and third positions ($n_2 = 26$, $n_3 = 26$). Since 0 cannot appear in the fourth position (the first position for a number), there are nine possibilities ($n_4 = 9$). Any digit from 0 through 9 may appear in the last two positions, so there are 10 possible digits for each position ($n_5 = 10$, $n_6 = 10$). By the Fundamental Counting Principle, the total number of license plates is the product

$$24 \cdot 26 \cdot 26 \cdot 9 \cdot 10 \cdot 10 = 14{,}601{,}600. \quad \blacksquare$$

EXAMPLE 3

A committee consisting of eight Democrats and six Republicans must choose a chairperson, a vice-chairperson, a secretary, and a treasurer. The chairperson must be a Democrat, and the vice-chairperson must be a Republican. The secretary and treasurer may belong to either party. If no person can hold more than one office, in how many different ways can these four officers be chosen?

SOLUTION   Consider the choice of a chairperson as the first event. Since there are eight Democrats, there are eight possible ways for this event to occur ($n_1 = 8$). The second event is the choice of a vice-chairperson. It can occur in six ways because there are six Republicans ($n_2 = 6$). Once the chairperson and vice-chairperson have been determined, any of the remaining 12 people may be chosen secretary. So that event can occur in 12 ways ($n_3 = 12$). The fourth and last event ($k = 4$) is the choice of a treasurer, who can be any of the 11 people remaining ($n_4 = 11$). By the Fundamental Counting Principle, the number of ways in which all four officers can be chosen is the product $n_1 n_2 n_3 n_4 = 8 \cdot 6 \cdot 12 \cdot 11 = 6336$.

$$\blacksquare$$

▉▉ PERMUTATIONS

A **permutation** of a set of elements is an ordering of the elements in the set. For example, each of the following is a permutation of the numbers 1, 2, 3.

$$1, 2, 3 \quad 1, 3, 2 \quad 2, 1, 3 \quad 2, 3 \: 1 \quad 3, 1, 2 \quad 3, 2, 1.$$

You should convince yourself that these six permutations are all of the possible orders for the numbers 1, 2, 3. We can use the Fundamental Counting Principle to determine the number of possible permutations of any set.

EXAMPLE 4

The number of possible batting orders for a nine-person baseball team is the number of permutations of a set of nine elements. There are nine possibilities for first batter, eight possibilities for second (after the first one is determined), seven possibilities for third (after the first two are determined), and so on, down to the eighth (two possibilities) and ninth batter (one possibility). So the number of possible batting orders is

$$9 \cdot 8 \cdot 7 \cdot 6 \cdot 5 \cdot 4 \cdot 3 \cdot 2 \cdot 1 = 362{,}880. \quad \blacksquare$$

The argument in the example carries over to the general case with n in place of 9: To determine a permutation of n elements, there are n possible choices for first position, $n - 1$ choices for second position, and so on. Therefore, we have the following.

Permutations of n Elements

The total number of permutations of a set of n elements is

$$n! = n(n - 1)(n - 2) \cdots 3 \cdot 2 \cdot 1.^*$$

EXAMPLE 5

In how many ways can a 52-card deck of cards be shuffled (that is, how many different orderings of the deck are possible)?

Figure 7–17

SOLUTION   The number of orderings is the number of permutations of a 52-element set, namely, 52!. Using the factorial key on a calculator (located in the MATH menu or its PROB submenu), we see that this is *enormous* (approximately 8.066×10^{67}, as shown in Figure 7–17).   ■

In many cases, we are interested in the order of some (but not necessarily all) of the elements of a set. For instance, in Example 1 (the short-story contest), we considered the possible ways of ordering three of the six contestants (the prize winners). Such an ordering is called a permutation of six things taken three at a time. More generally, if $r \le n$, an ordering of r elements of an n-element set is called a **permutation of n elements taken r at a time.**

The number of permutations of n elements taken r at a time is denoted $_nP_r$. Example 1 shows that $_6P_3$, the number of permutations of six elements taken three at a time, is

$$_6P_3 = 6 \times 5 \times 4 = 120.$$

Note that $_6P_3$ is the product of three consecutive integers, beginning with 6 and working downward.

A similar pattern holds in other cases. For example, to compute $_8P_5$, the number of permutations of eight elements taken five at a time, there are eight choices for the first position, seven choices for the second, six choices for third, five choices for fourth, and four choices for the fifth position. By the Fundamental Counting Principle,

$$_8P_5 = 8 \cdot 7 \cdot 6 \cdot 5 \cdot 4 = 6720.$$

Thus, $_8P_5$ is the product of five consecutive integers, beginning with 8 and working downward. The same argument works in the general case.

Permutations of n Elements Taken r at a Time

Let r and n be positive integers with $r \le n$. Then

$_nP_r$ is the product of r consecutive integers, beginning with n and working downward.

---

*Factorial notation ($n!$) is explained on page 583.

EXAMPLE 6

Find $_{20}P_4$ and $_9P_6$.

SOLUTION For $_{20}P_4$, we start with 20 and go down 4 factors:

$$_{20}P_4 = \underbrace{20 \cdot 19 \cdot 18 \cdot 17}_{\text{4 factors}} = 116{,}280.$$

For $_9P_6$, we start with 9 and go down 6 factors.

$$_9P_6 = \underbrace{9 \cdot 8 \cdot 7 \cdot 6 \cdot 5 \cdot 4}_{\text{6 factors}} = 60{,}480 \qquad \blacksquare$$

By using an algebraic trick, we can find another useful formula for computing $_nP_r$. For example, since $\dfrac{3!}{3!} = 1$,

$$_8P_5 = 8 \cdot 7 \cdot 6 \cdot 5 \cdot 4 = \frac{8 \cdot 7 \cdot 6 \cdot 5 \cdot 4}{1} \cdot \frac{3!}{3!} = \frac{8 \cdot 7 \cdot 6 \cdot 5 \cdot 4 \cdot 3 \cdot 2 \cdot 1}{3!} = \frac{8!}{3!} = \frac{8!}{(8-5)!}.$$

A similar computation works in the general case (with n in place of 8 and r in place of 5) and leads to the following description.

$_nP_r$

> The number of permutations of n elements taken r at a time is
>
> $$_nP_r = \frac{n!}{(n-r)!}.$$

Although it is sometimes necessary to use the two formulas for $_nP_r$ directly, most numerical computations can be readily done on a calculator by using the *nPr* key (which is in the MATH menu or its PROB submenu).

EXAMPLE 7

Use a calculator to find $_{12}P_5$, $_{20}P_7$, and $_{15}P_9$.

SOLUTION See Figure 7–18. $\qquad \blacksquare$

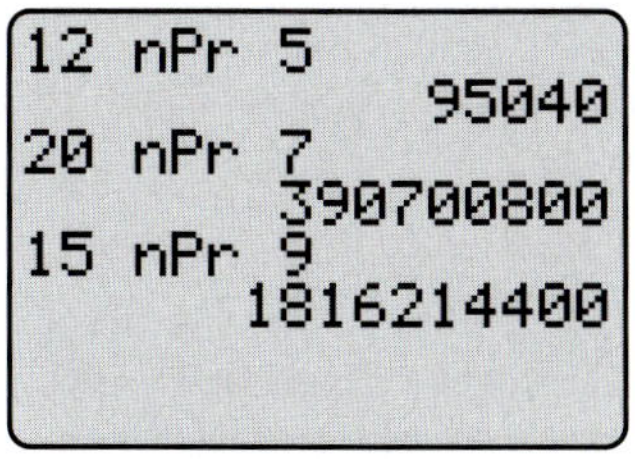

```
12 nPr 5
            95040
20 nPr 7
         390700800
15 nPr 9
        1816214400
```

Figure 7–18

▪▪ COMBINATIONS

To play the Ohio state lottery, you select six different integers from 1 to 44. The order in which you list them doesn't matter. If these six numbers are drawn in *any* order, you win. Each possible choice of six numbers is called a combination of 44 elements taken 6 at a time. More generally, a **combination of *n* elements taken *r* at a time** is any collection of r distinct objects in a set of n objects, without regard to the order in which the r objects might be chosen.

EXAMPLE 8

The alphabet has 26 letters. So the letters B, Q, T form a combination of 26 elements taken 3 at a time. If we list these three letters as B, T, Q, or Q, B, T, or Q, T, B, or T, B, Q, or T, Q, B, we still have the *same combination* because the order doesn't matter. The situation with permutations is quite different. There are six different permutations of these three letters (BQT, BTQ, QBT, QTB, TBQ, TQB) because order *does* matter for permutations.   ∎

The number of combinations of n elements taken r at a time is denoted $_nC_r$. We can compute this number by considering the relationship between combinations and permutations. A *permutation* of n elements taken r at a time is determined by two events:

1. A choice of r elements in the n-element set, that is, a combination of n elements taken r at a time; and

2. A specific ordering of these r elements, that is, a permutation of r elements.

So the number $_nP_r$ of permutations of n elements taken r at a time is the same as the total number of ways in which events 1 and 2 can occur. By the Fundamental Counting Principle,

$$_nP_r = \begin{pmatrix} \text{Number of ways} \\ \text{event 1 can occur} \end{pmatrix} \times \begin{pmatrix} \text{Number of ways} \\ \text{event 2 can occur} \end{pmatrix}$$

$$_nP_r = \begin{pmatrix} \text{Number of combinations of} \\ n \text{ elements taken } r \text{ at a time} \end{pmatrix} \times \begin{pmatrix} \text{Number of permutations} \\ \text{of } r \text{ elements} \end{pmatrix}$$

$$\frac{n!}{(n-r)!} = {_nC_r} \times r!$$

Dividing both sides of this last equation by $r!$ we have the following result.

Combinations of n Elements Taken r at a Time

The number of combinations of n elements taken r at a time is

$$_nC_r = \frac{n!}{r!(n-r)!}.^*$$

---

*If you have read Section 7.4, you see that the number of combinations of n elements taken r at a time is precisely the binomial coefficient, which is why the same notation $_nC_r$ is used for both.

EXAMPLE 9

The number of three-person committees that can be selected from a group of ten people is the number of combinations of ten elements taken three at a time:

$$_{10}C_3 = \frac{10!}{3!(10-3)!} = \frac{10!}{3! \cdot 7!} = 120.$$ ∎

EXAMPLE 10

The number of possible five-card poker hands is the number of combinations of 52 elements taken 5 at a time.

$$_{52}C_5 = \frac{52!}{5!(52-5)!} = \frac{52!}{5! \cdot 47!} = 2{,}598{,}960.$$ ∎

The easiest way to evaluate the number $_nC_r$ is to use the *nCr* key on a calculator (in the MATH menu or its PROB submenu).

EXAMPLE 11

To play the Ohio state lottery, which has a minimum prize of $4 million, you select six different integers from 1 to 44. At $1 per ticket (one choice), how much would you have to spend to guarantee that one of your tickets would be a winner?

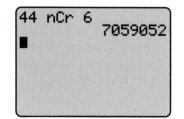

Figure 7–19

SOLUTION   To guarantee that you have a winning ticket, you must have a ticket for every possible six-number choice. The number of choices is the number of combinations of 44 elements taken 6 at a time, namely, $_{44}C_6$. At $1 per ticket, this would cost $7,059,052, as shown in Figure 7–19. Not a very good investment if the prize is $4 million (even worse if someone else also has a winning ticket and you have to split the prize). ∎

EXAMPLE 12

The city council consists of 18 Democrats, 20 Republicans, and 7 Independents. In how many ways can you select a committee of three Democrats, four Republicans, and two Independents?

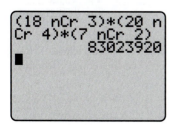

Figure 7–20

SOLUTION   The number of ways of choosing 3 of the 18 Democrats is $_{18}C_3$, the number of combinations of 18 elements taken 3 at a time. Similarly, the number of ways of choosing four Republicans is $_{20}C_4$, and the number of ways of choosing two Independents is $_7C_2$. By the Fundamental Counting Principle, the number of ways in which all these choices can be made is the product

$$_{18}C_3 \cdot {}_{20}C_4 \cdot {}_7C_2 = 83{,}023{,}920,$$

as shown in Figure 7–20. ∎

EXERCISES 7.5

In Exercises 1–16, compute the number.

1. $_4P_3$
2. $_7P_5$
3. $_8P_8$
4. $_5P_1$

5. $_{12}P_2$
6. $_{11}P_3$
7. $_{14}C_2$
8. $_{20}C_3$

9. $_{99}C_{95}$
10. $_{65}C_{60}$
11. $_nP_2$
12. $_nP_{n-1}$

13. $_nC_{n-1}$
14. $_nC_1$
15. $_nC_2$
16. $_nP_1$

In Exercises 17–22, find the number n that makes the statement true.

17. $_nP_4 = 8(_nP_3)$
18. $_nP_5 = 7(_nP_4)$

19. $_nP_6 = 9(_{n-1}P_5)$
20. $_nP_7 = 11(_{n-1}P_6)$

21. $_nP_5 = 21(_{n-1}P_3)$
22. $_nP_6 = 30(_{n-2}P_4)$

In Exercises 23–32, use the Fundamental Counting Principle to answer the question.

23. Every Social Security number has nine digits. How many possible Social Security numbers are there?

24. (a) How many five-digit zip codes are possible?
    (b) How many nine-digit zip codes are possible?

25. How many four-letter radio station call letters are possible if each must start with either W or K?

26. How many seven-digit phone numbers can be formed from the numerals 0 through 9 if none of the first three digits can be 0?

27. An early-bird restaurant menu offers 6 appetizers, 3 salads, 12 entrees, and 4 desserts. How many different meals can be ordered by choosing one item from each category?

28. A man has four pairs of pants, six sports coats, and eight ties. How many different outfits can he wear?

29. How many five-letter identification codes can be formed from the first 20 letters of the alphabet if no repetitions are allowed?

30. How many five-digit numbers can be formed from the numerals 2, 3, 4, 5, 6 if
    (a) no repeated digits are allowed?
    (b) repeated digits are allowed?

31. In how many different ways can an eight-question true-false test be answered?

32. An exam consists of ten multiple-choice questions with four choices for each question. In how many ways can this exam be answered?

In Exercises 33–38, use permutations to answer the question.

33. In how many different orders can eight people be seated in a row?

34. In how many orders can six pictures be arranged in a vertical line?

35. How many outcomes are possible in a four-team tournament (in which there are no ties)?

36. In how many ways can six mathematics texts be stacked in one pile on your desk?

37. In how many ways can the first four people in the batting order of a nine-player baseball team be chosen?

38. In how many different ways can first through third prizes be awarded in a ten-person race (assuming no ties)?

In Exercises 39–44, use combinations to answer the question.

39. A student must answer any five questions on an eight-question exam. In how many ways can this be done?

40. How many different straight lines are determined by eight points in the plane, no three of which lie on the same straight line? [*Hint:* A straight line is determined by two points.]

41. How many different six-card cribbage hands are possible from a 52-card deck?

42. Six pens are to be randomly selected from 100 newly manufactured pens to estimate the defect rate. How many six-pen samples are possible?

43. A committee of five students is to be chosen from a class of 22. How many different committees are possible?

44. An ice-cream store has 31 flavors. How many different two-scoop cones can they offer (assuming two different flavors for each cone)?

In the remaining exercises, use any appropriate technique (or several of them) to answer the question.

45. In how many ways can four men and eight women be seated in a row of 12 chairs if men sit at both ends of the row?

46. In how many ways can six men and six women be seated in a row of 12 chairs if the women sit in even-numbered seats?

47. How many different positive integers can be formed from the digits 3, 5, 7, 9 if no repeated digits are allowed? [*Hint:* How many one-digit numbers can be formed? How many two-digit ones?]

48. How many games must be played in an eight-team league if each team is to play every other team exactly twice?

49. How many batting orders are possible for a nine-player baseball team if the four infielders bat first and the pitcher bats last?

50. An electronic lock has five buttons. To unlock it, you press two buttons simultaneously, then two more, then two more. For instance, you might have to press 1-5, then 2-3, then 4-2. How many codes are possible for this lock?

51. (a) A three-digit code is needed to open the combination lock in the figure. To open it, rotate the dial to the right until the first digit of the code is at the marker at the top of the dial; then rotate the dial to the left until the second digit of the code is at the marker; then rotate the dial to the right until the third digit is at the marker. How many possible "combinations" are there for this lock?

   (b) Explain why combination locks should really be called permutation locks.

52. A dish holds six yellow, eight red, and four brown M&Ms and you randomly pick four of them. How many such samples are possible in which the M&Ms are

   (a) all red?   (b) all yellow?

   (c) 2 yellow, 2 brown?

   (d) 2 red, 1 yellow, 1 brown?

53. In the Michigan state lottery, you must pick 6 integers from 1 to 49. If a ticket (one choice) costs $1, how much would it cost to guarantee that you held a winning ticket?

54. Do Exercise 53 for the Powerball lottery that is played in 15 different states. You must select five distinct integers from 1 to 45 and one "Powerball number," also an integer from 1 to 45 (possibly the same as one of the five selected first). You win if your first five numbers and your Powerball number match the ones drawn in the lottery.

In Exercises 55 and 56, you may want to refer to the figure below, which shows a standard deck of cards. It consists of four suits (spades, hearts, diamonds, clubs), each with 13 cards (Ace, 2, 3, 4, 5, 6, 7, 8, 9, 10, Jack, Queen, King).

55. How many five-card poker hands consist of three aces and two kings?

Spades   Hearts   Diamonds   Clubs

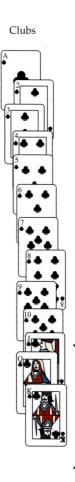

Face cards

56. How many five-card poker hands are flushes (five cards of the same suit)? [*Hint:* The number of club flushes is the number of ways in which five cards can be chosen from the 13 clubs.]

57. How many eight-digit numbers can be formed using three 6's and five 7's?

58. How many different five-person committees consisting of three women and two men can be chosen from a group of ten women and eight men?

59. How many ways can two committees, one of four people and one of three people, be chosen from a group of 11 people if no person serves on both committees?

60. How many six-digit numbers can be formed using only the numerals 5 and 9?

61. (a) A basketball squad consists of five people who can play center or forward and seven people who can play only guard. In how many ways can a team consisting of a center, two forwards, and two guards be chosen?

(b) How many teams are possible if the squad consists of two centers, four forwards, and six people who can play either guard or forward?

62. (a) How many different pairs are possible from a standard 52-card deck? [*Hint:* How many pairs can be formed from the four aces? How many from the four kings?]

(b) How many five-card poker hands consist of a pair of aces and three other cards, none of which are aces?

(c) How many five-card poker hands contain at least two aces?

(d) How many five-card poker hands are full houses (three of a kind and a pair of another kind)?

Thinkers

63. (a) In how many different orders can three objects be placed around a circle? (Note that ABC, BCA, and CAB are different orders when the three letters are in a line but the *same* order in a circular arrangement. However, ABC is not the same circular arrangement as ACB, because B is to the left of A in one case and to the right of A in the other.)

(b) Do part (a) for four objects.

(c) Do part (a) for five objects.

(d) Let *n* be a positive integer and do part (a) for *n* objects.

64. In how many ways can six different foods be arranged around the edge of a circular table? (See Exercise 63.)

65. In how many different orders could King Arthur and 12 knights be seated at the Round Table? (See Exercise 63.)

66. All students at a certain college are required to take year-long courses in economics, mathematics, and history during their first year. The economics classes meet at 9, 11, 1, and 3 o'clock; mathematics classes meet at 10, 12, 2, and 4 o'clock; history classes meet at 8, 10, 12, and 5 o'clock. How many different schedules are possible for a student? Assume that each class lasts 50 minutes and that a student has the same schedule for the entire year.

67. A promoter wants to make up a program consisting of ten acts. He has seven singing acts and nine instrumental acts available. If singing and instrumental acts are to be alternated, how many different program orders are possible? [*Hint:* First consider the possibilities when a singing act begins the program.]

7.5.A   *SPECIAL TOPICS*   Distinguishable Permutations

Figure 7–21

Suppose you have three identical red marbles, three identical white marbles, and two identical blue ones as in Figure 7–21. How many different color patterns can be formed by placing the eight marbles in a row? Each way of placing the marbles in a row is a permutation, and we know that there are 8! permutations of eight marbles. But the *same* color pattern may result from *different* permutations. To see how this can happen, mentally label the red marbles R_1, R_2, R_3, the white ones W_1, W_2, W_3, and the blue ones B_1, B_2. Here are several different permutations of the marbles that all produce the *same* color pattern shown in Figure 7–21.

$$B_1R_3B_2R_1W_1W_2R_2W_3; \qquad B_1R_2B_2R_3W_2W_1R_1W_3;$$

$$B_2R_1B_1R_2W_3W_2R_3W_1.$$

Permutations that produce different color pattern, such as the three just listed, are said to be *indistinguishable*. Observe that any one of these three indistinguishable permutations may be obtained from any of the others simply by rearranging the red marbles among themselves, the white marbles among themselves, and the blue marbles among themselves.

Permutations that produce different color patterns are said to be *distinguishable*. If two permutations are distinguishable, then one *cannot* be obtained from the other simply by rearranging the marbles of the same color. Finding the total number of different patterns is the same as finding the total number of distinguishable permutations of the eight marbles.

To find the total number of different color patterns (distinguishable permutations), we shall proceed indirectly. We shall examine the way in which a specific permutation of the marbles is determined. Any *permutation* of the marbles (for instance, $B_2 R_2 B_1 R_1 W_1 W_3 R_3 W_2$) is determined by these four things:

(i) A color pattern (in the example, the pattern B R B R W W R W).

(ii) The order in which the red marbles appear in the red positions of the pattern (in the example, $R_2 R_1 R_3$), that is, a permutation of the three red marbles.

(iii) The order in which the white marbles appear in the white positions of the pattern (in the example, $W_1 W_3 W_2$), that is, a permutation of the three white marbles.

(iv) The order in which the blue marbles appear in the blue positions of the pattern (in the example, $B_2 B_1$), that is, a permutation of the two blue marbles.

A specific choice of one possibility for *each* of items (i) to (iv) will lead to exactly one permutation of the eight marbles. Conversely, every permutation of the eight marbles uniquely determines a choice in each one of items (i) to (iv), as shown by the example above. So the total number of permutations of the eight marbles, namely, 8!, is the same as the total number of ways that items (i) to (iv) can all occur. According to the Fundamental Counting Principle, this number is the product of the numbers of ways each of the four items can occur. Therefore,

$$8! = \begin{pmatrix} \text{Number of} \\ \text{color} \\ \text{patterns} \end{pmatrix} \times \begin{pmatrix} \text{Number of} \\ \text{permutations of} \\ \text{three red marbles,} \\ \text{namely, 3!} \end{pmatrix} \times \begin{pmatrix} \text{Number of} \\ \text{permutations of} \\ \text{three white marbles,} \\ \text{namely, 3!} \end{pmatrix} \times \begin{pmatrix} \text{Number of} \\ \text{permutations of} \\ \text{two blue marbles,} \\ \text{namely, 2!} \end{pmatrix}.$$

If we let N denote the number of color patterns, this statement becomes

$$8! = N \cdot 3! \cdot 3! \cdot 2!.$$

Solving this equation for N, we see that

$$N = \frac{8!}{3! \cdot 3! \cdot 2!} = 560.$$

More generally, suppose that n and $k_1, k_2, \ldots, k_t$ are positive integers such that $n = k_1 + k_2 + \cdots + k_t$. Suppose that a set consists of n objects and that k_1 of these objects are all of one kind (such as red marbles), that k_2 of the objects are all of another kind (such as white marbles), that k_3 of the objects are all of a third kind, and so on. We say that two permutations of this set are **distinguishable** if one cannot be obtained from the other simply by rearranging objects of the same kind. The total number of distinguishable permutations can be found by using the same method that was used to find the total number of color patterns in the marble example (where we had $n = 8$ and $k_1 = 3$, $k_2 = 3$, $k_3 = 2$).

Distinguishable Permutations

Given a set of n objects in which k_1 are of one kind, k_2 are of a second kind, k_3 are of a third kind, and so on, then the number of distinguishable permutations of the set is

$$\frac{n!}{k_1! \cdot k_2! \cdot k_3! \cdots k_t!}$$

EXAMPLE 1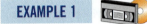

The number of distinguishable ways that the letters in the word TENNESSEE can be arranged is just the number of distinguishable permutations of the set consisting of the nine symbols T, E, N, N, E, S, S, E, E. There are four E's, two N's, two S's, and one T. So we apply the formula in the preceding box with $n = 9$; $k_1 = 4$; $k_2 = 2$; $k_3 = 2$; $k_4 = 1$; and find that the number of distinguishable permutations is

$$\frac{9!}{4!\,2!\,2!\,1!} = \frac{9 \cdot 8 \cdot 7 \cdot 6 \cdot 5 \cdot 4 \cdot 3 \cdot 2 \cdot 1}{4 \cdot 3 \cdot 2 \cdot 1 \cdot 2 \cdot 1 \cdot 2 \cdot 1 \cdot 1} = 3780. \quad \blacksquare$$

✓ **EXERCISES 7.5.A**

In Exercises 1–4, determine the number of distinguishable ways the letters in the word can be arranged.

1. LOOK

2. MISSISSIPPI

3. CINCINNATI

4. BOOKKEEPER

5. How many color patterns can be obtained by placing five red, seven black, three white, and four orange disks in a row?

6. In how many different ways can you write the algebraic expression $x^4 y^2 z^3$ without using exponents or fractions?

7.6 Introduction to Probability

Probability theory was first developed in the seventeenth century to analyze games of chance. It provides the mathematical tools for dealing with questions that involve uncertainty, such as these:

What is the chance of rain tomorrow?

How does smoking affect your chances of getting cancer?

If you drink and drive, what is the probability that you will have an accident?

We begin with some basic definitions. An **experiment** is any activity with an observable result; this result is called an **outcome** of the experiment.

EXAMPLE 1

What are the possible outcomes for each of the following experiments?

(a) Flip a coin.

(b) Draw a card from a standard deck.

(c) Test an item on an assembly line for satisfactory quality.

SOLUTION

(a) There are two possible outcomes: heads or tails.

(b) Since a standard deck has 52 cards, there are 52 possible outcomes (see the picture on page 599).

(c) Presumably, the outcomes of quality testing are "satisfactory" or "defective."
 ■

We first consider experiments with **equally likely** outcomes. Flipping a fair coin (one that is not weighted or shaved to favor one side over the other) is an example of such an experiment, as is rolling a fair die (one that isn't loaded so that one number comes up more often than the others).[*] A well-run factory, however, will produce far more satisfactory than defective items, so the two outcomes of a quality control test will not be equally likely.

The set of all possible outcomes of an experiment is called the **sample space** of the experiment.

EXAMPLE 2

Find the sample space for each of these experiments.

(a) A single die is rolled.

(b) Two coins are flipped, one after the other.

---

[*]Hereafter, all dice and coins are assumed to be fair.

SOLUTION

(a) The sample space is

$$\{1, 2, 3, 4, 5, 6\},$$

since these numbers are the possible outcomes of rolling the die.

(b) There are four possible outcomes.

Heads on both coins (HH)

Heads on the first coin and tails on the second (HT)

Tails on the first coin and heads on the second (TH)

Tails on both coins (TT)

Using the preceding labels, the sample space is the set

$$\{HH, HT, TH, TT\}. \quad \blacksquare$$

Any subset of the sample space (that is, any collection of some of the possible outcomes) is called an **event.**

EXAMPLE 3

When rolling a die, the sample space is $\{1, 2, 3, 4, 5, 6\}$. What events (subsets of the sample space) correspond to the following?

(a) An even number is rolled.

(b) A number greater than 2 is rolled.

(c) A number less than 7 is rolled.

SOLUTION

(a) $\{2, 4, 6\}$

(b) $\{3, 4, 5, 6\}$

(c) Rolling a die always results in a number less than 7, so this event is the entire sample space $\{1, 2, 3, 4, 5, 6\}$. $\quad \blacksquare$

The basic idea of probability theory is to assign to each event a number between 0 and 1 that indicates its likelihood.

Probability of an Event

If S is the sample space of an experiment with equally likely outcomes and E is an event, then the *probability* of E is denoted $P(E)$ and defined by

$$P(E) = \frac{\text{Number of outcomes in } E}{\text{Number of outcomes in } S}.$$

EXAMPLE 4

In the experiment of rolling a single die, find the probability of rolling

(a) 5;　　　　　　　　　　　　　　(b) an even number;

(c) a number greater than 2;　　　　(d) a number less than 7;

(e) a number greater than 6.

SOLUTION

(a) The sample space $S = \{1, 2, 3, 4, 5, 6\}$ contains six possible outcomes. The event E of "rolling a 5" has just one outcome. So the probability of rolling a 5 is

$$P(E) = \frac{\text{Number of outcomes in } E}{\text{Number of outcomes in } S} = \frac{1}{6}.$$

(b) The event F of "rolling an even number" has three possible outcomes (2, 4, and 6). Therefore, the probability of rolling an even number is

$$P(F) = \frac{\text{Number of outcomes in } F}{\text{Number of outcomes in } S} = \frac{3}{6} = \frac{1}{2}.$$

(c) The event G of "rolling a number greater than 2" has four possible outcomes (3, 4, 5, and 6). Hence,

$$P(G) = \frac{\text{Number of outcomes in } G}{\text{Number of outcomes in } S} = \frac{4}{6} = \frac{2}{3}.$$

(d) Rolling a number less than 7 is an event that *always* occurs because the set of possible outcomes (1, 2, 3, 4, 5, and 6) is the entire sample space S. Its probability is

$$P(S) = \frac{\text{Number of outcomes in } S}{\text{Number of outcomes in } S} = \frac{6}{6} = 1.$$

(e) The event T of rolling a number greater than 6 *never* occurs, so there are no outcomes for this event and its probability is

$$P(T) = \frac{\text{Number of outcomes in } T}{\text{Number of outcomes in } S} = \frac{0}{6} = 0. \quad \blacksquare$$

As Example 4 illustrates, the numerator of the fraction

$$P(E) = \frac{\text{Number of outcomes in } E}{\text{Number of outcomes in } S}$$

is always less than or equal to the denominator (because the event E cannot have more outcomes than the entire sample space). Therefore, we have the following.

Properties of Probability

For any event E,

$$0 \leq P(E) \leq 1.$$

If an event E must always occur, then $P(E) = 1$.

If an event E can never occur, the $P(E) = 0$.

EXAMPLE 5

Assuming that the probability of a girl being born is the same as that of a boy, find the probability that a family with three children has

(a) at least two girls;   (b) exactly two girls.

SOLUTION   The possible ways for three children to be born lead to the following sample space, in which b stands for boy and g for girl.

$$S = \{bbb, bbg, bgb, bgg, gbb, gbg, ggb, ggg\}$$

(a) The event of having at least two girls is

$$E = \{bgg, gbg, ggb, ggg\},$$

and

$$P(E) = \frac{\text{Number of outcomes in } E}{\text{Number of outcomes in } S} = \frac{4}{8} = \frac{1}{2}.$$

(b) The event of having exactly two girls is

$$F = \{bgg, gbg, ggb\},$$

whose probability is

$$P(F) = \frac{\text{Number of outcomes in } F}{\text{Number of outcomes in } S} = \frac{3}{8}. \quad \blacksquare$$

EXAMPLE 6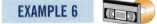

If a pair of dice are rolled as in Figure 7–22, find the probability that the total is
(a) 7;   (b) 11.

Figure 7–22

SOLUTION   Each outcome in the sample space can be denoted by an ordered pair of numbers: The first coordinate is the number showing on the first die and the second coordinate is the number showing on the second die. The sample space consists of all 36 such pairs, as shown here.

(1, 6)	(2, 6)	(3, 6)	(4, 6)	(5, 6)	(6, 6)
(1, 5)	(2, 5)	(3, 5)	(4, 5)	(5, 5)	(6, 5)
(1, 4)	(2, 4)	(3, 4)	(4, 4)	(5, 4)	(6, 4)
(1, 3)	(2, 3)	(3, 3)	(4, 3)	(5, 3)	(6, 3)
(1, 2)	(2, 2)	(3, 2)	(4, 2)	(5, 2)	(6, 2)
(1, 1)	(2, 1)	(3, 1)	(4, 1)	(5, 1)	(6, 1)

(a) A total of 7 can occur in six ways:

$$(1, 6), (2, 5), (3, 4), (4, 3), (5, 2), (6, 1).$$

So the probability of a total of 7 is $6/36 = 1/6$.

(b) A total of 11 can occur in only two ways: (5, 6) and (6, 5). So the probability of a total of 11 is $2/36 = 1/18$. ■

CAUTION

The set

$$\{2, 3, 4, 5, 6, 7, 8, 9, 10, 11, 12\}$$

consists of the possible totals when rolling two dice, but it cannot be used as the sample space in Example 6 bcause the outcomes in this set are *not* equally likely. For instance, the total 2 only occurs in one way (namely, (1, 1)) and therefore is less likely than a total of 7, which can occur in six ways.

The solution of many probability problems depends on the counting techniques of Section 7.5.

EXAMPLE 7

A committee of three people is chosen at random[*] from a group of four Republicans and six Democrats. Find the probability that the committee consists of

(a) three Democrats;   (b) one Republican and two Democrats.

SOLUTION   The sample space S consists of all the possible three-person committees. The number of outcomes in S is the number of ways of choosing three people from a group of ten, namely,

$$_{10}C_3 = \frac{10!}{3!7!} = 120.$$

(a) The number of ways of choosing three Democrats from the six available is

$$_6C_3 = \frac{6!}{3!3!} = 20,$$

so the probability that the committee consists entirely of Democrats is $20/120 = 1/6$.

(b) There are four choices for the Republican member. The number of ways of choosing two of the six Democrats is $_6C_2 = \frac{6!}{2!4!} = 15$. By the Fundamental Counting Principle, the number of ways of choosing the committee is

$$\begin{pmatrix} \text{Number of ways} \\ \text{of choosing one} \\ \text{Republican} \end{pmatrix} \times \begin{pmatrix} \text{Number of ways} \\ \text{of choosing two} \\ \text{Democrats} \end{pmatrix} = 4 \times {}_6C_2 = 4 \cdot 15 = 60.$$

The probability of such a committee being chosen is $60/120 = 1/2$. ■

---

[*]This means that each person is equally likely to be chosen.

▚ PROBABILITY OF TWO EVENTS

The next example illustrates a basic probability relationship when two events are involved.

EXAMPLE 8

In the experiment of rolling a die, consider these events:

$$E = \text{an even number is rolled} = \{2, 4, 6\}$$

$$F = \text{a number larger than 4 is rolled} = \{5, 6\}.$$

(a) Find $P(E)$ and $P(F)$.

(b) Find $P(E \text{ and } F)$, that is, the probability of rolling an even number that is greater than 4.

(c) Find $P(E \text{ or } F)$, that is, the probability of rolling an even number or one greater than 4.

(d) How do these four probabilities seem to be related?

SOLUTION

(a) Since the sample space $\{1, 2, 3, 4, 5, 6\}$ has six outcomes.

$$P(E) = \frac{3}{6} \qquad \text{and} \qquad P(F) = \frac{2}{6}.$$

(b) There is only one outcome in both E and F, namely, 6. So

$$P(E \text{ and } F) = \frac{1}{6}.$$

(c) The outcomes that are in E or F are $\{2, 4, 5, 6\}$. Hence,

$$P(E \text{ or } F) = \frac{4}{6}.$$

(d) Examination of these four fractions shows that

$$P(E) + P(F) - P(E \text{ and } F) = \frac{3}{6} + \frac{2}{6} - \frac{1}{6} = \frac{4}{6} = P(E \text{ or } F).$$

To understand why this works, look at the numerators.

$$P(E) + P(F) - P(E \text{ and } F) = P(E \text{ or } F)$$

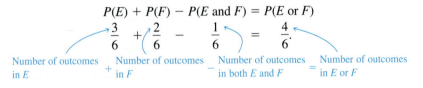

When you add the number of outcomes in E to the number of outcomes in F, you count the outcome in both E and F (namely, 6) twice. Subtracting it gives the corrrect total of outcomes in E or F. ■

By extending the argument in Example 8(d), it can be shown that the equation developed there holds in the general case.

Union Rule

> If E and F are events in the same sample space, then
>
> $$P(E \text{ or } F) = P(E) + P(F) - P(E \text{ and } F).$$

EXAMPLE 9

A pair of dice are rolled, one after the other. Find the probability that the first die shows 3 or the sum of the dice is 7 or 8.

SOLUTION Let E = the first die shows 3, and F = the sum is 7 or 8. Figure 7–23 shows the sample space and the events of E and F.

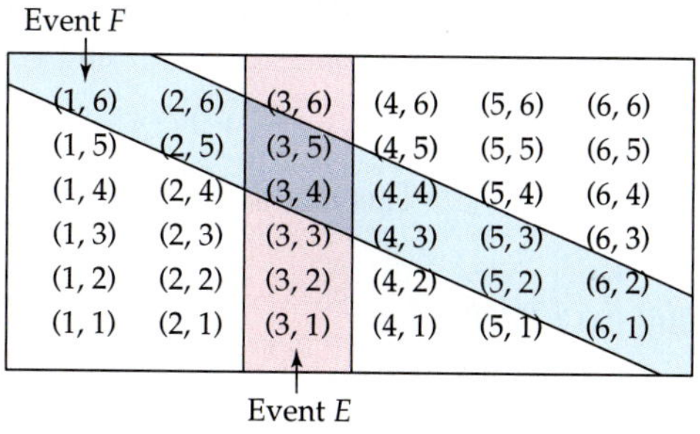

Figure 7–23

As you can see, the sample space has 36 outcomes. There are 6 outcomes in E, 11 outcomes in F, and 2 outcomes in both E and F. Therefore,

$$P(E) = \frac{6}{36}, \qquad P(F) = \frac{11}{36}, \qquad \text{and} \qquad P(E \text{ and } F) = \frac{2}{36}.$$

By the union rule,

$$P(E \text{ or } F) = P(E) + P(F) - P(E \text{ and } F)$$

$$= \frac{6}{36} + \frac{11}{36} - \frac{2}{36} = \frac{15}{36} = \frac{5}{12}.$$

So the probability of the first die being 3 or the total being 7 or 8 is $5/12$. ∎

⠶ MUTUALLY EXCLUSIVE EVENTS

Two events in the same sample space are said to be **mutually exclusive** if they cannot occur simultaneously.

EXAMPLE 10

In the experiment of rolling a single die, which of these pairs of events are mutually exclusive?

(a) "Rolling a 1" and "rolling an even number."

(b) "Rolling an odd number" and "rolling a number larger than 4."

SOLUTION

(a) Rolling a 1 is the event $\{1\}$, and rolling an even number is the event $\{2, 4, 6\}$. Since they have no outcomes in common, these events cannot occur simultaneously and hence are mutually exclusive.

(b) Rolling an odd number is the event $\{1, 3, 5\}$, and rolling a number larger than 4 is the event $\{5, 6\}$. The outcome 5 is in both events, meaning that both events occur when 5 is rolled. Therefore, these events are not mutually exclusive. ■

If E and F are mutually exclusive, then $P(E \text{ and } F) = 0$ because there are no outcomes in both E and F. Consequently, the union rule becomes

$$P(E \text{ or } F) = P(E) + P(F) - P(E \text{ and } F) = P(E) + P(F) - 0 = P(E) + P(F).$$

This fact is worth recording.

Mutually Exclusive Events

If E and F are mutually exclusive events in the same sample space, then

$$P(E \text{ or } F) = P(E) + P(F).$$

EXAMPLE 11

A five-card hand is dealt from a well-shuffled standard 52-card deck.[*] What is the probability that it will consist of five hearts or five spades?

SOLUTION   The sample space consists of the possible five-card hands. The number of such hands is the number of ways in which five cards can be chosen from 52 without regard to order, that is,

$$_{52}C_5 = \frac{52!}{5!47!} = 2,598,960.$$

The event E of getting five hearts can occur in as many ways as five cards can be chosen from the 13 hearts in the deck, namely,

$$_{13}C_5 = \frac{13!}{5!8!} = 1287.$$

---

*See the picture on page 599.

Similarly, the event F of getting five spades can occur in 1287 ways. Since E and F are mutually exclusive,

$$P(E \text{ or } F) = P(E) + P(F) = \frac{1287}{2,598,960} + \frac{1287}{2,598,960} = \frac{2574}{2,598,960}$$

$$\approx .00099$$

Rounded to three decimal places, this probability is .001. So the chances of being dealt five hearts or five spades are about 1 in 1000.   ■

▦ COMPLEMENTS

The **complement** of an event consists of all the outcomes in the sample space that are *not* in the event. In rolling a die, for instance, "rolling a 3 or 4" is the set $\{3, 4\}$. The complement of this event is the set $\{1, 2, 5, 6\}$, which is the event "rolling a 1, 2, 5, or 6." Note that either the event $\{3, 4\}$ or its complement *must* occur whenever the die is rolled.

The complement of an event E is denoted E'. In every case, E or E' must occur (since every possible outcome is in one or the other). Thus, $P(E \text{ or } E') = 1$. Since E and E' are mutually exclusive (no outcome is in both E and E'), we have

$$P(E \text{ or } E') = P(E) + P(E')$$

$$1 = P(E) + P(E').$$

Rearranging this last equation produces the following.

Complements

> If E is an event, then the probability of its complement E' occurring is
>
> $$P(E') = 1 - P(E).$$

EXAMPLE 12

If a committee of three people is chosen at random from a group of four Republicans and six Democrats, what is the probability that it will have at least one Republican member?

SOLUTION   The event "at least one Republican member" is the complement of the event E, "all three committee members are Democrats." In Example 7(a), we saw that $P(E) = 1/6$. Therefore,

$$P(E') = 1 - P(E) = 1 - \frac{1}{6} = \frac{5}{6}.   ■$$

EXAMPLE 13

You are one of 35 people at a party. A man offers to bet $100 that at least two people in the room have the same birthday (same month and day, not necessarily the same year). Should you take the bet?

SOLUTION*   The event "at least two people in the room have the same birthday" is the complement of the event $E =$ "no people in the room have the same birthday." So we first find the probability of E, which is

$$\frac{\text{Number of ways 35 people can have different birthdays}}{\text{Number of ways 35 people can have birthdays}}.$$

To find the denominator, we note that there are 365 possible birthdays for the first person in the room, 365 for the second, 365 for the third, and so on. By the Fundamental Counting Principle, the number of ways for all 35 birthdays to occur is

$$\underbrace{365 \cdot 365 \cdot 365 \cdots 365}_{\text{35 factors}} = 365^{35}.$$

To find the numerator, note that there are 365 possible birthdays for the first person in the room, 364 for the second (since they must have different birthdays), 363 for the third, and so on. By the Fundamental Counting Principle the number of ways in which 35 people can all have different birthdays is

$$365 \cdot 364 \cdot 363 \cdots \qquad \text{(35 factors working downward from 365),}$$

which is precisely the number $_{365}P_{35}$ (see the box on page 594). Thus, the probability that no people in the room have the same birthday is

$$\frac{_{365}P_{35}}{365^{35}} \approx .1856,$$

as shown in Figure 7–24. Therefore, the probability of the complement (at least two people have the same birthday) is

$$1 - .1856 = .8144.$$

So the man who wants to bet has a better than 81% probability of winning. Don't take the bet!   ■

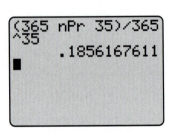

Figure 7–24

PROBABILITY DISTRIBUTIONS

In many real-life situations, it isn't possible to determine the probability of particular outcomes with the kind of theoretical analysis that we have used with dice, cards, etc. In such cases, probabilities are determined empirically, by running experiments, constructing models, or analyzing previous data. To predict the chance of rain tomorrow, for example, a weather forecaster might note that

---

*This solution ignores leap years and assumes that every day of the year is equally likely as a birthday.

there were 3800 days in the past century with atmospheric conditions similar to those today and that 1600 of these days were followed by rain. It would be reasonable to conclude that the probability of rain tomorrow is approximately $1600/3800 \approx .42$.

In a sample space with n equally likely outcomes, such as those studied earlier, each outcome has a probability of $1/n$, and the sum of these n probabilities is 1. This idea can be generalized to deal with experiments in which the outcomes are not all equally likely. Each outcome in the sample space is assigned a number (its probability) by some reasonable means (using theoretical analysis or an empirical approach), subject to the following conditions:

1. The probability of each outcome is a number between 0 and 1.

2. The sum of the probabilities of all the outcomes is 1.

Such an assignment of probabilities is called a **probability distribution.** It can be shown that the formulas developed above for mutually exclusive events and complements are valid for any probability distribution.

EXAMPLE 14

On the basis of survey data, a pollster concludes that Smith has a 42% chance of winning an upcoming election and that Jones has a 38% chance of winning. Assuming that these figures are accurate, find the probability that

(a) Smith or Jones will win;

(b) the third candidate, Brown, will win.

SOLUTION   The sample space consists of three outcomes:

$$S = \text{Smith wins}, \quad J = \text{Jones wins}, \quad B = \text{Brown wins}.$$

We are given that $P(S) = .42$ and $P(J) = .38$.

(a) Since Smith and Jones can't both win, S and J are mutually exclusive events; hence,

$$P(S \text{ or } J) = P(S) + P(J) = .42 + .38 = .80.$$

Thus, there is an 80% chance that Smith or Jones will win.

(b) The complement of the event "S or J" is B (the only other outcome in the sample space). Consequently,

$$P(B) = 1 - P(S \text{ or } J) = 1 - .80 = .20.$$

Alternatively, $P(B) = .20$ because the sum $P(S) + P(J) + P(B)$ must be 1. In any case, we conclude that Brown has a 20% chance of winning the election.   ■

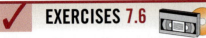

EXERCISES 7.6

In Exercises 1–6, list or describe a sample space for the experiment and state the number of outcomes in the sample space.

1. Three coins are flipped.

2. A day in March is chosen at random.

3. A marble is drawn from a jar containing two red and two blue marbles.

4. A coin is flipped and a die is rolled.

5. Two cards are drawn from a stack consisting of four jacks, four queens, four kings, and four aces.

6. Spin both spinners shown here.

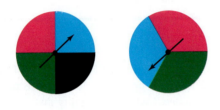

In Exercises 7–10, a single die is rolled. Find the probability of rolling

7. An odd number.

8. A number greater than 3.

9. A number other than 5.

10. Any number except 5 or 6.

In Exercises 11–14, three coins are flipped. Find the probability that

11. Exactly one is heads.   12. At least one is heads.

13. At least two are tails.   14. All three are tails.

In Exercises 15–18, a letter is chosen at random from the word

ABRACADABRA.

Find the probability that the chosen letter is

15. A   16. B

17. C or D   18. E

In Exercises 19–22, the 12-sided die shown here is rolled. Each of the numbers 1–6 appears twice (on opposite faces of the die). Find the probability of rolling

19. 6

20. An even number

21. A number greater than 3

22. 3 or 5

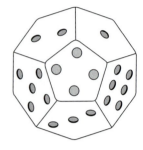

In Exercises 23–28, two dice are rolled, as in Example 6. Find the probability that the sum is

23. Less than 9   24. Greater than 7   25. 2 or 11

26. 8, 9, or 10   27. Even   28. Odd

In Exercises 29–36, a marble is drawn from an urn containing three red, four white, and eight blue marbles. Find the probability that this marble is

29. Red   30. White

31. Blue or white   32. Black

33. Not blue   34. Not white

35. Red or white   36. Not black

In Exercises 37–44, a card is drawn from a well-shuffled standard 52-card deck (see the picture on page 599). Find the probability that this card is

37. An ace

38. A heart

39. A red queen

40. A face card (king, queen, or jack)

41. A red face card

42. A red card that is not a face card

43. Not a face card and greater than 5 (ace is 1)

44. A black card less than 6 (ace is 1)

In Exercises 45–48, determine whether the events are mutually exclusive.

45. *Experiment:* Checking the attendance records for this class. *Events:* Finding a person with perfect attendance and finding a person with two absences.

46. *Experiment:* Drawing a card from a standard deck. *Events:* Drawing a spade and drawing a diamond.

47. *Experiment:* Drawing a card from a standard deck. *Events:* Drawing a heart and drawing an ace.

48. *Experiment:* Rolling a pair of dice. *Events:* One die shows 6, and the total of both dice is 9.

In Exercises 49–52, a pair of dice are rolled. Use the union rule to find the probability of the event.

49. The second die shows 5 or the sum of both dice is 9 or 10.

50. The first die shows 1 or the sum of both dice is 4 or 5.

51. The sum of both dice is 8 or both dice show the same number.

52. The sum of both dice is 9 or the difference (larger − smaller) of the dice is 3.

In Exercises 53–56, a card is drawn from a standard 52-card deck (see the picture on page 599). Find the probability that the card is

53. A spade or a face card

54. A heart or a non-face card

55. Black or an ace

56. Red or a face card

In Exercises 57–60, a jar contains eight green balls, numbered 1–8 and twelve red balls numbered 1–12. If one ball is taken from the jar, find the probability that this ball is

57. Red or even numbered

58. Green or odd numbered

59. Even or has a number larger than 4

60. Odd or has a number less than 6

In Exercises 61–70, a five-card hand is dealt from a well-shuffled standard 52-card deck (picture on page 599). Find the probability that it contains

61. Four aces

62. Four of a kind

63. Two aces and three kings

64. Five face cards

65. At least one ace

66. At least one spade

67. Four aces or four of a kind

68. At least one ace or at least one spade

69. Five red cards or five face cards

70. Only clubs or diamonds, or only non-face cards.

71. If a student taking a 10-question true-false quiz guesses randomly, find the probability that the student will get
(a) Ten questions correct
(b) Five questions correct and five wrong
(c) At least seven questions correct

72. A 5-question exam is to be given, with the questions taken from a list of 12 questions. If a student knows the answer to 7 of the 12 questions and does not know the answers to the other 5, what is the probability that the student will correctly answer
(a) All five questions on the exam?
(b) Exactly four questions on the exam?
(c) No more than two questions on the exam?

73. A committee of six people is to be randomly chosen from a group of five men and ten women. Find the probability that the committee consists of
(a) Three women and three men
(b) Four women and two men
(c) Six women

74. A gift package of 3 items is to be chosen randomly from a collection of 12 books, 6 boxes of candy, and 8 movie tickets. Find the probability that the gift package contains
(a) All books
(b) No movie tickets
(c) Two books
(d) One book, one box of candy, and one movie ticket

75. A shipment of 20 stereos contains 5 defective ones. If five stereos in this shipment are chosen at random, what is the probability that exactly two of them will be defective?

76. A package contains 14 fasteners, 2 of which are defective. If three fasteners are chosen at random from the package, what is the probability that all three are not defective?

77. In the Ohio state lottery, the bettor chooses six numbers between 1 and 44. If these six numbers are drawn in the lottery, the player wins a prize of several million dollars. What is the probability that a bettor who buys only one ticket will win the lottery?

78. In another state lottery (see Exercise 77), the bettor chooses six numbers from 1 to 50, but gets two such choices for each ticket bought. What is the probability that a bettor who buys only one ticket will win this lottery?

Exercises 79–82, deal with the birthday problem in Example 13. The same argument used there shows that the probability that no people in a group of n people have the same birthday is $\frac{365P_n}{365^n}$.

79. Fill the blanks in the following table. [*Hint:* Use the table feature of a calculator.]

Number of People	Probability That at Least Two Have the Same Birthday
10	
20	
25	
30	
40	
50	

80. What is the smallest number of people for which the probability that at least two of them have the same birthday is at least 1/2?

81. What is the probability that at least two American presidents have had the same birthday? [Through 2004, there were 42 different presidents.]

82. (a) What is the probability that two members of the U.S. Senate have the same birthday?
(b) What is the probability that two members of the U.S. House of Representatives have the same birthday? [Think!]

83. On the basis of data going back for several generations, the probability that a baby born to a certain family is a girl is .55. If such a family has four children, find the probability that they have
(a) Two boys and two girls
(b) All boys
(c) At least one boy

84. A coin is weighted so that when it is flipped it comes up heads 60% of the time. If this coin is flipped three times, find the probability of getting
(a) Three heads
(b) At least one head
(c) At least two tails

85. If a dart is thrown at the dart-board shown in the figure, find the probability that it hits
(a) Red
(b) Blue
(c) Green
(d) White

[*Hint:* What is the area of each color on the board? What is the area of the entire board?]

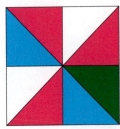

86. In the dartboard shown in the figure, the red inner circle is 3 inches in diameter; the outer edge of the green ring is 10 inches in diameter; the outer edge of the white ring is 18 inches in diameter; and the outer edge of the blue ring is 24 inches in diameter. If a dart is thrown at this board, find the probability that it hits
(a) Red
(b) Blue
(c) Green
(d) White

In Exercises 87–92, use the following table, which shows the number of alcohol-related automobile crashes in Ohio on New Year's Eve over a 10-year period.[*]

Time	Crashes
11 P.M.–Midnight	104
Midnight–1 A.M.	168
1–2 A.M.	201
2–3 A.M.	195
3–4 A.M.	117
4–5 A.M.	87
5–6 A.M.	44

Assume that no crashes occurred exactly on the hour. If a crash is chosen at random, what is the probability that it occurred between

87. Midnight and 1 A.M.   **88.** 11 P.M. and 4 A.M.

89. After 4 A.M.

90. Between 11 P.M. and 1 A.M. or between 2 A.M. and 6 A.M.

91. Between 11 P.M. and 1 A.M. or between midnight and 4 A.M.

92. Between 11 P.M. and 3 A.M. or between midnight and 5 A.M.

In Exercises 93–98, use the following population projections (in thousands) for the year 2005.[†]

Age	Male	Female
Less than 20	40,941	39,016
20–34	28,559	28,532
35–54	41,592	42,913
More than 54	29,505	36,555

---

*State of Ohio Records complied by the Cleveland *Plain Dealer*.
†Bureau of the Census, U.S. Department of Commerce.

If a person is randomly selected in 2005, find the probability that this person is

93. Female

94. Less than 20 years old

96. A female less than 35 years old

96. A male at least 20 years old

97. A male or in the 20–34 age bracket

98. A female or less than 35 years old

In Exercises 99–104, use the following results of a survey of 800 consumers, which shows the number of people in each income class who watch TV for the stated number of hours each week.

Annual Income	Hours of TV Watched per week		
	<10	10–20	>20
Less than $20,000	40	120	100
20,000–39,999	70	160	50
40,000–59,999	80	60	20
60,000 or more	40	50	10

If a person from the survey group is chosen at random, find the probability that this person

99. Has an income of at least $40,000

100. Watches TV at least 10 hours per week

101. Has an income under $40,000 and watches TV at least 10 hours per week

102. Has an income of at least $20,000 and watches TV less than 10 hours per week

103. Has an income of at least $60,000 or watches TV more than 20 hours per week

104. Watches TV between 10 and 20 hours per week and has an income under $40,000

7.7 Mathematical Induction

Mathematical induction is a method of proof that can be used to prove a wide variety of mathematical facts, including the Binomial Theorem and statements such as the following.

The sum of the first n positive integers is the number $\dfrac{n(n+1)}{2}$.

$2^n > n$ for every positive integer n.

For each positive integer n, 4 is a factor of $7^n - 3^n$.

All of the preceding statements have a common property. For example, a statement such as

The sum of the first n positive integers is the number $\dfrac{n(n+1)}{2}$

or, in symbols,

$$1 + 2 + 3 + \cdots + n = \frac{n(n+1)}{2},$$

is really an infinite sequence of statements, one for each possible value of n.

$$n = 1: \qquad\qquad 1 = \frac{1(2)}{2}$$

$$n = 2: \qquad 1 + 2 = \frac{2(3)}{2}$$

$$n = 3: \qquad 1 + 2 + 3 = \frac{3(4)}{2},$$

and so on. Obviously, there isn't time enough to verify every one of the statements on this list, one at a time. But we can find a workable method of proof by examining how each statement on the list is *related* to the *next* statement on the list.

For example, for $n = 50$, the statement is

$$1 + 2 + 3 + \cdots + 50 = \frac{50(51)}{2}.$$

At the moment, we don't know whether or not this statement is true. But just *suppose* that it were true. What could then be said about the next statement, the one for $n = 51$?

$$1 + 2 + 3 + \cdots + 50 + 51 = \frac{51(52)}{2}.$$

Well, *if* it is true that

$$1 + 2 + 3 + \cdots + 50 = \frac{50(51)}{2},$$

then adding 51 to both sides and simplifying the right-hand side would yield these equalities.

$$1 + 2 + 3 + \cdots + 50 + 51 = \frac{50(51)}{2} + 51$$

$$1 + 2 + 3 + \cdots + 50 + 51 = \frac{50(51)}{2} + \frac{2(51)}{2} = \frac{50(51) + 2(51)}{2}$$

$$1 + 2 + 3 + \cdots + 50 + 51 = \frac{(50 + 2)51}{2}$$

$$1 + 2 + 3 + \cdots + 50 + 51 = \frac{51(52)}{2}.$$

Since this last equality is just the original statement for $n = 51$, we conclude that

If the statement is true for $n = 50$, *then* it is also true for $n = 51$.

We have *not* proved that the statement actually *is* true for $n = 50$, but only that *if* it is, then it is also true for $n = 51$.

We claim that this same conditional relationship holds for any two consecutive values of n. In other words, we claim that for any positive integer k,

(1) **If** the statement is true for *n* = *k*, *then* it is also true for *n* = *k* + **1.**

The proof of this claim is the same argument used earlier (with *k* and *k* + 1 in place of 50 and 51): *If* it is true that

$$1 + 2 + 3 + \cdots + k = \frac{k(k+1)}{2}, \qquad \text{[Original statement for } n = k\text{]}$$

then adding *k* + 1 to both sides and simplifying the right side produces these equalities.

$$1 + 2 + 3 + \cdots + k + (k+1) = \frac{k(k+1)}{2} + (k+1)$$

$$1 + 2 + 3 + \cdots + k + (k+1)$$

$$= \frac{k(k+1)}{2} + \frac{2(k+1)}{2} = \frac{k(k+1) + 2(k+1)}{2}$$

$$1 + 2 + 3 + \cdots + k + (k+1) = \frac{(k+2)(k+1)}{2}$$

$$1 + 2 + 3 + \cdots + k + (k+1) = \frac{(k+1)[(k+1)+1]}{2}.$$

[Original statement for *n* = *k* + 1]

We have proved that claim (1) is valid for each positive integer *k*. We have *not* proved that the original statement is true for any value of *n*, but only that *if* it is true for *n* = *k*, then it is also true for *n* = *k* + 1. Applying this fact when *k* = 1, 2, 3, . . . , we see that

(2)
⎧
If the statement is true for *n* = 1,   *then* it is also true for *n* = 1 + 1 = 2;

If the statement is true for *n* = 2,   *then* it is also true for *n* = 2 + 1 = 3;

If the statement is true for *n* = 3,   *then* it is also true for *n* = 3 + 1 = 4;
.
.
.
If the statement is true for *n* = 50,   *then* it is also true for *n* = 50 + 1 = 51;

If the statement is true for *n* = 51,   *then* it is also true for *n* = 51 + 1 = 52;
.
.
.
⎩

and so on.

We are finally in a position to *prove* the original statement:

$$1 + 2 + 3 + \cdots + n = n(n+1)/2.$$

Obviously, it *is true* for *n* = 1, since 1 = 1(2)/2. Now apply in turn each of the propositions on list (2). Since the statement *is* true for *n* = 1, it must also be true for *n* = 2, and hence for *n* = 3, and hence for *n* = 4, and so on, for every value of *n*. Therefore, the original statement is true for *every* positive integer *n*.

The preceding proof is an illustration of the following principle.

Principle of Mathematical Induction

Suppose there is given a statement involving the positive integer n and that

(i) The statement is true for $n = 1$.

(ii) If the statement is true for $n = k$ (where k is any positive integer), then the statement is also true for $n = k + 1$.

Then the statement is true for every positive integer n.

Property (i) is simply a statement of fact. To verify that it holds, you must prove the given statement is true for $n = 1$. This is usually easy, as in the preceding example.

Property (ii) is a *conditional* property. It does not assert that the given statement *is* true for $n = k$, but only that *if* it is true for $n = k$, then it is also true for $n = k + 1$. So to verify that property (ii) holds, you need only prove this conditional proposition.

If the statement is true for $n = k$, *then* it is also true for $n = k + 1$.

To prove this or any conditional proposition, you must proceed as in the previous example: Assume the "if" part, and use this assumption to prove the "then" part. As we saw earlier, the same argument will usually work for any possible k. Once this conditional proposition has been proved, you can use it *together* with property (i) to conclude that the given statement is necessarily true for every n, just as in the preceding example.

Thus, proof by mathematical induction reduces to two steps.

Step 1   Prove that the given statement is true for $n = 1$.

Step 2   Let k be a positive integer. Assume that the given statement is true for $n = k$. Use this assumption to prove that the statement is true for $n = k + 1$.

Step 2 may be performed before step 1 if you wish. Step 2 is sometimes referred to as the **inductive step.** The assumption that the given statement is true for $n = k$ in this inductive step is called the **induction hypothesis.**

EXAMPLE 1

Prove that $2^n > n$ for every positive integer n.

SOLUTION   Here, the statement involving n is $2^n > n$.

Step 1   When $n = 1$, we have the statement $2^1 > 1$. This is obviously true.

Step 2   Let k be any positive integer. We assume that the statement is true for $n = k$; that is, we assume that $2^k > k$. We shall use this assumption to prove that the statement is true for $n = k + 1$, that is, that $2^{k+1} > k + 1$.

We begin with the induction hypothesis:* $2^k > k$. Multiplying both sides of this inequality by 2 yields

③
$$2 \cdot 2^k > 2k$$
$$2^{k+1} > 2k.$$

Since k is a positive integer, we know that $k \geq 1$. Adding k to each side of the inequality $k \geq 1$, we have

$$k + k \geq k + 1$$
$$2k \geq k + 1.$$

Combining this result with inequality ③, we see that

$$2^{k+1} > 2k \geq k + 1.$$

The first and last terms of this inequality show that $2^{k+1} > k + 1$. Therefore, the statement is true for $n = k + 1$. This argument works for any positive integer k. Thus, we have completed the inductive step. By the Principle of Mathematical Induction, we conclude that $2^n > n$ for every positive integer n. ∎

EXAMPLE 2

Simple arithmetic shows that

$$7^2 - 3^2 = 49 - 9 = 40 = 4 \cdot 10$$

and

$$7^3 - 3^3 = 343 - 27 = 316 = 4 \cdot 79.$$

In each case, 4 is a factor. These examples suggest that

For each positive integer n, 4 is a factor of $7^n - 3^n$.

This conjecture can be proved by induction as follows.

Step 1 When $n = 1$, the statement is "4 is a factor of $7^1 - 3^1$." Since $7^1 - 3^1 = 4 = 4 \cdot 1$, the statement is true for $n = 1$.

Step 2 Let k be a positive integer and assume that the statement is true for $n = k$, that is, that 4 is a factor of $7^k - 3^k$. Let us denote the other factor by D, so that the induction hypothesis is: $7^k - 3^k = 4D$. We must use this assumption to prove that the statement is true for $n = k + 1$, that is, that 4 is a factor of $7^{k+1} - 3^{k+1}$. Here is the proof.

$$7^{k+1} - 3^{k+1} = 7^{k+1} - 7 \cdot 3^k + 7 \cdot 3^k - 3^{k+1} \qquad \text{[Since } -7 \cdot 3^k + 7 \cdot 3^k = 0\text{]}$$
$$= 7(7^k - 3^k) + (7 - 3)3^k \qquad \text{[Factor]}$$
$$= 7(4D) + (7 - 3)3^k \qquad \text{[Induction hypothesis]}$$
$$= 7(4D) + 4 \cdot 3^k \qquad \text{[7 - 3 = 4]}$$
$$= 4(7D + 3^k). \qquad \text{[Factor out 4]}$$

---

*This is the point at which you usually must do some work. Remember that what follows is the "finished proof." It does not include all the thought, scratch work, false starts, and so on that were done before this proof was actually found.

From this last line, we see that 4 is a factor of $7^{k+1} - 3^{k+1}$. Thus, the statement is true for $n = k + 1$, and the inductive step is complete. Therefore, by the Principle of Mathematical Induction, the conjecture is actually true for every positive integer n. ■

Another example of mathematical induction, the proof of the Binomial Theorem, is given at the end of this section.

Sometimes a statement involving the integer n may be false for $n = 1$ and (possibly) other small values of n but true for all values of n beyond a particular number. For instance, the statement $2^n > n^2$ is false for $n = 1, 2, 3$, and 4. But it is true for $n = 5$ and all larger values of n. A variation on the Principle of Mathematical Induction can be used to prove this fact and similar statements. See Exercise 28 for details.

▋▋ A COMMON MISTAKE WITH INDUCTION

It is sometimes tempting to omit step 2 of an inductive proof when the given statement can easily be verified for small values of n, especially if a clear pattern seems to be developing. As the next example shows, however, *omitting step 2 may lead to error.*

EXAMPLE 3

An integer (> 1) is said to be *prime* if its only positive integer factors are itself and 1. For instance, 11 is prime, since its only positive integer factors are 11 and 1. But 15 is not prime because it has factors other than 15 and 1 (namely, 3 and 5). For each positive integer n, consider the number

$$f(n) = n^2 - n + 11.$$

You can readily verify that

$$f(1) = 11, \qquad f(2) = 13, \qquad f(3) = 17, \qquad f(4) = 23, \qquad f(5) = 31$$

and that *each of these numbers is prime.* Furthermore, there is a clear pattern: The first two numbers (11 and 13) differ by 2; the next two (13 and 17) differ by 4; the next two (17 and 23) differ by 6; and so on. On the basis of this evidence, we might conjecture that

For each positive integer n, the number $f(n) = n^2 - n + 11$ is prime.

We have seen that this conjecture is true for $n = 1, 2, 3, 4, 5$. Unfortunately, however, it is *false* for some values of n. For instance, when $n = 11$,

$$f(11) = 11^2 - 11 + 11 + 11^2 = 121.$$

But 121 is obviously *not* prime, since it has a factor other than 121 and 1, namely, 11. You can verify that the statement is also false for $n = 12$ but true for $n = 13$. ■

In the preceding example, the proposition

If the statement is true for $n = k$, then it is true for $n = k + 1$

is false when $k = 10$ and $k + 1 = 11$. If you were not aware of this and tried to complete step 2 of an inductive proof, you would not have been able to find a valid proof for it. Of course, the fact that you can't find a proof of a proposition doesn't always mean that no proof exists. But when you are unable to complete step 2, you are warned that there is a possibility that the given statement may be false for some values of n. This warning should prevent you from drawing any wrong conclusions.

██ PROOF OF THE BINOMIAL THEOREM

We shall use induction to prove that for every positive integer n,

$$(x + y)^n = x^n + \binom{n}{1}x^{n-1}y$$

$$+ \binom{n}{2}x^{n-2}y^2 + \binom{n}{3}x^{n-3}y^3 + \cdots + \binom{n}{n-1}xy^{n-1} + y^n.$$

This theorem was discussed and its notation explained in Section 7.4.

Step 1   When $n = 1$, there are only two terms on the right-hand side of the preceding equation, and the statement reads $(x + y)^1 = x^1 + y^1$. This is certainly true.

Step 2   Let k be any positive integer, and assume that the theorem is true for $n = k$, that is, that

$$(x + y)^k = x^k + \binom{k}{1}x^{k-1}y + \binom{k}{2}x^{k-2}y^2 + \cdots$$

$$+ \binom{k}{r}x^{k-r}y^r + \cdots + \binom{k}{k-1}xy^{k-1} + y^k.$$

[On the right-hand side of this equation, we have included a typical middle term $\binom{k}{r}x^{k-r}y^r$. The sum of the exponents is k, and the bottom part of the binomial coefficient is the same as the y exponent.] We shall use this assumption to prove that the theorem is true for $n = k + 1$, that is, that

$$(x + y)^{k+1} = x^{k+1} + \binom{k+1}{1}x^ky + \binom{k+1}{2}x^{k-1}y^2 + \cdots$$

$$+ \binom{k+1}{r+1}x^{k-r}y^{r+1} + \cdots + \binom{k+1}{k}xy^k + y^{k+1}.$$

We have simplified some of the terms on the right-hand side; for instance, $(k + 1) - 1 = k$ and $(k + 1) - (r + 1) = k - r$. But this is the correct statement for $n = k + 1$: The coefficients of the middle terms are $\binom{k+1}{1}$, $\binom{k+1}{2}$, $\binom{k+1}{3}$, and so on; the sum of the exponents of each middle term is $k + 1$, and the bottom part of each binomial coefficient is the same as the y exponent.

To prove the theorem for $n = k + 1$, we shall need this fact about binomial coefficients: For any integers r and k with $0 \leq r < k$,

④
$$\binom{k}{r+1} + \binom{k}{r} = \binom{k+1}{r+1}.$$

A proof of this fact is outlined in Exercise 67 on page 591.

To prove the theorem for $n = k + 1$, we first note that

$$(x + y)^{k+1} = (x + y)(x + y)^k.$$

Applying the induction hypothesis to $(x + y)^k$, we see that

$$(x + y)^{k+1} = (x + y)\left[x^k + \binom{k}{1}x^{k-1}y + \binom{k}{2}x^{k-2}y^2 + \cdots + \binom{k}{r}x^{k-r}y^r \right.$$

$$\left. + \binom{k}{r+1}x^{k-(r+1)}y^{r+1} + \cdots + \binom{k}{k-1}xy^{k-1} + y^k \right]$$

$$= x\left[x^k + \binom{k}{1}x^{k-1}y + \cdots + y^k \right] + y\left[x^k + \binom{k}{1}x^{k-1}y + \cdots + y^k \right].$$

Next we multiply out the right-hand side. Remember that multiplying by x increases the x exponent by 1 and multiplying by y increases the y exponent by 1.

$$(x + y)^{k+1} = \left[x^{k+1} + \binom{k}{1}x^k y + \binom{k}{2}x^{k-1}y^2 + \cdots + \binom{k}{r}x^{k-r+1}y^r \right.$$

$$\left. + \binom{k}{r+1}x^{k-r}y^{r+1} + \cdots + \binom{k}{k-1}x^2 y^{k-1} + xy^k \right]$$

$$+ \left[x^k y + \binom{k}{1}x^{k-1}y^2 + \binom{k}{2}x^{k-2}y^3 + \cdots + \binom{k}{r}x^{k-r}y^{r+1} \right.$$

$$\left. + \binom{k}{r+1}x^{k-(r+1)}y^{r+2} + \cdots + \binom{k}{k-1}xy^k + y^{k+1} \right]$$

$$= x^{k+1} + \left[\binom{k}{1} + 1 \right]x^k y + \left[\binom{k}{2} + \binom{k}{1} \right]x^{k-1}y^2 + \cdots$$

$$+ \left[\binom{k}{r+1} + \binom{k}{r} \right]x^{k-r}y^{r+1} + \cdots + \left[1 + \binom{k}{k-1} \right]xy^k + y^{k+1}.$$

Now apply statement ④ to each of the coefficients of the middle terms. For instance, with $r = 1$, statement ④ shows that

$$\binom{k}{2} + \binom{k}{1} = \binom{k+1}{2}.$$

Similarly, with $r = 0$,

$$\binom{k}{1} + 1 = \binom{k}{1} + \binom{k}{0} = \binom{k+1}{1},$$

and so on. Then the expression above for $(x + y)^{k+1}$ becomes

$$(x + y)^{k+1} = x^{k+1} + \binom{k+1}{1}x^k y + \binom{k+1}{2}x^{k-1}y^2 + \cdots$$

$$+ \binom{k+1}{r+1}x^{k-r}y^{r+1} + \cdots + \binom{k+1}{k}xy^k + y^{k+1}.$$

Since this last statement says that the theorem is true for $n = k + 1$, the inductive step is complete. By the Principle of Mathematical Induction, the theorem is true for every positive integer n.

✓ EXERCISES 7.7

In Exercises 1–18, use mathematical induction to prove that each of the given statements is true for every positive integer n.

1. $1 + 2 + 2^2 + 2^3 + 2^4 + \cdots + 2^{n-1} = 2^n - 1$

2. $1 + 3 + 3^2 + 3^3 + 3^4 + \cdots + 3^{n-1} = \dfrac{3^n - 1}{2}$

3. $1 + 3 + 5 + 7 + \cdots + (2n - 1) = n^2$

4. $2 + 4 + 6 + 8 + \cdots + 2n = n^2 + n$

5. $1^2 + 2^2 + 3^2 + \cdots + n^2 = \dfrac{n(n + 1)(2n + 1)}{6}$

6. $\dfrac{1}{2} + \dfrac{1}{4} + \dfrac{1}{8} + \cdots + \dfrac{1}{2^n} = 1 - \dfrac{1}{2^n}$

7. $\dfrac{1}{1 \cdot 2} + \dfrac{1}{2 \cdot 3} + \dfrac{1}{3 \cdot 4} + \cdots + \dfrac{1}{n(n + 1)} = \dfrac{n}{n + 1}$

8. $\left(1 + \dfrac{1}{1}\right)\left(1 + \dfrac{1}{2}\right)\left(1 + \dfrac{1}{3}\right) \cdots \left(1 + \dfrac{1}{n}\right) = n + 1$

9. $n + 2 > n$

10. $2n + 2 > n$

11. $3^n \geq 3n$

12. $3^n \geq 1 + 2n$

13. $3n > n + 1$

14. $\left(\dfrac{3}{2}\right)^n > n$

15. 3 is a factor of $2^{2n+1} + 1$

16. 5 is a factor of $2^{4n-2} + 1$

17. 64 is a factor of $3^{2n+2} - 8n - 9$

18. 64 is a factor of $9^n - 8n - 1$

19. Let c and d be fixed real numbers. Prove that
$$c + (c + d) + (c + 2d) + (c + 3d) + \cdots$$
$$+ [c + (n - 1)d] = \dfrac{n[2c + (n - 1)d]}{2}$$

20. Let r be a fixed real number with $r \neq 1$. Prove that
$$1 + r + r^2 + r^3 + \cdots + r^{n-1} = \dfrac{r^n - 1}{r - 1}.$$
[*Remember:* $1 = r^0$; so when $n = 1$, the left side reduces to $r^0 = 1$.]

21. (a) Write *each* of $x^2 - y^2$, $x^3 - y^3$, and $x^4 - y^4$ as a product of $x - y$ and another factor.
    (b) Make a conjecture as to how $x^n - y^n$ can be written as a product of $x - y$ and another factor. Use induction to prove your conjecture.

22. Let $x_1 = \sqrt{2}$; $\quad x_2 = \sqrt{2 + \sqrt{2}}$;
$x_3 = \sqrt{2 + \sqrt{2 + \sqrt{2}}}$; and so on. Prove that $x_n < 2$ for every positive integer n.

In Exercises 23–27, if the given statement is true, prove it. If it is false, give a counterexample.

23. Every odd positive integer is prime.

24. The number $n^2 + n + 17$ is prime for every positive integer n.

25. $(n + 1)^2 > n^2 + 1$ for every positive integer n.

26. 3 is a factor of the number $n^3 - n + 3$ for every positive integer n.

27. 4 is a factor of the number $n^4 - n + 4$ for every positive integer n.

28. Let q be a *fixed* integer. Suppose a statement involving the integer n has the following two properties:
    (i) The statement is true for $n = q$.
    (ii) *If* the statement is true for $n = k$ (where k is any integer with $k \geq q$), *then* the statement is also true for $n = k + 1$.
    Then we claim that the statement is true for every integer n greater than or equal to q.
    (a) Give an informal explanation that shows why this claim should be valid. Note that when $q = 1$, this claim is precisely the Principle of Mathematical Induction.
    (b) The claim made before part (a) will be called the *Extended Principle of Mathematical Induction*. State the two steps necessary to use this principle to prove that a given statement is true for all $n \geq q$. (See the discussion on page 620.)

In Exercises 29–34, use the Extended Principle of Mathematical Induction (Exercise 28) to prove the given statement.

29. $2n - 4 > n$ for every $n \geq 5$. (Use 5 for q here.)

30. Let r be a fixed real number with $r > 1$. Then $(1 + r)^n > 1 + nr$ for every integer $n \geq 2$. (Use 2 for q here.)

31. $n^2 > n$ for all $n \geq 2$

32. $2^n > n^2$ for all $n \geq 5$

33. $3^n > 2^n + 10n$ for all $n \geq 4$

34. $2n < n!$ for all $n \geq 4$

Thinkers

35. Let n be a positive integer. Suppose that there are three pegs and on one of them, n rings are stacked, each ring being smaller in diameter than the one below it (see the figure). We want to transfer the stack of rings to another peg according to these rules: (i) Only one ring may be moved at a time; (ii) a ring can be moved to any peg, provided that it is never placed on top of a smaller ring; (iii) the final order of the rings on the new peg must be the same as the original order on the first peg.

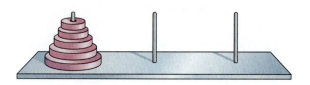

(a) What is the smallest possible number of moves when $n = 2$? when $n = 3$? when $n = 4$?

(b) Make a conjecture as to the smallest possible number of moves required for any n. Prove your conjecture by induction.

36. The basic formula for compound interest $T(x) = P(1 + r)^x$ was discussed on page 381. Prove by induction that the formula is valid whenever x is a positive integer. [*Note: P* and r are assumed to be constant.]

Chapter 7 Review

IMPORTANT FACTS & FORMULAS

- In an arithmetic sequence $\{a_n\}$ with common difference d,

$$a_n = a_1 + (n-1)d, \qquad \sum_{n=1}^{k} a_n = \frac{k}{2}(a_1 + a_k),$$

$$\sum_{n=1}^{k} a_n = ka_1 + \frac{k(k-1)}{2}\,d.$$

- In a geometric sequence $\{a_n\}$ with common ratio $r \neq 1$,

$$a_n = r^{n-1}a_1, \qquad \sum_{n=1}^{k} a_n = a_1\left(\frac{1 - r^k}{1 - r}\right).$$

- $n! = 1 \cdot 2 \cdot 3 \cdots (n-2)(n-1)n$

- $\displaystyle \binom{n}{r} = \frac{n!}{r!(n-r)!} = {}_nC_r$

■ *The Binomial Theorem:*

$$(x + y)^n = x^n + \binom{n}{1}x^{n-1}y + \binom{n}{2}x^{n-2}y^2 + \binom{n}{3}x^{n-3}y^3 + \cdots + \binom{n}{n-1}xy^{n-1} + y^n$$

$$= \sum_{j=0}^{n}\binom{n}{j}x^{n-j}y^j$$

■ The number of permutations of n elements is $n!$

■ $_nP_r = \dfrac{n!}{(n-r)!} = \left(\begin{array}{l}\text{The product of } r \text{ consecutive integers}\\ \text{starting with } n \text{ and working down}\end{array}\right)$

REVIEW QUESTIONS

In Questions 1–4, find the first four terms of the sequence $\{a_n\}$.

1. $a_n = 2n - 5$

2. $a_n = 3^n - 27$

3. $a_n = \left(\dfrac{-1}{n}\right)^2$

4. $a_n = (-1)^{n+1}(n-1)$

5. The number of people employed at Lucent Technologies over a three-year period can be approximated by the sequence $\{a_n\}$, where n is measured in months (with $n = 1$ corresponding to September 1999) and

$$a_n = -.153n^4 + 18.09n^3 - 668.19n^2 + 4980.5n + 148,270.^*$$

(a) About how many were employed in September 1999? In July 2001? In August 2002?

(b) Approximately when did employment drop below 60,000?

6. Find the fifth partial sum of the sequence $\{a_n\}$, where $a_1 = -4$ and $a_n = 3a_{n-1} + 2$.

7. Find the fourth partial sum of the sequence $\{a_n\}$, where $a_1 = 1/9$ and $a_n = 3a_{n-1}$.

8. $\displaystyle\sum_{n=0}^{4} 2^n(n+1) = ?$

9. $\displaystyle\sum_{n=2}^{4} (3n^2 - n + 1) = ?$

10. Average tuition and fees each year in private colleges can be approximated by the sequence given by

$$b_n = 9401.63 \cdot e^{.05467n},$$

where $n = 0$ corresponds to the 1990–1991 school year.[†]

(a) What were the average tuition and fees in 2000–2001? What are they this year?

(b) If you started at a private college with average tuition and fees in Fall 1990 and graduated after four years, what was the total that you (or your parents) spent on tuition and fees?

(c) Do part (b), assuming that you start in Fall 2002.

In Questions 11–14, find a formula for a_n; *assume that the sequence is arithmetic.*

11. $a_1 = 3$ and the common difference is -6.

12. $a_2 = 4$ and the common difference is 3.

13. $a_1 = -5$ and $a_3 = 7$.

14. $a_3 = 2$ and $a_7 = -1$.

15. Find the 11th partial sum of the arithmetic sequence with $a_1 = 5$ and common difference -2.

16. Find the 12th partial sum of the arithmetic sequence with $a_1 = -3$ and $a_{12} = 16$.

In Questions 17–20, find a formula for a_n; *assume that the sequence is geometric.*

17. $a_1 = 2$ and the common ratio is 3.

18. $a_1 = 5$ and the common ratio is $-1/2$.

19. $a_2 = 192$ and $a_7 = 6$.

20. $a_3 = 9/2$ and $a_6 = -243/16$.

21. Find the fifth partial sum of the geometric sequence with $a_1 = 1/4$ and common ratio 3.

22. Find the sixth partial sum of the geometric sequence with $a_1 = 5$ and common ratio $1/2$.

23. Find numbers b, c, d such that 4, b, c, d, 23 are the first five terms of an arithmetic sequence.

24. Find numbers c and d such that 8, c, d, 27 are the first four terms of a geometric sequence.

25. The amount spent on movie theater admissions (in billions of dollars) in year n is approximated by a geometric sequence $\{a_n\}$ in which $n = 1$ corresponds to 1994.[‡]

(a) If $5.123 billion was spent in 1994 and $5.5067 billion was spent in 1995, find a formula for a_n.

(b) How much was spent in 2002?

(c) What was the total amount spent from 1994 to 2002?

---

*Based on data from Lucent Technologies.
[†]Based on data from the College Board.

[‡]Based on data from the U.S. Bureau of Economic Analysis.

26. Attendance at college football games in year n is approximated by an arithmetic sequence $\{b_n\}$ in which $n = 1$ corresponds to 1995.[*]
    (a) If attendance was 35,410,000 in 1995 and 36,320,000 in 1996, find the attendance in 2003.
    (b) What was the total attendance from 1995 to 2000?

In Questions 27 and 28, find the sums, if they exist.

27. $\displaystyle\sum_{n=1}^{\infty} \frac{1}{2^{n-1}}$

28. $\displaystyle\sum_{n=1}^{\infty} \left(\frac{-1}{4^n}\right)$

29. Use the Binomial Theorem to show that $(1.02)^{51} > 2.5$.

30. What is the coefficient of $u^3 v^2$ in the expansion of $(u + 5v)^5$?

31. $\displaystyle\binom{15}{12} = ?$

32. $\displaystyle\binom{18}{3} = ?$

33. Let n be a positive integer. Simplify $\displaystyle\binom{n+1}{n}$.

34. Use the Binomial Theorem to expand $(\sqrt{x} + 1)^5$. Simplify your answer.

35. $\dfrac{20!5!}{6!17!} = ?$

36. $\dfrac{7! - 5!}{4!} = ?$

37. Find the coefficient of $x^2 y^4$ in the expansion of $(2y + x^2)^5$.

In Questions 38–42, compute the number.

38. $_7P_3$

39. $_{24}P_6$

40. $_7C_3$

41. $_{24}C_{20}$

42. $_kC_t$

43. How many three-digit numbers are there in which the first digit is even and the second is odd?

44. How many possible finishes (first, second, third, place) are there in an eight-horse race?

45. A woman has three skirts, four blouses, and two scarves. How many different outfits can she wear (an outfit being a skirt, blouse, and scarf)?

46. A bridge hand consists of 13 cards from a standard 52-card deck. How many different bridge hands are there?

47. How many different committees consisting of two men and two women can be formed from a group of eight women and five men?

48. List a sample space for the experiment of flipping a coin four times in succession.

49. If two dice are rolled, what is the probability that the sum is 8?

50. Which event is more likely: rolling a 3 with a single die or rolling a total of 6 with two dice?

51. Assuming that the probability of a girl being born is the same as that of a boy, find the probability that a family with four children has three girls and a boy.

52. If a card is drawn from a well-shuffled standard 52-card deck, what is the probability that it is a 6 or a king?

In Exercises 53–56, a card is drawn from a well-shuffled standard 52-card deck. Find the probability that this card is

53. An ace

54. A four or a queen

55. Black or a king

56. Red or even-numbered (2, 4, 6, 8, 10)

57. A committee of four people is to be randomly chosen from a group of 8 mathematicians and 12 chemists. What is the probability that the committee will consist of two mathematicians and two chemists?

58. Two stale candy bars are inadvertently dropped in a box with 18 fresh bars of the same brand. If you take three bars from the box, what is the probability that all of them will be fresh?

In Exercises 59–62, use the following table, which shows the 2002 U.S. population by race.[†]

Race[‡]	Population
African American	36,622,825
Asian	11,534,748
Native American[§]	3,460,424
White	232,713,539
Other	4,037,162

If a person is chosen at random, what is the probability that this person is

59. Native American

60. Nonwhite

[†]Bureau of the Census, U.S. Department of Commerce.
[‡]There were 38,641,406 people of Hispanic or Latino origin. They can be of any race and are included in the groups listed.
[§]Includes Native Alaskans and Native Hawaiians.

[*]Based on data from the NCAA.

61. African American or Asian

62. White or Other

63. Prove that for every positive integer n,

$$\frac{1}{3} + \frac{1}{3^2} + \frac{1}{3^3} + \cdots + \frac{1}{3^n} = \frac{3^n - 1}{2 \cdot 3^n}.$$

64. Prove that for every positive integer n,

$$1^3 + 2^3 + 3^3 + \cdots + n^3 = \frac{n^2(n + 1)^2}{4}.$$

65. Prove that for every positive integer n,

$$1 + 5 + 5^2 + 5^3 + \cdots + 5^{n-1} = \frac{5^n - 1}{4}.$$

66. Prove that $2^n \geq 2n$ for every positive integer n.

67. If x is a real number with $|x| < 1$, then prove that $|x^n| < 1$ for all $n \geq 1$.

68. Prove that for any positive integer n,

$$1 + 5 + 9 + \cdots + (4n - 3) = n(2n - 1).$$

69. Prove that for any positive integer n,

$$1 + 4 + 4^2 + 4^3 + \cdots + 4^{n-1} = \frac{1}{3}(4^n - 1).$$

70. Prove that $3n < n!$ for every $n \geq 4$.

71. Prove that for every positive integer n, 8 is a factor of $9^n - 8n - 1$.

DISCOVERY PROJECT 7   Racking, Stacking, and Packing

Charles Shoffner/Index Stock Imagery

Arranging objects in orderly space-filling patterns is a common occurrence throughout our experience; we see 50 stars in a rectangular field on the U.S. flag, 15 balls in a pool ball rack, and boxes of 88 oranges or 64 crayons. What drives these patterns? Often it is purely aesthetics—we like to have objects arranged in polygons, pyramids, cubes, or other regular objects. First, let's start with a historical pattern change.

In 1959, the number of states in the United States increased from 48 to 50. It's easy to arrange 48 stars in a roughly rectangular pattern; use six rows of eight. Forty-nine stars is also easy; use seven rows of seven, giving a square pattern. Unfortunately, the field of blue on the U.S. flag is not square; it's slightly longer than it is wide. The solution adopted was to offset the lines of stars to make them closer together, as shown below. This general kind of problem is called a packing problem: How do you optimize the organization of regular or irregular objects to fit the most or best into a fixed area?

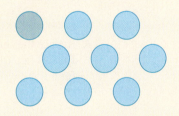

1. How would you arrange 50 objects in offset rows to fit a roughly rectangular pattern? (You can check your solution against any U.S. flag.)

Similar problems occur in gaming. In the game of eight-ball, played on a pool table, balls are prepared for the game by placing 15 numbered balls in an equilateral triangular pattern, called a "rack," with 5 balls on each side.

2. How many balls are there in a game of eight-ball?

3. What is the general formula for the number of pool balls in a rack with n balls on a side?

The exercises become more interesting as you move to three-dimensional forms. For example, spherical objects can be stacked with offset layers in the shape of trianglular or square pyramids as long as the bottom layer sits on some type of rack. This is the custom when stacking cannonballs for artillery emplacements. It turns out that triangular pyramids are more stable, so we will consider only those stacks.

4. In a triangular stack, each nested layer is an equilateral triangle. How many cannonballs are there in a triangular stack that is five layers high?

5. One approach to finding a general formula for the number of spheres in a triangular stack with n layers is to calculate $\sum_{j=1}^{n} \left(\sum_{k=1}^{j} k \right)$. Why does this formula work?

6. Reduce the sum in Exercise 5 to a polynomial that tells the number of spheres in a triangular stack. Does this formula produce the same result for a five-layer stack as you got in Exercise 4?

CHAPTER 8

Analytic Geometry

Peter Menzel/Stock, Boston Inc./PictureQuest

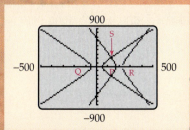

Calling all ships!

The planets travel in elliptical orbits around the sun, and satellites travel in elliptical orbits around the earth. Parabolic reflectors are used in spotlights, radar antennas, and satellite dishes. The long-range navigation system (loran) uses hyperbolas to enable a ship to determine its exact location. See Exercise 48 on page 659.

Chapter Outline

Interdependence of Sections

The first three sections of this chapter are essentially independent of one another, but a few exercises in Sections 8.2 and 8.3 assume knowledge of preceding section(s). Sections 8.1–8.3 are prerequisites for Section 8.4.

The discussion of analytic geometry that was begun in Secton 1.1 is continued here, with an examination of conic sections (which have played a significant role in mathematics since ancient times).

When a right circular cone is cut by a plane, the intersection is a curve called a **conic section,** as shown in Figure 8–1.[*] Conic sections were studied by the ancient Greeks and are still of interest. For instance, planets travel in elliptical orbits, parabolic mirrors are used in telescopes, and certain atomic particles follow hyperbolic paths.

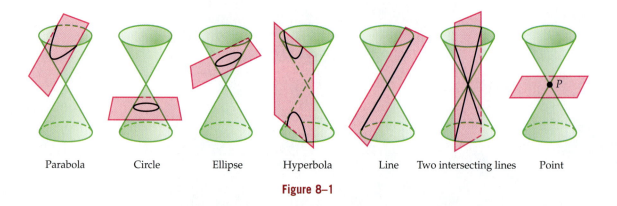

| Parabola | Circle | Ellipse | Hyperbola | Line | Two intersecting lines | Point |

Figure 8–1

Although the Greeks studied conic sections from a purely geometric point of view, the modern approach is to describe them in terms of the coordinate plane and distance, or as the graphs of certain types of equations. This was done for circles in Section 1.1 and will be done here for ellipses, hyperbolas, and parabolas.

---

[*]A point, a line, or two intersecting lines are sometimes called **degenerate** conic sections.

8.1 Circles and Ellipses

Let P be a point in the plane, and let r be a positive number. Then the **circle** with center P and radius r consists of all points X in the plane such that

$$\text{Distance from } X \text{ to } P = r,$$

as shown in Figure 8–2.

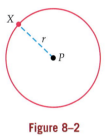

Figure 8–2

In Section 1.1, we proved the following result.

***Equation of
a Circle***

> The circle with center $(0, 0)$ and radius r is the graph of the equation
> $$x^2 + y^2 = r^2.$$
> The circle with center (h, k) and radius r is the graph of
> $$(x - h)^2 + (y - k)^2 = r^2.$$

EXAMPLE 1

Find the center and radius of the circle whose equation is given.

(a) $(x - 2)^2 + (y - 7)^2 = 25$
(b) $(x - 4)^2 + (y + 3)^2 = 36$

SOLUTION

(a) The right side of the equation is 5^2. So the center is at $(2, 7)$, and the radius is 5.

(b) First, we rewrite the equation to match the form in the box above:
$$(x - 4)^2 + \quad (y + 3)^2 = 36$$
$$(x - 4)^2 + [y - (-3)]^2 = \ 6^2.$$

Hence, the circle has center $(4, -3)$ and radius 6. ∎

The left-hand side of the equation of the circle can be thought of as the sum of two perfect squares [in Example 1(a), for instance, $(x - 2)^2 + (y - 7)^2$]. To find the center and radius of a circle whose equation is multiplied out, such as

$$x^2 + y^2 - 8x - 6y = -9,$$

we must first express the left-hand side as the sum of two perfect squares. The technique for doing this is based on a simple algebraic fact:

$$\left(x \pm \frac{b}{2}\right)^2 = x^2 \pm bx + \left(\frac{b}{2}\right)^2,$$

which you can easily verify by multiplying out the left-hand side. This equation says that if you add $\left(\frac{b}{2}\right)^2$ to $x^2 \pm bx$, the result is a perfect square. Note how $\left(\frac{b}{2}\right)^2$ is related to the expression $x^2 \pm bx$.

$\left(\frac{b}{2}\right)^2$ is the square of one-half the coefficient of x.

EXAMPLE 2

Show that the graph of

$$3x^2 + 3y^2 - 12x - 30y + 45 = 0$$

is a circle and find its center and radius.

SOLUTION   The technique to be used below requires that x^2 and y^2 each have coefficient 1. So we begin by dividing both sides of the equation by 3 and regrouping the terms.

$$x^2 + y^2 - 4x - 10y + 15 = 0$$
$$(x^2 - 4x) + (y^2 - 10y) = -15.$$

We want to add a constant to the expression $x^2 - 4x$ so that the result will be a perfect square. Using the fact discussed before the example, we take half the coefficient of x, namely, -2, and square it, obtaining 4. Adding 4 to both sides of the equation, we obtain

$$(x^2 - 4x + 4) + (y^2 - 10y) = -15 + 4,$$

which factors as

$$(x - 2)^2 + (y^2 - 10y) = -11.$$

Similarly, in the expression $y^2 - 10y$, we take half the coefficient of y, namely, -5, and square it, obtaining 25. Now add 25 to both sides of the equation and factor.

$$(x - 2)^2 + (y^2 - 10y + 25) = -11 + 25$$

$$(x - 2)^2 + (y - 5)^2 = 14$$

$$(x - 2)^2 + (y - 5)^2 = (\sqrt{14})^2$$

Now we see that the graph is a circle with center $(2, 5)$ and radius $\sqrt{14}$. ■

The technique of adding the square of half the coefficient of x to $x^2 - 4x$ to obtain the perfect square $(x - 2)^2$ in Example 2 is called **completing the square.** It will be used frequently in this chapter.

The preceding discussion of circles provides the model for our discussion of the other conic sections. In each case, the conic is defined in terms of points and distances, and its Cartesian equation is determined. The standard form of the equation of a conic includes the key information necessary for a rough sketch of its graph, just as the standard form of the equation of a circle tells you its center and radius.

▪▪ ELLIPSES

Definition. Let P and Q be points in the plane, and let r be a number greater than the distance from P to Q. The **ellipse** with **foci**[*] P and Q is the set of all points X such that

(Distance from X to P) + (Distance from X to Q) = r.

To draw this ellipse, take a piece of string of length r and pin its ends on P and Q. Put your pencil point against the string and move it, keeping the string taut. You will trace out the ellipse, as shown in Figure 8–3.

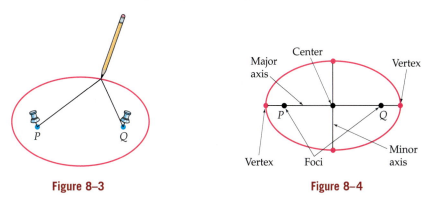

Figure 8–3   **Figure 8–4**

The midpoint of the line segment from P to Q is the **center** of the ellipse. The points where the straight line through the foci intersects the ellipse are its **vertices.** The **major axis** of the ellipse is the line segment joining the vertices; its **minor axis** is the line segment through the center, perpendicular to the major axis, as shown in Figure 8–4.

---

[*]"Foci" is the plural of "focus."

Equation.   Suppose that the foci P and Q are on the x-axis, with coordinates

$$P = (-c, 0) \quad \text{and} \quad Q = (c, 0) \quad \text{for some } c > 0.$$

Let $a = r/2$, so that $2a = r$. Then the point (x, y) is on the ellipse exactly when

[Distance from (x, y) to P] + [Distance from (x, y) to Q] = r

$$\sqrt{(x + c)^2 + (y - 0)^2} + \sqrt{(x - c)^2 + (y - 0)^2} = 2a$$

$$\sqrt{(x + c)^2 + y^2} = 2a - \sqrt{(x - c)^2 + y^2}.$$

Squaring both sides and simplifying (Exercise 76), we obtain

$$a\sqrt{(x - c)^2 + y^2} = a^2 - cx.$$

Again squaring both sides and simplifying, we have

$$(a^2 - c^2)x^2 + a^2 y^2 = a^2(a^2 - c^2).$$

To simplify the form of this equation, let $b = \sqrt{a^2 - c^2}$ * so that $b^2 = a^2 - c^2$ and the equation becomes

$$b^2 x^2 + a^2 y^2 = a^2 b^2.$$

Dividing both sides by $a^2 b^2$ shows that the coordinates of every point on the ellipse satisfy the equation

$$\frac{x^2}{a^2} + \frac{y^2}{b^2} = 1.$$

Conversely, it can be shown that every point whose coordinates satisfy this equation is on the ellipse. When the equation is in this form the x- and y-intercepts of the graph are easily found. For instance, to find the x-intercepts, we set $y = 0$ and solve

$$\frac{x^2}{a^2} + \frac{0^2}{b^2} = 1$$

$$x^2 = a^2$$

$$x = \pm a.$$

Similarly, the y-intercepts are $\pm b$.

A similar argument applies when the foci are on the y-axis and leads to this conclusion.

---

*The distance between the foci is $2c$. Since $r = 2a$ and $r > 2c$ by definition, we have $2a > 2c$, and hence, $a > c$. Therefore, $a^2 - c^2$ is a positive number and has a real square root.

Standard Equations of Ellipses Centered at the Origin

Let a and b be real numbers with $a > b > 0$. Then the graph of each of the following equations is an ellipse centered at the origin.

$\dfrac{x^2}{a^2} + \dfrac{y^2}{b^2} = 1$
$\begin{cases} x\text{-intercepts: } \pm a \qquad y\text{-intercepts: } \pm b \\[4pt] \text{major axis on the } x\text{-axis, with vertices } (a, 0) \text{ and } (-a, 0) \\[4pt] \text{foci: } (c, 0) \text{ and } (-c, 0), \text{ where } c = \sqrt{a^2 - b^2} \end{cases}$

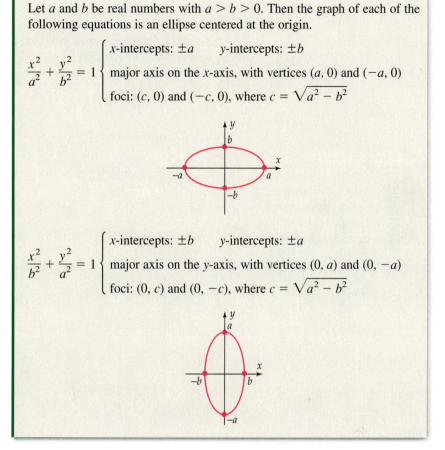

$\dfrac{x^2}{b^2} + \dfrac{y^2}{a^2} = 1$
$\begin{cases} x\text{-intercepts: } \pm b \qquad y\text{-intercepts: } \pm a \\[4pt] \text{major axis on the } y\text{-axis, with vertices } (0, a) \text{ and } (0, -a) \\[4pt] \text{foci: } (0, c) \text{ and } (0, -c), \text{ where } c = \sqrt{a^2 - b^2} \end{cases}$

In the preceding box $a > b$, but don't let all the letters confuse you: When the equation is in standard form, the denominator of the x term tells you the x-intercepts, the denominator of the y term tells you the y-intercepts, and the major axis is the longer one, as illustrated in the following examples.

EXAMPLE 3

Identify and sketch the graph of the equation $4x^2 + 9y^2 = 36$.

SOLUTION   To identify the graph, we put the equation in standard form.

$$4x^2 + 9y^2 = 36$$

Divide both sides by 36:
$$\frac{4x^2}{36} + \frac{9y^2}{36} = \frac{36}{36}$$

Simplify:
$$\frac{x^2}{9} + \frac{y^2}{4} = 1$$

$$\frac{x^2}{3^2} + \frac{y^2}{2^2} = 1$$

The graph is now in the form of the first equation in the preceding box, with $a = 3$ and $b = 2$. So its graph is an ellipse with x-intercepts ± 3 and y-intercepts ± 2. Its major axis and foci lie on the x-axis, as do its vertices $(3, 0)$ and $(-3, 0)$. A hand-sketched graph is shown in Figure 8–5. To graph this ellipse on a calculator, we first solve its equation for y.

$$4x^2 + 9y^2 = 36$$

Subtract $4x^2$ from both sides: $$9y^2 = 36 - 4x^2$$

Divide both sides by 9: $$y^2 = \frac{36 - 4x^2}{9}.$$

Taking square roots on both sides, we see that

$$y = \sqrt{\frac{36 - 4x^2}{9}} \quad \text{or} \quad y = -\sqrt{\frac{36 - 4x^2}{9}}.$$

Graphing both of these equations on the same screen, we obtain Figure 8–6. ■

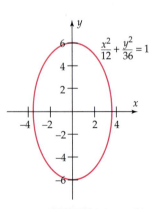

TIP

On most calculators, you can graph both equations in Example 3 simultaneously by keying in

$$y = \{-1, 1\} \sqrt{\frac{36 - 4x^2}{9}}.$$

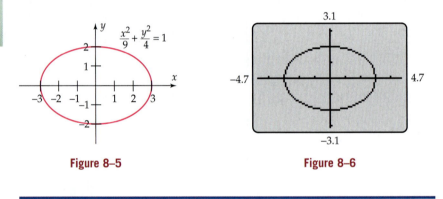

Figure 8–5

Figure 8–6

EXAMPLE 4

Find the equation of the ellipse with vertices $(0, \pm 6)$ and foci $(0, \pm 2\sqrt{6})$, and sketch its graph.

SOLUTION   Since the foci are $(0, 2\sqrt{6})$ and $(0, -2\sqrt{6})$, the center of the ellipse is $(0, 0)$, and its major axis lies on the y-axis. Hence, its equation is of the form

$$\frac{x^2}{b^2} + \frac{y^2}{a^2} = 1.$$

From the box on page 639, we see that $a = 6$ and $c = 2\sqrt{6}$. Since $c = \sqrt{a^2 - b^2}$, we have $c^2 = a^2 - b^2$, so

$$b^2 = a^2 - c^2 = 6^2 - (2\sqrt{6})^2 = 36 - 4 \cdot 6 = 12.$$

Hence, $b = \sqrt{12}$, and the equation of the ellipse is

$$\frac{x^2}{(\sqrt{12})^2} + \frac{y^2}{6^2} = 1 \quad \text{or, equivalently,} \quad \frac{x^2}{12} + \frac{y^2}{36} = 1.$$

The graph has x-intercepts $\pm\sqrt{12} \approx \pm 3.46$ and y-intercepts ± 6, as sketched in Figure 8–7. ■

Figure 8–7

◼ VERTICAL AND HORIZONTAL SHIFTS

We now examine ellipses that have foci on a line parallel to one of the coordinate axes. Recall that in Section 3.4, we saw that replacing the variable x by $x - 5$ in the rule of the function $y = f(x)$ shifts the graph horizontally 5 units to the right, whereas replacing x by $x + 5$ [that is, $x - (-5)$] shifts the graph horizontally 5 units to the left (see the box on page 234).

Similarly, if the rule of a function is given by $y = f(x)$, then replacing y by $y - 4$ shifts the graph 4 units vertically upward because

$$y - 4 = f(x) \qquad \text{is equivalent to} \qquad y = f(x) + 4$$

(see the box on page 232). For arbitrary equations, we have similar results.

Vertical and Horizontal Shifts

> Let h and k be constants. Replacing x by $x - h$ and y by $y - k$ in an equation shifts the graph of the equation
>
> $|h|$ units horizontally (right for positive h, left for negative h) and
>
> $|k|$ units vertically (upward for positive k, downward for negative k).

EXAMPLE 5

Identify and sketch the graph of

$$\frac{(x - 5)^2}{9} + \frac{(y + 4)^2}{36} = 1.$$

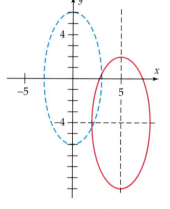

Figure 8–8

SOLUTION   This equation can be obtained from the equation $\dfrac{x^2}{9} + \dfrac{y^2}{36} = 1$ (whose graph is known to be an ellipse) as follows.

Replace x by $x - 5$ and   replace y by $y - (-4) = y + 4$.

This is the situation described in the previous box, with $h = 5$ and $k = -4$. Therefore, the graph is the ellipse $\dfrac{x^2}{9} + \dfrac{y^2}{36} = 1$ shifted horizontally 5 units to the right and vertically 4 units downward, as shown in Figure 8–8. The center of the ellipse is at $(5, -4)$. Its major axis (the longer one) lies on the vertical line $x = 5$, as do its foci. The minor axis is on the horizontal line $y = -4$.   ◼

EXAMPLE 6

Identify and sketch the graph of

$$4x^2 + 9y^2 - 32x - 90y + 253 = 0.$$

SOLUTION   We first rewrite the equation.

$$(4x^2 - 32x) + (9y^2 - 90y) = -253$$

$$4(x^2 - 8x) + 9(y^2 - 10y) = -253$$

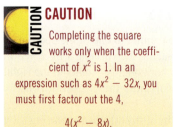

CAUTION

Completing the square works only when the coefficient of x^2 is 1. In an expression such as $4x^2 - 32x$, you must first factor out the 4,

$$4(x^2 - 8x),$$

and then complete the square on the expression in parentheses.

The factoring in the last equation was done in preparation for completing the square (see the Caution in the margin). To complete the square on $x^2 - 8x$, we add 16 (the square of half the coefficient of x), and to complete the square on $y^2 - 10y$, we add 25 (the square of half the coefficient of y).

$$4(x^2 - 8x + 16) + 9(y^2 - 10y + 25) = -253 + ? + ?.$$

Be careful here: On the left-hand side, we haven't just added 16 and 25. When the left-hand side is multiplied out, we have actually added $4 \cdot 16 = 64$ and $9 \cdot 25 = 225$. To leave the equation unchanged, we must add these numbers on the right.

$$4(x^2 - 8x + 16) + 9(y^2 - 10y + 25) = -253 + 64 + 225$$

Factor and simplify:
$$4(x - 4)^2 + 9(y - 5)^2 = 36$$

Divide both sides by 36:
$$\frac{4(x - 4)^2}{36} + \frac{9(y - 5)^2}{36} = \frac{36}{36}$$

Simplify:
$$\frac{(x - 4)^2}{9} + \frac{(y - 5)^2}{4} = 1.$$

The graph of this equation is the ellipse $\dfrac{x^2}{9} + \dfrac{y^2}{4} = 1$ shifted 4 units to the right and 5 units upward. Its center is at $(4, 5)$. Its major axis lies on the horizontal line $y = 5$, and its minor axis lies on the vertical line $x = 4$, as shown in Figure 8–9. ■

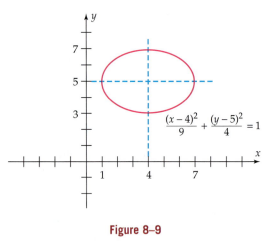

Figure 8–9

By translating the information about ellipses centered at the origin in the box on page 639, we obtain the following.

Standard Equations of Ellipses with Center at (h, k)

Let (h, k) be any point in the plane. If a and b are real numbers with $a > b > 0$, then the graph of each of the following equations is an ellipse with center (h, k).

$$\frac{(x - h)^2}{a^2} + \frac{(y - k)^2}{b^2} = 1 \begin{cases} \text{major axis on the horizontal line } y = k \\ \text{minor axis on the vertical line } x = h \\ \text{foci: } (h - c, k) \text{ and } (h + c, k), \text{ where} \\ \quad c = \sqrt{a^2 - b^2} \end{cases}$$

$$\frac{(x - h)^2}{b^2} + \frac{(y - k)^2}{a^2} = 1 \begin{cases} \text{major axis on the vertical line } x = h \\ \text{minor axis on the horizontal line } y = k \\ \text{foci: } (h, k - c) \text{ and } (h, k + c), \text{ where} \\ \quad c = \sqrt{a^2 - b^2} \end{cases}$$

TECHNOLOGY TIP

Casio 9850 and FX 2.0 have conic section graphers (on the main menu) that produce the graphs of equations in standard form when the various coefficients are entered. TI-83+ users can download a conic grapher from TI.

EXAMPLE 7

Find an appropriate viewing window and graph the ellipse

$$\frac{(x - 3)^2}{40} + \frac{(y + 2)^2}{120} = 1.$$

SOLUTION　To graph the ellipse, we first solve its equation for y.

Subtract $\dfrac{(x - 3)^2}{40}$ from both sides: $\qquad \dfrac{(y + 2)^2}{120} = 1 - \dfrac{(x - 3)^2}{40}$

Multiply both sides by 120: $\qquad (y + 2)^2 = 120\left[1 - \dfrac{(x - 3)^2}{40}\right]$

Multiply out right side: $\qquad (y + 2)^2 = 120 - 3(x - 3)^2$

Take square roots on both sides: $\qquad y + 2 = \pm\sqrt{120 - 3(x - 3)^2}$

$$y = \sqrt{120 - 3(x - 3)^2} - 2 \quad \text{or} \quad y = -\sqrt{120 - 3(x - 3)^2} - 2$$

So we should graph both of these last two equations on the same screen.

　　To determine an appropriate window, look at the original form of the equation:

$$\frac{(x - 3)^2}{40} + \frac{(y + 2)^2}{120} = 1.$$

The center of the ellipse is at $(3, -2)$. Its graph extends a distance of $\sqrt{40}$ to the left and right of the center and a distance of $\sqrt{120}$ above and below the center. Since $\sqrt{40}$ is a bit less than 7 and $\sqrt{120}$ is a bit less than 11 (why?), our window should include x-values from $3 - 7$ to $3 + 7$ (that is, $-4 \le x \le 10$) and y-values from $-2 - 11$ to $-2 + 11$ (that is, $-13 \le y \le 9$). So we first try the window in Figure 8–10 on the next page. In this window, the ellipse looks longer horizontally than vertically. We know from its equation, however, that the major (longer) axis is vertical because the larger constant 120 is the denominator of the y-term. So we change to a square window and obtain the more accurate graph in

Figure 8–11.* Even this graph has gaps that shouldn't be there (an ellipse is a connected figure), but this is unavoidable because of the limited resolution of the calculator screen. ■

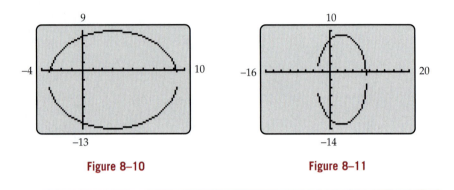

Figure 8–10   **Figure 8–11**

Graph the equation $x^2 + 8y^2 + 6x + 9y + 4 = 0$ without first putting it in standard form.

SOLUTION   Rewrite it like this:

$$8y^2 + 9y + (x^2 + 6x + 4) = 0.$$

This is a quadratic equation of the form $ay^2 + by + c = 0$, with

$$a = 8, \qquad b = 9, \qquad c = x^2 + 6x + 4$$

and hence can be solved by using the quadratic formula.

$$y = \frac{-b \pm \sqrt{b^2 - 4ac}}{2a}$$

$$y = \frac{-9 \pm \sqrt{9^2 - 4 \cdot 8 \cdot (x^2 + 6x + 4)}}{2 \cdot 8}$$

$$y = \frac{-9 \pm \sqrt{81 - 32(x^2 + 6x + 4)}}{16}$$

TECHNOLOGY TIP

TI users can save keystrokes by entering the first equation in Example 8 as y_1 and then using the RCL key to copy the text of y_1 to y_2. [On TI-89, use COPY and PASTE in place of RCL.] Then only one sign needs to be changed to make y_2 into the second equation of Example 8.

GRAPHING EXPLORATION

Find a complete graph of the original equation by graphing both of the following functions on the same screen. The Technology Tip in the margin may be helpful.

$$y = \frac{-9 + \sqrt{81 - 32(x^2 + 6x + 4)}}{16}$$

$$y = \frac{-9 - \sqrt{81 - 32(x^2 + 6x + 4)}}{16}. \qquad ■$$

---

*Figure 8–11 shows a square window for a TI-83+. On wide-screen calculators, a longer x-axis is needed for a square window.

■■ APPLICATIONS

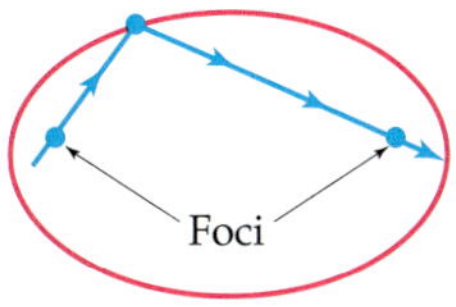

Elliptical surfaces have interesting reflective properties. If a sound or light ray passes through one focus and reflects off an ellipse, the ray will pass through the other focus, as shown in Figure 8–12. Exactly this situation occurs under the elliptical dome of the U.S. Capitol. A person who stands at one focus and whispers can be clearly heard by anyone at the other focus. Before this fact was widely known, when Congress used to sit under the dome, several political secrets were inadvertently revealed by congressmen to members of the other party.

Foci

Figure 8–12

The planets and many comets have elliptical orbits, with the sun as one focus. The moon travels in an elliptical orbit with the earth as one focus. Satellites are usually put into elliptical orbits around the earth.

EXAMPLE 9

The earth's orbit around the sun is an ellipse that is almost a circle. The sun is one focus, and the major and minor axes have lengths 186,000,000 miles and 185,974,062 miles, respectively. What are the minimum and maximum distances from the earth to the sun?

SOLUTION   The orbit is shown in Figure 8–13. If we use a coordinate system with the major axis on the *x*-axis and the sun having coordinates $(c, 0)$, then we obtain Figure 8–14.

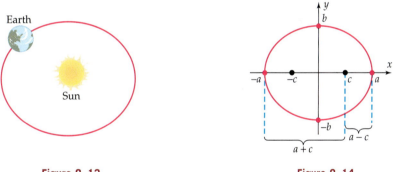

Earth

Sun

Figure 8–13

Figure 8–14

The length of the major axis is $2a = 186,000,000$, so that $a = 93,000,000$. Similarly, $2b = 185,974,062$, so $b = 92,987,031$. As was shown earlier, the equation of the orbit is

$$\frac{x^2}{a^2} + \frac{y^2}{b^2} = 1, \qquad \text{where}$$

$$c = \sqrt{a^2 - b^2} = \sqrt{(93,000,000)^2 - (92,987,031)^2} \approx 1,553,083.$$

Figure 8–14 suggests a fact that can also be proven algebraically: The minimum and maximum distances from a point on the ellipse to the focus $(c, 0)$ occur at the endpoints of the major axis:

Minimum distance $= a - c \approx 93,000,000 - 1,553,083 = 91,446,917$ miles

Maximum distance $= a + c \approx 93,000,000 + 1,553,083 = 94,553,083$ miles.

✓ EXERCISES 8.1

In Exercises 1–6, determine which of the following equations could possibly have the given graph.

$2x^2 + y^2 = 12,$
$x^2 + 6y^2 = 18,$
$(x + 3)^2 + y^2 = 9,$
$(x - 3)^2 + (y + 4)^2 = 5,$

$(x - 4)^2 + (y - 3)^2 = 4,$
$(x + 3)^2 + (y + 4)^2 = 6,$
$x^2 + (y - 3)^2 = 4,$
$(x + 2)^2 + (y - 3)^2 = 2$

1.

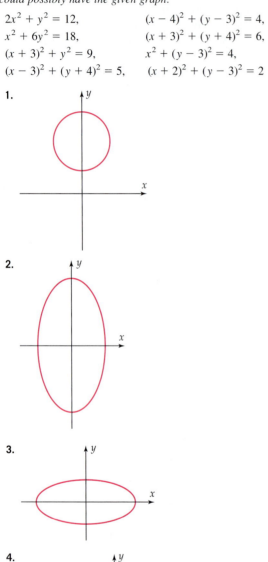

2.

3.

4.

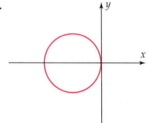

5.

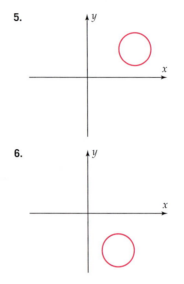

6.

In Exercises 7–12, find the center and radius of the circle whose equation is given, as in Example 2.

7. $x^2 + y^2 + 8x - 6y - 15 = 0$

8. $15x^2 + 15y^2 = 10$

9. $x^2 + y^2 + 6x - 4y - 15 = 0$

10. $x^2 + y^2 + 10x - 75 = 0$

11. $x^2 + y^2 + 25x + 10y = -12$

12. $3x^2 + 3y^2 + 12x + 12 = 18y$

In Exercises 13–20, identify the conic section whose equation is given and find a complete graph of the equation as in Example 3.

13. $\dfrac{x^2}{25} + \dfrac{y^2}{4} = 1$

14. $\dfrac{x^2}{6} + \dfrac{y^2}{16} = 1$

15. $4x^2 + 3y^2 = 12$

16. $9x^2 + 4y^2 = 72$

17. $\dfrac{x^2}{10} - 1 = \dfrac{-y^2}{36}$

18. $\dfrac{y^2}{49} + \dfrac{x^2}{81} = 1$

19. $4x^2 + 4y^2 = 1$

20. $x^2 + 4y^2 = 1$

In Exercises 21–26, find the equation of the ellipse that satisfies the given conditions.

21. Center $(0, 0)$; foci on x-axis; x-intercepts ± 7; y-intercepts ± 2.

22. Center $(0, 0)$; foci on y-axis; x-intercepts ± 1; y-intercepts ± 8.

23. Center $(0, 0)$; foci on x-axis; major axis of length 12; minor axis of length 8.

24. Center $(0, 0)$; foci on y-axis; major axis of length 20; minor axis of length 18.

25. Center $(0, 0)$; endpoints of major and minor axes: $(0, -7)$, $(0, 7)$, $(-3, 0)$, $(3, 0)$.

26. Center $(0, 0)$; vertices $(8, 0)$ and $(-8, 0)$; minor axis of length 8.

Calculus can be used to show that the area of the ellipse with equation $\dfrac{x^2}{a^2} + \dfrac{y^2}{b^2} = 1$ is $\pi a b$. Use this fact to find the area of each ellipse in Exercises 27–32.

27. $\dfrac{x^2}{16} + \dfrac{y^2}{4} = 1$ **28.** $\dfrac{x^2}{9} + \dfrac{y^2}{5} = 1$

29. $3x^2 + 4y^2 = 12$ **30.** $7x^2 + 5y^2 = 35$

31. $6x^2 + 2y^2 = 14$ **32.** $5x^2 + y^2 = 5$

In Exercises 33–44, identify the conic section whose equation is given, list its center, and find its graph.

33. $\dfrac{(x - 1)^2}{4} + \dfrac{(y - 5)^2}{9} = 1$

34. $\dfrac{(x - 2)^2}{16} + \dfrac{(y + 3)^2}{12} = 1$

35. $\dfrac{(x + 1)^2}{16} + \dfrac{(y - 4)^2}{8} = 1$

36. $\dfrac{(x + 5)^2}{4} + \dfrac{(y + 2)^2}{12} = 1$

37. $9x^2 + 4y^2 + 54x - 8y + 49 = 0$

38. $4x^2 + 5y^2 - 8x + 30y + 29 = 0$

39. $x^2 + y^2 + 6x - 8y + 5 = 0$

40. $x^2 + y^2 - 4x + 2y - 7 = 0$

41. $4x^2 + y^2 + 24x - 4y + 36 = 0$

42. $9x^2 + y^2 - 36x + 10y + 52 = 0$

43. $25x^2 + 16y^2 + 50x + 96y = 231$

44. $9x^2 + 25y^2 - 18x + 50y = 191$

In Exercises 45–50, find the equation of the ellipse that satisfies the given conditions.

45. Center $(2, 3)$; endpoints of major and minor axes: $(2, -1)$, $(0, 3)$, $(2, 7)$, $(4, 3)$.

46. Center $(-5, 2)$; endpoints of major and minor axes: $(0, 2)$, $(-5, 17)$, $(-10, 2)$, $(-5, -13)$.

47. Center $(7, -4)$; foci on the line $x = 7$; major axis of length 12; minor axis of length 5.

48. Center $(-3, -9)$; foci on the line $y = -9$; major axis of length 15; minor axis of length 7.

49. Center $(3, -2)$; passing through $(3, -6)$ and $(9, -2)$.

50. Center $(2, 5)$; passing through $(2, 4)$ and $(-3, 5)$.

In Exercises 51 and 52, find the equations of two distinct ellipses satisfying the given conditions.

51. Center at $(-5, 3)$; major axis of length 14; minor axis of length 8.

52. Center at $(2, -6)$; major axis of length 15; minor axis of length 6.

In Exercises 53–58, determine which of the following equations could possibly have the given graph.

$$\dfrac{(x + 3)^2}{4} + \dfrac{(y + 3)^2}{8} = 1, \qquad \dfrac{(x - 3)^2}{9} + \dfrac{(y + 4)^2}{4} = 1,$$

$$2x^2 + 2y^2 - 8 = 0, \qquad 4x^2 + 2y^2 - 8 = 0,$$

$$2x^2 + y^2 - 8x - 6y + 9 = 0,$$

$$x^2 + 3y^2 + 6x - 12y + 17 = 0.$$

53.

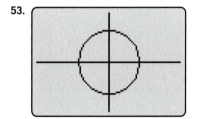

54.

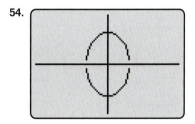

55.

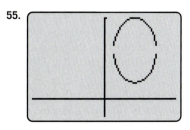

56.

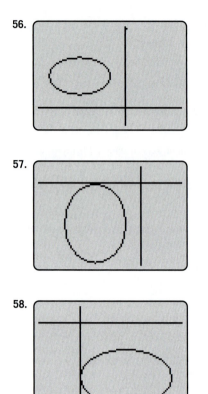

57.

58.

59. A circle is inscribed in the ellipse with equation

$$x^2 + 4y^2 = 64.$$

A diameter of the circle lies on the minor axis of the ellipse. Find the equation of the circle.

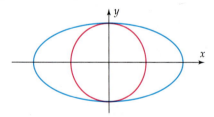

60. If (c, d) is a point on the ellipse in Exercise 59, prove that $(c/2, d)$ is a point on the circle.

61. The orbit of the moon around the earth is an ellipse with the earth as one focus. If the length of the major axis of the orbit is 477,736 miles and the length of the minor axis is 477,078 miles, find the minimum and maximum distances from the earth to the moon.

62. Halley's Comet has an elliptical orbit with the sun as one focus and a major axis that is 1,636,484,848 miles long. The closest the comet comes to the sun is 54,004,000 miles. What is the maximum distance from the comet to the sun?

63. A cross section of the ceiling of a "whispering room" at a museum is half of an ellipse. Two people stand so that their heads are approximately at the foci of the ellipse. If one whispers upward, the sound waves are reflected off the elliptical ceiling to the other person, as indicated in the figure. (For the reason why, see Figure 8–12 and the accompanying text.) How far from the center are the two people standing? [*Hint:* Use the rectangular coordinate system suggested in the figure to find the equation of the ellipse; then find the foci.]

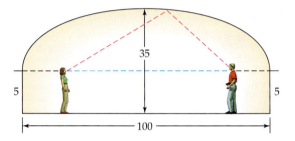

64. Suppose that the whispering room in Exercise 63 is 100 feet long and that the two people at the foci are 80 feet apart. How high is the center of the roof?

65. The bottom of a bridge is shaped like half an ellipse and is 20 feet above the river at the center, as shown in the figure. Find the height of the bridge bottom over a point on the river 25 feet from the center of the river. [*Hint:* Use a coordinate system, with the *x*-axis on the surface of the water and the *y*-axis running through the center of the bridge, to find an equation for the ellipse.]

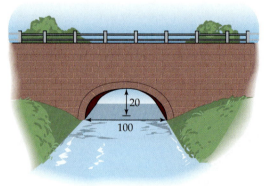

66. The stained glass window in the figure is shaped like the top half of an ellipse. The window is 10 feet wide, and the two figures in the window are located 3 feet from the center. If the figures are 1.6 feet high, find the height of the window at the center. [*Hint:* Use a coordinate system, with the bottom of the window as the *x*-axis and the vertical line through the center of the window as the *y*-axis.]

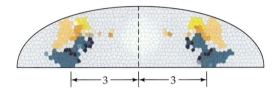

If $a > b > 0$, then the **eccentricity** of the ellipse

$$\frac{(x-h)^2}{a^2} + \frac{(y-k)^2}{b^2} = 1 \quad \text{or} \quad \frac{(x-h)^2}{b^2} + \frac{(y-k)^2}{a^2} = 1$$

is the number $\dfrac{\sqrt{a^2 - b^2}}{a}$. In Exercises 67–70, find the eccentricity of the ellipse whose equation is given.

67. $\dfrac{x^2}{100} + \dfrac{y^2}{99} = 1$

68. $\dfrac{x^2}{18} + \dfrac{y^2}{25} = 1$

69. $\dfrac{(x-3)^2}{10} + \dfrac{(y-9)^2}{40} = 1$

70. $\dfrac{(x+5)^2}{12} + \dfrac{(y-4)^2}{8} = 1$

71. On the basis of your answers to Exercises 67–70, how is the eccentricity of an ellipse related to its graph? [*Hint:* What is the shape of the graph when the eccentricity is close to 0? When it is close to 1?]

72. Assuming that the same scale was used for the axes, which of these ellipses has the larger eccentricity?

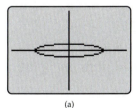

(a)

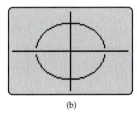

(b)

73. A satellite is to be placed in an elliptical orbit, with the center of the earth as one focus. The satellite's maximum distance from the surface of the earth is to be 22,380 km, and its minimum distance is to be 6540 km. Assume that the radius of the earth is 6400 km, and find the eccentricity of the satellite's orbit.

74. The first step in landing Apollo 11 on the moon was to place the spacecraft in an elliptical orbit such that the minimum distance from the *surface* of the moon to the spacecraft was 110 km and the maximum distance was 314 km. If the radius of the moon is 1740 km, find the eccentricity of the Apollo 11 orbit.

75. Consider the ellipse whose equation is $\dfrac{x^2}{a^2} + \dfrac{y^2}{b^2} = 1$. Show that if $a = b$, then the graph is actually a circle.

76. Complete the derivation of the equation of the ellipse on page 638 as follows.
(a) By squaring both sides, show that the equation

$$\sqrt{(x+c)^2 + y^2} = 2a - \sqrt{(x-c)^2 + y^2}$$

may be simplified as

$$a\sqrt{(x-c)^2 + y^2} = a^2 - cx.$$

(b) Show that the last equation in part (a) may be further simplified as

$$(a^2 - c^2)x^2 + a^2 y^2 = a^2(a^2 - c^2).$$

Thinker

77. The punch bowl and a table holding the punch cups are placed 50 feet apart at a garden party. A portable fence is then set up so that any guest inside the fence can walk straight to the table, then to the punch bowl, and then return to his or her starting point without traveling more than 150 feet. Describe the longest possible such fence that encloses the largest possible area.

8.2   Hyperbolas

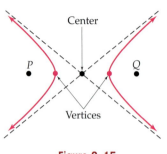

Figure 8–15

Definition.   Let P and Q be points in the plane, and let r be a positive number. The set of all points X such that

$$\big|(\text{Distance from } P \text{ to } X) - (\text{Distance from } Q \text{ to } X)\big| = r$$

is the **hyperbola** with **foci** P and Q; r will be called the **distance difference.** Every hyperbola has the general shape shown by the red curve in Figure 8–15. The dashed straight lines are the **asymptotes** of the hyperbola; it gets closer and closer to the asymptotes, but never touches them. The asymptotes intersect at the midpoint of the line segment from P to Q; this point is called the **center** of the hyperbola. The **vertices** of the hyperbola are the points where it intersects the line segment from P to Q. The line through P and Q is called the **focal axis.**

Equation.   Another complicated exercise in the use of the distance formula, which will be omitted here, leads to the following algebraic description.

Standard Equations of Hyperbolas Centered at the Origin

Let a and b be positive real numbers. Then the graph of each of the following equations is a hyperbola centered at the origin.

$$\frac{x^2}{a^2} - \frac{y^2}{b^2} = 1 \begin{cases} x\text{-intercepts: } \pm a \qquad y\text{-intercepts: none} \\ \text{focal axis on the } x\text{-axis, with vertices } (a, 0) \text{ and } (-a, 0) \\ \text{foci: } (c, 0) \text{ and } (-c, 0), \text{ where } c = \sqrt{a^2 + b^2}. \\ \text{asymptotes: } y = \frac{b}{a}x \text{ and } y = -\frac{b}{a}x \end{cases}$$

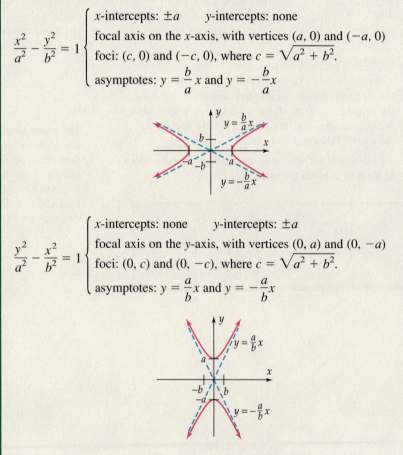

$$\frac{y^2}{a^2} - \frac{x^2}{b^2} = 1 \begin{cases} x\text{-intercepts: none} \qquad y\text{-intercepts: } \pm a \\ \text{focal axis on the } y\text{-axis, with vertices } (0, a) \text{ and } (0, -a) \\ \text{foci: } (0, c) \text{ and } (0, -c), \text{ where } c = \sqrt{a^2 + b^2}. \\ \text{asymptotes: } y = \frac{a}{b}x \text{ and } y = -\frac{a}{b}x \end{cases}$$

Once again, don't worry about all the letters in the box. When the equation is in standard form with the x term positive and y term negative, the hyperbola intersects the x-axis and opens from side to side. When the x term is negative and the y term is positive, the hyperbola intersects the y-axis and opens up and down.

EXAMPLE 1

Identify and sketch the graph of the equation $9x^2 - 4y^2 = 36$.

SOLUTION   We first put the equation in standard form.

$$9x^2 - 4y^2 = 36$$

Divide both sides by 36:
$$\frac{9x^2}{36} - \frac{4y^2}{36} = \frac{36}{36}$$

Simplify:
$$\frac{x^2}{4} - \frac{y^2}{9} = 1$$

$$\frac{x^2}{2^2} - \frac{y^2}{3^2} = 1$$

Applying the fact in the box with $a = 2$ and $b = 3$ shows that the graph is a hyperbola with vertices $(2, 0)$ and $(-2, 0)$ and asymptotes $y = \frac{3}{2}x$ and $y = -\frac{3}{2}x$. We first plot the vertices and sketch the rectangle determined by the vertical lines $x = \pm 2$ and the horizontal lines $y = \pm 3$. The asymptotes go through the origin and the corners of this rectangle, as shown on the left in Figure 8–16. It is then easy to sketch the hyperbola. ■

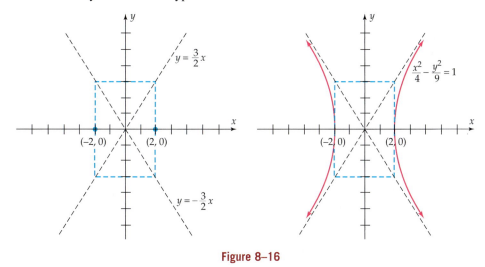

Figure 8–16

EXAMPLE 2

Find the equation of the hyperbola with vertices $(0, 1)$ and $(0, -1)$ that passes through the point $(3, \sqrt{2})$. Then sketch its graph.

SOLUTION   The vertices are on the y-axis, and the equation is of the form

$$\frac{y^2}{a^2} - \frac{x^2}{b^2} = 1$$

with $a = 1$. Since $(3, \sqrt{2})$ is on the graph, we have

$$\frac{(\sqrt{2})^2}{1^2} - \frac{3^2}{b^2} = 1$$

Simplify: $\qquad\qquad 2 - \dfrac{9}{b^2} = 1$

Subtract 2 from both sides: $\qquad\qquad -\dfrac{9}{b^2} = -1$

Multiply both sides by $-b^2$: $\qquad\qquad 9 = b^2.$

Therefore, $b = 3$, and the equation is

$$\frac{y^2}{1^2} - \frac{x^2}{3^2} = 1 \qquad \text{or, equivalently,} \qquad y^2 - \frac{x^2}{9} = 1.$$

The asymptotes of the hyperbola are the lines $y = \pm\frac{1}{3}x$. To sketch the graph, we first solve the equation for y.

$$y^2 - \frac{x^2}{9} = 1$$

Add $\dfrac{x^2}{9}$ to both sides: $\qquad\qquad\qquad y^2 = 1 + \dfrac{x^2}{9}$

Take square roots on both sides: $\quad y = \sqrt{1 + \dfrac{x^2}{9}} \qquad \text{or} \qquad y = -\sqrt{1 + \dfrac{x^2}{9}}$

Graphing these last two equations on the same screen, we obtain Figure 8–17. ∎

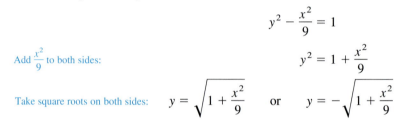

Figure 8–17

■■ VERTICAL AND HORIZONTAL SHIFTS

Let h and k be constants. As we saw on page 641, replacing x by $x - h$ and y by $y - k$ in an equation shifts the graph of the equation $|h|$ units horizontally and $|k|$ units vertically. If the graph of the equation is a hyperbola with center at the origin, then the shifted graph has its center at (h, k). Using this fact and translating the information in the box on page 650, we obtain the following.

Standard Equations
of Hyperbolas
with Center at (h, k)

If a and b are positive real numbers, then the graph of each of the following equations is a hyperbola with center (h, k).

$$\frac{(x - h)^2}{a^2} - \frac{(y - k)^2}{b^2} = 1 \begin{cases} \text{focal axis on the horizontal line } y = k \\ \text{foci: } (h - c, k) \text{ and } (h + c, k), \text{ where} \\ \qquad c = \sqrt{a^2 + b^2} \\ \text{vertices: } (h - a, k) \text{ and } (h + a, k) \\ \text{asymptotes: } y = \pm\dfrac{b}{a}(x - h) + k \end{cases}$$

$$\frac{(y - k)^2}{a^2} - \frac{(x - h)^2}{b^2} = 1 \begin{cases} \text{focal axis on the vertical line } x = h \\ \text{foci: } (h, k - c) \text{ and } (h, k + c), \text{ where} \\ \qquad c = \sqrt{a^2 + b^2} \\ \text{vertices: } (h, k - a) \text{ and } (h, k + a) \\ \text{asymptotes: } y = \pm\dfrac{a}{b}(x - h) + k \end{cases}$$

EXAMPLE 3

Identify and sketch the graph of

$$\frac{(x-3)^2}{4} - \frac{(y+2)^2}{9} = 1.$$

SOLUTION If we rewrite the equation as

$$\frac{(x-3)^2}{2^2} - \frac{(y-(-2))^2}{3^2} = 1,$$

then it has the first form in the preceding box with $a = 2$, $b = 3$, $h = 3$, and $k = -2$. Its graph is a hyperbola with center $(3, -2)$. There are several ways to obtain the graph.

Method 1. The equation of this hyperbola can be obtained from

(∗)
$$\frac{x^2}{4} - \frac{y^2}{9} = 1$$

by replacing x by $x - 3$ and y by $y + 2 = y - (-2)$. So its graph is just the graph of (∗) (see Figure 8–16) shifted 3 units to the right and 2 units downward, as shown in Figure 8–18.

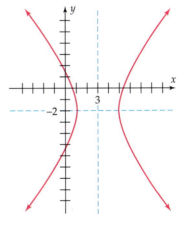

Figure 8–18

Method 2. Solve the original equation for y.

$$\frac{(x-3)^2}{4} - \frac{(y+2)^2}{9} = 1$$

Multiply both sides by 36: $\quad 9(x-3)^2 - 4(y+2)^2 = 36$

Rearrange terms: $\quad 4(y+2)^2 = 9(x-3)^2 - 36$

Divide both sides by 4: $\quad (y+2)^2 = \dfrac{9(x-3)^2 - 36}{4}$

Take square roots of both sides: $\quad y + 2 = \pm\sqrt{\dfrac{9(x-3)^2 - 36}{4}}$

$$y = -2 + \sqrt{\frac{9(x-3)^2 - 36}{4}} \quad \text{or} \quad y = -2 - \sqrt{\frac{9(x-3)^2 - 36}{4}}$$

EXAMPLE 4

Sketch the graph of $6y^2 - 8x^2 - 24y - 48x - 96 = 0$.

SOLUTION   We first solve the equation for y. Begin by rewriting it as

$$6y^2 - 24y + (-8x^2 - 48x - 96) = 0.$$

This is a quadratic equation of the form $ay^2 + by + c = 0$, with

$$a = 6, \qquad b = -24, \qquad c = -8x^2 - 48x - 96.$$

We use the quadratic formula to solve it.

$$y = \frac{-b \pm \sqrt{b^2 - 4ac}}{2a}$$

$$y = \frac{-(-24) \pm \sqrt{(-24)^2 - 4 \cdot 6(-8x^2 - 48x - 96)}}{2 \cdot 6}$$

$$y = \frac{24 \pm \sqrt{576 - 24(-8x^2 - 48x - 96)}}{12}$$

Now we graph both

$$y = \frac{24 + \sqrt{576 - 24(-8x^2 - 48x - 96)}}{12} \qquad \text{and}$$

$$y = \frac{24 - \sqrt{576 - 24(-8x^2 - 48x - 96)}}{12}$$

on the screen to obtain the hyperbola in Figure 8–19. ■

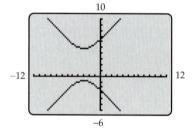

Figure 8–19

EXAMPLE 5

Find the center of the hyperbola in Example 4.

SOLUTION   Begin by rearranging the equation.

$$6y^2 - 8x^2 - 24y - 48x - 96 = 0$$

Add 96 to both sides: $\qquad 6y^2 - 24y - 8x^2 - 48x = 96$

Group x- and y-terms: $\qquad (6y^2 - 24y) - (8x^2 + 48x) = 96$

Factor out coefficients of y^2 and x^2: $\qquad 6(y^2 - 4y) - 8(x^2 + 6x) = 96.$

Complete the square in the expression $y^2 - 4y$ by adding 4 (the square of half the coefficient of y), and complete the square in $x^2 + 6x$ by adding 9 (the square of half the coefficient of x).

$$6(y^2 - 4y + 4) - 8(x^2 + 6x + 9) = 96 + ? + ?$$

CAUTION

Completing the square only works when the coefficient of y^2 is 1. In an expression such as $6y^2 - 24y$, you must first factor out the 6,

$$6(y^2 - 4y),$$

and then complete the square on the expression in parentheses.

On the left side, we have actually added $6 \cdot 4 = 24$ and $-8 \cdot 9 = -72$, so we must add these numbers on the right to keep the equation unchanged.

$$6(y^2 - 4y + 4) - 8(x^2 + 6x + 9) = 96 + 24 - 72$$

Factor and simplify:
$$6(y - 2)^2 - 8(x + 3)^2 = 48$$

Divide both sides by 48:
$$\frac{(y - 2)^2}{8} - \frac{(x + 3)^2}{6} = 1$$

$$\frac{(y - 2)^2}{8} - \frac{(x - (-3))^2}{6} = 1$$

In this form, we can see that the graph is a hyperbola with center at $(-3, 2)$. ∎

▖ APPLICATIONS

The reflective properties of hyperbolas are used in the design of camera and telescope lenses. If a light ray passes through one focus of a hyperbola and reflects off the hyperbola at a point P, then the reflected ray moves along the straight line determined by P and the other focus, as shown in Figure 8–20.

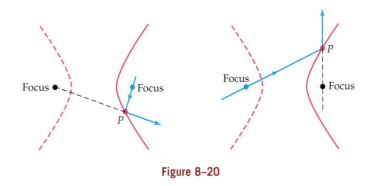

Figure 8–20

Hyperbolas are also the basis of the long-range navigation system (loran), which enables a ship to determine its exact location by radio, as illustrated in the next example.

EXAMPLE 6

Three loran transmitters Q, P, and R are located 200 miles apart along a straight shoreline and simultaneously transmit signals at regular intervals. These signals travel at a speed of 980 feet per microsecond. A ship S receives a signal from P and 305 microseconds later a signal from R. It also receives a signal from Q 528 microseconds after the one from P. Determine the ship's location.

SOLUTION   Take the line through the loran stations as the x-axis, with the origin located midway between Q and P, so that the situation looks like Figure 8–21.

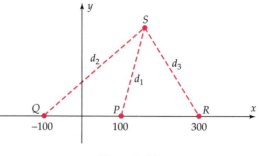

Figure 8–21

If the signal takes t microseconds to go from P to S and $t + 528$ microseconds to go from Q to S, then

$$d_1 = 980t \qquad \text{and} \qquad d_2 = 980(t + 528),$$

so

$$|d_1 - d_2| = |980t - 980(t + 528)| = 980 \cdot 528 = 517{,}440 \text{ feet.}$$

Since 1 mile is 5280 feet, this means that

$$|d_1 - d_2| = 517{,}440/5{,}280 \text{ miles} = 98 \text{ miles.}$$

In other words,

$$|(\text{Distance from } P \text{ to } S) - (\text{Distance from } Q \text{ to } S)| = |d_1 - d_2| = 98.$$

This is precisely the situation described in the definition of "hyperbola" on page 650: S is on the hyperbola with foci $P = (100, 0)$, $Q = (-100, 0)$, and distance difference $r = 98$. This hyperbola has an equation of the form

$$\frac{x^2}{a^2} - \frac{y^2}{b^2} = 1,$$

where $(\pm a, 0)$ are the vertices, $(\pm c, 0) = (\pm 100, 0)$ are the foci, and $c^2 = a^2 + b^2$. Figure 8–22 and the fact that the vertex $(a, 0)$ is on the hyperbola show that

$$|[\text{Distance from } P \text{ to } (a, 0)] - [\text{Distance from } Q \text{ to } (a, 0)]| = r = 98$$

$$|(100 - a) - (100 + a)| = 98$$

$$|-2a| = 98$$

$$|a| = 49.$$

Figure 8–22

Consequently, $a^2 = 49^2 = 2401$, and hence, $b^2 = c^2 - a^2 = 100^2 - 49^2 = 7599$. Thus, the ship lies on the hyperbola

$$\text{(*)} \qquad \frac{x^2}{2401} - \frac{y^2}{7599} = 1.$$

A similar argument using P and R as foci shows that the ship also lies on the hyperbola with foci $P = (100, 0)$ and $R = (300, 0)$ and center $(200, 0)$, whose distance difference r is

$$|d_1 - d_3| = 980 \cdot 305 = 298,900 \text{ feet} \approx 56.61 \text{ miles}.$$

As before, you can verify that $a = 56.61/2 = 28.305$, and hence, $a^2 = 28.305^2 = 801.17$. This hyperbola has center $(200, 0)$, and its foci are $(200 - c, k) = (100, 0)$ and $(200 + c, k) = (300, 0)$, which implies that $c = 100$. Hence, $b^2 = c^2 - a^2 = 100^2 - 801.17 = 9198.83$, and the ship also lies on the hyperbola

$$\text{(**)} \qquad \frac{(x - 200)^2}{801.17} - \frac{y^2}{9198.83} = 1.$$

Since the ship lies on both hyperbolas, its coordinates are solutions of both the equations (*) and (**). They can be found algebraically by solving each of the equations for y^2, setting the results equal, and solving for x. They can be found geometrically by graphing both hyperbolas and finding the intersection point. As shown in Figure 8–23, there are actually four points of intersection. However, the two below the x-axis represent points on land in our situation. Furthermore, since the signal from P was received first, the ship is closest to P. So it is located at the point S in Figure 8–23. A graphical intersection finder shows that this point is approximately $(130.48, 215.14)$, where the coordinates are in miles from the origin. ∎

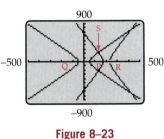

Figure 8–23

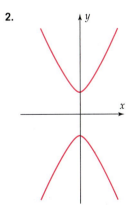

EXERCISES 8.2

In Exercises 1–6, determine which of the following equations could possibly have the given graph.

$$3x^2 + 3y^2 = 12, \qquad 6y^2 - x^2 = 6,$$
$$x^2 + 4y^2 = 1, \qquad 4x^2 + 4(y + 2)^2 = 12,$$
$$4(x + 4)^2 + 4y^2 = 12, \qquad 6x^2 + 2y^2 = 18,$$
$$2x^2 - y^2 = 8, \qquad 3x^2 - y = 6$$

1.

2.

3.

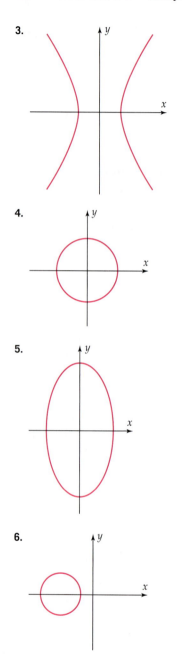

4.

5.

6.

In Exercises 7–14, identify the conic section whose equation is given and find a complete graph of the equation, as in Example 1.

7. $\dfrac{x^2}{6} - \dfrac{y^2}{16} = 1$ **8.** $\dfrac{x^2}{4} - y^2 = 1$

9. $4x^2 - y^2 = 16$ **10.** $3y^2 - 5x^2 = 15$

11. $\dfrac{x^2}{10} - \dfrac{y^2}{36} = 1$ **12.** $\dfrac{y^2}{9} - \dfrac{x^2}{16} = 1$

13. $x^2 - 4y^2 = 1$ **14.** $2x^2 - y^2 = 4$

In Exercises 15–18, find the equation of the hyperbola that satisfies the given conditions.

15. Center $(0, 0)$; x-intercepts ± 3; asymptote $y = 2x$.

16. Center $(0, 0)$; y-intercepts ± 12; asymptote $y = 3x/2$.

17. Center $(0, 0)$; vertex $(2, 0)$; passing through $(4, \sqrt{3})$.

18. Center $(0, 0)$; vertex $(0, \sqrt{12})$; passing through $(2\sqrt{3}, 6)$.

In Exercises 19–32, identify the conic section whose equation is given, list its center, and find its graph.

19. $\dfrac{(y + 3)^2}{25} - \dfrac{(x + 1)^2}{16} = 1$

20. $\dfrac{(y + 1)^2}{9} - \dfrac{(x - 1)^2}{25} = 1$

21. $\dfrac{(x + 3)^2}{1} - \dfrac{(y - 2)^2}{4} = 1$

22. $\dfrac{(y + 5)^2}{9} - \dfrac{(x - 2)^2}{1} = 1$

23. $(y + 4)^2 - 8(x - 1)^2 = 8$

24. $(x - 3)^2 + 12(y - 2)^2 = 24$

25. $4y^2 - x^2 + 6x - 24y + 11 = 0$

26. $x^2 - 16y^2 = 0$

27. $2x^2 + 2y^2 - 12x - 16y + 26 = 0$

28. $3x^2 + 3y^2 + 12x + 6y = 0$

29. $2x^2 + 3y^2 - 12x - 24y + 54 = 0$

30. $x^2 + 2y^2 + 4x - 4y = 8$

31. $x^2 - 3y^2 + 4x + 12y = 20$

32. $2x^2 + 16x = y^2 - 6y - 55$

In Exercises 33–36, find the equation of the hyperbola that satisfies the given conditions.

33. Center $(-2, 3)$; vertex $(-2, 1)$; passing through $(-2 + 3\sqrt{10}, 11)$.

34. Center $(-5, 1)$; vertex $(-3, 1)$; passing through $(-1, 1 - 4\sqrt{3})$.

35. Center $(4, 2)$; vertex $(7, 2)$; asymptote $3y = 4x - 10$.

36. Center $(-3, -5)$; vertex $(-3, 0)$; asymptote $6y = 5x - 15$.

In Exercises 37–42, determine which of the following equations could possibly have the given graph.

$\dfrac{(y - 2)^2}{4} - \dfrac{(x - 3)^2}{9} = 1,$ $\dfrac{(x + 3)^2}{3} - \dfrac{(y + 3)^2}{4} = 1,$

$4x^2 - 2y^2 = 8,$ $9(y - 2)^2 = 36 + 4(x + 3)^2,$

$3(y + 3)^2 = 4(x - 3)^2 - 12,$ $y^2 - 2x^2 = 6.$

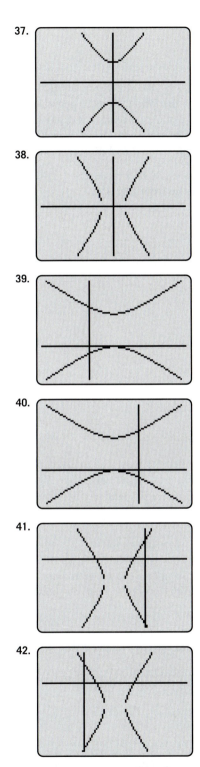

37.

38.

39.

40.

41.

42.

43. Sketch the graph of $\dfrac{y^2}{4} - \dfrac{x^2}{b^2} = 1$ for $b = 2$, $b = 4$, $b = 8$, $b = 12$, and $b = 20$. What happens to the hyperbola as b takes larger and larger values? Could the graph ever degenerate into a pair of horizontal lines?

44. Find a number k such that $(-2, 1)$, is on the graph of $3x^2 + ky^2 = 4$. Then graph the equation.

45. Show that the asymptotes of the hyperbola $\dfrac{x^2}{a^2} - \dfrac{y^2}{a^2} = 1$ are perpendicular to each other.

46. Find the approximate coordinates of the points where these hyperbolas intersect:

$$\frac{(x-1)^2}{4} - \frac{(y+1)^2}{8} = 1 \qquad \text{and} \qquad 4y^2 - x^2 = 1.$$

47. Two listening stations that are 1 mile apart record an explosion. One microphone receives the sound 2 seconds after the other does. Use the line through the microphones as the x-axis, with the origin midway between the microphones, and the fact that sound travels at 1100 feet per second to find the equation of a hyperbola on which the explosion is located. Can you determine the exact location of the explosion?

48. Two transmission stations P and Q are located 200 miles apart on a straight shoreline. A ship 50 miles from shore is moving parallel to the shoreline. A signal from Q reaches the ship 400 microseconds after a signal from P. If the signals travel at 980 feet per microsecond, find the location of the ship (in terms of miles) in the coordinate system with x-axis through P and Q and origin midway between them.

---

*If $a > 0$ and $b > 0$, then the **eccentricity** of the hyperbola*

$$\frac{(x-h)^2}{a^2} - \frac{(y-k)^2}{b^2} = 1 \qquad \text{or} \qquad \frac{(y-k)^2}{a^2} - \frac{(x-h)^2}{b^2} = 1$$

is the number $\dfrac{\sqrt{a^2 + b^2}}{a}$. In Exercises 49–53, find the eccentricity of the hyperbola whose equation is given.

49. $\dfrac{(x-6)^2}{10} - \dfrac{y^2}{40} = 1$ **50.** $\dfrac{y^2}{18} - \dfrac{x^2}{25} = 1$

51. $6(y-2)^2 = 18 + 3(x+2)^2$

52. $16x^2 - 9y^2 - 32x + 36y + 124 = 0$

53. $4x^2 - 5y^2 - 16x - 50y + 71 = 0$

54. (a) Graph these hyperbolas (on the same screen if possible).

$$\frac{y^2}{4} - \frac{x^2}{1} = 1, \qquad \frac{y^2}{4} - \frac{x^2}{12} = 1, \qquad \frac{y^2}{4} - \frac{x^2}{96} = 1.$$

(b) Compute the eccentricity of each hyperbola in part (a).

(c) On the basis of parts (a) and (b), how is the shape of a hyperbola related to its eccentricity?

8.3   Parabolas

Definition.   Parabolas appeared in Section 4.1 as the graphs of quadratic functions. Parabolas of this kind are a special case of the following more general definition. Let L be a line in the plane, and let P be a point not on L. If X is any point not on L, the distance from X to L is defined to be the length of the perpendicular line segment from X to L. The **parabola** with **focus** P and **directrix** L is the set of all points X such that

Distance from X to P = Distance from X to L

as shown in Figure 8–24.

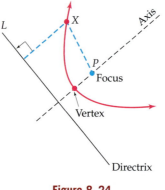

Figure 8–24

The line through P perpendicular to L is called the **axis**. The intersection of the axis with the parabola (the midpoint of the segment of the axis from P to L) is the **vertex** of the parabola, as illustrated in Figure 8–24. The parabola is symmetric with respect to its axis.

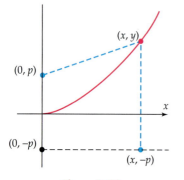

Figure 8–25

Equation.   Suppose that the focus is on the y-axis at the point $(0, p)$, where p is a nonzero constant, and that the directrix is the horizontal line $y = -p$. If (x, y) is any point on the parabola, then the distance from (x, y) to the horizontal line $y = -p$ is the length of the vertical segment from (x, y) to $(x, -p)$ as shown in Figure 8–25.

By the definition of the parabola,

Distance from (x, y) to $(0, p)$ = Distance from (x, y) to line $y = -p$

Distance from (x, y) to $(0, p)$ = Distance from (x, y) to $(x, -p)$

$$\sqrt{(x - 0)^2 + (y - p)^2} = \sqrt{(x - x)^2 + [y - (-p)]^2}.$$

Squaring both sides and simplifying, we have

$$(x - 0)^2 + (y - p)^2 = (x - x)^2 + (y + p)^2$$

$$x^2 + y^2 - 2py + p^2 = 0^2 + y^2 + 2py + p^2$$

$$x^2 = 4py.$$

Conversely, it can be shown that every point whose coordinates satisfy this equation is on the parabola.

A similar argument works for the parabola with focus $(p, 0)$ on the x-axis and directrix the vertical line $x = -p$, and leads to this conclusion.

Standard Equations of Parabolas with Vertex at the Origin

Let p be a nonzero real number. Then the graph of each of the following equations is a parabola with vertex at the origin.

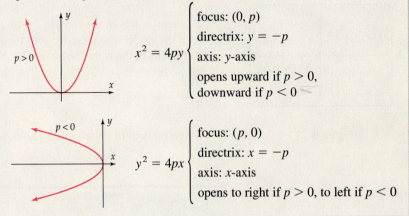

$x^2 = 4py$ $\begin{cases} \text{focus: } (0, p) \\ \text{directrix: } y = -p \\ \text{axis: } y\text{-axis} \\ \text{opens upward if } p > 0, \\ \text{downward if } p < 0 \end{cases}$

$y^2 = 4px$ $\begin{cases} \text{focus: } (p, 0) \\ \text{directrix: } x = -p \\ \text{axis: } x\text{-axis} \\ \text{opens to right if } p > 0, \text{ to left if } p < 0 \end{cases}$

EXAMPLE 1

Show that the graph of the equation is a parabola, and find its focus and directrix; then sketch the graph.

(a) $y = -x^2/8$ (b) $x = 3y^2$

SOLUTION

(a) We rewrite the equation so that it matches one of the forms in the preceding box.

$$y = -\frac{x^2}{8}$$

Multiply both sides by -8: $-8y = x^2$

The equation $x^2 = -8y$ is of the form $x^2 = 4py$, with $4p = -8$, so $p = -2$. Hence, the graph is a downward-opening parabola with focus $(0, p) = (0, -2)$ and directrix $y = -p = -(-2) = 2$, as shown in Figure 8–26.

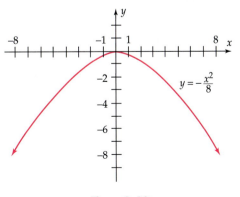

Figure 8–26

(b) Divide both sides of $x = 3y^2$ by 3 so that the equation becomes

$$y^2 = \frac{x}{3}$$

$$y^2 = \frac{1}{3}x.$$

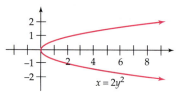

Figure 8–27

This equation is of the form $y^2 = 4px$, with $4p = 1/3$, so $p = 1/12$. Therefore, the graph is a parabola with focus $(1/12, 0)$ and directrix $x = -1/12$ that opens to the right. To sketch its graph, we solve the equation $y^2 = x/3$ for y.

$$y = \sqrt{\frac{x}{3}} \qquad \text{or} \qquad y = -\sqrt{\frac{x}{3}}$$

Graphing both of these equations on the same screen produces Figure 8–27.

■

EXAMPLE 2　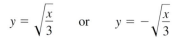

Find the focus, directrix, and equation of the parabola that passes through the point $(8, 2)$, has vertex $(0, 0)$, and focus on the x-axis.

SOLUTION　The equation is of the form $y^2 = 4px$. Since $(8, 2)$ is on the graph, we have $2^2 = 4p \cdot 8$, so $p = \frac{1}{8}$. Therefore, the focus is $\left(\frac{1}{8}, 0\right)$, and the directrix is the vertical line $x = -\frac{1}{8}$. The equation is $y^2 = 4\left(\frac{1}{8}\right)x = \frac{1}{2}x$ or, equivalently, $x = 2y^2$. Its graph is sketched in Figure 8–28.　■

Figure 8–28

▓ VERTICAL AND HORIZONTAL SHIFTS

Let h and k be constants. If we replace x by $x - h$ and y by $y - k$ in the equation of a parabola with vertex at the origin, then the graph of the new equation can be obtained by shifting the parabola vertically and horizontally so that its vertex is at (h, k), as explained on page 641. Using this fact and translating the information in the box on page 661, we obtain the following.

Standard Equations of Parabolas with Vertex at (h, k)

If p is a nonzero real number, then the graph of each of the following equations is a parabola with vertex (h, k).

$$(x - h)^2 = 4p(y - k) \begin{cases} \text{focus: } (h, k + p) \\ \text{directrix: the horizontal line } y = k - p \\ \text{axis: the vertical line } x = h \\ \text{opens upward if } p > 0, \text{ downward if } p < 0 \end{cases}$$

$$(y - k)^2 = 4p(x - h) \begin{cases} \text{focus: } (h + p, k) \\ \text{directrix: the vertical line } x = h - p \\ \text{axis: the horizontal line } y = k \\ \text{opens to right if } p > 0, \text{ to left if } p < 0 \end{cases}$$

EXAMPLE 3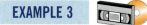

Identify and sketch the graph of $y = 2(x - 3)^2 + 1$.

SOLUTION　　We first rewrite the equation.

$$y = 2(x - 3)^2 + 1$$

Subtract 1 from both sides:　　$y - 1 = 2(x - 3)^2$

Divide both sides by 2:　　$\dfrac{1}{2}(y - 1) = (x - 3)^2$

$$(x - 3)^2 = \dfrac{1}{2}(y - 1)$$

This is the first form in the preceding box, with $h = 3$, $k = 1$, and $4p = 1/2$. Hence, $p = 1/8$, and the graph is an upward-opening parabola with vertex $(3, 1)$, focus $(3, 1 + 1/8) = (3, 9/8)$, and directrix the horizontal line $y = 1 - 1/8 = 7/8$. The graph of

$$(x - 3)^2 = \dfrac{1}{2}(y - 1) \qquad \text{or, equivalently,} \qquad y = 2(x - 3)^2 + 1$$

is the graph of $y = 2x^2$ shifted 3 units to the right and 1 unit upward, as shown in Figure 8–29. ■

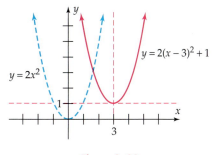

Figure 8–29

EXAMPLE 4

Graph the equation $x = 5y^2 + 30y + 41$ without putting it in standard form.

SOLUTION　　There are two methods of graphing this equation on a calculator.

Method 1. Rewrite the equation as

$$5y^2 + 30y + 41 - x = 0.$$

This is a quadratic equation of the form $ay^2 + by + c = 0$, with $a = 5$, $b = 30$, and $c = 41 - x$. It can be solved by using the quadratic formula.

$$y = \dfrac{-30 \pm \sqrt{30^2 - 4 \cdot 5(41 - x)}}{2 \cdot 5} = \dfrac{-30 \pm \sqrt{900 - 20(41 - x)}}{10}$$

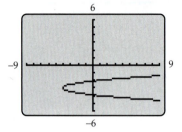

Figure 8–30

Now graph both

$$y = \frac{-30 + \sqrt{900 - 20(41 - x)}}{10} \quad \text{and} \quad y = \frac{-30 - \sqrt{900 - 20(41 - x)}}{10}$$

on the same screen to obtain the parabola in Figure 8–30.

Method 2. Since the equation $x = 5y^2 + 30y + 41$ defines x as a function of y, we can use parametric graphing, as explained in Special Topics 3.3.A on page 230. The parametric equations are

$$x = 5t^2 + 30t + 41$$

$$y = t.$$

Graphing these equations in parametric mode also produces Figure 8–30. ∎

The parametric graphing technique used in Example 4 can be applied to the parabola with equation $(y - k)^2 = 4p(x - h)$ by solving the equation for x and letting $y = t$.

Parametric Equations for Parabolas

The parabola with vertex (h, k) and equation

$$(y - k)^2 = 4p(x - h)$$

is given by the parametric equations

$$x = \frac{(t - k)^2}{4p} + h \quad \text{and} \quad y = t \quad (t \text{ any real number}).$$

EXAMPLE 5

Find the vertex, focus, and directrix of the parabola

$$x = 5y^2 + 30y + 41$$

that was graphed in Example 4.

SOLUTION   We first rewrite the equation.

$$5y^2 + 30y + 41 = x$$

Subtract 41 from both sides: $5y^2 + 30y = x - 41$

Factor out 5 on left-hand side: $5(y^2 + 6y) = x - 41$

Complete the square on the expression $y^2 + 6y$ by adding 9 (the square of half the coefficient of y).

$$5(y^2 + 6y + 9) = x - 41 + ?$$

On the left-hand side, we have actually added $5 \cdot 9 = 45$, so we must add the same amount to the right-hand side.

$$5(y^2 + 6y + 9) = x - 41 + 45$$

Factor left-hand side: $\qquad 5(y + 3)^2 = x + 4$

Divide both sides by 5: $\qquad (y + 3)^2 = \dfrac{1}{5}(x + 4)$

$$[y - (-3)]^2 = \dfrac{1}{5}[x - (-4)]$$

Thus, the graph is a parabola with vertex $(-4, -3)$. In this case, $4p = 1/5$, so $p = 1/20 = .05$. Hence, the focus is $(-4 + .05, -3) = (-3.95, -3)$, and the directrix is $x = -4 - .05 = -4.05$.

GRAPHING EXPLORATION

Use the parametric equations in the preceding box, with $h = -4$, $k = -3$, and $p = 1/20$, to graph this parabola in the window with

$$-9 \leq x \leq 9 \qquad \text{and} \qquad -6 \leq y \leq 6 \qquad (-6 \leq t \leq 6).$$

Your graph should be identical to Figure 8–30. ∎

■■ APPLICATIONS

Certain laws of physics show that sound waves or light rays from a source at the focus of a parabola will reflect off the parabola in rays parallel to the axis of the parabola, as shown in Figure 8–31. This is the reason that parabolic reflectors are used in automobile headlights and searchlights.

Conversely, a light ray coming toward a parabola will be reflected into the focus, as shown in Figure 8–32. This fact is used in the design of radar antennas, satellite dishes, and field microphones used at outdoor sporting events to pick up conversation on the field.

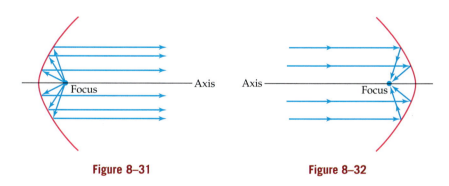

Figure 8–31   **Figure 8–32**

Projectiles follow a parabolic curve, a fact that is used in the design of water slides in which the rider slides down a sharp incline, then up and over a hill, before plunging downward into a pool. At the peak of the hill, the rider shoots up along a parabolic arc several inches above the slide, experiencing a sensation of weightlessness.

EXAMPLE 6

The radio telescope in Figure 8–33 has the shape of a parabolic dish (a cross section through the center of the dish is a parabola). It is 30 feet deep at the center and has a diameter of 200 feet. How far from the vertex of the parabolic dish should the receiver be placed to catch all the rays that hit the dish?

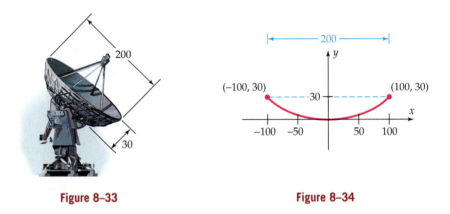

Figure 8–33   **Figure 8–34**

SOLUTION   Rays hitting the dish are reflected into the focus, as explained above. So the radio receiver must be located at the focus. To find the focus, draw a cross section of the dish, with vertex at the origin, as in Figure 8–34. The equation of this parabola is of the form $x^2 = 4py$. Since the point (100, 30) is on the parabola, we have

$$x^2 = 4py$$

Substitute: $$100^2 = 4p(30)$$

Simplify: $$120p = 100^2$$

Divide both sides by 120: $$p = \frac{100^2}{120}$$

Simplify: $$p = \frac{250}{3}.$$

As we saw in the box on page 661, the focus is the point $(0, p)$, which is p units from the vertex $(0, 0)$. Therefore, the receiver should be placed $250/3 \approx 83.33$ feet from the vertex. ■

✓ EXERCISES 8.3

In Exercises 1–6, determine which of the following equations could possibly have the given graph.

$y = x^2/4$, $x^2 = -8y$, $6x = y^2$, $y^2 = -4x$,

$2x^2 + y^2 = 12$, $x^2 + 6y^2 = 18$,

$6y^2 - x^2 = 6$, $2x^2 - y^2 = 8$

1.

2.

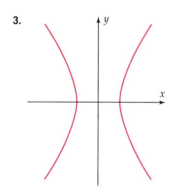

3.

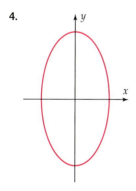

4.

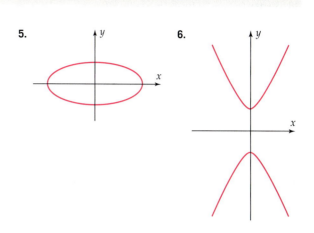

5.

6.

In Exercises 7–10, find the focus and directrix of the parabola.

7. $y = 3x^2$

8. $x = .5y^2$

9. $y = .25x^2$

10. $x = -6y^2$

In Exercises 11–22, determine the vertex, focus, and directrix of the parabola without graphing *and state whether it opens upward, downward, left, or right.*

11. $x - y^2 = 2$

12. $y - 3 = x^2$

13. $x + (y + 1)^2 = 2$

14. $y + (x + 2)^2 = 3$

15. $3x - 2 = (y + 3)^2$

16. $2x - 1 = -6(y + 1)^2$

17. $x = y^2 - 9y$

18. $x = y^2 + y + 1$

19. $y = 3x^2 + x - 4$

20. $y = -3x^2 + 4x - 1$

21. $y = -3x^2 + 4x + 5$

22. $y = 2x^2 - x - 1$

In Exercises 23–30, sketch the graph of the equation and label the vertex.

23. $y = 4(x - 1)^2 + 2$

24. $y = 3(x - 2)^2 - 3$

25. $x = 2(y - 2)^2$

26. $x = -3(y - 1)^2 - 2$

27. $y = x^2 - 4x - 1$

28. $y = x^2 + 8x + 6$

29. $y = x^2 + 2x$

30. $x = y^2 - 3y$

In Exercises 31–42, find the equation of the parabola satisfying the given conditions.

31. Vertex $(0, 0)$; axis $x = 0$; $(2, 12)$ on graph

32. Vertex $(0, 1)$; axis $x = 0$; $(2, -7)$ on graph

33. Vertex $(1, 0)$; axis $x = 1$; $(2, 13)$ on graph

34. Vertex $(-3, 0)$; axis $y = 0$; $(-1, 1)$ on graph

35. Vertex $(2, 1)$; axis $y = 1$; $(5, 0)$ on graph

36. Vertex $(1, -3)$; axis $y = -3$; $(-1, -4)$ on graph

37. Vertex $(-3, -2)$; focus $(-47/16, -2)$

38. Vertex $(-5, -5)$; focus $(-5, -99/20)$

39. Vertex $(1, 1)$; focus $(1, 9/8)$

40. Vertex $(-4, -3)$; $(-6, -2)$ and $(-6, -4)$ on graph

41. Vertex $(-1, 3)$; $(8, 0)$ and $(0, 4)$ on graph

42. Vertex $(1, -3)$; $(0, -1)$ and $(-1, 5)$ on graph

In Exercises 43–48, find parametric equations for the curve whose equation is given, and use them to find a complete graph of the curve.

43. $8x = 2y^2$ **44.** $4y = x^2$

45. $y = 4(x - 1)^2 + 2$ **46.** $y = 3(x - 2)^2 - 3$

47. $x = 2(y - 2)^2$ **48.** $x = -3(y - 1)^2 - 2$

In Exercises 49–56, identify the conic section whose equation is given, list its vertex, or vertices, if any, and find its graph.

49. $x^2 = 6x - y - 5$ **50.** $y^2 = x - 2y - 2$

51. $3y^2 = x - 1 + 2y$ **52.** $2y^2 = x - 4y - 5$

53. $3x^2 + 3y^2 - 6x - 12y - 6 = 0$

54. $2x^2 + 3y^2 + 12x - 6y + 9 = 0$

55. $2x^2 - y^2 + 16x + 4y + 24 = 0$

56. $4x^2 - 40x - 2y + 105 = 0$

In Exercises 57–62, determine which of the following equations could possibly have the given graph.

$$y = (x + 5)^2 - 3, \qquad x = (y + 3)^2 + 2,$$

$$y = (x - 4)^2 + 2, \qquad x = -(y - 3)^2 - 2.$$

$$y^2 = 4y + x - 1, \qquad y = -x^2 - 8x - 18$$

57.

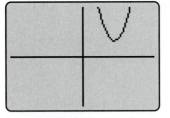

58.

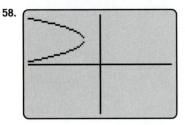

59.

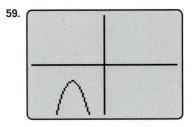

60.

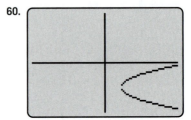

61.

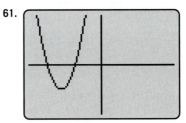

62.

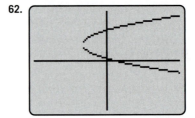

63. Find the number b such that the vertex of the parabola $y = x^2 + bx + c$ lies on the y-axis.

64. Find the number d such that the parabola $(y + 1)^2 = dx + 4$ passes through $(-6, 3)$.

65. Find the points of intersection of the parabola $4y^2 + 4y = 5x - 12$ and the line $x = 9$.

66. Find the points of intersection of the parabola $4x^2 - 8x = 2y + 5$ and the line $y = 15$.

67. A parabolic satellite dish is 4 feet in diameter and 1.5 feet deep. How far from the vertex should the receiver be placed to catch all the signals that hit the dish?

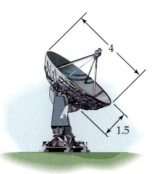

68. A flashlight has a parabolic reflector that is 3 inches in diameter and 1.5 inches deep. For the light from the bulb to reflect in beams that are parallel to the center axis of the flashlight, how far from the vertex of the reflector should the bulb be located? [*Hint:* See Figure 8–31 and the preceding discussion.]

69. A radio telescope has a parabolic dish with a diameter of 300 feet. Its receiver (focus) is located 130 feet from the vertex. How deep is the dish at its center? [*Hint:* Position the dish as in Figure 8–34, and find the equation of the parabola.]

70. The 6.5-meter MMT telescope on top of Mount Hopkins in Arizona has a parabolic mirror. The focus of the parabola is 8.125 meters from the vertex of the parabola. Find the depth of the mirror.

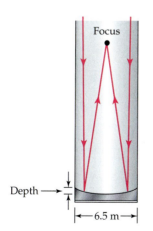

71. The Hale telescope at Mount Palomar in California also has a parabolic mirror, whose depth is .096 meter (see the figure for Exercise 70). The focus of the parabola is 16.75 meters from the vertex. Find the diameter of the mirror.

72. A large spotlight has a parabolic reflector that is 3 feet deep at its center. The light source is located $1\frac{1}{3}$ feet from the vertex. What is the diameter of the reflector?

73. The cables of a suspension bridge are shaped like parabolas. The cables are attached to the towers 100 feet from the bridge surface, and the towers are 420 feet apart. The cables touch the bridge surface at the center (midway between the towers). At a point on the bridge 100 feet from one of the towers, how far is the cable from the bridge surface?

74. At a point 120 feet from the center of a suspension bridge, the cables are 24 feet above the bridge surface. Assume that the cables are shaped like parabolas and touch the bridge surface at the center (which is midway between the towers). If the towers are 600 feet apart, how far above the surface of the bridge are the cables attached to the towers?

8.4   Rotations and Second-Degree Equations

A **second-degree equation** in x and y is one that can be written in the form

$$Ax^2 + Bxy + Cy^2 + Dx + Ey + F = 0$$

for some constants A, B, C, D, E, F, with at least one of A, B, C nonzero.

EXAMPLE 1

Show that each of the following conic sections is the graph of a second-degree equation.

(a) Ellipse: $\dfrac{x^2}{6} + \dfrac{y^2}{5} = 1$
(b) Hyperbola: $\dfrac{(x + 1)^2}{4} - \dfrac{(y - 3)^2}{6} = 1$

SOLUTION   We need only show that each of these equations is in fact a second-degree equation. In each case, eliminate denominators, multiply out all terms, and gather them on one side of the equal sign.

(a)

$$\frac{x^2}{6} + \frac{y^2}{5} = 1$$

Multiply both sides by 30: $5x^2 + 6y^2 = 30$

Rearrange terms: $5x^2 + 6y^2 - 30 = 0$

This equation is a second-degree equation because it has the form

$$Ax^2 + Bxy + Cy^2 + Dx + Ey + F = 0$$

with $A = 5, B = 0, C = 6, D = 0, E = 0$, and $F = -30$.

(b)

$$\frac{(x + 1)^2}{4} - \frac{(y - 3)^2}{6} = 1$$

Multiply both sides by 12: $3(x + 1)^2 - 2(y - 3)^2 = 12$

Multiply out left side: $3(x^2 + 2x + 1) - 2(y^2 - 6y + 9) = 12$

$3x^2 + 6x + 3 - 2y^2 + 12y - 18 = 12$

Rearrange terms: $3x^2 - 2y^2 + 6x + 12y - 27 = 0$

This is a second-degree equation with $A = 3, B = 0, C = -2, D = 6, E = 12$, and $F = -27$.   ■

Calculations like those in Example 1 can be used on the equation of any conic section to show that it is the graph of a second-degree equation. Conversely, it can be shown that

The graph of every second-degree equation is a conic section

(possibly degenerate). When the second-degree equation has no xy-term (that is, $B = 0$), as was the case in Example 1, the graph is a conic section in standard position (axis or axes parallel to the coordinate axes). When $B \neq 0$, however, the conic is rotated from standard position, so its axis or axes are not parallel to the coordinate axes.

EXAMPLE 2

Graph the equation

$$3x^2 + 6xy + y^2 + x - 2y + 7 = 0.$$

SOLUTION   We first rewrite it as

$$y^2 + 6xy - 2y + 3x^2 + x + 7 = 0$$

$$y^2 + (6x - 2)y + (3x^2 + x + 7) = 0.$$

This equation has the form $ay^2 + by + c = 0$, with $a = 1$, $b = 6x - 2$, and $c = 3x^2 + x + 7$. It can be solved by using the quadratic formula.

$$y = \frac{-b \pm \sqrt{b^2 - 4ac}}{2a} = \frac{-(6x - 2) \pm \sqrt{(6x - 2)^2 - 4 \cdot 1 \cdot (3x^2 + x + 7)}}{2 \cdot 1}$$

The top half of the graph is obtained by graphing

$$y = \frac{-6x + 2 + \sqrt{(6x - 2)^2 - 4(3x^2 + x + 7)}}{2},$$

and the bottom half is obtained by graphing

$$y = \frac{-6x + 2 - \sqrt{(6x - 2)^2 - 4(3x^2 + x + 7)}}{2}.$$

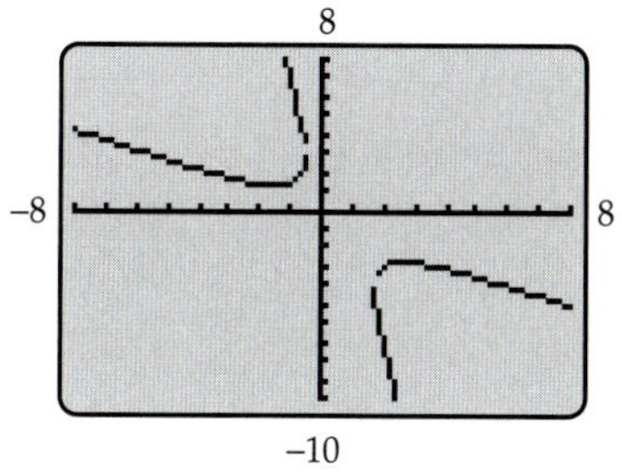

The graph is a hyperbola whose focal axis tilts upward to the left, as shown in Figure 8–35. ■

8

−8   8

−10

Figure 8–35

▝▝ THE DISCRIMINANT

The following fact, whose proof requires trigonometry, makes it easy to identify the graphs of second-degree equations without graphing them.

Graphs of Second-Degree Equations

If the equation

$$Ax^2 + Bxy + Cy^2 + Dx + Ey + F = 0 \qquad (A, B, C \text{ not all } 0)$$

has a graph, then that graph is

A circle or an ellipse (or a point), if $B^2 - 4AC < 0$;

A parabola (or a line or two parallel lines), if $B^2 - 4AC = 0$;

A hyperbola (or two intersecting lines), if $B^2 - 4AC > 0$.

The expression $B^2 - 4AC$ is called the **discriminant** of the equation.

EXAMPLE 3

Identify the graph of

$$2x^2 - 4xy + 3y^2 + 5x + 6y - 8 = 0$$

and sketch the graph.

SOLUTION   We compute the discriminant with $A = 2$, $B = -4$, and $C = 3$.

$$B^2 - 4AC = (-4)^2 - 4 \cdot 2 \cdot 3 = 16 - 24 = -8.$$

Hence, the graph is an ellipse (possibly a circle or a single point). To find the graph, we rewrite the equation as

$$3y^2 - 4xy + 6y + 2x^2 + 5x - 8 = 0$$

$$3y^2 + (-4x + 6)y + (2x^2 + 5x - 8) = 0.$$

The equation has the form $ay^2 + by + c = 0$ and can be solved by the quadratic formula.

$$y = \frac{-b \pm \sqrt{b^2 - 4ac}}{2a}$$

$$= \frac{-(-4x + 6) \pm \sqrt{(-4x + 6)^2 - 4 \cdot 3 \cdot (2x^2 + 5x - 8)}}{2 \cdot 3}.$$

The graph can now be found by graphing the last two equations on the same screen.

GRAPHING EXPLORATION

Find a viewing window that shows a complete graph of the equation. In what direction does the major axis run?   ■

EXAMPLE 4

Does Figure 8–36 show a complete graph of

$$3x^2 + 5xy + 2y^2 - 8y - 1 = 0?$$

SOLUTION   Although the graph in the figure looks like a parabola, appearances are deceiving. The discriminant of the equation is

$$B^2 - 4AC = 5^2 - 4 \cdot 3 \cdot 2 = 1,$$

which means that the graph is a hyperbola. So Figure 8–36 cannot possibly be a complete graph of the equation.[*]

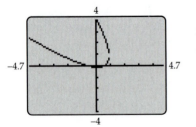

Figure 8–36

GRAPHING EXPLORATION

Solve the equation for y, as in Example 3. Then find a viewing window that clearly shows both branches of the hyperbola.   ■

---

[*]When I asked students to graph this equation on an examination, many of them did not bother to compute the discriminant and produced something similar to Figure 8–36. They didn't get much credit for this answer. *Moral:* Use the discriminant to identify the conic so that you know the shape of the graph you are looking for.

EXAMPLE 5

Sketch the graph of

$$3x^2 + 6xy + 3y^2 + 13x + 9y + 53 = 0.$$

SOLUTION   The discriminant is $B^2 - 4AC = 6^2 - 4 \cdot 3 \cdot 3 = 0$. Hence, the graph is a parabola (or a line or parallel lines in the degenerate case).

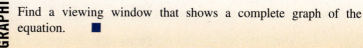

GRAPHING EXPLORATION

Find a viewing window that shows a complete graph of the equation. ■

EXERCISES 8.4

In Exercises 1–6, assume that the graph of the equation is a nondegenerate conic section. Without graphing, determine whether the graph is an ellipse, hyperbola, or parabola.

1. $x^2 - 2xy + 3y^2 - 1 = 0$

2. $xy - 1 = 0$

3. $x^2 + 2xy + y^2 + 2\sqrt{2}x - 2\sqrt{2}y = 0$

4. $2x^2 - 4xy + 5y^2 - 6 = 0$

5. $17x^2 - 48xy + 31y^2 + 50 = 0$

6. $2x^2 - 4xy - 2y^2 + 3x + 5y - 10 = 0$

In Exercises 7–24, use the discriminant to identify the conic section whose equation is given, and find a viewing window that shows a complete graph.

7. $9x^2 + 4y^2 + 54x - 8y + 49 = 0$

8. $4x^2 + 5y^2 - 8x + 30y + 29 = 0$

9. $4y^2 - x^2 + 6x - 24y + 11 = 0$

10. $x^2 - 16y^2 = 0$

11. $3y^2 - x - 2y + 1 = 0$

12. $x^2 - 6x + y + 5 = 0$

13. $41x^2 - 24xy + 34y^2 - 25 = 0$

14. $x^2 + 2\sqrt{3}xy + 3y^2 + 8\sqrt{3}x - 8y + 32 = 0$

15. $17x^2 - 48xy + 31y^2 + 49 = 0$

16. $52x^2 - 72xy + 73y^2 = 200$

17. $9x^2 + 24xy + 16y^2 + 90x - 130y = 0$

18. $x^2 + 10xy + y^2 + 1 = 0$

19. $23x^2 + 26\sqrt{3}xy - 3y^2 - 16x + 16\sqrt{3}y + 128 = 0$

20. $x^2 + 2xy + y^2 + 12\sqrt{2}x - 12\sqrt{2}y = 0$

21. $17x^2 - 12xy + 8y^2 - 80 = 0$

22. $11x^2 - 24xy + 4y^2 + 30x + 40y - 45 = 0$

23. $3x^2 + 2\sqrt{3}xy + y^2 + 4x - 4\sqrt{3}y - 16 = 0$

24. $3x^2 + 2\sqrt{2}xy + 2y^2 - 12 = 0$

Chapter 8 Review

IMPORTANT FACTS & FORMULAS

■ Equations of conic sections

Conic Section	Equation
Circle center (h, k) radius r	$(x - h)^2 + (y - k)^2 = r^2$
Ellipse center (h, k) axes on the lines $x = h, y = k$	$\dfrac{(x - h)^2}{a^2} + \dfrac{(y - k)^2}{b^2} = 1$
Hyperbola center (h, k) vertices on the line $y = k$	$\dfrac{(x - h)^2}{a^2} - \dfrac{(y - k)^2}{b^2} = 1$
Hyperbola center (h, k) vertices on the line $x = h$	$\dfrac{(y - k)^2}{a^2} - \dfrac{(x - h)^2}{b^2} = 1$
Parabola vertex (h, k) axis $x = h$	$(x - h)^2 = 4p(y - k)$

Parabola
vertex (h, k) $(y - k)^2 = 4p(x - h)$
axis $y = k$

■ The discriminant of the equation $Ax^2 + Bxy + Cy^2 + Dx + Ey + F = 0$ (where A, B, C are not all zero) is $B^2 - 4AC$.

If $B^2 - 4AC < 0$, the graph is a circle or ellipse (or a point).

If $B^2 - 4AC = 0$, the graph is a parabola (or a line or two parallel lines).

If $B^2 - 4AC > 0$, the graph is a hyperbola (or two intersecting lines).

REVIEW QUESTIONS 💿

In Questions 1–4, find the foci and vertices of the conic, and state whether it is an ellipse or a hyperbola.

1. $\dfrac{x^2}{16} + \dfrac{y^2}{20} = 1$ **2.** $\dfrac{x^2}{9} - \dfrac{y^2}{16} = 1$

3. $\dfrac{(x-1)^2}{7} + \dfrac{(y-3)^2}{16} = 1$ **4.** $3x^2 = 1 + 2y^2$

5. Find the focus and directrix of the parabola $10y = 7x^2$.

6. Find the focus and directrix of the parabola
$$3y^2 - x - 4y + 4 = 0.$$

In Questions 7–20, sketch the graph of the equation. If there are asymptotes, give their equations.

7. $\dfrac{x^2}{4} + \dfrac{y^2}{25} = 1$ **8.** $25x^2 + 4y^2 = 100$

9. $\dfrac{(x-3)^2}{9} + \dfrac{(y+5)^2}{4} = 1$ **10.** $\dfrac{x^2}{9} - \dfrac{y^2}{16} = 1$

11. $\dfrac{(y+4)^2}{25} - \dfrac{(x-1)^2}{4} = 1$ **12.** $4x^2 - 9y^2 = 144$

13. $x^2 + 4y^2 - 10x + 9 = 0$

14. $9x^2 - 4y^2 - 36x + 24y - 36 = 0$

15. $2y = 4(x - 3)^2 + 6$ **16.** $3y = 6(x + 1)^2 - 9$

17. $x = y^2 + 2y + 2$ **18.** $y = x^2 - 2x + 3$

19. $x^2 + y^2 - 6x + 5 = 0$

20. $x^2 + y^2 - 4x + 6y + 4 = 0$

21. Find the center and radius of the circle whose equation is
$$x^2 + y^2 + 8x + 10y + 33 = 0.$$

22. Find the equation of the circle with center $(-2, 3)$ that passes through the point $(1, 7)$.

23. What is the center of the ellipse
$$4x^2 + 3y^2 - 32x + 36y + 124 = 0?$$

24. Find the equation of the ellipse with center at the origin, one vertex at $(0, 4)$, passing through $(\sqrt{3}, 2\sqrt{3})$.

25. Find the equation of the ellipse with center at $(3, 1)$, one vertex at $(1, 1)$, passing through $(2, 1 + \sqrt{3}/2)$.

26. Find the equation of the hyperbola with center at the origin, one vertex at $(0, 5)$, passing through $(1, 3\sqrt{5})$.

27. Find the equation of the hyperbola with center at $(3, 0)$, one vertex at $(3, 2)$, passing through $(1, \sqrt{5})$.

28. Find the equation of the parabola with vertex $(2, 5)$, axis $x = 2$, and passing through $(3, 12)$.

29. Find the equation of the parabola with vertex $(3/2, -1/2)$, axis $y = -1/2$, and passing through $(-3, 1)$.

30. Find the equation of the parabola with vertex $(5, 2)$ that passes through the points $(7, 3)$ and $(9, 6)$.

31. The arch shown in the figure has the shape of half of an ellipse. How wide is the arch at a point 8 feet above the ground? [*Hint:* Think of the arch as sitting on the *x*-axis, with its center at the origin, and find its equation.]

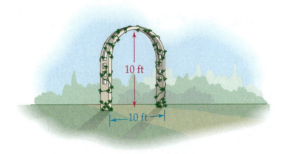

10 ft

10 ft

32. A bridge, whose bottom is shaped like half an ellipse, spans a 52-foot wide canal. The bottom of the bridge is 8 feet above the water at a point 20 feet from the center of the canal. Find the height of the bridge bottom above the water at the center of the canal.

33. A parabolic satellite dish is 3 feet in diameter and 1 foot deep at the vertex. How far from the vertex should the receiver be placed to catch all the signals that hit the dish?

3

1

34. A satellite dish with a parabolic cross section is 6 feet in diameter. The receiver is located on the axis, 1 foot from the base of the dish. How deep is the dish at its center?

In Questions 35–38, assume that the graph of the equation is a nondegenerate conic. Use the discriminant to identify the graph.

35. $3x^2 + 2\sqrt{2}xy + 2y^2 - 12 = 0$

36. $x^2 + y^2 - xy - 4y = 0$

37. $4xy - 3x^2 - 20 = 0$

38. $4x^2 - 4xy + y^2 - \sqrt{5}x - 2\sqrt{5}y = 0$

In Questions 39–44, find a viewing window that shows a complete graph of the equation.

39. $x^2 - xy + y^2 - 6 = 0$

40. $x^2 + xy + y^2 - 3y - 6 = 0$

41. $x^2 + xy - 2 = 0$

42. $x^2 - 4xy + y^2 + 5 = 0$

43. $x^2 + 3xy + y^2 - 2\sqrt{2}x + 2\sqrt{2}y = 0$

44. $x^2 + 2xy + y^2 - 4\sqrt{2}y = 0$

DISCOVERY PROJECT 8   Designing Light Fixtures

Steve Niedorf Photography/Getty Images

Lighting fixtures are often designed using a parabolic reflector that focuses the light into a relatively parallel beam. This kind of light is referred to as a spotlight, since the beam of light can create a bright circular pool of light on a wall or floor. The effect is created when the light source is placed at the focus deep inside the parabola, sending most of the rays of light in the forward direction.

When the focus is placed closer to the top of the reflector, lots of light leaks out the side, creating a light source with two intensities. This is a popular style with many design advantages; you can accent art objects or create pools of light with a bright center for reading and soft light to surround a chair. We'll explore the factors in designing these lighting fixtures.

To make the job simpler, orient the parabolic reflector so that the axis of symmetry is the y-axis and the vertex sits at the origin. Using the equation $x^2 = 4py$, we also know that the focus of the parabola is at the point $(0, p)$.

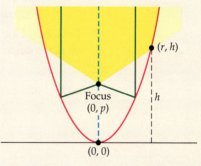

1. Explain why the equation for a nonvertical line through the focus of the parabola is of the form $y = mx + p$.

2. Find two expressions for the value of h, based on the equations of the two graphs that intersect at (r, h).

3. Suppose that you wanted a 4-inch-diameter light where the angle of the diffused light was 45°. This corresponds to a slope of $m = 1$ on the boundary line. How tall is the reflector? How far from the base should the light source (focal point) be?

4. A more focused light would have an angle of diffused light of 30°, corresponding to a slope of $m = 1.73$ on the boundary line. How tall is the reflector? How far from the base should the light source be?

Appendix 1 GEOMETRY REVIEW

An **angle** consists of two half-lines that begin at the same point P, as in Figure 1. The point P is called the **vertex** of the angle, and the half-lines are called the **sides** of the angle.

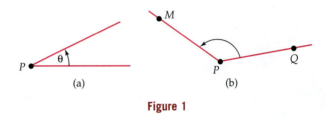

Figure 1

An angle may be labeled by a Greek letter, such as angle θ in Figure 1(a), or by listing three points (a point on one side, the vertex, a point on the other side), such as angle QPM in Figure 1(b).

To measure the size of an angle, we must assign a number to each angle. Here is the classical method for doing this.

1. Construct a circle whose center is the vertex of the angle.

2. Divide the circumference of the circle into 360 equal parts (called **degrees**) by marking 360 points on the circumference, beginning with the point where one side of the angle intersects the circle. Label these points $0°$, $1°$, $2°$, $3°$, and so on.

3. The label of the point where the second side of the angle intersects the circle is the degree measure of the angle.

For example, Figure 2 shows an angle θ of measure 25 degrees (in symbols, $25°$) and an angle β of measure $135°$.

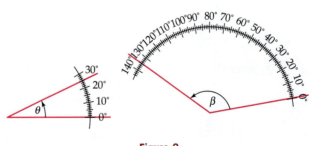

Figure 2

An **acute angle** is an angle whose measure is strictly between $0°$ and $90°$, such as angle θ in Figure 2. A **right angle** is an angle that measures $90°$. An **obtuse angle** is an angle whose measure is strictly between $90°$ and $180°$, such as angle β in Figure 2.

■■ TRIANGLES

A **triangle** has three sides (straight line segments) and three angles, formed at the points where the various sides meet. When angles are measured in degrees,

The sum of the measures of all three angles of a triangle is *always* $180°$.

For instance, see Figure 3. This fact is proved in Exercise 37.

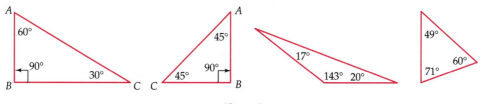

Figure 3

A **right triangle** is a triangle, one of whose angles is a right angle, such as the first two triangles shown in Figure 3. The side of a right triangle that lies opposite the right angle is called the **hypotenuse;** the sides of the right angle are sometimes called the **legs** of the triangle. In each of the right triangles in Figure 3, side *AC* is the hypotenuse. The legs are sides *AB* and *BC*.

Pythagorean Theorem

If the legs of a right triangle have lengths a and b and the hypotenuse has length c, then

$$c^2 = a^2 + b^2.$$

See Exercise 38 for a proof of the Pythagorean Theorem.

EXAMPLE 1

Consider the right triangle with legs of lengths 5 and 12, as shown in Figure 4.

According to the Pythagorean Theorem, the length c of the hypotenuse satisfies the equation $c^2 = 5^2 + 12^2 = 25 + 144 = 169$. Since $169 = 13^2$, we see that c must be 13. ∎

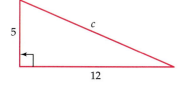

Figure 4

Isosceles Triangle Theorem

If two angles of a triangle are equal, then the two sides opposite these angles have the same length.

Conversely, if two sides of a triangle have the same length, then the two angles opposite these sides are equal.

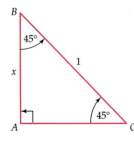

Figure 5

Suppose the hypotenuse of the right triangle shown in Figure 5 has length 1 and that angles B and C measure 45° each.

Then by the Isosceles Triangle Theorem, sides AB and AC have the same length. If x is the length of side AB, then by the Pythagorean Theorem,

$$x^2 + x^2 = 1^2$$
$$2x^2 = 1$$
$$x^2 = \frac{1}{2}$$
$$x = \sqrt{\frac{1}{2}} = \frac{1}{\sqrt{2}} = \frac{\sqrt{2}}{2}.$$

(We ignore the other solution of this equation, namely, $x = -\sqrt{1/2}$, since x represents a length here and therefore must be nonnegative.) Therefore, the legs of a 90°–45°–45° triangle with hypotenuse 1 are each of length $\sqrt{2}/2$. ∎

30-60-90 Triangle Theorem

In a right triangle that has an angle of 30°, the length of the side opposite the 30° angle is one-half the length of the hypotenuse.

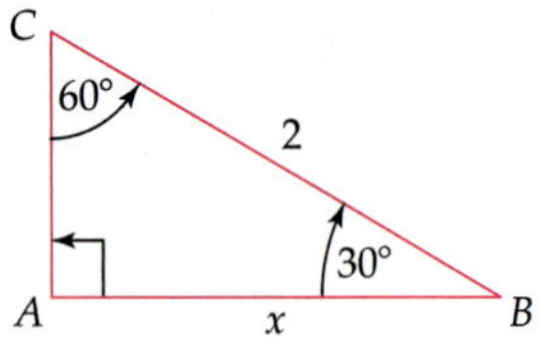

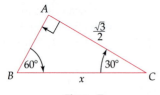

Figure 6

EXAMPLE 3

Suppose that in the right triangle shown in Figure 6 angle B is 30° and the length of hypotenuse BC is 2.

By the 30–60–90 Triangle Theorem, the side opposite the 30° angle, namely, side AC, has length 1. If x denotes the length of side AB, then by the Pythagorean Theorem,

$$1^2 + x^2 = 2^2$$
$$x^2 = 3$$
$$x = \sqrt{3}. ∎$$

Figure 7

EXAMPLE 4

The right triangle shown in Figure 7 has a 30° angle at C, and side AC has length $\sqrt{3}/2$.

Let x denote the length of the hypotenuse BC. By the 30–60–90 Triangle Theorem, side AB has length $\frac{1}{2}x$. By the Pythagorean Theorem,

$$\left(\frac{1}{2}x\right)^2 + \left(\frac{\sqrt{3}}{2}\right)^2 = x^2$$
$$\frac{x^2}{4} + \frac{3}{4} = x^2$$

$$\frac{3}{4} = \frac{3}{4}x^2$$

$$x^2 = 1$$

$$x = 1.$$

Therefore, the triangle has hypotenuse of length 1 and legs of lengths $1/2$ and $\sqrt{3}/2$. ∎

Figures 6 and 7 illustrate properties shared by all triangles.

The longest side of a triangle is always opposite its largest angle.

The shortest side of a triangle is always opposite its smallest angle.

You may find these facts helpful when checking solutions to triangle problems.

▟ CONGRUENT TRIANGLES

Two angles are said to be **congruent** if they have the same measure, and two line segments are **congruent** if they have the same length. We say that two triangles are **congruent** if the three sides and three angles of one are congruent respectively to the corresponding sides and angles of the other, as illustrated in Figure 8.

Congruent Triangles

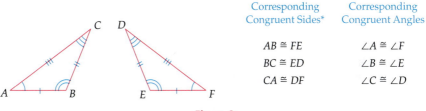

	Corresponding Congruent Sides*	Corresponding Congruent Angles
	$AB \cong FE$	$\angle A \cong \angle F$
	$BC \cong ED$	$\angle B \cong \angle E$
	$CA \cong DF$	$\angle C \cong \angle D$

Figure 8

When two triangles are congruent, you can place one on top of the other so that they coincide. (You might have to rotate or flip one of the triangles over to do this, as is the case in Figure 8.) Thus,

Congruent triangles have the same size and same shape.

The key facts about congruent triangles are proved in high school geometry.

Congruent Triangles Theorem

If two triangles are given, then any one of the following conditions guarantees that they are congruent.

1. SAS: Two sides and the angle between them in one triangle are congruent to the corresponding sides and angle in the other triangle.

2. ASA: Two angles and the side between them in one triangle are congruent to the corresponding angles and side in the other triangle.

3. SSS: The three sides of one triangle are congruent to the corresponding sides of the other triangle.

---

*The symbol $\cong$ means "is congruent to."

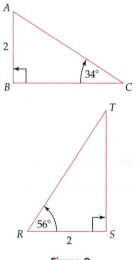

Figure 9

EXAMPLE 5

Show that the right triangles in Figure 9 are congruent.

SOLUTION   Triangle ABC has angles of 90° and 34°. Since the sum of all three angles is 180°, angle A must measure 56°. Hence, we have these congruences.

$$\angle A \cong \angle R \quad \text{[Both measure 56°.]}$$

$$\angle B \cong \angle S \quad \text{[Both measure 90°.]}$$

$$AB \cong RS. \quad \text{[Both have length 2.]}$$

Since two angles and the side between them in triangle ABC are congruent to the corresponding angles and side in triangle DEF, the triangles are congruent by ASA.   ■

▓ SIMILAR TRIANGLES

Two triangles are said to be **similar** if the three angles of one are congruent respectively to the three angles of the other, as illustrated in Figure 10.

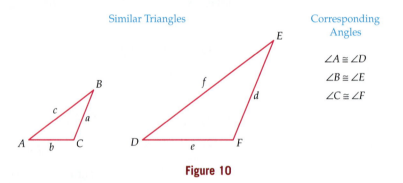

Figure 10

Thus,

> **Similar triangles have the same shape, but not necessarily the same size.**

In Figure 10, for example, the sides of triangle DEF are twice as long as the corresponding sides of triangle ABC:

$$d = 2a, \quad e = 2b, \quad f = 2c,$$

which is equivalent to:

$$\frac{d}{a} = 2, \quad \frac{e}{b} = 2, \quad \frac{f}{c} = 2.$$

Hence,

$$\frac{d}{a} = \frac{e}{b} = \frac{f}{c}.$$

A similar fact holds in the general case.

Ratios Theorem

Suppose triangle *ABC* is similar to triangle *DEF* (with $\angle A \cong \angle D$; $\angle B \cong \angle E$; $\angle C \cong \angle F$).

Then the ratios of corresponding sides are equal, that is,

$$\frac{d}{a} = \frac{e}{b} = \frac{f}{c}.$$

NOTE: There are many ways to rewrite the conclusions of the Ratios Theorem. For example,

$$\frac{d}{a} = \frac{f}{c} \quad \text{is equivalent to} \quad \frac{a}{c} = \frac{d}{f}.$$

To see this, multiply the first equation by $\frac{a}{f}$. Similarly, multiplying by $\frac{b}{f}$ shows that

$$\frac{e}{b} = \frac{f}{c} \quad \text{is equivalent to} \quad \frac{b}{c} = \frac{e}{f}.$$

EXAMPLE 6

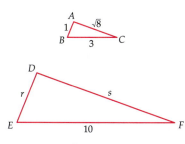

Figure 11

Suppose the triangles in Figure 11 are similar and that the sides have the lengths indicated. Find *r* and *s*.

SOLUTION   By the Ratios Theorem,

$$\frac{\text{Length } DF}{\text{Length } AC} = \frac{\text{Length } EF}{\text{Length } BC}.$$

In other words,

$$\frac{s}{\sqrt{8}} = \frac{10}{3},$$

so

$$3s = 10\sqrt{8}$$

$$s = \left(\frac{10}{3}\right)\sqrt{8}.$$

Similarly,

$$\frac{\text{Length } DE}{\text{Length } AB} = \frac{\text{Length } EF}{\text{Length } BC},$$

so

$$\frac{r}{1} = \frac{10}{3}$$

$$r = \frac{10}{3}.$$

Therefore, the sides of triangle DEF are of lengths 10, $\frac{10}{3}$, and $\frac{10}{3}\sqrt{8}$. ■

EXAMPLE 7

Tom, who is 6 feet tall, stands in the sunlight near the main library on campus, and Anne determines that his shadow is 7 feet long. At the same time, their friend Mary Alice finds that the library casts a 100-foot-long shadow. How high is the main library?

SOLUTION   Both Tom and the library are perpendicular to the ground. So we have the right triangles in Figure 12 (which is not to scale).

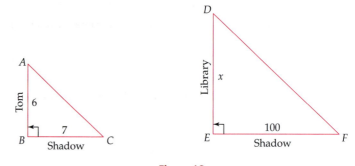

Figure 12

Angles B and E each measure 90°; hence,

$$\angle B \cong \angle E.$$

Because the sun hits both Tom and the library at the same angle,

$$\angle A \cong \angle D.$$

If each of these angles measures $k°$, then angle C measures $180° - 90° - k°$ (because the sum of the angles in triangle ABC must be 180°). Similarly, angle F must also measure $180° - 90° - k°$, so

$$\angle C \cong \angle F.$$

Therefore, triangle ABC is similar to triangle DEF. By the Ratios Theorem,

$$\frac{\text{Length } DE}{\text{Length } AB} = \frac{\text{Length } EF}{\text{Length } BC}$$

$$\frac{x}{6} = \frac{100}{7}$$

$$x = 6\left(\frac{100}{7}\right) \approx 85.7.$$

So the library is about 85.7 feet high. ■

In addition to the definition, there are several other ways to show that two triangles are similar.

Similar Triangles Theorem

Given triangles *ABC* and *DEF*,

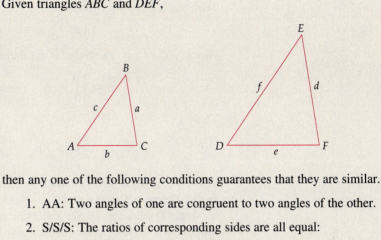

then any one of the following conditions guarantees that they are similar.

1. AA: Two angles of one are congruent to two angles of the other.

2. S/S/S: The ratios of corresponding sides are all equal:

$$\frac{d}{a} = \frac{e}{b} = \frac{f}{c}.$$

3. S/A/S: An angle in one is congruent to an angle in the other, and the ratios of the corresponding sides of these angles are the same:

$$\angle A \cong \angle D \qquad \text{or} \qquad \angle B \cong \angle E \qquad \text{or} \qquad \angle C \cong \angle F$$

$$\frac{e}{b} = \frac{f}{c} \qquad\qquad \frac{d}{a} = \frac{f}{c} \qquad\qquad \frac{d}{a} = \frac{e}{b}.$$

Example 7 shows why the first condition guarantees similarity: If two pairs of angles are congruent, then the third angles are also congruent because the sum of the angles in each triangle is 180°. The proofs of the other conditions will be omitted.

EXAMPLE 8

Show that the triangles in Figure 13 are similar, and find angles *C*, *D*, and *E* and side *f*.

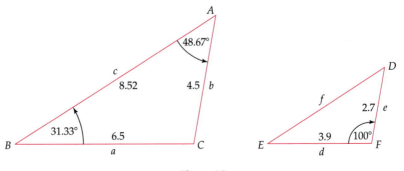

Figure 13

SOLUTION   Since the sum of the measures of the angles in triangle ABC is 180°, angle C must measure

$$180° - 31.33° - 48.67° = 100°.$$

Hence, $\angle C \cong \angle F$. The ratios of the corresponding sides of these angles are

$$\frac{d}{a} = \frac{3.9}{6.5} = .6 \qquad \text{and} \qquad \frac{e}{b} = \frac{2.7}{4.5} = .6$$

Thus,

$$\angle C \cong \angle F \qquad \text{and} \qquad \frac{d}{a} = \frac{e}{b}.$$

Therefore, triangle ABC is similar to triangle DEF by S/A/S of the Similar Triangles Theorem. Consequently, corresponding angles are congruent. In particular, angle D measures 48.67° and angle E measures 31.33° because $\angle D \cong \angle A$ and $\angle E \cong \angle B$. Since corresponding side ratios are equal,

$$\frac{f}{c} = \frac{d}{a}$$

$$\frac{f}{8.52} = \frac{3.9}{6.5} = .6$$

$$f = .6(8.52) = 5.112. \quad \blacksquare$$

✓ EXERCISES 1

In Exercises 1–8, find the missing side(s) of the triangle.

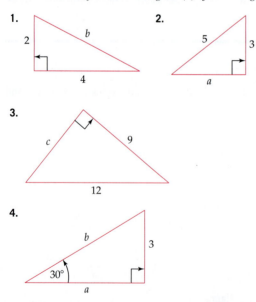

1.

2.

3.

4.

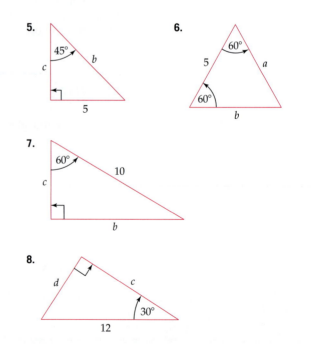

5.

6.

7.

8.

In Exercises 9 and 10, two congruent triangles are given. Find the missing sides and angles.

9.

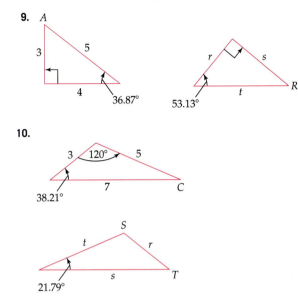

10.

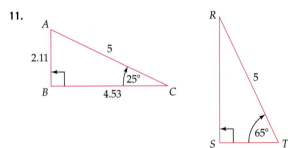

In Exercises 11 and 12, prove that the two triangles are congruent.

11.

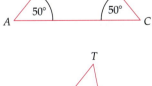

12.

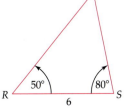

13. If $\angle u \cong \angle v$ and $OP \cong OQ$, show that $PD \cong DQ$ and that PQ is perpendicular to OD. [*Hint:* Prove that triangles *POD* and *DOQ* are congruent.]

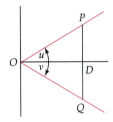

14. If $AB \cong AD$, $BC \cong DC$, $\angle x \cong \angle y$, and $\angle u \cong \angle v$, prove that triangles *ADC* and *ABC* are congruent.

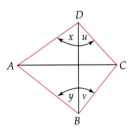

15. If $AB \cong DC$ and $\angle u \cong \angle v$, prove that triangles *ABC* and *ACD* are congruent.

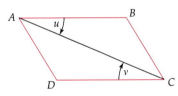

16. If $AB \cong CB$ and $\angle u \cong \angle v$, prove that triangles *ABD* and *BCD* are congruent.

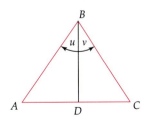

17. If $\angle x \cong \angle y$ and $\angle u \cong \angle v$, prove that triangles ABC and ADC are congruent.

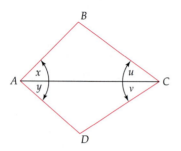

18. If $\angle x \cong \angle y$ and $\angle u \cong \angle v$, prove that triangles ABD and DBC are congruent.

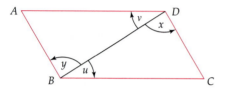

In Exercises 19–22, triangle ABC is similar to triangle RST. The following notation is used: The side opposite angle A is denoted a, the side opposite angle B is denoted b, the side opposite angle R is denoted r, and so on.

19. If $\angle B = 37°$, $\angle C = 90°$, $a = 6$, $b = 4.5$ and $\angle R = 53°$, $\angle T = 90°$, $s = 21$, find c, r, t.

20. If $\angle A = 90°$, $c = 220$ and $\angle T = 90°$, $\angle S \cong \angle B$, $r = 100$, $s = 50$, find a, b, t.

21. If $a = 7$, $b = 5$, $c = 6$, and $\angle A \cong \angle T$, $\angle C \cong \angle R$, $s = 2.5$, find r and t.

22. If $a = 12$, $b = 9$, $c = 4$, and $\angle B \cong \angle T$, $\angle C \cong \angle R$, $t = 6.3$, find r and s.

In Exercises 23–26, prove that the triangles are similar, and find the missing sides and angles.

23.

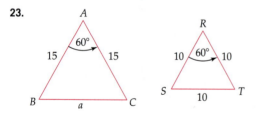

24.

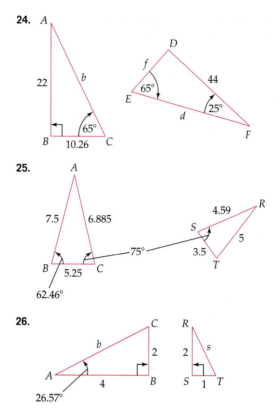

25.

26.

In Exercises 27–30, prove that triangles ABC and RST are similar, and find the missing sides and angles. The notation is the same as in Exercises 19–22.

27. $\angle C = 90°$, $a = 5$, $b = 5$ and $\angle T = 90°$, $r = 17.5$, $s = 17.5$.

28. $\angle B = (160/3)°$, $\angle C = 90°$, $a = 3.5$, $b = 4.7$ and $\angle R = (110/3)°$, $\angle T = 90°$, $r = 7$, $s = 9.4$

29. $\angle A = 105°$, $\angle B = 33.5°$, $a = 7$, $b = 4$, $c = 4.8$ and $\angle R = 105°$, $\angle T = 33.5°$, $r = 3.5$, $t = 2$.

30. $\angle B = 25°$, $\angle C = 142.8°$, $a = 5$, $b = 10$, $c = 14.3$, and $\angle S = 25°$, $\angle T = 12.2°$, $t = 10$

31. A flagpole casts a 99-foot shadow at the same time as a 5-foot-long stick (perpendicular to the ground) casts a 9-foot shadow. How high is the flagpole?

32. If a 6-foot-tall person casts a 10-foot shadow, how long is the shadow cast by a 28.3-foot-high tree that is next to the person?

33. A scale model of a pyramid is 1/20 the size of the original. If the model pyramid is 4 feet tall, how tall is the original pyramid?

34. A scale model of a ship is 8 inches long, and its mast is 4.5 inches high. If the real ship is 65 feet long, how high is the real mast?

35. On a map of Ohio, Columbus is 9.2 centimeters from Cincinnati and 8.5 centimeters from Marietta. On the map, Cincinnati is 15.75 centimeters from Marietta. If the actual distance from Columbus to Cincinnati is 100 miles, how far is Marietta from Columbus and from Cincinnati?

36. A small mirror is placed on level ground 4 feet from the base of a flagpole. A person whose eyes are 66 inches above the ground sees the top of the pole in the mirror when he is 2 feet from the mirror. The base of the pole, the mirror, and the person are aligned on a straight line. How high is the flagpole?

Thinkers

37. If *ABC* is a triangle, prove that the sum of its angles is 180°. [*Hint:* In the figure below, use congruent triangles to show that $\angle u \cong \angle x$ and $\angle v \cong \angle y$. You may assume that opposite sides of a rectangle have the same length.]

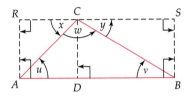

38. Prove the Pythagorean Theorem. [*Hint:* Use the figure below, in which triangle *ABC* has a right angle at *C*. Find three similar triangles in the figure, and use the properties of similar triangles to show that $a^2 + b^2 = c^2$.]

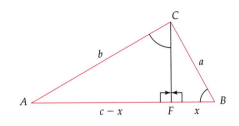

Appendix 2 PROGRAMS

Each program is preceded by a *Description,* which describes, in general terms, how the program operates and what it does. Some programs require that certain things be done before the program is run (such as entering a function in the function memory); these requirements are listed as *Preliminaries.* Occasionally, italic remarks appear in brackets after a program step; they are *not* part of the program but are intended to provide assistance when you are entering the program into your calculator. A remark such as "[*MATH NUM menu*]" means that the symbols or commands needed for that step of the program are in the NUM submenu of the MATH menu.

Fraction Conversion for Casio (Built-in on other calculators)

Description: Enter a positive repeating decimal; the program converts it into a fraction.

Note: When two submenus are given, the first is for 9850 and the second is for FX 2.0.

Fix 7   [*SETUP DISPLAY menu*]

"N ="?→N   [*PROG REL or LOGIC menu*]

0→D

Lbl 1   [*PROG JUMP menu*]

D + 1→D

N × D

Rnd   [*OPTN NUM menu*]

If (Frac Ans) ≠ 0   [*PROG COM or IF; OPTN NUM; PROG REL or LOGIC menus*]

Then Goto 1   [*PROG COM or IF; PROG JUMP menus*]

If End   [*PROG COM or IF menu*]

(Ans + .5)→N

Norm   [*SETUP DISP menu*]

Locate 3, 4, Int N   [*PROG IO; OPTN NUM menus*]

Locate 3, 5, "—"

Locate 3, 6, D◢

TI-82/83+ Quadratic Formula (Built-in on other calculators)

Description: Enter the coefficients of the quadratic equation $ax^2 + bx + c = 0$; the program finds all real solutions.

:ClrHome [*Optional*]

:Disp "AX² + BX + C = 0" [*Optional*]

:Prompt A

:Prompt B

:Prompt C

:(B² − 4AC) → S

:If S < 0

:Goto 1

:Disp $(-B + \sqrt{S})/(2A)$

:Disp $(-B - \sqrt{S})/(2A)$

:Stop

:Lbl 1

:Disp "NO REAL ROOTS"

TI-85 Table Maker (Built-in on other calculators)

Preliminaries: Enter the function to be evaluated in the function memory as Y_1.

Description: Select a starting point and an increment (the amount by which adjacent x entries differ); the program displays a table of function values. To scroll through the table a page at a time, press "down" or "up." Press "quit" to end the program. [*Note:* An error message will occur if the calculator attempts to evaluate the function at a number for which it is not defined. In this case, change the starting point or increment to avoid the undefined point.]

:Lbl SETUP
:ClLCD [*Optional*]
Disp " "TABLE SETUP"
:Input "TblMin =", tblmin
:Input "ΔTbl =", dtbl   [*CHAR GREEK menu*]
:tblmin → x   [*Use the x-var key for x.*]
:Lbl CONTD
:ClLCD
:Output(1,1,"x")   [*I/O menu*]
:Output(1,10,"y_1")
:For (cnt,2,7,1)
:Output(cnt,1,x)
Output(cnt,8," ")

:Output(cnt,9,y_1)
:x + dtbl → x
:End
:Menu(1, "Down",CONTD, 2,"UP",CONTU, 4, "Setup," SETUP, 5, "quit", TQUIT)
:Lbl CONTU
:x − 12*dtbl → x
:Goto CONTD
:Lbl TQUIT
:ClLCD
:Stop

Synthetic Division (Built-in on TI-89/92)

Preliminaries: Enter the coefficients of the dividend $F(x)$ (in order of decreasing powers of x, putting in zeros for missing coefficients) as list L_1 (or List 1). If the coefficients are 1, 2, 3, for example, key in {1, 2, 3,} and store it in L_1. The symbols { } are on the keyboard or in the LIST menu. The list name L_1 is on the keyboard of TI-82/83+. On TI-85/86 and HP-39+, type in L1. On Casio, use "List" in the LIST submenu of the OPTN menu to type in List 1.

Description: Write the divisor in the form $x − a$ and enter a. The program displays the degree of the quotient $Q(x)$, the coefficients of $Q(x)$ (in order of decreasing powers of x), and the remainder. If the program pauses before it has displayed all these items, you can use the arrow keys to scroll through the display; then press ENTER (or OK) to continue.

TI

:ClrHome [ClLCD on TI-85/86]

:Disp "DIVISOR IS X − A"

:Prompt A

:$L_1 \to L_2$ [*See Preliminaries for how to enter list names*]

:dim $L_1 \to N$ [dimL L_1 on TI-85/86]   [*LIST OPS menu*]

:For (K, 2, N)

:$(L_1(K) + A \times L_2(K − 1)) \to L_2(K)$

:End

:round($L_2(N)$,9) $\to$ R [*MATH NUM menu*]

:$(N − 1) \to$ dim L_2

:Disp "DEGREE OF QUOTIENT"

:Disp dim $L_2 − 1$

:Disp "COEFFICIENTS"

:Pause L_2

:Disp "REMAINDER"

:Disp R

Casio

"DIVISOR IS X − A"

"A = "? $\to$ A

List 1 $\to$ List 2   [*See Preliminaries for how to enter list names*]

dim List 1 $\to$ N   [*OPTN LIST menu*]

2 $\to$ K

Lbl 1

(List 1[K] + A × List 2 [K − 1]) $\to$ List 2 [K]

K + 1 $\to$ K

If K ≤ N

Then Goto 1

If End

Fix 9   [*SETUP DISP menu*]

List 2 [N]

Rnd   [*OPTN MATH NUM menu*]

Ans $\to$ R

Norm   [*SETUP DISP menu*]

Seq(List 2 [X], X, 1, N − 1, 1) $\to$ List 2

    [*OPTN LIST menu*]

"DEGREE OF QUOTIENT"

dim List 2 − 1◢

"COEFFICIENTS"

List 2◢

"REMAINDER"

R

HP-39+

Input A; "SYNDIV", "X − A"; "ENTER A"; 0:

$L_1 \to L_2$:   [*See Preliminaries for how to enter list names*]

Size(L_2) $\to$ N: [*MATH LIST menu*]

For C = 1 to (N − 1)   Step 1; $L_2(C) \times A + L_2(C + 1) \to L_2(C + 1)$ End:

Makelist ($L_2(J)$, J, 1, N − 1, 1) $\to L_3$:   [*MATH LIST menu*]

Msgbox "DEGREE OF QUOTIENT" Size(L_3) − 1:

Msgbox "COEFFICIENTS" L_3:

Msgbox "REMAINDER" $L_2(N)$:

RREF Program for TI-82 and Casio (Built-in on other calculators)

Preliminaries: Enter a matrix in the matrix memory as matrix A.

Description: When the program is run, it uses row operations to put matrix A in reduced row echelon form. The reduced matrix is displayed; you may scroll through it, using the arrow keys. When the program is finished, the original matrix is in the matrix memory as matrix A, and the reduced matrix is in the matrix memory as matrix B.

TI

```
:ClrHome
:[A] → [B]
:dim([B]) → L₁
:L₁(1) → M
:L₁(2) → N
:1 → R
:1 → J
:Lbl 6
:abs(seq([B](I, J), I, R, M, 1)) → L₂     [LIST menu]
:max(L₂) → T
:If T = 0
:Goto 3
:R → I
:Lbl 1
:If abs([B](I, J)) = T
:Goto 2
:I + 1 → I
:Goto 1
:Lbl 2
:*row([B](I, J)⁻¹, [B], I) → [B]    [MATRIX MATH menu]
:For(K, 1, I − 1, 1)
:*row+(−[B](K, J), [B], I, K) → [B]
:End
:For(K, I + 1, M, 1)
:*row+(−[B](K, J), [B], I, K) → [B]
:End
:rowSwap([B], I, R) → [B]
:round([B], 9) → [B]
:If R = min(M, N)
:Then
:Lbl 7
:Pause round([B], 9)▶Frac
```

```
:Stop
:End
:R + 1 → R
:J + 1 → J
:If J = N + 1
:Goto 7
:Goto 6
:Lbl 3
:J + 1 → J
:If J = N + 1
:Goto 7
:Goto 6
```

Casio

```
Mat A → Mat B
"NUMBER OF ROWS"? → M
"NUMBER OF COLUMNS"? → N
{M, N} → List 3
1 → R
1 → J
Lbl 6
Mat → List(Mat B, J) → List 1
Seq(List 1[X], X, R, M, 1) → List 1 [OPTN LIST menu]
Abs(List 1) → List 1
Max(List 1) → T   [OPTN LIST menu]
If T = 0
Then Goto 3
If End
R → I
Lbl 1
Abs(Mat B[I, J]) → S
If S = T
Then Goto 2
If End
```

I + 1 → I

Goto 1

Lbl 2

*Row Mat B[I, J]$^{-1}$, B, I   [(*MENU*) *MAT menu*]

For 1 → K To I − 1

*Row+ −Mat B[K, J], B, I, K

Next

For I + 1 → K To M

*Row+ −Mat B[K, J], B, I, K

Next

Swap B, I, R

If R = Min(List 3)

Then Goto 7

Else Goto 8

If End

Lbl 7

Mat B ◢

Stop

Lbl 8

R + 1 → R

J + 1 → J

If J = N + 1

Then Goto 7

If End

Goto 6

Lbl 3

J + 1 → J

If J = N + 1

Then Goto 7

If End

Goto 6

HP-39 Rectangular/Polar Conversion Program

Description: Enter the rectangular coordinates of a point in the plane; the program displays the polar coordinates of the point.

Input X; "RECTANGULAR TO POLAR"; "X ="; "ENTER X"; 0:

Input Y: "RECTANGULAR TO POLAR"; "Y ="; "ENTER Y"; 0:

If X > 0

Then (ATAN(Y/X)) → C

Else (ATAN(X, Y) + π) → C:

End:

MSGBOX "R =" $\sqrt{X^2 + Y^2}$ "θ =" C:

HP-39 Polar/Rectangular Conversion Program

Description: Enter the polar coordinates of a point in the plane; the program displays the rectangular coordinates of the point.

Input R; "POLAR TO RECTANGULAR"; "R = "; "ENTER R"; 0:

Input θ; "POLAR TO RECTANGULAR"; "θ = "; "ENTER θ"; 0: [*CHARS menu*]

MSGBOX "X = " R(cos θ) "Y = " R(sin θ):

CHAPTER 0

Section 0.1, page 10

1.

3. $12 > 9$ **5.** $-4 > -8$ **7.** $\pi < 100$

9. $y \le 7.5$ **11.** $t > 0$ **13.** $c \le 3$

15. $-6 < -2$ **17.** $3/4 = .75$ **19.** $1/3 > .33$

21. 5   **23.** 6   **25.** 0

27. $\dfrac{2040}{523}, \dfrac{189}{37}, \sqrt{27}, \dfrac{4587}{691}, 6.735, \sqrt{47}$

29. $a = b + c$ **31.** a lies to the right of b.

33. $b - a > 0$ or $a < b$.

35. 0   **37.** 10   **39.** 169   **41.** $\pi - 2$

43. $|-2| < |-5|$ **45.** $|3| > -|4|$ **47.** $-7 < |-1|$

49. 7   **51.** $29/2$

53. $\pi - 3$ **55.** $|316.3 - 408.6|$

57. For 5 ft 8 in., $|x - 143| \le 21$; for 6 ft 0 in., $|x - 163| \le 26$.

59. $|21 - (-12)| = 33$ **61.** 2000, 2003, 2004

63. t^2 **65.** $(-3 - y)^2$

67. $b - 3$ **69.** $|x - 5| < 4$

71. $|x + 4| \le 17$ **73.** $|c| < |b|$

75. The distance from x to 3 is less than 2.

77. x is at most 3 units from -7.

79. $x = \pm 1$ **81.** $x = 1$ or $x = 3$

83. $x = -\pi \pm 4$

85. $(c - d)^2 \ge 0$, so $|(c - d)^2| = (c - d)^2$ by the definition. But $(c - d)^2 = c^2 - 2cd + d^2$.

87. If all of a, b, and c were zero, then by property 1, $|a| = 0$, $|b| = 0$, and $|c| = 0$, so $|a| + |b| + |c|$ would be zero.

Special Topics 0.1.A, page 14

1. $.777\cdots$ **3.** $1.6428571428571\cdots$

5. $.0526315789473684210526315789473684210526315789473684421\cdots$

7. $37/99$

9. $\dfrac{758{,}679}{9900} = \dfrac{252{,}893}{3300}$

11. $5/37$ **13.** $\dfrac{14{,}386}{275}$

15. no   **17.** yes   **19.** no   **21.** yes

23. Let $a = .74999\cdots$; then $100a = 74.999\cdots$ and $10{,}000a = 7499.999\cdots$, so $9900a = 7425$, so $a = 7425/9900 = 3/4$. Let $b = .75000$; then $100b = 75$ and $b = 75/100 = 3/4$.

25. $.05882352941176470588235294117647\cdots$

27. $.034482758620689655172413793103448275862068965\,517241379310344827586\cdots$

29. $6.0212765957446808510638297872340425531914893617\,70212765957446808510638297872340425531914893617\cdots$

31. Answers will vary. Examples follow.
   (a) To each terminating decimal, associate the nonterminating, nonrepeating decimal obtained by appending the string of digits $010010001\cdots$ (one more zero in each successive block).
   (b) With each repeating decimal, associate the nonrepeating decimal obtained by inserting longer and longer strings of zeros between blocks of repeated digits. An example other than the one in the hint would be to associate with $0.27272727\cdots$ the decimal $0.270270027000\cdots$.

Section 0.2, page 21

1. 36   **3.** 73   **5.** -5 **7.** -112

9. $\dfrac{81}{16}$ **11.** $\dfrac{129}{8}$ **13.** x^{10} **15.** $.03y^9$

17. $24x^7$ **19.** $9x^4 y^2$ **21.** $384w^6$ **23.** ab^3

25. $8x^{-1} y^3$ **27.** $3xy$ **29.** 2^{12}

31. 2^{-12} **33.** x^7 **35.** $e^9 c$

37. $a^{12} b^8$ **39.** $\dfrac{1}{c^{10} d^6}$ **41.** $\dfrac{a^7 c}{b^6}$

43. $c^3 d^6$ **45.** negative   **47.** negative

49. negative   **51.** 3^s **53.** $a^{6t} b^{4t}$

55. 6.275×10^9 **57.** 5.91×10^{12} m

59. 2×10^{-9} m   **61.** 299,790,000 mi/s

63. .00000000000000000000000000001602 coulomb
(26 zeros between decimal point and 1)

65. (a) 5.874602×10^{12} mi   (b) 3.994729×10^{15} mi

67. 1×10^5 genes   **69.** about 5.147×10^8 km^2

71. Elizabeth's friend

73.

r	$1000(1 + r)^5$
$3\frac{1}{2}\%$	1187.69
5%	1276.28
7.5%	1435.63
10%	1610.51

75–77. Many examples exist, including the following.

75. If $a = 2$, $r = 3$, $s = 4$, then $a^r a^s = 2^3 2^4 = (8)(16) = 128$, but $a^{rs} = 2^{12} = 4096$.

77. If $c = 2$, $r = 2$, then $c^{-r} = 2^{-2} = 1/4$ but $-c^r = -2^2 = -4$.

Section 0.3, page 28

1. $4\sqrt{5}$ **3.** $7\sqrt{3}$ **5.** $6\sqrt{2}$

7. $\dfrac{-2 + \sqrt{11}}{5}$ **9.** $-\sqrt{2}$ **11.** $15\sqrt{5}$

13. $4a^4|b^{-1}|$ **15.** $\dfrac{|d^5|}{2\sqrt{c}}$ or $\dfrac{|d^5|\sqrt{c}}{c}$

17. $3a^3b^{-1}$ **19.** $>$ **21.** $=$

23. $(a^2 + b^2)^{1/3}$ **25.** $a^{3/16}$ **27.** $4t^{27/10}$

29. $1000k^{13/4}$ **31.** $x^{9/2}$ **33.** $\dfrac{1}{3y^{2/3}}$

35. $\dfrac{a^{1/2}}{49b^{5/2}}$ **37.** $x^{7/6} - x^{11/6}$ **39.** $x - y$

41. $x + y - (x + y)^{3/2}$ **43.** $2684.84

45. $3166.75   **47.** about 35,863,131 mi

49. about 886,781,537 mi   **51.** about 5,312,985

53. about 20,003,970   **55.** 172.3

57. 187.4

59. (a)

L	C	$Q = L^{1/4}C^{3/4}$
10	7	7.65
20	14	15.31
30	21	22.96
40	28	30.61
60	42	45.92

(b) If both labor and capital are doubled, output is doubled. If both are tripled, output is tripled.

61. (a)

L	C	$Q = L^{1/2}C^{3/4}$
10	7	13.61
20	14	32.37
30	21	53.73
40	28	76.98
60	42	127.79

(b) If both labor and capital are doubled, output is multiplied by $2^{5/4}$. If both are tripled, output is multiplied by $3^{5/4}$.

63. (a) The square (or any even power) of a real number is never negative.

(b) When $c = -8$, $c^{1/3} = \sqrt[3]{-8} = -2$, but $c^{2/6} = \sqrt[6]{(-8)^2} = \sqrt[6]{64} = 2$.

65. (a) (i) gives $3\frac{1}{7} \approx 3.142857$; (ii) gives $3\frac{1}{6} \approx 3.16666$;

(iii) gives $3\frac{35}{216} \approx 3.162037$; (iii) is best;

(b) (i) gives $4\frac{3}{9} \approx 4.33333$; (ii) gives $4\frac{3}{8} = 4.375$;

(iii) gives $4\frac{183}{512} \approx 4.35742$; (iii) is best.

67. $\sqrt{3^2 + 4^2} = \sqrt{9 + 16} = \sqrt{25} = 5$, wheras $3 + 4 = 7$

69. $\sqrt{64} \cdot \sqrt[3]{64} = 8 \cdot 4 = 32$, whereas $\sqrt[4]{64} = 2\sqrt{2} \approx 2.828$

Section 0.4, page 34

1. $8x$ **3.** $-2a^2b$

5. $-x^3 + 4x^2 + 2x - 3$ **7.** $5u^3 + u - 4$

9. $6z^3w^2 - zw^3 - 12z^2w + 4z + 8$

11. $-3x^3 + 15x + 8$ **13.** $-5xy - x$

15. $15y^3 - 5y$

17. $12a^2x^2 - 6a^3xy + 6a^2xy$

19. $12z^4 + 30z^3$ **21.** $12a^2b - 18ab^2 + 6a^3b^2$

23. $x^2 - x - 2$ **25.** $2x^2 + 2x - 12$

27. $y^2 + 7y + 12$ **29.** $-6x^2 + x + 35$

31. $3y^3 - 9y^2 + 4y - 12$ **33.** $x^2 - 16$

35. $16a^2 - 25b^2$ **37.** $y^2 - 22y + 121$

39. $25x^2 - 10bx + b^2$ **41.** $16x^6 - 8x^3y^4 + y^8$

43. $9x^4 - 12x^2y^4 + 4y^8$ **45.** $2y^3 + 9y^2 + 7y - 3$

47. $-15w^3 + 2w^2 + 9w - 18$

49. $24x^3 - 4x^2 - 4x$ **51.** $x - 25$

53. $9 + 6\sqrt{y} + y$ **55.** $\sqrt{3}x^2 + 4x + \sqrt{3}$

57. $3ax^2 + (2a + 3b)x + 2b$

59. $abx^2 + (a^2 + b^2)x + ab$

61. $x^3 - (a + b + c)x^2 + (ab + ac + bc)x - abc$

63. $12x^2 - 7x - 10$

	$3x$	2
$4x$	$12x^2$	$8x$
-5	$-15x$	-10

65. $8x^3 + 125$

	$4x^2$	$-10x$	25
$2x$	$8x^3$	$-20x^2$	$50x$
5	$20x^2$	$-50x$	125

67. (a) length: $20 - 2x$; width: $15 - 2x$;
(b) volume is $x(20 - 2x)(15 - 2x) =$
$4x^3 - 70x^2 + 300x$;
(c) 234 in.^{3}, 352 in.^{3}, 336 in.^{3}

69. (a) $x(16 - 2x) = 16x - 2x^2$;
(b) $14, 24, 30, 32, 30, 24, 14, 0$;
(c) $x = 4$ appears to produce the largest cross-sectional area.

71. $4x^2 + 32x$

73. If x represents the number chosen, the values at each step are $x, x + 1, (x + 1)^2 = x^2 + 2x + 1, x - 1, (x - 1)^2 = x^2 - 2x + 1, (x^2 + 2x + 1) - (x^2 - 2x + 1) = 4x$;
$\dfrac{4x}{x} = 4$.

75. Answers will vary.

Section 0.5, page 41

1. $(x + 2)(x - 2)$

3. $(3y + 5)(3y - 5)$

5. $(9x + 2)^2$

7. $(\sqrt{5} + x)(\sqrt{5} - x)$

9. $(7 + 2z)^2$

11. $(x^2 + y^2)(x + y)(x - y)$

13. $(x + 3)(x - 2)$

15. $(z + 3)(z + 1)$

17. $(y + 9)(y - 4)$

19. $(x - 3)^2$

21. $(x + 5)(x + 2)$

23. $(x + 9)(x + 2)$

25. $(3x + 1)(x + 1)$

27. $(2z + 3)(z + 4)$

29. $9x(x - 8)$

31. $2(5x + 1)(x - 1)$

33. $(4u - 3)(2u + 3)$

35. $(2x + 5y)^2$

37. $(x - 5)(x^2 + 5x + 25)$

39. $(x + 2)^3$

41. $(2 + x)(4 - 2x + x^2)$

43. $(x + 1)(x^2 - x + 1)$

45. $(2x - y)(4x^2 + 2xy + y^2)$

47. $(x + 2)(x^2 - 2x + 4)(x - 2)(x^2 + 2x + 4)$

49. $(y^2 + 5)(y^2 + 2)$

51. $(x + z)(x - y)$

53. $(a + 2b)(a^2 - b)$

55. $(x + \sqrt{8})(x - \sqrt{8})(x + 4)$

57. $(x^2 + x + 1)(x^2 - x + 1)$.

59. $(x^2 + x + 3)(x^2 - x + 3)$.

61. (a) $(x^2 + 5 + x)(x^2 + 5 - x)$ and
$(x^2 + 6 + x)(x^2 + 6 - x)$;
(b) $(x^2 + 5 + x)(x^2 + 5 - x) = x^2(x^2 + 5 - x) +$
$5(x^2 + 5 - x) + x(x^2 + 5 - x) = x^4 + 5x^2 - x^3 +$
$5x^2 + 25 - 5x + x^3 + 5x - x^2 = x^4 + 9x^2 + 25$;
$(x^2 + 6 + x)(x^2 + 6 - x) = x^2(x^2 + 6 - x) +$
$6(x^2 + 6 - x) + x(x^2 + 6 - x) = x^4 + 6x^2 - x^3 +$
$6x^2 + 36 - 6x + x^3 + 6x - x^2 = x^4 + 11x^2 + 36$.

63. The statement $3(y + 2) = 3y + 2$ is false when $y = 1$, since $3(1 + 2) = 9$ but $3(1) + 2 = 5$. The mistake is the value 2. The correct statement is $3(y + 2) = 3y + 6$.

65. The statement $(x + y)^2 = x + y^2$ is false when $x = 1$ and $y = 2$, since $(1 + 2)^2 = 9$ but $1 + 2^2 = 5$. The mistake is the first part of the expression $x + y^2$. The correct statement is $(x + y)^2 = x^2 + 2xy + y^2$.

67. The statement $y + y + y = y^3$ is false when $y = 1$, since $1 + 1 + 1 = 3$ but $1^3 = 1$. The mistake is the use of 3 as an exponent. The correct statement is $y + y + y = 3y$.

69. The statement $(x - 3)(x - 2) = x^2 - 5x - 6$ is false when $x = 3$, since $(3 - 3)(3 - 2) = 0$ but $3^2 - 5(3) - 6 = -12$. The mistake is the sign on the 6. The correct statement is $(x - 3)(x - 2) = x^2 - 5x + 6$.

Section 0.6, page 51

1. $\dfrac{9}{7}$

3. $\dfrac{195}{8}$

5. $\dfrac{x - 2}{x + 1}$

7. $\dfrac{a + b}{a^2 + ab + b^2}$

9. $\dfrac{1}{x}$

11. $\dfrac{29}{35}$

13. $\dfrac{121}{42}$

15. $\dfrac{ce + 3cd}{de}$

17. $\dfrac{b^2 - c^2}{bc}$

19. $\dfrac{-1}{x(x + 1)}$

21. $\dfrac{x + 3}{(x + 4)^2}$

23. $\dfrac{2x - 4}{x(3x - 4)}$

25. $\dfrac{x^2 - xy + y^2 + x + y}{x^3 + y^3}$

27. $\dfrac{-6x^5 - 38x^4 - 84x^3 - 71x^2 - 14x + 1}{4x(x + 1)^3(x + 2)^3}$

29. 2

31. $\dfrac{2}{3c}$

33. $\dfrac{3y}{x^2}$

35. $\dfrac{12x}{x - 3}$

37. $\dfrac{5y^2}{3(y + 5)}$

39. $\dfrac{u + 1}{u}$

41. $\dfrac{35}{24}$

43. $\dfrac{u^2}{vw}$

45. $\dfrac{x + 3}{2x}$

47. $\dfrac{cd(c + d)}{c - d}$

49. $\dfrac{-3(y - 1)}{y}$

51. $\dfrac{-2x - h}{(x + h)^2 x^2}$

53. $\dfrac{3\sqrt{2}}{4}$

55. $\dfrac{\sqrt{3} - 3}{4}$

57. $\dfrac{2\sqrt{x} - 4}{x - 4}$

59. $\dfrac{1}{\sqrt{x + h + 1} + \sqrt{x + 1}}$

61. $\dfrac{2x + h}{\sqrt{(x + h)^2 + 1} + \sqrt{x^2 + 1}}$

63. (a) \$298.62   (b) \$2333.76

65. (a) \$503.43   (b) \$109,234.80

67. $\pi/4$ **69.** $1/25$

71. (a) .25
(b) .25, .333 · · ·, .321, .286, .25, .220
(c) at $t = 1$;
(d) The concentration continues to decrease.

73. The statement $\dfrac{1}{a} + \dfrac{1}{b} = \dfrac{1}{a + b}$ is false when $a = 1$ and

$b = 1$, since $1/1 + 1/1 = 2$ but $\dfrac{1}{1 + 1} = \dfrac{1}{2}$. The correct

statement is $\dfrac{1}{a} + \dfrac{1}{b} = \dfrac{a + b}{ab}$.

75. The statement $\left(\dfrac{1}{\sqrt{a} + \sqrt{b}}\right)^2 = \dfrac{1}{a + b}$ is false when $a = 1$

and $b = 1$, since $\left(\dfrac{1}{\sqrt{1} + \sqrt{1}}\right)^2 = \dfrac{1}{4}$ but $\dfrac{1}{1 + 1} = \dfrac{1}{2}$.

The correct statement is $\left(\dfrac{1}{\sqrt{a} + \sqrt{b}}\right)^2 = \dfrac{1}{a + 2\sqrt{ab} + b}$.

77. The statement $\dfrac{x}{2 + x} = \dfrac{1}{2}$ is false when $x = 1$, since

$\dfrac{1}{2 + 1} = \dfrac{1}{3}$. The expression $\dfrac{x}{2 + x}$ is already in simplest

form.

79. The statement $\left(\sqrt{x} + \sqrt{y}\right) \cdot \dfrac{1}{\left(\sqrt{x} + \sqrt{y}\right)} = x + y$ is false

when $x = 1$ and $y = 1$, since $(1 + 1) \cdot \dfrac{1}{(1 + 1)} = 1$ but

$1 + 1 = 2$. The correct statement is

$\left(\sqrt{x} + \sqrt{y}\right) \cdot \dfrac{1}{\left(\sqrt{x} + \sqrt{y}\right)} = 1$.

Chapter 0 Review, page 54

1. (a) >   (b) <   (c) <   (d) >   (e) =.

3. (a) $-10 < y < 10$ (b) $0 \le x \le 10$

5. (a) $|x + 7| < 3$ (b) $|y| > |x - 3|$

7. $x = 2$ or $x = 8$ **9.** $x = -1/2$ or $x = -11/2$

11. (a) $7 - \pi$ (b) $\sqrt{23} - \sqrt{3}$

13. (c)

15. $\dfrac{28}{99}$

17. $\dfrac{c^{15}}{d^{15}e^5}$

19. $\dfrac{2}{u}$

21. 1.232×10^{16}; 7.89×10^{-11}

23. c^2

25. $a^{10/3}b^{42/5}$

27. $u^{1/2} - v^{1/2}$

29. $\dfrac{c^2 d^4}{2}$

31. (d)

33. $-2p^3 - 4p^2 + 3p + 9$ **35.** $12z^3 + 14z^2 - 7z + 5$

37. $12k^2 - 20k + 3$

39. $12x^3 - 24x^2 + 12x = 12x(x - 1)^2$

41. $-.0027x^4 + .25x^3 - 6.2x^2 + 70x - 520$

43. $(x + 4)(x + 1)$ **45.** $(x - 4)(x + 2)$

47. $(2x + 3)(2x - 3)$ **49.** $(2x - 5)(2x + 3)$

51. $x(3x - 1)(x + 2)$ **53.** $(x^2 + 1)(x + 1)(x - 1)$

55. (e)   **57.** $1 + 2a$

59. $\dfrac{6x - 11}{24x}$

61. $\dfrac{-y}{x - y}$

63. $\dfrac{x - 1}{x - 3}$

65. $\dfrac{4}{9}$

67. $\dfrac{2}{r + 2}$

69. $\dfrac{x + 1}{x - 1}$

71. $\dfrac{2}{\sqrt{2x + 2h + 1} + \sqrt{2x + 1}}$

CHAPTER 1

Section 1.1, page 68

1. $A(-3, 3)$; $B(-1.5, 3)$; $C(-2.5, 0)$; $D(-1.5, -3)$; $E(0, 2)$;
$F(0, 0)$; $G(2, 0)$; $H(3, 1)$; $I(3, -1)$

3. $(-6, 3)$ **5.** $(4, 2)$

7.

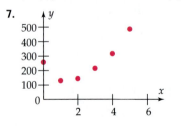

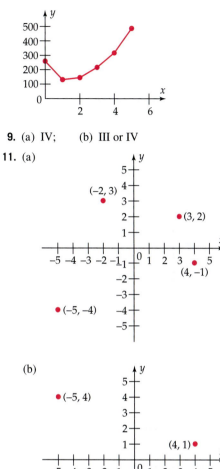

9. (a) IV;   (b) III or IV

11. (a)

(b)

(c) The points (a, b) and $(a, -b)$ are on opposite sides of, and the same distance from, the x-axis.

13. 13; $(-1/2, -1)$ **15.** $\sqrt{17}$; $(3/2, -3)$

17. $\sqrt{6 - 2\sqrt{6}}$; $\left(\dfrac{\sqrt{2} + \sqrt{3}}{2}, \dfrac{3}{2}\right)$

19. $|a - b|\sqrt{2}$; $\left(\dfrac{a + b}{2}, \dfrac{a + b}{2}\right)$

21. $13 + 2\sqrt{2} + \sqrt{13} \approx 19.434$

23. The distances are $(0, 0)$ to $(1, 1)$: $\sqrt{2}$; $(0, 0)$ to $(2, -2)$: $2\sqrt{2}$; $(1, 1)$ to $(2, -2)$: $\sqrt{10}$. But $\left(\sqrt{2}\right)^2 + \left(2\sqrt{2}\right)^2 = 2 + 8 = 10$ and $\left(\sqrt{10}\right)^2 = 10$, so the triangle is a right triangle with hypotenuse $\sqrt{10}$.

25. (a)

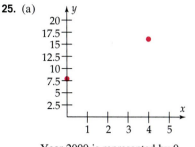

Year 2000 is represented by 0

(b) $(2, 11.9)$

(c) The midpoint might be interpreted to project sales of organically grown food in the United States to be \$11.9 billion in 2002. This requires the assumption of linear growth.

27. (a) QB$(20, 10)$; R$\left(48\frac{1}{3}, 45\right)$; distance ≈ 45.03 yd
    (b) $\left(34\frac{1}{6}, 27\frac{1}{2}\right)$

29. yes   **31.** yes   **33.** no

35. $(x + 3)^2 + (y - 4)^2 = 4$ **37.** $x^2 + y^2 = 2$

39. $(x - 5)^2 + (y + 2)^2 = 5$

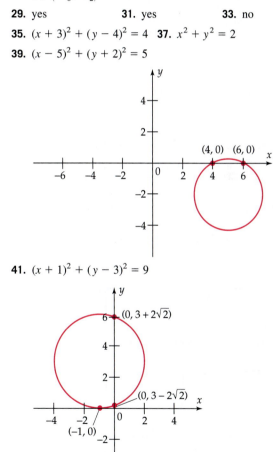

41. $(x + 1)^2 + (y - 3)^2 = 9$

43. $(x - 2)^2 + (y - 2)^2 = 8$

45. $(x - 1)^2 + (y - 2)^2 = 8$

47. $(x + 5)^2 + (y - 4)^2 = 16$

49. $(x - 2)^2 + (y - 1)^2 = 5$

51. $(-3, -4)$ and $(2, 1)$ **53.** $x = 6$

55. (a) about 7,250,000
(b) about 2007 and 7,750,000
(c) before 2003 and after 2011

57. (a) under 56;   (b) age 56;
(c) retiring at age 60: about \$30,000; retiring at age 65: about \$35,000.

59. B   **61.** C

63. Using the coordinates suggested by the hint: The coordinates of M are $\left(\dfrac{0 + s}{2}, \dfrac{r + 0}{2}\right) = \left(\dfrac{s}{2}, \dfrac{r}{2}\right)$. The distances from M to the two endpoints are $\sqrt{\left(0 - \dfrac{s}{2}\right)^2 + \left(r - \dfrac{r}{2}\right)^2}$
$= \dfrac{\sqrt{r^2 + s^2}}{2}$ and $\sqrt{\left(\dfrac{s}{2} - s\right)^2 + \left(\dfrac{r}{2} - 0\right)^2} = \dfrac{\sqrt{r^2 + s^2}}{2}$.
The distance from M to the origin is
$\sqrt{\left(0 - \dfrac{s}{2}\right)^2 + \left(0 - \dfrac{r}{2}\right)^2} = \dfrac{\sqrt{r^2 + s^2}}{2}$. The distances are identical.

65. The circles are all those that are tangent to the y-axis and have center on the x-axis.

67. The midpoint of the segment joining (c, d) and $(-c, -d)$ is the point $\left(\dfrac{c + (-c)}{2}, \dfrac{d + (-d)}{2}\right)$, which is $(0, 0)$, the origin. The points (c, d) and $(-c, -d)$ thus lie on a straight line through the origin and equidistant from the origin. Since the x- and y-coordinates both have opposite signs, the points are on opposite sides of the origin.

Section 1.2, page 83

1. $P(3, 5), Q(-6, 2)$ **3.** $P(-12, 8), Q(-3, -9)$

5.

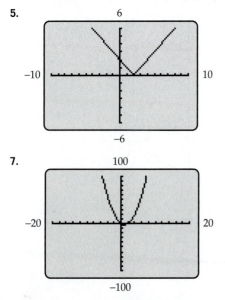

7.

9.

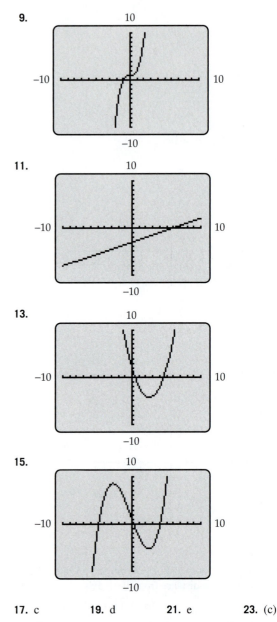

11.

13.

15.

17. c   **19.** d   **21.** e   **23.** (c)

25. $0 \le x \le 26, 0 \le y \le 120,000$, x-scl $= 5$, y-scl $= 10,000$. The deer population starts at 4000, climbs to a maximum of about 100,000 in year 20, then dies out after about 25 years.

27. $0 \le x \le 48, 0 \le y \le 35$, x-scl $= 6$, y-scl $= 5$. The concentration starts at 0, rises to about 31 after about 6 seconds, and then drops. We are told that the amount after 48 hours has no effect.

29. (a) 1917, 1929;   (b) 1925; 104,000

31. (a) at about 6.3 hours
(b) at about 18.7 hours

33. no

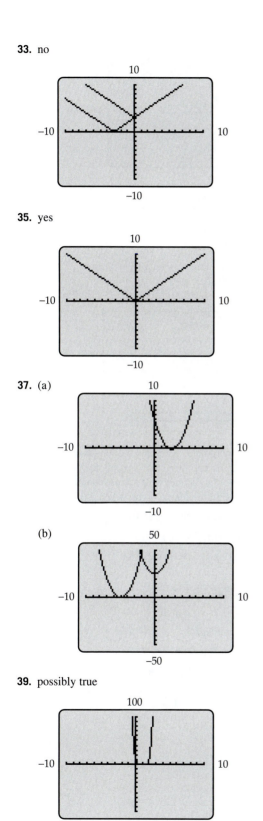

35. yes

37. (a)

(b)

39. possibly true

41. (a)

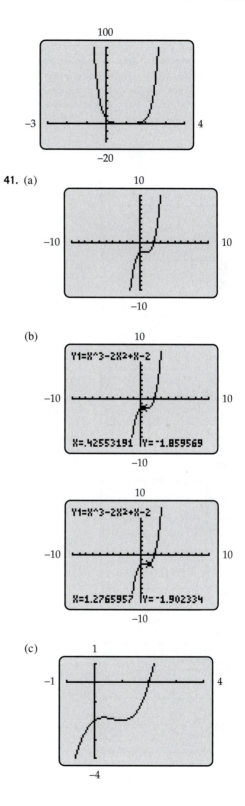

(b)

(c)

There are many possible square windows for Exercises 43–51; these were made on a TI-83+. To have a square window with the same y-axis shown here on wider screen calculators (such as TI-86/89), the x-axis should be longer than the one shown here. Because of the calculator's limited resolution, some graphs that should be connected (such as circles) may appear to have breaks in them.

43.

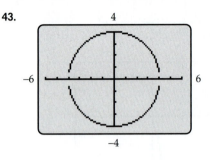

45.

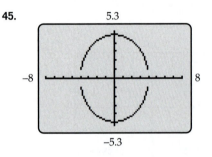

47.

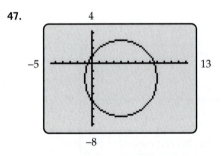

49.

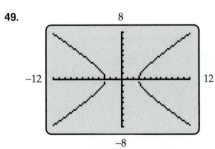

51.

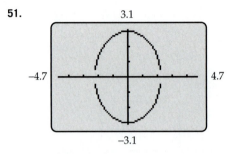

53. Maximum at about (.7922, 4.4849);
Minimum at about (4.2078, −3.4849)

55. The model suggests a maximum population of 613,230 in the year 1961 or 1962.

57. The maximum number of subscribers, about 22.5 million, occurred in 2004.

59. (a)

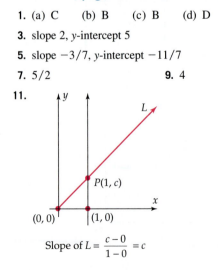

(b) The x-coordinate represents the length of the side of the squares cut from the corners; the y-coordinates represent the volumes of the corresponding boxes.

(c) The maximum volume occurs when x is about 4.182.

61. $-5 \le x \le 5$ and $-100 \le y \le 100$

63. $-5 \le x \le 5$ and $-10 \le y \le 10$

65. $-6 \le x \le 12$ and $-50 \le y \le 250$

Section 1.3, page 95

1. (a) C   (b) B   (c) B   (d) D

3. slope 2, y-intercept 5

5. slope $-3/7$, y-intercept $-11/7$

7. 5/2   **9.** 4

11.

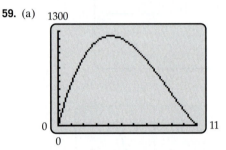

Slope of $L = \dfrac{c - 0}{1 - 0} = c$

13. $y = x + 2$

15. $y = -x + 8$

17. $y = -x - 5$

19. $y = (-7/3)x + 34/9$

21. perpendicular   **23.** parallel   **25.** parallel

27. The line through $(-5, -2)$ and $(3, 0)$ and the line through $(-3, 1)$ and $(5, 3)$ both have slope $1/4$. The line through $(-5, -2)$ and $(-3, 1)$ and the line through $(3, 0)$ and $(5, 3)$ both have slope $3/2$. Thus both pairs of opposite sides are parallel.

29. The line through $(-1, 2)$ and $(9, 6)$ has slope $2/5$. The line through $(-1, 2)$ and $(1, -3)$ has slope $-5/2$. The product of the slopes is -1, so the lines are perpendicular and the triangle has a right angle at the vertex $(-1, 2)$.

31. $y = 3x + 7$

33. $y = 1.5x$

35. $y = x - 5$

37. $y = -x + 2$

39. $y = (-3/4)x + 25/4$

41. $y = (-1/2)x + 6$

43. Since B is nonzero, we can solve both equations for y to get $y = (-A/B)x - C/B$ and $y = (-A/B)x - D/B$. The lines have the same slope, so they are parallel (unless $C = D$, in which case both equations give the same line).

45. (a) $y = (9/8)x + 54$
(b) about 57.375 million

47. (a) $y = 3.97x + 144.5$
(b) $200.08 \approx 200.1$
(c) 3.97 per year

49. $2,750$ per year; $4,750$

51. (a) $y = 60.4x + 696$
(b) for 1996: $1058.4 billion, for 2006: $1662.4 billion

53. (a) $c = 110,000 + 50x$
(b) $r = 72x$
(c) $p = 22x - 110,000$
(d) 5,000 books

55. (a) $y = 8.50x + 50,000$
(b) $11, $9.50, $9.00

57. (a) $y = (30/7)x + 1$
(b) about 18.1 million, about 43.8 million

59. (a) 12.5 gallons per minute, $8\frac{1}{3}$ gallons per minute, 25 gallons per minute
(b)

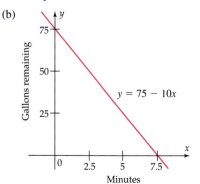

61. (a) 10   (b) 30

63. (a) $y = 5x + 115$
(b) 125 pounds, 185 pounds
(c) 6 ft 7 in.

65. If $b = c$, then the equations would be the same. Using the hint, if (x_1, y_1) satisfied both equations, then $y_1 = mx_1 + b$ and $y_1 = mx_1 + c$. Thus, $mx_1 + b = mx_1 + c$ and $b = c$. Since $b \neq c$, this is a contradiction, and no point can lie on both lines, that is, the lines are parallel.

67. Sketch a figure:

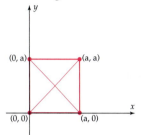

The slopes of the diagonals are given by $\dfrac{a - 0}{a - 0} = 1$ and $\dfrac{0 - a}{a - 0} = -1$. Since $1(-1) = -1$, the diagonals are perpendicular.

Section 1.4, page 107

1. (a) The data points are $(1, 2)$, $(2, 2)$, $(3, 3)$, $(4, 3)$ and $(5, 5)$. For $y = x$, the residuals are $1, 0, 0, -1, 0$; their sum is 0. For $y = .5x + 1.5$, the residuals are $0, -.5, 0, -.5$ and 1; their sum is 0.
(b) The sum is 2 for $y = x$ and 1.5 for $y = .5x + 1.5$.
(c) The second model is a better fit.

3. (a) The data points are $(0, 1.4)$, $(5, 2.2)$, $(10, 3.2)$ and $(15, 4)$. For $y = .17x + 1.4$, the residuals are $0, -.05, .1$ and $.05$; their sum is $.1$. For $y = .18x + 1.38$, the residuals are $.02, -.08, .02$ and $-.08$; their sum is $-.12$.
(b) The sum is .015 for $y = .17x + 1.4$ and .0136 for $y = .18x + 1.38$.
(c) The second model is a better fit.

5. positive correlation   **7.** very little correlation

9.

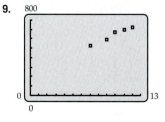

(a) The data appears linear.
(b) a positive correlation

11.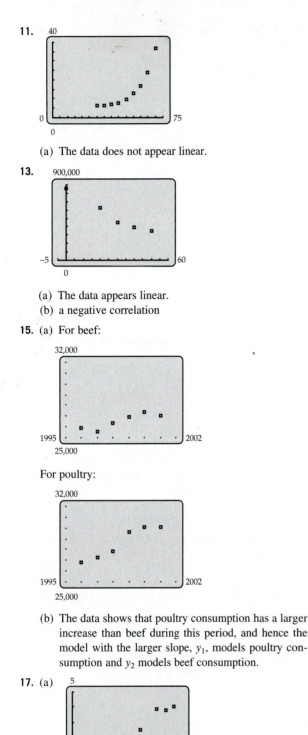

(a) The data does not appear linear.

13.

(a) The data appears linear.
(b) a negative correlation

15. (a) For beef:

For poultry:

(b) The data shows that poultry consumption has a larger increase than beef during this period, and hence the model with the larger slope, y_1, models poultry consumption and y_2 models beef consumption.

17. (a)

(b) $y = .393x - .487$
(c) In 1997, $2.3 billion; in 1999, $3.0 billion
(d) $5.4 billion

19. (a) $y = 67.7x + 660.6$
(b) The slope of 67.7 indicates that national expenditures on health care increase by about $67.7 billion each year. The y-intercept of 660.6 reflects the model's prediction of $660.6 billion spent in 1990.
(c) If the model remains accurate in the future, the national health care expenditures in 2007 will be approximately $1811.2 billion.

21. (a) $y = 16.54x + 86.5$; the number of Internet users in 2009 will be about 268.45 million.
(b) This model will certainly be inaccurate by 2013, since it predicts about 335 million users when the population will be only 324 million. More likely, it will become inaccurate much sooner. In 2010, for example it predicts a number that would include *every* person over five years old, which is unlikely.

23. (a) $y = 230.5x + 1118$
(b) about 2007

Chapter 1 Review, page 112

1. $\sqrt{58}$ **3.** $\sqrt{c^2 + d^2}$ **5.** $\left(d, d + \dfrac{c}{2}\right)$

7. (a) $\sqrt{17}$ (b) $(x - 2)^2 + (y + 3)^2 = 17$

9.

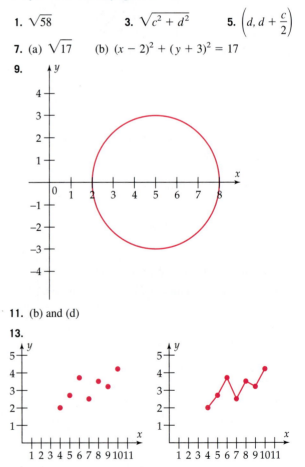

11. (b) and (d)

13.

15. (a) a, d
   (b) b and c do not show the peaks and valleys; e shows only a vertical segment
   (c) $-4 \le x \le 7$, $-6 \le y \le 8$

17. (a) None of the windows shows a complete graph though c comes close
   (b) a, b and d do not show any peaks or valleys; c does not show the valleys; no graph appears in window e
   (c) $-8 \le x \le 12$, $-1000 \le y \le 500$

19. (a) b, c
   (b) a does not show peaks or valleys; d shows what appears to be several vertical lines; no graph appears in window e
   (c) $-9 \le x \le 9$, $-150 \le y \le 150$

21. (a) 1999   (b) 1992   (c) 2003

23. definitely false

25.

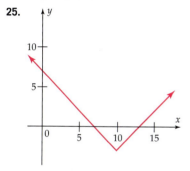

27.

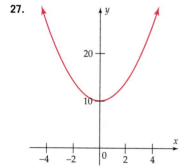

29. (a) y-intercept is 1
   (b) slope is 4/5.

31. $y = 3x - 7$ **33.** $y = (1/5)x + 29/5$

35. 25,000 ft   **37.** false   **39.** false

41. false   **43.** false   **45.** (d)

47. (e)

49. (a) $y = .2x + 70.8$
   (b) 73.6 years
   (c) in 2016

51. (c)   **53.** (d)

55. (a)

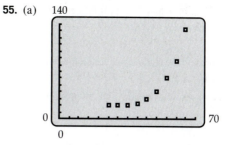

   (b) The data do not appear to be linear.

57. (a)

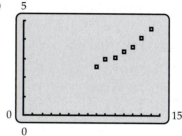

   (b) $y = .31x + .068$
   (c) about $5.0 billion

59. (a) $y = .39x + 10.48$
   (b) for 1995, $12.43; for 2001, $14.77; the estimate for 1995 is a little high; for 2001 the estimate is a little low.
   (c) For 2004, $15.94.

61. (a) $y = 14.055x + 62.959$
   (b) In 2004, $259.7 billion; in 2007, $301.9 billion
   (c) about 2009.

CHAPTER 2

Section 2.1, page 129

1. 1.5   **3.** -5 **5.** 8   **7.** $-5/6$

9. -32 **11.** $-1/13$ **13.** -3 **15.** 5/6

17. no solution   **19.** 1   **21.** no solution

23. $-5/2$ **25.** $-1474/1189 \approx -1.2397$

27. 3.765   **29.** $2831/9774 \approx .2436$

31. $\dfrac{x + 5}{3}$ **33.** $\dfrac{3x - 72}{4}$

35. $\dfrac{2A}{h} - c \quad (h \ne 0)$

37. $\dfrac{4V}{\pi d^2} \quad (d \ne 0)$

39. -3 **41.** 1500

43. (a) 1998   (b) 2004

45. (a) 2005   (b) 2010

47. (a) $y = 6.25x - 5$, where y is in billions of dollars and $x = 0$ corresponds to 2000
(b) 2006

49. (a) Let x represent the student's score on the fourth exam.
(b) x, $88 + 62 + 79 + x$, $(88 + 62 + 79 + x)/4$

51. (a) Let l represent the length, and let w represent the width.
(b) l, w, $l + w + l + w = 2l + 2w$, $l = 2w$.

53. Let x represent the worker's salary before the raise. Then $x + (.08)x = 1593$.

55. Let x represent the number. Then $x + 9 = 3x - 1$.

57. $17.88   **59.** $14.40

61. $366.67 at 12% and $733.33 at 6%

63. 14%   **65.** 96   **67.** 30 pounds

69. 8/3 qt   **71.** 65 mph   **73.** 4:40 P.M.

75. 60 mph   **77.** 45 hr   **79.** 30 min

81. 90 min   **83.** 60 m by 20 m

85. $1437.50   **87.** 84 years

Special Topics 2.1.A, page 136

1. A varies directly as the square of r; the constant of variation is π.

3. A varies jointly as l and w; the constant of variation is 1.

5. V varies jointly as the square of r and h; the constant of variation is $\dfrac{\pi}{3}$.

7. $a = \dfrac{k}{b}$ **9.** $z = kxyw$

11. $d = k\sqrt{h}$ **13.** $v = ku$, $k = 4$

15. $v = \dfrac{k}{u}$, $k = 16$ **17.** $t = krs$, $k = 4$

19. $w = kxy^2$, $k = 2$ **21.** $T = \dfrac{kpv^3}{u^2}$, $k = 16$

23. 4   **25.** 9/4   **27.** 50

29. 3   **31.** $1638   **33.** 12 pounds

35. 80 kilograms per square centimeter

37. 400 ft   **39.** 30 inches

41. 3750 kilograms

43. (a) 2400 pounds
(b) no more than 4 ft

Section 2.2, page 148

1. 3 or 5   **3.** −2 or 7   **5.** −3 or 1/2

7. −2 or −1/4   **9.** −4/3 or 1

11. −4/3 or 1/4   **13.** $2 \pm \sqrt{3}$ **15.** $-3 \pm \sqrt{2}$

17. no real number solutions

19. $\dfrac{1}{2} \pm \sqrt{2}$ **21.** $1 \pm \dfrac{\sqrt{3}}{2}$

23. 2   **25.** 2   **27.** 1

29. −5 or 8   **31.** $\dfrac{-5 \pm \sqrt{57}}{8}$

33. −6 or −3   **35.** $\dfrac{-1 \pm \sqrt{2}}{2}$

37. −3/2 or 5

39. no real number solutions

41. no real number solutions

43. approximately 1.8240 or .4701

45. (a) 2116.1 pounds per square foot, 670.5229 pounds per square foot
(b) 14,410 feet

47. 1960   **49.** ± 10 **51.** 16

53. rational

55. $c = \pm\sqrt{\dfrac{E}{m}}$, $\dfrac{E}{m} \geq 0$

57. $r = \dfrac{-\pi h \pm \sqrt{\pi^2 h^2 + 4\pi A}}{2\pi}$, $\pi^2 h^2 + 4\pi A \geq 0$

59. $k = 4$

61. x, y, $2x + 2y$, $2x + 2y = 45$, xy, $xy = 112.5$; 7.5 cm by 15 cm

63. x, y, xy, $xy = 5.7$, $x = y + 2.3$; width is 1.5 ft, length is 3.8 ft

65. 3 cm, 4 cm, and 5 cm   **67.** 4 cm

69. about 1.753 ft   **71.** 2 m

73. 2.5 yd   **75.** 2 in. by 2 in.

77. 1.48 in.   **79.** 9 mph and 16 mph

81. 176 mph   **83.** about 132.7 ft

85. 12 hr

87. about 6.32 seconds. The time in Example 7 is less because in that example the ball is thrown down.

89. (a) about 4.38 seconds
(b) after 50 seconds

91. 2000 meters

Section 2.3, page 159

1. 3   **3.** 3   **5.** 2

7. −2.4265   **9.** 1.1640   **11.** 1.2372

13. 1.1921   **15.** −1.3794   **17.** −1.60051

19. 1.8608   **21.** 2.1017

23. -1.7521 **25.** $.9505$ **27.** $0, 2.2074$

29. 2.3901 **31.** $-.6514, 1.1514$

33. 7.0334 **35.** $2/3$

37. $1/12$ **39.** $\sqrt{3}$

41. $1.4528, -112.0000$ **43.** $1.9097, .0499$

45. 2006

47. (a) 1999 (b) 2003

Section 2.4, page 167

1. $\pm 1, \pm\sqrt{6}$ **3.** $\pm\sqrt{7}$ **5.** $\pm\dfrac{\sqrt{2}}{2}, \pm 2$

7. $\pm\dfrac{\sqrt{5}}{5}$ **9.** 7 **11.** 4

13. -2 **15.** ± 3 **17.** $2, -1$

19. 9 **21.** $1/2$ **23.** $1/2, -4$

25. 6 **27.** $3, 7$

29. $b = \sqrt{\dfrac{a^2}{A^2 - 1}} = \dfrac{a}{\sqrt{A^2 - 1}}$

31. $u = \sqrt{\dfrac{x^2}{1 - K^2}} = \dfrac{x}{\sqrt{1 - K^2}}$

33. $3, -6$ **35.** $3/2$ **37.** 2

39. $3/2$ **41.** $-5, -3, -1, 1$

43. 50 cm by 120 cm

45. 4.658 in.   **47.** 8.02 in.

49. either 11.47 ft or 29.91 ft

51. 6.205 miles

53. 2.2 ft by 4.4 ft by 4 ft high

Section 2.5, page 175

1. $x \le \dfrac{3}{2}$

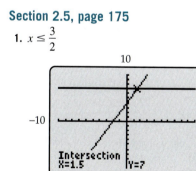

3. $x > -2$

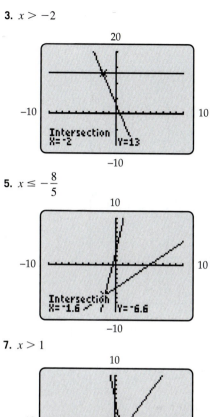

5. $x \le -\dfrac{8}{5}$

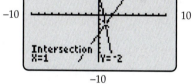

7. $x > 1$

9. $2 < x < 4$

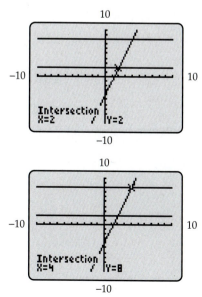

11. $-3 \le x < \dfrac{5}{2}$

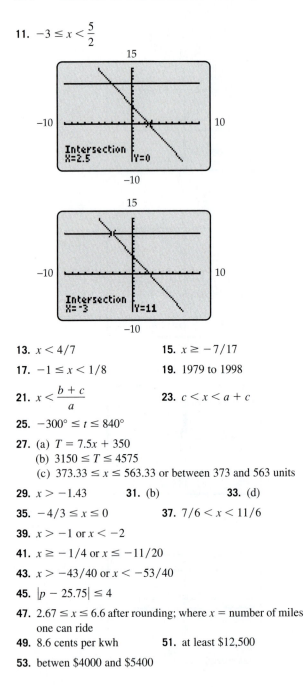

13. $x < 4/7$

15. $x \ge -7/17$

17. $-1 \le x < 1/8$

19. 1979 to 1998

21. $x < \dfrac{b+c}{a}$

23. $c < x < a + c$

25. $-300° \le t \le 840°$

27. (a) $T = 7.5x + 350$
(b) $3150 \le T \le 4575$
(c) $373.33 \le x \le 563.33$ or between 373 and 563 units

29. $x > -1.43$

31. (b)

33. (d)

35. $-4/3 \le x \le 0$

37. $7/6 < x < 11/6$

39. $x > -1$ or $x < -2$

41. $x \ge -1/4$ or $x \le -11/20$

43. $x > -43/40$ or $x < -53/40$

45. $|p - 25.75| \le 4$

47. $2.67 \le x \le 6.6$ after rounding; where $x =$ number of miles one can ride

49. 8.6 cents per kwh

51. at least $12,500

53. betwen $4000 and $5400

Section 2.6, page 184

1. $1 \le x \le 3$

3. $x \le -7$ or $x \ge -2$

5. $x \le -3$ or $x \ge 3$

7. $x \le -2$ or $x \ge 3$

9. $x \le 1/2$ or $x \ge 2/3$

11. $x < \dfrac{-11 - \sqrt{97}}{2}$ or $x > \dfrac{-11 + \sqrt{97}}{2}$

13. $\dfrac{5 - 3\sqrt{5}}{2} < x < \dfrac{5 + 3\sqrt{5}}{2}$

15. all real numbers

17. $-1 \le x \le 0$ or $x \ge 1$

19. $x < -1$ or $0 < x < 3$

21. $x \le -7$ or $x \ge -3$

23. $x \le -1$ or $x = 3$

25. $-3 - \sqrt{3} < x < -3 + \sqrt{3}$ or $x > 0$

27. $\dfrac{3 - \sqrt{29}}{2} \le x \le \dfrac{3 + \sqrt{29}}{2}$

29. $-2.26354 \le x \le .75564$ or $x \ge 3.50790$

31. $.5 < x < .83929$

33. $-3.79 \le x \le -.6003$ or $.442 \le x \le 8.9486$

35. $x < -3.8694$, $-1.5588 < x < -1.1783$, or $.06130 < x < 2.2952$

37. $x < -1/3$ or $x > 2$

39. $-2 < x < -1$ or $1 < x < 3$

41. $x > 1$

43. $x \le -9/2$ or $x > -3$

45. $-\sqrt{7} < x < \sqrt{7}$ or $x > 5.3445$

47. $9.6 < x < 12.4$ (July 1999 to May 2002)

49. $0 < x < 16.8$ (2000 to 2017) and $x > 59.2$ (after 2059)

51. any two numbers between 1 and 19 whose sum is 20 (2 and 18, 3.7 and 16.3, for example)

53. between 10 and 35 medallions

55. $1 \le t \le 4$ where t is seconds after liftoff

57. $2 < t < 2.25$, where t is seconds after dropping the ball

59. $x < -1.326$, $.0315 < x < 4.053$, or $x > 47.242$

61. $x < -3$ or $1/2 < x < 5$

Chapter 2 Review, page 186

1. $44/7$

3. 5

5. $r = \dfrac{b - a}{2Q}$

7. 2012

9. 3/11 ounce of gold and 8/11 ounce of silver

11. $2\dfrac{2}{9}$ hours

13. 9.6 ft

15. 6

17. $k = 1$

19. no real solution

21. $5/3$ or -1

23. no real solutions

25. 1984

27. No. If x represents one number then $2 - x$ would be the other number. If their product is 2, then $x(2 - x) = 2$ gives a quadratic equation with negative discriminant, and hence, no real solutions.

29. (a) 211.68 square inches
(b) 94.08 square inches
(c) 2.25 times as large

31. $x = 3.2678$

33. $x = -3.2843$

35. $x = 1.6511$

37. 1999

39. $\pm\sqrt{2}, \pm 3$

41. $5/3, -1$

43. $\pm\sqrt{5}$

45. $1, -2$

47. $\dfrac{5 - \sqrt{5}}{2}$

49. no solution

51. $s = \dfrac{gt^2}{2}$

53. $-9/2 < x < 2$

55. $x \le -4/3$ or $x \ge 0$

57. $x < -2$ or $x > -1/3$

59. incomes less than \$360

61. $x \le -1$ or $0 \le x \le 1$

63. (e)

65. $x \le 0$ or $x \ge 4/5$

67. $x \le -7$ or $x > -4$

69. $x < -2\sqrt{3}$ or $-3 < x < 2\sqrt{3}$

71. $x < \dfrac{1 - \sqrt{13}}{6}$ or $x > \dfrac{1 + \sqrt{13}}{6}$

73. 2000 to 2003

CHAPTER 3

Section 3.1, page 199

1. This could be a table of values of a function, because each input determines one and only one output.

3. This could not be a table of values of a function, because two output values are associated with the input -5.

5. 6

7. -2

9. -17

11. This defines y as a function of x.

13. This defines x as a function of y.

15. This defines both y as a function of x and x as a function of y.

17. Neither

19.
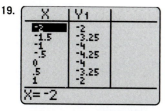

21.
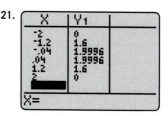

23. \$0, \$0, \$0, \$45.66, \$162.50, \$2483.02

25. A function may assign the same output to many different inputs.

27. Postage is a function of weight, but weight is not a function of postage; for example, all letters less than one ounce use the same postage amount.

29. (a) Average man:

Drinks in 1 hour	2	3	4	5
Blood alcohol content	.03	.05	.07	.10

Average woman:

Drinks in 1 hour	2	3
Blood alcohol content	.05	.08

(b) These tables define functions. For the average man, the domain consists of all x such that $2 \le x \le 5$, and the range of all y such that $.03 \le y \le .10$. For the average woman, the domain consists of all x such that $2 \le x \le 3$, and the range of all y such that $.05 \le y \le .08$.

31. (a) BMI is a function of weight in this case, because one BMI is associated with each weight.
(b) Weight is a function of BMI according to the data in the table, because one weight is associated with each BMI.
(c) For this data, BMI is still a function of weight, but weight is not a function of BMI because there are, for example, two weights associated with BMI 19.

33. (a) $A = \pi r^2$ (b) $A = \dfrac{\pi d^2}{4}$

35. $V = 4x^3$

37. $D = 400 - 16t^2, 0 \le D \le 400$

39. (a)
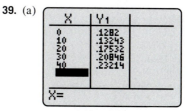

(b) They are the same after rounding. Yes
(c) 1995: 22.0%, 2005: 25.4%
(d) 2010 or 2011

41. (a) $x > 0$
(b) $-1 \le y \le 1$ where $y = \cos(\ln x)$

43. (a) For a nonnegative number, the part of the number to the left of the decimal point (the integer part) is the closest integer to the left of the number on the number line.
(b) the negative integers
(c) negative numbers that are not integers.

Section 3.2, page 209

1. (a) $-.8$ (b) $-.75$ (c) $-.4$ (d) $-.125$ (e) 0

3. $\sqrt{3} + 1$

5. $\sqrt{\sqrt{2} + 3} - \sqrt{2} + 1$

7. 4 **9.** $34/3$ **11.** $59/12$

13. $(a + k)^2 + \dfrac{1}{a + k} + 2$ **15.** $6 - 4x + x^2 + \dfrac{1}{2 - x}$

17. 8 **19.** -1

21. $s^2 + 2s$ **23.** $t^2 - 1$

25. -2 **27.** 2

29. $4 \pm \sqrt{26}$ **31.** $-7/2$ or 3 **33.** 1

35. 3 **37.** $1 - 2x - h$

39. $\dfrac{1}{\sqrt{x + h} + \sqrt{x}}$ **41.** $2x + h$

43. (a) $f(a) = a^2, f(b) = b^2, f(a + b) = (a + b)^2 = a^2 + 2ab + b^2$, so $(a + b)^2 \neq a^2 + b^2$, that is, $f(a + b) \neq f(a) + f(b)$
(b) $f(a) = 3a, f(b) = 3b, f(a + b) = 3(a + b) = 3a + 3b = f(a) + f(b)$.

45. (a) $x \leq 20$ (b) 3 (c) -1 (d) 1 (e) 2

47. (iii) and (v) **49.** all real numbers

51. all real numbers **53.** $x \geq 0$

55. all real numbers except 0 **57.** all real numbers

59. all real numbers except 3 and -2

61. $6 \leq x \leq 12$

63. Many examples including $f(x) = x^2$ and $g(x) = x^4$.

65. $f(x) = \sqrt{2x - 5}$ **67.** $f(x) = \dfrac{x^3 + 6}{5}$

69. (a)

$$T(x) \begin{cases} .1x & \text{if } 0 < x \leq 7150 \\ .15(x - 7150) + 715 & \text{if } 7150 < x \leq 29{,}050 \\ .25(x - 29{,}050) + 4000 & \text{if } 29{,}050 < x \leq 70{,}350 \\ .28(x - 70{,}350) + 14{,}325 & \text{if } 70{,}350 < x \leq 146{,}750 \\ .33(x - 146{,}750) + 35{,}717 & \text{if } 146{,}750 < x \leq 319{,}100 \\ .35(x - 319{,}100) + 92{,}529.50 & \text{if } x > 319{,}100 \end{cases}$$

(b) $3242.50, $5487.50, $22,627$

71. (a) Here are tables of the residuals for each model.

$g(x) = 1.1x + 39$

Data Point (x, p)	Model Point (x, y)	Residual $p - y$	Squared Residual $(p - y)^2$
$(5, 43.3)$	$(5, 44.5)$	-1.2	1.44
$(10, 50.9)$	$(10, 50)$	$.9$	$.81$
$(15, 56.3)$	$(15, 55.5)$	$.8$	$.64$

The sum of the squares of the residuals for this model is 2.89.

$g(x) = -.005x^2 + 1.1x + 44$

Data Point (x, p)	Model Point (x, y)	Residual $p - y$	Squared Residual $(p - y)^2$
$(5, 43.3)$	$(5, 49.375)$	-6.075	36.905625
$(10, 50.9)$	$(10, 54.5)$	-3.6	12.96
$(15, 56.3)$	$(15, 59.375)$	3.075	9.455625

The sum of the squares of the residuals for this model is 59.32125.

$g(x) = .07x^3 - 2x^2 + 20x - 14.8$

Data Point (x, p)	Model Point (x, y)	Residual $p - y$	Squared Residual $(p - y)^2$
$(5, 43.3)$	$(5, 43.95)$	$-.65$	$.4225$
$(10, 50.9)$	$(10, 55.2)$	-4.3	18.49
$(15, 56.3)$	$(15, 71.45)$	-15.15	229.5225

The sum of the squares of the residuals for this model is 248.435.

Clearly, the first model is by far the best for the given data.
(b) 56.6 million

73. (a) $f(x) = 200 + 5x$ $g(x) = 10x$
(b)

x	20	35	50
$f(x)$	300	375	450
$g(x)$	200	350	500

(c) 45 suits

75. $C(x) = \dfrac{45{,}000}{x} + 5.75x$

77. (a) $y = 108 - 4x$
(b) $V(x) = 108x^2 - 4x^3$

79. (a) $f(x) = 916.17x + 10,075.21$
(b) $16,487; $18,319; $20,151; the first two figures are lower, the last is higher.
(c) $23,815.

Section 3.3, page 223

1. yes, 0   **3.** no

5.

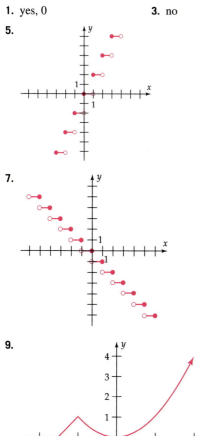

7.

9.

11.

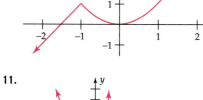

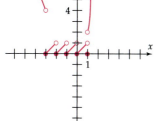

13. (a) one possible answer is $p(x) = -[-x], x > 0$

(b)

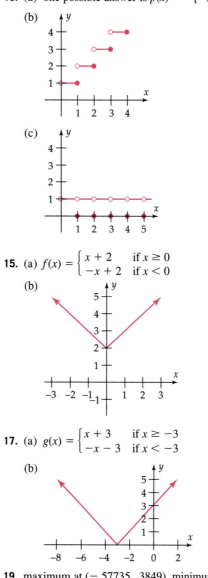

(c)

15. (a) $f(x) = \begin{cases} x + 2 & \text{if } x \geq 0 \\ -x + 2 & \text{if } x < 0 \end{cases}$

(b)

17. (a) $g(x) = \begin{cases} x + 3 & \text{if } x \geq -3 \\ -x - 3 & \text{if } x < -3 \end{cases}$

(b)

19. maximum at $(-.57735, .3849)$, minimum at $(.57735, -.3849)$

21. maximum at $(1, .5)$, minimum at $(-1, -.5)$

23. maximum at $(.4367, 2.1767)$, minimum at $(.7633, 2.1593)$

25. increasing when $-2.5 < x < 0$ and when $1.7 < x < 4$; decreasing when $-6 < x < -2.5$ and when $0 < x < 1.7$

27. constant when $x < -1$ and when $x > 1$; decreasing when $-1 < x < 1$

29. decreasing when $x < -5.7936$ and when $x > .4603$; increasing when $-5.7936 < x < .4603$

31. decreasing when $x < 0$ and when $.867 < x < 2.883$; increasing when $0 < x < .867$ and when $x > 2.883$

33. One of many possible graphs is shown.

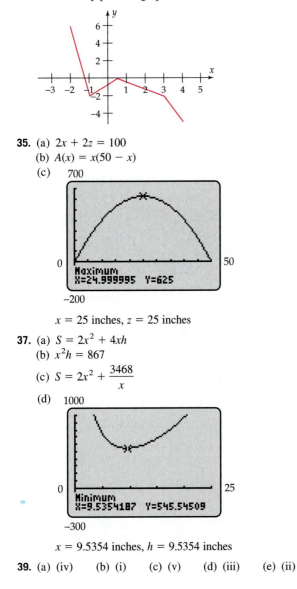

35. (a) $2x + 2z = 100$
(b) $A(x) = x(50 - x)$
(c)

700

Maximum
X=24.999995 Y=625

0 ······ 50

−200

$x = 25$ inches, $z = 25$ inches

37. (a) $S = 2x^2 + 4xh$
(b) $x^2h = 867$
(c) $S = 2x^2 + \dfrac{3468}{x}$
(d)

1000

Minimum
X=9.5354187 Y=545.54509

0 ······ 25

−300

$x = 9.5354$ inches, $h = 9.5354$ inches

39. (a) (iv)   (b) (i)   (c) (v)   (d) (iii)   (e) (ii)

In Exercises 41–43, many correct answers are possible, one of which is given here.

41.

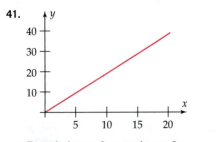

Domain is $x \geq 0$, range is $y \geq 0$.

43.

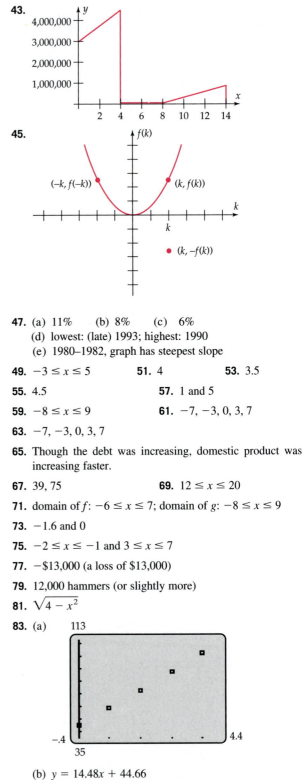

45.

$f(k)$

$(-k, f(-k))$ $(k, f(k))$

k

$(k, -f(k))$

47. (a) 11%   (b) 8%   (c) 6%
(d) lowest: (late) 1993; highest: 1990
(e) 1980–1982, graph has steepest slope

49. $-3 \leq x \leq 5$ **51.** 4   **53.** 3.5

55. 4.5   **57.** 1 and 5

59. $-8 \leq x \leq 9$ **61.** $-7, -3, 0, 3, 7$

63. $-7, -3, 0, 3, 7$

65. Though the debt was increasing, domestic product was increasing faster.

67. 39, 75   **69.** $12 \leq x \leq 20$

71. domain of f: $-6 \leq x \leq 7$; domain of g: $-8 \leq x \leq 9$

73. -1.6 and 0

75. $-2 \leq x \leq -1$ and $3 \leq x \leq 7$

77. $-\$13,000$ (a loss of \$13,000)

79. 12,000 hammers (or slightly more)

81. $\sqrt{4 - x^2}$

83. (a)

113

−.4 ······ 4.4

35

(b) $y = 14.48x + 44.66$
(c) \$131.54 billion   (d) 2010

85. During the next five minutes, she rested, then continued her run at her original pace for ten minutes. She then ran home at this pace, a 25-minute run.

Special Topics 3.3.A, page 231

1. $-10 \leq x \leq 30$ and $-8 \leq y \leq 3$ $(-10 \leq t \leq 10)$

3. $-10 \leq x \leq 10$ and $-10 \leq y \leq 10$ $(-10 \leq t \leq 10)$

5. $-30 \leq x \leq 0$ and $0 \leq y \leq 10$ $(0 \leq t \leq 10)$

7. $-5 \leq x \leq 45$ and $-65 \leq y \leq 65$

9. $-2 \leq x \leq 32$ and $-10 \leq y \leq 75$; Near y-axis: $-2 \leq x \leq 5$ and $-10 \leq y \leq 10$

11. $-15 \leq x \leq 2$ and $-25 \leq y \leq 75$; Near y-axis: $-2 \leq x \leq 2$ and $-4 \leq y \leq 2$

13. seven times

Section 3.4, page 239

1. H   **3.** F   **5.** K   **7.** C

9.

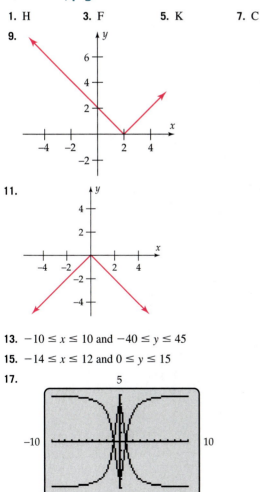

11.

13. $-10 \leq x \leq 10$ and $-40 \leq y \leq 45$

15. $-14 \leq x \leq 12$ and $0 \leq y \leq 15$

17.

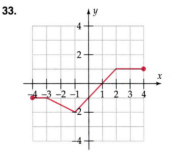

19. Shift 3 units to the right then 2 units up.

21. Reflect across the x-axis, then shrink vertically by a factor of $1/2$, and then shift 6 units down.

23. $g(x) = (x + 5)^2 + 6$ **25.** $g(x) = 2\sqrt{x - 6} - 3$

27. (a) $g(x) = x^2 + 3x + 2$
    (b) both difference quotients are $2x + h + 3$

29.

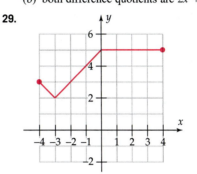

31.

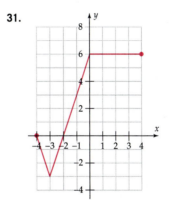

33.

35.

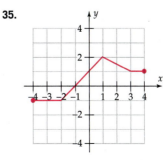

37.

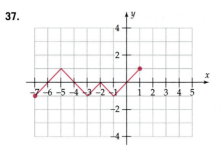

39.

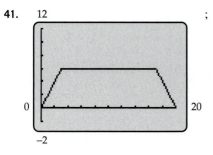

41.

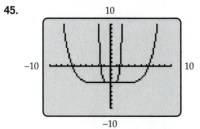

$$g(x) = |x - 3| + |x - 17| - 8$$

43.

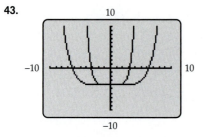

45.

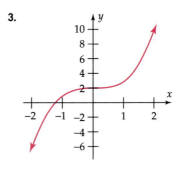

47.

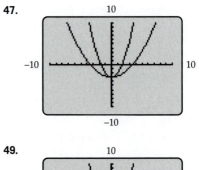

49.

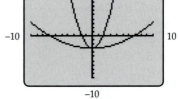

Special Topics 3.4.A, page 245

1.

Graph has y-axis symmetry

3.

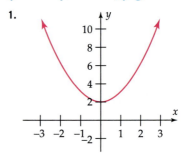

5. odd **7.** even **9.** even

11. even **13.** neither **15.** Yes

17. No **19.** Origin

21. Origin **23.** y-axis

25.

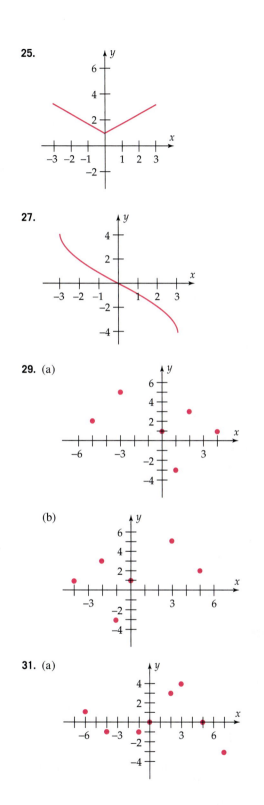

27.

29. (a)

(b)

31. (a)

(b)

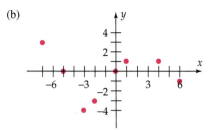

(c) Many correct answers, including this one.

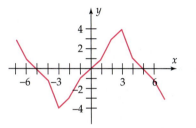

33. If replacing x by $-x$ does not change the graph and replacing y by $-y$ does not change the graph, then doing both does not change the graph. Thus, x-axis plus y-axis symmetry guarantees origin symmetry. If replacing x by $-x$ does not change the graph and replacing y by $-y$ *does* change the graph, then doing both must change the graph. Thus, y-axis symmetry and origin symmetry without x-axis symmetry is impossible. The argument is analogous for x-axis and origin symmetry.

Section 3.5, page 352

1. $x^3 - 3x + 2$; $-x^3 - 3x + 2$; $x^3 + 3x - 2$

3. $x^2 + 2x - 5 + \dfrac{1}{x}$; $\dfrac{1}{x} - x^2 - 2x + 5$; $x^2 + 2x - 5 - \dfrac{1}{x}$

5. $-3x^4 + 2x^3$; $\dfrac{-3x + 2}{x^3}$; $\dfrac{x^3}{-3x + 2}$

7. $(x^2 - 3)\sqrt{x - 3}$; $\dfrac{x^2 - 3}{\sqrt{x - 3}}$; $\dfrac{\sqrt{x - 3}}{x^2 - 3}$

9. all real numbers except 0; all real numbers except 0

11. $-4/3 \le x \le 2$; $-4/3 < x \le 2$

13. 0   **15.** 30

17. 49; 1; -8 **19.** -3; -3; 0

21. $(x + 3)^2$, all real numbers; $x^2 + 3$, all real numbers

23. $\dfrac{1}{\sqrt{x}}$, $x > 0$; $\dfrac{1}{\sqrt{x}}$, $x > 0$

25. x^6; x^9 **27.** $\dfrac{1}{x^2}$; x

29. $(f \circ g)(x) = f(g(x)) = 9g(x) + 2 = 9\left(\dfrac{x-2}{9}\right) + 2 =$

$x - 2 + 2 = x; (g \circ f)(x) = g(f(x)) = \dfrac{f(x) - 2}{9} =$

$\dfrac{9x + 2 - 2}{9} = \dfrac{9x}{9} = x$

31. $(f \circ g)(x) = f(g(x)) = \sqrt[3]{g(x)} + 2 = \sqrt[3]{(x-2)^3} + 2 =$

$x - 2 + 2 = x; (g \circ f)(x) = g(f(x)) = (f(x) - 2)^3 =$

$\left(\sqrt[3]{x} + 2 - 2\right)^3 = \left(\sqrt[3]{x}\right)^3 = x$

33.

x	$f(x)$	$g(x) = f(f(x))$
-4	-3	-1
-3	-1	$1/2$
-2	0	1
-1	$1/2$	$5/4$
0	1	$3/2$
1	$3/2$	2
2	1	$3/2$
3	-2	0
4	-2	0

35.

x	$(g \circ f)(x)$
1	4
2	2
3	5
4	4
5	4

37.

x	$(f \circ f)(x)$
1	1
2	3
3	3
4	5
5	1

There may be correct answers to Exercises 39–43, other than those given here.

39. Let $g(x) = x^2 + 2$ and $h(t) = \sqrt[3]{t}$. Then $(h \circ g)(x) = h(g(x))$ $= \sqrt[3]{g(x)} = \sqrt[3]{x^2 + 2}$.

41. Let $g(x) = 7x^3 - 10x + 17$ and $f(t) = t^7$. Then $(f \circ g)(x)$ $= f(g(x)) = (7x^3 - 10x + 17)^7$.

43. Let $g(x) = 3x^2 + 5x - 7$ and $h(t) = 1/t$. Then $(h \circ g)(x)$ $= h(g(x)) = \dfrac{1}{g(x)} = \dfrac{1}{3x^2 + 5x - 7}$.

45. Both yield the function f.

47.

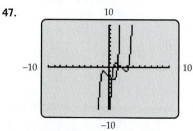

Functions are not the same.

49. $x^2 + 6x + 10; 2x + h + 6$

51. $\dfrac{1}{x+1}; \dfrac{-1}{(x+1)(x+h+1)}$

53. (a) about 1.22×10^{-4} square inches, about 2.5×10^{-6} square inches, about 1.36×10^{-7} square inches
(b) no, no, over a certain time period

55. (a) $A = \dfrac{\pi}{4}d^2 = \dfrac{\pi}{4}\left(6 - \dfrac{50}{t^2 + 10}\right)^2$
(b) about .7854 square inches, about 22.265 square inches
(c) about 11.4 weeks

57. $V(t) = \dfrac{256\pi t^3}{3}$, about 17,157 cm^3

59. $s(t) = \dfrac{10t}{3}$

Section 3.6, page 259

1. (a) 14 ft per sec
(b) 54 ft per sec
(c) 112 ft per sec
(d) 93.3 ft per sec

3. (a) Increasing at about 2086 people per year
(b) decreasing at about 2756 people per year

5. (a) decreasing at 245,800 per year
(b) increasing at 552,400 per year
(c) increasing at 296,200 per year
(d) decreasing at 89,600 per year
(e) fastest from 1985 to 1995, slowest from 1995 to 2005
(f) fastest from 1980 to 1985, slowest from 2005 to 2010

7. (a) increasing at 16.6 billion shares per year
(b) increasing at 29.175 billion shares per year
(c) increasing at 53.1 billion shares per year
(d) increasing at 32.9 billion shares per year
(e) 1999–2002

9. (a) $5000 per page
(b) $1875 per page
(c) $625 per page
(d) $1750 per page
(e) no, no, no

11. (a) 10 ft per sec
(b) 30 ft per sec
(c) 80 ft per sec

13. -72 **15.** -2 **17.** -1 **19.** 1.5858

21. (a) Car C: 62.5 ft per sec, Car D: 75 ft per sec
(b) $4 \le t \le 9$
(c) The slopes are about 50 ft per sec for Car C and 85 ft per sec for Car D.

23. (a) day 0 to day 50
(b) day 0 to day 100
(c) -26
(d) $20, -20, 0$

25. (a) $y = 11.8399x + 232.72779$
(b) 11.84
(c) 11, 13.2, the first is smaller, the second is larger
(d) 2011

Special Topics 3.6.A, page 267

1. 1   **3.** $2x + h$

5. $-8000 + 2t + h$ **7.** $2\pi r + \pi h$

9. (a) -7979.9 gallons per minute
(b) -7979.99 gallons per minute
(c) -7980 gallons per minute

11. (a) 6.5π (b) 6.2π (c) 6.1π (d) 6π
(e) They are the same.

Chapter 3 Review, page 269

1. (a) -3 (b) 1755   (c) 2   (d) -14

3.

x	0	1	2	-4	t	k	$b-1$	$1-b$	$6-2u$
$f(x)$	7	5	3	15	$7-2t$	$7-2k$	$9-2b$	$5+2b$	$4u-5$

5. Many possible answers, including these.
(a) If $f(x) = x + 1$, $a = 1$, $b = 2$, then $f(a + b) = f(1 + 2) = f(3) = 4$, but $f(a) + f(b) = f(1) + f(2) = 2 + 3 = 5$
(b) If $f(x) = x + 1$, $a = 1$, $b = 2$, then $f(ab) = f(1 \cdot 2) = f(2) = 3$, but $f(a) \cdot f(b) = f(1) \cdot f(2) = 2 \cdot 3 = 6$.

7. $r \geq 4$ **9.** $t^2 + t - 2$

11. $\dfrac{x^3 + 2x + 4}{4}$

13. (a) $f(t) = 50\sqrt{t}$
(b) $g(t) = 2500\pi t$
(c) radius is 150 meters, area is $22{,}500\pi$ square meters
(d) after approximately 12.7 hours

15.

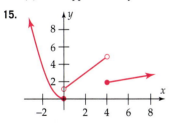

17. (b)

19. No local maximum, local minimum at $-.5$, increasing when $x > -.5$, decreasing when $x < -.5$

21. Local maximum at -5.0737, local minimum at $-.263$, increasing when $x < -5.073$ and when $x > -.263$, decreasing when $-5.0737 < x < -.263$.

23.

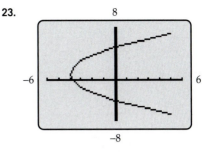

25. Here is one possibility:

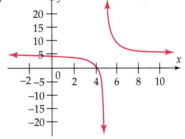

27. Graph has symmetry with respect to x-axis, y-axis, and the origin.

29. even   **31.** odd

33. Symmetric with respect to y-axis.

35.

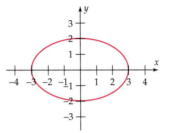

37. $-3 \leq y \leq 4$

39. Many correct answers, including: any number from 2 to 3.5 or from 4.5 to 6

41. 1   **43.** -3 **45.** true

47. 4   **49.** $x \leq 3$ **51.** $x < 3$

53. (a) King Richard
(b) King Richard
(c) Fireball Bob

55. Shrink vertically by a factor of .25, then shift 2 units up.

57. Shift 7 units right, then stretch vertically by a factor of 3, then reflect in the x-axis, and then shift 2 units up.

59. (e)

61. (a) $1/3$

(b) $(x - 1) \sqrt{x^2 + 5}$ $(x \neq 1)$

(c) $\dfrac{\sqrt{(c + 1)^2 + 5}}{c}$ $(c \neq 0)$

63.

x	-4	-3	-2	-1	0	1	2	3	4
$g(x)$	1	4	3	1	-1	-3	-2	-4	-3
$h(x) = g(g(x))$	-3	-3	-4	-3	1	4	3	1	4

65. $82/27$ **67.** $\dfrac{1}{x^3} + 3$ **69.** $1/4$

71. (a) $\dfrac{1}{x^2 - 1}$ (b) $\dfrac{1}{x^2} - 1$

73. $x \geq 0, x \neq 1$ **75.** (a) $-1/3$ (b) $5/8$

77. 6   **79.** 3   **81.** $2x + h$

83. (a) $290 per ton
(b) $230 per ton
(c) $212 per ton

85. (a) approximately 45 to 50
(b) approximately 25 to 35
(c) approximately 30 to 44

87. (a) $4400 per year, $10,480 per year, about $6738 per year
(b) $259,120

CHAPTER 4

Section 4.1, page 288

1. $(5, 2)$, upward   **3.** $(1, 2)$, downward

5. $(3, -6)$, upward   **7.** $(-3/2, 15/4)$, upward

9. I   **11.** K   **13.** J

15. F   **17.** $f(x) = 3x^2$

19. $f(x) = 3(x - 2)^2 + 5$ **21.** $f(x) = 2(x + 3)^2 + 4$

23. $f(x) = \frac{1}{2}(x - 4)^2 + 3$

25. vertex: $(-3, -21)$, x-intercepts: $\dfrac{-6 \pm \sqrt{42}}{2}$

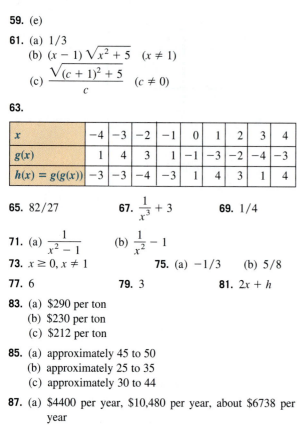

27. vertex: $(4, 14)$, x-intercepts: $4 \pm \sqrt{14}$

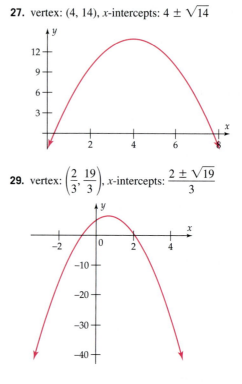

29. vertex: $\left(\dfrac{2}{3}, \dfrac{19}{3}\right)$, x-intercepts: $\dfrac{2 \pm \sqrt{19}}{3}$

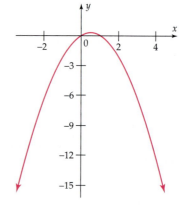

31. vertex: $(1/2, 1/4)$, x-intercepts: 0, 1

33. $g(x) = 2x^2 - 5$; $(0, -5)$

35. $h(x) = 2(x - 3)^2 + 4$; $(3, 4)$

37. (a) $-2x - h + 1$ (b) $(1/2, 1/4)$ (c) $-h$

39. (a) $-4x - 2h + 2$ (b) $(1/2, -1/2)$ (c) $-2h$

41. 2.5 sec; 196 ft   **43.** 22 ft

45. (a) $20,000, $100,000
(b) 200, $550

47. (a) 3906.25 ft at 15.625 sec
(b) 31.25 sec

49. 842 ft high, 13,987 ft or 2.65 miles away

51. (a) 30 m, 170 m   (b) 50 kph

53. 62.5 ft by 62.5 ft

55. 704.167 square feet, approximately

57. 50 ft (perpendicular to river) by 100 ft (parallel to river)

59. (a) $y = \left(\dfrac{1}{1152}\right)x^2 - 2$

(b) $\sqrt{1152} \approx 33.94$ inches

61. $3.50

63. $3.67; But if tickets must be priced in multiples of $0.20, $3.60 is best.

65. (a) $f(x) = 15.679(x - 2)^2 + 1546$
(b) $1938 billion
(c) 2014

67. $f(x) = 4.96875(x - 8)^2 + 268.5$; 2006

Section 4.2, page 300

1. $3x^3 - 3x^2 + 5x - 11$; 12

3. $x^2 + 2x - 6$; $-7x + 7$

5. $5x^2 + 5x + 5$; 0

7. no   **9.** yes   **11.** 2 and 0

13. $2\sqrt{2}$ and -1 **15.** 2   **17.** 6

19. -30 **21.** 170, 802   **23.** no

25. yes   **27.** $(x + 4)(2x - 7)(3x - 5)$

29. $(x - 3)(x + 3)(2x + 1)^2$

31. $f(x) = (x + 2)(x + 1)(x - 1)(x - 2)(x - 3)$
$= x^5 - 3x^4 - 5x^3 + 15x^2 + 4x - 12$

33. $f(x) = (x + 1)x(x - 1)(x - 2)(x - 3)$
$= x^5 - 5x^4 + 5x^3 + 5x^2 - 6x$

35. Many correct answers, including $f(x) =$
$(x - 1)(x - 7)(x + 4) = x^3 - 4x^2 - 25x + 28$

37. Many correct answers, including $f(x) =$
$(x - 1)^4(x - 2)(x - \pi)$ and $f(x) =$
$(x - 1)^2(x - 2)^2(x - \pi)^2$

39. $f(x) = .17(x - 8)(x - 5)x$

41. $-3, -1, 1$ **43.** $-5, -1, 1$

45. $-4, 0, 1/2, 1$ **47.** $-3, 2$

49. 2   **51.** $-5, 2, 3$

53. $(x - 2)(2x^2 + 1)$ **55.** $x^3(x + 2)(x^2 + 3)$

57. $(x - 1)^2(x - 2)(x^2 + 3)$

59. $-1/2, 1, 2$ **61.** $1/3, 1/2, 1$

63. (a) By the Rational Root Test, the only possible rational roots of $x^2 - 2$ are ± 1 and ± 2. Since $\sqrt{2}$ is a root of the polynomial and it is not one of those four numbers, it is not rational.

(b) By the Rational Root Test, the only possible rational roots of $x^2 - 3$ are ± 1 and ± 3. Since $\sqrt{3}$ is a root of the polynomial and it is not one of those four numbers, it is not rational.

65. (a) 8.2 per 100,000; 5.5 per 100,000
(b) 1996   (c) 1991

67. (a) 77.7 bacteria per hour
(b) -29.8875 bacteria per hour. The number of bacteria is declining—they are dying.
(c) $t = 140.485$ or about 140 hours into the experiment
(d) $t = 147.765$ or about 148 hours into the experiment

69. 3.5

Special Topics 4.2.A, page 306

1. Quotient: $3x^3 - 2x^2 - 4x + 1$
Remainder: 7

3. Quotient: $2x^3 - x^2 + 3x - 11$
Remainder: 25

5. Quotient: $5x^3 + 35x^2 + 242x + 1690$
Remainder: 11,836

7. Quotient: $x^3 - 4x^2 - 4x - 6$
Remainder: -19

9. Quotient: $3x^3 + \dfrac{3}{4}x^2 - \dfrac{29}{16}x - \dfrac{29}{64}$, remainder: $\dfrac{483}{256}$

11. Quotient: $2x^3 - 6x^2 + 2x + 2$, remainder: 1

13. $(x + 4)(3x^2 - 3x + 1)$

15. $\left(x - \dfrac{1}{2}\right)(2x^4 - 6x^3 + 12x^2 - 10)$

17. Quotient: $x^2 - 2.15x + 4$, remainder: 2.25

19. -4

Section 4.3, page 312

1. yes   **3.** yes   **5.** no

7. possibly degree 3 or 5   **9.** not a polynomial function

11. possibly degree 5

13. $-50 \le x \le 50$, $-100,000 \le y \le 1,000,000$ or similar

15. $-2, 1$, and 3, each odd

17. -2 odd, -1 odd, 2 even

19. (e)   **21.** (f)   **23.** (c)

25.

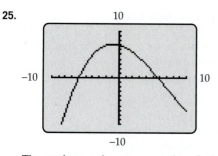

The graph must shoot up somewhere farther to the right.

27.

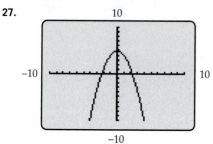

The graph must shoot back up somewhere to the left and somewhere to the right.

29. $-9 \le x \le 3$ and $-20 \le y \le 40$

31. $-6 \le x \le 6$ and $-60 \le y \le 320$

33. $-3 \le x \le 4$ and $-35 \le y \le 20$

35.

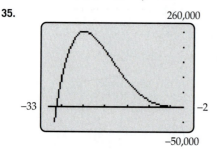

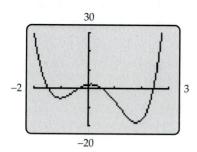

37.

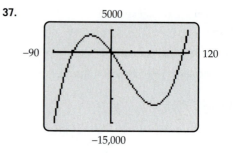

39. (a) The graph of a cubic polynomial can have no more than two local extrema. If it had only one, the ends would go in the same direction. Hence, it can have two or none.

(b) When the behavior for $|x|$ large and the number of extrema are both accounted for, these four shapes are the only possible ones.

41. (a) odd   (b) positive
(c) -2, 0, 4, and 6   (d) 5

43. (d)

45.

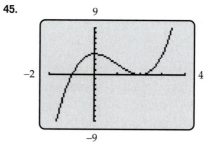

local maximum at (0, 4), local minimum at (2, 0); x-intercepts at -1 and 2.

47.

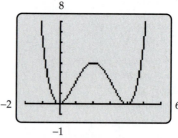

local maximum at (2, 4), local minima at (0, 0) and (4, 0); x-intercepts 0 and 4

49.

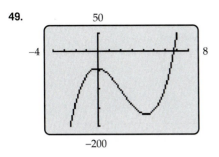

local maximum at $(-.12, -44.73)$, local minimum at $(4.23, -167.99)$; x-intercept 6.72

51. (a) Since the leading coefficient is positive and the degree is even, the graph will eventualy shoot upward.

(b)

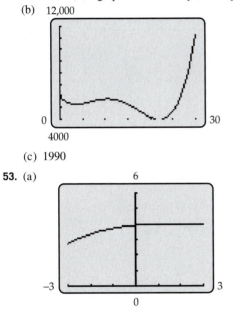

(c) 1990

53. (a)

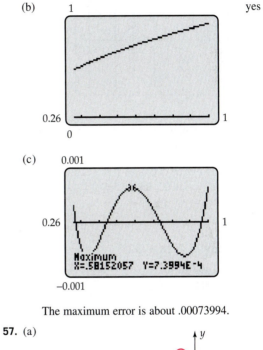

If the graph had the horizontal portion appearing in the window, then the equation $g(x) = 4$ would have infinitely many solutions, which is impossible, since a polynomial equation of degree 3 can have at most three solutions.

(b) $1 \le x \le 3$ and $3.99 \le y \le 4.01$

(c) If the graph of a polynomial of degree n had a horizontal portion appearing in the window, it would be a portion also of the line $y = k$. Then the equation $f(x) = k$ would have infinitely many solutions, which is impossible, since a polynomial equation of degree n can have at most n solutions.

55. (a)

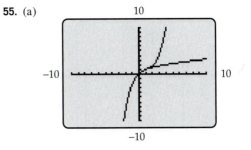

In the standard window, the graphs appear similar only in the interval from 0 to 2.

(b) yes

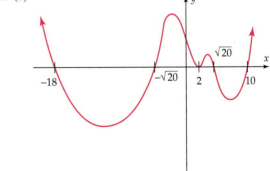

(c)

The maximum error is about .00073994.

57. (a)

(b)

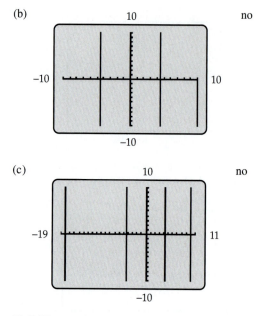

no

(c)

no

(d) Different windows would be needed for different portions of the graph. $-20 \leq x \leq -3$ and $-5 \times 10^6 \leq y \leq 10^6$; Then $-3 \leq x \leq 2$ and $-5000 \leq y \leq 60,000$; then $1 \leq x \leq 5$ and $-5000 \leq y \leq 5000$; finally $5 \leq x \leq 11$ and $-100,000 \leq y \leq 100,000$

Special Topics 4.3.A, page 321

1. (a) Volume was lowest at the right-hand end of the interval, in 1994. This probably is a result of the business cycle.
   (b) 1989

3. (a) $P(x) = -.01x^2 + 275x - 600,000$
   (b) 13,750 radiators, $1,290,625

5. (a) $R(x) = 29x$
   (b) $P(x) = -.001x^3 - .06x^2 + 30.5x$
   (c) 82,794 tags for a profit of $1546.39

7. (a) about 4.43 by 4.43 inches
   (b) 4 by 4 inches

9.

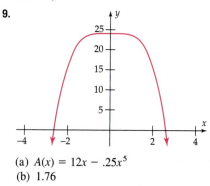

(a) $A(x) = 12x - .25x^5$
(b) 1.76

11. 34.20 cm by 34.20 cm by 17.10 cm

13. 1.25 miles down the road

15. (a)

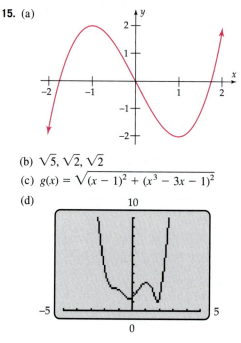

(b) $\sqrt{5}$, $\sqrt{2}$, $\sqrt{2}$
(c) $g(x) = \sqrt{(x-1)^2 + (x^3 - 3x - 1)^2}$
(d)

(e) (1.86388, .88359)

17. about 2.975

19. (a) 206 units
    (b) 269 units, $577 per unit

Section 4.4, page 327

1. cubic   3. quadratic or quartic

5. (a) $y = .7221x^3 - 39.0114x^2 + 529.8468x + 2939.4581$
   (b) 1987: 4984.5 per 100,000; 1993: 4820.9 per 100,000
   (c) 4546.6 per 100,000
   (d) Not clear since the crime rate rose slightly in 2001, after declining for the preceding decade.

7. (a)

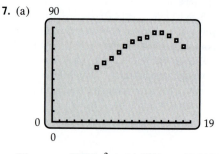

(b) $y = -.51798x^2 + 14.6568x - 20.8871$
(c) 80.4°, 69°, 82.8°

9. (a)   7000

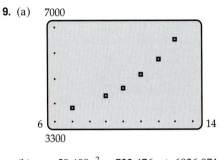

6   3300   14

(b) $y = 58.488x^2 - 722.476x + 6026.071$
(c) \$3990   (d) \$9439

11. (a) $y = -.00503x^3 + .19062x^2 - .93687x + 33.27203$
(b) 1998: 48,844; 2001: 51,094
(c) The model predicts a very small decrease in this period.

13. (a) $y = -.06019x^3 + .97321x^2 + 9.65146x + 99.61905$ and $y = .104167x^4 - 3.81019x^3 + 48.77083x^2 - 243.20569x + 562.47619$; For both models, $x = 0$ corresponds to 1990, and expenditures are in billions of dollars. The quartic model seems better.
(b) Neither model can be accurate indefinitely; the quartic model shows sharp increases soon, while the cubic model has a negative leading coefficient indicating eventual decline to zero.

Section 4.5, page 341

1. all real numbers except -2.5

3. all real numbers except $3 \pm \sqrt{5}$

5. all real numbers except 1 and $\pm\sqrt{2}$

7. F   **9.** A   **11.** $x = 6, x = -1$

13. $x = -1, x = 0$ **15.** $x = -2, x = 2$

17. $y = 3$, one window is $-100 \le x \le 100$ and $-8 \le y \le 8$

19. $y = -1$, one window is $-40 \le x \le 40$ and $-8 \le y \le 8$

21. $y = 5/2$, one window is $-40 \le x \le 40$ and $-8 \le y \le 8$

23. vertical asymptote: $x = -5$, horizontal asymptote: $y = 0$;

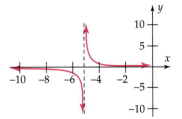

25. vertical asymptote: $x = -5/2$; horizontal asymptote: $y = 0$;

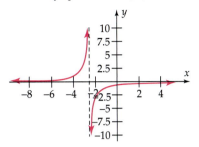

27. vertical asymptote: $x = 1$; horizontal asymptote: $y = 3$;

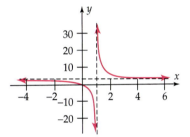

29. vertical asymptote: $x = 3$; horizontal asymptote: $y = -1$;

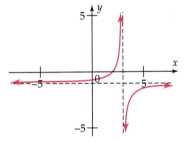

31. vertical asymptotes: $x = 0, x = -1$; horizontal asymptote: $y = 0$;

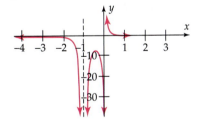

33. vertical asymptotes: $x = -2, x = 1$; horizontal asymptote: $y = 0$;

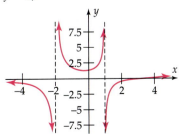

35. vertical asymptotes: $x = -5, x = 1$; horizontal asymptote: $y = 0$;

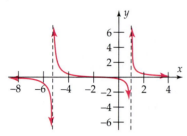

37. vertical asymptote: $x = 0$; horizontal asymptote: $y = -4$;

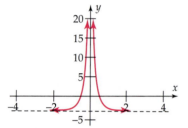

39. vertical asymptotes: $x = 5, x = -1$; horizontal asymptote: $y = 1$;

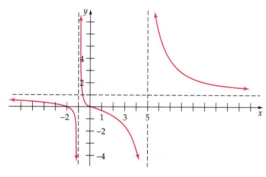

In Exercises 41–47, many correct answers are possible. The ones given here work well for TI-83+.

41. Three windows are needed: $-5 \le x \le 4.4$ and $-8 \le y \le 4$; then $-2 \le x \le 2$ and $-.5 \le y \le 5$; finally, $-15 \le x \le -3$ and $-.07 \le y \le .02$

43. $-9.4 \le x \le 9.4$ and $-4 \le y \le 4$

45. Two windows are needed: $-4.7 \le x \le 4.7$ and $-8 \le y \le 8$; then $-40 \le x \le 35$ and $-2 \le y \le 3$

47. Two windows are needed: $-4.7 \le x \le 4.7$ and $-2 \le y \le 2$; then $3 \le x \le 14$ and $-.02 \le y \le .01$

49. $\dfrac{-1}{x(x + h)}$

51. $\dfrac{-2x - h}{x^2(x + h)^2}$

53. $.2 \le x \le 1$ and $-1 \le y \le .5$

55. (a) 9 weeks
(b) 28.996 mm
(c) yes, 30 mm

57. (a) 98 degrees
(b) 4 hours after the infection began
(c) 55.3 hours after the infection began.

59. (a) \$259,000; \$273,000; \$407,000
(b) 250 bikes, \$1240 per bike
(c) between 167 and 338 bikes

61. 8.4343 in. by 8.4343 in. by 14.057 in.

63. (a) $c(x) = \dfrac{2800 + 3.5x^2}{x}$

(b) 13.9 mph and 57.5 mph
(c) 28.3 mph

65. (a) $P(x) = x + \dfrac{500}{x}$

(b) if $10 < x < 50$, then $P(x) < 60$
(c) $x \approx 22.4$, 22.4 by 11.2 meters

67. (a) $h_1 = h - 2$

(b) $h_1 = \dfrac{150}{\pi r^2} - 2$

(c) $V = \pi(r - 1)^2\left(\dfrac{150}{\pi r^2} - 2\right)$

(d) Since the walls are 1 foot thick, r must be greater than 1.
(e) $r \approx 2.88$ ft, $h \approx 5.76$ ft

Special Topics 4.5.A, page 347

1. $y = x$; $-14 \le x \le 14$ and $-14 \le y \le 14$

3. $y = x^2 - x$; $-15 \le x \le 6$ and $-40 \le y \le 240$

5. Vertical asymptote: $x = 2$. nonvertical asymptote: $y = x + 1$,

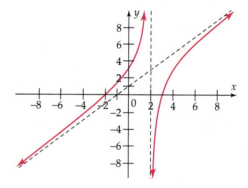

7. Vertical asymptote: $x = 5/2$, nonvertical asymptote: $y = 2x + 7$.

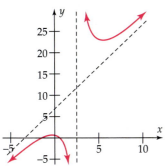

9. Vertical asymptote: $x = 1$, nonvertical asymptote: $y = x^2 + x + 1$

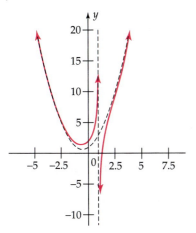

11. Vertical asymptote: $x = 2$; nonvertical asymptote: $y = x^2 + 2x + 4$

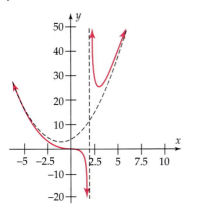

13. $-15.5 \le x \le 8.5$ and $-16 \le y \le 8$

15. $-4.7 \le x \le 4.7$ and $-12 \le y \le 8$

17. Two windows are needed: $-13 \le x \le 7$ and $-20 \le y \le 20$; Then $-2.5 \le x \le 1$ and $-.02 \le y \le .02$

19. (a)

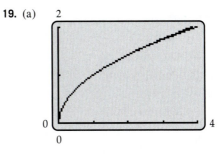

(b) Approximately $.06 \le x \le 2.78$

Section 4.6, page 353

1. $8 + 2i$ **3.** $-2 - 10i$

5. $-1/2 - 2i$ **7.** $\dfrac{\sqrt{2} - \sqrt{3}}{2} + 2i$

9. $1 + 13i$ **11.** $-10 + 11i$

13. $-21 - 20i$ **15.** 4

17. $-i$ **19.** i **21.** i

23. $\dfrac{5}{29} + \dfrac{2}{29}i$ **25.** $-\dfrac{1}{3}i$ **27.** $\dfrac{12}{41} - \dfrac{15}{41}i$

29. $-\dfrac{5}{41} - \dfrac{4}{41}i$ **31.** $\dfrac{10}{17} - \dfrac{11}{17}i$ **33.** $\dfrac{7}{10} + \dfrac{11}{10}i$

35. $-\dfrac{113}{170} + \dfrac{41}{170}i$ **37.** $6i$ **39.** $\sqrt{14}i$

41. $-4i$ **43.** $11i$

45. $\left(\sqrt{15} + 3\sqrt{2}\right)i$ **47.** $2/3$

49. $-41 - i$ **51.** $\left(2 + 5\sqrt{2}\right) + \left(\sqrt{5} - 2\sqrt{10}\right)i$

53. $\dfrac{1}{3} - \dfrac{\sqrt{2}}{3}i$ **55.** $x = 2,\, y = -2$

57. $x = -3/4,\, y = 3/2$

59. $\dfrac{1}{3} + \dfrac{\sqrt{14}}{3}i,\, \dfrac{1}{3} - \dfrac{\sqrt{14}}{3}i$

61. $-\dfrac{1}{2} + \dfrac{\sqrt{7}}{2}i,\, -\dfrac{1}{2} - \dfrac{\sqrt{7}}{2}i$

63. $\dfrac{1}{4} + \dfrac{\sqrt{31}}{4}i,\, \dfrac{1}{4} - \dfrac{\sqrt{31}}{4}i$

65. $\dfrac{3 + \sqrt{3}}{2} + 0i,\, \dfrac{3 - \sqrt{3}}{2} + 0i$

67. $2 + 0i,\, -1 + \sqrt{3}i,\, -1 - \sqrt{3}i$

69. $1 + 0i,\, -1 + 0i,\, 0 + i,\, 0 - i$

71. -1

73. $\overline{\overline{z}} = \overline{\overline{a + bi}} = \overline{a - bi} = a + bi = z$

75. $\dfrac{1}{z} = \dfrac{a}{a^2 + b^2} - \dfrac{b}{a^2 + b^2}i$

Section 4.7, page 358

1. 2   **3.** 6   **5.** -30

7. 0: multiplicity 54, $-4/5$: multiplicity 1

9. 0: multiplicity 15; π: multiplicity 14;
$\pi + 1$: multiplicity 13

11. $1 \pm 2i$, $[x - (1 + 2i)][x - (1 - 2i)]$

13. $\dfrac{-1}{3} \pm \dfrac{2\sqrt{5}}{3}i$, $3\left(x - \dfrac{-1 + 2\sqrt{5}i}{3}\right)\left(x - \dfrac{-1 - 2\sqrt{5}i}{3}\right)$

15. $3, \dfrac{-3}{2} + \dfrac{3\sqrt{3}}{2}i, \dfrac{-3}{2} - \dfrac{3\sqrt{3}}{2}i,$
$(x - 3)\left(x - \dfrac{-3 + 3\sqrt{3}i}{2}\right)\left(x - \dfrac{-3 - 3\sqrt{3}i}{2}\right)$

17. $-2, 1 + \sqrt{3}i, 1 - \sqrt{3}i,$
$(x + 2)\left[x - \left(1 + \sqrt{3}i\right)\right]\left[x - \left(1 - \sqrt{3}i\right)\right]$

19. $\pm 1, \pm i$, $(x - 1)(x + 1)(x - i)(x + i)$

21. $\pm\sqrt{5}, \pm\sqrt{2}i$, $\left(x - \sqrt{5}\right)\left(x + \sqrt{5}\right)\left(x - \sqrt{2}i\right)\left(x + \sqrt{2}i\right)$

In Exercises 23–47, there may be correct answers other than the ones shown here.

23. $(x - 1)(x - 7)(x + 4)$

25. $(x - 1)^2(x - 2)^2(x - \pi)^2$

27. $f(x) = 2x(x + 3)(x - 4)$

29. $f(x) = x^2 - 4x + 5$

31. $f(x) = x^3 - 6x^2 + 13x - 10$

33. $f(x) = x^5 - x^4 - x^3 + 19x^2 - 32x + 30$

35. $f(x) = x^2 - 2x + 5$

37. $f(x) = x^4 - 14x^3 + 74x^2 - 176x + 160$

39. $f(x) = x^6 - 5x^5 + 8x^4 - 6x^3$

41. $f(x) = 3x^2 - 6x + 6$

43. $f(x) = -2x^3 + 2x^2 - 2x + 2$

45. $x^2 + (-1 + i)x + (2 + i)$

47. $x^3 - 5x^2 + (7 + 2i)x + (-3 - 6i)$

49. $3, -\dfrac{1}{2} + \dfrac{\sqrt{3}}{2}i, -\dfrac{1}{2} - \dfrac{\sqrt{3}}{2}i$

51. $i, -i, -1, -2$ **53.** $1, 2i, -2i$

55. $2 - i, 2 + i, i, -i$

57. (a) $\overline{z + w} = \overline{(a + bi) + (c + di)}$
$= \overline{(a + c) + (b + d)i}$
$= (a + c) - (b + d)i$
$= (a - bi) + (c - di)$
$= \bar{z} + \bar{w}$

(b) $\overline{z \cdot w} = \overline{(a + bi) \cdot (c + di)}$
$= \overline{(ac - bd) + (bc + ad)i}$
$= (ac - bd) - (bc + ad)i$
$= (a - bi) \cdot (c - di)$
$= \bar{z} \cdot \bar{w}$

59. (a) If z is a root of $f(x), f(z) = 0$. Then $\overline{f(z)} = \bar{0} = 0$. Also
$$\overline{f(z)} = \overline{az^3 + bz^2 + cz + d}$$
$$= \overline{az^3} + \overline{bz^2} + \overline{cz} + \overline{d}$$
$$= \bar{a}\overline{z^3} + \bar{b}\overline{z^2} + \bar{c}\bar{z} + \bar{d}$$
$$= a\bar{z}^3 + b\bar{z}^2 + c\bar{z} + d = f(\bar{z})$$

(b) Since $\overline{f(z)} = 0, a\bar{z}^3 + b\bar{z}^2 + c\bar{z} + d = 0$; therefore, $\bar{z}$ is a root of $f(x)$.

61. For each nonreal complex root z, there must be two factors of the polynomial: $(x - z)$ and $(x - \bar{z})$. This yields an even number of factors. There will remain at least one factor, hence at least one root, which must be real.

Chapter 4 Review, page 361

1. $(2, 3)$ **3.** $(4, -4)$ **5.** $(1.5, -5.75)$

7. (a) $y = 260 - x$
(b) $A(x) = -x^2 + 260x - 3500$
(c) 130 ft by 130 ft

9. 30 ft perpendicular to the building by 60 ft parallel to the building

11. (a), (c), (e), and (f)   **13.** 0

15.
$$
\begin{array}{r|rrrrrrr}
2 & 1 & -5 & 8 & 1 & -17 & 16 & -4 \\
  &   & 2 & -6 & 4 & 10 & -14 & 4 \\
\hline
  & 1 & -3 & 2 & 5 & -7 & 2 & 0 \\
\end{array}
$$
the other factor is $x^5 - 3x^4 + 2x^3 + 5x^2 - 7x + 2$

17. $f(x) = x^3 - 5x^2 - x + 5$ is one of many correct answers

19. $5/3$ and -1 **21.** $-2, \dfrac{4 \pm \sqrt{3}}{3}$

23. (a) $1, -1, 3, -3, \dfrac{1}{2}, -\dfrac{1}{2}, \dfrac{3}{2}, -\dfrac{3}{2}$
(b) 3   (c) $x = \dfrac{1 \pm \sqrt{3}}{2}$ and $x = 3$

25. (d)

27. $2x + h + 1$

29.

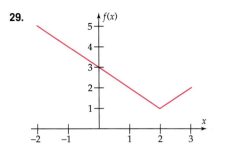

This graph could not possibly be the graph of a polynomial function because of the sharp corner. Many other graphs are possible.

31. (c)

33. $-2 \le x \le 9$ and $-35 \le y \le 10$

35. $-2 \le x \le 18$ and $-500 \le y \le 1200$

37. (a)

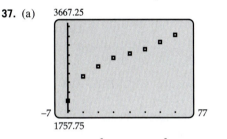

(b) $y = .00852x^3 - 1.10945x^2 + 56.2583x + 2017.2576$
(c) $C(71) - C(70) = \$26.79$
(d) about $\$85.49$, about $\$47.84$

39.

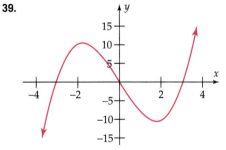

x-intercepts: $-3, 0, 3$; local maximum: $(-1.732, 10.392)$; local minimum: $(1.732, -10.392)$

41.

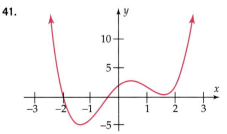

x-intercepts: $-1.908, -.376$; local maximum: $(.469, 2.941)$; local minima: $(-1.326, -4.914)$, $(1.607, .617)$

43.

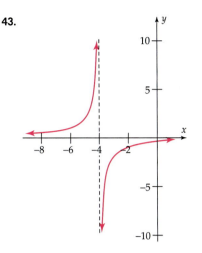

x-intercepts: none; vertical asymptote: $x = -4$; horizontal asymptote: $y = 0$

45.

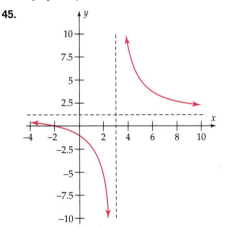

x-intercept: -2.5; vertical asymptote: $x = 3$; horizontal asymptote: $y = 4/3$

47. vertical asymptotes: $x = -2, x = 3$; horizontal asymptote: $y = 0$

49. $-4.7 \le x \le 4.7$ and $-5 \le y \le 5$; then $2 \le x \le 20$ and $-.2 \le y \le .1$

51. $-18.8 \le x \le 18.8$ and $-8 \le y \le 8$

53. between 400 and 2500 bags; between $\$1.75$ (at 400 bags) and $\$0.70$ (at 2500 bags)

55. (a) $t(x) = \dfrac{40}{x} + \dfrac{110}{x + 25}$, $0 \le x \le 55$
(b) at least a little over 44 mph

57. $\dfrac{1}{(x + 1)(x + h + 1)}$

59. $\left(\sqrt{6} + 2\right) + \left(\sqrt{3} - 2\sqrt{2}\right)i$

61. $-i$

63. $-\dfrac{3}{2} \pm \dfrac{\sqrt{31}}{2}i$

65. $\dfrac{3}{10} \pm \dfrac{\sqrt{31}}{10}i$ **67.** $\pm\dfrac{\sqrt{6}}{3}, \pm i$

69. $-2, 1 \pm \sqrt{3}i$ **71.** $i, -i, 2, -1$

73. $x^4 - 2x^3 + 2x^2$ is one possibility

CHAPTER 5

Section 5.1, page 377

1.

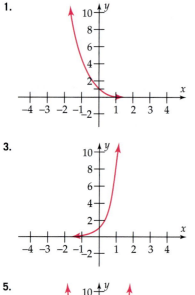

3.

5.

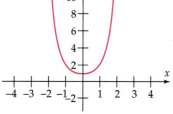

7. Shift vertically 5 units down.

9. Stretch away from the x-axis by a factor of 3.

11. Shift horizontally 2 units left, then shift vertically 5 units down.

13. $f(x)$ has graph C, $g(x)$ has graph A, $h(x)$ has graph B.

15. Neither   **17.** Even   **19.** Even

21. 11   **23.** 4/5

25. $\dfrac{10^{x+h} - 10^x}{h}$

27. $\dfrac{2^{x+h} + 2^{-x-h} - 2^x - 2^{-x}}{h}$

29. $-4 \le x \le 4, -1 \le y \le 10$

31. $-4 \le x \le 4, -1 \le y \le 10$

33. $-10 \le x \le 10, -10 \le y \le 10$

35. $-5 \le x \le 10, -1 \le y \le 6$

37. The negative x-axis is an asymptote; $(-1.443, -.531)$ is a local minimum.

39. No asymptote; $(0, 1)$ is a local minimum.

41. The x-axis is an asymptote; $(0, 1)$ is a local maximum.

43. (a)

Time (hours)	Number of Cells
0	1
.25	2
.5	4
.75	8
1	16

(b) $C(t) = 2^{4t}$

45. $g(x) = -315(e^{.00418x} + e^{-.00418x}); C = 1260$

47. (a) About 1.984 billion, about 3.998 billion.
(b) 2008

49. (a) About 286.4 million, about 300.1 million.
(b) About 2041

51. (a) $k = 15$ (b) About 12.13 psi
(c) About .0167 psi

53. (a) About 74.1 years, about 76.3 years
(b) 1930

55. (a) About 62.2 million, about 73.5 million.
(b) 2000
(c) No. 76.7 million is an upper bound.

57. (a)

Fold	0	1	2	3	4
Thickness (inches)	.002	.004	.008	.016	.032

(b) $f(x) = .002(2^x)$
(c) 2097.152 inches
(d) 43 folds

59. $f(x) = 2^x$ is one example.

61. (a)

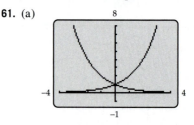

Each graph is the reflection of the other in the y-axis.
(b) Each graph is the reflection of the other in the y-axis.

Section 5.2, page 388

1. (a) $1469.33
   (b) $1485.95
   (c) $1489.85
   (d) $1491.37

3. $585.83   **5.** $610.40   **7.** $639.76

9. $563.75   **11.** $582.02   **13.** $3325.29

15. $3359.59   **17.** $6351.16   **19.** $568.59

21. Fund C   **23.** $385.18

25. $1,162,003.14   **27.** $4000 after 4 years

29. About 5%   **31.** About 5.92%

33. (a) 9 years
   (b) 9 years
   (c) 9 years
   (d) Doubling time is independent of investment amount.

35. About 9.9 years   **37.** $7835.26

39. $21,303.78

41. (a) About 12.55%
   (b) Quarterly: 12.55%, monthly: 12.6825%, daily: 12.7475%.

43. (a) $f(x) = 6(3^x)$
   (b) 3
   (c) In 12 weeks no, in 14 weeks yes.

45. (a) $f(x) = 100.4(1.014^x)$
   (b) About 115.375 million
   (c) In 2015

47. (a) $f(x) = 4966\,(1.039^x)$
   (b) $8484.46
   (c) 2005–2006

49. (a) $f(x) = 32.44(1.0224^x)$
   (b) In 2010: 40.485 million, in 2025: 56.442 million
   (c) 2023

51. 256 bacteria after 10 hours, 654 bacteria after 2 days

53. (a) $f(x) = .75^x$
   (b) 8 feet

55. 103 months

57. (a) $f(x) = 100(.5^{x/1620})$
   (b) After 800 years: 71.0 milligrams, after 1600 years: 50.4 milligrams, after 3200 years: 25.4 milligrams.

59. About 7171 years

61. (a) After the third 50-year period: $c\left(\dfrac{1}{2}\right)^3$, after the fourth 50-year period: $c\left(\dfrac{1}{2}\right)^4$.

   (b) $c\left(\dfrac{1}{2}\right)^t$

   (c) $c\left(\dfrac{1}{2}\right)^{x/50}$ becomes $c\left(\dfrac{1}{2}\right)^{x/h}$.

Section 5.3, page 400

1. 4   **3.** -2.5 **5.** $10^3 = 1000$

7. $10^{2.88} = 750$ **9.** $e^{1.0986} = 3$

11. $e^{-4.6052} = .01$ **13.** $e^{z+w} = x^2 + 2y$

15. $\log .01 = -2$ **17.** $\log 3 = .4771$

19. $\ln 25.79 = 3.25$ **21.** $\ln 5.5527 = 12/7$

23. $\ln w = 2/r$ **25.** $\sqrt{43}$ **27.** 15

29. .5   **31.** 931   **33.** $x + y$

35. x^2 **37.** $f(x) = 4e^{3.2188x}$

39. $g(x) = -16e^{3.4177x}$ **41.** $x > -1$

43. $x < 0$ **45.** $x > 0$

47. (a) For all $x > 0$.
   (b) According to the fourth property of logarithms, $e^{\ln x} = x$ for every $x > 0$.

49. Stretch vertically by a factor of 2 away from the x-axis.

51. Shift horizontally 4 units to the right.

53. Shift horizontally 3 units to the left, then shift vertically 4 units down.

55.
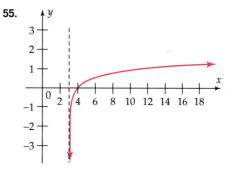

57.
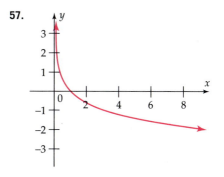

59. $0 \le x \le 20$ and $-10 \le y \le 10$; then $0 \le x \le 2$ and $-20 \le y \le 20$

61. $-10 \le x \le 10$ and $-3 \le y \le 3$

63. $0 \le x \le 20$ and $-6 \le x \le 3$

65. .5493   **67.** $-.2386$

69. $\dfrac{\ln (3 + h) - \ln 3}{h}$ (b) $h = 2.2008$

71. (a) 4%: 17.7 years, 6%: 11.9 years, 8%: 9 years, 12%: 6.1 years, 18%: 4.2 years, 24%: 3.2 years, 36%: 2.3 years

(b) 18, 12, 9, 6, 4, 3, 2; they are the same. The doubling time for an investment at $p\%$ is approximately $72/p$ years.

73. (a) 77.7
(b) 2012

75. (a) 44.4%
(b) 2010

77. $A = -9, B = 10$

79. About 4392 meters

81. (a) In 1999: about 26.55 billion pounds, in 2002: about 27.72 billion pounds.
(b) 2008

83. (a) About 9.9 days
(b) 6986 people

85. (a) No ads: about 120 bikes, $1000: about 299 bikes, $10,000: about 513 bikes.
(b) $1000: yes, $10,000: no.
(c) $1000: yes, $10,000: yes.

87. $n = 30$ gives an approximation with a maximum error of .00001 when $-.7 \le x \le .7$.

Section 5.4, page 408

1. $\ln (x^2 y^3)$ **3.** $\log (x - 3)$ **5.** $\ln \left(\dfrac{1}{x^7}\right)$

7. $\ln (e - 1)^3$ **9.** $\log (20\,xy)$ **11.** $\log (x^2 - 9)$

13. $\log \left(\dfrac{1}{(x + 1)(x - 2)}\right)^2$ **15.** $\ln \sqrt{\dfrac{x}{3x + 5}}$

17. $\log x$ **19.** $2u + 5v$ **21.** $\frac{1}{2}u + 2v$

23. $\dfrac{2}{3}u + \dfrac{1}{6}v$ **25.** False   **27.** True

29. False

31. $a = 100, b = 10$, among many examples

33. $b = e$

35. The graphs are not the same. Negative real numbers are in the domain of $v = \log x^2$ but not in the domain of $y = 2 \log x$.

37. The graph of $f(x) = \log (5x) = \log 5 + \log x$ is the graph of $g(x) = \log x$ shifted upward by $\log 5$ units.

39. The graph of $f(x) = \ln (x^5) = 5 \ln x$ is the graph of $g(x) = \ln x$ stretched away from the x-axis by a factor of 5.

41. $\log \left(\dfrac{x + h}{x}\right)^{\frac{1}{h}}$

43. (a) About 5.38 million vehicles   (b) 2009

45. 2   **47.** 2.5   **49.** 20

51. 66   **53.** Twice as loud

55. (a) Both are approximately 1.255.
(b) Both are approximately 2.659.
(c) Both are approximately 3.952.
(d) These results suggest that $\log c = \dfrac{\ln c}{\ln 10}$.

57. (a) .9422
(b) 1.9422
(c) 2.9422
(d) 3.9422
(e) 4.9422
(f) The numbers 8.753, 87.53, . . ., 87,530 are related in that each is 10 times the previous number. Their logarithms are related in that each is 1 more than the previous logarithm. If the numbers a and b are related as follows. If $b = 10a$, then $\log b = 1 + \log a$.

Special Topics 5.4.A, page 417

1.

x	0	1	2	4
$f(x) = \log_4 x$	Not defined	0	.5	1

3.

x	1/36	1/6	1	216
$h(x) = \log_6 x$	-2	-1	0	3

5.

x	0	1/7	$\sqrt{7}$	49
$f(x) = 2 \log_7 x$	Not defined	-2	1	4

7.

x	-2.75	-1	1	29
$h(x) = 3 \log_2(x + 3)$	-6	3	6	15

9. $\log .01 = -2$ **11.** $\log \sqrt[3]{10} = \dfrac{1}{3}$

13. $\log r = 7k$ **15.** $\log_7 5{,}764{,}801 = 8$

17. $\log_3 (1/9) = -2$ **19.** $10^4 = 10{,}000$

21. $10^{2.88} \approx 750$ **23.** $5^3 = 125$

25. $2^{-2} = 1/4$ **27.** $10^{z+w} = x^2 + 2y$

29. $\sqrt{97}$ **31.** $x^2 + y^2$ **33.** .5

35. 6   **37.** $b = 3$ **39.** $b = 20$

41. 5   **43.** 3   **45.** 4

47. $\log \left(\dfrac{x^2 y^3}{z^6}\right)$ **49.** $\log [x(x - 3)]$

51. $\log_2 [5|c|]$ **53.** $\log_4 \left(\dfrac{1}{49c^2}\right)$

55. $\ln \left(\dfrac{(x + 1)^2}{x + 2}\right)$ **57.** $\log_2 x$

59. $\ln (e - 1)^2$

61. 3.3219   **63.** .8271   **65.** 1.1115   **67.** 1.6199

69. 1.5497, more diverse

71. For Community 1, $H = 1.5850$; for Community 2, $H = .8118$.

73. True by the Quotient Law.

75. True by the Power Law.

77. False. By the Product Law, the logarithm of $5x$ is the logarithm of 5 plus the logarithm of x, not 5 times the logarithm of x.

79. 397^{398} is larger.

81. $\log_b u = \dfrac{\log_a u}{\log_a b}$

83. $\log_{10} u = 2 \log_{100} u$

85. $\log_b x = \frac{1}{2}\log_b v + 3 = \frac{1}{2}\log_b v + 3 \log_b b = \log_b v^{1/2} + \log_b b^3 = \log_b \left(b^3 \sqrt{v} \right)$, so $x = b^{\log_b x} = b^{\log_b \left(b^3 \sqrt{v} \right)} = b^3 \sqrt{v}$.

87.

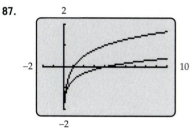

The graphs are not the same; they intersect only when $x = 10^{\log 4/(1-1/\log 4)}$, or approximately 0.1228. Statements such as the one given are generally false.

89.

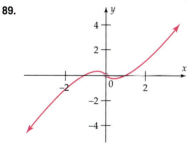

Local minimum at approximately $(.3679, -.3196)$. Local maximum at approximately $(-.3679, .3196)$. No asymptotes. There is a hole at $x = 0$ because $h(0)$ is not defined.

Section 5.5, page 426

1. $x = 4$

3. $x = 1/9$

5. $x = 1/2$ or -3

7. $x = -2$ or $-1/2$

9. $x = \ln 5/\ln 3 \approx 1.4650$

11. $x = \ln 3/\ln 1.5 \approx 2.7095$

13. $x = \dfrac{\ln 3 - 5 \ln 5}{\ln 5 + 2 \ln 3} \approx -1.8254$

15. $x = \dfrac{\ln 2 - \ln 3}{\ln 3 + 3 \ln 2} \approx .1276$

17. $x = (\ln 5)/2 \approx .8047$

19. $x = (\ln 2/7)/1.4 \approx -.8948$

21. $x = \dfrac{2 \ln (5/2.1)}{\ln 3} \approx 1.5793$

23. If $\ln u = \ln v$, then $u = e^{\ln u} = e^{\ln v} = v$.

25. $x = 9$

27. $x = 5$

29. $x = 6$

31. $x = 3$

33. $x = \dfrac{-5 + \sqrt{37}}{2}$

35. $x = 9/(e - 1)$

37. $x = 5$

39. $x = \pm\sqrt{10,001}$

41. $x = \sqrt{\dfrac{e + 1}{e - 1}}$

43. (a) About \$4.75 billion
    (b) 2009

45. (a) About 72.2 million families
    (b) In 2008

47. About .444 billion years
49. About 14.95 days

51. About 3.813 days
53. About 2534 years ago

55. About 3689 years
57. 6.99%

59. (a) About 22.52 years
    (b) About 22.11 years

61. \$3197.05
63. About 79.4 years

65. (a) 1.36%
    (b) 2027

67. (a) $k = 21.459$
    (b) .182

69. (a) At the beginning, 20 bacteria; three hours later, 2500 bacteria
    (b) About .43 hours

71. (a) 200 people; 2795 people
    (b) After 6 weeks

73. (a) $k \approx .2290$, $c \approx 83.33$
    (b) About 12.4 weeks

Section 5.6, page 435

1. Logistic or cubic

3. Exponential, quadratic, cubic, or power

5. Exponential or power
7. Quadratic or cubic

9. Quadratic or cubic

11. Ratios are approximately 5.07, 5.06, 5.06, 5.08, and 5.05. An exponential model might be appropriate.

13. (a) As x increases, $e^{-.0216x}$ decreases from 1 toward 0. Thus, $(1 + 56.33e^{-.0216x})$ decreases toward 1, so the ratio increases toward (but never reaches) 442.1.

(b)

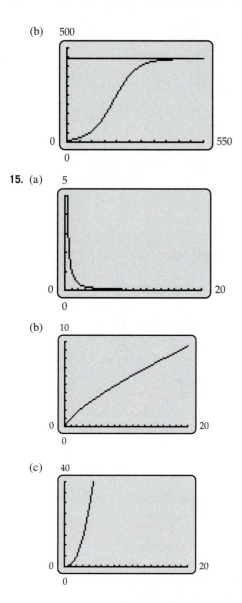

15. (a)

(b)

(c)

17. {(ln x, ln y)} is approximately linear, so a power model is appropriate.

19. {(ln x, ln y)} is approximately linear, so a power model is appropriate.

21. (a)

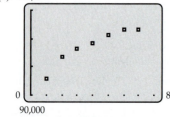

(b) Logarithmic model

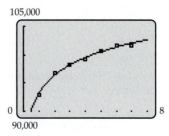

Logistic model

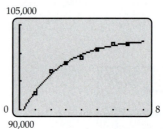

(c)

Y_1 is the logarithmic model and Y_2 is the logistic model.

(d) The logarithmic model is better.

23. (a)

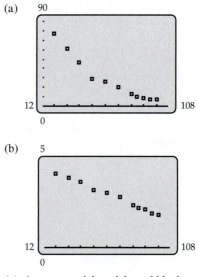

(b)

(c) An exponential model would be best:
$y = 15.225(.9695^x)$

25. (a)

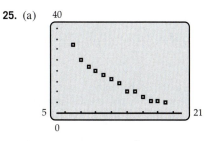

(b) $y = 91.1630(.85919^x)$ or equivalently,
$y = 91.1630e^{-.15176x}$

(c) About 2.8 students per computer

(d) 2009

(e) Any projection of future trends becomes increasingly less reliable as the dates recede from those for which data is known.

27. (a)

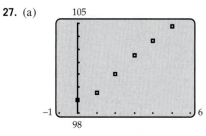

(b) Successive ratios are all 1.01 (rounded to two decimal places); an exponential model.

(c) $y = 98.7395(1.012^x)$ or equivalently, $y = 98.74e^{.012x}$

(d)

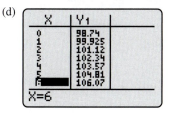

The model fits the data well.

(e) The model predicts about 108.6 million households in 2003.

29. (a) $y = 17.594 + 13.4239 \ln x$

(b) 77.4 years

(c) The model predicts 80.9 years in 2012 and 81.1 years in 2013.

Section 5.7, page 448

1. Not one-to-one.

3. One-to-one.

5. One-to-one.

7. Not one-to-one.

9. $g(x) = -x$

11. $g(x) = \dfrac{x + 4}{5}$

13. $g(x) = \sqrt[3]{\dfrac{5 - x}{2}}$

15. $g(x) = \dfrac{x^2 + 7}{4} \quad (x \geq 0)$

17. $g(x) = \dfrac{1}{x}$

19. $g(x) = \dfrac{1 - x}{2x}$

21. $g(x) = \sqrt[3]{\dfrac{5x + 1}{1 - x}}$

23. $g(f(x)) = g(x + 1) = x + 1 - 1 = x$
$f(g(x)) = f(x - 1) = x - 1 + 1 = x$

25. $g(f(x)) = g\left(\dfrac{1}{x + 1}\right) = \dfrac{1 - \dfrac{1}{x + 1}}{\dfrac{1}{x + 1}} = \dfrac{x + 1 - 1}{1} = x$

$f(g(x)) = f\left(\dfrac{1 - x}{x}\right) = \dfrac{1}{\dfrac{1 - x}{x} + 1} = \dfrac{x}{1 - x + x} = x$

27. $g(f(x)) = g(x^5) = \sqrt[5]{x^5} = x$
$f(g(x)) = f\left(\sqrt[5]{x}\right) = \left(\sqrt[5]{x}\right)^5 = x$

29. $f(f(x)) = \dfrac{2f(x) + 1}{3f(x) - 2} = \dfrac{2\left(\dfrac{2x + 1}{3x - 2}\right) + 1}{3\left(\dfrac{2x + 1}{3x - 2}\right) - 2} =$

$\dfrac{4x + 2 + 3x - 2}{6x + 3 - 6x + 4} = \dfrac{7x}{7} = x$

31.

33.

35.

37.

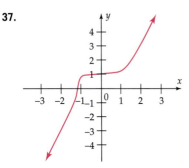

39. One possibility is $h(x) = |x|$ with $x \geq 0$; this function is its own inverse.

41. One possibility is $h(x) = -x^2$ with $x \geq 0$; the inverse is $g(x) = \sqrt{-x}$ with $x \leq 0$.

43. One possibility is $h(x) = \dfrac{x^2 + 6}{2}$ with $x \geq 0$; the inverse is $g(x) = \sqrt{2x - 6}$ with $x \geq 3$.

45. One possibility is $h(x) = \dfrac{1}{x^2 + 1}$ with $x \geq 0$; the inverse is $g(x) = \sqrt{\dfrac{1}{x} - 1}$ with $0 < x \leq 1$.

47. (a) $f^{-1}(x) = \dfrac{x - 2}{3}$

(b) $f^{-1}(1) = -\dfrac{1}{3}$; $\dfrac{1}{f(1)} = \dfrac{1}{5}$

49. First note that f is one-to-one, since if $c \neq d$, then $mc \neq md$ and $mc + b \neq md + b$. The inverse is $g(x) = (x - b)/m$.

51. (a) Slope $= (a - b)/(b - a) = -1$

(b) The slopes are 1 and -1; their product is -1, so the lines are perpendicular.

(c) Length of $PR = \sqrt{(c - a)^2 + (c - b)^2}$, length of $RQ = \sqrt{(b - c)^2 + (a - c)^2}$. Clearly, these are equal, since $(b - c)^2 = (c - b)^2$ and $(c - a)^2 = (a - c)^2$. Thus, the line $y = x$ is perpendicular to PQ and bisects PQ, so the line is the perpendicular bisector of PQ, and thus P and Q are symmetric with respect to the line $y = x$.

53. (a) Suppose $a \neq b$. If $f(a) = f(b)$, then $g(f(a)) = g(f(b))$. But $a = g(f(a))$ by (1) and $b = g(f(b))$. Hence, $a = b$, contrary to our hypothesis. Therefore, it cannot happen that $f(a) = f(b)$, that is, $f(a) \neq f(b)$. Hence, f is one-to-one.

(b) If $g(y) = x$, then $f(g(y)) = f(x)$. But $f(g(y)) = y$ by (2). Hence, $y = f(g(y)) = f(x)$.

(c) If $f(x) = y$, then $g(f(x)) = g(y)$. But $g(f(x)) = x$ by (1). Hence, $x = g(f(x)) = g(y)$.

Chapter 5 Review, page 452

1. $-3 \leq x \leq 3$ and $0 \leq y \leq 10$.

3. $-3 \leq x \leq 3$ and $-2 \leq y \leq 8$.

5. $-3 \leq x \leq 3$ and $0 \leq y \leq 2$.

7. $-5 \leq x \leq 10$ and $-5 \leq y \leq 2$.

9. (a) About 2.03%

(b) About 31.97%

(c) From the sixth to the tenth month

(d) Never. The line $y = 100$ is a horizontal asymptote, so $P(t)$ never reaches 100; but after 17 months, P is over 99%.

11. (a) $1341.68

(b) $541.68

13. (a) $2357.90

(b) After 392 months (32 years, 8 months)

15. $S(x) = 56{,}000(1.065)^x$

17. About 3.747 grams

19. $\ln 756 = 6.628$

21. $\ln(u + v) = r^2 - 1$

23. $\log 756 = 2.8785$

25. $e^{7.118} = 1234$

27. $e^t = rs$

29. $5^u = cd - k$

31. 3

33. $3/4$

35. $2 \ln x$ or $\ln x^2$

37. $\ln\left(\dfrac{9y}{x^2}\right)$

39. $\ln x^{-10}$

41. 2

43. (c)

45. (c)

47. $\pm\sqrt{2}$

49. $\dfrac{3 \pm \sqrt{57}}{4}$

51. $-1/2$

53. $e^{(u-c)/d}$

55. 2

57. 101

59. About 1.64 milligrams

61. About 12.05 years

63. $452.89

65. 7.6

67. (a) $11°F$

(b)

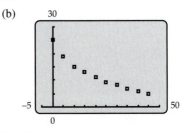

(c) The plotted data strongly suggests exponential decay. One might also argue that the ratios of successive values are approximately equal.

(d) $y = 22.42157(.9666^x)$

(e) About $10.3°F$

69. $g(x) = \dfrac{x - 1}{2}$

71. $g(x) = \sqrt[3]{x^5 - 1}$

73.

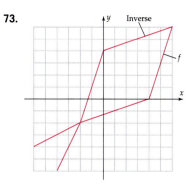

75. Yes. The function is one-to-one; it is its own inverse.

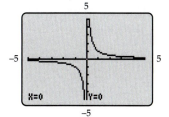

77. No. The function is not one-to-one.

CHAPTER 6

Section 6.1, page 470

1. The values are a solution.

3. The values are a solution.

5. The values are not a solution.

7. $x = 11/5, y = -7/5$ **9.** $x = 2/7, y = -11/7$

11. $r = 5/2, s = -5/2$

13. $x = \dfrac{3}{2}c, y = d - \dfrac{1}{2}c$

15. $x = 28, y = 22$ **17.** $x = 2, y = -1$

19. No solution

21. $x = b, y = \dfrac{3b - 4}{2}$, where b is any real number.

23. $x = b, y = \dfrac{3b - 2}{4}$, where b is any real number.

25. No solution   **27.** $x = -6, y = 2$

29. $x = 13.2, y = 3.6$ **31.** $x = 7/11, y = -7$

33. $x = 1, y = 1/2$

35. (a) Solve each equation for y, and observe that a positive slope represents an increasing population; a negative slope represents a decreasing population.
    (b) 1987

37. Never

39. $x = \dfrac{rd - sb}{ad - bc}, y = \dfrac{as - rc}{ad - bc}$

41. Since the two lines have slopes $-\frac{1}{2}$ and 2, they are perpendicular and intersect at one point, regardless of c. Therefore, there is exactly one solution for the system. Alternatively, solve the system to find that the only solution is $x = \dfrac{3c + 8}{15}, y = \dfrac{6c - 4}{15}$.

43. $c = -3, d = 1/2$

45. Equilibrium quantity is 8000; equilibrium price is $6.

47. Equilibrium quantity is 4000; equilibrium price is $180.

49. $x = 120$ units

51. (a) $R = 60x$
    (b) 400 hedge trimmers

53. $x = 500$. Since the break-even point is above the maximum number that can be sold, the product should not be produced.

55. $x = 1400$. Since the break-even point is below the maximum number that can be sold, the product should be produced.

57. 140 adults; 60 children

59. Boat: 18 mph; current: 2 mph

61. 3/4 pounds of cashews; $2\frac{1}{4}$ pounds of peanuts

63. 24 grams of 50% alloy; 16 grams of 75% alloy

65. (a) Electric: $y = 2000 + 960x$, solar: $y = 14,000 + 114x$
    (b) Electric: $6800, solar: $14,570
    (c) The costs will be the same when $x = 14.2$. Electric heating will be cheaper before that, solar afterwards.

67. (a) $R = 250,500x$
    (b) $C = 1,295,000 + 206,500x$
    (c) $P = R - C = 44,000x - 1,295,000$
    (d) $R = 109,000x, C = 440,000 + 82,000x,$
       $P = R - C = 27,000x - 440,000$
    (e) 50.3 weeks, $918,000
    (f) It would be better to open off Broadway if a run of 50 weeks or less is expected.

69. $14,450 at 9%, $6,450 at 11%

71. (a) $y = .373x + 27.02$
    (b) $y = .624x + 23.65$ (c) 2003

73. 80 bowls and 120 plates

Special Topics 6.1.A, page 479

1. $x = 3, y = 9$ or $x = -1, y = 1.$

3. $x = \dfrac{-1 - \sqrt{73}}{6}, y = \dfrac{37 + \sqrt{73}}{18}$ or
    $x = \dfrac{-1 + \sqrt{73}}{6}, y = \dfrac{37 - \sqrt{73}}{18}.$

5. $x = 7, y = 3$ or $x = 3, y = 7.$

7. $x = 6, y = 1$ or $x = 0, y = -2.$

9. $x = 2, y = 0$ or $x = 4, y = 2$.

11. $x = 4, y = 3$ or $x = -4, y = 3$ or $x = \sqrt{21}, y = -2$ or $x = -\sqrt{21}, y = -2$

13. $x = -1.6237, y = -8.1891$ or $x = 1.3163, y = 1.0826$ or $x = 2.8073, y = 2.4814$

15. $x = -1.9493, y = .4412$ or $x = .3634, y = .9578$ or $x = 1.4184, y = .5986$

17. No real solutions

19. $x = \dfrac{13 + \sqrt{105}}{8}, y = \dfrac{-3 + \sqrt{105}}{8}$ or $x = \dfrac{13 - \sqrt{105}}{8}, y = \dfrac{-3 - \sqrt{105}}{8}$

21. The intersection point nearest the origin is $(2.4, 3.2)$.

23. The two graphs do not intersect; hence, there is no solution.

25. Two possible boxes: One is 2 by 2 by 4 meters and the other is approximately 3.1231 by 3.1231 by 1.6404 meters.

27. -4 and -12 **29.** 15 and -12

31. 12 feet by 17 feet **33.** 8 inches by 15 inches

35. $y = 6x - 9$

Section 6.2, page 486

1. $x = -68, y = 13, w = 3, z = -6$

3. $x = 20, y = -9, w = 3, z = 15/2$

5. $\begin{aligned} x \quad\quad - 3z &= 2 \\ 2x - 4y + 5z &= 1 \\ 5x - 8y + 7z &= 6 \\ 3x - 4y + 2z &= 3 \end{aligned}$

7. $\begin{aligned} 3x \quad\quad + z + 2w + 18v &= 0 \\ -4x + y \quad\quad - w - 24v &= 0 \\ 7x - y + z + 3w + 42v &= 0 \\ 4x \quad\quad + z + 2w + 24v &= 0 \end{aligned}$

9. $\begin{aligned} x + y + 2z + 3w &= 1 \\ - y - z - 2w &= -1 \\ 3x + y + 4z + 5w &= 2 \end{aligned}$

11. $\begin{aligned} x + 12y - 3z + 4w &= 10 \\ 2y + 3z + w &= 4 \\ - z &= -7 \\ 6y - 2z - 3w &= 0 \end{aligned}$

13. $x = 3/2, y = 3/2, z = -3/2$

15. $x = -14, y = -6, z = 2$

17. $x = 100, y = 50, z = 50$

19. $x = 1, y = 2$

21. $x = 2.4, y = 1.8$

23. The system has no solution.

25. \$10,000 should be invested in AAA bonds. \$20,000 in A bonds, and \$5000 in B bonds.

27. 40 pounds of corn chips, $46\frac{2}{3}$ pounds of nuts, and $13\frac{1}{3}$ pounds of pretzels

29. 400 box seats, 5400 grandstand seats and 1200 bleacher seats

Section 6.3, page 497

1. $\begin{pmatrix} 2 & -3 & 4 & | & 1 \\ 1 & 2 & -6 & | & 0 \\ 3 & -7 & 4 & | & -3 \end{pmatrix}$

3. $\begin{pmatrix} 1 & -\frac{1}{2} & \frac{7}{4} & | & 0 \\ 2 & -\frac{3}{2} & 5 & | & 0 \\ 0 & -2 & \frac{1}{3} & | & 0 \end{pmatrix}$

5. $\begin{aligned} 2x - 3y &= 1 \\ 4x + 7y &= 2 \end{aligned}$

7. $\begin{aligned} x + \quad\quad z \quad\quad &= 1 \\ x - y + 4z - 2w &= 3 \\ 4x + 2y + 5z \quad\quad &= 2 \end{aligned}$

9. $x = 3/2, y = 5, z = -2, w = 0$

11. $x = 2 - w, y = -3 - 2w, z = 4, w$ any real number

13. $y = 3/2, x = 3/2, z = -3/2$

15. $z = t, y = -1 + \frac{1}{3}t, x = 2 - \frac{4}{3}t$, where t is any real number.

17. $x = -14, y = -6, z = 2$

19. $x = 100, y = 50, z = 50$

21. $z = t, y = -2t + 1/2, x = t$, where t is any real number.

23. $x = 1, y = 2$ **25.** No solution

27. $z = t, y = t - 1, x = -t + 2$, where t is any real number.

29. $x = y = z = 0$

31. $x = -1, y = 1, z = -3, w = -2$

33. $x = 7/31, y = 6/31, z = 1/31, w = 29/31$

35. $x = 1/2, y = 1/3, z = -1/4$

37. $x = 3/5, y = 1/5$

39. $x = -1, y = 2$

41. $A = -1, B = 2$

43. $A = -3/25, B = 3/25, C = 7/5$

45. $A = 2, B = 3, C = -1$

47. \$3000 from friend, \$6000 from the bank, \$1000 from the insurance company

49. \$15,000 in the mutual fund, \$30,000 in bonds, \$25,000 in the fast-food franchise

51. 6 cups of Progresso, 9 cups of Healthy Choice, 2 cups of Campbell's. The serving size is 1.7 cups.

53. Three possible solutions: 18 bedroom models, 13 living-room models, 0 whole-house models, or 16 bedroom models, 8 living-room models, 2 whole-house models, or 14 bedroom models, 3 living-room models, 4 whole-house models

55. (a) X-ray 1: $a + b$. X-ray 2: $a + c$.
(b) $a = .405$, $b = .345$, $c = .195$
(c) A is bone, B is tumorous, C is healthy.

57. 2000 chairs, 1600 chests, and 2500 tables.

59. (a) $x + t = 1000$
$x + y = 1100$
$y + z = 700$
$t + z = 600$
(b) $z = 600 - t$, $y = 100 + t$, $x = 1000 - t$
(c) The smallest number of cars leaving A on 4th Avenue is 0; the largest is 600. The smallest number of cars leaving A on Euclid is 400; the largest is 1000. The smallest number of cars leaving C on 5th Avenue is 100; the largest is 700. The smallest number of cars leaving C on Chester is 0; the largest is 600.

61. Tom: 8 hours, Dick: 24 hours, Harry: 12 hours

Section 6.4, page 512

1. AB is a 2×4 matrix, but BA is not defined.

3. AB is a 3×3 matrix, BA is a 2×2 matrix.

5. AB is a 3×2 matrix, but BA is not defined.

7. $\begin{pmatrix} 3 & 0 & 11 \\ 2 & 8 & 10 \end{pmatrix}$

9. $\begin{pmatrix} 1 & -3 \\ 2 & -1 \\ 5 & 6 \end{pmatrix}$

11. $\begin{pmatrix} 1 & -1 & 1 & 2 \\ 4 & 3 & 3 & 2 \\ -1 & -1 & -3 & 2 \\ 5 & 3 & 2 & 5 \end{pmatrix}$

13. $AB = \begin{pmatrix} 17 & -3 \\ 33 & -19 \end{pmatrix}$ $BA = \begin{pmatrix} -4 & 9 \\ 24 & 2 \end{pmatrix}$

15. $AB = \begin{pmatrix} 8 & 24 & -8 \\ 2 & -2 & 6 \\ -3 & -21 & 15 \end{pmatrix}$ $BA = \begin{pmatrix} 19 & 9 & 8 \\ -10 & 2 & 0 \\ 0 & 0 & 0 \end{pmatrix}$

17. $\begin{pmatrix} -2 & 1 \\ 3/2 & -1/2 \end{pmatrix}$

19. The matrix has no inverse.

21. The matrix has no inverse.

23. $\begin{pmatrix} -3 & 2 & -4 \\ -1 & 1 & -1 \\ 8 & -5 & 10 \end{pmatrix}$

25. $x = -1$, $y = 0$, $z = -3$ **27.** $x = -8$, $y = 16$, $z = 5$

29. $x = -.5$, $y = -2.1$, $z = 6.7$, $w = 2.8$

31. $x = 10.5$, $y = 5$, $z = -13$, $v = 32$, $w = 2.5$

33. $x = -1149/161$, $y = 426/161$, $z = -1124/161$, $w = 579/161$

35. $x = y = z = v = w = 0$

37. The system is inconsistent and has no solution.

39. $w = t$, $z = s$, $y = \dfrac{5}{4} - \dfrac{1}{4}s$, $x = -2 - 3t$, where s and t are any real numbers.

41. The equation of the parabola is $y = \dfrac{35}{8}x^2 + \dfrac{13}{4}x - \dfrac{117}{8}$.

43. (a) $a = \dfrac{115}{9933}$, $b = \dfrac{7991}{9933}$, $c = 315$
(b) 1983: 342.35 ppm; 1993: 357.34 ppm; 2003: 374.65 ppm

45. (a) $y = -\dfrac{31}{32}x^2 + \dfrac{61}{2}x + \dfrac{895}{8}$
(b) 1993: $194.66; 1998: $293.88; 2002: $338.38

47. $a = 1$, $b = -4$, $c = 1$

49. One pair of jeans is $34.50, one jacket is $72, one sweater is $44, and one shirt is $21.75

51. Box A: 15,000, box B: 18,000, box C: 54,000

Special Topics 6.4.A, page 522

1. $A + B = \begin{pmatrix} 10 & -3 \\ 3 & 7 \end{pmatrix}$; $AB = \begin{pmatrix} 17 & -3 \\ 33 & -19 \end{pmatrix}$;
$BA = \begin{pmatrix} -4 & 9 \\ 24 & 2 \end{pmatrix}$; $2A - 3B = \begin{pmatrix} -15 & 19 \\ 16 & -16 \end{pmatrix}$

3. $A + B = \begin{pmatrix} 2 & 1/2 \\ 13/2 & 9/2 \end{pmatrix}$; $AB = \begin{pmatrix} 23/4 & -1/4 \\ 43/4 & -9/2 \end{pmatrix}$;
$BA = \begin{pmatrix} -21/4 & -17/4 \\ 31/4 & 17/2 \end{pmatrix}$; $2A - 3B = \begin{pmatrix} 3/2 & 17/2 \\ 1/2 & 4 \end{pmatrix}$

5. $A + B = \begin{pmatrix} 3 & 6 \\ 8 & 9 \end{pmatrix}$; $AB = \begin{pmatrix} 7 & 9 \\ 3 & 5 \end{pmatrix}$;
$BA = \begin{pmatrix} 5 & 3 \\ 9 & 7 \end{pmatrix}$; $2A - 3B = \begin{pmatrix} -9 & -13 \\ -19 & -27 \end{pmatrix}$

7. $A + B = \begin{pmatrix} 4 & 4 \\ 6 & -5 \end{pmatrix}$; $AB = \begin{pmatrix} 1 & 2 \\ -6 & 10 \end{pmatrix}$;
$BA = \begin{pmatrix} 1 & -4 \\ 3 & 10 \end{pmatrix}$; $2A - 3B = \begin{pmatrix} 13 & -2 \\ -3 & 10 \end{pmatrix}$

9. $\begin{pmatrix} -2 & 0 \\ 1 & 1 \end{pmatrix}$ **11.** $\begin{pmatrix} 7 & 0 \\ -3 & -2 \end{pmatrix}$

13. $\begin{pmatrix} 0 & -8 & -2 \\ 4 & 16 & 18 \end{pmatrix}$ **15.** $\begin{pmatrix} 8 & -8 \\ 5 & 2 \\ -3 & 25 \end{pmatrix}$

17. $\begin{pmatrix} -6 & 11 \\ 10 & 3 \end{pmatrix}$ **19.** $\begin{pmatrix} 4 & -7/2 \\ 4 & 21/2 \end{pmatrix}$

21. $\begin{pmatrix} 5/2 & 1 \\ -6 & 11/2 \end{pmatrix}$

23. $A + B = \begin{pmatrix} 1 & 2 \\ 3 & 0 \end{pmatrix} + \begin{pmatrix} -1 & 2 \\ 3 & 4 \end{pmatrix} = \begin{pmatrix} 0 & 4 \\ 6 & 4 \end{pmatrix}$ and

$B + A = \begin{pmatrix} -1 & 2 \\ 3 & 4 \end{pmatrix} + \begin{pmatrix} 1 & 2 \\ 3 & 0 \end{pmatrix} = \begin{pmatrix} 0 & 4 \\ 6 & 4 \end{pmatrix}$

25. $c(A + B) = 2\begin{pmatrix} 0 & 4 \\ 6 & 4 \end{pmatrix} = \begin{pmatrix} 0 & 8 \\ 12 & 8 \end{pmatrix}$ and

$cA + cB = 2\begin{pmatrix} 1 & 2 \\ 3 & 0 \end{pmatrix} + 2\begin{pmatrix} -1 & 2 \\ 3 & 4 \end{pmatrix}$

$= \begin{pmatrix} 2 & 4 \\ 6 & 0 \end{pmatrix} + \begin{pmatrix} -2 & 4 \\ 6 & 8 \end{pmatrix} = \begin{pmatrix} 0 & 8 \\ 12 & 8 \end{pmatrix}$

27. $(cd)A = (2 \cdot 3)\begin{pmatrix} 1 & 2 \\ 3 & 0 \end{pmatrix} = 6\begin{pmatrix} 1 & 2 \\ 3 & 0 \end{pmatrix} = \begin{pmatrix} 6 & 12 \\ 18 & 0 \end{pmatrix}$ and

$c(dA) = 2\left[3\begin{pmatrix} 1 & 2 \\ 3 & 0 \end{pmatrix}\right] = 2\begin{pmatrix} 3 & 6 \\ 9 & 0 \end{pmatrix} = \begin{pmatrix} 6 & 12 \\ 18 & 0 \end{pmatrix}$

29. $A(B + C) = \begin{pmatrix} 1 & 2 \\ 3 & 0 \end{pmatrix}\begin{pmatrix} 1 & 5 \\ 4 & 6 \end{pmatrix} = \begin{pmatrix} 9 & 17 \\ 3 & 15 \end{pmatrix};$

$AB + AC = \begin{pmatrix} 5 & 10 \\ -3 & 6 \end{pmatrix} + \begin{pmatrix} 4 & 7 \\ 6 & 9 \end{pmatrix} = \begin{pmatrix} 9 & 17 \\ 3 & 15 \end{pmatrix}$

31. For AB to be defined, $k = r$. For BA also to be defined, $n = t$.

33. $AB = \begin{pmatrix} 0 & 0 \\ 0 & 0 \end{pmatrix} = AC$, but clearly, $B \neq C$.

35. $(A + B)(A - B) = \begin{pmatrix} 5 & 0 \\ 7 & -1 \end{pmatrix}\begin{pmatrix} 1 & 2 \\ -3 & -7 \end{pmatrix} =$

$\begin{pmatrix} 5 & 10 \\ 10 & 21 \end{pmatrix};$ but

$A^2 - B^2 = \begin{pmatrix} 11 & -1 \\ -2 & 18 \end{pmatrix} - \begin{pmatrix} -1 & -5 \\ 25 & 4 \end{pmatrix} = \begin{pmatrix} 12 & 4 \\ -27 & 14 \end{pmatrix}$

37. $A^t = \begin{pmatrix} a & c \\ b & d \end{pmatrix}; B^t = \begin{pmatrix} r & u \\ s & v \end{pmatrix}$

39. $(A^t)^t = \begin{pmatrix} a & c \\ b & d \end{pmatrix}^t = \begin{pmatrix} a & b \\ c & d \end{pmatrix} = A$

41. A possible matrix would be

	Heart	Lung	Liver	Kidney
2000	4164	3732	17,132	50,426
2001	4148	3821	17,546	52,216
2003	3568	3927	17,567	59,333

The transpose of this matrix would also serve.

43. $\begin{pmatrix} 220 & 400 & 280 \\ 360 & 300 & 225 \\ 180 & 380 & 300 \end{pmatrix}\begin{pmatrix} 2.75 \\ 3.25 \\ 2.20 \end{pmatrix} = \begin{pmatrix} \$2521 \\ \$2460 \\ \$2390 \end{pmatrix}$

45. $\begin{pmatrix} 340 & 60 \\ 108 & 292 \end{pmatrix}.$

47. 158, −125, 103, 151, 174, −166, 115, 116, 145, −159, 89, 18, 131, −86, 77, 115, 130, −89, 58, 11, 138, −51, 32, −51

49. "Tom has a new job"

51. 222, 104, 41, 128, 120, 279, 172, 227, 102, 47, 199, 10, 15, −10, 218, 160, 60, −23, 80, 150, 374, 174, 165, 114, 36

53. (a) 81, 189, 166, 113
  (b) Decoding leads to an error message. A matrix must have an inverse to be used in this coding scheme.

Section 6.5, page 533

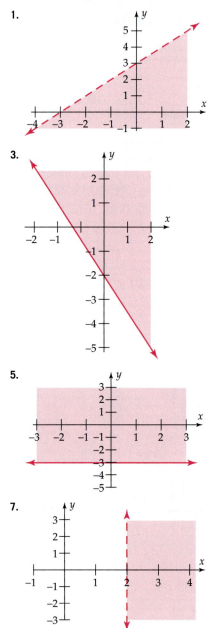

1.

3.

5.

7.

9.

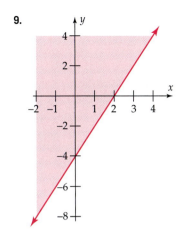

11.

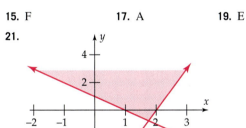

13.

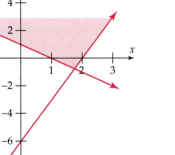

15. F   **17.** A   **19.** E

21.

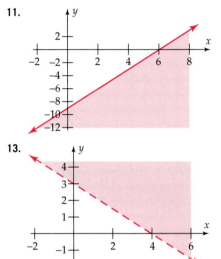

23.

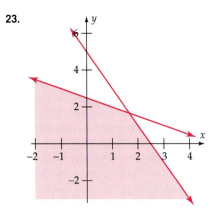

25.

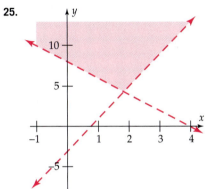

27.

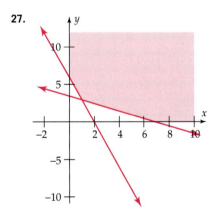

29.

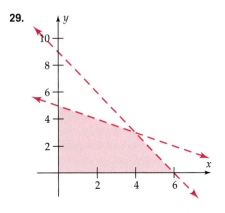

31.

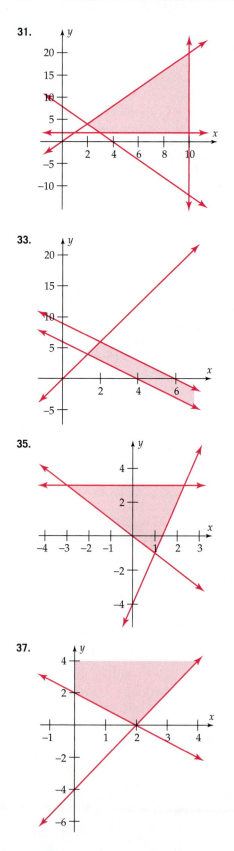

33.

35.

37.

39.

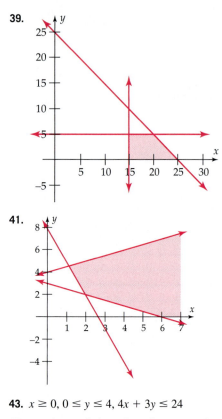

41.

43. $x \geq 0$, $0 \leq y \leq 4$, $4x + 3y \leq 24$

45. $y \geq 0$, $x \leq 4$, $4y \leq 3x + 8$, $x + y \geq 2$

47. $2 < x < 7$, $-1 < y < 3$

49. $x < 2$, $3y < 2x + 8$, $x + 6y > -4$

51. Let x = number of tanks of regular, and let y = number of tanks of premium.

$$2x + 3y \leq 12$$
$$3x + 3.5y \leq 16$$
$$x \geq 0, y \geq 0$$

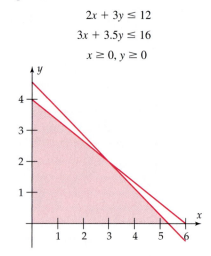

53. Let x = number of toy chests, and let y = number of silverware cases.

$$4x + 2y \leq 20$$

$$3x + 4y \leq 20$$

$$x \geq 0, y \geq 0$$

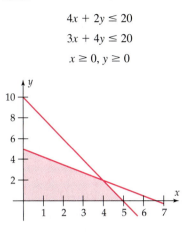

55. Let x = number of grams of grain, and let y = number of grams of meat by-products.

$$2x + 3y \leq 12$$

$$2x + y \geq 2$$

$$y \geq 1$$

$$x \geq 0$$

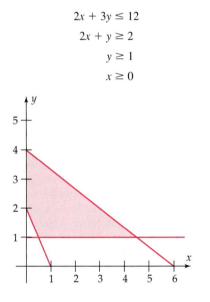

Section 6.6, page 542

1. Minimum: 19, maximum: 49

3. Minimum: 11, maximum: 27

5. Minimum: 5, maximum: 19

7. Minimum: 7, maximum: 16

9. Minimum: 0, maximum: 128/3

11. Minimum: 0, maximum: 42

13. Minimum: 0, maximum: 15

15. Minimum: 23, no maximum (unbounded)

17. Minimum: 15, maximum: 28

19. Minimum: 5, maximum: 27

21. Minimum: 0, maximum: 22

23. A graph will show that there are no points that satisfy these conditions.

25. Maximum profit of $780 when 6 four-person rafts and 18 two-person rafts are made.

27. Maximum revenue of $480 when no boxes of Choco-Mix and 30 boxes of Minto-Mix are made.

29. Minimum cost of $11.40 when 30 Supervites pills and 120 Vitahealth pills are taken.

31. Minimum cost of $2.80 occurs when 8 pounds of feed A and 3 pounds of feed B are used.

33. Minimum cost of $6,300 occurs when 0 tons of bran and 28 tons of oats are used.

35. Maximum number of quality points, 268, occurs when 8 humanities courses and 12 science courses are taken.

37. A graph will show that the conditions are contradictory. There are no points satisfying all the conditions. Adding $2x + y \leq 10$ and $x + 2y \leq 8$ gives $3x + 3y \leq 18$, or equivalently, $x + y \leq 6$, which contradicts $x + y \geq 8$.

39. The GPA is 2.89 at vertex (4, 4), 2.95 at (4, 12), 2.91 at (8, 12), and 2.80 at (18, 4). However, the total quality points are 268 at (8, 12) and 224 at (4, 12). So the distribution of courses that produces the highest number of quality points does not yield the highest grade point average. This is not a contradiction, however; to maximize the grade point average would require a different objective function, one that, moreover, would not be linear and therefore would not be revealed through linear programming.

Chapter 6 Review, page 546

1. $x = -5, y = -7$ **3.** $x = 0, y = -2$

5. 2000; about 69 days

7. $33\frac{1}{3}$ pounds of 70% alloy and $16\frac{2}{3}$ pounds of 40% alloy

9. $x = 3, y = 9$ or $x = -1, y = 1$

11. $x = 1 + \sqrt{7}, y = 1 - \sqrt{7}$ or $x = 1 - \sqrt{7}, y = 1 + \sqrt{7}$

13. $\begin{pmatrix} 1 & -2 & 3 & | & 4 \\ 2 & 1 & -4 & | & 3 \\ -3 & 4 & -1 & | & -2 \end{pmatrix}$

15. $\begin{pmatrix} 2 & -1 & -2 & 2 \\ 1 & 3 & -2 & 1 \\ -1 & 4 & 2 & -3 \end{pmatrix}$

17. $x = -35, y = 140, z = 22$

19. $x = 2, y = 4, z = 6$

21. $z = t, y = -2t, x = 1 - t$, where t is any real number.

23. (c)   **25.** $A = 1, B = 3$

27. $\begin{pmatrix} -2 & 3 \\ -4 & 1 \end{pmatrix}$

29. The product AE does not exist.

31. $\begin{pmatrix} -9 & 7 \\ -4 & 3 \end{pmatrix}$ **33.** $\begin{pmatrix} 1 & 2 & -2 \\ -1 & 3 & 0 \\ 0 & -2 & 1 \end{pmatrix}$

35. $x = -1/85, y = -14/85, z = -21/34, w = 46/85$

37. $y = 5x^2 - 2x + 1$

39. (a) $y = \dfrac{17}{16}x^2 - \dfrac{17}{4}x + \dfrac{49}{4}$

(b) 1998: 46.25 hours, 2002: 114.25 hours, 2006: 216.25 hours

41. 100 faculty members

43. 30 pounds of corn, 15 pounds of soybeans, 40 pounds of by-products

45. $\begin{pmatrix} 4 & 5 \\ 24 & 1 \\ 18 & -7 \end{pmatrix}$ **47.** $\begin{pmatrix} -5 & 1 \\ -6 & -5 \end{pmatrix}$

49. Not defined **51.** $\begin{pmatrix} 227 & 124 & 49 \\ 266 & 40 & 94 \end{pmatrix}$

53.

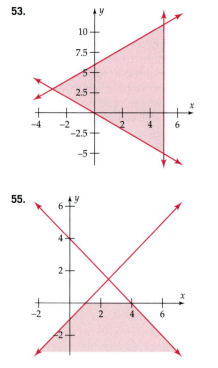

55.

57. Minimum: .72, for $x = 4$, $y = 8$; no maximum (unbounded)

59. Maximum revenue of \$10,900 when 15 containers and 35 barrels are loaded. There will then be no unused space in the truck.

CHAPTER 7

Section 7.1, page 562

1. $a_1 = 8, a_2 = 10, a_3 = 12, a_4 = 14, a_5 = 16$

3. $a_1 = 1, a_2 = \dfrac{1}{8}, a_3 = \dfrac{1}{27}, a_4 = \dfrac{1}{64}, a_5 = \dfrac{1}{125}$

5. $a_1 = \dfrac{1}{2}, a_2 = \dfrac{1}{2}, a_3 = \dfrac{3}{8}, a_4 = \dfrac{1}{4}, a_5 = \dfrac{5}{32}$

7. $a_1 = -\sqrt{3}, a_2 = 2, a_3 = -\sqrt{5}, a_4 = \sqrt{6}, a_5 = -\sqrt{7}$

9. $a_1 = 3.9, a_2 = 4.01, a_3 = 3.999, a_4 = 4.0001,$ $a_5 = 3.99999$

11. $a_1 = 2, a_2 = 7, a_3 = 8, a_4 = 13, a_5 = 14$

13. $a_1 = 3, a_2 = 1, a_3 = 4, a_4 = 1, a_5 = 5$

15. $a_n = (-1)^n$ **17.** $a_n = \dfrac{n}{n+1}$

19. $a_n = 5n - 3$ **21.** $a_n = 3 \cdot 2^{n-1}$

23. $a_n = 4\sqrt{n}$

25. $a_1 = 4, a_2 = 11, a_3 = 25, a_4 = 53, a_5 = 109$

27. $a_1 = -16, a_2 = -8, a_3 = -4, a_4 = -2, a_5 = -1$

29. $a_1 = 2, a_2 = 0, a_3 = 3, a_4 = 1, a_5 = 4$

31. $a_1 = 1, a_2 = -2, a_3 = 3, a_4 = 2, a_5 = 3$

33. $a_0 = 2, a_1 = 3, a_2 = 3, a_3 = 9/2, a_4 = 27/4$

35. $\displaystyle\sum_{i=1}^{11} i$ **37.** $\displaystyle\sum_{i=1}^{7} \dfrac{1}{2^{i+6}}$ or $\displaystyle\sum_{i=7}^{13} \dfrac{1}{2^i}$

39. $\displaystyle\sum_{i=1}^{7} \dfrac{2^i}{i}$ **41.** 28 **43.** 34

45. 45 **47.** 224 **49.** 15,015

51. $a_1 + a_2 + a_3 = -10$
$a_1 + a_2 + a_3 + a_4 + a_5 + a_6 = -2$

53. $a_1 + a_2 + a_3 = 5$
$a_1 + a_2 + a_3 + a_4 + a_5 + a_6 = 0$

55. $\displaystyle\sum_{n=1}^{6} \dfrac{1}{2n+1} \approx .9551$ **57.** $\displaystyle\sum_{n=1}^{5} \dfrac{(-1)^{n+1}n}{n+7} \approx .2558$

59. 2.613035 **61.** 33.465

63. 1.5759958

65. $a_{12} = \$7299.70, a_{36} = \$2555.50$

67. (a) 2000:979.1 hours, 2003:1005.1 hours
(b) 5874.1 hours

69. (a) 2002: \$41.67, 2004: \$58.17
(b) \$234.25

71. (b) $a_{17} = 59, a_{18} = 61, a_{19} = 67, a_{20} = 71$

73. $a_1 = 4, a_2 = 9, a_3 = 25, a_4 = 49, a_5 = 121$

75. $a_1 = 3, a_2 = 7, a_3 = 13, a_4 = 19, a_5 = 23$

77. (a) 1, 1, 2, 3, 5, 8, 13, 21, 34, 55
(b) 1, 2, 4, 7, 12, 20, 33, 54, 88, 143
(c) The nth partial sum is $a_{n+2} - 1$.

79. $5(1)^2 + 4(-1)^1 = 1 = 1^2$
$5(1)^2 + 4(-1)^2 = 9 = 3^2$
$5(2)^2 + 4(-1)^3 = 16 = 4^2$
$5(3)^2 + 4(-1)^4 = 49 = 7^2$
$5(5)^2 + 4(-1)^5 = 121 = 11^2$
$5(8)^2 + 4(-1)^6 = 324 = 18^2$
$5(13)^2 + 4(-1)^7 = 841 = 29^2$
$5(21)^2 + 4(-1)^8 = 2209 = 47^2$
$5(34)^2 + 4(-1)^9 = 5776 = 76^2$
$5(55)^2 + 4(-1)^{10} = 15129 = 123^2$

81. Since $a_n = a_{n-1} + a_{n-2}, a_n - a_{n-1} = a_{n-2}$.
Thus, $a_3 - a_2 = a_1$
$a_4 - a_3 = a_2$
$a_5 - a_4 = a_3$
$\cdots\cdots\cdots$
$a_{k+2} - a_{k+1} = a_k$

Adding: $a_{k+2} - a_2 = \sum_{n=1}^{k} a_n$

Since $a_2 = 1$, $a_{k+2} - 1 = \sum_{n=1}^{k} a_n$.

Section 7.2, page 570

1. The sequence is arithmetic, with common difference 2.

3. The sequence is not arithmetic.

5. The sequence is arithmetic, with common difference log 2.

7. The sequence is arithmetic, with common difference -1.

9. $a_1 = 9, a_2 = 13, a_3 = 17, a_4 = 21, a_5 = 25$
The sequence is arithmetic, with common difference 4.

11. $c_1 = -1, c_2 = 1, c_3 = -1, c_4 = 1, c_5 = -1$
The sequence is not arithmetic.

13. $a_1 = 1, a_2 = 4, a_3 = -1, a_4 = 6, a_5 = -3$
The sequence is not arithmetic.

15. $a_1 = 1, a_2 = 2, a_3 = 3, a_4 = 4, a_5 = 5,$
The sequence is arithmetic, with common difference 1.

17. $a_{n+1} - a_n = [3 - 2(n + 1)] - (3 - 2n) = -2$
The sequence is arithmetic, with common difference -2.

19. $a_{n+1} - a_n = \left(4 + \dfrac{n+1}{3}\right) - \left(4 + \dfrac{n}{3}\right) = \dfrac{1}{3}$
The sequence is arithmetic, with common difference $\frac{1}{3}$.

21. $a_{n+1} - a_n = \dfrac{5 + 3(n+1)}{2} - \dfrac{5 + 3n}{2}$
$= \dfrac{5 + 3n + 3 - 5 - 3n}{2} = \dfrac{3}{2}$
The sequence is arithmetic, with common difference $\frac{3}{2}$.

23. $a_{n+1} - a_n = [c + 2(n+1)] - (c + 2n) = 2$
The sequence is arithmetic, with common difference 2.

25. $a_5 = 13;\ a_n = 2n + 3$

27. $a_5 = 5;\ a_n = \dfrac{n+15}{4}$

29. $a_5 = 8;\ a_n = \dfrac{21-n}{2}$ **31.** $a_5 = 8.4;\ a_n = 7.9 + .1n$

33. $a_1 = 6;\ a_n = 2n + 4$ **35.** $a_1 = -7;\ a_n = 5n - 12$

37. $a_1 = -3;\ a_n = 7n - 10$

39. $a_1 = -6;\ a_n = \dfrac{3n - 15}{2}$

41. 87   **43.** $-21/4$ **45.** 30   **47.** -81

49. 710   **51.** $-285/2$ **53.** $470/3$

55. (a) $a_n = 1049.8 + 173.7n$
(b) 2003: \$5044.90; 2005: \$5392.30

57. 428 seats

59. 23.25, 22.5, 21.75, 21, 20.25, 19.5, 18.75 inches

61. 2550   **63.** 20,100

65. \$77,500 in tenth year; \$437,500 over 10 years

67. (a) $a_n = 127.4 + 9n$
(b) \$235.4 billion
(c) \$1277.4 billion

69. (a) $c_n = 15,100.8 + 728n$
(b) \$614,120

Section 7.3, page 576

1. Arithmetic   **3.** Geometric   **5.** Arithmetic

7. Geometric   **9.** Geometric

11. Both arithmetic ($d = 0$) and geometric ($r = 1$)

13. $a_6 = 160;\ a_n = 5 \cdot 2^{n-1}$

15. $a_6 = 1/256;\ a_n = 4^{2-n}$

17. $a_6 = -5/16;\ a_n = \dfrac{5(-1)^{n-1}}{2^{n-2}}$

19. $a_6 = 4/27;\ a_n = 36(1/3)^{n-1}$

21. $a_6 = -16/125;\ a_n = -\dfrac{2^{n-2}}{5^{n-3}}$

23. $a_n \div a_{n-1} = (-1/2)^n \div (-1/2)^{n-1} = -1/2$. The sequence is geometric, with common ratio $-1/2$.

25. $a_n \div a_{n-1} = 5^{n+2} \div 5^{n-1+2} = 5$. The sequence is geometric, with common ratio 5.

27. $a_n \div a_{n-1} = \left(\sqrt{5}\right)^n \div \left(\sqrt{5}\right)^{n-1} = \sqrt{5}$. The sequence is geometric, with common ratio $\sqrt{5}$.

29. $a_5 = 1;\ a_n = (-1/4)^{n-5}$

31. $a_5 = 5000;\ a_n = \dfrac{1}{2}(10)^{n-1}$

33. $a_5 = 1/4;\ a_n = 4^{4-n}$

35. $a_5 = 10\sqrt[3]{2};\ a_n = 5\left(\sqrt[3]{2}\right)^{n-1}$

37. $\dfrac{315}{32}$ **39.** 381

41. 254 **43.** $-\dfrac{4921}{19{,}683}$ **45.** $\dfrac{665}{8}$

47. (a) $a_n = 44.4313(.8556)^{n-1}$
(b) 2.6826 students per computer
(c) 2009 or early 2010.

49. (a) $c_n = 3.9631(1.0515)^{n-1}$
(b) 1998: \$5.632 billion
2002: \$6.886 billion
2004: \$7.613 billion

51. (a) $b_n = 130.76(1.0760)^{n-1}$
(b) \$1605.85

53. 23.75 ft **55.** \$21,474,836.47

57. \$1898.44

59. Since $a_n = a_1 r^{n-1}$, $\log a_n = \log a_1 + (n-1)\log r$. Therefore, $\log a_n - \log a_{n-1} = [\log a_1 + (n-1)\log r] - [\log a_1 + (n-2)\log r] = \log r$. This shows that the sequence $\{\log a_n\}$ is arithmetic with common difference $\log r$.

61. $a_k = 2^{k-1}$ and $r = 2$. The sum of the preceding terms is
$$\sum_{n=1}^{k-1} a_n = a_1\left(\frac{1-r^{k-1}}{1-r}\right) = 1\left(\frac{1-2^{k-1}}{1-2}\right) = 2^{k-1} - 1$$
Thus, each term $a_k = 2^{k-1}$ equals 1 plus the sum of the preceding terms.

63. 3 years and 2 months

Special Topics 7.3.A, page 582

1. 1 **3.** 3/47 **5.** $833\frac{1}{3}$

7. $4 + 2\sqrt{2}$ **9.** 2/9 **11.** 597/110

13. $\dfrac{10{,}702}{4995}$ **15.** $\dfrac{174{,}067}{99{,}900}$

17. (a) Since $\displaystyle\sum_{n=1}^{\infty} 2(1.5)^n = 2(1.5)^1 + 2(1.5)^2 + 2(1.5)^3 + \cdots = 3 + 3(1.5)^1 + 3(1.5)^2 + \cdots$, this is a geometric series with $a_1 = 3$ and $r = 1.5$.
(b) $f(x) = 6(1.5^x - 1)$.
(c) The function increases faster and faster as x gets larger; the graph does not approach a horizontal line, and the series does not converge.

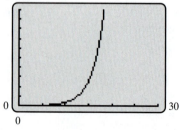

19. We may conjecture that the sum is 1.

Section 7.4, page 590

1. 720 **3.** 56 **5.** 220 **7.** 15

9. 100 **11.** 40 **13.** 0 **15.** 64

17. 3,921,225

19. $x^4 + 4x^3 + 6x^2 + 4x + 1$

21. $x^4 + 8x^3 + 24x^2 + 32x + 16$

23. $x^5 + 5x^4 y + 10x^3 y^2 + 10x^2 y^3 + 5xy^4 + y^5$

25. $a^5 - 5a^4 b + 10a^3 b^2 - 10a^2 b^3 + 5ab^4 - b^5$

27. $32x^5 + 80x^4 y^2 + 80x^3 y^4 + 40x^2 y^6 + 10xy^8 + y^{10}$

29. $x^3 + 6x^2\sqrt{x} + 15x^2 + 20x\sqrt{x} + 15x + 6\sqrt{x} + 1$

31. $1 - 10c + 45c^2 - 120c^3 + 210c^4 - 252c^5 + 210c^6 - 120c^7 + 45c^8 - 10c^9 + c^{10}$

33. $x^{-12} + 4x^{-8} + 6x^{-4} + 4 + x^4$

35. 56 **37.** $-8i$

39. $g(x) = x^3 + x^2 - 4$

41. $g(x) = x^4 - 8x^3 + 24x^2 - 29x + 15$

43. $10x^3 y^2$ **45.** $35c^3 d^4$ **47.** $\dfrac{35}{8u^5}$

49. 4032 **51.** 160 **53.** $(x+1)^5$

55. $(2z+1)^4$

57. (a) $\dbinom{9}{1} = \dfrac{9!}{1!8!} = \dfrac{9 \cdot 8!}{8!} = 9$ $\dbinom{9}{8} = \dfrac{9!}{8!1!} = 9$

(b) $\dbinom{n}{1} = \dfrac{n!}{1!(n-1)!} = \dfrac{n(n-1)!}{(n-1)!} = n$

$\dbinom{n}{n-1} = \dfrac{n!}{(n-1)!1!} = n$

59. Since $2 = 1 + 1$,

$$2^n = (1+1)^n = \binom{n}{0}1^n + \binom{n}{1}1^{n-1}\cdot 1$$
$$+ \binom{n}{2}1^{n-2}\cdot 1^2 + \cdots + \binom{n}{n}1^n = \binom{n}{0} + \binom{n}{1} +$$
$$\binom{n}{2} + \cdots + \binom{n}{n}$$

61. (a) $f(x+h) - f(x) = \dbinom{5}{1}x^4 h + \dbinom{5}{2}x^3 h^2 + \dbinom{5}{3}x^2 h^3 + \dbinom{5}{4}xh^4 + \dbinom{5}{5}h^5$

(b) $\dfrac{f(x+h) - f(x)}{h}$

$$= \frac{h\left[\binom{5}{1}x^4 + \binom{5}{2}x^3 h + \binom{5}{3}x^2 h^2 + \binom{5}{4}xh^3 + \binom{5}{5}h^4\right]}{h}$$

$$= \binom{5}{1}x^4 + \binom{5}{2}x^3 h + \binom{5}{3}x^2 h^2 + \binom{5}{4}xh^3 + \binom{5}{5}h^4$$

(c) $\dbinom{5}{1}x^4$ or $5x^4$

63. (a) $f(x + h) - f(x)$

$$= \binom{12}{1}x^{11}h + \binom{12}{2}x^{10}h^2 + \cdots + \binom{12}{12}h^{12}$$

(b) $\dfrac{f(x + h) - f(x)}{h}$

$$= \frac{h\left[\binom{12}{1}x^{11} + \binom{12}{2}x^{10}h + \cdots + \binom{12}{12}h^{11}\right]}{h}$$

$$= \binom{12}{1}x^{11} + \binom{12}{2}x^{10}h + \cdots + \binom{12}{12}h^{11}$$

(c) $\binom{12}{1}x^{11}$ or $12x^{11}$

65. (a) $f(x + 10)$ adds 10 to each input; hence, if $x = 0$, $f(x + 10) = f(10)$, that is, if $x = 0$, the value of the function for $1980 + 10 = 1990$ is calculated.

(b) The graph of $g(x) = f(x + 10)$ is the graph of $f(x)$ shifted 10 units to the left.

(c) $g(x) = .0021x^3 + .953x^2 + 27.53x + 212.31$

67. (a) $(n - r)! =$
$(n - r)[(n - r) - 1]! = (n - r)[n - (r + 1)]!$

(b) $(n - r)! = (n + 1 - r - 1)! = [(n + 1) - (r + 1)]!$

(c) $\dbinom{n}{r + 1} + \dbinom{n}{r} = \dfrac{n!}{(r + 1)![n - (r + 1)]!} +$

$$\frac{n!}{r!(n - r)!} = \frac{(n - r)n!}{(r + 1)![(n - r)[n - (r + 1)]!} +$$

$$\frac{(r + 1)n!}{(r + 1)r!(n - r)!} = \frac{(n - r)n!}{(r + 1)!(n - r)!} +$$

$$\frac{(r + 1)n!}{(r + 1)!(n - r)!} = \frac{[(n - r) + (r + 1)]n!}{(r + 1)!(n - r)!} =$$

$$\frac{(n + 1)n!}{(r + 1)![(n + 1) - (r + 1)]!} =$$

$$\frac{(n + 1)!}{(r + 1)![(n + 1) - (r + 1)]!} = \binom{n + 1}{r + 1}$$

(d) Since each entry in row n has form $\binom{n}{k}$, the sum of two adjacent entries is $\binom{n}{r + 1} + \binom{n}{r}$. These are the entries above and to the right and left of $\binom{n + 1}{r + 1}$. This explains the assertion.

Section 7.5, page 598

1. 24   **3.** 40,320   **5.** 132

7. 91   **9.** 3,764,376   **11.** $n(n - 1)$

13. n **15.** $\dfrac{n(n - 1)}{2}$ **17.** 11

19. 9   **21.** 7 (positive root only)

23. 1,000,000,000 possible Social Security numbers

25. 35,152 possible call letters

27. 864 possible meals

29. 1,860,480 possible identification codes

31. 256 possible ways to answer the test

33. 40,320   **35.** 24   **37.** 3024   **39.** 56

41. 20,358,520   **43.** 26,334

45. 43,545,600 ways to fill the row

47. 64 possible integers to be formed

49. 576 ways to choose the lineup

51. (a) 59,280 possible "combinations"
(b) Since the order in which the numbers are chosen matters, permutations are being used and not combinations.

53. \$13,983,816

55. 24   **57.** 56   **59.** 11,550

61. (a) 630   (b) 840

63. (a) 2 ways   (b) 6 ways   (c) 24 ways
(d) $(n - 1)!$ ways

65. 479,001,600 ways   **67.** 76,204,800

Special Topics 7.5.A, page 602

1. 12   **3.** 50,400

5. 1,396,755,360

Section 7.6, page 614

1. {HHH, HHT, HTH, HTT, THH, THT, TTH, TTT}. There are eight outcomes.

3. The sample space consists of the possible drawings, that is, $\{r, b\}$. There are two outcomes.

5. The sample space consists of the number of two-card hands that can be drawn from a 16-card pack. Therefore, there are $_{16}C_2 = 120$ possible outcomes.

7. 1/2   **9.** 5/6   **11.** 3/8

13. 1/2   **15.** 5/11   **17.** 2/11

19. 1/6   **21.** 1/2   **23.** 13/18

25. 1/12   **27.** 1/2   **29.** 1/5

31. 4/5   **33.** 7/15   **35.** 7/15

37. 1/13   **39.** 1/26   **41.** 3/26

43. 5/13   **45.** Mutually exclusive

47. Not mutually exclusive

49. 11/36   **51.** 5/18   **53.** 11/26

55. 7/13   **57.** 4/5   **59.** 4/5

61. $\dfrac{1}{54{,}145}$ **63.** $\dfrac{1}{108{,}290}$

65. .3412   **67.** 1/4165   **69.** .0256

71. (a) 1/1024   (b) .2461   (c) .1719

73. (a) .2398   (b) .4196   (c) .0420

75. .2935   **77.** $\dfrac{1}{7{,}059{,}052}$

79.

Number of People	Probability That at Least Two Have the Same Birthday
10	$1 - \dfrac{_{365}P_{10}}{365^{10}} = .1169$
20	$1 - \dfrac{_{365}P_{20}}{365^{20}} = .4114$
25	$1 - \dfrac{_{365}P_{25}}{365^{25}} = .5687$
30	$1 - \dfrac{_{365}P_{30}}{365^{30}} = .7063$
40	$1 - \dfrac{_{365}P_{40}}{365^{40}} = .8912$
50	$1 - \dfrac{_{365}P_{50}}{365^{50}} = .97$

81. .9159

83. (a) .3675   (b) .0410   (c) .9085

85. (a) 3/8   (b) 1/4   (c) 1/8   (d) 1/4

87. .1834   **89.** .1430   **91.** .8570

93. .5112   **95.** .2348   **97.** .5880

99. .325   **101.** .5375   **103.** .3375

Section 7.7, page 625

1. When $n = 1$, both sides of the equation are equal to 1; thus the statement is true. Assume that the statement is true for $n = k$. Then $1 + 2 + 2^2 + 2^3 + \cdots + 2^{k-1} = 2^k - 1$. Add 2^k to both sides of the equation to get

$$1 + 2 + 2^2 + 2^3 + \cdots + 2^{k-1} + 2^k = 2^k - 1 + 2^k = 2(2^k) - 1 = 2^{k+1} - 1.$$

Thus, the statement is true for $n = k + 1$. Therefore, by the Principle of Mathematical Induction, the statement is true for all positive integers n.

3. When $n = 1$, both sides of the equation are equal to 1; thus, the statement is true. Assume that the statement is true for $n = k$. Then

$$1 + 3 + 5 + 7 + \cdots + (2k - 1) = k^2.$$

Add the next odd integer, $(2k + 1)$ to both sides:

$$1 + 3 + 5 + 7 + \cdots + (2k - 1) + (2k + 1) = k^2 + (2k + 1) = (k + 1)^2.$$

Thus, the statement is true for $n = k + 1$. Therefore, by the Principle of Mathematical Induction, the statement is true for all positive integers n. That is, the sum of the first n odd integers is n^2.

5. When $n = 1$, both sides of the equation are equal to 1; thus, the statement is true. Assume that the statement is true for $n = k$. Then

$$1^2 + 2^2 + 3^2 + \cdots + k^2 = \frac{k(k + 1)(2k + 1)}{6}.$$

Add the next square $(k + 1)^2$ to both sides of the equation:

$$1^2 + 2^2 + 3^2 + \cdots + k^2 + (k + 1)^2$$

$$= \frac{k(k + 1)(2k + 1)}{6} + (k + 1)^2$$

$$= (k + 1)\left[\frac{k(2k + 1)}{6} + (k + 1)\right]$$

$$= (k + 1)\left[\frac{2k^2 + 7k + 6}{6}\right]$$

$$= (k + 1)\frac{(k + 2)(2k + 3)}{6}$$

$$= \frac{(k + 1)[(k + 1) + 1][2(k + 1) + 1]}{6}.$$

Thus, the statement is true for $n = k + 1$. Therefore, the statement is true for all positive integers n.

7. When $n = 1$, both sides of the equation are equal to $\frac{1}{2}$; thus, thus, the statement is true. Assume that the statement is true for $n = k$. Then

$$\frac{1}{1 \cdot 2} + \frac{1}{2 \cdot 3} + \frac{1}{3 \cdot 4} + \cdots + \frac{1}{k(k + 1)} = \frac{k}{k + 1}.$$

Add the next term $\dfrac{1}{(k + 1)(k + 2)}$ to both sides of the equation:

$$\frac{1}{1 \cdot 2} + \frac{1}{2 \cdot 3} + \frac{1}{3 \cdot 4} + \cdots + \frac{1}{k(k + 1)} +$$

$$\frac{1}{(k + 1)(k + 2)} = \frac{k}{k + 1} + \frac{1}{(k + 1)(k + 2)}$$

$$= \frac{(k + 1)^2}{(k + 1)(k + 2)} = \frac{k + 1}{k + 2}.$$

Thus, the statement is true for $n = k + 1$. Therefore, the statement is true for all positive integers n.

9. When $n = 1$, we have $1 + 2 > 1$, which is true. Assume that the statement is true for $n = k$. Then $k + 2 > k$. Add 1 to both sides of the inequality to get $k + 2 + 1 > k + 1$, $(k + 1) + 2 > (k + 1)$, and the statement is true for $n = k + 1$. Therefore, the statement $n + 2 > n$ is true for all positive integers n.

11. When $n = 1$, we have $3^1 \geq 3(1)$, which is true. Assume that the statement is true for $n = k$. Then $3^k \geq 3(k)$. Multiply both sides of the inequality by 3 to get $3^{k+1} \geq 9k$ or equivalently, $3^{k+1} \geq 3k + 6k$. Since k is a positive integer, $6k > 3$, hence, $3^{k+1} \geq 3k + 6k \geq 3k + 3 = 3(k + 1)$.

Thus, the statement is true for $n = k + 1$. Therefore, $3^n \geq 3n$ is true for all positive integers n.

13. When $n = 1$, we have $3(1) > 1 + 1$, which is true. Assume that the statement is true for $n = k$. Then $3k > k + 1$. Add 3 to both sides of the inequality to get $3k + 3 > k + 4$, and since $k + 4 > k + 2$, we have $3k + 3 > k + 2$ which is equivalent to $3(k + 1) > (k + 1) + 1$, and the statement is true for $n = k + 1$. Therefore, the statement $3n > n + 1$ is true for all positive integers n.

15. When $n = 1$, $2^{2n+1} + 1 = 2^3 + 1 = 9$, and the statement is true, since 3 is a factor of 9. Assume that the statement is true for $n = k$. Then 3 is a factor of $2^{2k+1} + 1$. Thus, there is an integer p that we can multiply by 3 to get $2^{2k+1} + 1$. That is, $3p = 2^{2k+1} + 1$, or $3p - 1 = 2^{2k+1}$. Now consider

$$2^{2(k+1)+1} + 1 = 2^{2k+3} + 1 = 2^{2k+1}2^2 + 1 =$$
$$(3p - 1)2^2 + 1 = 12p - 3 = 3(4p - 1).$$

Thus, 3 is a factor of $2^{2(k+1)+1} + 1$, and the statement is true for $n = k + 1$. Thus, 3 is a factor of $2^{2n+1} + 1$ for all positive integers n.

17. When $n = 1$, $3^{2n+2} - 8n - 9 = 64$, and the statement is true, since 64 is a factor of 64. Assume that the statement is true for $n = k$. Then 64 is a factor of $3^{2k+2} - 8k - 9$. Thus, there is an integer p that we can multiply by 64 to get $3^{2k+2} - 8k - 9$. That is, $64p = 3^{2k+2} - 8k - 9$ or $3^{2k+2} = 64p + 8k + 9$. Now consider

$$3^{2(k+1)+2} - 8(k + 1) - 9 = 3^{2k+4} - 8k - 17$$

$$= 9(3^{2k+2}) - 8k - 17$$

$$= 9(64p + 8k + 9) - 8k - 17$$

$$= 9 \cdot 64p + 64k + 64$$

$$= 64(9p + k + 1).$$

Therefore, 64 is a factor of $3^{2(k+1)+2} - 8(k + 1) - 9$, and the statement is true for $n = k + 1$. Thus, 64 is a factor of $3^{2n+2} - 8n - 9$ for all positive integers n.

19. When $n = 1$, both sides of the equation are equal to c, and the statement is true. Assume that the statement is true for $n = k$. Then

$$c + (c + d) + (c + 2d) + (c + 3d) + \cdots +$$

$$(c + (k - 1)d) = \frac{k(2c + (k - 1)d)}{2}.$$

Add $c + kd$ to both sides of the equation to obtain

$$c + (c + d) + (c + 2d) + \cdots + (c + (k - 1)d) +$$

$$(c + kd) = \frac{k(2c + (k - 1)d)}{2} + (c + kd)$$

$$= \frac{2kc + k(k - 1)d + 2c + 2kd}{2}$$

$$= \frac{2c(k + 1) + (k + 1)kd}{2}$$

$$= \frac{(k + 1)(2c + kd)}{2},$$

and the statement is true for $n = k + 1$. Thus, it is true for all positive integers n.

21. (a) $x^2 - y^2 = (x - y)(x + y)$
$x^3 - y^3 = (x - y)(x^2 + xy + y^2)$
$x^4 - y^4 = (x - y)(x^3 + x^2y + xy^2 + y^3)$

(b) *Conjecture:*

$$x^n - y^n =$$
$$(x - y)(x^{n-1} + x^{n-2}y + x^{n-3}y^2 + \cdots + y^{n-1})$$

From part (a) we see that the statement is true for $n = 2$. Assume that the statement is true for $n = k$. Then
$$x^k - y^k =$$
$$(x - y)(x^{k-1} + x^{k-2}y + x^{k-3}y^2 + \cdots + y^{k-1}).$$
Consider

$$x^{k+1} - y^{k+1} = xx^k - yy^k = xx^k - xy^k + xy^k - yy^k$$
$$= x(x^k - y^k) + (x - y)y^k$$
$$= x(x - y)(x^{k-1} + x^{k-2}y + x^{k-3}y^2$$
$$+ \cdots + y^{k-1}) + (x - y)y^k$$
$$= (x - y)[x(x^{k-1} + x^{k-2}y + x^{k-3}y^2$$
$$+ \cdots + y^{k-1}) + y^k]$$
$$= (x - y)(x^k + x^{k-1}y + x^{k-2}y^2$$
$$+ \cdots + xy^{k-1} + y^k),$$

and the statement is true for $n = k + 1$. Thus, it is true for all positive integers n.

23. This statement is false. For example, 9 is an odd positive integer, and it is not a prime.

25. When $n = 1$, the statement is true. Assume that the statement is true for $n = k$. Then $(k + 1)^2 > k^2 + 1$. Then

$$(k + 2)^2 = k^2 + 4k + 4 = k^2 + 2k + 1 + 2k + 3$$
$$= (k + 1)^2 + (2k + 3)$$
$$> k^2 + 1 + (2k + 3) = k^2 + 2k + 1 + 3$$
$$= (k + 1)^2 + 1 + 3$$
$$> (k + 1)^2 + 1,$$

and the statement is true for $n = k + 1$. Thus, $(n + 1)^2 > n^2 + 1$ is true for all positive integers n.

27. This statement is false; counterexample: $n = 2$, $n^4 - n + 4 = 18$, and 4 is not a factor of 18.

29. When $n = 5$, we have $2(5) - 4 > 5$, which is true. Assume that the statement is true for $n = k$, where $k \geq 5$. Then $2k - 4 > k$. And $2(k + 1) - 4 = 2k - 4 + 2 > k + 2 > k + 1$. Thus, the statement is true for $n = k + 1$. Therefore, by induction, the statement is true for all $n \geq 5$.

31. When $n = 2$, we have $2^2 > 2$, which is true. Assume that the statement is true for $n = k$, where $k \geq 2$. Then $k^2 > k$. Thus $(k + 1)^2 = k^2 + 2k + 1 > k + 2k + 1 > k + 1$. So the statement is true for $n = k + 1$. Thus, $n^2 > n$ for all $n \geq 2$.

33. When $n = 4$, we have $3^4 > 2^4 + 10(4)$, which is true. Assume that the statement is true for $n = k$, where $k \geq 4$. Then $3^k > 2^k + 10k$. So we have

$$3^{k+1} = 3 \cdot 3^k > 3(2^k + 10k) = 3 \cdot 2^k + 30k >$$
$$2 \cdot 2^k + 30k > 2^{k+1} + 10k + 10$$
$$= 2^{k+1} + 10(k + 1).$$

Therefore, the statement is true for $n = k + 1$. Thus, $3^n > 2^n + 10n$ for all $n \geq 4$.

35. (a) When $n = 2$, you can move the stack in three moves (that is, $2^2 - 1$ moves). When $n = 3$, you can move the stack in seven moves (that is, $2^3 - 1$ moves). When $n = 4$, it takes 15 moves (that is, $2^4 - 1$ moves).
 (b) We conjecture that it takes $2^n - 1$ moves to move n rings. Clearly, this holds true for $n = 1$. Assume that the statement is true for $n = k$. That is, it takes $2^k - 1$ moves to move k rings. Now consider $k + 1$ rings. You can move the top k rings in $2^k - 1$ moves, take one move to move the bottom ring, and take $2^k - 1$ moves to move the top k rings onto the bottom. This gives a total of $2^k - 1 + 1 + 2^k - 1 = 2(2^k) - 1 = 2^{k+1} - 1$ moves. Thus the conjecture is true for $n = k + 1$. It will take $2^n - 1$ moves to move n rings for all positive integers n.

Chapter 7 Review, page 628

1. $a_1 = -3; a_2 = -1; a_3 = 1; a_4 = 3$

3. $a_1 = 1; a_2 = 1/4; a_3 = 1/9; a_4 = 1/16$

5. (a) September 1999: 152,600
  July 2001: 86,634
  August 2002: 48,620
 (b) February 2002

7. $40/9$ **9.** 81

11. $a_n = 9 - 6n$ **13.** $a_n = 6n - 11$

15. -55 **17.** $a_n = 2 \cdot 3^{n-1}$

19. $a_n = \dfrac{3}{2^{n-8}}$ **21.** $121/4$

23. $b = 35/4, c = 27/2, d = 73/4$

25. (a) $a_n = 5.123(1.0749)^{n-1}$
 (b) 2002: $9.1298 billion
 (c) $62.6274 billion

27. 2

29. $(1.02)^{51} = (1 + .02)^{51} = 1^{51} + 51(1)^{50}(.02) + \dfrac{51 \cdot 50}{2 \cdot 1}(1)^{49}(.02)^2 + \cdots$. The first three terms add to 2.53, and the rest are positive; hence, $(1.02)^{51} > 2.5$.

31. 455 **33.** $n + 1$ **35.** 1140 **37.** 80

39. $96,909,120$ **41.** $10,626$

43. 200 **45.** 24

47. 280 **49.** $5/36$

51. $.25$ **53.** $1/13$ **55.** $7/13$

57. $.3814$ **59.** $.012$ **61.** $.167$

63. When $n = 1$, both sides of the equation are equal to $1/3$, thus, the statement is true. Assume that the statement is true for $n = k$. Then

$$\frac{1}{3} + \frac{1}{3^2} + \frac{1}{3^3} + \cdots + \frac{1}{3^k} = \frac{3^k - 1}{2 \cdot 3^k}.$$

Add the $\dfrac{1}{3^{k+1}}$ to both sides of the equation:

$$\frac{1}{3} + \frac{1}{3^2} + \frac{1}{3^3} + \cdots + \frac{1}{3^k} + \frac{1}{3^{k+1}} = \frac{3^k - 1}{2 \cdot 3^k} + \frac{1}{3^{k+1}}$$

$$\frac{1}{3} + \frac{1}{3^2} + \frac{1}{3^3} + \cdots + \frac{1}{3^k} + \frac{1}{3^{k+1}} = \frac{3(3^k - 1)}{2 \cdot 3^{k+1}} + \frac{2}{2 \cdot 3^{k+1}}$$

$$\frac{1}{3} + \frac{1}{3^2} + \frac{1}{3^3} + \cdots + \frac{1}{3^k} + \frac{1}{3^{k+1}} =$$

$$\frac{3(3^k - 1) + 2}{2 \cdot 3^{k+1}} = \frac{3^{k+1} - 3 + 2}{2 \cdot 3^{k+1}} = \frac{3^{k+1} - 1}{2 \cdot 3^{k+1}}.$$

Thus, the statement is true for $n = k + 1$. And by induction, it is true for all positive integers n.

65. When $n = 1$, both sides of the equation are equal to 1; thus, the statement is true. Assume that the statement is true for $n = k$. Then

$$1 + 5^1 + 5^2 + \cdots + 5^{k-1} = \frac{5^k - 1}{4}.$$

Add 5^k to both sides of the equation:

$$1 + 5^1 + 5^2 + \cdots + 5^{k-1} + 5^k = \frac{5^k - 1}{4} + 5^k$$

$$= \frac{5^k - 1 + 4 \cdot 5^k}{4} = \frac{5 \cdot 5^k - 1}{4} = \frac{5^{k+1} - 1}{4}.$$

Thus, the statement is true for $n = k + 1$. And by induction, it is true for all positive integers n.

67. Clearly, the statement is true for $x = 0$; hence, assume that $x \neq 0$. Then the statement is true for $n = 1$. Assume that the statement is true for $n = k$; then $|x^k| < 1$. Multiply both sides of this inequality by $|x|$ to obtain $|x^k| \cdot |x| < 1 \cdot |x|$. Thus, $|x^{k+1}| = |x^k| \cdot |x| < |x| < 1$. Thus, the statement is true for $n = k + 1$. Thus, it is true for all positive integers n.

69. When $n = 1$, both sides of the equation are equal to 1; thus, the statement is true. Assume that the statement is true for $n = k$. Then

$$1 + 4^1 + 4^2 + \cdots + 4^{k-1} = \frac{4^k - 1}{3}.$$

Add 4^k to both sides:

$$1 + 4^1 + 4^2 + \cdots + 4^{k-1} + 4^k = \frac{4^k - 1}{3} + 4^k$$

$$= \frac{4^k - 1}{3} + \frac{3 \cdot 4^k}{3} = \frac{4 \cdot 4^k - 1}{3} = \frac{4^{k+1} - 1}{3}.$$

Thus, the statement is true for $n = k + 1$. And by induction, it is true for all positive integers n.

71. When $n = 1$, $9^n - 8n - 1 = 0$, and the statement is true, since 8 is a factor of 0. Assume that the statement is true for $n = k$. Then 8 is a factor of $9^k - 8k - 1$. That is, there is an integer p that we can multiply by 8 to get $9^k - 8k - 1$. Thus, $8p = 9^k - 8k - 1$ or $8p + 8k + 1 = 9^k$. Consider $9^{k+1} - 8(k + 1) - 1$. Rearrange this expression to get $9(9^k) - 8k - 9 = 9(8p + 8k + 1) - 8k - 9 = 72p + 64k = 8(9p + 8k)$, so 8 is a factor of $9^{k+1} - 8(k + 1) - 1$. Thus, the statement is true for $n = k + 1$. Therefore, by induction, the statement is true for all positive integers n.

CHAPTER 8

Section 8.1, page 646

1. $x^2 + (y - 3)^2 = 4$ **3.** $x^2 + 6y^2 = 18$

5. $(x - 4)^2 + (y - 3)^2 = 4$

7. Center $(-4, 3)$, radius $2\sqrt{10}$

9. Center $(-3, 2)$, radius $2\sqrt{7}$

11. Center $(-5/2, -5)$, radius $\dfrac{\sqrt{677}}{2} \approx 13.01$

13. Ellipse

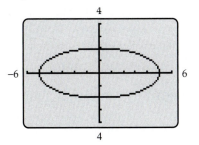

15. Ellipse

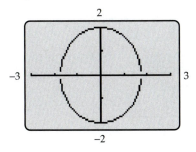

17. Ellipse

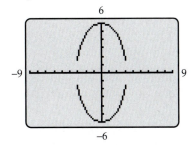

19. Circle

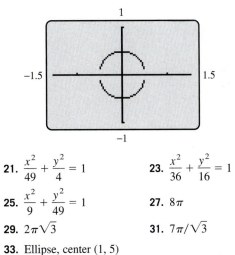

21. $\dfrac{x^2}{49} + \dfrac{y^2}{4} = 1$ **23.** $\dfrac{x^2}{36} + \dfrac{y^2}{16} = 1$

25. $\dfrac{x^2}{9} + \dfrac{y^2}{49} = 1$ **27.** 8π

29. $2\pi\sqrt{3}$ **31.** $7\pi/\sqrt{3}$

33. Ellipse, center $(1, 5)$

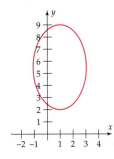

35. Ellipse, center $(-1, 4)$

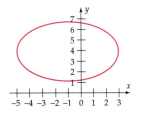

37. Ellipse, center $(-3, 1)$

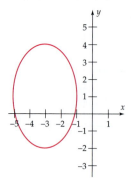

39. Circle, center $(-3, 4)$, radius $\sqrt{20}$

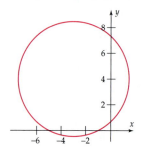

41. Ellipse, center $(-3, 2)$

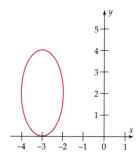

43. Ellipse, center $(-1, -3)$

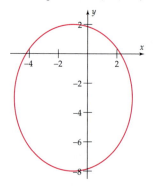

45. $\dfrac{(x-2)^2}{4} + \dfrac{(y-3)^2}{16} = 1$

47. $\dfrac{(x-7)^2}{25/4} + \dfrac{(y+4)^2}{36} = 1$

49. $\dfrac{(x-3)^2}{36} + \dfrac{(y+2)^2}{16} = 1$

51. $\dfrac{(x+5)^2}{49} + \dfrac{(y-3)^2}{16} = 1$ or $\dfrac{(x+5)^2}{16} + \dfrac{(y-3)^2}{49} = 1$

53. $2x^2 + 2y^2 - 8 = 0$

55. $2x^2 + y^2 - 8x - 6y + 9 = 0$

57. $\dfrac{(x+3)^2}{4} + \dfrac{(y+3)^2}{8} = 1$

59. The ellipse has y-intercepts $(0, \pm 4)$. Therefore, the circle, which has center at the origin, has radius 4. The equation of the circle must be $x^2 + y^2 = 16$.

61. The minimum distance is 226,335 miles. The maximum distance is 251,401 miles.

63. 80 feet   **65.** 17.3 feet

67. Eccentricity = .1

69. Eccentricity = $\dfrac{\sqrt{3}}{2} \approx .87$

71. The closer the eccentricity is to zero, the more the ellipse resembles a circle. The closer the eccentricity is to 1, the more elongated the ellipse becomes.

73. Eccentricity $\approx .38$.

75. If $a = b$, the equation becomes

$$\frac{x^2}{a^2} + \frac{y^2}{a^2} = 1,$$
$$x^2 + y^2 = a^2.$$

This is the equation of a circle with center at the origin, radius a.

77. The fence is an ellipse with major axis 100 ft, minor axis $50\sqrt{3} \approx 86.6$ ft.

Section 8.2, page 657

1. $x^2 + 4y^2 = 1$ **3.** $2x^2 - y^2 = 8$

5. $6x^2 + 2y^2 = 18$

7. Hyperbola

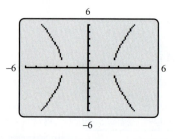

Note: Because of limited resolution, this calculator graph does not show that the top and bottom halves of the graph are connected.

9. Hyperbola (See Note in Exercise 7 answer.)

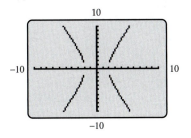

11. Hyperbola

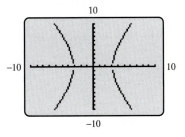

13. Hyperbola

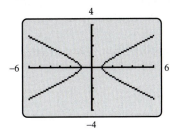

15. $\dfrac{x^2}{9} - \dfrac{y^2}{36} = 1$ **17.** $\dfrac{x^2}{4} - \dfrac{y^2}{1} = 1$

19. Hyperbola, center $(-1, -3)$

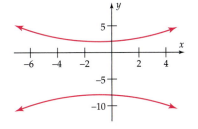

21. Hyperbola, center $(-3, 2)$

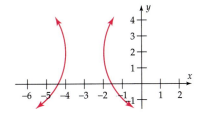

23. Hyperbola, center $(1, -4)$

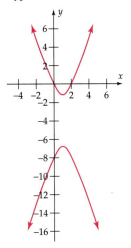

25. Hyperbola, center $(3, 3)$

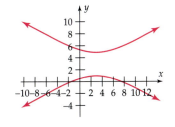

27. Circle, center $(3, 4)$

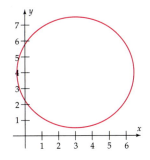

29. Ellipse, center (3, 4)

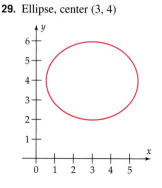

31. Hyperbola, center $(-2, 2)$

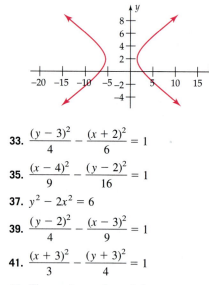

33. $\dfrac{(y-3)^2}{4} - \dfrac{(x+2)^2}{6} = 1$

35. $\dfrac{(x-4)^2}{9} - \dfrac{(y-2)^2}{16} = 1$

37. $y^2 - 2x^2 = 6$

39. $\dfrac{(y-2)^2}{4} - \dfrac{(x-3)^2}{9} = 1$

41. $\dfrac{(x+3)^2}{3} - \dfrac{(y+3)^2}{4} = 1$

43. The graphs are shown below on one set of coordinate axes.

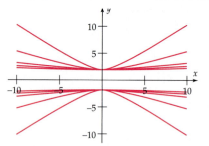

As b increases, the graphs get flatter. Although the hyperbola may look like two horizontal lines for very large b, it is still a hyperbola, with asymptotes $y = \pm 2x/b$ that have nonzero slopes.

45. The asymptotes of $\dfrac{x^2}{a^2} - \dfrac{y^2}{a^2} = 1$ are $y = \pm\dfrac{a}{a}x$ or $y = \pm x$. Since they have slopes 1 and -1 and $1(-1) = -1$, they are perpendicular.

47. The equation is $\dfrac{x^2}{1,210,000} - \dfrac{y^2}{5,759,600} = 1$ (measurement in feet). The exact location cannot be determined from only the given information.

49. $\sqrt{5} \approx 2.24$ **51.** $\sqrt{3}$ **53.** $3/2$

Section 8.3, page 667

1. $6x = y^2$ **3.** $2x^2 - y^2 = 8$

5. $x^2 + 6y^2 = 18$

7. Focus: $(0, 1/12)$, directrix $y = -1/12$

9. Focus: $(0, 1)$, directrix $y = -1$

11. Opens to the right.
Vertex: $(2, 0)$, focus: $(9/4, 0)$, directrix: $x = 7/4$

13. Opens to the left.
Vertex: $(2, -1)$, focus: $(7/4, -1)$, directrix: $x = 9/4$

15. Opens to the right.
Vertex: $(2/3, -3)$, focus: $(17/12, -3)$, directrix: $x = -1/12$

17. Opens to right.
Vertex: $(-81/4, 9/2)$, focus: $(-20, 9/2)$, directrix: $x = -41/2$

19. Opens upward.
Vertex: $(-1/6, -49/12)$, focus: $(-1/6, -4)$, directrix: $x = -25/6$

21. Opens downward.
Vertex: $(2/3, 19/3)$, focus: $(2/3, 25/4)$, directrix: $y = 77/12$

23.

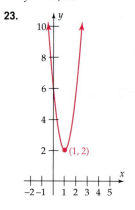

25.

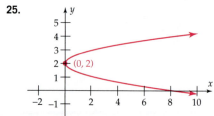

27.

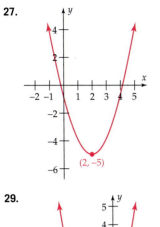

(2, –5)

29.

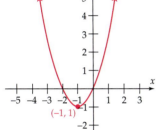

(–1, 1)

31. $y = 3x^2$

33. $y = 13(x - 1)^2$

35. $x - 2 = 3(y - 1)^2$

37. $x + 3 = 4(y + 2)^2$

39. $y - 1 = 2(x - 1)^2$

41. $(y - 3)^2 = x + 1$

43. $x = t^2/4, \ y = t$

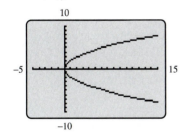

45. $x = t, \ y = 4(t - 1)^2 + 2$

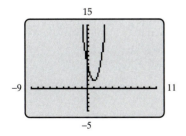

47. $x = 2(t - 2)^2, \ y = t$

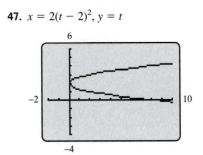

49. Parabola, vertex $(3, 4)$

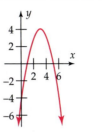

51. Parabola, vertex $(2/3, 1/3)$

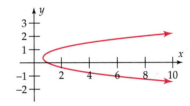

53. Circle, center $(1, 2)$

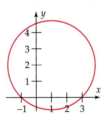

55. Hyperbola, center $(-4, 2)$, vertices $\left(-4 \pm \sqrt{2}, \, 2\right)$

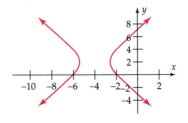

57. $y = (x - 4)^2 + 2$

59. $y = -x^2 - 8x - 18$

61. $y = (x + 5)^2 - 3$

63. $b = 0$

65. $\left(9, \dfrac{-1 \pm \sqrt{34}}{2}\right)$

67. The receiver should be placed at the focus, $\frac{2}{3}$ feet or 8 inches from the vertex.

69. 43.3 feet   **71.** 5.072 meters

73. 27.44 feet

Section 8.4, page 673

1. Ellise   **3.** Parabola   **5.** Hyperbola

7. Ellipse. Window $-6 \le x \le 3$ and $-2 \le y \le 4$.

9. Hyperbola. Window $-7 \le x \le 13$ and $-3 \le y \le 9$.

11. Parabola. Window $-1 \le x \le 8$ and $-3 \le y \le 3$.

13. Ellipse. Window $-1.5 \le x \le 1.5$ and $-1 \le y \le 1$.

15. Hyperbola. Window $-15 \le x \le 15$ and $-10 \le y \le 10$.

17. Parabola. Window $-19 \le x \le 2$ and $-1 \le y \le 13$.

19. Hyperbola. Window $-15 \le x \le 15$ and $-15 \le y \le 15$.

21. Ellipse. Window $-6 \le x \le 6$ and $-4 \le y \le 4$.

23. Parabola. Window $-9 \le x \le 4$ and $-2 \le y \le 10$.

Chapter 8 Review, page 675

1. Ellipse, vertices $\left(0, \pm 2\sqrt{5}\right)$, foci $(0, \pm 2)$

3. Ellipse, vertices $(1, -1)$ and $(1, 7)$, foci $(1, 0)$ and $(1, 6)$

5. Focus: $(0, 5/14)$, directrix: $y = -5/14$

7.

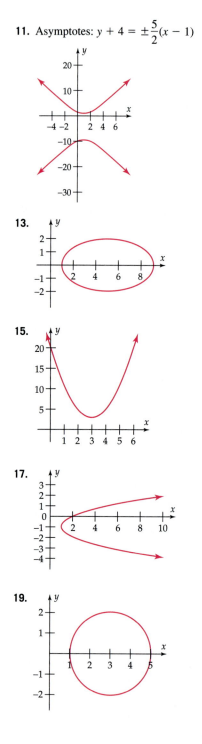

9.

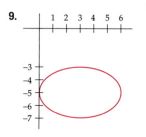

11. Asymptotes: $y + 4 = \pm\dfrac{5}{2}(x - 1)$

13.

15.

17.

19.

21. Center: $(-4, -5)$, radius: $2\sqrt{2}$

23. $(4, -6)$

25. $\dfrac{(x-3)^2}{4} + \dfrac{(y-1)^2}{2} = 1$

27. $\dfrac{y^2}{4} - \dfrac{(x-3)^2}{16} = 1$

29. $\left(y + \dfrac{1}{2}\right)^2 = -\dfrac{1}{2}\left(x - \dfrac{3}{2}\right)$

31. 6 feet

33. The receiver should be placed at the focus, $\frac{9}{16}$ feet or 6.25 inches from the vertex.

35. Ellipse   **37.** Hyperbola

39. $-6 \le x \le 6$ and $-4 \le y \le 4$

41. $-9 \le x \le 9$ and $-6 \le y \le 6$

43. $-15 \le x \le 10$ and $-10 \le y \le 20$

APPENDIX 1, page 678

1. $b = \sqrt{20} = 2\sqrt{5}$ **3.** $c = \sqrt{63} = 3\sqrt{7}$

5. $b = \sqrt{50} = 5\sqrt{2}, c = 5$

7. $b = \sqrt{75} = 5\sqrt{3}, c = 5$

9. Angle $A = 53.13°$
Angle $R = 36.87°$
$r = 3, s = 4, t = 5$

11. Angle $A = 65°$ and Angle $R = 25°$.
Thus, $\angle A \cong \angle T$, $\angle C \cong \angle R$, $AC \cong TR$.
By ASA, the triangles are congruent.

13. $\angle u \cong \angle v$, that is, $\angle POD \cong \angle QOD$. $\angle PDO \cong \angle QDO$, since all right angles are congruent. $OD \cong OD$. By ASA, $PDO \cong QDO$. Hence, corresponding parts are congruent and $PD \cong QD$.

15. $\angle u \cong \angle v$, that is, $\angle BAC \cong \angle DCA$. $AB \cong DC$ and $AC \cong AC$. By SAS, the triangles ABC and ACD are congruent.

17. $\angle x \cong \angle y$, that is, $\angle BAC \cong \angle DAC$. $\angle u \cong \angle v$, that is, $\angle BCA \cong \angle DCA$. $AC \cong AC$. By ASA, the triangles ABC and ADC are congruent.

19. $c = 7.5, r = 28, t = 35$

21. $r = 3, t = 3.5$

23. Since $\dfrac{AB}{RS} = \dfrac{15}{10} = \dfrac{AC}{RT}$ and $\angle A \cong \angle R$, the triangles are similar by S/A/S. $a = 15$, $\angle S \cong \angle T$, and both have measure $60°$, since the triangles are equilateral.

25. Since $\dfrac{AC}{RS} = \dfrac{6.885}{4.59} = \dfrac{5.25}{3.5} = \dfrac{BC}{TS}$ and $\angle C \cong \angle S$, the triangles are similar by S/A/S. (S/S/S can also be used.) $\angle T \cong \angle B$, and both have measure $62.46°$. $\angle A \cong \angle R$, and both have measure $42.54°$.

27. Since $\dfrac{a}{r} = \dfrac{5}{17.5} = \dfrac{b}{s}$ and $\angle C \cong \angle T$, the triangles are similar by S/A/S. Since the triangles are isosceles right triangles, $\angle A$, $\angle B$, $\angle R$, $\angle S$ all have measure $45°$. $c = 5\sqrt{2}$ and $t = 17.5\sqrt{2}$.

29. Since $\angle A \cong \angle R$ and $\angle B \cong \angle T$, the triangles are similar by AA. $\angle C \cong \angle S$ and both have measure $41.5°$; $s = 2.4$.

31. 55 ft   **33.** 80 ft

35. Distance from Marietta to Columbus: 92.4 mi., distance from Marietta to Cincinnati: 171.2 mi.

37. In the figure, $AD \cong CR$, $CD \cong AR$, $AC \cong CA$. Hence, by SSS, triangle $ADC \cong$ triangle CRA, and so $\angle x \cong \angle u$. Similarly, $\angle v \cong \angle y$. Since $\angle x + \angle w + \angle y = 180°$, it follows that $\angle u + \angle w + \angle y = 180°$.

INDEX
of Applications

Athletics

Baseball batting orders, permutations of, 593–594
Darts, 616
Field dimensions, 64–65, 70
Field house construction, 189
Game attendance, 629
Hit ball, distance of, 189–190, 284–285, 510–511
Seating, 568–569
Snowboarding, 112, 270–271
Swimming, winning times, 115
Teams, permutations of, 600
Thrown ball, distance of, 290
Ticket sales, 98, 291, 467–468, 487

Biology and Life Sciences.
See also Medical; Population Growth

Accidental deaths, 104, 438
Air quality standards, 546–547
Blood alcohol content, 199
Breast-feeding mothers, 302
Dinosaur age, 367
Diversity of species, 416, 418
Fish, length of, 342
Genes, weight of, 22
Height and weight, 199–200
Inhibited population growth, 422
Multiple births, 402, 436
Population density, 22
Seedlings, survival rate of, 364
Sex, probability of birth, 606, 616
Size of an animal, 55

Business and Manufacturing.
See also Income

Advertising expenditures, 107, 259, 260, 328, 402
Average profits, 261
Beef consumption, 402
Box dimensions, 34, 35, 150, 169, 211, 224–225, 291, 318–320, 339–340, 343, 480
Break-even points, 177, 208, 472–473
Budget discussions, 134

Business and Manufacturing
(*continued*)

Budget variance, 12
Combined work, 129, 132, 187
Consumer confidence survey, 71
Cost-benefit function, 342
Cost per unit, 90, 97, 176, 183–184, 185, 200, 207–208, 210, 241, 323, 338–339, 343, 363, 487
Costs of production, 53, 55
Cylindrical containers, 150, 322
Debt, 496
Employees, approximating number of, 628
Expenses and sales income, 228
Filtering, 391
Foreign tourists, 210
Gutters, 290
Imports, value of, 390
Input-output analysis, 550–551
Mail costs, 270
Measuring, 125
Mixtures, 130, 132, 499, 514, 532–533, 534–535, 538–541, 544, 548, 549
Postage rates, 214, 223
Poultry consumption, 108
Prescription drug marketing, 115
Price of specific goods, 131
Price per unit, 362, 364
Production costs, 500
Profit graphs, 99–100
Profit maximization, 79–80, 83, 84, 291, 321
Profit rate of change, 273
Rent increases, 286
Research and development expenditures, 109
Resource allocation, 29, 458
Revenue, 197, 240, 520
Sales, holiday, 571
Sales of specific goods, 69, 130, 427, 514
Sales tax function, 209
Sequences, 571
Shared work, 151
Shipping charges, 487
Telephone cards, prepaid, 115
Theft rate, 85
Transistor growth, 374–375, 391
Triangular display case project, 57
Tubes, 480
Vehicle production, 96

Business and Manufacturing
(*continued*)

Web use, prevalence of, 60
Women in labor force, 287–288
Workers' compensation costs, 323–324

Chemistry and Physics.
See also Temperature

Atmospheric pressure, 148, 379, 402, 428
Barometric pressure, 402
Boiling water, 96
Carbon dioxide (CO_2) emissions, 184
Concentrations, 52–53, 96, 291, 401, 427, 513
Earth's dimensions, 22, 23
Electrical resistance, 135–136, 137
Evaporation, 251–252, 254
Exponential decay, 379, 427
Exponential growth, 378
Falling objects, 145, 151, 185, 254–255
Filling pools, 72
Filtering tap water, 386–387
Folded paper, 169, 379
Gas pressure variation, 137
Gravitational acceleration, 344
Inequalities, 177
Listening stations, 659
Loran transmitters, 655–657
Metals, averaging, 187
Mixtures, 127–128, 132, 343, 469–470, 487, 499, 547
Natural gas, 260, 302
Period of a pendulum, 11
Planetary orbits, 22, 29, 432–434, 645, 648
Probability of, 615
Projectiles, parabolic path of, 666
Radioactive decay, 373–374, 387–388, 391, 420–421, 452, 454
Radio waves, speed of, 22
Reflective properties, elliptical surfaces, 677
Rocket paths, 83, 151, 185, 290, 362
Satellite dish, 676
Satellite orbit, 649
Shadow length, 254
Sound waves, 22, 145, 409
Speed of light, 22
Traffic, 364
Transmission stations, 659

Equations and Graphs

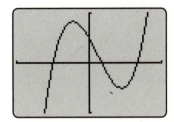

The solutions of the equation $f(x) = 0$ are the x-intercepts of the graph of $y = f(x)$.

Distance Formula

Length of segment $PQ = \sqrt{(x_1 - x_2)^2 + (y_1 - y_2)^2}$

Midpoint Formula

Midpoint M of segment $PQ = \left(\dfrac{x_1 + x_2}{2}, \dfrac{y_1 + y_2}{2} \right)$

Slope of nonvertical line through (x_1, y_1) and (x_2, y_2)

$\dfrac{y_2 - y_1}{x_2 - x_1}$

Equation of line with slope m through (x_1, y_1)

$y - y_1 = m(x - x_1)$

Equation of line with slope m and y-intercept b

$y = mx + b$

Exponential Growth Functions

$f(x) = Pa^x \quad (a > 1)$

$f(x) = Pe^{kx} \quad (k > 0)$

$f(x) = P(1 + r)^x \quad (0 < r < 1)$

Exponential Decay Functions

$f(x) = Pa^x \quad (0 < a < 1)$

$f(x) = Pe^{kx} \quad (k < 0)$

$f(x) = P(1 - r)^x \quad (0 < r < 1)$

Compound Interest Formula

$A = P(1 + r)^t$

Natural Logarithms

For $v, w > 0$ and any u, k:

$\ln v = u$ means $e^u = v$

$\ln (vw) = \ln v + \ln w$

$\ln \left(\dfrac{v}{w} \right) = \ln v - \ln w$

$\ln(v^k) = k(\ln v)$

Logarithms to Base b

For $v, w > 0$ and any u, k:

$\log_b v = u$ means $b^u = v$

$\log_b(vw) = \log_b v + \log_b w$

$\log_b \left(\dfrac{v}{w} \right) = \log_b v - \log_b w$

$\log_b(v^k) = k(\log_b v)$

Special Notation

$\ln v$ means $\log_e v$

$\log v$ means $\log_{10} v$

Change of Base Formula

$\log_b v = \dfrac{\ln v}{\ln b}$